THE PESTICIDE MANUAL

A WORLD COMPENDIUM

THE
PESTICIDE
MANUAL

A World Compendium

Published by The British Crop Protection
Council

Editor
CHARLES R. WORTHING
B.Sc., M.A., D. Phil.

Glasshouse Crops Research Institute

BCPC BRITISH
CROP
PROTECTION
COUNCIL

6th EDITION
1979

First published   *1968*

Second edition   *1971*

Third edition   *1972*

Fourth edition   *1974*

Fifth edition   *1977*

Sixth edition   *1979*

Reprint   *1980*

This and other publications of The British Crop Protection Council can be obtained from any major bookseller or direct from BCPC Publications Sales, 'Shirley', Westfields, Cradley, Malvern, Worcestershire WR13 5LP, England.

ISBN 0-901436-44-5

Managing Editor: Dr. E. K. Woodford, O.B.E.

Every effort has been made to ensure that the statements made in this Manual are correct but neither The British Crop Protection Council nor the Editor accept responsibility for any loss, damage or other accident arising from any error in the Manual.

Printed by The Lavenham Press Limited, Lavenham, Suffolk

PREFACE

The continuing introduction of pesticidal chemicals and microbial agents of proven or potential usefulness, the withdrawal of some earlier compounds and the unprecedented demand for the 5th edition, made this an opportune time to update what has widely become known as 'Hubert Martin's *Pesticide Manual*'.

Dr. Martin, who with Mr. A. W. Billitt as Technical Officer had been associated with the project from the beginning, no longer wished to be an editor and the British Crop Protection Council invited me, as joint editor of the 4th and 5th editions, to undertake the preparation of a 6th edition. I trust the new edition follows the high standard set by its predecessors and that Dr. Martin will regard it as a suitable memento for his 80th birthday in June 1979.

The intention has been to include all chemicals and microbial agents used as active components of products to control crop pests and diseases, pests in public health and animal ectoparasites. Details of herbicides, plant growth regulators and of pest repellents are also included. Manufacturers were asked to update entries from previous editions and to add new compounds which had reached the stage of submission to outside laboratories for field tests. Compounds now only of historic interest or which have been superseded by others of greater potential have been placed, by name (molecular formula and Wiswesser Line-Formula Notation) in an appendix (Appendix I).

The collection of the basic technical information on these chemicals is a task requiring the collaboration of the industrial laboratories in which they were developed or are being manufactured. No attempt has been made to include specific details, such as residue tolerance levels, minimum intervals between treatment and harvest or allowed uses that vary from country to country. The response from manufacturers has been excellent, as may be judged by the uniformity of the information provided for each compound. It is particularly pleasing that the new entries include several from a wider spread of countries than hitherto.

It is impossible to list here the many friends, spread over several continents, who have helped in supplying details or in placing me in contact with firms that have changed their address since earlier editions. To them and to the readers who notified me of errors in previous editions I say a very sincere 'Thank you'.

In particular I wish to thank the Advisory Editorial Board, set up by The British Crop Protection Council, for guidance and help, especially in enabling me to contact suitable specialists on selected topics. I also wish to thank Dr. D. Rudd-Jones, Chairman of the Council's Publications Committee, Dr. E. K. Woodford, its Managing Editor, and Mr. A. W. Billitt, its Technical Officer, for their constant support and guidance throughout the operation, and the Printing Department of The Boots Company Limited for their co-operation and understanding.

I am indebted to Mr. J. O'G. Tatton for advice in selecting preferred chemical names within the rules of the International Union of Pure and Applied Chemistry. These will be in agreement with those in BS 1831, ISO 1750, ISO 765, and addenda when new editions of these documents are published shortly.

Two important changes have been introduced in this edition. Firstly, sub-headings make it simpler to locate specific information within a given entry. Secondly, four indexes have been included. They are based on: the Wiswesser Line-Formula Notation; molecular formula; code numbers given to the compounds by either the manufacturers or official bodies such as WHO or USDA; and a name index, covering chemical names, recognised common names and trade marks (*i.e.* similar to that used previously). I am much indebted to Mr. S. B. Walker, of Imperial Chemical Industries Limited, Plant Protection Division, for assigning the Wiswesser Line-Formula Notations and to Mr. R. W. Parsons for providing computer-generated indexes.

I must stress, however, that I accept the responsibility for any factual errors arising from transcription of the information supplied by the various firms. I would be grateful if readers could bring to my attention any obvious errors or omissions so that they can be corrected in the next edition.

Charles R. Worthing, *Editor*

Glasshouse Crops Research Institute

Member Organisations

BCPC BRITISH CROP PROTECTION COUNCIL

The corporate members are

Association of Applied Biologists,
National Vegetable Research Station,
Wellesbourne,
Warwick.   CB35 9EF

Agricultural Research Council,
160 Great Portland Street,
London.   W1N 6DT

Agricultural Engineers Association Ltd,
6 Buckingham Gate,
London.   SW1E 6JU

British Agrochemicals Association,
Alembic House,
93 Albert Embankment,
London.   SE1 7TU

National Association of Agricultural
   Contractors,
Huts Corner,
Tilford Road,
Hindhead,
Surrey.   GU26 6SF

National Farmers' Union,
Agriculture House,
25/31 Knightsbridge,
London.   SW1X 7NJ

Natural Environment Research Council,
Polaris House,
North Star Avenue,
Swindon,
Wilts.   SN2 1EU

Society of Chemical Industry,
Pesticides Group,
14 Belgrave Square,
London.   SW1X 8PS

UK Agricultural Supply Trade
   Association Limited,
3 Whitehall Court,
London.   SW1A 2EQ

Department of Agriculture and
   Fisheries for Scotland,
St. Andrew's House,
Edinburgh.   EH1 3D4

Department of Agriculture,
Northern Ireland,
Dundonald House,
Upper Newtonwards Road,
Belfast.   BT4 3SB
Northern Ireland.

Department of the Environment
   (Great Britain),
2 Marsham Street,
London.   SW1 3EB

Ministry of Agriculture, Fisheries and
      Food,
   Pesticides Branch,
   Great Westminster House,
   Horseferry Road,
   London.   SW1P 2AE

Agricultural Development & Advisory
   Service (Headquarters),
Great Westminster House,
Horseferry Road,
London.   SW1P 2AE

Agricultural Science Service,
The Harpenden Laboratory,
Hatching Green,
Harpenden,
Hertfordshire.   AL5 2BD

Ministry of Overseas Development,
Eland House,
Stag Place,
London.   SW1E 5DH

Further information about the BCPC, its organisation and work can be obtained from:
The Administrative Secretary, The British Crop Protection Council, 144-150 London
Road, Croydon. CR0 2TB.

CONTENTS

GUIDE TO USE OF THE MANUAL

The abbreviations used are listed in Appendix II

Compounds
Each compound is described on a separate page or pages. Entries are arranged alphabetically. The required compound is best located from the indexes at the end of the book; they are by Wiswesser Line-Formula Notation (explained in Appendix VI), molecular formula, manufacturer's or official code number and common name, trade mark or chemical name.

For ease of reference the information about each compound is grouped under sub-heads as follows. (See example on the opposite page, in which the numbers refer to the paragraph numbers below).

Heading
1 The name used is preferably the BSI Common name. If the compound has no BSI common name then the common name given by ISO or ANSI or WSSA or BPC or ESA is used in that order of priority. Otherwise a well known trivial name (sulfoxide) or chemical name (copper oxychloride) or IUPAC name (1,3-dichloropropene) is used.

2 If the BSI name differs from the ISO name, the latter is given in brackets below the BSI name.

3 Chemical structure.

4 Wiswesser Line-Formula Notation.

5 Molecular formula and weight.

Nomenclature and development.
6 Common names: BSI, ISO, ANSI, WSSA, BPC and JMAF are stated; national name or major spelling variations (excluding the addition or absence of a terminal 'e', accents or use of 't' for 'th') are listed. The phrase 'proposed by ISO' refers to names that have passed ISO's preliminary enquiry stage, and so are likely to be adopted, but have not yet been the subject of voting by member bodies in a letter ballot. BSI and ISO names are correct up to July 1979.

7 Code numbers used by USDA (prefix ENT, AI3 or AN4) and WHO (prefix OMS).

8 Preferred and alternative IUPAC names.

9 Current and former *C.A.* name(s) and Registry Number(s).

10 Discovering organisation or person; developing company, code number, trade mark(s) and protecting patents (which may have expired); well-known trade marks of withdrawn formulations are included in *italics*.

11 Early important scientific reference(s) using *Chemical Abstracts* style journal abbreviations—see *Chemical Abstracts Service Source Index*.

Manufacture and properties.
12 Manufacturing process.

13 Physical and chemical properties of the active ingredient and/or the technical product.

Uses.
14 Principal uses.

Toxicology.
15 Oral and dermal LD50 values. Chronic toxicity. Toxicity to wild life.

Formulations.
16 Principal formulations.

Analysis.
17 Methods, generally with relevant references, for product and residue analysis.

18 The addresses of the leading chemical manufacturers or suppliers are given in Appendix IV.

Note: this specimen entry has been deliberately shortened so that it illustrates the principles without unnecessary duplication. It is *not* the complete entry on trichloronate.

(1)→ **Trichloronate**

(2)→ (Trichloronat)

(4)→ GR BG DG EOPS&2&O2 $C_{10}H_{12}Cl_3O_2PS$ (333·6) ←(5)

(6)→ { **Nomenclature and development.** The common name trichloronat is approved ←(7)
by ISO—exception BSI (trichloronate); also known as ENT 25712. The
(8)→ IUPAC and *C.A.* name is *O*-ethyl *O*-2,4,5-trichlorophenyl ethylphosphonothioate, Registry No. *[327-98-0]*. It was introduced in 1960 by Bayer AG under
(9,10)→ { the code number 'Bayer 37289', the trade mark 'Agrisil' and protected by
DBP 1 099 530. Its insecticidal properties were described by R. O. Drummond, ←(11)
J. Econ. Entomol., 1963, **56**, 831.

(12)→ { **Manufacture and development.** Produced by the condensation of 2,4,5-
trichlorophenol with *O*-ethyl ethylphosphonochloridothioate, it is an amber-
coloured liquid, b.p. 108°/0·01 mmHg, d_4^{20} 1·365. Its solubility at 20° is 50 mg/l ←(13)
water; soluble in acetone. It is hydrolysed by alkali.

(14)→ { **Uses.** Trichloronate is a non-systemic insecticide recommended for the
control of root maggots.

(15)→ { **Toxicology.** The acute oral LD50 is: for rats 16-37·5 mg/kg; for rabbits 25-50
mg/kg. The acute dermal LD50 for male rats is 135-341 mg/kg. In 2-y feeding
trials rats receiving 3 mg/kg diet showed no symptom of poisoning.

(16)→ **Formulations.** Granules and seed dressings of various a.i. content.

(17)→ { **Analysis.** Product analysis is by u.v. spectroscopy at 313 nm after alkaline
hydrolysis to 2,4,5-trichlorophenol. Residues may be determined by GLC
(E. Möllhoff, *Pflanzenschutz-Nachr. (Am. Ed.),* 1968, **21**, 331). Details of both ←(18)
methods may be obtained from Bayer AG.

Acephate

$$O$$
$$\|$$
$$MeO.P.SMe$$
$$|$$
$$NH.C.Me$$
$$\|$$
$$O$$

1VMPO&S1&O1 $C_4H_{10}NO_3PS$ (183·2)

Nomenclature and development. The common name acephate is approved by BSI, ISO, ANSI and JMAF; also known as ENT 27822. The IUPAC and *C.A.* name is *O,S*-dimethyl acetylphosphoramidothioate, Registry No. *[30560-19-1]*. It was introduced in 1971, after field trials begun in 1969, by the Chevron Chemical Co. under the code number 'Ortho 12 420', trade mark 'Orthene' and protected by USP 3 716 600; 3 845 172. Its insecticidal properties were described by J. M. Grayson, *Pest Control,* 1972, **40**, 30.

Manufacture and properties. Acephate is produced by the acetylation of *O,O*-dimethyl phosphoramidothioate followed by isomerisation to the *O,S*-dimethyl analogue. The technical product (purity 80-90%) is a colourless solid, m.p. 82-89°, v.p. 1·7 x 10⁻⁶ mmHg at 24°, *d* 1·35. Its solubility at room temperature is: *c.* 650 g/l water; < 50 g/l aromatic solvents; > 100 g/l acetone or ethanol. It is relatively stable.

Uses. Acephate is a systemic insecticide of moderate persistence with residual activity lasting, at the foliar rates recommended, *c.* 10-15 d. At rates of *c.* 50-100 g a.i./100 l it is effective against a wide range of aphids, thrips, leaf miners, sawflies and lepidopterous larvae, and is safe on many crop plants. Chemical structure-biological activity relationships in analogues were summarised; P. S. Magee, *Residue Rev.,* 1974, **53**, 3.

Toxicology. The acute oral LD50 of the technical product is: for rats 866-945 mg/kg; for mice 361 mg/kg; for mallard ducks 350 mg/kg; for chickens 852 mg/kg; for ringneck pheasants 140 mg/kg. The minimum emetic dose for beagle dogs is 215 mg/kg and the minimum lethal dose 681 mg/kg. The acute dermal LD50 for rabbits is > 2000 mg/kg; no irritation or sensitisation was observed in skin tests on guinea-pigs. In 2-y feeding trials: dogs showed no significant differences at levels up to 100 mg/kg diet (except for depression of cholinesterase at the highest dose level); rats showed no effect on weight gain at 30 mg/kg diet and slightly depressed body weight at 100 mg/kg diet, slight to moderate cholinesterase depression was noted. There were no significant differences between treated and untreated animals for other parameters, including pathology or tumour incidence or classification. No teratogenic effects were noted in rats or rabbits, nor mutagenic effects in the dominant lethal gene study in mice. The LC50 (96 h) is: for rainbow trout > 1000 mg/l; for bluegill 2050 mg/l; for large-mouth black bass 1725 mg/kg; for channel catfish 2230 mg/l; for mosquito fish 6650 mg/l; for goldfish 9550 mg/l.

Formulations. These include: w.s.p. (250, 500 and 750 g a.i./kg); pressurised sprays (2·5 and 10 g/l); granules.

Analysis. Product and residue analysis is by GLC: J. B. Leary, *Anal. Methods Pestic. Plant Growth Regul.,* 1973, **7**, 363. Particulars are available from the Chevron Chemical Co.

1

Acrolein

$$CH_2{=}CH.CHO$$

VH1U1   C_3H_4O (56.06)

Nomenclature and development. The trivial name acrolein (I) is accepted in lieu of a common name by BSI and WSSA. The IUPAC name is prop-2-enal or acrylaldehyde, in *C.A.* usage 2-propenal *[107-02-8]* formerly (I). It was introduced as an aquatic herbicide and algicide by the Shell Chemical Co. under the trade name 'Aqualin' or *'Aqualine'* (a name no longer used for this product in some countries) and protected by USP 2042220; 2959476; 2978475.

Manufacture and properties. Produced by passing glycerol vapour over magnesium sulphate at 330-340° or by the oxidation of allyl alcohol or propene, it is a mobile liquid, with a pungent odour, b.p. 52.5°, f.p. —87.7°, v.p. 210 mmHg at 20°, d^{20} 0.8410. Its solubility at 20° is 208 g/kg water; miscible with most organic solvents. It is highly reactive chemically, polymerising slowly on storage and violently in the presence of alkali or concentrated acids. It is flammable with a flash point (closed cup) < —17.8°. It is transported in oxygen-free atmospheres in the presence of quinolin-8-ol as polymerisation inhibitor.

Uses. Acrolein is used as an aquatic herbicide, being injected below the water surface by special equipment. It is also an algicide.

Toxicology. The acute oral LD50 is: for rats 46 mg/kg; for rabbits 7.1 mg/kg. It is lachrymatory and skin contact causes burns. A concentration of 2.5 mg/m³ air causes eye and mucous membrane irritation in 2-3 min and is intolerable in 5 min. It is highly toxic to fish.

Formulation. Solution (850 and 920 g/l).

Analysis. Product analysis is by reaction with hydroxylamine hydrochloride and titration of the acid liberated. Residues in water may be determined by reaction with 2,4-dinitrophenylhydrazine and colorimetry of the product. Details of these methods are available from the Shell Chemical Co.

Alachlor

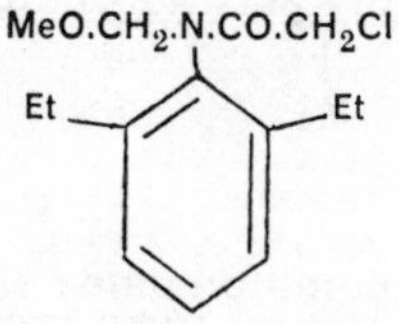

MeO.CH₂.N.CO.CH₂Cl

$$\text{MeO.CH}_2\text{.N.CO.CH}_2\text{Cl}$$

Et

Et

G1VN1O1&R   B2   F2 $C_{14}H_{20}ClNO_2$ (269·8)

Nomenclature and development. The common name alachlor is approved by BSI, ISO, ANSI, WSSA and JMAF. The IUPAC name is 2-chloro-2′,6′-diethyl-N-(methoxy-methyl)acetanilide (I) formerly α-chloro-2′,6′-diethyl-N-methoxymethylacetanilide, in *C.A.* usage 2-chloro-N-(2,6-diethylphenyl)-N-(methoxymethyl)acetamide *[15972-60-8]* formerly (I). It was introduced in 1966 by the Monsanto Co. under the code number 'CP 50144', trade mark 'Lasso' and protected by USP 3442945. Herbicidal compositions and method of their use are protected by USP 3547620. Its herbicidal properties were first described by R. F. Husted *et al., Proc. North Cent. Weed Control Conf.,* 1966, **21,** 44.

Manufacture and properties. Produced by the reaction of chloroacetyl chloride with the azomethine of 2,6-diethylaniline and formaldehyde followed by treatment with methanol, it is a non-volatile crystalline solid, m.p. 40-41°, $d_{15.6}^{25}$ 1·133. Its solubility at 23° is 240 mg/l water; it is soluble in acetone, benzene, ethanol, ethyl acetate. It is hydrolysed under strongly acid or alkaline conditions.

Uses. Alachlor is a selective pre-em. herbicide of similar action to propachlor, but with a longer residual action, lasting some 70-84 d; it is used at 2 kg a.i./ha to control annual grasses and many broad-leaved weeds in maize, soyabeans, cotton, sugarcane, peanuts, radish, rape-seed and other brassicas.

Toxicology. The acute oral LD50 for rats is 1200 mg/kg; the lethal dose by dermal application is > 2000 mg/kg; non-irritant.

Formulations. An e.c. (480 g a.i./l); granules (150 g/kg).

Analysis. Isothermal GLC is used: Monsanto Co. (AQC No. 118-70). See also: R. A. Conkin, *Anal. Methods Pestic. Plant Growth Regul.,* 1978, **10,** 255.

Aldicarb

Me
|
MeS.C.CH:N.O.CO.NHMe
|
Me

1SX1&1&1UNOVM1

$C_7H_{14}N_2O_2S$ (190·3)

Nomenclature and development. The common name aldicarb is approved by BSI, ISO and ANSI—exception Germany; also known as ENT 27093, OMS 771, USDA AI3-27093. Its IUPAC name is 2-methyl-2-(methylthio)propionaldehyde O-methylcarbamoyloxime (I), in *C.A.* usage 2-methyl-2-(methylthio)propanal O-[(methylamino)carbonyl]oxime *[116-06-3]* formerly (I). Its insecticidal properties were first described by M. H. J. Weiden *et al., J. Econ. Entomol.,* 1965, **58**, 154, and it was introduced in 1965 by the Union Carbide Corp. under the code number 'UC 21 149', trade mark 'Temik' and protected by USP 3 217 037.

Manufacture and properties. Produced by the reaction of methyl isocyanate with 2-methyl-2-(methylthio)propionaldehyde (L. K. Payne & M. H. J. Weiden, *J. Agric. Food Chem.,* 1966, **14**, 356), it forms colourless crystals, m.p. 98-100°, v.p. 1 x 10⁻⁴ mmHg at 25°. Its solubility at room temperature is 6 g/l water; it is practically insoluble in heptane but soluble in most organic solvents. It is stable, except to concentrated alkali, is non-corrosive to metal containers and equipment and is non-flammable.

It is a cholinesterase inhibitor and was specifically designed to resemble O-acetylcholine structurally: Payne & Weiden, *loc. cit.* The sulphur atom of the molecule is readily oxidised to form the sulphoxide which is more slowly oxidised to the sulphone (aldoxycarb, p.5). The sulphoxide has solubility > 330 g/l water (D. L. Bull *et al., ibid.,* 1967, **15**, 610).

Uses. Aldicarb is a soil-applied systemic pesticide used against certain mites, nematodes and insects (especially aphids, whiteflies, leaf miners). Seed furrow, band or broadcast treatments either pre-plant or at planting as well as post-em. sidedress treatments at rates of 0·56-11·25 kg a.i./ha are used. Soil moisture is required to release the active chemical from the granules, so irrigation or rainfall should follow application. Uptake by the roots is rapid and the plants are protected for up to 84 d. The persistence of aldicarb and its oxidation products in soil has been studied: J. H. Smelt *et al., Pestic. Sci.,* 1978, **9**, 279, 286, 293.

Toxicology. The acute oral LD50 for male rats is 0·93 mg tech./kg. The acute dermal LD50 for male rabbits is 5·0 mg/kg. Rats were killed within 5 min by a dust concentration of 200 mg/m³ air. In 2-y feeding trials rats receiving 0·3 mg/kg daily were unaffected. The chronic LD50 (7 d) for 56-d-old bobwhite quail is 2400 mg/kg diet. The LC50 (96 h) for rainbow trout is 8·8 mg/l.

Formulations. Because of handling hazards only granular formulations in the 0·42-1·41 mm mesh size are available: on corn cob grits (50, 100 and 150 g a.i./kg); in the UK only, on coal (100 g/kg). Formulations based on gypsum have also been developed (50 and 100 g/kg). The formulation *'Ambush'* (20 g/kg corn cob) has been withdrawn.

Analysis. Product analysis is by i.r. spectroscopy: *CIPAC Handbook,* 1979, **1A**, in press. Residues in plants, animal tissues, soil and water are determined by GLC with FPD after oxidation to the sulphone (aldoxycarb) by peracetic acid or 3-chloroperbenzoic acid and a column chromatographic cleanup: R. R. Romine, *Anal. Methods Pestic. Plant Growth Regul.,* 1973, **7**, 147; J. C. Maitlan *et al., J. Agric. Food Chem.,* 1968, **16**, 549; J. H. Smelt *et al., loc. cit.,* pp.279, 293. Colorimetric methods have been used based on hydrolysis to hydroxylamine which is oxidised to nitrous acid, the latter used to diazotise suphanilic acid and the resulting diazonium salt coupled with a suitable component: D. P. Johnson *et al., J. Assoc. Off. Anal. Chem.,* 1966, **49**, 399; D. F. Lee & J. A. Rougham, *Analyst (London),* 1971, **96**, 798.

Aldoxycarb

O Me   O
‖ | ‖
Me.S.C.CH=N.O.C.NH.Me
‖ |
O Me

WS1&X1&1&1UNOVM1 $C_7H_{14}N_2O_4S$ (222·3)

Nomenclature and development. The common name aldoxycarb is approved by BSI and proposed by ISO; also known as ENT 29 261, USDA AI3-29 261 and AN4-9. The IUPAC name is 2-mesyl-2-methylpropionaldehyde *O*-methylcarbamoyloxime formerly 2-methyl-2-methylsulphonylpropionaldehyde *O*-methylcarbamoyloxime, in *C.A.* usage 2-methyl-2-(methylsulfonyl)propanal *O*-[(methylamino)carbonyl]oxime *[1646-88-4]* formerly 2-methyl-2-methylsulfonylpropionaldehyde *O*-methylcarbamoyloxime. Its insecticidal and nematicidal properties were first described by M. H. J. Weiden *et al., J. Econ. Entomol.,* 1965, **58,** 154 and it was introduced by Union Carbide Corp. under the code number 'UC 21 865', proposed trade mark 'Standak' and protected by USP 3 217 037. It is one of the oxidation products of aldicarb formed metabolically or chemically.

Manufacture and properties. Produced by the oxidation of aldicarb (p.4) with a strong oxidising agent, *e.g.* peracetic acid, aldoxycarb is a colourless crystalline solid, with a slightly pungent odour, m.p. 140-142°, v.p. 9 x 10^{-5} mmHg at 25°. Its solubility at 25° is: *c.* 9 g/l water; 50 g/l acetone; 75 g/l acetonitrile; 41 g/l dichloromethane. It is stable to light and heat but not to concentrated alkali. It is non-corrosive to alloys and non-flammable.

Uses. Aldoxycarb is a nematicide and systemic insecticide. It is a potent inhibitor of cholinesterase. Applied to the soil, sprayed overall or as a band, followed by incorporation before sowing or planting, it is effective depending on pest species and crop at 0·5-11·25 kg a.i./ha. Uptake through the root system protects plants against pests such as aphids, thrips and plant bugs for 25-56 d. Applied directly in the water at transplant at 2-3 kg/ha it controls *Meloidogyne, Pratylenchus* and *Trichodorus* spp. on, for example, tobacco without crop damage. Nematicidal and systemic insecticidal activity has been noted with cereals and cotton when used as a dressing at 0·5-2·0 kg/100 kg seed. It can be applied as a foliar spray at 3-4 kg/ha. Its persistence in soil has been studied: J. H. Smelt *et al., Pestic Sci.,* 1978, **9,** 279.

Toxicology. The acute oral LD50 is: for male rats 26·8 mg tech./kg; for mallard duck 33·5 mg/kg. The acute dermal LD50 (4 h) on intact skin for male rats is 700-1400 mg/kg. Inhalation LC50 (4-h dust exposure) for rats is 120 mg/m³ air. There was no skin irritation or eye injury in standard tests. The 'no effect' level from life-span feeding trials was: for mice 9·6 mg/kg; for rats 2·4 mg/kg. Teratology, reproduction and mutagenic studies were all negative. The LC50 (96 h) is: for trout 40 mg/l; for bluegill 55·5 mg/l. The 8-d chronic LD50 is: for mallard duck > 10000 mg/kg diet; for bobwhite quail 5706 mg/kg diet.

Formulations. These include: w.p. and low concentration granules.

Analysis. Product analysis is by i.r. spectroscopy: c.f. *CIPAC Handbook,* 1979, **1A,** in press. Residues in plants, animal tissues, soil and water is made by GLC with FPD after column chromatographic clean-up: R. R. Romine, *Anal. Methods Pestic. Plant Growth Regul.,* 1973, **7,** 147; J. C. Maitlan *et al., J. Agric. Food Chem.,* 1968, **16,** 549; J. H. Smelt *et al., loc. cit.*

Aldrin

(HHDN)

L D5 C555 A D- EU JUTJ AG AG BG IG JG KG

$C_{12}H_8Cl_6$ (364·9)

Nomenclature and development. The common name HHDN is approved by BSI and ISO for the pure compound—exceptions Canada, Denmark, USSR and JMAF (aldrin), and USA. The common name aldrin is approved by BSI and ISO for a product containing ⩾95% HHDN—exceptions Canada, Denmark and USSR (see above); also known as ENT 15949. The IUPAC name is (1R,4S,5S,8R)-1,2,3,4,10,10-hexachloro-1,4,4a,5,8,8a-hexahydro-1,4:5,8-dimethanonaphthalene formerly 1,2,3,4,10,10-hexachloro-1,4,4a,5,8,8a-hexahydro-exo-1,4-$endo$-5,8-dimethanonaphthalene, in *C.A.* usage 1,2,3,4,10,10-hexa-chloro-1α,4α,4aβ,5α,8α,8aβ-hexahydro-1,4:5,8-dimethanonaphthalene *[309-00-2]* formerly 1,2,3,4,10,10-hexachloro-1,4,4a,5,8,8a-hexahydro-$endo$-1,4-exo-5,8-dimethanonaphthalene. It was introduced in 1948 by J. Hyman & Co. as 'Compound 118' under the trade mark 'Octalene' and protected by USP 2635977. Its insecticidal action was first described by C. W. Kearns *et al., J. Econ. Entomol.,* 1949, **42,** 127.

Manufacture and properties. Produced by condensing hexachlorocyclopentadiene with the Diels-Alder adduct of cyclopentadiene and acetylene, it is a colourless crystalline solid, m.p. 104-104·5°, v.p. 7·5 x 10^{-5} mmHg at 20°, 1·4 x 10^{-4} mmHg at 25°. Its solubility at 27° is 27 µg/l water; moderately soluble in petroleum oils; readily soluble in acetone, benzene, xylene. The technical grade is a tan to dark brown solid, m.p. 49-60°, containing ⩾85% HHDN. Aldrin is stable to heat, to alkali and dilute acids, but oxidising agents and concentrated acids attack the unchlorinated ring. It is compatible with most pesticides and fertilisers, but is corrosive because of the slow formation of hydrogen chloride on storage.

Uses. Aldrin is a non-systemic and persistent insecticide, effective against soil insects at rates of 0·5-5·0 kg/ha and is non-phytotoxic. It is readily oxidised to dieldrin (see p.185).

Toxicology. The acute oral LD50 for rats is 67 mg/kg; it is absorbed through the skin. In 2-y feeding trials rats receiving 5 mg/kg diet suffered no ill-effect, but liver changes resulted at 25 mg/kg diet.

Formulations. These include: e.c. (240-480 g/l); w.p. (400-700 g/kg), urea may be added to prevent dehydrochlorination by certain carriers; dusts (25-50 g/kg); seed dressings; granules.

Analysis. Technical product and formulation analysis is by i.r. spectrometry (*CIPAC Handbook,* 1970, **1,** 428) or by GLC (details available from the Shell International Co. Ltd). Residues may be determined by GLC with ECD (K. Elgar *et al., Analyst (London),* 1966, **91,** 143; *Anal. Methods Pestic. Plant Growth Regul.,* 1972, **6,** 268).

Allethrin

Me$_2$C=CH—⟨cyclopropane ring with Me, Me⟩—C(=O)—O—⟨cyclopentenone ring with Me⟩—CH$_2$CH=CH$_2$

L5V BUTJ B2U1 C1 DOV- BL3TJ A1 A1 C1UY1&1   C$_{19}$H$_{26}$O$_3$ (302·4)

Nomenclature and development. The common name allethrin is approved by BSI, ISO and JMAF—exceptions France (palléthrine) and Germany; also known as ENT 17510. The IUPAC name is (RS)-3-allyl-2-methyl-4-oxocyclopent-2-enyl (1RS)-cis,trans-chrysanthemate, in C.A. usage 2-methyl-4-oxo-3-(2-propenyl)-2-cyclopenten-1-yl 2,2-dimethyl-3-(2-methyl-1-propenyl)cyclopropanecarboxylate [584-79-2] (formerly [137-98-4]) formerly (±)-2-allyl-4-hydroxy-3-methyl-2-cyclopenten-1-one ester of 2,2-dimethyl-3-(2-methyl-propenyl)cyclopropanecarboxylic acid. It was introduced by the Sumitomo Chemical Co. under the trade mark 'Pynamin'. It was first described by M. S. Schechter *et al., J. Am. Chem. Soc.,* 1949, **71,** 3165. Its development and properties have been reviewed (W. Barthel, *Wld. Rev. Pest Control,* 1967, **6,** 59). Bioallethrin, a stereoisomer of allethrin, is listed separately (p.42).

Manufacture and properties. Produced by the esterification of chrysanthemic acid with 2-allyl-4-hydroxy-3-methylcyclopent-2-en-1-one (H. J. Sanders & A. W. Taff, *Ind. Eng. Chem.,* 1954, **46,** 414), technical allethrin is a pale yellow oil containing 75-95% allethrin isomers, b.p. *c.* 140°/0·1 mmHg, d_4^{25} 1·005, n_D^{21} 1·5070. It is sparingly soluble in water; soluble in most organic solvents; miscible with petroleum oils. Its chemical properties are similar to those of the pyrethrins (p.459) but, having a more stable side chain, it is somewhat more persistent (P. Granett *et al., J. Econ. Entomol.,* 1951, **44,** 552).

Uses. Allethrin is a contact insecticide as effective as natural pyrethrins against *Musca domestica* but less effective against household pests and most other species of insect. Its activity is enhanced by pyrethrin synergists though less markedly by piperonyl butoxide.

Toxicology. The acute oral LD50 is: for mice 480 mg/kg; for rats 920 mg/kg. In 1-y feeding trials rats tolerated 2000 mg/kg diet without ill-effect.

Formulation. Usually it is combined with synergists and other insecticides as fly sprays in kerosine, aerosol concentrates (1-6 g allethrin/l), or dusts.

Analysis. Product analysis is by GLC (*cf* bioallethrin) or by i.r. spectroscopy (S. K. Freeman, *Anal. Chem.,* 1955, **27,** 1261). The proportions of stereoisomers may be determined by GLC of suitable derivatives (F. E. Rickett, *Analyst (London),* 1973, **98,** 687; A. Horiba *et al., Agric. Biol. Chem.,* 1977, **41,** 2003; 1978, **42,** 671) or by NMR (F. E. Rickett & P. B. Henry, *Analyst (London),* 1974, **99,** 330). Residues may be determined by GLC—details from McLaughlin Gormley King Co. See also: D. B. McClellan, *Anal. Methods Pestic., Plant Growth Regul. Food Addit.,* 1964, **2,** 25; *Anal. Methods Pestic. Plant Growth Regul.,* 1972, **6,** 283; J. Sherma, *ibid.,* 1976, **8,** 117.

Alloxydim-sodium

L6V BUTJ BY3&UNO2U1 CQ E1 E1 FVO1 *alloxydim* $C_{17}H_{25}NO_5$ (323·4)

alloxydim-sodium $C_{17}H_{24}NNaO_5$ (345·4)

L6V BUTJ BY3&UNO2U1 CO E1 E1 FVO1 &-NA-

Nomenclature and development. The common name alloxydim is approved by BSI and proposed by ISO. The IUPAC name is methyl 3-[1-(allyloxyimino)butyl]-4-hydroxy-6,6-dimethyl-2-oxocyclohex-3-enecarboxylate, in *C.A.* usage methyl 2,2-dimethyl-4,6-dioxo-5-[1-[(2-propenyloxy)amino]butylidene]cyclohexanecarboxylate *[55634-91-8]*; it is also known by the code number 'NP-48' (Nippon Soda Co. Ltd) and protected by JapanPA 77-95 636. Its sodium salt, alloxydim-sodium, in *C.A.* usage methyl 4-hydroxy-6,6-dimethyl-2-oxo-3-[1-[(2-propenyloxy)imino]butyl]-3-cyclohexene-1-carboxylate sodium salt *[66003-55-2]* was introduced by Nippon Soda Co. Ltd as an experimental herbicide under the code number 'NP-48 Na', also known as 'BAS 90 210H' (BASF AG). Its properties were first reported by Y. Horono *et al., Meet. Pestic. Sci. Soc. Jpn., 1st,* 1976; A. Formigoni & Y. Horono, *Meded. Fac. Landbouwwet. Rijksuniv. Gent,* 1977, **42**, 1597.

Properties. Alloxydim-sodium is a colourless crystalline solid, m.p. 185·5° (decomp.). Its solubility at 30° is: >2 kg/kg water; 1 kg/kg dimethylformamide; 619 g/kg methanol; 50 g/kg ethanol; 15 g/kg butanone; 14 g/kg acetone; <4 g/kg cyclohexanone, ethyl acetate, xylene. It is very hygroscopic.

Uses. Alloxydim-sodium is a selective herbicide effective post-em. against grass weeds in sugar beet, vegetables and broad-leaved crops at 0·5-1·0 kg a.i./ha. Tank mixes with other herbicides or split applications with herbicides effective against broad-leaved weeds are recommended to increase the range of herbicidal activity.

Toxicology. The acute oral LD50 is: for rats 2260-2322 mg/kg; for mice 3000-3200 mg/kg.

Formulation. 'NP-48 Na 75% SP', w.s.p. (750 g alloxydim-sodium/kg).

Allyl alcohol

$$CH_2=CH.CH_2OH$$

Q2U1 C_3H_6O (58.08)

Nomenclature and development. The IUPAC name allyl alcohol or prop-2-en-1-ol is accepted in lieu of a common name by BSI and ISO—in France (alcool allylique)—in *C.A.* usage 2-propen-1-ol *[107-18-6]*. Its experimental use as a herbicide was first reported in 1950.

Manufacture and properties. Produced by the high temperature chlorination of propene forming 3-chloroprop-1-ene which is then hydrolysed, it is a colourless mobile liquid, with a pungent odour, b.p. 96.9°, v.p. 17.3 mmHg at 20°, d^{20} 0.8535, n_D^{20} 1.4133. It forms a glass at —190°. It is miscible with water, forming a constant boiling mixture (b.p. 88.9°) containing 723 g allyl alcohol/kg mixture, and with common organic solvents. On storage it polymerises to a thick syrup that is insoluble in water. Its flash point (open cup) is 32.2° and it is flammable in air in mixtures of 2.5-18% *m/m*. It is not appreciably corrosive to metals.

Uses. Allyl alcohol is highly phytotoxic and is used for weeding forest nursery beds at 85-170 g in 50000-100000 l water/ha. Also for the partial sterilisation of glasshouse soil (E. M. Emmert & J. E. Klinker, *Ky. Agric. Exp. Stn. Prog. Rep.,* 1950) at 18.4 g in 6.1 l water/m² of soil, which should be > 16° or preferably ⩾ 21°. Planting may be made 3-10 d after application.

Toxicology. The acute oral LD50 is: for rats 64 mg/kg; for mice 85 mg/kg. The acute dermal LD50 for rabbits is 89 mg/kg. It is lachrymatory and intensely irritating to mucous membranes and skin.

Formulation. No formulation of the technical material is necessary.

Analysis. Product analysis is by measuring the addition of bromine or acetylation of the hydroxy group. It may be detected by trapping in water or sulphuric acid and oxidising to acrylaldehyde by chromic-sulphuric acid.

Aluminium phosphide

.AL..P   *aluminium phosphide*   AlP   (57·96)

PHHH   *phosphine* H_3P (34·00)

Nomenclature and development. There is no common name. The IUPAC name is aluminium phosphide, in *C.A.* usage aluminum phosphide *[20859-73-8]*. It was introduced as a pesticide in the early 1930's by the Dr. Werner Freyberg Chemische Fabrik under the protection of BP 461 997; USP 2 117 158.

Properties. It forms dark grey or yellowish crystals, m.p. > 1000°. Though stable when dry, it reacts with moist air, violently with acids, producing phosphine and is used to generate the latter toxicant.

Phosphine. Phosphine, PH_3, is a gas, with an objectionable ammoniacal odour, b.p. —87·4°, f.p. —132·5°. It is sparingly soluble in water. It is spontaneously flammable (due to the presence of traces of other hydrides of phosphorus) in air, with a lower explosion limit of 26·1-27·1 g/m³.

Uses. Phosphine is highly insecticidal and is used to fumigate: silos using 1-3 g/t grain; sacked goods at 0·75-1·5 g/m³ sackpile. Dosages vary with the fumigation conditions: *Trop. Stored Prod. Inf.*, 1972, No. 23, p.6. Fumigation takes 3-10 d and should be undertaken only by trained personnel. Adequate airing after this period is checked by a gas detector. The residue is non-poisonous and is removed in the subsequent screening of the grain.

Toxicology. Phosphine is a potent mammalian poison, MAC value accepted in Germany is 0·18 mg/m³, in USA 0·54 mg/m³.

Formulations. Aluminium phosphide formulations are normally based on products containing *c.* 570 g/kg. They include: 'Detia Gas-Ex-B' (W. Freyberg Chemische Fabrik): crepe paper bags (34 g powder, to produce 11 g phosphine); 'Detia Gas-Ex-T', 'Delecia Gastoxin' (Messrs Ernst Freyberg), 'Phostoxin' (Degesch AG), 'Celphos' (Excel Ind.): tablets (3 g, producing 1 g phosphine); 'Phostoxin', pellets (0·6 g, containing 55% aluminium phosphide, 40% ammonium carbamate and 5% aluminium oxide). These formulations are inserted by probe into the material to be fumigated, *e.g.* grain, the moisture content of which should be > 10%. 'Phostoxin' evolves a non-flammable mixture of phosphine, ammonia and carbon dioxide.

Analysis. Product and residue analysis depends upon determining the phosphine liberated by acid treatment. Measurement is: by GLC (B. Berck *et al., J. Agric. Food Chem.*, 1970, **18**, 143; T. Dumas, *J. Assoc. Off. Anal. Chem.*, 1978, **61**, 51); or by trapping the phosphine in potassium permanganate solution or bromine water and determining the phosphate produced (J. W. Elmore & F. R. Roth, *J. Assoc. Off. Agric. Chem.*, 1943, **26**, 559; 1947, **30**, 312; B. L. Giswold *et al., Anal. Chem.*, 1951, **23**, 192; R. B. Bruce *et al., J. Agric. Food Chem.*, 1962, **10**, 18). Phosphine in air may be determined by aspiration through dimercury dichloride solution and noting the change in pH (R. W. D. Taylor, *Chem. Ind. (London)*, 1968, p.1116).

Aluminium tris(ethyl phosphonate)

$$\left[\text{EtO} - \overset{\overset{\text{O}}{\|}}{\underset{\underset{\text{H}}{|}}{\text{P}}} - \text{O}^- \right]_3 \quad \text{Al}^{3+}$$

OPHO&O2 3 &-AL-   *aluminium tris(ethyl phosphonate)* $C_6H_{18}AlO_9P_3$ (354·1)

Nomenclature and development. No common name has yet been agreed for this compound. The IUPAC and *C.A.* name is aluminium tris(ethyl phosphonate) (IUPAC), in *C.A.* usage aluminium tris(ethyl phosphonate) *[39148-24-8]*. It was introduced in 1978 by Rhône-Poulenc Phytosanitaire under the code number 'LS 74 783', trade marks 'Aliette' and 'Mikal' and protected by FP 2 254 276. Its fungicidal properties were first reported by D. Horrière *et al., Phytiatr.-Phytopharm.,* 1977, **26**, 3.

Manufacture and properties. Produced by precipitation from solutions of the water-soluble sodium ethyl phosphonate and an aluminium salt, it is a colourless powder, decomposing > 200°, of negligible v.p. at room temperature. Its solubility at room temperature is 122 g/l water; practically insoluble in most organic solvents. It is stable under normal storage conditions, but unstable in acidic or alkaline solutions and oxidised by strong oxidising agents. It is compatible with most other pesticides and non-corrosive.

Uses. It is a fungicide with systemic activity, being translocated upwards as well as downwards in the plant. It is mainly effective against Phycomycete diseases.

Toxicology. The acute oral LD50 is: for rats 5800 mg a.i./kg; for mice 3700 mg/kg. The acute dermal LD50 for rats is > 3200 mg/kg. It is non-irritant to the skin. In 90-d feeding trials the 'no effect' level was: for rats 5000 mg/kg diet; for dogs 50000 mg/kg diet. Toxicity to honey bees, birds and fish is low. The LC50 (96 h) for rainbow trout is 428 mg/l.

Formulations. These include: 'Aliette', w.p. (800 mg a.i./kg); 'Mikal', w.p. (500 g aluminium salt + 250 g folpet/kg).

Analysis. Technical product and formulation analysis is by iodometric titration. Residue analysis is by GLC after methylation, using a phosphorus-specific detector. Details of the methods are available from Rhône-Poulenc Phytosanitaire.

Ametryne
(Ametryn)

T6N CN ENJ BS1 DMY1&1 FM2 $\qquad$ $C_9H_{17}N_5S$ (227·3)

Nomenclature and development. The common name ametryn is approved by ISO, ANSI, WSSA and JMAF; spellings by BSI (ametryne) and France (amétryne). The IUPAC name is 2-ethylamino-4-isopropylamino-6-methylthio-1,3,5-triazine, in *C.A.* usage *N*-ethyl-*N'*-(1-methylethyl)-6-(methylthio)-1,3,5-triazine-2,4-diamine *[834-12-8]* formerly 2-(ethylamino)-4-(isopropylamino)-6-(methylthio)-*s*-triazine. It was introduced in 1964 by J. R. Geigy S.A. (now Ciba-Geigy AG) under the code number 'G 34162', trade marks 'Gesapax' and 'Evik' (USA only) and protected by SwissP 337019; BP 814948; USP 2909420. Its herbicidal properties were described by H. Gysin & E. Knüsli, *Adv. Pest Control Res.,* 1960, **3**, 289.

Manufacture and properties. Produced by reacting atrazine with methanethiol in the presence of an equivalent of NaOH or by reacting 4-ethylamino-6-isopropylamino-1,3,5-triazine-2-thiol with a methylating agent in the presence of NaOH, ametryne forms colourless crystals, m.p. 84-86° (recryst. from light petroleum), v.p. 8·4 x 10^{-7} mmHg at 20°. Its solubility at 20° is 190 mg/l water; it is readily soluble in organic solvents. It is stable in slightly acidic or alkaline media, but is converted into the herbicidally inactive 6-hydroxy derivative in strongly acidic or alkaline media.

Uses. Ametryne is a selective herbicide used pre- and post-em. to control broad-leaved and grassy weeds in pineapple and sugarcane at 2-4 kg a.i./ha; in banana, citrus, potatoes, corn (directed spray), coffee, tea and oil palms at 1-5 kg/ha. In pineapple, sugarcane and sisal it is also used in combination with atrazine ('Gesapax combi'). The degradation of ametryne and other triazine herbicides is discussed by E. Knüsli *et al.* in *Degradation of Herbicides,* p.51.

Toxicology. The acute oral LD50 for rats is 1100-1750 mg/kg; the acute dermal LD50 for rats is > 3100 mg/kg. No difference was observed between treated and control groups when fed for 2 y to rats and dogs at 1000 mg/kg diet. It is slightly toxic to fish.

Formulations. These include: 'Gesapax 50' w.p. (500 g a.i./kg), 'Gesapax 80' w.p. (800 g/kg); 'Gesapax 500 FW', flowable. Also in combination with other herbicides, e.g. with atrazine ('Gesapax combi') or with simazine ('Gesatop Z', for tropical crops).

Analysis. Product analysis is by titration with perchloric acid in acetic acid or by GLC with internal standard: A. H. Hofberg *et al., J. Assoc. Off. Anal. Chem.,* 1973, **56**, 586; *CIPAC Handbook,* 1978, **1A**, in press; FAO Specification CP/61. Residues are determined by GLC, TLC or by spectrophotometry: particulars from Ciba-Geigy AG; see also: A. M. Mattson *et al., J. Agric. Food Chem.,* 1965, **13**, 120; K. Ramsteiner *et al., J. Assoc. Off. Anal. Chem.,* 1974, **57**, 192; E. Knüsli, *Anal. Methods Pestic., Plant Growth Regul. Food Addit.,* 1964, **4**, 13; T. H. Byast *et al., Agric. Res. Counc. (G.B.) Weed Res. Organ., Tech. Rep.,* No. 15 (2nd Ed.), p.40.

Aminocarb

$$Me_2N{-}\underset{\underset{}{Me}}{\bigcirc}{-}O.\overset{\overset{O}{\parallel}}{C}.NHMe$$

1N1&R B1 DOVM1 $C_{11}H_{16}N_2O_2$ (208·3)

Nomenclature and development. The common name aminocarb is approved by BSI and ISO; also known as ENT 25 784. The IUPAC name is 4-dimethylamino-*m*-tolyl methyl-carbamate (I) also known as 4-dimethylamino-3-methylphenyl methylcarbamate (II), in *C.A.* usage (II) *[2032-59-9]* formerly (I). It was introduced in 1963 by Bayer AG under code numbers 'Bayer 44 646' and 'A363', trade mark 'Matacil', and protected by DBP 1 145 162. Its development was first described by G. Unterstenhöfer, *Meded. Landbouwhogesch. Opzoekingsstn. Staat Gent,* 1963, **28,** 758.

Properties. It is a colourless crystalline solid, m.p. 93-94° and practically non-volatile. It is slightly soluble in water; moderately soluble in aromatic solvents; soluble in polar organic solvents. It is non-compatible with highly alkaline pesticides.

Uses. Aminocarb is a non-systemic insecticide, main use against lepidopterous larvae and other biting insects, but has some acaricidal activity and is an effective molluscicide; generally recommended at 75 g/100 l.

Toxicology. The acute oral LD50 for rats is about 50 mg/kg; the acute intraperitoneal LD50 for rats is 21 mg/kg; the acute dermal LD50 for rats is 275 mg/kg. Rats fed 2 y at 200 mg/kg diet showed no symptoms of poisoning. Hazardous to honey bees.

Formulation. A w.p. (500-750 g a.i./kg).

Analysis. Product and residue analysis is by u.v. spectroscopy: particulars from Bayer AG; see also: H. Niessen & H. Frehse, *Pflanzenschutz-Nachr. (Am. Ed.),* 1963, **16,** 205. Residues may be determined by GLC after hydrolysis and acylation or reaction with 1-fluoro-2,4-dinitrobenzene: R. J. Argauer, *J. Agric. Food Chem.,* 1969, **17,** 888; E. R. Holden *et al., ibid.,* p.56; L. I. Butler & L. M. McDonough, *ibid.,* 1968, **16,** 403.

Aminotriazole

(Amitrole)

T5MN DNJ CZ $C_2H_4N_4$ (84·08)

Nomenclature and development. The trivial name aminotriazole is accepted in lieu of a common name by BSI, in France and USSR; the common name amitrole is approved by ISO, ANSI and WSSA—exception JMAF (ATA); also known as ENT 25445. The IUPAC name is 1,2,4-triazol-3-ylamine formerly 3-amino-1,2,4-triazole, in *C.A.* usage 1*H*-1,2,4-triazol-3-amine *[61-82-5]* formerly 3-amino-*s*-triazole. It was introduced as a herbicide in 1955 by Amchem Products, Inc. under the trade mark 'Weedazol' and protected by USP 2670282. Its herbicidal properties were first reported by R. Behrens, *Proc. North Cent. Weed Control Conf.,* 1953, p.61.

Manufacture and properties. Produced by the condensation of formic acid with aminoguanidine, it is a colourless crystalline powder, m.p. 157-159°. Its solubility at 25° is: 280 g/l water; insoluble in non-polar solvents, diethyl ether, acetone; at 75°, 260 g/kg ethanol. It forms salts with most acids or bases and is a powerful chelating agent. It is somewhat corrosive to iron, aluminium and copper.

Uses. Aminotriazole is a non-selective herbicide absorbed by roots and leaves and translocated, inhibiting chlorophyll formation and regrowth from buds. It is used around established apple and pear trees between harvest and the following summer; as a non-selective herbicide before planting wheat, oats, maize, potatoes, kale and oilseed rape, on fallow land and in other non-crop situations. Its activity is enhanced by the addition of ammonium thiocyanate. General reviews have been published (E. Kröller, *Residue Rev.,* 1966, **12,** 163; M. C. Carter, *Herbicides: Chemistry, Degradation and Mode of Action,* 1975, **1,** p.377).

Toxicology. The acute oral LD50 for rats is 1100-24600 mg/kg; the acute dermal LD50 for rats is > 10000 mg/kg. In 476-d feeding trials rats receiving 50 mg/kg diet suffered no effect on growth or food intake, but the male rats developed an enlarged thyroid after 90 d. Rats fed 120 d at 500 mg/kg diet and returnèd to normal aminotriazole-free diet 14 d before sacrifice, appeared to have normal thyroids (T. H. Jukes & C. B. Schaeffer, *Science,* 1960, **132,** 296); it is doubtful whether this antithyroid action is carcinogenic (E. B. Ashwood, *J. Am.Med. Assoc.,* 1960, **172,** 1319; T. H. Jukes & C. B. Shaeffer, *loc. cit.).*

Formulations. These include: 'Weedazol', w.s.p. (500 g a.i./kg); 'Weedazol TL', a mixture of aminotriazole + ammonium thiocyanate. There are numerous mixtures with one (or more) other herbicides.

Analysis. Product analysis is by acid-base titration of a dimethylformamide extract *(CIPAC Handbook,* in press); other components of mixtures being determined by the appropriate methods *(loc. cit.).* Residues in soils may be determined by extraction with water, methanol or aqueous ethanol and, after clean-up, diazotising and coupling with 4-amino-5-hydroxynaphthalene-2,6-disulphonic acid and colorimetry at 538 nm (R. A. Herrett & A. J. Linck, *J. Agric. Food Chem.,* 1961, **9,** 466) or with *N*-(1-naphthyl)ethylenediamine with colorimetry at 455 nm (J. Burke & R. W. Storherr, *J. Assoc. Off. Agric. Chem.,* 1961, **44,** 196). See also: G. L. Sutherland, *Anal. Methods Pestic., Plant Growth Regul. Food Addit.,* 1964, **4,** 17.

Amitraz

1R C1 DNU1N1&1UNR B1 D1 $C_{19}H_{23}N_3$ (293·4)

Nomenclature and development. The common name amitraz is approved by BSI, ISO, ANSI and BPC; also known as ENT 27967. The IUPAC name is *N*-methylbis(2,4-xylyliminomethyl)amine, previously known as *N,N*-di-(2,4-xylyliminomethyl)methyl-amine and 2-methyl-1,3-di-(2,4-xylylimino)-2-azapropane, in *C.A.* usage *N'*-(2,4-dimethyl-phenyl)-*N*-[[(2,4-dimethylphenyl)imino]methyl]-*N*-methylmethaniminamide *[33089-61-1]* formerly *N*-methyl-*N'*-2,4-xylyl-*N*-(*N*-2,4-xylylformimidoyl)formamidine. Its acaricidal properties were first described by B. H. Palmer *et al., Proc. Int. Congr. Acarol., 3rd,* 1971, p.687 (for veterinary use) and by D. M. Weighton *et al., Meded. Fac. Landbouwwet. Rijksuniv. Gent,* 1972, **37,** 765 (for crop use). It was introduced in 1973 by The Boots Co. Ltd, under the code number 'BTS 27419', trade marks 'Taktic', 'Mitac' (The Boots Co. Ltd) and 'Triatox' (Wellcome). It is protected by BP 1 327935.

Manufacture and properties. Produced by reaction of 2,4-xylidine, ethyl orthoformate and methylamine, it forms colourless monoclinic needles, m.p. 86-87°: I. R. Harrison *et al., Pestic Sci.,* 1972, **3,** 679; 1973, **4,** 901. Its solubility at room temperature is: *c.* 1 mg/l water; > 300 g/l acetone or toluene. It is unstable at pH < 7 and a slow deterioration of the *moist* compound occurs on prolonged standing. It is compatible with commonly-used pesticides.

Uses. Amitraz is an insecticide and acaricide effective against a wide range of phytophagous mites and insects. Against mites, it is effective at 20-60 g a.i./100 l depending on the species. All stages of mites are susceptible. It is effective against *Psylla pyricola, Bemisia tabaci* and the eggs of various species of Lepidoptera such as *Heliothis zea, Spodoptera littoralis* and *Pectinophora gossypiella* at 300-700 g/ha. It is relatively non-toxic to predatory insects. Its main veterinary uses are against the ticks and mites of cattle and sheep, including those strains resistant to other chemical classes of ixodicide, and the compound persists long enough on hair and wool to control all stages of the parasite. Amitraz has also been shown to possess activity against sarcoptic mange mite in pigs: A. J. Griffiths, *Proc. Br. Insectic. Fungic. Conf., 8th.,* 1975, **2,** 557. Used as a spray or dip, its effective concentration lies between 5-100 g/100 l depending on species of parasite and treatment intervals. It penetrates *Boophilus microplus* rapidly, amitraz and *N'*-methyl-*N'*-2,6-xylylformamidine formed from it *in vivo* being toxic to the pest: C. A. Schunter & P. G. Thompson, *Aust. J. Biol. Sci.,* 1978, **31,** 141.

Toxicology. The acute oral LD50 is: for rats 800 mg/kg; for mice *c.* 1600 mg/kg. The acute dermal LD50 for rats is > 1600 mg/kg. No pathological effect was observed in 2-y trials on rats receiving 200 mg/kg diet, or in dogs dosed at 1 mg/kg daily. Relatively non-toxic to honey bees. The LC50 (48 h) for trout is 2·7-4·0 mg/l, for Japanese carp 1·17 mg/l; the LC50 (96 h) for bluegill is 1·3 mg/l, for harlequin 3·2-4·2 mg/l. The LC50 (8 d) for mallard duck is 7000 mg/l, for Japanese quail 1800 mg/l.

Formulations. These include: e.c. (200 g a.i./l) for crop protection, (125 g/l) for animal use; d.p. (250 and 500 g a.i./kg).

Analysis. Product analysis is by GLC, details from The Boots Co. Ltd.

15

Ammonium sulphamate

$$H_2N.SO_2.O^- \quad NH_4^+$$

ZSWQ &ZH

$H_6N_2O_3S$ (114·1)

Nomenclature and development. The IUPAC name ammonium sulphamate is accepted in lieu of a common name by ISO—exception WSSA (AMS). The *C.A.* name is monoammonium sulfamate *[7773-06-0]*. It was introduced as a herbicide in 1945 by E. I. duPont de Nemours & Co. Inc. under the trade mark 'Ammate' and protected by USP 2277744, and later by Albright Wilson (Mfg) Ltd under the trade mark 'Amcide'.

Manufacture and properties. Produced by converting sulphamic acid, made by the action of fuming sulphuric acid on urea, to the ammonium salt (USP 2102350; 2487480) or by the action of non-gaseous sulphur trioxide on liquid ammonia (USP 2426420), it forms colourless plates, m.p. 130°. Its solubility at 25° is 2·16 kg/kg water; soluble in formamide, glycerol. A solution of 50 g/kg water has pH 5·2. The technical product, m.p. 131-132°, is $\geqslant 97\%$ pure. It is decomposed at 160° to non-flammable gases and acts as a flame retardant. It forms addition products with aldehydes and is readily oxidised by bromine and chlorine. It is somewhat corrosive to mild steel.

Uses. Ammonium sulphamate is a non-selective herbicide of particular value for the control of most woody plants when used as an overall spray at 60 g a.i./l and for the control of poison ivy in fruit orchards by localised application. For basal bark, cut stump, or frill-girdle treatment it is used at 400 g/l, or by direct application of crystals.

Toxicology. The acute oral LD50 for rats is 3900 mg/kg. There was no influence on the circulation or respiration of rats by the intravenous injection of 100 mg/kg (A. M. Ambrose, *J. Ind. Hyg. Toxicol.*, 1943, **25**, 26).

Formulation. 'Ammate', w.s.p. ($\geqslant 800$ g a.i./kg); 'Amicide', w.s.p. (970 g/kg).

Analysis. Product analysis is by titration with sodium nitrate (W. W. Bowler & E. A. Arnold, *Anal. Chem.*, 1947, **19**, 336). Residues may be determined by colorimetry (H. L. Pease *et al., J. Agric. Food Chem.*, 1966, **14**, 140) or measurement of the nitrogen evolved on reaction with nitrite (W. N. Carson, *Anal. Chem.*, 1951, **23**, 1016).

Ancymidol

T6N CNJ EXQR DO1&- AL3TJ· $C_{15}H_{16}N_2O_2$ (256·3)

Nomenclature and development. The common name ancymidol is approved by BSI, ANSI and by ISO. The IUPAC name is α-cyclopropyl-4-methoxy-α-(pyrimidin-5-yl)benzyl alcohol, in *C.A.* usage α-cyclopropyl-α-(4-methoxyphenyl)-5-pyrimidinemethanol *[12771-68-5]*. Ancymidol was introduced in 1971 by Elanco Products, a division of Eli Lilly & Co., under the code number 'EL-531', the trade marks 'A-Rest' and 'Reducymol' and protected by BP 1 218 623.

Manufacture and properties. Produced by the condensation of pyrimidin-5-yllithium with cyclopropyl 4-methoxyphenyl ketone, it is a colourless crystalline solid m.p. 110-111°. Its solubility at 25° is *c.* 650 mg/l water and the solution is stable except when acidified (⩽ pH 4). Ancymidol is freely soluble in acetone, methanol, ethyl acetate, chloroform, 2-methoxyethanol, acetonitrile; moderately soluble in aromatic hydrocarbons; but only slightly soluble in saturated hydrocarbon solvents. The technical material has v.p. $< 1 \times 10^{-6}$ mmHg at 50°.

Uses. Ancymidol is a plant growth regulator which reduces internode elongation in a wide range of species: E. E. Tschabold *et al., Plant Physiol.,* 1971, **48,** 519; the growth inhibition may be reversed by gibberellic acid: A. C. Leopold, *ibid.,* p.537. Ancymidol has been evaluated on a range of commercial greenhouse plants including many varieties of chrysanthemum, poinsettia, dahlia, tulip, Easter lily and other ornamentals with excellent results. Delay of 1-3 d in flower maturation may occur at high application rates, but flower development is not otherwise affected.

Toxicology. The acute oral LD50 is: for adult rats 4500 mg/kg; for mice 5000 mg/kg; for chickens, dogs and monkeys > 500 mg/kg. There was no skin irritation but eye irritation was moderate. In 90-d feeding trials rats and dogs receiving 8000 mg/kg diet suffered no ill-effect. The 'no effect' level in teratology studies in rats was 8000 mg/kg diet. LC50 values were for fingerlings of: bluegill 146 mg/l; rainbow trout 55 mg/l; goldfish > 100 mg/l.

Formulation. An aqueous solution (0·25 g/l).

Analysis. Product and residue analysis is by GLC with FID: R. Frank & E. Q. Day, *Anal. Methods Pestic. Plant Growth Regul.,* 1976, **8,** 475. Methods for the determination of ancymidol in formulations and in soil are available from Elanco Products.

Anthraquinone

L   C666   BV   IVJ $C_{14}H_8O_2$ (208·2)

Nomenclature and development. The IUPAC name anthraquinone (I) is accepted in lieu of a common name by BSI and ISO; in *C.A.* usage 9,10-anthracenedione *[84-65-1]* formerly (I). It has long been in use as a bird repellant (*Hoefchen-Briefe (Engl. Ed.)*, 1951, **4**, 227; *Chem. Eng. News*, 1955, **33**, 4998) and is marketed by Bayer AG under the trade mark 'Morkit'.

Manufacture and properties. Produced by the oxidation of anthracene or the condensation of phthalic anhydride with benzene, it forms yellow-green crystals, m.p. 285°, and sublimes, d_4^{40} 1·42-1·44. It is practically insoluble in water; solubility at 20°: 1 g/kg ethanol; 2·6 g/kg benzene; 6·1 g/kg chloroform; 1 g/kg diethyl ether.

Uses. Anthraquinone is used as a seed treatment to deter attack by birds, in particular rooks, being formulated as seed dressings, w.p. or in conjunction with other seed protectants.

Toxicology. It is non-poisonous, mice being unharmed by single oral doses of 5000 mg/kg.

Formulations. These include: 'Morkit' seed dressing (250 g a.i./kg); w.p. (500 g/kg).

Analysis. Particulars of a gravimetric method are available from Bayer AG.

18

Antu

NH.CS.NH$_2$

L66J BMYZUS   C$_{11}$H$_{10}$N$_2$S (202·3)

Nomenclature and development. The common name antu is approved by BSI, ISO and JMAF. The IUPAC name is 1-(1-naphthyl)-2-thiourea (I), in *C.A.* usage 1-naphthalenyl-thiourea *[86-88-4]* formerly (I); also known as σ-naphthylthiourea. Its toxicity to rodents was discovered during a search for a derivative of thiourea (used in genetic studies) more acceptable to rats (C. F. Richter, *J. Am. Med. Assoc.,* 1945, **129**, 927; *Proc. Soc. Exp. Biol. Med.,* 1946, **63**, 364), a discovery protected by USP 2390848.

Manufacture and properties. Produced by the reaction of 1-naphthyl isothiocyanate with ethanolic ammonia, it forms colourless crystals, m.p. 198°. The technical product is a blue-grey powder, its solubility at room temperature is: 600 mg/l water; 24·3 g/l acetone; 86 g/l triethyleneglycol. It is stable on exposure to sun and air.

Uses. Antu is a rodenticide specific for adult *Rattus norvegicus* for which the toxic dose is 6-8 mg/kg. It is less toxic to other *Rattus* spp. and tolerance is developed in rats by repeated administration of sub-lethal doses.

Toxicology. It induces vomiting in dogs. It has been withdrawn from use, in some countries, because of the carcinogenicity of naphthylamines present as impurities.

Formulations. These include: baits (10-30 g/kg) in suitable protein- or carbohydrate-rich materials; tracking powder (200 g/kg).

Analysis. Product analysis is by extraction with hot ethanol, reaction with silver nitrate and titration of the liberated nitric acid. (*CIPAC Handbook,* 1970, **1**, 16). Residues may be determined by addition of bromine to an acetic acid solution, and the violet colour produced in alkali measured colorimetrically (E. Bremais & K. G. Bergner, *Pharm. Zentralhalle,* 1950, **89,** 115).

Arsenous oxide

$$As_2O_3$$

.AS2.O3 As_2O_3 (197·8)

Nomenclature and development. The traditional chemical name arsenous oxide is accepted by ISO in lieu of a common name; also known as arsenic trioxide, arsenious oxide and by the trivial name white arsenic. The IUPAC name is diarsenic trioxide, in *C.A.* usage arsenic oxide (As_2O_3) *[1327-53-3]*. It has been used as a rodenticide since the 16th century.

Manufacture and properties. Produced by roasting diarsenic trisulphide and purified by sublimation, it is a colourless solid which exists in 3 allotropic forms. The amorphous form is unstable, reverting to the octahedral form, m.p. 272°, subliming 125-150°; the rhombic form has m.p. 312°, v.p. 66·1 mmHg at 312°. Its solubility at 16° is 17 g/l water; practically insoluble in ethanol, chloroform, diethyl ether. It is stable in air but slowly oxidised in acid media. It dissolves in alkali to form arsenites.

Uses. Arsenous oxide is used as a rodenticide in baits, containing 100 g/kg soaked wheat or cereal offals, to control *Rattus norvegicus, R. rattus* and *Mus musculus*. It is also used as a sheep dip to control ectoparasites.

Toxicology. The acute oral LD50 is: for rats 180-200 mg a.i. (in carbohydrate or protein)/kg, 300 mg (in bacon fat)/kg, 20 mg (in aqueous solution)/kg (E. W. Packman *et al., J. Agric. Food Chem.,* 1961, **9**, 271); for mice 34·4-63·5 mg/kg (J. W. E. Harrison, *AMA Arch. Ind. Health,* 1958, **17**, 118). It is extremely toxic to man, the minimum lethal dose is 2 mg/kg (H. O. Calvery, *J. Am. Med. Assoc.,* 1938, **111**, 1722). It is non-cumulative and is eliminated from the animal body in 7-42 d.

Formulation. Paste (103 g/kg).

Analysis. Product analysis is by titration with iodine (*WHO Specifications Insectic.* (2nd Ed.)) or by AOAC methods. Residues may be determined by the Gutzeit method (AOAC Methods) or by titration with bromate (AOAC Methods).

Asulam

$$H_2N—\!\!\!\overset{\displaystyle\bigcirc}{\underset{\displaystyle\bigcirc}{S}}\!\!\!—NH.CO_2Me$$

ZR   DSWMVO1 $C_8H_{10}N_2O_4S$ (230·2)

Nomenclature and development. The common name asulam is approved by BSI, ISO, ANSI, WSSA and JMAF—exception Germany. The IUPAC name is methyl 4-amino-phenylsulphonylcarbamate formerly methyl 4-aminobenzenesulphonylcarbamate, in *C.A.* usage methyl [(4-aminophenyl)sulfonyl]carbamate *[3337-71-1]* formerly methyl sulfanilylcarbamate. It should be stated which salt is present, *e.g.* the potassium, *C.A.* Registry No. *[24951-45-9]*, or sodium *[2302-17-2]* salts. It was introduced in 1968 by May & Baker Ltd under the code number 'M & B 9057', trade mark 'Asulox' and protected by BP 1 040 541. Its herbicidal properties were described by H. J. Cottrell & B. J. Heywood, *Nature (London)*, 1965, **207**, 655.

Manufacture and properties. Produced by the reaction of 4-aminobenzenesulphonamide, after protection of the amino group, with methyl chloroformate in the presence of an acid-binding agent, it forms colourless crystals m.p. 143-144°. Its solubility at 20-25° is: 4 g/l water; 300 g/l acetone; 290 g/l methanol; < 20 g/l hydrocarbons or chlorinated hydrocarbons. The presence of the sulphonyl and and ester groups enhances the acidic nature of the NH group pK_a 4·82 and water-soluble salts are formed by treatment with alkali hydrogen carbonates. The calcium and magnesium salts are not precipitated by hard water. Its stability in storage is good and its corrosive properties are slight or absent.

Uses. Asulam is absorbed by leaves and roots causing, in susceptible plants, a slow chlorosis. It does not inhibit the Hill reaction at low concentrations but appears to interfere with cell division and expansion. It is used for the post-em. control of *Rumex* spp. in pasture and deciduous fruit orchards at 1·0-1·7 kg a.i./ha; for pre- and post-em. control of weed grasses in sugarcane at 3·36-3·64 kg/ha; for post-em. control of grasses in tropical tree-plantation crops at 2·7-3·36 kg/ha; for post-em. control of *Pteridium aquilinum* in pastures and forestry at 4-6 kg/ha; for post-em. control of *Avena fatua* in linseed at 1 kg/ha; and for control of broad-leaved weeds and grasses pre- and post-em. of the crop oilseed poppy at 3-7 kg/ha.

Toxicology. The acute oral LD50 of the sodium salt for mice, rats and rabbits is > 4000 mg/kg; the acute dermal LD50 for rats is > 1200 mg/kg. In 90-d feeding trials rats receiving 400 mg asulam-sodium/kg diet suffered no significant effects. The acute oral LD50 for chickens, pigeon and quail is > 2000 mg/kg; the LC50 (96 h) for rainbow trout, channel catfish and goldfish is > 5000 mg/l.

Formulation. A solution of the sodium salt (400 g asulam/l).

Analysis. Product and residue analysis is by hydrolysis, diazotisation and coupling with a suitable component to produce a dye that can be estimated colorimetrically; alternatively titration with sodium nitrite plus lithium methoxide can be used. See: C. H. Brockelsby & D. F. Muggleton, *Anal. Methods Pestic. Plant Growth Regul.*, 1973, **7**, 497.

Atrazine

T6N CN ENJ BMY1&1 DM2 FG $C_8H_{14}ClN_5$ (215·7)

Nomenclature and development. The common name atrazine is approved by BSI, ISO, ANSI, WSSA and JMAF. The IUPAC name is 2-chloro-4-ethylamino-6-isopropyl-amino-1,3,5-triazine, in *C.A.* usage 6-chloro-*N*-ethyl-*N'*-(1-methylethyl)-1,3,5-triazine-2,4-diamine *[1912-24-9]* formerly 2-chloro-4-(ethylamino)-6-(isopropylamino)-*s*-triazine. It was introduced in 1958 by J. R. Geigy S.A. (now Ciba-Geigy AG) under code number 'G 30027', trade marks 'Gesaprim', 'Primatol' and protected by SwissP 329 277; BP 814 947; USP 2 891 855. Its herbicidal properties were first recorded by H. Gysin & E. Knüsli, *Proc. Int. Congr. Crop Prot., 4th,* Hamburg, 1957.

Manufacture and properties. Produced by the reaction, with 2,4,6-trichloro-1,3,5-triazine, of one equivalent of ethylamine, followed by one equivalent of isopropyl-amine in the presence of an acid-binding agent, it forms colourless crystals, m.p. 175-177°, v.p. 3·0 x 10^{-7} mmHg at 20°, Its solubility at 20° is 30 mg/l water; readily soluble in methanol, chloroform. It is stable in neutral, slightly acidic or basic media, but alkali or mineral acids at higher temperatures hydrolyse it to the herbicidally-inactive hydroxy derivative.

Uses. Atrazine is used as a selective pre- and post-em. herbicide on many crops including maize, sorghum, sugarcane, pineapple and asparagus at 1-4 kg a.i./ha. Higher rates serve for general weed control.

Toxicology. The acute oral LD50 for rats is 1859 to *c.* 3080 mg/kg; the acute dermal LD50 for rabbits is 7500 mg/kg. It is slightly toxic to fish.

Formulations. These include: w.p., *e.g.* 'Gesaprim 50', 'Primatol A 50' (500 g a.i./kg), and 'Gesaprim 80' and 'Primatol A 80' (800 g/kg); 'Gesaprim 500 FW' (flowable). Combinations include: 'Fanoprim' (atrazine and bromofenoxim); 'Gesapax combi' (atrazine + ametryne); 'Gesaprim combi' (atrazine + terbutryne); 'Gesaprim S' (atrazine + simazine); 'Primagram' and 'Primetextra' (atrazine + metolachlor).

Analysis. Product analysis is by titration of ionic chloride liberated by treatment with morpholine (H. P. Bosshardt, *et al., J. Assoc. Off. Anal. Chem.,* 1971, **54**, 749); or GLC with internal standard: R. T. Murphy *et al., ibid.,* p.697; *CIPAC Handbook,* 1979, **1A,** in press; FAO Specification (CP/61). Residues are determined by GLC, TLC or spectrophotometry: particulars from Ciba-Geigy AG. See also: K. Ramsteiner *et al., J. Assoc. Off. Anal. Chem.,* 1974, **57,** 92; E. Knüsli, *Anal. Methods Pestic., Plant Growth Regul. Food Addit.,* 1964, **4,** 33; *Anal. Methods Pestic. Plant Growth Regul.,* 1972, **6,** 600; T. H. Byast *et al., Agric. Res. Counc. (G.B.) Weed Res. Organ., Tech. Rep.,* No. 15 (2nd Ed.), p.40.

Azinphos-ethyl

$$(EtO)_2P.S.CH_2-N$$

T66   BNNNVJ   D1SPS&O2&O2 $C_{12}H_{16}N_3O_3PS_2$ (345·4)

Nomenclature and development. The common name azinphos-ethyl is approved by BSI and ISO—exception USSR (triazotion); also known as ENT 22 014. The IUPAC name is S-(3,4-dihydro-4-oxobenzo[d]-[1,2,3]-triazin-3-ylmethyl)   O,O-diethyl   phosphorodithioate, in C.A. usage O,O-diethyl S-[(4-oxo-1,2,3-benzotriazin-3(4H)-yl)methyl] phosphoro-dithioate [2642-71-9] formerly O,O-diethyl phosphorodithioate S-ester with 3-(mercapto-methyl)-1,2,3-benzotriazin-4(3H)-one. It was developed by W. Lorenz and introduced in 1953 by Bayer AG under the code numbers 'Bayer 16 259' and 'R 1513'; the trade mark 'Gusathion A' (in Federal Republic of Germany: 'Gusathion H', 'Gusathion K forte') and protected by USP 2 758 115; DBP 927 270. Its insecticidal activity was described by E. E. Ivy et al., J. Econ. Entomol., 1955, **48**, 293.

Manufacture and properties. Produced by the reaction of 3,4-dihydrobenzo[d]-[1,2,3]-triazin-4-one with paraldehyde and thionyl chloride, and the resulting 3-chloro-methyl derivative with O,O-diethyl hydrogen phosphorodithioate in the presence of an acid-binding agent such as sodium hydrogen carbonate, it forms colourless needles, m.p. 53°, b.p. 111°/1 x 10^{-3} mmHg, v.p. 2·2 x 10^{-7} mmHg at 20°, d_4^{20} 1·284, n_D^{53} 1·5928. It is of negligible solubility in water, but is soluble in organic solvents except light petroleum and aliphatic hydrocarbons. It is thermo-stable but readily hydrolysed by alkali.

Uses. Azinphos-ethyl is a non-systemic insecticide and acaricide with good ovicidal properties and long persistence. It is used on a large range of crops, including cotton, top fruit, citrus, vegetables, potatoes, beets, grapes, hops, tobacco, coffee, rice, cereals, oilseed crops, forage crops and some ornamentals. Rates are generally 400 g a.i./ha. Metabolism in beans has been studied (W. Steffens & J. Wieneke, Pflanzenschutz-Nachr. (Am. Ed.), 1976, **29**, 35).

Toxicology. The acute oral LD50 for rats is 12·5-17·5 mg/kg; the acute dermal LD50 for rats is 250 mg/kg (2 h); the intraperitoneal LD50 for rats is > 7·5 mg/kg. In 90-d feeding trials rats receiving 2 mg/kg diet showed no symptom of poisoning.

Formulations. These include: e.c. (200-400 g a.i./l); w.p. (250-400 g/kg); ULV (500 g/l).

Analysis. Product analysis is by alkaline hydrolysis, and the liberated O,O-diethyl hydrogen phosphorodithioate complexed with copper(II) ions, extracted and measured at 420 nm (CIPAC Handbook, 1970, **1**, 18; FAO Specification (CP/41)). Residues are measured by GLC, details are available from Bayer AG.

Azinphos-methyl

$(MeO)_2 \overset{\underset{\displaystyle \|}{S}}{P}.S.CH_2$—[structure: 3,4-dihydro-4-oxobenzo[d]-[1,2,3]-triazin-3-yl]

T66 BNNNVJ D1SPS&O1&O1 $C_{10}H_{12}N_3O_3PS_2$ (317·1)

Nomenclature and development. The common name azinphos-methyl is approved by BSI and ISO—exception USSR (metiltriazotion); also known as ENT 23 233. The IUPAC name is *S*-(3,4-dihydro-4-oxobenzo[*d*]-[1,2,3]-triazin-3-ylmethyl) *O,O*-dimethyl phosphorodithioate, in *C.A.* usage *O,O*-dimethyl *S*-[(4-oxo-1,2,3-benzotriazin-3(4*H*)-yl)-methyl] phosphorodithioate *[86-50-0]* formerly *O,O*-dimethyl phosphorodithioate *S*-ester with 3-(mercaptomethyl)-1,2,3-benzotriazin-4(3*H*)-one. It was developed by W. Lorenz and introduced in 1953 by Bayer AG under the code numbers 'Bayer 17 147' and 'R 1582', trade marks 'Gusathion M' and 'Guthion' (Bayer AG), and protected by USP 2758115; DBP 927270. It was first described by E. E. Ivy *et al., J. Econ. Entomol.*, 1955, **48**, 293.

Manufacture and properties. In its production the process used is as for azinphos-ethyl (p.23) but substituting *O,O*-dimethyl hydrogen phosphorodithioate for the diethyl ester. It forms colourless crystals, m.p. 73-74°, v.p. negligible at 20°, d_4^{20} 1·44, n_D^{76} 1·6115. Its solubility at room temperature is 33 mg/l water; it is soluble in most organic solvents. It is unstable at temperatures above 200° and is rapidly hydrolysed by cold alkali and acid.

Uses. Azinphos-methyl is a non-systemic insecticide and acaricide of long persistence, chiefly effective against biting and sucking insect pests. It is mainly used on top fruit, vegetables, grapes, maize, cotton, citrus and some ornamentals. Residues in plants, soil and water have been reviewed: C. A. Anderson *et al., Residue Rev.,* 1974, **51**, 123; see also W. Steffens & J. Wieneke, *Pflanzenschutz-Nachr. (Am. Ed.),* 1976, **29**, 1; J. Wieneke & W. Steffens, *ibid.,* p.18.

Toxicology. The acute oral LD50 is: for female rats 16·4 mg/kg; for male guinea-pigs 80 mg/kg. No symptom of poisoning occurred in rats receiving 2·5 mg/kg diet for 2 y.

Formulations. These include: e.c. (200 g a.i./l); w.p. (200, 250, 400 and 500 g/kg); dust (50 g/kg); w.p. with added systemic insecticides (*e.g.* demeton-S-methyl sulphone).

Analysis. Product analysis is by alkaline hydrolysis, reacting the liberated *O,O*-dimethyl hydrogen phosphorodithioate with copper(II) ions, extracting the resulting complex whose absorption is measured at 420 nm: M. V. Norris *et al., J. Agric. Food Chem.,* 1954, **2**, 570; *CIPAC Handbook,* 1970, **1**, 25; FAO Specification (CP/41). Residues may be determined by hydrolysis with KOH in propan-2-ol giving anthranilic acid, which is diazotised and coupled with *N*-(1-naphthyl)ethylenediamine and the colour measured at 555 nm: W. R. Meagher *et al., J. Agric. Food Chem.,* 1960, **8**, 282; see also: D. MacDougall, *Anal. Methods Pestic., Plant Growth Regul. Food Addit.,* 1964, **2**, 231; *Anal. Methods Pestic. Plant Growth Regul.,* 1972, **6**, 397.

Aziprotryne

$$N_3$$

MeS—[triazine ring]—NHPr^i

T6N CN ENJ BS1 DNNN FMY1&1 $C_7H_{11}N_7S$ (225·3)

Nomenclature and development. The common name aziprotryne is approved by BSI, ISO and JMAF—spelling in USA, aziprotryn. The IUPAC name is 2-azido-4-isopropylamino-6-methylthio-1,3,5-triazine, in *C.A.* usage 4-azido-*N*-(1-methylethyl)-6-methylthio-1,3,5-triazin-2-amine *[4658-28-0]* formerly 2-azido-4-(isopropylamino)-6-(methylthio)-*s*-triazine. It was introduced in 1967 by Ciba AG (now Ciba-Geigy AG) as an experimental herbicide under the code number 'C 7019', trade mark 'Mesoranil' and protected by BP 1 093 376; FP 1 537 312. It was first reported by D. H. Green *et al., C. R. Journ. Etud. Herbic. Conf. COLUMA, 4th,* 1967, **1**, 1.

Manufacture and properties. Produced by the reaction of 2,4,6-trichloro-1,3,5-triazine with one equivalent of isopropylamine in the presence of sodium hydroxide, followed by sodium methanethiolate and, finally, sodium azide, it is a colourless crystalline powder, m.p. 95°, v.p. 2×10^{-8} mmHg at 20°. Its solubility at 20° is 55 mg/l water.

Uses. Aziprotryne is a pre-em. herbicide effective against a wide range of broad-leaved annual weeds and some grasses. It is used in transplanted brassica crops (excluding cauliflowers) at 1·5-2·5 kg a.i./ha after emergence of the weeds. It is also applied to seeded onions and leeks at 1 kg/ha.

Toxicology. The acute oral LD50 for rats is 3600-5833 mg/kg; the acute dermal LD50 for rats is > 3000 mg/kg. In 90-d feeding trials on rats and dogs the 'no effect' level was > 50 mg/kg daily. It is slightly toxic to fish.

Formulation. A w.p. (500 g a.i./kg).

Analysis. Product analysis is by reaction with sodium iodide in acid solution and titration of the liberated iodine. Residues are determined by GLC using MCD. Details of the methods are available from Ciba-Geigy AG.

Nomenclature and development. The gram-positive bacterium *Bacillus thuringiensis* Berliner, in *C.A.* usage *Bacillus thuringiensis,* was detected in 1902 in dying larvae of *Bombyx mori* Lim. by S. Ishiwata (cited by K. Ishikawa, *Pathology of the Silkworm*) and was first studied and named after isolation from larvae of *Ephestia kuehniella* Zell. by E. Berliner, *Z. Angew. Entomol.,* 1915, **2**, 29. It was first used as a microbial insecticide 'Sporeine' against lepidopterous larvae in 1938 (S. E. Jacobs, *Proc. Soc. Appl. Bacteriol.,* 1950, **13**, 83) and has been developed for the control of these pests by several companies, of which 3 currently market products under the trade names 'Bactospeine' (distributor, Biochem Products Ltd), 'Dipel' (Abbott Laboratories) and 'Thuricide' (Sandoz AG).

Manufacture and properties. Produced by fermentation in undefined media containing molasses, fishmeal, plus essential trace elements, in well-controlled conditions; at sporulation each rod-shaped bacterium also produces a glycoprotein crystal. The dormant spores and crystals are harvested by centrifugation, low temperature evaporation and spray drying. The crystals are insoluble in water and are unstable in alkaline media or in the presence of certain enzymes.

The high molecular weight glycoprotein is broken down by the alkaline gut juices of susceptible insect species, forming smaller units that attack the lining of the larval gut, disrupting osmotic balance, paralysing mouthparts and gut, thus stopping feeding. In some species the spores germinate in the gut and play an important role in pathengenicity, the larvae eventually dying of septicaemia.

An adenine nucleotide, the β-toxin, is produced by many *B. thuringiensis* strains during the logarithmic phase of bacterial growth, but not by the strains in commercial production.

Formulated products are compatible with most insecticides, acaricides, fungicides and plant growth regulators but not with captafol, dinocap, alkaline sprays, *e.g.* Bordeaux mixture, or, under some conditions, with foliar nutrients.

Uses. *B. thuringiensis* is a microbial insecticide effective only against larvae of many lepidopterous species. Its specificity to Lepidoptera and safety to man and the natural enemies of many crop pests make it ideal for use in pest management. It is important to treat the parts of the plant normally attacked by the larvae because the bacteria do not spread. It is non-phytotoxic and is used in cotton, tobacco, soyabeans, pome fruits and ornamentals.

Toxicology. There is no evidence of acute or chronic toxicity in rats, guinea-pigs, mice, dogs, other test mammals, man, fish or birds (R. Fisher & L. Rosner, *J. Agric. Food Chem.,* 1959, **7**, 687). Very slight inhalation and dermal irritation has been observed in test animals, probably due to the physical rather than the biological properties of the *B. thuringiensis* formulation.

Formulations. The products are not markedly different and taking 'Dipel' as a typical example, they include: 'Dipel WP', w.p. (16000 i.u./mg) of spores and glycoprotein crystals, stable at 30% r.h. for ⩾ 3 y; 'Dipel LC', liquid concentrate (4000 i.u./mg; with spray adjuvants) and 'Dipel SC', spray concentrate (9000 i.u./mg; without adjuvants) are intended for aerial LV application and have a shelf life of *c.* 0·5 y; 'Dipel' dusts (160-800 i.u./mg); 'Dipel' baits (320 i.u./mg) used on high value crops where placement is critical.

Analysis. The activity of *B. thuringiensis* is measured in i.u. relative to that of a standard product against *Trichoplusia ni* in a standard bioassay. Assays based on the number of spores are not satisfactory.

Barban

$$NH.CO.O.CH_2C \equiv C.CH_2Cl$$

G2UU2OVMR   CG $C_{11}H_9Cl_2NO_2$ (258·1)

Nomenclature and development. The common name barban is approved by BSI, ISO, ANSI and WSSA—exceptions: Republic of South Africa (barbamate), USSR (chlorinat), JMAF (CBN) and Italy. The IUPAC name is 4-chlorobut-2-ynyl 3-chlorocarbanilate, also known as 4-chlorobut-2-ynyl 3-chlorophenylcarbamate, in *C.A.* usage 4-chloro-2-butynyl 3-chlorophenylcarbamate *[101-27-9]* formerly 4-chloro-2-butynyl *m*-chlorocarbanilate. It was introduced in 1958 by the Spencer Chemical Co. under the code number 'CS-847', the trade mark 'Carbyne' and protected by USP 2906614. Its herbicidal properties were first described by A. D. Brown, *Proc. North Cent. Weed Control Conf.,* 1958, **15**, 98.

Manufacture and properties. Produced by the reaction of 4-chlorobut-2-yn-1-ol with 3-chlorophenyl isocyanate in the presence of pyridine in an inert solvent, it is a crystalline solid, m.p. 75-76°. Its solubility at 25° is 11 mg/l water; readily soluble in benzene, 1,2-dichloroethane; slightly soluble in hexane. It is rapidly hydrolysed by alkali with displacement of the terminal chlorine (50% loss occurs in 58 s at 25° in 1M sodium hydroxide). Hydrolysis in acid gives 3-chloroacrylic acid.

Uses. Barban is a selective post-em. herbicide controlling *Avena fatua* at the 1·5-2·5 leaf stage using 420-840 g a.i./ha without serious damage to seedling wheat, barley, rapeseed, lucerne, clovers, *Elymus junceus* and other grasses, field and broad beans, sugar beet and vining peas.

Toxicology. The acute oral LD50 for rats and mice is 1300-1500 mg/kg. The acute dermal LD50 is: for rats > 1600 mg/kg; for rabbits > 23000 mg/kg. In 2-y feeding trials adverse effects on rats were apparent only at the highest level tested (5000 mg/kg diet). Possible allergic reaction in man by skin contact should be avoided by the use of plastic protective clothing.

Formulations. An e.c. (125 and 250 g a.i./l).

Analysis. Product analysis is by u.v. spectrometry at 277·5 nm after column chromatographic separation from impurities (K. J. Bombaugh & W. C. Bull, *J. Agric. Food Chem.,* 1961, **9**, 386). Residues are determined by hydrolysis to 3-chloroaniline which is measured by GLC with ECD after bromination (R. J. Harris & R. J. Whiteoak, *Analyst (London),* 1972, **97**, 294) or by diazotisation, coupling with *N*-(1-naphthyl)ethylenediamine and colorimetry at 550 nm (J. R. Ridden & T. R. Hopkins, *J. Agric. Food Chem.,* 1961, **9**, 47). See also: T. R. Hopkins, *Anal. Methods Pestic., Plant Growth Regul. Food Addit.,* 1964, **4**, 37.

Benazolin

T56	BNVSJ	B1VQ	IG		*benazolin*	$C_9H_6ClNO_3S$	(243·7)
T56	BNVSJ	B1VO	IG	&-KA-	*benazolin-potassium*	$C_9H_5ClKNO_3S$	(281·8)
T56	BNVSJ	B1VO2	IG		*benazolin-ethyl*	$C_{11}H_{10}ClNO_3S$	(275·4)

Nomenclature and development. The common name benazolin is approved by BSI and ISO—exception Germany. The IUPAC name is 4-chloro-2-oxobenzothiazolin-3-ylacetic acid, in *C.A.* usage 4-chloro-2-oxo-3(*2H*)-benzothiazoleacetic acid *[3813-05-6]* formerly 4-chloro-2-oxo-3-benzothiazolineacetic acid. It was introduced in 1965 by The Boots Co. Ltd as 'Cornox CWK' in New Zealand, and in 1966 under the code number 'RD 7693' in the UK, protected by BP 862226. Its herbicidal properties were first described by E. L. Leafe, *Proc. Br. Weed Control Conf., 7th,* 1964, p.32. Plant growth substances containing the benzothiazole nucleus were first described in 1949 by D. J. Jayne *et al.,* USP 2 468 075, but the assigned structures were shown to be incorrect by R. F. Brookes & E. L. Leafe, *Nature (London),* 1963, **198,** 589.

Manufacture and properties. Cyclisation of 1-(2-chlorophenyl)thiourea yields 2-amino-4-chlorobenzothiazole which is converted to the 2-hydroxyanalogue, reacted with ethyl chloroacetate and hydrolysed to benazolin. It is a colourless crystalline solid, m.p. 193°, with negligible v.p. at 20°. The solubility of the acid at 20° is 600 mg/l water, but alkali metal and diethanolamine salts are readily soluble (600 g benazolin-potassium/l). Technical grade benazolin is *c.* 90% pure, m.p. 189°.

Benazolin is stable except to concentrated alkali; its salts are compatible with similar salt formulations of the phenoxyalkanoic acids. Solutions of the alkali metal salts at pH 8·5-10·0 are non-corrosive to mild steel, but are mildly corrosive to galvanised iron, tin-plate and aluminium.

Uses. Benazolin is a highly specific post-em. translocated growth regulator herbicide effective against *Stellaria media* at 140-210 g a.i./ha, *Galium aparine* at 450 g/ha and *Sinapis arvensis* at 700 g/ha. Synergism with dicamba has been shown, particularly against *Matricaria* spp. and *Anthemis* spp. in cereals: BP 1 243 006. In non-undersown cereals, it is combined with other herbicides such as dicamba + MCPA, dicamba + dichlorprop. In cereals undersown with grass/clover mixtures, it is combined with MCPA, MCPA + MCPB, MCPA + 2,4-DB. For grassland uses, where clover safety is not important, it is combined with dicamba and dichlorprop and, where clover safety is required, with MCPA + 2,4-DB. Benazolin is also used for selective weed control in oilseed rape either alone or in mixture with 3,6-dichloropicolinic acid.

Toxicology. The acute oral LD50 is: for mice 3200 mg/kg; for rats > 4800 mg/kg. In 90-d feeding tests the 'no effect' level is between 300 and 1000 mg/kg daily for rats and *c.* 300 mg/kg daily for dogs. A 30% solution of the potassium salt causes a mild irritation of the skin and eyes of rabbits, while a 1 in 24 dilution is non-irritant.

Formulations. An aqueous solution of the potassium salt, Registry No. *[67338-65-2]* (250 g a.i./l) and as the ethyl ester, Registry No. *[25059-80-7]* (280 g a.i./kg). In mixtures, benazolin is formulated as ethyl ester for use in oilseed rape; as ester, alkali metal or amine salts in cereals, and as alkali metal salts in grassland. Potassium or amine salt formulations are typically aqueous solutions; the ethyl ester can be utilised as either w.p. or e.c.

Analysis. Product and residue analysis is by GLC after conversion to the methyl ester: details available from The Boots Co. Ltd.

Bendiocarb

T56 BO DO CHJ Cl Cl FOVM1 $C_{11}H_{13}NO_4$ (223·2)

Nomenclature and development. The common name bendiocarb is approved by BSI and ISO. The IUPAC name is 2,3-isopropylidenedioxyphenyl methylcarbamate (I), also known as 2,2-dimethylbenzo-1,3-dioxol-4-yl methylcarbamate, in *C.A.* usage 2,2-dimethyl-1,3-benzodioxol-4-yl methylcarbamate *[22781-23-3]* formerly (I). It was introduced in 1971 by Fisons Ltd Agrochemical Division under the code number 'NC 6897', the trade mark 'Ficam' and protected by 'BP 1 220056. Its insecticidal properties and synthesis were described by R. W. Lemon, *Proc. Br. Insectic. Fungic. Conf., 6th,* 1971, **2**, 570; P. J. Brooker *et al., Pestic. Sci.,* 1972, **3**, 735.

Manufacture and properties. Produced by the reaction of 2,3-isopropylidenedioxyphenol with methyl isocyanate, it is a colourless solid, m.p. 129-130°, v.p. 5 x 10⁻⁶ mmHg at 25°. Its solubility at 25° is 40 mg/l water; 350 mg/kg hexane; < 1 g/kg odourless kerosine; 200 g/kg acetone, dioxane, chloroform; 40 g/kg ethanol, benzene. The partition ratio at 25° is 1 : 9 for water/hexane. It is stable to hydrolysis in water at pH 5; 50% decomposition (tested under EPA guidelines) occurs in 4 d at 25° and pH 7, the products being 2,3-isopropylidenedioxyphenol, methylamine and carbon dioxide.

Uses. Bendiocarb is an insecticide acting by cholinesterase inhibition, effective as a contact and stomach poison. It is active against mosquitoes, flies, wasps, ants, fleas, cockroaches and many other industrial and storage pests. In agriculture it is active against Lepidoptera, Coleoptera and Collembola, especially soil pests. It is used as a seed treatment on sugar beet and maize. Granular formulations are used against *Oscinella frit, Agriotes* spp. in Europe and corn rootworm and a variety of turf pests in N. America. A sprayable formulation is also used against turf pests. Other crops where it is being used experimentally include rice, cotton, tobacco, small grains and potatoes. Residues of bendiocarb in sugar beet and maize have been below the limits of detection.

Toxicology. The acute LD50 for most mammals lies in the range 34-64 mg/kg. The acute dermal LD50 for rats is 566-800 mg/kg. The dietary and cumulative toxicity to rats is low. It is very toxic to honey bees. The LC50 for a range of fish species is 0·4-1·8 mg a.i.(as w.p.)/l.

Formulations. These include: 'Ficam W', w.p. (800 g a.i./kg); granules (30 and 100 g/kg); 'Ficam D', dust (10 g/kg); residual spray with sucrose; paint-on bait for fly control; granules for pest control on turf.

Analysis. Product analysis is by GLC with internal standard. Analysis of residues in foodstuffs is based on hydrolysis and reaction with 1-fluoro-2,4-dinitrobenzene, the derivative being determined by GLC with ECD. See also: R. J. Whiteoak *et al., Anal. Methods Pestic. Plant Growth Regul.,* 1978, **10**, 3.

Benfluralin

Et.N.Bu

O_2N — ⬡ — NO_2

CF_3

FXFFR   CNW   ENW   DN4&2

$C_{13}H_{16}F_3N_3O_4$ (335·3)

Nomenclature and development. The common name benfluralin is approved by BSI and ISO—exceptions WSSA (benefin), JMAF (bethrodine). The IUPAC name is *N*-butyl-*N*-ethyl-α,α,α-trifluoro-2,6-dinitro-*p*-toluidine (I) formerly *N*-butyl-*N*-ethyl-2,6-dinitro-4-trifluoromethylaniline, in *C.A.* usage *N*-butyl-*N*-ethyl-2,6-dinitro-4-(trifluoromethyl)-benzenamine *[1861-40-1]* formerly (I). Its herbicidal properties were first described by J. F. Schwer, *Proc. North Cent. Weed Control Conf., 1965* and it was introduced in 1965 by Eli Lilly & Co. under the code number 'EL-110', trade marks 'Balan' 'Bonalan', and protected by USP 3 257 190.

Manufacture and properties. Produced by the nitration of 4-chloro-α,α,α-trifluoro-toluene and reaction with *N*-ethylbutylamine, it is a yellow-orange crystalline solid, m.p. 65-66·5°, v.p. 3·89 x 10^{-4} mmHg at 30°. Its solubility at 25° is < 1 mg/l water; it is readily soluble in most organic solvents, less so in ethanol. It is stable but susceptible to decomposition by u.v. light and is compatible with most other pesticides.

Uses. Benfluralin is incorporated into the soil as a pre-em. herbicide effective for the control of annual grasses and broad-leaved weeds in lettuce, groundnuts, tobacco, chicory, cucumbers, endives, lucerne and other forage crops at rates of 1·0-1·35 kg a.i./ha. At 1·5 kg/ha it controls annual grass weeds in turf. It is of low to moderate persistence.

Toxicology. The acute oral LD50 is: for rats > 10000 mg/kg; for mice > 5000 mg/kg; for rabbits, dogs and chickens > 2000 mg/kg. Skin application of 200 mg/kg and eye application produced no irritation. In 90-d feeding trials the 'no effect' level was: in rats 1250 mg/kg diet; in dogs 500 mg/kg diet. The LC50 to bluegill fingerlings is 0·37 mg/l.

Formulation. An e.c. (180 g a.i./l); granules (25 g/kg).

Analysis. Benfluralin is isolated from products by column chromatography and determined spectroscopically. Residues are analysed by GLC with ECD. See also: T. H. Byast *et al., Agric. Res. Counc. (G.B.) Weed Res. Organ., Tech. Rep.,* No. 15 (2nd Ed.), p.11; W. S. Johnson & R. Frank, *Anal. Methods Pestic. Plant Growth Regul.,* 1976, **8**, 335. Details of analytical procedures and other information are available from Eli Lilly & Co.

Benodanil

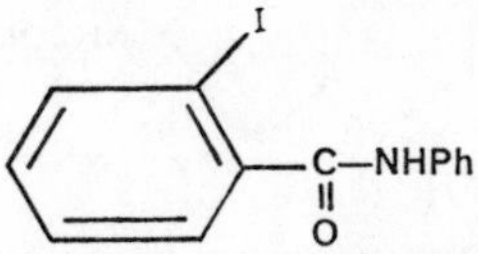

I
C—NHPh
O

IR   BVMR $C_{13}H_{10}INO$ (323·1)

Nomenclature and development. The common name benodanil is approved by BSI and ISO. The IUPAC name is 2-iodobenzanilide (I), in *C.A.* usage 2-iodo-*N*-phenylbenzamide *[15310-01-7]* formerly (I). Its fungicidal properties were first reported by F. Löcher *et al., Meded. Fac. Landbouwwet. Rijksuniv. Gent,* 1974, **39,** 1079; and by E. -H. Pommer *et al.,* at the 39th Deutsche Pflanzenschutztagung, Stuttgart, 1973. It was introduced in 1974 by BASF AG under the code number 'BAS 3170 F'.

Manufacture and properties. Produced by the reaction of 2-iodobenzoic acid, carbonyl chloride and aniline, it is a colourless crystalline solid, m.p. 137° and of negligible v.p. at 20°. Its solubility at 20° is: 20 mg/kg water; 401 g/kg acetone; 77 g/kg chloroform; 93 g/kg ethanol; 120 g/kg ethyl acetate. It is stable at temperatures up to 50° and no hydrolysis was observed in water, acid or alkalies.

Uses. Benodanil is a systemic fungicide used at 1·1 kg a.i./ha to control *Puccinia striiformis* on wheat and barley and *P. hordei* on barley; it is effective against rust diseases of plums, coffee, tobacco, vegetables, carnations, roses and other ornamentals.

Toxicology. The acute oral LD50 for rats and guinea-pigs is > 6400 mg/kg; it is mildly irritating to rabbits' skin and eyes. A 90-d feeding trial indicated that the 'no effect' level is ≥ 100 mg/kg diet.

Formulation. 'Calirus', w.p. (500 g a.i./kg).

Analysis. Products are extracted with tetrahydrofuran and benodanil measured by i.r. spectroscopy. Residues may be determined, after extraction and purification, by GLC with ECD.

Benomyl

$$O = C.NHBu$$

(chemical structure: benzimidazole ring with N–C(=O)NHBu substituent and 2-NHCOMe group)

—NHCOMe

T56   BN   DNJ   BVM4   CMVO1 $C_{14}H_{18}N_4O_3$ (290·3)

Nomenclature and development. The common name benomyl is approved by BSI, ISO, ANSI and JMAF. The IUPAC name is methyl 1-(butylcarbamoyl)benzimidazol-2-ylcarbamate, in *C.A.* usage methyl 1-[(butylamino)carbonyl]-1*H*-benzimidazol-2-ylcarbamate *[17804-35-2]* formerly methyl 1-(butylcarbamoyl)-2-benzimidazolecarbamate. It was introduced in 1967 by E. I. du Pont de Nemours & Co. (Inc.) as an experimental fungicide, code number 'DuPont 1991', trade mark 'Benlate' and protected by DutchP 6 706 331; USP 3 631 176. It was first described by C. J. Delp & H. L. Klopping, *Plant Dis. Rep.*, 1968, **52**, 95.

Manufacture and properties. Produced by condensing *o*-phenylenediamine with NCNHCOOCH$_3$ and then reacting with butyl isocyanate, it is a colourless crystalline solid with a faint acrid odour. It decomposes before melting and is non-volatile at room temperature. Its solubility at pH 7 and 20° is 3·8 mg/l water; readily soluble in chloroform (*c.* 90 g/kg); significantly less soluble in other common organic solvents and oils. In some solvents dissociation occurs to form carbendazim and butyl isocyanate (M. Chiba & E. A. Cherniak, *J. Agric. Food Chem.*, 1978, **26**, 573). The mechanism of the acid-catalysed decomposition in aqueous solvents has been studied (J. P. Calmon & D. R. Sayag, *ibid.*, 1976, **24**, 314, 317). It is non-corrosive to metals.

Uses. Benomyl is a protective and eradicant fungicide with systemic activity, effective against a wide range of fungi affecting fruits, nuts, vegetables, field crops, turf and ornamentals. It is also effective against mites, primarily as an ovicide. Its systemic activity is sometimes enhanced by surfactants. It is also used as pre- or post-harvest sprays or dips for the control of storage rots of fruits and vegetables. Typical rates are: on field and vegetable crops, 140-550 g a.i./ha; on tree crops 550-1100 kg/ha; for post-harvest use 25-200 g/100 l. Benomyl *per se* is stable on the surface of banana skins (J. Cox *et al.*, *Pestic. Sci.*, 1974, **5**, 135; J. Cox & J. A. Pinegar, *ibid.*, 1976, **7**, 193). Factors affecting the duration of the fungicidal effect from soil treatments have been studied (D. J. Austin & G. G. Briggs, *ibid.*, p.201; P. M. Smith & C. R. Worthing, *Proc. Br. Insectic. Fungic. Conf., 7th*, 1972, **1**, 200; *Rep. Glasshouse Crops Res. Inst., 1974*, 1975, pp.63, 106).

Toxicology. The acute oral LD50 for rats is > 10 000 mg/kg. It is mildly irritant to the skin but without dermatitis hazard. In 90-d feeding trials there was no evidence of toxicity for rats receiving 2500 mg/kg diet. In 2-y feeding trials with rats and dogs a low order of toxicity occurred. The LC50 for mallard ducks and quail was > 5000 mg/kg.

Formulations. These include: 'Benlate' fungicide, w.p. (500 g a.i./kg); 'Benlate' OD fungicide, oil dispersible (500 g/kg); 'Tersan' 1991 turf fungicide, w.p. (500 g/kg); 'Benlate' T 20 fungicide (200 g benomyl + 200 g thiram/kg).

Analysis. The preferred method for determination of benomyl-derived residues in soils and plant tissue is cation-exchange HPLC (J. J. Kirkland *et al.*, *J. Agric. Food Chem.*, 1973, **21**, 368) and in cows' milk, urine, faeces and tissues (J. J. Kirkland, *ibid.*, p. 171). Residues in soils and plant tissues may be determined, after extraction and purification, by converting to 2-aminobenzimidazole which is determined by fluorimetry or colorimetry following bromination (H. L. Pease & J. A. Gardiner, *ibid.*, 1969, **17**, 267; H. L. Pease & R. F. Holt, *J. Assoc. Off. Anal. Chem.*, 1971, **54**, 1399). See also: H. L. Pease *et al.*, *Anal. Methods Pestic. Plant Growth Regul.*, 1973, **7**, 674; W. E. Bleidner *et al.*, *ibid.*, 1978, **10**, 157.

Bensulide

$$\text{(Me}_2\text{CHO)}_2\overset{\displaystyle S}{\overset{\displaystyle \|}{\text{P}}}\text{.S. CH}_2\text{CH}_2\text{NH.}\overset{\displaystyle O}{\underset{\displaystyle O}{\overset{\displaystyle \|}{\underset{\displaystyle \|}{\text{S}}}}}\text{.Ph}$$

WSR&M2SPS&OY1&1&OY1&1 $\qquad\qquad\qquad$ $C_{14}H_{24}NO_4PS_3$ (397·5)

Nomenclature and development. The common name bensulide is approved by BSI, WSSA and ISO—exception JMAF (SAP). The IUPAC name is *O,O*-di-isopropyl *S*-2-phenyl-sulphonylaminoethyl phosphorodithioate, in *C.A.* usage *O,O*-bis(1-methylethyl) *S*-[2-[(phenylsulfonyl)amino]ethyl] phosphorodithioate *[741-58-2]* formerly *O,O*-diisopropyl phosphorodithioate *S*-ester with *N*-(2-mercaptoethyl)benzenesulfonamide. It was introduced about 1964 by the Stauffer Chemical Co. under the code number 'R-4461', trade marks 'Betasan' (for turf use) and 'Prefar' (for crop use), and protected by USP 3 205 253. Its herbicidal properties were described by D. D. Hemphill, *Res. Rep. North Cent. Weed Control Conf.,* 1962, pp.104, 111.

Properties. It is an amber liquid forming a solid, m.p. 34·4°, d^{22} 1·25, n_D^{30} 1·5438. Its solubility at room temperature is 25 mg/l water; slightly soluble in kerosine; moderately soluble in xylene; readily soluble in acetone, methanol. It is stable at 80° for 50 h but decomposes at 200° in 18-40 h. It is non-corrosive.

Uses. Bensulide is a pre-em. herbicide suitable for pre-plant use on crops such as cucurbits, brassicas, lettuces and cotton, at 2·3-7·0 kg a.i./ha., or for use on established turf at 11-22 kg/ha, its effects persisting for 0·33-1·0 y according to rate of application. When applied to the roots, bensulide is not translocated to the leaves except as metabolites.

Toxicology. The acute oral LD50 for male albino rats is 770 mg/kg; the acute dermal LD50 for albino rats is 3950 mg/kg. In 90-d feeding trials it was well tolerated by rats at up to 250 mg/kg diet and dogs at up to 625 mg/kg diet.

Formulations. These include: e.c. (480 g a.i./l); granules (100 g/kg) on vermiculite.

Analysis. Product analysis is by HPLC. Crop residues are similarly determined but isolated by TLC; particulars of these methods and of GLC methods are available from the Stauffer Chemical Co. See also: G. G. Patchett *et al., Anal. Methods Pestic., Plant Growth Regul. Food Addit.,* 1967, **5**, 483; R. W. Buxton *et al., Anal. Methods Pestic. Plant Growth Regul.,* 1972, **6**, 672.

33

Bentazone

T66 BMSWNVJ DY1&1 $C_{10}H_{12}N_2O_3S$ (240·3)

Nomenclature and development. The common name bentazone is approved by BSI and ISO, the spelling by ANSI, WSSA (bentazon)—exception the Republic of South Africa (bendioxide). The IUPAC name is 3-isopropyl-(1*H*)-benzo-2,1,3-thiadiazin-4-one 2,2-dioxide, in *C.A.* usage 3-(1-methylethyl)-(1*H*)-2,1,3-benzothiadiazin-4(3*H*)-one 2,2-dioxide *[25057-89-0]* formerly 3-isopropyl-(1*H*)-2,1,3-benzothiadiazin-4(3*H*)-one 2,2-dioxide. It was introduced in 1968 by BASF AG under the code number 'BAS 351H'. Its herbicidal properties were first described by A. Fischer, *Proc. Br. Weed Control Conf., 9th,* 1968, p.1042.

Manufacture and properties. Produced by reacting anthranilic acid with isopropyl-sulphamoyl chloride and cyclisation of the intermediate *N*-(isopropylsulphamoyl)-anthranilic acid with carbonyl chloride, it is a colourless crystalline powder, m.p. 137-139°. Its solubility at 20° is: 500 mg/kg water; 1·51 kg/kg acetone; 33 g/kg benzene; 180 g/kg chloroform; 861 g/kg ethanol.

Uses. Bentazone is a contact herbicide for the control of *Matricaria* and *Anthemis* spp., *Galium aparine, Stellaria media, Chrysanthemum segetum* and *Lapsana communis* in winter and spring cereals when applied at 1·1-2·2 kg a.i. in 225-450 l water/ha after the cereal has reached the 3-4 leaf stage. Other crops are, *e.g.* soyabeans, peas, rice, onions. It is absorbed by the leaves and, having little effect on germinating seeds, it cannot be used as pre-em. herbicide. Herbicidal activity is brief, giving no residual problems.

Toxicology. The acute oral LD50 for rats is *c.* 1100 mg/kg; the dermal LD50 for rats is > 2500 mg/kg; applications to the shaved skin of rabbits caused no irritation but severe eye irritation healed after a week.

Formulations. These include 'Basagran' ('BAS 35 100H'; w.p., 500 g a.i./kg); 'Basagran liquid' ('BAS 35 107H'; 480 g/l); 'Basagran DP' ('BAS 35 801H'); solution 250 g bentazone + 350 g dichlorprop-ethylammonium/l); 'Basagran MCPB', solution (bentazone + MCPB); 'Basagran M' ('BAS 43 300H'); 250 g bentazone + 125 g MCPA/l).

Analysis. Product analysis is by isolation of the a.i. by chromatography and estimation either by *(a)* titration with tetrabutylammonium hydroxide in anhydrous media or *(b)* by u.v. absorption at 296 nm. For residue analysis convert bentazone to its 1-methyl derivative with diazomethane and estimate by GLC using ECD (T. H. Byast *et al., Agric. Res. Counc. (G.B.) Weed Res. Organ., Tech. Rep.,* No. 15 (2nd Ed.), p.59).

Benzoximate

2ONUYOVR&R CG BO1 FO1 $C_{18}H_{18}ClNO_5$ (363·8)

Nomenclature and development. The common name benzoximate is approved by BSI and ISO—exception JMAF (benzomate). The IUPAC name is 3-chloro-α-ethoxyimino-2,6-dimethoxybenzyl benzoate also known as ethyl *O*-benzoyl 3-chloro-2,6-dimethoxybenzohydroximate, in *C.A.* usage benzoic acid anhydride with 3-chloro-*N*-ethoxy-2,6-dimethoxybenzenecarboximidic acid *[29104-30-1]* formerly benzoic acid anhydride with 3-chloro-*N*-ethoxy-2,6-dimethoxybenzimidic acid. It was introduced in 1972 by Nippon Soda Co. Ltd under the trade mark 'Citrazon' and protected by BP 1 247 817; USP 3 821 402; DAS 2 012 973. In some European countries the trade mark 'Aazomate' is used.

Properties. It is a colourless crystalline solid, m.p. 73°. Solubilities at 20°: practically insoluble in water; 650 g/l benzene, 80 g/l hexane, 710 g/l xylene, 1·46 kg/l dimethylformamide. It is stable in acid media but decomposed by strong alkali.

Uses. Benzoximate is a non-systemic acaricide, effective against *Panonychus ulmi* and *P. citri* on apples and citrus at 10-13 g a.i./100 l. See also: A. Formigoni *et al., Atti Giornate Fitopathol.,* 1973, p.95; K. K. Siccama *et al., Meded. Fac. Landbouwwet. Rijksuniv. Gent,* 1977, **42**, 1479.

Toxicology. The acute oral LD50 is: for Wistar rats > 15 000 mg/kg; for mice 12 000-14 500 mg/kg; the acute dermal LD50 for rats and mice is > 15 000 mg/kg. Rats fed 2 y at 400 mg/kg diet were unharmed. The LC50 (48 h) for carp is 1·75 mg tech./l.

Formulation. 'Citrazon 20 EC', e.c. (200 g a.i./l).

Analysis. Spectrophotometry at 258 nm is used after chromatographic separation.

35

Benzoylprop-ethyl

GR   BG   DNVR&Yl&VO2 $C_{18}H_{17}Cl_2NO_3$ (366·2)

Nomenclature and development. The common name benzoylprop-ethyl is approved by BSI and ISO. The IUPAC and *C.A.* name is ethyl *N*-benzoyl-*N*-(3,4-dichlorophenyl)- DL-alaninate *[22212-55-1]*; formerly known as ethyl (±)-2-[*N*-(3,4-dichlorophenyl)benzamido]-propionate and ethyl (±)-2-(*N*-benzoyl-3,4-dichloroanilino)propionate. It was introduced in 1969 by Shell Research Ltd under the code number 'WL 17731', the trade mark 'Suffix' and protected by BP 1 164 160. Its herbicidal properties were first described by T. Chapman *et al., Symp. New Herbic., 3rd*, 1969, p.40.

Properties. The technical material is an off-white crystalline powder, m.p. 70-71°, v.p. 3·5 x 10⁻⁸ mmHg at 20°. Its solubility at 25° is: *c.* 20 mg/l water; 700-750 g/l acetone; 250-300 g/l 'Shellsol A'. It is photochemically stable and hydrolytically stable at intermediate pH.

Uses. Benzoylprop-ethyl, applied at 1·0-1·5 kg a.i./ha to wheat between the end of tillering and second node stage of the crop, gives an 85-95% control of wild oat *(Avena fatua, A. ludoviciana, A. sterilis, A. barbata)*. Wheat tolerates applications of 4 kg/ha. Crop competition ensures the effective control of wild oat, the stem elongation of which is inhibited (B. Jeffcoat & W. N. Harries, *Pestic Sci.*, 1973, **4**, 891). The fate of the compound in plants, soils and animals has been reported (K. I. Beynon *et al., ibid.,* 1974, **5**, 429, 443, 451; J. V. Crayford *et al., ibid.,* 1976, **7**, 559).

Toxicology. The acute oral LD50 is: for rats 1555 mg/kg; for mice 716 mg/kg; for domestic fowl > 1000 mg/kg. The acute dermal LD50 for rats is > 1000 mg/kg. In 90-d feeding trials no change of toxicological significance was seen in rats receiving 1000 mg/kg diet or dogs 300 mg/kg diet. The LC50 (96 h) for rainbow trout is 2·2 mg/l.

Formulations. An e.c. (200 and 250 g a.i./l).

Analysis. Product analysis is by GLC (*CIPAC Handbook,* in press) or i.r. spectroscopy. Residues may be determined by GLC with ECD: details of the methods can be obtained from Shell International Chemical Co. Ltd. See also: A. N. Wright & B. L. Mathews, *Pestic. Sci.,* 1976, **7**, 339.

Benzthiazuron

MeNH.C.NH— (structure)

T56 BN DSJ CMVM1 $C_9H_9N_3OS$ (207·2)

Nomenclature and development. The common name benzthiazuron is approved by BSI and ISO. The IUPAC name is 1-(benzothiazol-2-yl)-3-methylurea, in *C.A.* usage *N*-2-benzothiazolyl-*N'*-methylurea *[1929-88-0]* formerly 1-(2-benzothiazolyl)-3-methylurea. It was introduced in 1966 by Bayer AG as a herbicide under the code number 'Bayer 60 618', trade mark 'Gatnon' and protected by BelgP 647 740; FP 1 401 483; BP 1 004 469; see also USP 2 756 135.

Manufacture and properties. Produced by the reaction of benzothiazol-2-ylamine and methyl isocyanate in benzene, it is a colourless powder which decomposes, with sublimation, at 287°, v.p. 1 x 10⁻⁵ mmHg at 90°. Its solubility at 20° is: 12 mg/l water; 5-10 mg/l acetone, chlorobenzene, xylene. It is of satisfactory stability and is non-corrosive.

Uses. Benzthiazuron is used as a pre-em. herbicide on sugar beet at rates of 3·2-6·4 kg a.i./ha. No exact data on persistence are available, but wheat sown in October after pre-em. use in April on sugar beet was not affected.

Toxicology. The acute oral LD50 for rats is 1280 mg/kg; dermal applications of 500 mg/kg gave no symptom. In 60-d feeding trials all rats receiving 130 mg/kg daily survived.

Formulations. A w.p. (800 g/kg); 'Merpelan', w.p. (600 g benzthiazuron + 125 g lenacil/kg).

Analysis. Product analysis is by i.r. spectroscopy, details are available from Bayer AG. Residue analysis is by u.v. spectroscopy: H. J. Jarczyk & K. Vogeler, *Pflanzenschutz-Nachr. (Am. Ed.)*, 1967, **20**, 575.

S-Benzyl di-sec-butylthiocarbamate

$$\text{Me} \quad \text{O}$$
$$| \qquad ||$$
$$(EtCH)_2.N.C.S.CH_2Ph$$

2Y1&NVS1R&Y2&1 $C_{16}H_{25}NOS$ (279·4)

Nomenclature and development. The common name tiocarbazil is proposed by BSI. The IUPAC name is *S*-benzyl di-*sec*-butylthiocarbamate, in *C.A.* usage *S*-phenylmethyl bis(1-methylpropyl)carbamothioate *[36756-79-3]*. It was introduced in 1972 by Montedison S.p.A. under the trade name 'Drepamon' and protected by Italian P 907 710; DBP 2 144 700. Its herbicidal properties were first described by N. Caracalli *et al., Proc. Congr. Risicultura, 8th,* 1973, p.446; P. Bergamaschi *et al., Atti Giornate Fitopatol.,* 1975, p.1139.

Properties. It is a colourless liquid of aromatic odour, v.p. 7 x 10^{-4} mmHg at 50°, d_4^{20} 1·023; n_D^{20} 1·535. Its solubility at 30° is 2·5 mg/l water; miscible with most organic solvents. It suffered no decomposition in storage for 60 d at 40° nor in aqueous solutions exposed to sunlight for 100 h. It is stable to hydrolysis and is only slightly decomposed in aqueous ethanolic solution at pH 1·5 after 30 d at 40°. The technical product is $\geqslant 95\%$ pure.

Uses. It is a pre- and post-em. herbicide effective against weeds in rice, in particular *Echinochloa crusgalli, E. colonum, Leptochloa fascicularis, Lolium perenne* and *Cyperus* spp., at rates of 4·0 kg a.i./ha, provided that the rice field remains submerged for $\geqslant 20$ d after application. No ill-effect was seen in the rice varieties tested.

Toxicology. The acute oral LD50 is: for rats, guinea-pigs, rabbits, chickens, quails, pheasants and hares > 10 000 mg tech./kg; for mice 8000 mg/kg. There was no mortality to rats or rabbits treated percutaneously with 1200 mg/kg. In 2-y feeding trials albino rats and beagle dogs receiving 1000 mg tech./kg diet suffered no ill-effect except for a slight weight loss in male dogs. There was no significant difference between controls and albino rats receiving 300 mg/kg diet for 3 generations. The LC50 for the fish species tested is $\geqslant 8$ mg/l; for the mollusc *Australorbis glabratus* > 60 mg/l.

Formulations. These include: w.s.c. (500 and 700 g a.i./l); seed treatment (700 g/l); granules (50 and 75 g/kg).

Analysis. Product analysis is by GLC. Residue analysis is by GLC after suitable clean-up. Details of methods are available from Montedison S.p.A. See also R. Fabbrini *et al., Anal. Methods Pestic. Plant Growth Regul.,* **11,** in press.

38

S-Benzyl O,O-di-isopropyl phosphorothioate

$$\text{(Me}_2\text{CHO)}_2\overset{\overset{\displaystyle O}{\|}}{\text{P}}.\text{S.CH}_2\text{Ph}$$

1Y1&OPO&S1R&OY1&1 $C_{13}H_{21}O_3PS$ (288·3)

Nomenclature and development. The common name IBP is used by JMAF; there is no BSI or ISO name. The IUPAC name is S-benzyl O,O-di-isopropyl phosphorothioate (I), in C.A. usage O,O-bis(1-methylethyl) S-phenylmethyl phosphorothioate [26087-47-8] formerly (I). The O,O-diethyl analogue was introduced in 1965 by the Ihara Chemical Co. under the trade mark 'Kitazin', but it was replaced in 1968 by the di-isopropyl homologue by Kumiai Chemical Industry Co. Ltd under the trade mark 'Kitazin P'. Patents are filed in more than 30 countries.

Manufacture and properties. Produced by the direct phosphorylation of benzyl alcohol using phosphorous trichloride, sulphur and propan-2-ol, it is a yellow oil, b.p. 126°/0·04 mmHg, n_D 1·5106. Its solubility at 18° is 1 g/l water; soluble in most organic solvents. The technical product is c. 94% pure.

Uses. It is a systemic fungicide used to control *Piricularia oryzae* in rice. It is applied at 400-600 g a.i. (as e.c.) in 1000 l/ha as soon as the blast lesions appear. One or 2 sprays may be needed during the head-sprouting season. Its translocation and metabolism in rice plants has been reported (H. Yamamoto *et al., Agric. Biol. Chem.*, 1973, **37**, 1553).

Toxicology. The acute oral LD50 for mice is 600 mg/kg; the acute dermal LD50 for mice is 4000 mg/kg. The LC50 for carp is 5·1 mg/l.

Formulations. These include: e.c. (480 g a.i./l); dust (20 g/kg); granules (170 g/kg).

Analysis. Product analysis is by TLC and colorimetry. Residues are analysed by GLC with FPD.

Bifenox

$$O_2N \text{—} \underset{O=COMe}{\overset{}{\bigcirc}} \text{—O—} \underset{Cl}{\overset{Cl}{\bigcirc}}$$

GR CG DOR DNW CVO1 $C_{14}H_9Cl_2NO_5$ (342·1)

Nomenclature and development. The common name bifenox is approved by ANSI and WSSA; there is no BSI or ISO name. The IUPAC and *C.A.* name is methyl 5-(2,4-dichlorophenoxy)-2-nitrobenzoate, *C.A.* Registry No. *[42576-02-3]* (formerly *[12680-11-4]*). It was introduced in 1970 by Mobil Chemical Co. as an experimental herbicide under the code number 'MC-4379', the trade mark 'Modown' and protected by BP 1232368; USP 3652645; 3776715. Its use as a herbicide is reported by W. M. Dest *et al., Proc. Northeast. Weed Sci. Conf.,* 1973, **27,** 31.

Manufacture and properties. Produced by the reaction of potassium 2,4-dichlorophenoxide with methyl 5-chloro-2-nitrobenzoate, it forms yellow crystals, m.p. 84-86°, v.p. 2·4 x 10^{-6} mmHg at 30°. Its solubility is: 0·35 mg/l water; 400 g/kg acetone; 300 g/kg xylene; 400 g/kg chlorobenzene; < 50 g/kg ethanol; < 10 g/kg aliphatic hydrocarbons.

Uses. Bifenox is used for pre-em. and directed post-em. treatment to control important broad-leaved weeds and some grasses in soyabeans, corn, sorghum, rice and small grains. Pre-em. applications of 1·68-2·24 kg a.i./ha are recommended to provide effective weed control over a range of soil types and a variety of climatic conditions. Post-em. applications, as directed sprays, have shown good activity at rates of 1·12-1·68 kg/ha. Residual control usually persists for 35-70 d depending upon rainfall, soil type and plant species. Tests indicate that broad-leaved weed control is not adversely affected by rainfall extremes; there is no need for incorporation and no carry-over or residue problems.

Toxicology. The acute oral LD50 for rats is > 6400 mg tech./kg. The acute dermal LD50 for rabbits is > 20000 mg/kg. There is no serious inhalation toxicity or eye irritation. High dosages in the diets of rats and dogs produced no pathological or histological changes.

Formulation. 'Modown Herbicide' as: e.c. (210 g/l); w.p. (800 g/kg); flowable (440 g/kg); granules (100 g/kg).

Analysis. Product analysis is by GLC, comparing with an internal standard. Residues may be determined by GLC using halogen-specific MCD.

Binapacryl

$$O = C.CH:CMe_2$$

2Y1&R   CNW   ENW   BOV1UY1&1 $C_{15}H_{18}N_2O_6$ (322·4)

Nomenclature and development. The common name binapacryl is approved by BSI, ISO, ANSI and JMAF; also known as ENT 25 793. The IUPAC name is 2-*sec*-butyl-4,6-dinitrophenyl 3-methylcrotonate (I), in *C.A.* usage 2-(1-methylpropyl)-4,6-dinitrophenyl 3-methyl-2-butenoate *[485-31-4]* formerly (I). Its biological properties were first described by L. Emmel & M. Czech, *Anz. Schaedlingskd.*, 1960, **33**, 145, and it was introduced in 1960 by Hoescht AG under the code number 'Hoe 2784', trade marks 'Acricid', 'Endosan', 'Morocide' and protected by BP 855 736; DBP 1 099 787; USP 3 123 522.

Manufacture and properties. Produced by the esterification of dinoseb (p.209) with 3-methylcrotonoyl chloride, binapacryl is a colourless crystalline powder. The technical product has m.p. 65-69°, d_4^{20} 1·25-1·28. Pure binapacryl has v.p. 1 x 10^{-4} mmHg at 60° and is practically insoluble in water but soluble in most organic solvents: 780 g/kg acetone, 110 g/kg ethanol, 570 g/kg isophorone, 107 g/kg kerosine, 700 g/kg xylene, 570 g/kg heavy aromatic naphtha.

It is unstable in alkalies and concentrated acids, suffers slight hydrolysis on long contact with water and is slowly decomposed by u.v. light. It is non-corrosive and is compatible with w.p. formulations of insecticides and non-alkaline fungicides. It may be phytotoxic when mixed with organophosphorus compounds.

Uses. Binapacryl is a non-systemic acaricide which is effective against powdery mildews, used mainly against all stages of spider mites and powdery mildews of top fruit at 25-50 g a.i./100 l. Non-phytotoxic to a wide range of apples, pears, cotton and citrus; some risk of damage to young tomatoes, grapes and roses.

Toxicology. The acute oral LD50 is: for rats 150-225 mg/kg; for male mice 1600-3200 mg/kg; for female guinea-pigs 300 mg/kg; for dogs 450-640 mg/kg. The acute dermal LD50 (in acetone solution) for rabbits and mice is 750 mg/kg; it is slightly irritating to the eyes. In 2-y feeding studies rats tolerated without symptom 200-500 mg/kg diet and dogs 20-50 mg/kg diet. The maximum tolerated dose is: for guppies 0·5 mg/l; for carp 1 mg/l; for trout 2 mg/l.

Formulations. These include: w.p. (242·5 and 485 g a.i./kg); e.c. (388 g/l); dispersion (485 g/l); dust (38·8 g/kg).

Analysis. Product analysis is by extracting with heptane, treating the extract with sodium ethoxide, steam distilling the dinoseb so formed and measuring its absorption at 420 nm in dilute NaOH; alternatively the nitro groups are determined by the titanium trichloride method. Residues are analysed by conversion to dinoseb, as above, which is purified and measured colorimetrically at 380 nm; particulars of these methods may be obtained from Hoescht AG. See also R. W. Buxton & T. A. Mohr, *Anal. Methods Pestic., Plant Growth Regul. Food Addit.*, 1967, **5**, 235; *Anal. Methods Pestic. Plant Growth Regul.*, 1972, **6**, 314.

Bioallethrin

(i)　　　　　　　　　　　　　　　　　　　(ii)

L5V BUTJ B2U1 C1 DOV- BL3TJ A1 A1 C1UY1&1 $C_{19}H_{26}O_3$ (302·4)

Nomenclature and development. The common name bioallethrin is approved by BSI and in New Zealand; there is no ISO name; the name *d-trans*-allethrin is used in USA; also known as ENT 16275. The IUPAC name is (*RS*)-3-allyl-2-methyl-4-oxocyclopent-2-enyl (1*R*)-*trans*-chrysanthemate, in *C.A.* usage 2-methyl-4-oxo-3-(2-propenyl)-2-cyclopenten-1-yl 2,2-dimethyl-3-(2-methyl-1-propenyl)cyclopropanecarboxylate *[584-79-2]* (formerly *[22431-63-6]*) formerly *dl*-2-allyl-4-hydroxy-3-methyl-2-cyclopenten-1-one ester of *d-trans*-2,2-dimethyl-3-(2-methylpropenyl)cyclopropanecarboxylic acid. The commercial availability of (1*R*)-*trans*-chrysanthemic acid enabled the preparation of bioallethrin (i), which is a more potent insecticide than allethrin (J. Lhoste *et al., C. R. Seances Acad. Agric. Fr.,* 1967, **53**, 686). It was introduced in 1969 by Roussel-Uclaf under the trade mark, in some countries, 'Bioallethrine' and by the McLaughlin Gormley King Co. under the trade mark 'D-Trans' and protected by USP 3159535.

Further research by Roussel-Uclaf resulted in the production of the corresponding ester of the (*S*)-alcohol, with faster 'knockdown' and insecticidal activity than bioallethrin (F. Rauch *et al., Meded. Fac. Landbouwwet. Rijksuniv. Gent,* 1972, **37**, 755). This isomer, in *C.A.* usage [1*R*-[1α(*S**),3β]]-2-methyl-4-oxo-3-(2-propenyl)-2-cyclopenten-1-yl 2,2-dimethyl-3-(2-methyl-1-propenyl)cyclopropanecarboxylate *[28434-00-6]*, is known as (*S*)-bioallethrin (ii) and AI3-29 024, the technical product having the trade mark 'Esbiol'.

Properties. Bioallethrin, as 'D-Trans', is an amber viscous liquid, d 0·997, n_D^{25} 1·500, flash point 65·6° (tag open cup) and contains 90% *m/m* bioallethrin. 'Esbiol' is a yellow viscous liquid, d^{20} 0·980, and contains up to 90% *m/m* (*S*)-bioallethrin. These esters are practically insoluble in water but miscible with organic solvents including refined kerosine.

Uses. Bioallethrin and (*S*)-bioallethrin are potent non-systemic insecticides causing a rapid paralysis or 'knockdown'. Their metabolic detoxication in insects is delayed by the addition of synergists such as piperonyl butoxide or 'MGK 264' (see p.217), so increasing the effectiveness of such formulations.

Toxicology. The acute oral LD50 of 'D-Trans' for male rats is 425-575 mg/kg, for female rats 845-875 mg/kg. The acute oral LD50 of (*S*)-bioallethrin for male rats is 784 mg/kg, for female rats 1545 mg/kg; the acute dermal LD50 for rabbits is 1545 mg/kg. The LC50 (96 h) for (*S*)-bioallethrin is: for rainbow trout 0·0105 mg/l; for bluegill 0·0332 mg/l.

Formulations. Usually these compounds are combined with synergists and other insecticides as fly sprays in kerosine, aerosol concentrates or dusts.

Analysis. Product analysis is by GLC (*CIPAC Handbook,* 1979, **1A**, in press). The stereoisomer composition may be established by GLC of suitable derivatives (F. E. Rickett, *Analyst (London),* 1973, **98**, 687; A. Horiba *et al., Agric. Biol. Chem.,* 1977, **41**, 2003; 1978, **42**, 671) or NMR (F. E. Rickett & P. B. Henry, *Analyst (London),* 1974, **99**, 330). Residue analysis is by GLC.

Bioresmethrin

T5OJ B1R& D1OV- BL3TJ A1 A1 C1UY1&1 $C_{22}H_{26}O_3$ (338·4)

Nomenclature and development. The common name bioresmethrin is approved by BSI and ISO; also known as ENT 27622, AI3-27 622, OMS 1206. The IUPAC name is 5-benzyl-3-furylmethyl (1*R*)-*trans*-chrysanthemate (I), in *C.A.* usage (5-phenylmethyl-3-furanyl)methyl (1*R-trans*)-2,2-dimethyl-3-(2-methyl-1-propenyl)cyclopropanecarboxylate *[28434-01-7]* (formerly *[10453-54-0]*) formerly (I). It was first described by M. Elliott *et al., Nature (London),* 1967, **213,** 493. It has been called 'NRDC 107', 'FMC 18 739', 'RU 11 484', 'biobenzyfuroline' and is a stereoisomer of resmethrin (p.467).

Manufacture and properties. Produced by esterification of 5-benzyl-3-furylmethanol (BP 1 168 797; 1 168 798; 1 168 799), it is a viscous yellow to reddish brown liquid which partially solidifies on standing, m.p. 30-35°, b.p. 180°/0·01 mmHg, d^{20} 1·050 $[\alpha]_D^{20}$ —5 to —8° (ethanol). It is insoluble in water; soluble in all common organic solvents. It is more stable than the pyrethrins, but it is fairly rapidly decomposed on exposure of films to air and sunlight (M. Elliott *et al., Proc. Insectic. Fungic. Conf., 7th,* 1973, **2,** 721). Readily hydrolysed in alkaline media.

Uses. Bioresmethrin is a powerful contact insecticide effective against a wide range of insects including flies, mosquitoes, cockroaches and plant pests (I. C. Brookes *et al., Soap Chem. Spec.,* 1969, **45** (3), 62); and as a grain protectant. The toxicity to normal *Musca domestica* is 50 x that of pyrethrins (M. Elliott, *Nature (London), loc. cit.),* but synergistic factors are lower than those of pyrethrins with methylenedioxyphenyl synergists. Toxicity to plants is low.

Toxicology. The acute oral LD50 for rats is: 7070-8000 mg/kg; for chicks > 10 000 mg/kg. The acute dermal LD50 for female rats is > 10 000 mg/kg. In 90-d feeding trials the 'no effect' level was: for rats 1200 mg/kg diet; for dogs > 500 mg/kg diet. Rats tolerated 4000 mg/kg diet for 60 d. In rats dosed daily with 80 mg/kg from d 8-16 of pregnancy, no teratogenic effects were observed; there were foetal deaths at this dose but not at 40 mg/kg. It is very toxic to honey bees (oral LD50 3 ng/bee; topical LD50 6·2 ng/bee). The LC50 (96 h) for harlequin fish is 0·014 mg/l; LC50 (48 h) for harlequin fish 0·018 mg/l, for guppies 0·5-1·0 kg/l.

Formulations. These include: aerosol concentrates (1 g a.i./l) with bioallethrin or related pyrethroids; ULV; ready-to-use liquids (2·5 g/l) for oil- or water-based sprays. Some formulations include a synergist (piperonyl butoxide).

Analysis. Product analysis is by GLC with FID using an internal standard such as dicyclohexyl phthalate or dioctyl phthalate (B. B. Brown, *Anal. Methods Pestic. Plant Growth Regul.,* 1973, **7,** 441).

Biphenyl

Ph.Ph

RR $C_{12}H_{10}$ (154·2)

Nomenclature and development. The IUPAC name biphenyl (I) is accepted by BSI in lieu of a common name, in *C.A.* usage 1,1'-biphenyl *[92-52-4]* formerly (I). It was introduced in 1944 for the impregnation of citrus fruit wraps against rot fungi (G. B. Ramsey *et al., Bot. Gaz.*, 1944, **106**, 74).

Manufacture and properties. Produced by the pyrolytic dehydrogenation of benzene, it forms colourless leaflets, m.p. 70·5°, b.p. 256·1°, *d* 1·041, n_D^{77} 1·588. It is practically insoluble in water; soluble in most organic solvents.

Uses. Biphenyl is used to inhibit the mycelial growth and spore formation of citrus fruit rots.

Toxicology. The acute oral LD50 for rats is 3280 mg/kg. Prolonged exposure of humans to vapour concentrations > 5 mg/m³ is considered dangerous (J. Deichmann *et al., J. Ind. Hyg. Toxicol.*, 1947, **29**, 1).

Analysis. Residues may be determined by i.r. spectroscopy (W. F. Newhall *et al., Anal. Chem.*, 1954, **26**, 1234) or by colorimetric methods (J. B. Davenport, *Analyst (London)*, 1953, **78**, 558; A. Rajzman, *ibid.*, 1960, **85**, 116). See also: A. Rajzman, *Residue Rev.*, 1964, **8**, 1; S. W. Souci & G. Maier-Haarländer, *ibid.*, 1966, **16**, 103.

Bis(tributyltin) oxide

Bu₃Sn.O.SnBu₃

4-SN-4&4&O-SN-4&4&4 $C_{24}H_{54}OSn_2$ (596·1)

Nomenclature and development. Bis(tributyltin) oxide (I) is also known as ENT 24 979. The IUPAC name is (I), in *C.A.* usage hexabutyldistannoxane *[56-35-9]*. It was introduced in 1966 by M & T Chemicals Ltd under the trade mark 'TBTO'. It was first described by J. G. A. Luijten & G. J. M. van der Kerk, *Investigations in the field of organotin chemistry,* and is not protected by patent except for some specialised applications.

Manufacture and properties. Produced by hydrolysis of tributyltin chloride, which is obtained by the reaction of butylmagnesium chloride on tin tetrachloride, it is a colourless to light yellow liquid, b.p. 180°/2 mmHg, f.p. < —45°, of negligible v.p. at 20°, d_4^{20} 1·14, n_D 1·4870, viscosity 4·8 x 10⁻⁶ m²/s at 25°. The solubility of the commercial grade at room temperature is *c.* 100 mg/l water (60 mg/l in sea water); miscible with most organic solvents. It is stable in storage in dark, closed containers but usually forms tributyltin carbonate on exposure to air; is compatible with most fungicides and bactericides and is non-corrosive to copper, brass, stainless steel, galvanised or black iron.

Uses. Bis(tributyltin) oxide is a fungicide, bactericide, algicide and molluscicide. It is mainly used as a toxicant in marine antifouling paints, as an industrial cooling water algicide and as a wood preservative. It is phytotoxic and unsuitable for use on growing crops or on cultivated soil.

Toxicology. The acute oral LD50 for rats is 194 mg a.i. (as aqueous suspension)/kg (J. R. Elsea & O. E. Paynter, *AMA Arch. Ind. Health,* 1958, **18**, 214). It is absorbed by skin and may cause irritation.

Formulations. It is readily emulsified with anionic, cationic or non-ionic surfactants.

Analysis. Product analysis involves wet oxidation, reduction to the tin(II) state and determination by titration with potassium iodate. Residue analysis is by hydrolysis to tin(II) salts with colorimetric determination with dithiol (R. E. D. Clark, *Analyst (London),* 1936, **61**, 242; 1937, **62**, 661).

Blasticidin-S

T6O CUTJ EMV1YZ2N1&YZUM FVQ B- AT6NVNJ DZ $C_{17}H_{26}N_8O_5$ (422·4)

Nomenclature and development. The common name blasticidin-S is approved by JMAF. In *C.A.* usage (*S*)-4[[3-amino-5[(aminoiminomethyl)methylamino]-1-oxopentyl]amino]-1-[4-amino-2-oxo-1(2*H*)-pyrimidinyl]-1,2,3,4-tetradeoxy-β-D-erythrohex-2-enopyranuronic acid *[2079-00-7]* (formerly *[11002-92-9]*, *[12767-55-4]*) formerly 4-[3-amino-5-(1-methyl-guanidino)valeramido-1-[4-amino-2-oxo-1(2*H*)-pyrimidinyl]-1,2,3,4-tetradeoxy-β-D-erythro]hex-2-enopyranuronic acid. This antibiotic was discovered in 1955 by K. Fukunaga *et al.*, *Bull. Agric. Chem. Soc. Jpn.*, 1955, **19,** 181; S. Takeuchi *et al.*, *J. Antibiot. Ser. A,* 1958, **11,** 1; and its structure was elucidated by H. Yonehara & N. Otake, *Tetrahedron Lett.*, 1966, p.3785; H. Seto *et al., ibid.,* p.3793.

Manufacture and properties. Produced by the fermentation of *Streptomyces griseochromogenes,* it forms colourless needles, m.p. 253-255°, [α]$_D$ + 108·4°. It is soluble in water, acetic acid; insoluble in other common organic solvents.

Uses. Blasticidin-S is a contact fungicide used mainly for the control of *Piricularia oryzae* on rice at *c.* 10 g a.i. in 100 l water/ha. Its range of use is limited by phytotoxicity. Of its derivatives, the benzylaminobenzenesulphonate is the least phytotoxic and is widely used as the a.i. in formulations.

Toxicology. The acute oral LD50 for mice is 39·5 mg/kg. The acute dermal LD50 is: for rats 3100 mg/kg; for mice 220 mg/kg. Precautions are necessary in handling as it can produce conjunctivitis, but it is broken down rapidly after application. Carp are killed in water containing 8700 mg/l.

Formulations. These include: e.c. (20 g a.i./l); w.p. (40 g/kg); dust (1·6 g/kg).

Borax

$$Na_2B_4O_7 \cdot 10H_2O$$

.NA2.B4-O7.QH1∅ $B_4H_{20}Na_2O_{17}$ (381·4)

Nomenclature and development. Borax, *C.A.* Registry No. *[1303-96-4]* (formerly *[1344-90-7]*), is the trivial name recognised by BSI and ISO—exception New Zealand (sodium borate). The IUPAC name is disodium tetraborate decahydrate. Borax has long been used as a mild antiseptic and fungicide

Manufacture and properties. Produced by crystallisation from concentrated liquors after dissolving of sodium borate ores, borax is a colourless crystalline solid, efflorescent in dry air, losing 5 mol of water of crystallisation at 100°, 4 or more at 160° and becomes anhydrous at 320°. On rapid heating it melts at 75°. Its solubility at 20° is 51·4 g/l water and it is soluble in glycerol and ethylene glycol but insoluble in ethanol. Its aqueous solution is alkaline, but it is hydrolysed by mild alkali and it is not compatible with certain herbicides. Borax is used as a flame retardant and as a corrosion inhibitor for ferrous alloys.

Uses. Borax has been recommended for mould prevention on citrus (H. R. Fulton & J. J. Bowman, *J. Agric. Res.,* 1924, **28**, 961; F. Lauriol, *Fruits,* 1954, **9**, 3). As it is strongly phytotoxic, its main use is for weed control on non-crop sites at rates equivalent to 1·25-3·0 t B_2O_3/ha, and for the eradication of St. John's Wort (S. C. Litzenberger *et al., Bull. Mont. Agric. Exp. Stn.,* No. 426, 1945) and poison ivy (E. M. Stodard, *Circ. Conn. Agric. Exp. Stn.,* No. 160, 1944). In soil, at the above doses, it persists for 2 y, depending on rainfall and soil texture. Borax is used in insect baits for food stores and as 'Nippon' in the control of ants.

Toxicology. The acute oral LD50 for rats is 4500-6000 mg/kg (R. J. Weir & R. S. Fisher, *Toxicol. Appl. Pharmacol.,* 1972, **23**, 351); the lethal dose to human infants is 5-6 g.

Formulation. Borax, in addition to its use alone, is formulated in mixture with sodium chlorate, reducing the fire hazard of the latter, and with organic herbicides such as bromacil for use on industrial sites.

Analysis. Product analysis is by the mannitol-NaOH titration: AOAC Methods. Residues in plant material may be analysed *(a)* by extracting the ash with dilute sulphuric acid and estimating colorimetrically using quinalizarin: AOAC Methods, or *(b)* by ashing with barium hydroxide, extracting the ash with dilute acetic acid and determining spectrophotometrically with 4,5-dihydroxy-1-(4-nitrobenzeneazo)naphthalene-2,7-disulphonic acid: C. M. Austin & J. S. McHargue, *J. Assoc. Off. Agric. Chem.,* 1948, **31**, 427.

47

Nomenclature, development and preparation. The trivial name Bordeaux mixture, *C.A.* Registry No. *[8011-63-0]* is used. The mixture was introduced in 1885 (A. Millardet, *J. Agric. Prat. (Paris),* 1885, **49,** 513). It used to be prepared as a tank-mix from aqueous copper sulphate solutions and calcium oxide (later calcium dihydroxide) suspensions, giving a flocculent blue amorphous precipitate which, on standing, tends to become blue, crystalline and purplish. As ordinarily prepared the precipitate consists of copper dihydroxide stabilised by absorbed calcium sulphate (H. Martin, *Ann. Appl. Biol.,* 1932, **19,** 98). Pre-formed formulations, ready to mix with water, have naturally largely replaced the tank-mix product.

Properties. It is a pale blue powder. It is insoluble in water or common organic solvents. It is decomposed by acids. It dissolves in ammonium hydroxide forming a cuprammonium complex. It is incompatible with alkali-sensitive pesticides such as organophosphorus compounds or carbamates. It must not be prepared in metal vessels.

Uses. Bordeaux mixture is used as a protective fungicide for foliage applications, the freshly prepared precipitate having a high tenacity. Its use is limited to crop plants at stages of growth when its phytotoxicity is low. Major uses include the control of *Phytophthora infestans* on potatoes, *Venturia inaequalis* on apples and *Pseudoperonospora humuli* on hops.

Formulations. These include: w.p.; flowable suspensions. It can be prepared as a tank-mix using copper sulphate pentahydrate (1·0 kg) with calcium dihydroxide (1·25 kg) in water (100 l) for HV application, or (4 kg + 2 kg + 100 l) for LV application (R. Gair & D. J. Yarham, *Insecticide and Fungicide Handbook,* (5th Ed.), p.128).

Analysis. Product analysis is by digestion with sulphuric acid and determination of the copper content by electrolytic, volumetric, colorimetric or atomic absorption spectroscopic methods (AOAC Methods; *CIPAC Handbook,* 1970, **1,** 226; 1979, **1A,** in press; later volume, in press). Residues are determined by digestion with sulphuric acid and perchloric acid, extraction from acidic potassium iodide solution with dithizone and measurement as diethyldithiocarbamate (AOAC Methods).

Brodifacoum

T66 BOVJ EQ D- GL66&TJ IR DR DE $C_{31}H_{23}BrO_3$ (523·4)

Nomenclature and development. The common name brodifacoum is approved by BSI and proposed by ISO. The IUPAC name is 3-[3-(4'-bromobiphenyl-4-yl)-1,2,3,4-tetrahydro-1-naphthyl]-4-hydroxycoumarin, in *C.A.* usage 3-[3-(4'-bromo[1,1'-biphenyl]-4-yl)-1,2,3,4-tetrahydro-1-naphthalenyl]-4-hydroxy-2*H*-1-benzopyran-2-one *[56073-10-0]*. It was introduced in 1978 as a rodenticide by Sorex (London) Ltd under the code number 'WBA 8119' and developed by ICI Ltd, Plant Protection Division, under the code number 'PP 581' and trade marks 'Talon' and 'Ratak+'. Its rodenticidal properties were first described by R. Redfern *et al., J. Hyg.*, 1976, **77**, 419.

Manufacture and properties. Produced by condensing 4-hydroxycoumarin with 3-(4'-bromobiphenyl-4-yl)-1,2,3,4-tetrahydro-1-naphthol, it is an off-white to fawn powder, m.p. 228-235°. It is almost insoluble in water; slightly soluble in benzene, alcohols; soluble in acetone, chloroform and other chlorinated solvents. It does not readily form water-soluble alkali metal salts, but does form certain amine salts of limited solubility in water.

Uses. Brodifacoum is an indirect anticoagulant active against rats and mice including strains resistant to other anticoagulants, and against rodent species, such as hamsters, that are difficult to control with other anticoagulants. Its potency is such that, unlike other anticoagulants, a rodent may absorb a lethal dose by taking only a 50 mg/kg bait as part of its food intake on only one occasion. This makes brodifacoum particularly promising for non-commensal rodent control.

Toxicology. The acute oral LD50 is: for male rats 0·27 mg a.i./kg; for male rabbits 0·3 mg/kg; for male mice 0·4 mg/kg; for female guinea-pigs 2·8 mg/kg; for cats *c.* 25 mg/kg; for dogs 0·25-1·0 mg/kg.

Formulations. Sorex brodifacoum Bait (20 and 50 mg a.i./kg); 'Talon', bait (50 mg/kg).

Analysis. Products are analysed by u.v. spectroscopy—particulars from Sorex (London) Ltd. HPLC may also be used—details from ICI Ltd, Plant Protection Division.

Bromacil

T6MVNVJ CY2&1 EE F1 $C_9H_{13}BrN_2O_2$ (261·1)

Nomenclature and development. The common name bromacil is approved by BSI, ISO, ANSI, WSSA and JMAF—exception Germany. The IUPAC name is 5-bromo-3-*sec*-butyl-6-methyluracil (I), in *C.A.* usage 5-bromo-6-methyl-3-(1-methylpropyl)-2,4(1*H,3H*)-pyrimidinedione *[314-40-9]* formerly (I). The herbicidal activity of certain substituted uracils was reported by H. C. Bucha *et al., Science,* 1962, **137**, 537. Bromacil was introduced, in 1963, by E. I. du Pont de Nemours & Co. (Inc.) under the code number 'Du Pont Herbicide 976', trade mark 'Hyvar' X and the protection of USP 3 235 357; 3 352 862; BP 968 661; 968 663; 968 664; 968 665; 968 666; BelgP 625 897.

Manufacture and properties. Produced by the bromination of 3-*sec*-butyl-6-methyluracil, bromacil is a colourless crystalline solid, m.p. 158-159°, v.p. 8 x 10⁻⁴ mmHg at 100°. Its solubility at 25° is 815 mg/l water; moderately soluble in strong aqueous bases, acetone, acetonitrile, ethanol. It is stable up to m.p. and slowly sublimes at temperatures just below its m.p. It is decomposed slowly by concentrated acids, and is non-corrosive. The water-soluble formulation is incompatible with ammonium sulphamate, weed oils and liquid formulations of aminotriazole and may be precipitated by pesticides containing soluble calcium salts.

Uses. Bromacil is a non-selective inhibitor of photosynthesis absorbed mainly through the roots and recommended for general weed control on non-crop land at 5-15 kg a.i./ha; for subsequent annual maintenance 2-4 kg/ha is enough. It may also be used for annual weed control in established citrus plantations at 1·6-3·2 kg/ha; for perennial grasses and annual weeds at 3·2-8·0 kg/ha; in pineapple plantations at 1·8-5·5 kg/ha. At the higher rates its effect usually persists for more than one season.

Toxicology. The acute oral LD50 for rats is 5200 mg/kg. In tests on intact and abraded skin of guinea-pigs, a 50% suspension of the w.p. was mildly irritating to young animals, more so to older ones but did not produce sensitisation. In 2-y feeding trials the 'no effect' level for rats and dogs was 250 mg/kg diet.The LC50 (96 h) for mallard ducklings and bobwhite quail was 10 000 mg/l. The LC50 for bluegill and rainbow trout was *c.* 80 mg a.i. (as w.p.)/l (24 h) and 56-60 mg a.i. (as w.p.)/l (48 h). See H. Sherman & A. M. Kaplan, *Toxicol. Appl. Pharmacol.,* 1975, **34**, 189.

Formulations. 'Hyvar' X, w.p. (800 g a.i./kg); 'Hyvar' X-L, w.s.c. (240 g a.e./l; 219 g bromacil-lithium/kg); 'Krovar I', w.p. (400 g bromacil + 400 g diuron/kg); 'Krovar II', w.p. (530 g bromacil + 270 g diuron/kg).

Analysis. Residue analysis is by GLC (H. L. Pease, *J. Agric. Food Chem.,* 1966, **14**, 94; 1968, **16**, 54: H. Jarczyk, *Pflanzenschutz-Nachr. (Am. Ed.),* 1975, **28**, 319). See also: H. L. Pease & J. F. Deye, *Anal. Methods Pestic., Plant Growth Regul. Food Addit.,* 1967, **5**, 335; *Anal. Methods Pestic. Plant Growth Regul.,* 1972, **6**, 663; for residues in soils, T. H. Byast *et al., Agric. Res. Counc. (G.B.) Weed Res. Organ., Tech. Rep.,* No. 15 (2nd Ed.), p.47.

Bromofenoxim

WNR CNW DONU1R DQ CE EE $\qquad$ $C_{13}H_7Br_2N_3O_6$ (461·0)

Nomenclature and development. The common name bromofenoxim is approved by BSI and ISO. The IUPAC name is 3,5-dibromo-4-hydroxybenzaldehyde 2,4-dinitrophenyl-oxime, in *C.A.* usage 3,5-dibromo-4-hydroxybenzaldehyde *O*-(2,4-dinintrophenyl)oxime *[13181-17-4]* (formerly *[37273-85-1]*). It was introduced in 1969 by Ciba AG (now Ciba-Geigy AG) as an experimental herbicide under the code number 'C 9122', trade mark 'Faneron' and protected by BP 1096037. Its herbicidal properties were first described by D. H. Green *et al., Symp. New Herbic., 3rd,* 1969, p.177.

Manufacture and properties. Made by the reaction of sodium salt of 3,5-dibromo-4-hydroxybenzaldehyde oxime with 1-chloro-2,4-dinitrobenzene, it is a cream-coloured, crystalline powder, m.p. 196-197°, v.p. $< 10^{-8}$ mmHg at 20°. Its solubility at 20° is 0·1 mg/l water.

Uses. Bromofenoxin is a foliar-acting herbicide with a strong contact activity on annual dicotyledons. It has negligible activity through the soil. Suggested rates of application are: on winter-sown cereals, 1·5-2·5 kg a.i./ha when the weeds have emerged; on spring-sown cereals, 1·0-2·0 kg/ha to emerged broad-leaved weeds (up to 6 leaves). It is also used in combination (see below) with terbuthylazine to broaden its spectrum of weed control and extend its period of activity at rates of 0·8-1·25 kg mixed herbicide/ha.

Toxicology. The acute oral LD50 for rats of the technical material is 1217 mg/kg; the acute dermal LD50 for rats is > 3000 mg/kg. In 90-d feeding trials on rats and dogs the 'no effect' level was 300 mg/kg diet. The toxic response to fish varies with the species from moderate to high. It is slightly toxic to birds.

Formulations. These include a w.p. (500 g a.i./kg) and mixtures: 'Faneron multi 500 FW and 50 WP' (flowable and w.p., bromofenoxim + terbuthylazine, 3 : 1); 'Faneron combi 500 FW and 50 WP' (flowable and w.p., bromofenoxim + terbuthylazine, 2 : 1); 'Faneron GB 500 FW and WP' (flowable and w.p., bromofenoxim + terbuthylazine, 1·65 : 1).

Analysis. Product analysis is by titration of the phenolic hydroxy-group with sodium hydroxide, or by colorimetric determination of the 2,4-dinitrophenol liberated after hydrolysis: particulars from Ciba-Geigy AG. Residues are determined by GLC with ECD.

Bromophos

Cl

(MeO)₂PS.O—[benzene ring]—Br

Cl

GR   DG   BE   EOPS&O1&O1 $C_8H_8BrCl_2O_3PS$ (366·0)

Nomenclature and development. The common name bromophos is approved by BSI and by ISO; also known as ENT 27 126, OMS 658. The IUPAC and *C.A.* name is *O*-(4-bromo-2,5-dichlorophenyl) *O,O*-dimethyl phosphorothioate, Registry No. *[2104-96-3]*. Its insecticidal properties were first described by R. Immel & G. Geisthardt, *Meded. Landbouwhogesch. Opzoekingsstn. Staat Gent,* 1964, **29,** 1242. It was introduced in 1964 by C. H. Boehringer Sohn/Cela GmbH under code number 'S1942', trade mark 'Nexion' and protected by DBP 1 174 104; BP 956 343; USP 3 275 718.

Manufacture and properties. Produced by the reaction of *O,O*-dimethyl phosphoro-chloridothioate with 4-bromo-2,5-dichlorophenol, it forms yellowish crystals, m.p. 53-54°, v.p. 1·3 x 10⁻⁴ mmHg at 20°. Its solubility at room temperature is 40 mg/l water; soluble in most organic solvents, particularly carbon tetrachloride, diethyl ether, toluene. The technical product is 90% pure (m.p. ≥ 51°). It is stable in media up to pH 9, is non-corrosive and is compatible with all pesticides except sulphur and the organometal fungicides.

Uses. Bromophos is a non-systemic contact and stomach insecticide, active against Hemiptera, Diptera, certain Lepidoptera, Coleoptera and other insects. It is recommended for crop protection at 25-75 g a.i./100 l; for the control of flies, mosquitoes at 0·5 g/m². It is non-phytotoxic at insecticidal concentrations though suspect of damage under glass. It persists on sprayed foliage for 7-10 d. For general review, see D. Eichler, *Residue Rev.,* 1972, **41,** 65.

Toxicology. The acute oral LD50 is: for rats 3750-7700 mg/kg; for mice 2829-5850 mg/kg; for rabbits 720 mg/kg; for hens 9700 mg/kg; for honey bees 18·8-19·6 mg/kg. The acute dermal LD50 for rabbits is 2181 mg/kg. In 2-y feeding trials there was no clinical symptom for rats receiving 350 mg/kg daily or dogs 44 mg/kg daily. The LC50 for guppies was 0·5 mg/l; concentrations of 0·5-1·0 mg/l were non-fatal to mosquito fish in natural surroundings.

Formulations. These include: e.c. (250 and 400 g a.i./l); w.p. (250 g/kg); dusts (20-50 g/kg); granules (50-100 g/kg); atomising concentrate (400 g/l); dip (200 g/l); coarse powder (30 g/kg).

Analysis. Product analysis is by i.r. spectroscopy (FAO specification, CP/70). Residue analysis by GLC. Determination of the halogenated phenol produced by hydrolysis can be used for both product and residue analysis. See also: R. D. Weeren & D. Eichler, *Anal. Methods Pestic. Plant Growth Regul.,* 1978, **10,** 31.

Bromophos-ethyl

GR   DG   BE   EOPS&O2&O2 $C_{10}H_{12}BrCl_2O_3PS$ (394·0)

Nomenclature and development. The common name bromophos-ethyl is approved by BSI and ISO; also known as ENT 27 258, OMS 659. The IUPAC and *C.A.* name is *O*-(4-bromo-2,5-dichlorophenyl) *O,O*-diethyl phosphorothioate, Registry No. *[4824-78-6]*. Its insecticidal properties were first described by M. S. Mulla *et al., Mosq. News,* 1964, **24**, 312. It was introduced by C. H. Boehringer Sohn/Cela GmbH, under the code number 'S 2225', trade mark 'Nexagan' and protected by DBP 1 174 104; BP 956 343; USP 3 275 718.

Manufacture and properties. Produced by the reaction of *O,O*-diethyl phosphoro-chloridothioate with 4-bromo-2,5-dichlorophenol, it is a colourless to pale yellow liquid, b.p. 122-133°/0·001 mmHg, v.p. 4·6 x 10^{-5} mmHg at 30°. The technical grade as d^{20} 1·52-1·55. The solubility of the pure compound at room temperature is 2 mg/l water; miscible with all common organic solvents. It is stable in aqueous suspension but is slowly hydrolysed at pH 9; in homogeneous aqueous alcoholic solution at this pH de-ethylation occurs though at higher pH the phenol is removed as anion. It is non-corrosive and compatible with other pesticides except sulphur and organometal fungicides.

Uses. Bromophos-ethyl is a non-systemic contact and stomach insecticide, active against Hemiptera, Diptera, certain Lepidoptera, Coleoptera and other insects, and with some acaricidal activity. It is recommended for crop protection at 40-60 g a.i./100 l; against mites at 40-80 g/100 l; for cattle treatment against ticks at 50-100 g/100 l; against flies and mosquitoes at 0·4-0·8 ml/m². Phytotoxicity has not been reported. For general review see D. Eichler, *Residue Rev.,* 1972, **41,** 65.

Toxicology. The acute oral LD50 is: for rats 71-127 mg/kg, for mice 225-550 mg/kg; single oral doses of 100 mg/kg were lethal to guinea-pigs, of 125 mg/kg to sheep. Acute dermal LD50 (24-h patch test) on albino rabbits was 1366 mg/kg. No ill-effect was observed in rats fed 12 d at 1·5 mg/kg daily or dogs fed 42 d at 1 mg/kg daily.

Formulations. These include: e.c. (400 and 800 g a.i./l); w.p. (250 g/kg); granules (50 g/kg); atomising concentrate (800-900 g/l); dips (400 g/l).

Analysis. Product analysis is by i.r. spectroscopy. Residue analysis is by GLC. Determination of the halogenated phenol produced by hydrolysis can be used for both product and residue analysis. See also: R. D. Weeren & D. Eichler, *Anal. Methods Pestic. Plant Growth Regul.,* 1978, **10**, 41.

Bromopropylate

ER   DXQR   DE&VOY1&1 $C_{17}H_{16}Br_2O_3$ (428·1)

Nomenclature and development. The common name bromopropylate is approved by BSI, ANSI and ISO—exception JMAF (phenisobromolate); also known as ENT 27552. The IUPAC name is isopropyl 4,4'-dibromobenzilate (I), in *C.A.* usage 1-methylethyl 4-bromo-α-(4-bromophenyl)-α-hydroxybenzeneacetate *[18181-80-1]* formerly (I). It was introduced in 1966 by J. R. Geigy S.A. (now Ciba-Geigy AG) as an experimental acaricide under the code number 'GS 19851', trade marks 'Neoron' and 'Acarol' and protected by BelgP 691 105; SwissP 471 065. It was described by H. Grob *et al., Abstr. Int. Plant Prot. Congr., 6th,* 1967, p.198.

Manufacture and properties. Produced by the reaction of 4,4'-dibromobenzilic acid with propan-2-ol, it is a crystalline solid, m.p. 77°, v.p. 5·1 x 10^{-8} mmHg at 20°. Its solubility at 20° is < 5 mg/l water; it is readily soluble in most organic solvents.

Uses. Bromopropylate is a contact acaricide with residual activity, recommended for use on pome and stone fruit, citrus, grapes, cotton, vegetables, soyabeans, strawberries and ornamentals at 37·5-60 g a.i./100 l; on field crops at 0·5-1·0 kg/ha. It is also effective against mites resistant to organophosphorus compounds.

Toxicology. The acute oral LD50 for rats is > 5000 mg/kg. In 2-y feeding trials the 'no effect' level was: for rats 100 mg/kg diet; for dogs 250 mg/kg diet. It is highly toxic to fish and slightly toxic to birds.

Formulation. An e.c.: 'Neoron 250'/'Acarol 250' and 'Neoron 500'/'Acarol 500' (250 and 500 g a.i./l, respectively).

Analysis. Product analysis may be by determination of total bromine, by estimation of the ester group by saponification, or by GLC. Residue analysis is by GLC after column chromatographic clean-up. Particulars from Ciba-Geigy AG.

Bromoxynil

QR BE FE DCN   *bromoxynil* $C_7H_3Br_2NO$ (276·9)

OR BE FE DCN &-KA-   *bromoxynil-potassium* $C_7H_2Br_2KNO$ (315·0)

Nomenclature and development. The common name bromoxynil is approved by BSI, ISO, ANSI and WSSA; also known as ENT 20852. The IUPAC name is 3,5-dibromo-4-hydroxybenzonitrile (I) formerly 3,5-dibromo-4-hydroxyphenyl cyanide, in *C.A.* usage (I) *[1089-84-5]*. Its herbicidal properties were reported independently by R. L. Wain, *Nature (London)*, 1963, **200**, 28, by K. Carpenter & B. J. Heywood, *ibid.*, p.28, and by Amchem Products Inc. It was introduced in 1963 by Amchem Products, Inc. under the trade mark 'Brominil' and by May & Baker Ltd under the code number 'MB 10064' and the trade mark 'Buctril'.

Manufacture and properties. Produced by bromination of 4-hydroxybenzonitrile, it is a colourless solid, m.p. 194-195°, subliming *c.* 135°/0·15 mmHg and is slightly volatile in steam. Its solubility at 25° is: 130 mg/l water; 170 g/l acetone; 90 g/l methanol; < 20 g/l petroleum oils; 410 g/l tetrahydrofuran. The technical grade is *c.* 95% pure, m.p. 189-192°.

It is acidic, pK_a 4·06, and forms alkali metal and amine salts, typical solubilities at 20-25° being: bromoxynil-sodium, 42 g a.e./l water; bromoxynil-potassium (*C.A.* Registry No. *[2961-68-4]*), 61 g a.e./l water. Certain esters are oil-soluble—*e.g.* the octanoate (see following entry). Bromoxynil and its salts are stable and non-corrosive.

Uses. Bromoxynil is a contact herbicide with some translocated activity, and is used mainly for post-em. control of autumn and spring broad-leaved weeds in cereal crops at 280-560 g alkali salt/l, and in newly-sown turf. It has little residual activity and its fate in soils has been reported (G. H. Ingram & E. M. Pullin, *Pestic. Sci.*, 1974, **5**, 287).

Toxicology. The acute oral LD50 for rats is 190-270 mg/kg.

Formulations. These include w.s.c. of alkali metal salts, usually in combination with other herbicides, *e.g.* 'Tetroxone' (ICI Ltd, Plant Protection Division) (bromoxynil + ioxynil + dichlorprop + MCPA) as potassium salts. See next entry for the octanoate ester.

Analysis. Product analysis is by GLC. Residues may be determined by i.r. spectroscopy of bromoxynil or by GLC after conversion to the methyl ether (H. S. Segal & M. L. Sutherland, *Anal. Methods Pestic., Plant Growth Regul. Food Addit.*, 1967, **5**, 347; *Anal. Methods Pestic. Plant Growth Regul.*, 1972, **6**, 605; T. H. Byast *et al.*, *Agric. Res. Counc. (G.B.) Weed Res. Organ., Tech. Rep.*, No. 15 (2nd Ed.), p.13).

Bromoxynil octanoate

$$CN$$

(structure: 4-cyanophenyl ring with Br at 2,6 positions and OC(CH₂)₆Me ester at 4-position)

NCR CE EE DOV7 $C_{15}H_{17}Br_2NO_2$ (403·0)

Nomenclature and development. The common name bromoxynil is approved by BSI, ISO, ANSI and WSSA for 3,5-dibromo-4-hydroxybenzonitrile, the introduction of which as a herbicide is described in the previous entry. Bromoxynil octanoate is 2,6-dibromo-4-cyanophenyl octanoate (I) (IUPAC), in *C.A.* usage (I) *[1689-99-2]* formerly 3,5-dibromo-4-hydroxybenzonitrile octanoate ester. It was introduced in 1963 by May & Baker Ltd under the code numbers 'MB 10731' and '16272 RP', the trade mark 'Buctril' and protected by BP 1 067 033, and by Amchem Products, Inc. under the trade name *'Brominil'* later changed to 'Brominal'.

Manufacture and properties. Produced by the esterification of bromoxynil with octanoyl chloride, it is a cream waxy solid, m.p. 45-46°, of low volatility though it sublimes at 90° and 0·1 mmHg. The technical material has a slight ester odour and m.p. 40-44°. Its solubility at 25° is: practically insoluble in water; > 100 g/l acetone; 100 g/l methanol; 700 g/l xylene. It is stable on storage, does not react with most other pesticides and is only slightly corrosive. It is readily hydrolysed by dilute alkali.

Uses. Bromoxynil octanoate is a contact herbicide used for the post-em. control of seedling broad-leaved weeds in cereals at 280-560 g a.i./ha, and to control perennial weeds in industrial and other non-crop situations at 1·15 kg/ha. It is mainly used in combination with other herbicides. It is rapidly degraded in soil by micro-organisms and chemical reactions, 50% loss occurring in *c.* 10 d (R. F. Collins, *Pestic. Sci.,* 1973, **4**, 181; G. H. Ingram & E. M. Pullin, *ibid.,* 1974, **5**, 287). In plants and animals it is hydrolysed to bromoxynil, and the nitrile group is hydrolysed to the corresponding amide and carboxylic acid with some dehalogenation (J. H. Buckland, *ibid.,* 1973, **4**, 149, 689).

Toxicology. The acute oral LD50 is: for rats 250 mg/kg; for mice 245 mg/kg; for rabbits 325 mg/kg; for dogs > 50 mg/kg. In 90-d feeding trials: rats receiving 312 mg/kg diet suffered no ill-effect, but at 781 mg/kg diet there was a significant decrease in growth rate; dogs receiving 5 mg/kg daily suffered no ill-effect, but at 25 mg/kg daily there was a reduction in body weight without anorexia. The sub-acute LC50 (8 d) for pheasant chicks is 4400 mg/kg diet. The LC50 (96 h) for rainbow trout is 0·05 mg/l.

Formulations. These include e.c. such as: 'Brominal' (240 g a.i./l); 'Buctril M' (200 g bromoxynil octanoate + 200 g MCPA-iso-octyl/l); 'Oxytril CM' (bromoxynil octanoate + ioxynil octanoate; 400 g total a.e./l); 'Oxytril P' (bromoxynil octanoate + ioxynil octanoate + dichlorprop-iso-octyl; 500 g total a.e./l); 'Brittox' (bromoxynil octanoate + ioxynil octanoate + mecoprop-iso-octyl; 525 total a.e./l).

Analysis. Product analysis is by GLC or by bromine determination. Residues may be determined by GLC or by i.r. spectroscopy using the intense band at 4·5 μm due to aromatic nitrile. See also bromoxynil (previous entry).

Bronopol

$$\begin{array}{c} \text{Br} \\ | \\ \text{HOH}_2\text{C}-\text{C}-\text{CH}_2\text{OH} \\ | \\ \text{NO}_2 \end{array}$$

WNXE1Q1Q

$C_3H_6BrNO_4$ (200·0)

Nomenclature and development. The common name bronopol is approved by BPC. The IUPAC name is 2-bromo-2-nitropropane-1,3-diol, in *C.A.* usage 2-bromo-2-nitro-1,3-propanediol *[52-51-7]*. It was introduced in 1964 by The Boots Co. Ltd as an ingredient in a medical product and in 1970 for agricultural uses under the protection of BP 1 193 954. Its use in agriculture was first described by D. F. Spooner & S. B. Wakerley, *Proc. Br. Insectic. Fungic. Conf., 6th,* 1971, **1,** 201.

Manufacture and properties. Produced by bromination of the sodium salt of 2-nitropropane-1,3-diol, it is a colourless to pale brown-yellow solid, m.p. 130°, v.p. 1·26 x 10⁻⁵ mmHg· at 20°. Its solubility at 22° is 250 g/l water; it is soluble in 2 vol. acetone, 6 vol. 2-methoxyethanol and > 500 vol. toluene. The technical product is > 90% pure. It is slightly hygroscopic but stable under normal storage conditions; unstable in aluminium containers but stable in tin. It has been mixed successfully with aldrin, captan, dieldrin, heptachlor and thiram.

Uses. Bronopol is a bacteriostat active against a wide range of plant pathogenic bacteria, especially *Xanthomonas malvacearum* causing Blackarm disease of cotton, against which it is applied as a seed dressing. It is non-phytotoxic on a wide range of crops including cotton.

Toxicology. The acute oral LD50 is: for rats 180-400 mg/kg; for mice 270-400 mg/kg; for dogs 250 mg/kg. The acute dermal LD50 for rats is > 1600 mg/kg. The acute inhalation LC50 (6 h) for rats is > 5 g/m³ air. A 0·5% aqueous solution is moderately irritant to rabbit skin on repeated applications, and a single application of 1% to rabbits' eyes is slightly irritant. Rats fed for 72 d at 1000 mg/kg diet showed no clinical or pathological sign of toxicity.

Formulations. Seed dressing (120 g a.i./kg); dust with captan, for dressing cotton seed.

Analysis. Product analysis is by GLC; particulars from The Boots Co. Ltd.

Bufencarb

3Y1&R	COVM1	*(I)*	$C_{13}H_{19}NO_2$ (221·3)
2Y2&R	COVM1	*(II)*	

Nomenclature and development. The common name bufencarb is approved by BSI, ANSI, New Zealand and proposed by ISO for a reaction product; also known as ENT 27 127. The insecticidally active components of the product, *C.A.* Registry No. *[8065-36-9]*, are: **(I)** in IUPAC and current *C.A.* usage 3-(1-methylbutyl)phenyl methylcarbamate *[2282-34-0]*, former *C.A.* name *m*-(1-methylbutyl)phenyl methylcarbamate; and similarly 3-(1-ethylpropyl)phenyl methylcarbamate *[672-04-8]* formerly *m*-(1-ethylpropyl)phenyl carbamate. It was introduced in 1968 by the Chevron Chemical Co. as an experimental insecticide under the code number 'Ortho 5353' and trade mark 'Bux'. Its insecticidal properties were first disclosed in 1962: USP 3 062 864; 3 062 867.

Manufacture and properties. Produced by the reaction of the mixture of alkylated phenols with methyl isocyanate, the technical product consists of *c.* 65% of a 3 : 1 mixture of 3-(1-methylbutyl)phenyl methylcarbamate **(I)** and 3-(1-ethylpropyl)phenyl methyl-carbamate **(II)** and 35% of structurally related insecticidally inactive material (mainly the corresponding 4- and 2-alkyl isomers). The technical product is a yellow to amber coloured solid, m.p. 26-39°, b.p. *c.* 125°/0·04 mmHg, v.p. 3 x 10^{-5} mmHg at 30°, d^{26} 1·024. Its solubility at room temperature is: < 50 mg/l water; it is very soluble in methanol or xylene; less soluble in aliphatic hydrocarbons such as hexane. It is stable in neutral or acid solutions, but the rate of hydrolysis increases with rise in either pH or temperature. It is degraded fairly rapidly in soil and no carry-over would be expected from one season to the next with granules.

Uses. Bufencarb is active, at rates of 0·5-2·0 kg a.i./ha, against a range of soil and foliage insects, particularly corn rootworm larvae, rice stem borers, rice green leafhopper, rice water weevil, rice planthoppers and pineapple root mealybug.

Toxicology. The acute oral LD50 of the technical material for rats is 87 mg/kg; the acute dermal LD50 for rabbits is 680 mg/kg. Toxic if ingested or absorbed through the skin, it inhibits cholinesterase activity in red blood cells; atrophine sulphate is an effective antidote, 2-PAM is *not* effective. Eye irritation for rabbits is minimal. In 90-d feeding trials no effects were noted in beagle dogs or albino rats receiving levels up to 500 mg/kg diet. In an 8-d feeding trial the oral LD50 for pheasants and ducks was > 27 000 mg/kg diet. The LC50 (96 h) was: for goldfish 0·56 mg/l; for catfish 1·95 mg/l; for trout 0·064 mg/l.

Formulations. These include: 'Bux Ten Granular', granules (100 mg a.i./kg); 'Bux 2 Emulsive', e.c. (240 g/l); 'Bux 360 Emulsive', e.c. (360 g/l); dusts (20 and 40 g/kg).

Analysis. Product analysis is by GLC; particulars of the method and those for residues are available from the Chevron Chemical Co. See also, B. Tucker, *Anal. Methods Pestic. Plant Growth Regul.*, 1973, **7**, 179.

Bupirimate

T6N CNJ BM2 DOSWN1&1 E4 F1 $C_{13}H_{24}N_4O_3S$ (316·4)

Nomenclature and development. The common name bupirimate is approved by BSI, ANSI and New Zealand and proposed by ISO. The IUPAC name is 5-butyl-2-ethylamino-6-methyl-pyrimidin-4-yl dimethylsulphamate, in *C.A.* usage 5-butyl-2-(ethylamino)-6-methyl-4-pyrimidinyl dimethylsulfamate *[41483-43-6]*. It was introduced in 1975 by ICI Ltd, Plant Protection Division under the code number 'PP 588', trade marks 'Nimrod' and 'Nimrod T' and protected by BP 1400710. Its fungicidal properties were described by J. R. Finney *et al., Proc. Br. Insecticide. Fungic. Conf., 8th,* 1975, **2,** 667.

Manufacture and properties. Produced by the sulphamoylation of ethirimol (p.245), it is a pale tan waxy solid, m.p. 50-51°, v.p. 0·5 x 10^{-6} mmHg at 20°; the technical material has m.p. *c.* 40-45°. Its solubility at room temperature is 22 mg/l water; soluble in most organic solvents except paraffins. It is easily hydrolysed by dilute acids and is unstable at elevated temperatures. It is non-corrosive to metals.

Uses. Bupirimate is a systemic fungicide specifically effective against powdery mildews, especially those of apple and glasshouse rose, at rates of 5-15 g a.i./100 l HV. This specificity enhances its usefulness in integrated control programmes.

Toxicology. The acute oral LD50 for female rats, mice, rabbits and male guinea-pigs is *c.* 4000 mg/kg; no clinical signs were noted in rats after 10 daily treatments dermally at 500 mg/kg; mild irritation was caused to the eyes of rabbits. In 90-d feeding trials the 'no effect' level was: for rats 1000 mg/kg diet; for dogs 15 mg/kg daily.

Formulations. These include: 'Nimrod', e.c. (250 g. a.i./l); w.d.p. (250 g/kg); 'Nimrod T', e.c. (62·5 g bupirimate + 62·5 g triforine/l).

Analysis. Product analysis is by TLC on silica gel. The bupirimate is recovered from the plate with methanol and the extinction measured at 241 nm. Residues may be determined by extracting with methanol, hydrolysis to ethirimol which is converted by diazomethane to its methyl ether and measured by GLC. Details are available from ICI Ltd, Plant Protection Division.

Butachlor

$$BuO.CH_2.N.CO.CH_2Cl$$

Et———Et

G1VN1O4&R   B2   F2 $C_{17}H_{26}ClNO_2$ (311·9)

Nomenclature and development. The common name butachlor is approved by BSI, ANSI, WSSA and JMAF, and proposed by ISO. The IUPAC name is *N*-(butoxymethyl)-2-chloro-2′,6′-diethylacetanilide (I) formerly *N*-butoxymethyl-α-chloro-2′,6′-diethylacet-anilide, in *C.A.* usage *N*-(butoxymethyl)-2-chloro-*N*-(2,6-diethylphenyl)acetamide *[23184-66-9]* formerly (I). It was introduced in 1969 by the Monsanto Co. under the code number 'CP 53619', trade mark 'Machete' and protected by USP 3 442 945. Herbicidal compositions and method of use thereof are protected by USP 3 547 620. Its herbicidal properties were described by D. D. Baird & R. P. Upchurch, *Proc. South. Weed Control Conf., 23rd,* 1970, p.101.

Manufacture and properties. Produced by the reaction of chloroacetyl chloride with the azomethine of 2,6-diethylaniline and formaldehyde, followed by treatment with butan-1-ol, pure butachlor is a light yellow oil, b.p. 196°/0·5 mmHg, v.p. negligible at ordinary temperature, d_4^{30} 1·0695. Its solubility at 20° is 20 mg/l water; it is soluble in most organic solvents.

Uses. Butachlor is a pre-em. herbicide for the control of annual grasses and certain broad-leaved weeds in rice, both seeded and transplanted. It shows selectivity in wheat, barley, sugar beet, cotton, peanuts and several brassica crops. Effective rates range from 1-3 kg a.i./ha. Activity is dependent on water availability such as rainfall following treatment, overhead irrigation or applications to standing water as in rice culture.

Toxicology. The acute oral LD50 for rats is 3300 mg of the e.c./kg; the percutaneous LD50 for rabbits is 4000 mg/kg. It is a mild skin and eye irritant.

Formulations. An e.c. (600 g a.i./l); granules (50 g/kg).

Analysis. Isothermal GLC is used: Monsanto Co. (AQC 118-70).

Butam

$$PhCH_2.N.CCMe_3$$

(with O double-bonded above the second C, and CHMe_2 below the N)

1Y1&N1R&VX1&1&1 $C_{15}H_{23}NO$ (233·4)

Nomenclature and development. The common name butam is approved by ANSI and WSSA; there is no BSI or ISO name. The IUPAC name is *N*-benzyl-*N*-isopropylpival-amide, in *C.A.* usage 2,2-dimethyl-*N*-(1-methylethyl)-*N*-(phenylmethyl)propanamide *[35256-85-0]*. It was introduced by the Gulf Oil Chemicals Co. as an experimental herbicide under the code number 'GPC-5544' protected by USP 3 974 218; 3 707 366. It was reported by R. A. Schwartzbeck, *Proc. 1976 Br. Crop Prot. Conf., Weeds,* 1976, p.739.

Manufacture and properties. Produced by the action of pivaloyl chloride on *N*-benzylisopropylamine in benzene in the presence of triethylamine, it is a colourless oil, b.p. 86-87°/0·07 mmHg, n_D^{25} 1·5074. It is insoluble in water; very soluble in ethanol, benzene, toluene.

Uses. Butam is effective as a pre-em. herbicide controlling annual grass weeds, *Amaranthus retroflexus* and *Chenopodium album* in soyabeans, peanuts, cotton, rapeseed, beans, peas and potatoes at 2·3-3·4 kg a.i./ha. The weeds emerge but are stunted and die within a few weeks.

Toxicology. The acute oral LD50 is: for albino rats 6210 mg a.i./kg; for guinea-pigs 2025 mg/kg. The acute dermal LD50 for albino rabbits is > 2000 mg/kg; the e.c. is considered an eye but not a skin irritant to albino rabbits.

Formulations. 'GPC-5544-6EC', e.c. (720 g a.i./l).

Butamifos

WNR   D1   BOPS&O2&MY2&1 $C_{13}H_{21}N_2O_4PS$ (332·3)

Nomenclature and development. The common name butamifos is approved by BSI and proposed by ISO. The IUPAC name is *O*-ethyl *O*-(6-nitro-*m*-tolyl) *sec*-butylphosphorami-dothioate, in *C.A.* usage *O*-ethyl *O*-(5-methyl-2-nitrophenyl) 1-methylpropylphosphorami-dothioate *[36335-67-8]*. It was first described by M. Ueda, *Jpn. Pestic. Inf.,* 1975, No. 23, p.23 and was introduced in 1970 by Sumitomo Chemical Co. under the code number 'S-2846', trade name 'Cremart', and protected by BP 1 359 727; USP 3 936 433.

Manufacture and properties. Produced by the reaction of *O*-ethyl *O*-(6-nitro-*m*-tolyl) phosphorochoridodithioate with *sec*-butylamine, it is a brownish liquid, b.p. 160°/0·1 mmHg, v.p. 6·3 x 10^{-4} mmHg at 27°, d_4^{25} 1·188, n_D^{25} 1·5340. It is practically insoluble in water, but soluble in most organic solvents.

Uses. Butamifos is a contact herbicide used as a pre-em. treatment, effective at 1-2 kg a.i./ha against annual weeds and especially graminaceous weeds.

Toxicology. The acute oral LD50 is: for rats 630-790 mg/kg; for mice 400-430 mg/kg. Neither skin nor eye irritation was observed. The LC50 (48 h) for carp is > 1 mg/l.

Formulation. An e.c. (500 g a.i./l).

Analysis. Product analysis is by GLC.

Buthidazole

T5NVNTJ A1 DQ C- CT5NN DSJ EX1&1&1 $C_{10}H_{16}N_4O_2S$ (256·3)

Nomenclature and development. The common name buthidazole is approved by BSI, ANSI, WSSA and proposed by ISO. The IUPAC name is 3-(5-*tert*-butyl-1,3,4-thiadiazol-2-yl)-4-hydroxy-1-methyl-2-imidazolidone, in *C.A.* usage, 3-[5-(1,1-dimethylethyl)-1,3,4-thiadiazol-2-yl]-4-hydroxy-1-methyl-2-imidazolidinone *[55511-98-3]*. It was introduced in 1974 by the Veliscol Chemical Co. under the code number 'Vel-5026', trade mark 'Ravage' and protected by USP 3904640. Its herbicidal properties were described by R. F. Anderson, *Proc. Int. Velisicol Symp., 8th,* 1974.

Manufacture and properties. Produced by the reaction of carbonyl chloride with 2-amino-5-*tert*-butyl-1,3,4-thiadiazole followed by reaction with methyl(2,2-dimethoxy-ethyl)amine and the product heated with conc. HCl, buthidazole is a colourless crystalline solid, m.p. 133-134°. Its solubility at 25° is: 3·4 g/kg water; 144 g/kg acetone; 208 g/kg dimethylformamide; 44 g/kg toluene; 17 g/kg xylene.

Uses. Buthidazole is a residual herbicide of high persistence, suitable for use in non-crop areas at rates of 6·5 kg a.i./ha in regions of > 50 cm annual rainfall; at 4·5 kg/ha in regions of rainfall < 50 cm. At lower rates it has selective herbicidal properties against annual grass and broad-leaved weeds in maize, sugarcane and pineapple.

Toxicology. The acute oral LD50 for albino rats is 1483-1581 mg/kg. Non-toxic when applied dermally and non-irritant to the eyes of albino rats. The LC50 (96 h) is: for bluegill 122 mg/l; for rainbow trout 74·7 mg/l; for channel catfish 239 mg/l.

Formulations. These include: 'Ravage', w.p. (750 g a.i./kg); w.p. (500 g/kg); granules and pellets (50 and 100 g/kg).

Analysis. Product analysis is by HPLC with detection at 254 nm. Soil residues are extracted with methanol and determined by HPLC on a reversed phase column. Details of these methods are available from Veliscol Chemical Corp.

63

Buthiobate

T6NJ CNUYS4&S1R DX1&1&1 $C_{21}H_{28}N_2S_2$ (372·6)

Nomenclature and development. The common name buthiobate is approved by BSI and proposed by ISO. The IUPAC name is butyl 4-*tert*-butylbenzyl *N*-(3-pyridyl)dithiocarbonimidate, in *C.A.* usage butyl [4-(1,1-dimethylethyl)phenyl]methyl 3-pyridinylcarbonimidodithioate *[51308-54-4]*. It was first described by T. Kato *et al., Agric. Biol. Chem.,* 1975, **39**, 169; S. Tanaka *et al., ibid.,* 1977, **41**, 1627, and was introduced in 1977 by the Sumitomo Chemical Co. under the code number 'S-1358', trade name 'Denmert', and protected by BP 1 335 617; USP 3 832 351.

Manufacture and properties. Produced by the reaction of butyl (3-pyridyl)dithiocarbamate with a 4-*tert*-butylbenzyl halide, it is a yellowish oily liquid, m.p. 31-33°, v.p. 4·52 x 10⁻⁷ mmHg at 20°, d_{25}^{25} 1·0865, $n_D^{26.5}$ 1·596. It is practically insoluble in water but soluble in most organic solvents.

Uses. It is a preventive, curative and persistent fungicide, effective against powdery mildews of vegetables, beans, fruit and other crops. It is usually applied at 15-250 mg/l.

Toxicology. The acute oral LD50 is: for rats 2700-4900 mg tech./kg; for mice 4500-4550 mg/kg. Neither skin nor eye irritation was observed in rabbits. The LC50 (48 h) for carp is 6·4 mg/l.

Formulations. An e.c. (100 g a.i./l); w.p. (200 g/kg).

Analysis. Product analysis is by GLC.

Butocarboxim

Me
|
MeS.CH.C.Me
||
N.O.C.NHMe
||
O
(E)

Me
|
MeS.CH.C.Me
||
MeNH.C.O.N
||
O
(Z)

1SY1&Y1&UNOVM1 $C_7H_{14}N_2O_2S$ (190·3)

Nomenclature and development. The common name butocarboxim is approved by BSI and ISO. The IUPAC name is 3-(methylthio)butanone *O*-methylcarbamoyloxime, in *C.A.* usage 3-methylthio-2-butanone *O*-[(methylamino)carbonyl]oxime *[34681-10-2]*. It was introduced in 1973 by Wacker-Chemie GmbH under the code number 'Co 755', trade mark 'Drawin 755' and protected by DBP 2036491; USP 3816532; BP 1353202. Its insecticidal properties were first described by M. Vulić *et al., Meded. Fac. Landbouwwet. Rijksuniv. Gent,* 1973, **38,** 1175.

Manufacture and properties. Produced by oximation of 3-(methylthio)butanone and subsequent reaction with methyl isocyanate. The technical product is an 85% solution in xylene and contains the *(E)-* and *(Z)*-isomers in a ratio of 85:15. The pure *(E)*-isomer has m.p. 37°, the isomeric mixture v.p. *c.* 8 x 10^{-5} mmHg at 20°. Solubility is good in aromatic hydrocarbons, ketones, esters; poor in aliphatic hydrocarbons, carbon tetrachloride.

Uses. Butocarboxim is a systemic insecticide with moderate persistence against sucking insects. It has an excellent activity against *Aleurothrixus floccosus* in citrus cultures (100 g a.i./100 l) (M. Vulić & J. L. Beltran, *Z. Pflanzenkr. Pflanzenschutz,* 1977, **84,** 202); good activity (50-75 g/100 l) against aphids on vegetable, fruit and ornamental crops and moderate activity against mites. Structure/activity of analogues is reported by T. A. Magee & L. E. Limpel, *J. Agric. Food Chem.,* 1977, **25,** 1376.

Toxicology. The acute oral LD50 for rats is 158-240 mg/kg. The subcutaneous LD50 for rats is 188 mg/kg. The dermal LD50 for albino rabbits is 360 mg/kg. The 'no effect' level for beagle dogs (90 d) and rats (2 y) was 100 mg/kg diet; there was no carcinogenic effect at the highest dose (300 mg/kg diet) in rats during 2 y, nor any effect on fertility, growth rate or mortality. The LC50 (48 h) was: for rainbow trout 35 mg/l; for ide 55 mg/l; for Japanese quail 1180 mg/kg diet. No mutagenicity was observed with *Salmonella typhimurium* (TA 1535, TA 1537, TA 98, TA 100).

Formulation. 'Drawin 755' e.c. (500 g a.i./l) for agricultural and horticultural use; 'Drawin 75-5' solution (50 g/l) and spray (0·8 g/l) for garden and domestic use.

Analysis. Product analysis is by i.r. spectroscopy in chloroform at 5·78 µm or by HPLC. Residue analysis for butocarboxim, its sulphoxide and sulphone (butoxycarboxim) is by GLC with FID, details from Wacker-Chemie GmbH.

Butopyronoxyl

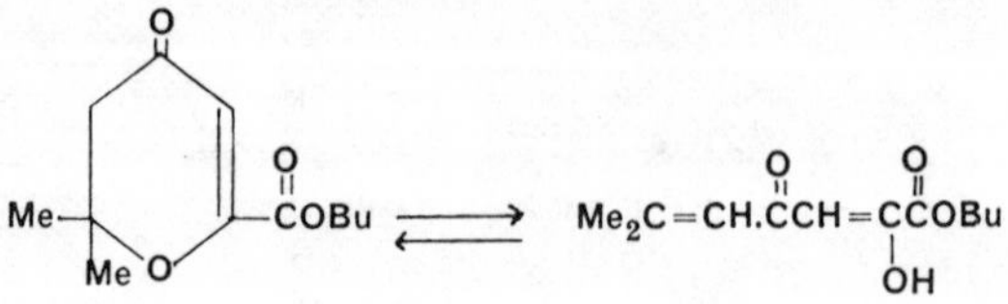

(II)

$$C_{12}H_{18}O_4 \quad (226\cdot3)$$

T6O DV BUTJ BVO4 F1 F1   4OVYQU1V1UY1&1

Nomenclature and development. The common name butopyronoxyl was adopted by US Pharm.; also known as ENT 9 and by the trivial names dihydropyrone, butyl mesityloxide oxalate. The IUPAC name is butyl dihydro-6,6-dimethyl-4-oxopyran-2-carboxylate, in *C.A.* usage butyl 3,4-dihydro-2,2-dimethyl-4-oxo-*2H*-pyran-6-carboxylate *[532-34-3]*. It was introduced as an insect repellent during the 1939-45 war by US Industrial Chemicals Inc. and Kilgore Chemicals under the protection of USP 2138540. The trade mark 'Indalone' (FMC Corp.) has been assigned to the Aldrich Chemical Co. Inc.

Manufacture and properties. Produced by the condensation of 4-methylpent-3-en-2-one and dibutyl oxalate in the presence of sodium ethoxide, it is a yellow to pale red liquid of mild aromatic odour; the technical product contains $\geqslant 85\%$ butopyronoxyl and $\leqslant 2\%$ mesityl oxide, the remainder being largely dibutyl oxalate and butanol; 90% distils between 256-270°; d_4^{20} 1·052-1·060; n_D^{20} 1·4750. It is practically insoluble in water, glycerol; slightly soluble in ethylene glycol, refined petroleum oils; miscible with chloroform, ethanol, diethyl ether, glacial acetic acid. It exists mainly as the dihydropyrone (I) but with the open chain enol (II) in equilibrium (S. A. Hall *et al., J. Am. Chem. Soc.,* 1945, **67**, 1224).

Uses. Butopyronoxyl is an insect repellent with little insecticidal activity.

Toxicology. The acute oral LD50 is: for rats 7840 mg/kg; for guinea-pigs 3200 mg/kg. The acute dermal LD50 for rabbits is 10000 mg/kg. In 2-y feeding trials some growth retardation was observed in rats receiving 80 g/kg diet. It is non-irritant though causing a slight stinging sensation to some men when applied to the skin.

Formulation. It is generally used in admixture (2 parts) with dimethyl phthalate (6 parts) and 2-ethylhexane-1,3-diol (2 parts).

66

Butoxycarboxim

Me
|
Me.SO₂.CH.C.Me
‖
N.O.C.NHMe
‖
O

(E)

Me
|
Me.SO₂.CH.C.Me
‖
MeNH.C.O.N
‖
O

(Z)

WS1&Y1&Y1&UNOVM1 $\qquad$ $C_7H_{14}N_2O_4S$ (222·3)

Nomenclature and development. The common name butoxycarboxim is approved by BSI and ISO. The IUPAC name is 3-(methylsulphonyl)butanone *O*-methylcarbamoyloxime, in *C.A.* usage 3-methylsulfonyl-2-butanone *O*-[(methylamino)carbonyl]oxime *[34681-23-7]*. It was introduced in 1972 by Wacker-Chemie GmbH under the code number 'Co 859', trade mark 'Plant Pin' and protected by DRP 2036491; USP 3816532; BP 2353202. Its insecticidal properties and the special application method were first described by M. Vulić & H. Bräunling, *Meded. Fac. Landbouwwet. Rijksuniv. Gent,* 1974, **39**, 847.

Manufacture and properties. Produced by oxidation of butocarboxim (p.65) with hydrogen peroxide, technical butoxycarboxim contains the (*E*)- and (*Z*)-isomers in a ratio of 85:15. The pure (*E*)-isomer has m.p. 83°; the isomeric mixture, v.p. 2 x 10⁻⁶ mmHg, is very soluble in water and polar solvents, moderately in aromatic hydrocarbons, almost insoluble in light petroleum, carbon tetrachloride.

Uses. Butoxycarboxim is a systemic insecticide effective against aphids and mites with special formulation and activity properties for soil application for potted plants. Structure/activity of analogues is reported by T. A. Magee & L. E. Limpel, *J. Agric. Food. Chem.,* 1977, **25**, 1376.

Toxicology. The acute oral LD50 is: for rats 458 mg/kg; for rabbits 275 mg/kg; for hens 367 mg/kg. The acute subcutaneous LD50 for female rats 288 mg/kg. The 'no effect' level in rats in 90-d trials was 300 mg/kg diet, whilst at 1000 mg/kg diet only slight inhibition of erythrocyte and plasma cholinesterase were noted. The oral LD50 for rats of the pasteboard stick formulation is > 5000 mg/kg. Butoxycarboxim is a metabolite of butocarboxim in plant and animal tissue, therefore the toxicological tests on the latter (p.65) partly include butoxycarboxim.

Formulation. Butoxycarboxim is formulated for the special use of soil application on potted ornamental domestic plants. It is incorporated in 'Plant Pin' (pasteboard-pins, 40 mm x 8 mm; containing 50 mg a.i. each).

Analysis. Product analysis is by i.r. spectroscopy in chloroform at 5·74 µm; residue analysis is by GLC with FID after conversion to a volatile derivative, details available from Wacker-Chemie GmbH.

Butralin

2Y1&MR   BNW   FNW   DX1&1&1 $C_{14}H_{21}N_3O_4$ (295·3)

Nomenclature and development. The common name butralin is approved by BSI, ANSI, New Zealand and WSSA, and proposed by ISO. The IUPAC name is *N-sec*-butyl-4-*tert*-butyl-2,6-dinitroaniline (I), in *C.A.* usage 4-(1,1-dimethylethyl)-*N*-(1-methylpropyl)-2,6-dinitrobenzenamine *[33629-47-9]* formerly (I). It was introduced by Amchem Products, Inc. under the code numbers 'Amchem 70-25', Amchem 'A-820', the trade marks 'Amex' and 'Tamex' and protected by USP 3672866. Its herbicidal properties were reported by J. R. Bishop *et al.*, *Pestic. Chem.*, *Proc. Int. Congr. Pestic. Chem.*, *2nd*, 1972, **5**, 197; S. R. McLane *et al.*, *Proc. South Weed Sci. Soc.*, 1971, **24**, 58.

Manufacture and properties. Produced by the nitration of 1-*tert*-butyl-4-chlorobenzene and subsequent reaction with *sec*-butylamine, it forms yellow-orange crystals, with a slight aromatic odour, m.p. 60-61°, b.p. 134-136°/0·5 mmHg, v.p. 1·3 x 10^{-5} mmHg at 25°, 2·8 x 10^{-5} mmHg at 30°. Its solubility at 24-26° is: 1 mg/l water; 125 g/kg methanol; 4·48 kg/kg acetone; 2·7 kg/kg benzene; 3·88 kg/kg xylene; 9·55 kg/kg butanone; 1·46 kg/kg carbon tetrachloride. Its flash point (Tag open cup) is 36° and it decomposes at 265°. Concentrates are stable on storage > 3 y but should not be stored < —5° nor allowed to freeze. No case of incompatibility is known. It is not corrosive to metals, but will permeate certain plastics and soften or swell certain types of rubber. It is stable to u.v. light.

Uses. Butralin is a pre-em. herbicide, which should be incorporated with the soil soon after application, used at 1·12-3·4 kg a.i./ha (depending on soil type) for weed control in soyabeans and cotton. It shows promise for weed control in lima beans, groundnuts, seedling alfalfa, flax, potatoes, *Phaseolus vulgaris* and cucumbers. Mixtures with vernolate or prometryne are also used on groundnuts and cotton, respectively. It is also used to control suckers on tobacco, being applied to stems at 125 mg/plant. It is not persistent in soil.

Toxicology. The acute oral LD50 for albino rats is 12600 mg tech./kg; the acute dermal LD50 for albino rabbits is 10200 mg/kg. The sub-chronic LC50 (8 d) for bobwhite quail and mallard duck was 10000 mg/kg diet. The LC50 (48 h) is: for bluegill 4·2 mg/l; for rainbow trout 3·4 mg/l.

Formulations. 'Amex', e.c. (480 g a.i./l).

Analysis. Product analysis is by GLC: details are available from Amchem Products, Inc.

Buturon

GR   DMVN1&Y1&1UU1 $C_{12}H_{13}ClN_2O$ (236·7)

Nomenclature and development. The common name buturon is approved by BSI and ISO—exception Portugal. The IUPAC name is 3-(4-chlorophenyl)-1-methyl-1-(1-methylprop-2-ynyl)urea, in *C.A.* usage N'-(4-chlorophenyl)-N-methyl-N-(1-methyl-2-propynyl)urea *[3766-60-7]* formerly 3-(*p*-chlorophenyl)-1-methyl-1-(1-methyl-2-propynyl)urea. Its herbicidal properties were first described by A. Fischer, *Meded. Landbouwhogesch. Opzoekingsstn. Staat Gent,* 1964, **29,** 719, and it was introduced in 1966 by BASF AG under the code number 'H 95', trade mark 'Eptapur' and protected by DBP 1 108 977.

Manufacture and properties. Produced by the reaction of 4-chlorophenyl isocyanate with methyl(1-methylprop-2-ynyl)amine, it is a colourless solid, m.p. 145-146°. Its solubility at 20° is: 30 mg/l water; 279 g/kg acetone; 9·8 g/kg benzene; 128 g/kg methanol. The technical grade has m.p. 132-142°. It is stable under normal conditions, but slowly decomposes in boiling water; it is compatible with other herbicides and is non-corrosive.

Uses. Buturon is a pre- and post-em. herbicide, absorbed mainly by the roots, and recommended for use at 0·5-1·5 kg a.i./ha in cereals and maize for the control of shallow-germinating grasses and broad-leaved weeds. It is rapidly degraded in plants.

Toxicology. The acute oral LD50 for rats is 3000 mg/kg; 20-h applications to the backs of rabbits produced a slight erythema but the ears were unaffected. In 120-d feeding trials, rats receiving 500 mg/kg diet showed no ill-effect.

Formulation. A w.p. (500 g a.i./kg).

Analysis. Product analysis is by hydrolysis and potentiometric titration of the 4-chloroaniline produced; residue analysis is by hydrolysis and colorimetric estimation of the 4-chloroaniline—particulars of both methods are available from BASF AG.

(±)-*sec*-**Butylamine**

$$\text{Me}$$
$$|$$
$$\text{Et.CH.NH}_2$$

ZY2&1 $\qquad\qquad\qquad\qquad\qquad\qquad\qquad\qquad$ $C_4H_{11}N$ (73·14)

Nomenclature and development. The IUPAC name is (±)-*sec*-butylamine (I), also known as (±)-2-aminobutane, in *C.A.* usage 2-butanamine *[13952-84-6]* formerly (I). It was introduced in 1962 by the University of California (Riverside) as a fungicidal fumigant and was first described by J. W. Eckert & M. J. Kolbezen, *Nature (London)*, 1962, **194**, 188. 'Tutane' is registered as a trade mark by Eli Lilly & Co., and 'Butafume' by BASF AG.

Properties. It is a colourless liquid, with an ammoniacal odour, b.p. 63°, v.p. 135 mmHg at 20°, n_D^{20} 1·394, d_4^{20} 0·724. It is miscible with water and most organic solvents. It is an organic base forming water-soluble salts with acids. Having an asymmetric carbon, it exists as optical isomers. It is stable but corrosive to tin, aluminium and some steels.

Uses. (±)-*sec*-Butylamine is a fungicide used for the control of many fruit-rotting fungi; aqueous solutions of its salts, containing 5-20 g amine/l, are used as dips or sprays on harvested fruit to prevent decay in transport or storage. The amine may be used to fumigate harvested fruit at 327 mg/m³ for 4 h, or its equivalent; potatoes (seed or ware) at 280 ml/t for 2·5 h to control gangrene and skin spot—see *Code of Practice for Fumigation of Potato Tubers with 2-Aminobutane* (sec-*Butylamine*). In neutral aqueous solution, the hydrochlorides of the enantiomorphs show marked differences in fungicidal acitivity; the control of decay of oranges due to *Penicillium digitatum* is largely by the *(R)*-(—)-isomer Registry No. *[13250-12-9]*: J. W. Eckert & M. J. Kolbezen, *Phytopathology*, 1967, **57**, 98. It is non-phytotoxic at 10 times the recommended concentration.

Toxicology. The acute oral LD50 is: for rats 380 mg amine/kg; for dogs 225 mg/kg; for hens 250 mg/kg. It is strongly irritant, but the dermal toxicity for rabbits is > 2500 mg/kg. In 2-y feeding trials rats and dogs receiving 2500 mg/kg diet suffered no ill-effect. In teratology studies, 'no effect' levels were: for rats 2500 mg/kg diet; for rabbits 5000 mg/kg diet. In reproduction studies in rats the 'no effect' level was 2500 mg/kg diet. The LC50 for bluegill fingerlings is > 50 mg/l.

Formulations. Concentrated aqueous solutions of the appropriate salts, preferably the acetate or phosphate; or as the free amine, 'Butafume', 99%.

Analysis. Product analysis is by the steam distillation of the amine into standard acid and back titration. Residues may be determined by the steam distillation of the amine into dilute sulphuric acid and the amine, after purification, treated with 1-fluoro-2,4-dinitrobenzene and the derivative estimated by GLC: E. W. Day *et al., J. Assoc. Off. Anal. Chem.*, 1968, **51**, 39; *Anal. Methods Pestic. Plant Growth Regul.*, 1976, **8**, 251.

Butylate

EtS.CO.N(CH$_2$CHMe$_2$)$_2$

1Y1&1NVS2&1Y1&1   C$_{11}$H$_{23}$NOS   (217·4)

Nomenclature and development. The common name butylate is approved by BSI, WSSA, ISO and in France (butilate)—exception Germany. The IUPAC name is *S*-ethyl di-isobutylthiocarbamate (I), in *C.A.* usage *S*-ethyl bis(2-methylpropyl)carbamothioate *[2008-41-5]*, formerly (I). Butylate is one of a series of thiolcarbamates introduced from 1954 onwards by the Stauffer Chemical Co. as herbicides. Its code number is 'R-1910' and trade mark 'Sutan'; protected by USP 2913327. Described by R. A. Gray *et al., Proc. North Cent. Weed Control Conf.,* 1962, **19**, 19.

Manufacture and properties. Produced by reaction of di-isobutylamine with carbonyl chloride giving the carbamoyl chloride which is reacted with ethanethiol, it is a clear liquid, with an aromatic odour, b.p. 138°/21 mmHg, v.p. 1·3 x 10^{-3} mmHg at 25°, *d* 0·9417, n_D^{30} 1·4701. Its solubility at room temperature is 45 mg/l water; miscible with kerosine, 4-methylpentan-2-one, xylene. It is non-corrosive.

Uses. It is toxic to germinating seeds and is incorporated with the soil immediately prior to sowing for the control of broad-leaved and grass weeds in maize. The rates recommended are: for the control of annual grasses and nut grass 3 kg a.i./ha; for the control of broad-leaved weeds in addition to grasses 4 kg/ha, or 3 kg butylate + 1 kg 2,4-D/ha. It is non-persistent.

Toxicology. The acute oral LD50 of the technical product for rats is 4000-4660 mg/kg; the acute dermal LD50 for rabbits is > 2000 mg/kg. In 90-d feeding trials 40 mg/kg daily was well tolerated by rats and dogs.

Formulations. These include: e.c. (720 g a.i./l); granules (100 g/kg). Also in combination with 2,4-D in the ratio 3:1; *e.g.,* e.c. (360 g butylate + 120 g 2,4-D/l), granules (60 g butylate + 20 g 2,4-D/kg; 120 g butylate + 40 g 2,4-D/kg).

Analysis. Product analysis is by GLC. Residues may be determined by hydrolysis, steam distillation of the amine and its estimation as the cupric dithiocarbamate complex. See also, J. E. Barney *et al., Anal. Methods Pestic. Plant Growth Regul.,* 1973, **7**, 641.

2-*sec*-Butylphenyl methylcarbamate

2Y1&R   BOVM1 $C_{12}H_{17}NO_2$ (207·3)

Nomenclature and development. The initials BPMC are approved as common name by JMAF; there is no BSI or ISO name. The IUPAC name is 2-*sec*-butylphenyl methylcarbamate (I), in *C.A.* usage 2-(1-methylpropyl)phenyl methylcarbamate *[3766-81-2]* formerly (I). It was introduced by the Sumitomo Chemical Co. under the trade mark 'Osbac', also by the Mitsubishi Chemical Industries Ltd and by Kumiai Chemical Industries Co. Ltd under the trade mark 'Bassa'. Its insecticidal properties were first described by R. L. Metcalf *et al., J. Econ. Entomol.,* 1962, **55,** 889.

Manufacture and properties. Produced by reacting 2-*sec*-butylphenyl chloroformate with methylamine, it is a crystalline solid, m.p. 32°, b.p. 112-113°/0·2 mmHg, d_4^{20} 1·050, $n_D^{27.7}$ 1·5115. Its solubility at 30° is 660 mg/l water; readily soluble in acetone, benzene, toluene, xylene. The technical grade is *c.* 95% pure. It is unstable to alkali and to concentrated acid.

Uses. It is an insecticide used to control hoppers on rice at 500 g a.i./100 l.

Toxicology. The acute oral LD50 is: for rats 410 mg/kg; for mice 340 mg/kg. The acute dermal LD50 for mice is 4200 mg/kg. The LC50 (48 h) for carp is 12·6 mg/l.

Formulations. An e.c. (500 mg a.i./l); dust (20 g/kg); microgranules (30 g/kg).

Analysis. Product analysis is by u.v. spectroscopy after purification by TLC.

Calciferol

$$C_{28}H_{44}O \quad (396{\cdot}7)$$

L56 FYTJ A1 BY1&1U1Y1&Y1&1 FU2U- BL6YYTJ AU1 DQ

Nomenclature and devlopment. The common name calciferol is approved by BPC, ergocalciferol by US Pharm.; also known as Vitamin D_2. The IUPAC name is ($3\beta,5Z,7E,22E$)-9,10-secoergosta-5,7,10(19),22-tetraen-3-ol (I), in *C.A.* usage (I) *[50-14-6]* formerly ergocalciferol. It was first prepared by u.v. irradiation of ergosterol (F. A. Askew *et al., Proc. Roy. Soc. Lond. B. Biol. Sci.,* 1932, **109,** 488) and its stereochemistry established (D. Crowfoot & J. D. Dunitz, *Nature (London),* 1948, **162,** 608; *Chem. Ind. (London),* 1957, p.1149) and its properties reviewed (H. H. Inhoffen, *Angew. Chem.,* 1960, **72,** 875). Its rodenticidal properties were discovered by Sorex Ltd and it was introduced as a commercial rodenticide in 1974 under the trade mark 'Sorexa C.R.' for a combination of calciferol and warfarin, protected by BP 1 371 135.

Manufacture and properties. Produced by u.v. irradiation of ergosterol which is extracted from yeast, it forms colourless crystals, m.p. 115-118°. It is insoluble in water; soluble in most organic solvents; solubility at 7° is 69·5 g/l acetone. Its stability on storage has been discussed (W. Huber & O. W. Barlow, *J. Biol. Chem.,* 1942, **149,** 125).

Uses. Calciferol is an essential natural vitamin, but in high doses produces a lethal hypervitaminosis characterised by hypercalcaemia and increase in serum cholesterol. Evidence suggests that admixture with warfarin increases the rodenticidal efficacy in many species. The mixture is semi-acute in action, 1-2 feeds being sufficient to cause death. This advantage, combined with excellent acceptability, is responsible for its success in the control of mice.

Formulations. Ready-to-use bait on canary seed (1 g calciferol + 250 mg warfarin/kg); oil concentrate (20 g/l).

Analysis. Product analysis is by colorimetry or by u.v. spectroscopy: details are available from Sorex Ltd.

Calcium arsenate

Nomenclature and development. The traditional name calcium arsenate is accepted by ISO and JMAF in lieu of a common name. The precise nature of the compounds present in calcium arsenate for crop protection use is unknown but phase rule studies indicate the presence of several basic calcium arsenates, of which $3Ca_3(AsO_4)_2.Ca(OH)_2$ is thought to be that most suitable for foliage use (G. W. Pearce & L. B. Norton, *J. Am. Chem. Soc.*, 1936, **58**, 1104; G. W. Pearce & A. W. Avens, *ibid.*, 1937, **59**, 1258). It came into insecticidal use about 1906.

Manufacture and properties. Produced by the interaction of arsenic acid and calcium hydroxide in proportions that yield arsenates more basic than calcium hydrogen arsenate (*cf.* lead arsenate, p.316), it is a white flocculent powder. It is practically insoluble in water; soluble in dilute mineral acids. It is decomposed by carbon dioxide yielding calcium carbonate and calcium hydrogen arsenate, the latter being appreciably soluble in cold water.

Uses. Calcium arsenate is a stomach poison, at one time widely used against leaf-eating insects, and still used on tobacco. Its liability to cause phytotoxicity is ascribed to the formation of water-soluble arsenic compounds.

Toxicology. It is of high mammalian toxicity by ingestion, the lethal dose being 35-100 mg/kg.

Formulation. Dusts.

Analysis. Product analysis is by determination of calcium, total and water-soluble arsenic content by standard methods (AOAC Methods; G. W. Pearce *et al., N.Y. St. Agric. Exp. Stn., Tech. Bull.,* No. 234, 1935). Residues may be determined by the Gutzeit method (AOAC Methods).

Calcium cyanide

Ca(CN)$_2$

. CA . . CN2 $\qquad\qquad\qquad\qquad\qquad\qquad\qquad\qquad\qquad$ C$_2$CaN$_2$ (92·11)

Nomenclature and development. It is known by the traditional chemical name calcium cyanide (I). The IUPAC name is calcium dicyanide, in *C.A.* usage (I) *[529-01-8]*. Its first use as a fumigant against insects was reported by H. J. Quayle, *J. Econ Entomol.*, 1923, **16,** 327. It was formerly marketed by American Cyanamid Co. under the trade mark *'Cyangas'*. It is now produced by other manufacturers and is used in countries other than the USA.

Manufacture and properties. Produced by the fusion of sodium chloride and calcium cyanamide, or by the reaction of liquid hydrogen cyanide with calcium carbide [a process which yields a product more akin to the acid cyanide CaH$_2$(CN)$_4$ than to Ca(CN)$_2$], it is a grey powder or a fine dust. Calcium cyanide is decomposed by moisture to give calcium hydroxide and hydrocyanic acid (which is highly toxic to animals and man), a decomposition proceeding at humidities as low as 25% r.h. (F. J. Metzger, *Ind. Eng. Chem.*, 1926, **18,** 161).

Uses. Its biological properties as a fumigant against insects and rodents are due wholly to the hydrocyanic acid produced by decomposition, these are discussed under hydrogen cyanide (p.300).

Analysis. Product analysis is by dissolving in water, precipitating sulphur compounds with lead carbonate and titrate with standard aqueous silver nitrate (AOAC Methods).

Camphechlor

Approximately $C_{10}H_{10}Cl_8$ (413·8)

Nomenclature and development. The common name camphechlor is approved by BSI and ISO—exceptions Belgium, Canada, France, India and USA (toxaphene) and USSR (polychlorcamphene)—also known as ENT 9735, for a reaction mixture of chlorinated camphenes containing 67-69% chlorine. 'Toxaphene' is the registered trade mark of Hercules Inc. but, in some countries, has been dedicated to the public and is there used as a common name; toxaphene is also used in *C.A.*, Registry No. *[8001-35-2]*. Camphechlor was first described by W. LeRoy Parker & J. R. Beacher, *Del. Univ. Agric. Exp. Stn. Bull.*, No. 264, 1947, and it was introduced in 1948 by Hercules Inc. under the code number 'Hercules 3956' and protected by USP 2 565 471; 2 657 164.

Properties. It is a yellow wax of mild terpene odour, softening in the range of 70-95°, v.p. 0·2-0·4 mmHg at 25°, d^{25} 1·65. Its solubility at room temperature is *c.* 3 mg/l water; it is readily soluble in organic solvents including petroleum oils. It is dehydrochlorinated by heat, by strong sunlight and by certain catalysts such as iron. It is incompatible with strongly alkaline pesticides and is non-corrosive in the absence of moisture.

The compounds present have been examined by J. E. Casida *et al., J. Agric. Food Chem.*, 1974, **22**, 653; 1975, **23**, 991; 26 components account for about 40% of the product; a heptachlorobornane and a mixture of 2 octachlorobornanes are more toxic to insects and mice than other components and are similarly biodegradable.

Uses. Camphechlor is a non-systemic contact and stomach insecticide with some acaricidal action. It is non-phytotoxic, except to cucurbits, and is used in the control of many insects on cotton, corn, small grains, vegetables and fruit. Also used for the control of animal ectoparasites.

Toxicology. The acute oral LD50 for rats is 80-90 mg/kg; the acute dermal LD50 for rats is 780-1075 mg/kg. The 'no effect' level from 2-y feeding tests on rats was 25 mg/kg diet. There was an accumulation in the body fat proportional to the dose fed but elimination was rapid when the intake was stopped.

Formulations. These include e.c. (480, 720 or 960 g a.i./l), the addition of 1-chloro-2,3-epoxypropane (5 g/l) to improve stability is restricted to countries where this use is permitted; w.p. (400 g/kg); dust (100 g/kg); granules (100 and 200 g/kg). Admixture with other insecticides is frequent.

Analysis. Product analysis is by total chlorine: AOAC Methods; *CIPAC Handbook,* 1970, **1**, 132; FAO Specifications (CP/35; CP/68). For residues use a spectrometric method; A. J. Grauper & C. L. Dunn, *J. Agric. Food Chem.,* 1960, **8**, 286, C. L. Dunn, *Anal. Methods Pestic., Plant Growth Regul. Food Addit.,* 1964, **2**, 523; or GLC using EDC after clean-up and dehydrochlorination: *Anal. Methods Pestic. Plant Growth Regul.,* 1972, **6**, 514.

Captafol

T56 BVNV GUTJ CSXGGYGG $C_{10}H_9Cl_4NO_2S$ (349·1)

Nomenclature and development. The common name captafol is approved by BSI, ISO and ANSI—exception JMAF (difolatan). The IUPAC name is 1,2,3,6-tetrahydro-*N*-(1,1,2,2-tetrachloroethylthio)phthalimide or *N*-(1,1,2,2-tetrachloroethylthio)cyclohex-4-ene-1,2-dicarboximide formerly 3a,4,7,7a-tetrahydo-*N*-(1,1,2,2-tetrachloroethanesulphenyl)-phthalimide, in *C.A.* usage 3a,4,7,7a-tetrahydro-2-[(1,1,2,2-tetrachoroethyl)thio]-1*H*-isoindole-1,3(2*H*)-dione *[2425-06-1]* formerly *N*-[(1,1,2,2-tetrachloroethyl)thio]-4-cyclohexane-1,2-dicarboximide. It was introduced in 1961 by the Chevron Chemical Co. under the code number 'Ortho-5865' and trade mark 'Difolatan'. Its fungicidal properties were first described by W. D. Thomas *et al., Phytopathology,* 1962, **52**, 754.

Manufacture and properties. Produced by the action of 1,1,2,2-tetrachloroethanesulphenyl chloride with 1,2,3,6-tetrahydrophthalimide, it is a colourless crystalline solid, m.p. 160-161°, v.p. negligible at room temperature. Its solubility is 1·4 mg/l water; slightly soluble in most organic solvents. The technical material is a light tan powder with a characteristic odour. Captafol is stable except under stongly alkaline conditions and it slowly decomposes at its melting point. It is compatible with most pesticides.

Uses. Captafol is a protective non-systemic fungicide applied to foliage at *c.* 200 g a.i./100 l. The stability of deposits which retain their fungicidal activity has led to spray programmes requiring fewer applications/season when applied at higher rates during the dormant season on pome fruits. It is widely used to control foliage and fruit disease of tomatoes, potato blight, coffee berry disease, tapping panel disease of hevea and diseases of other agricultural, horticultural and plantation crops. It is also used in the lumber and timber industries to reduce losses from wood rot fungi in logs and wood products.

Toxicology. The acute oral LD50 for rats is: 5000-6200 mg a.i./kg; 2500 mg w.p. (administered as an aqueous suspension)/kg. The acute dermal LD50 for rabbits is > 15400 mg/kg; some people develop an allergy to captafol. In 2-y feeding trials no apparent effect was observed in rats receiving 500 mg/kg diet and in dogs receiving 10 mg/kg daily. The chronic LD50 (10 d) is: for pheasants > 23070 mg/kg diet; for ducks 101700 mg/kg diet. The LC50 (96 h) is: for rainbow trout 0·50 mg/l; for goldfish 3·0 mg/l; for bluegill 0·15 mg/l.

Formulations. These include 'Ortho Difolatan 80W', w.p. (800 g a.i./kg); 'Ortho Difolatan 4 Flowable', liquid suspension (480 g/l).

Analysis. Technical product, formulation and residue analysis is by GLC: details available from the Chevron Chemical Co. See also: D. E. Pack, *Anal. Methods Pestic., Plant Growth Regul. Food Addit.,* 1967, **5**, 293; J. Crossley, *Anal. Methods Pestic. Plant Growth Regul.,* 1972, **6**, 556; A. A. Carlstrom & J. B. Leary, *ibid.,* 1978, **10**, 173.

Captan

T56 BVNV GUTJ CSXGGG $\qquad$ $C_9H_8Cl_3NO_2S$ (300·6)

Nomenclature and development. The common name captan is approved by BSI and ISO—exception Republic of South Africa; also known as ENT 26 538. The IUPAC name is 1,2,3,6-tetrahydro-*N*-(trichloromethylthio)phthalimide or *N*-(trichloromethylthio)-cyclohex-4-ene-1,2-dicarboximide formerly 3a,4,7,7a-tetrahydro-*N*-(trichloromethane-sulphenyl)phthalimide, in *C.A.* usage 3a,4,7,7a-tetrahydro-2-[(trichloromethyl)thio]-1*H*-isoindole-1,3(2*H*)-dione *[133-06-2]* formerly *N*-[(trichloromethyl)thio]-4-cyclohexene-1,2-dicarboximide. Its fungicidal properties were first described by A. R. Kittleston, *Science,* 1952, **115**, 84, and it was introduced by the Standard Oil Development Co., later by the Chevron Chemical Co. under the code number 'SR 406', trade marks 'Orthocide 406' and 'Orthocide' and protected by USP 2 553 770; 2 553 771; 2 553 776.

Manufacture and properties. It is produced by the reaction of trichloromethanesulphenyl chloride with 1,2,3,6-tetrahydrophthalimide, the latter being obtained by the action of ammonia on the condensation product of maleic anhydride and buta-1,3-diene. Captan forms colourless crystals, m.p. 178°, v.p. $< 1 \times 10^{-5}$ mmHg at 25°. Its solubility at 25° is: 3·3 mg/l water; 20 g/kg xylene; 70 g/kg chloroform; 21 g/kg acetone; 23 g/kg cyclohexanone; 1·7 g/kg propan-2-ol; it is insoluble in petroleum oils. The technical product is a colourless to beige amorphous solid, with a pungent odour, 90-95% pure, m.p. 160-170°. It is stable except under alkaline conditions and decomposes at or near its m.p. It is compatible with most other pesticides and is itself non-corrosive though its decomposition products are corrosive.

Uses. Captan is a fungicide used to control fungal diseases of many fruit, vegetable and ornamental crops including *Venturia inaequalis* of apple and *V. pirina* of pear, generally at 120 g a.i./100 l. It is not phytotoxic but should not be mixed with oil sprays. It is also used as a spray, root dip or seed dressing, to protect young plants against rots and damping-off.

Toxicology. The acute oral LD50 for rats is 9000 mg/kg. It may cause skin irritation. In 2-y feeding trials the 'no effect' level for rats was 1000 mg/kg diet. No teratogenic or mutagenic effects have been observed.

Formulations. These include: 'Orthocide 50 WP', w.p. (500 g a.i./kg; also 800 and 830 g/kg); dusts (50 and 100 g/kg); also dusts (600-750 g/kg) and w.p. for seed treatment, frequently when combined with other fungicides, or with insecticides.

Analysis. Product anlaysis is by i.r. spectrometry or by total chlorine content after alkaline hydrolysis: *CIPAC Handbook,* 1970, **1**, 171; FAO Specification (CP/57). Residues may be determined by reaction with resorcinol: A. R. Kittleson, *Anal. Chem.,* 1952, **24**, 1173. See also, J. N. Ospenson, *Anal. Methods Pestic., Plant Growth Regul. Food Addit.,* 1964, **3**, 7; *Anal. Methods Pestic. Plant Growth Regul.,* 1972, **6**, 546.

Carbaryl

O.CO.NHMe

L66J BOVM1

$C_{12}H_{11}NO_2$ (201·2)

Nomenclature and development. The common name carbaryl is approved by BSI, ISO and ANSI—exceptions JMAF (NAC), USSR (sevin) and Sweden; also known as ENT 23 969 and OMS 29. The IUPAC name is 1-naphthyl methylcarbamate (I), in *C.A.* usage 1-naphthalenyl methylcarbamate *[63-25-2]* formerly (I). Its insecticidal properties were first described by H. L. Haynes *et al., Contrib. Boyce Thompson Inst.,* 1957, **18**, 507. It was introduced in 1956 by the Union Carbide Corp. under the code number 'Experimental Insecticide 7744', the trade mark 'Sevin' and protected by USP 2 903 478.

Manufacture and development. Produced by the reaction of 1-naphthol with methyl isocyanate or with carbonyl chloride and methylamine, it is a colourless crystalline solid, m.p. 142°, v.p. $< 4 \times 10^{-5}$ mmHg at 25°, d_{20}^{20} 1·232. Its solubility at 30° is 120 mg/l water; it is soluble in most polar organic solvents such as dimethylformamide and dimethyl sulphoxide. The technical product is $\geqslant 99\%$ pure. It is stable to light, heat and hydrolysis under normal storage conditions and is non-corrosive. It is compatible with most other pesticides except those strongly alkaline, such as Bordeaux mixture or lime sulphur, which hydrolyse it to 1-naphthol.

Uses. Carbaryl is a contact insecticide with slight systemic properties recommended for use at 0·25-2·0 kg a.i./ha against many insect pests of fruit, vegetables, cotton and other crops. There is no evidence of phytotoxicity at these rates. It is also used to reduce the number of fruits on heavily laden apple trees.

Toxicology. The acute oral LD50 for male rats is 850 mg/kg. The acute dermal LD50 is: for rats > 4000 mg/kg; for rabbits > 2000 mg/kg. In 2-y feeding trials rats receiving 200 mg/kg diet suffered no ill-effect.

Formulations. These include: w.p. (500, 800 and 850 g a.i./kg); granules (50 g/kg); dusts (50 and 100 g/kg); bait pellets (50 g/kg); micronised suspensions in molasses, in non-phytotoxic oil or in aqueous media (480 g/l) and as true solutions in organic solvents.

Analysis. Product analysis is by i.r. spectroscopy at 5·75 μm (*CIPAC Handbook,* 1970, **1**, 185; 1979, **1A**, in press; FAO Specification (CP/55)) hydrolysis and the determination of the methylamine so produced has also been used. Residues may be determined by GLC (*Anal. Methods Pestic. Plant Growth Regul.,* 1972, **6**, 478); or by hydrolysis to 1-naphthol which is coupled with 4-nitrobenzenediazonium fluoroborate and the colour measured at 590 nm (R. Miskus *et al., J. Agric. Food Chem.,* 1959, **7**, 613; F. A. Gunther, *ibid.,* 1962, **10**, 222; see also H. A. Stansbury & R. Miskus, *Anal. Methods Pestic., Plant Growth Regul. Food Addit.,* 1964, **2**, 437).

Carbendazim

T56   BM   DNJ   CMVO1 $C_9H_9N_3O_2$ (191·2)

Nomenclature and development. The common name carbendazim is approved by BSI and ISO—exception USSR (BMC); the initials MBC or BCM were sometimes used before a common name was available. The IUPAC name is methyl benzimidazol-2-ylcarbamate, in *C.A.* usage methyl 1*H*-benzimidazol-2-ylcarbamate *[10605-21-7]* formerly methyl 2-benzimidazolecarbamate. Its fungicidal properties were described by H. Hampel & F. Löcher, *Proc. Br. Insectic. Fungic. Conf., 7th,* 1973, pp.127, 301. It was introduced by Hoechst AG under the code number 'Hoe 17411 OF' and trade mark 'Derosal'; by BASF AG under the code number 'BAS 346F' and trade mark 'Bavistin'; and by Du Pont under the trade mark 'Delsene' and protected by USP 3657443; BP 1190164.

Properties. Pure carbendazim is a light grey powder, m.p. 302-307° (decomp.). Its solubility at 20° is 5·8 mg/l water at pH 7 (D. J. Austin *et al., Pestic Sci.,* 1976, **7**, 211), at 24°: 0·5 mg/l hexane; 36 mg/l benzene; 68 mg/l dichloromethane; 300 mg/l ethanol; readily soluble in dimethylformamide. It slowly decomposes in alkaline solution; it has pK_a 4·48 (D. J. Austin & G. G. Briggs, *ibid.,* p. 201), is stable in acid forming water-soluble salts, *e.g.* carbendazim hydrochloride, *C.A.* Registry No. *[23454-47-7].*

Uses. Carbendazim is a systemic fungicide controlling a wide range of pathogens of fruit, vegetables, cereals, ornamentals and grapes. It is absorbed by the roots and green tissues of plants. Injections of solutions of salts, especially the hydrochloride and hypophosphite, into the trunks have given some control of Dutch elm disease: D. J. Clifford *et al., ibid.,* p.91.

Toxicology. The acute oral LD50 is: for rats > 15 000 mg/kg; for dogs > 2500 mg/kg; for quail > 10000 mg/kg. The acute dermal LD50 for rats is > 2000 mg/kg; the acute intraperitoneal LD50 for rats is > 15000 mg/kg. The 'no effect' level in 2-y feeding studies on rats and dogs was 300 mg/kg diet. The LC100 for ide and mirror carp is > 1000 mg/l. It is not toxic to honey bees.

Formulations. These include: 'Derosal', w.p. (594 g a.i./kg), dispersion (188 g/l); 'Bavistin', w.p. (500 g/kg); 'Lignasan', prepared injections of the hydrochloride for elm trees; 'Delsene' M, w.p. (100 g carbendazim + 640 g maneb/kg); 'Delsene' M-200, w.p. (100 g carbendazim + 640 g mancozeb/kg); 'Delsene' MX, w.p. (62 g carbendazim + 738 g maneb/kg); 'Delsene' MX-200 (62 g carbendazim + 738 g mancozeb/kg); 'Granosan' seed fungicide, powder (150 g carbendazim + 600 g maneb/kg); 'Granosan' 200 seed fungicide, powder (150 g carbendazim + 600 g mancozeb/kg).

Analysis. Product analysis is by extraction with glacial acetic acid and titration against perchloric acid. Residues in crops may be determined by u.v. spectroscopy after extraction and purification; or by HPLC: D. J. Austin & G. G. Briggs, *loc. cit.,* D. J. Austin *et al., loc. cit.*

Carbetamide

O   Me O
‖ | ‖
Ph.NH.C.O.CH.C.NHEt

2MVY1&OVMR $C_{12}H_{16}N_2O_3$ (236·3)

Nomenclature and development. The common name carbetamide is approved by BSI, ISO, ANSI and WSSA—exception Germany. The IUPAC name is (R)-(-)-1-(ethyl-carbamoyl)ethyl phenylcarbamate formerly D-(-)-N-ethyl-2-(phenylcarbamoyloxy)pro-pionamide, in *C.A.* usage N-ethyl-2-[[(phenylamino)carbonyl]oxy]propanamide (R)-isomer *[16118-45-0]* formerly D-N-ethyllactamide carbanilate ester. It was introduced in 1966 by Rhône-Poulenc Phytosanitaire under the code number '11 561 RP', the trade mark 'Legurame' and protected by BP 959 204; BelgP 597 035; USP 3 177 061. Its herbicidal properties were first described by J. Desmoras *et al., C. R. Journ. Etud. Herbic. Conf. COLUMA, 2nd,* 1963, p.14.

Manufacture and properties. Produced by the condensation of phenyl isocyanate with N-ethyllactamide, which is prepared by the reaction of ethylamine with methyl (R)-lactate, it forms colourless crystals, m.p. 119°, of negligible v.p. at 20°. The technical grade has m.p. > 110°. Its solubility at 20° is *c.* 3·5 g/l water; soluble in acetone, dimethylformamide, ethanol, methanol, dichloromethane. It is stable under normal storage conditions, compatible with most other pesticides and non-corrosive.

Uses. Carbetamide is a selective herbicide effective against grasses and some broad-leaved weeds. It is used for weed control in lucerne, red clover, oilseed rape, brassicas, chicory and endive at 2 kg a.i./ha. It persists in soil *c.* 60 d under normal field conditions.

Toxicology. The acute oral LD50 is: for rats 11 000 mg/kg; for mice 1250 mg/kg; for dogs 1000 mg/kg. Dermal applications of 500 mg/kg are non-toxic to rabbits. In 90-d feeding trials no effect was observed on: rats receiving 3200 mg/kg diet; dogs 12 800 mg/kg diet.

Formulations. 'Legurame liquide', e.c. (300 g a.i./l); 'Legurame P.M.' and 'Carbetamex', w.p. (700 g/kg).

Analysis. Product analysis is by titration of the ethylamine liberated on hydrolysis and separated by steam distillation. Residues may be determined by hydrolysis, isolation of the aniline produced, which is diazotised and measured by colorimetry of the dye obtained on coupling with N-(1-naphthyl)ethylenediamine (J. Desmoras *et al., Anal. Methods Pestic. Plant Growth Regul.,* 1973, 7, 509).

81

Carbofuran

MeNHCO.O

T56   BOT&J   Cl   Cl   IOVM1 $C_{12}H_{15}NO_3$ (221·3)

Nomenclature and development. The common name carbofuran is approved by BSI, ISO and ANSI; also known as ENT 27 164. The IUPAC name is 2,3-dihydro-2,2-dimethyl-benzofuran-7-yl methylcarbamate, in *C.A.* usage 2,3-dihydro-2,2-dimethyl-7-benzofuranyl methylcarbamate *[1563-66-2]*. It was introduced in 1967 by the Agricultural Chemical Div. of the FMC Corp. under the code number 'FMC 10242', trade mark 'Furadan' and protected by USP 3 474 170; 3 474 171. Bayer AG use the code number 'BAY 70 143', trade mark 'Curaterr'. Early references to the compound are F. L. McEwen & A. C. Davis, *J. Econ. Entomol.,* 1965, **58,** 369; E. J. Armburst & G. C. Gyrisco, *ibid.,* p.940.

Manufacture and properties. It is produced by the thermal rearrangement and cyclisation of 1-methylallyloxy-2-nitrobenzene to 2,3-dihydro-2,2-dimethyl-7-nitrobenzofuran, this is reduced to the amine, which is then diazotised and converted to 2,3-dihydro-2,2-dimethylbenzofuran-7-ol which in turn is esterified with methyl isocyanate. Carbofuran is a colourless crystalline solid, m.p. 150-152°, v.p. 2×10^{-5} mmHg at 33°, d_{20}^{20} 1·180. Its solubility at 25° is: 700 mg/l water; 150 g/kg acetone; 140 k/kg acetonitrile; 40 g/kg benzene; 90 g/kg cyclohexanone; 270 g/kg dimethylformamide; 250 g/kg dimethyl sulphoxide; 300 g/kg 1-methyl-2-pyrrolidone; it is essentially insoluble in conventional formulation solvents used in agriculture. It is non-corrosive, non-flammable but unstable in alkaline media.

Uses. Carbofuran is a systemic insecticide, acaricide and nematicide, applied to foliage at 0·25-1·0 kg a.i./ha for the control of insects and mites, or applied to the seed furrow at 0·5-4·0 kg/ha for the control of soil and foliar-feeding insects, or broadcast at 6-10 kg/ha for the control of nematodes. Its activity against these pests has been summarised (B. Homeyer, *Pflanzenschutz-Nachr. (Am. Ed.),* 1975, **28,** 3) and those of maize (P. Villeroy & P. Pourcharesse, *ibid.,* p.55; K. Küthe, *ibid.,* p.67; A. D. Cohick, *ibid.,* p.80; W. Kolbe, *ibid.,* p.144), brassicas (T. J. Martin & D. B. Morris, *ibid.,* p.92) and rice (K. Iwaya & G. Kollmer, *ibid.,* p.137). It is metabolised in the liver and excreted in the urine of animals, 50% being lost in 6-12 h; in soils 50% is lost in 30-60 d. 2,3-Dihydro-3-hydroxy-2,2-dimethylbenzofuran-7-yl methylcarbamate, which is of low toxicity to insects and mammals, is one of the products formed.

Toxicology. The acute oral LD50 is: for rats 8-14 mg a.i. (in corn oil)/kg; for dogs 19 mg a.i. (dry powder)/kg. The acute dermal LD50 for rabbits is 2550 mg a.i. (as w.p.)/kg. In 2-y feeding trials no effect was observed on rats receiving 25 mg/kg diet, on dogs receiving 20 mg/kg diet, nor on rats receiving 10 mg/kg diet for 3 generations nor on dogs receiving 50 mg/kg diet for one generation. The 10-d LD50 for pheasants was 960 mg a.i. (as 10% granule)/kg diet. The LC50 (96 h) for trout is 0·28 mg/l.

Formulations. These include: w.p. (750 g a.i./kg); flowable paste (480 g/l); granules (20, 30, 50 and 100 g/kg).

Analysis. Product analysis is by i.r. spectroscopy using absorbance in chloroform at 5·73 μm. Residues are determined by GLC: R. F. Cooke *et al., J. Agric. Food Chem.,* 1969, **17,** 277: E. Möllhoff, *Pflanzenschutz-Nachr. (Am. Ed.),* 1975, **28,** 370; R. F. Cooke, *Anal. Methods Pestic. Plant Growth Regul.,* 1973, **7,** 187.

Carbon disulphide

$$CS_2$$

SCS CS_2 (76·13)

Nomenclature and development. The IUPAC name carbon disulphide, in *C.A.* usage carbon disulfide *[75-15-0]*, is accepted in lieu of a common name by BSI and ISO. It was first used as an insecticide by Garreau in 1854 (see *Science*, 1926, **64**, 326).

Manufacture and properties. Produced by the direct reaction of sulphur and carbon in an electric furnace, it is a colourless to yellowish mobile liquid, b.p. 46·3°, m.p. —108·6°, v.p. 357·1 mmHg at 25°, d_4^{20} 1·2628, n_D^{18} 1·6295. Its solubility at 32° is 2·2 g/l water; miscible with ethanol, diethyl ether, chloroform. Its vapour is 2·63 times as dense as air, and is extremely flammable with flash point *c.* 20° and it ignites spontaneously *c.* 125-135°. Its impurities have an unpleasant odour.

Uses. Carbon disulphide is an insecticide used for fumigation of nursery stock and for soil treatment against insects and nematodes. It is also used in some countries in mixtures with carbon tetrachloride (to reduce fire hazard) for fumigating stored grain.

Toxicology. The vapour is highly poisonous, producing giddiness and vomiting in 30 min at 6·8 g/m³; repeated daily exposures to 227 mg/m³ caused ill health.

Formulations. Soil applications are made using the compound alone or as emulsions or solutions in alkali (thiocarbonates).

Analysis. Residues in grain may be determined by GLC (S. G. Heuser & K. A. Scudamore, *J. Sci. Food Agric.,* 1969, **20**, 566) or measuring the yellow colour of copper(II) dimethyldithiocarbamate produced with Viles reagent (C. L. Dunning, *J. Assoc. Off. Agric. Chem.,* 1957, **40**, 168). The concentration in air may be measured by drawing the air through an ethanolic solution of diethylamine and copper acetate and measuring the yellow colour produced.

CCl$_4$

GXGGG   CCl$_4$ (153·8)

Nomenclature and development. The traditional chemical name, carbon tetrachloride (I), is accepted in lieu of a common name by BSI and ISO; also known as ENT 27 164. The IUPAC name is tetrachloromethane (II), in *C.A.* usage (II) *[56-23-5]* formerly (I). It was used in 1908 for the fumigation of nursery stock: W. E. Britton, *Conn. Agric. Exp. Stn. Rep.,* No. 31, 1908.

Manufacture and properties. Produced by the reaction of chlorine with carbon disulphide in the presence of a catalyst, it is a colourless liquid, b.p. 76°, m.p. —23°, v.p. 114·5 mmHg at 25°, d_{25}^{25} 1·588, n_D^{20} 1·4607. The vapour is dense, 5·32 times that of air. Its solubility at 25° is 280 mg/kg water; miscible with most organic solvents. It is non-flammable and non-explosive; though generally inert, it is decomposed by water at high temperatures.

Uses. Carbon tetrachloride is of low insecticidal activity, but is used for grain disinfestation when long exposures are possible, its main advantage being low absorption by treated grain. It is often used in mixtures with more potent fumigants—*e.g.* ethylene dichloride, to reduce fire hazard of the latter. It has been used as a veterinary anthelmintic.

Toxicology. It is a general anaesthetic, prolonged exposure causing irritation of the mucous membranes, headache and nausea; repeated exposure to high concentrations causes liver damage. The acute oral LD50 is: for rats 5730-9770 mg/kg; for mice 12 800 mg/kg; for rabbits 6380-9975 mg/kg.

Analysis. Residues in cereals are determined by GLC (S. G. Heuser & K. A. Scudamore, *J. Sci. Food Agric.,* 1969, **20**, 566; *Analyst, (London),* 1974, **99**, 570; *Pestic. Sci.,* 1973, **4**, 1), or by the Fujiwara reaction (L. L. Ramsey, *J. Assoc. Off. Agric. Chem.,* 1957, **40**, 175). Concentrations in air after absorption, *e.g.* on silica gel, may be determined by hydrolysis to chloride ion and titration (F. F. Morehead, *Ind. Eng. Chem. Anal. Ed.,* 1940, **12**, 373; F. P. W. Winteringham, *J. Soc. Chem. Ind.,* 1942, **61**, 186).

Carbophenothion

$(EtO)_2PS.S.CH_2.S-\langle\bigcirc\rangle-Cl$

GR   DS1SPS&O2&O2 $C_{11}H_{16}ClO_2PS_3$ (342·9)

Nomenclature and development. The common name carbophenthion is approved by BSI, ISO and ANSI; also known as ENT 23708. The IUPAC name is S-(4-chlorophenyl-thio)methyl O,O-diethyl phosphorodithioate, in *C.A.* usage S-[[(4-chlorophenyl)thio]-methyl] O,O-diethyl phosphorodithioate *[786-19-6]*. It was introduced in 1955 by the Stauffer Chemical Co. under the code number 'R-1303', and trade marks 'Trithion' and 'Garrathion' and protected by USP 2793224. Its development is described in *Agric. Chem.*, 1956, **11**(11), 91.

Manufacture and properties. Produced by the condensation of sodium O,O-diethyl phosphorodithioate with chloromethyl 4-chlorophenyl sulphide, which is obtained by the interaction of 4-chloro(thiophenol), formaldehyde and hydrochloric acid, it is an almost colourless to amber liquid, with a mercaptan-like odour, b.p. 82°/0·01 mmHg, v.p. 3 x 10^{-7} mmHg at 20°, n_D^{25} 1·597. Its solubility is < 40 mg/l water; miscible with most organic solvents. The technical grade is *c.* 95% pure, d_{20}^{20} 1·285. It is relatively stable to hydrolysis and is oxidised, on the leaf surface, to the corresponding phosphorothioate (B. J. Luberoff *et al.*, *Agric. Chem.*, 1958, **13**(3), 83). It is compatible with most pesticides and is non-corrosive to mild steel.

Uses. Carbophenothion is a non-systemic acaricide and insecticide with a long residual action. It is used, in combination with petroleum oil, as a spray to control overwintering mites, aphids and scale insects on dormant deciduous fruit trees; as an acaricide on citrus trees; and as a cereal seed dressing to control wheat bulb fly. It is phytotoxic at high concentrations on some plants. Depending on soil type, 50% degradation occurs in soils in $\geqslant$ 100 d.

Toxicology. The acute oral LD50 for male albino rats is 32·3 mg/kg, 91 mg/kg has also been quoted (E. C. Hagen *et al.*, *Fed. Proc.*, 1961, **20**, 432). The acute dermal LD50 for rabbits is *c.* 1270 mg/kg.

Formulations. These include: e.c. (240, 480, 720 and 960 g a.i./l); w.p. (250 g/kg); dusts (10, 20 and 30 g/kg).

Analysis. Product analysis is by u.v. spectroscopy at 263 nm. Residues may be determined by GLC. Details of these methods are available from Stauffer Chemical Co. See also: J. J. Menn *et al.*, *Anal. Methods Pestic., Plant Growth Regul. Food Addit.*, 1964, **2**, 545; R. W. Buxton *et al.*, *Anal. Methods Pestic. Plant Growth Regul.*, 1972, **6**, 519.

Carboxin

T6O DS BUTJ B1 CVMR
$C_{12}H_{13}NO_2S$ (235·3)

Nomenclature and development. The common name carboxin is approved by BSI, ANSI and ISO—exceptions Canada, Denmark and Germany. The IUPAC name is 5,6-dihydro-2-methyl-1,4-oxathiin-3-carboxanilide (I) formerly 2,3-dihydro-6-methyl-5-phenylcarbamoyl-1,4-oxathiin, in *C.A.* usage 5,6-dihydro-2-methyl-*N*-phenyl-1,4-oxathiin-3-carboxamide *[5234-68-5]* formerly (I). It was introduced in 1966 by Uniroyal Inc. under the code number 'D 735', trade mark 'Vitavax' and protected by USP 3 249 499; 3 393 202; 3 454 391. Its fungicidal properties were described by B. von Schmeling & M. Kulka, *Science,* 1966, **152,** 659.

Manufacture and properties. Produced by the reaction of 2-chloro-3-oxobutyranilide with 2-mercaptoethanol, followed by cyclisation, it is a colourless solid, m.p. 91·5-92·5°; a dimorphic form has m.p. 98-100°. Its solubility at 25° is: 170 mg/l water; 600 g/kg acetone; 150 g/kg benzene; 1·5 kg/kg dimethyl sulphoxide; 100 g/kg ethanol; 210 g/kg methanol. The technical product is at least 97% pure. It is compatible with all except highly alkaline or acidic pesticides.

Uses. Carboxin is a systemic fungicide used for seed treatments of cereals against smuts and bunts and with co-fungicides for the control of most other soil-borne seedling diseases; also against *Rhizoctonia* spp. of cotton, groundnuts and vegetables. The dimorphic forms do not differ in fungicidal activity.

Toxicology. The acute oral LD50 for rats is 3820 mg/kg; the acute dermal LD50 for rabbits is > 8000 mg/kg. In 2-y feeding trials albino rats receiving 600 mg/kg diet suffered no detectable symptom.

Formulations. These include: a w.p. (750 g a.i./kg) for use on cereal seed at 40-120/100 kg seed; also combinations with organomercury fungicides, maneb and thiram for use on cereals, corn, cotton, rice; with captan for use on groundnuts (80-160 g/100 kg seed); flowable powders (170 and 340 g/kg) for cereals, corn, cotton, rice and in combination with thiram.

Analysis. Residue analysis is by hydrolysis and the determination of the aniline so formed either by coupling with 4-dimethylaminobenzaldehyde (green tissue) (J. R. Lane, *J. Agric. Food Chem.,* 1970, **18,** 409) or by GLC using a specific nitrogen detector (seed) (H. R. Siskin & J. E. Newell, *ibid.,* 1971, **19,** 738). See also, G. M. Stone, *Anal. Methods Pestic. Plant Growth Regul.,* 1976, **8,** 319.

Cartap

$$
\begin{array}{c}
O \\
\parallel \\
CH_2SCNH_2 \\
\mid \\
Me_2N-CH \\
\mid \\
CH_2SCNH_2 \\
\parallel \\
O
\end{array}
$$

ZVS1Y1SVZN1&1   *cartap* $C_7H_{15}N_3O_2S_2$ (237·3)

ZVS1Y1SVZN1&1   &GH   *cartap hydrochloride* $C_7H_{16}ClN_3O_2S_2$ (273·8)

Nomenclature and development. The common name cartap is approved by BSI, JMAF and ISO—exception France. The IUPAC name is *S,S'*-2-dimethylaminotrimethylene bis(thiocarbamate) (I) formerly 1,3-di(carbamoylthio)-2-dimethylaminopropane, in *C.A.* usage *S,S'*-[2-(dimethylamino)-1,3-propanediyl] dicarbamothioate *[15263-53-3]* formerly (I). Its hydrochloride was introduced in 1965 by Takeda Chemical Industries, Ltd, under the code number 'TI-1258', *C.A.* Registry Nos. *[22042-59-7]* (hydrochloride), *[15623-52-2]* (monohydrochloride), trade marks 'Padan', 'Cadan', 'Patap', 'Sanvex', 'Thiobel' and 'Vegetox', and protected by BP 1 126 204; USP 3 332 943; FP 1 452 338. Its properties were described by M. Sakai *et al., Jpn. J. Appl. Entomol. Zool.,* 1967, **11**, 125; its action and structure-activity relationships were reviewed: K. Konishi, *Pestic. Chem. (Congr. Pestic. Chem., 2nd, 1971),* 1972, **1**, 179; M. Sakai & Y. Sato, *ibid.,* p.445. Y. Kono, *Jpn. Pestic. Inf.,* 1978, No. 34, p.22.

Manufacture and properties. Cartap is produced by the thiocyanation of dimethyl-[2-chloro-1-(chloromethyl)ethyl]amine followed by hydration in the presence of hydrochloric acid. Its hydrochloride is a colourless crystalline solid, m.p. 179-181°, solubility at 25° *c.* 200 g/l water, slightly soluble in ethanol and methanol. The technical grade is 97% pure. It is stable under acidic conditions, but hydrolysed in neutral or alkaline solution. It is compatible with most pesticides and is virtually non-corrosive though susceptible to moisture.

Uses. Cartap hydrochloride paralyses the insect by a ganglionic blocking action on the central nervous system. It is used against rice stem borers at 600 g a.i./ha; Colorado potato beetles at 375-550 g/ha; Mexican bean beetles at 560 g/ha; cabbage caterpillars at 550 g/ha; diamond-back moths at 500 g/ha; boll weevils at 560 and 1120 g/ha. It is tolerated by a wide range of crops.

Toxicology. The acute oral LD50 for rats is 325-345 mg/kg; the acute dermal LD50 for mice is > 1000 mg/kg. The LC50 (48 h) for carp is 1·3 mg/l. It is moderately toxic to honey bees and is non-persistent.

Formulations. These include: w.s.p. (250 and 500 g a.i./kg); dust (20 g/kg); granules (40 and 100 g/kg).

Analysis. Product and residue analysis is by GLC or by oscillopolarography, K. Nishi *et al., Anal. Methods Pestic. Plant Growth Regul.,* 1973, **7**, 371.

Nomenclature and development. Cheshunt compound is the traditional name for a fungicide introduced by W. F. Bewley, *Rep. Exp. Res. Stn., Cheshunt, 1921,* 1922, p.38. It comprises mainly tetra-amminecopper sulphate, in *C.A.* usage diammonium carbonate mixture with copper (2+) sulfate *[55632-67-2].* (Cheshunt mixture).

Manufacture and properties. Produced by mixing copper sulphate pentahydrate (2 parts *m/m*) with technical ammonium carbonate (11 parts *m/m*). The latter largely comprises ammonium carbamate (H_2NCOO^- NH_4^+) and ammonium hydrogen carbonate (NH_4^+ HCO_3^-), and on exposure to air forms the hydrogen carbonate. Cheshunt compound is a deep violet powder, soluble in water.

Uses. Cheshunt compound is a typical copper fungicide, somewhat phytotoxic, used mainly for soil applications to control *Pythium, Phytophthora* spp. and other soil-borne fungi. It forms basic ammonium copper sulphate(s) after application.

Formulation. A w.s.p.

Analysis. Product analysis is by estimation of the copper content by standard methods (*CIPAC Handbook,* 1970, **1,** 226).

T55 BO DO FOTJ CXGGG GYQ1Q HQ $C_8H_{11}Cl_3O_6$ (309·5)

Nomenclature and development. The names chloralose and glucochloralose are accepted in lieu of a common name by BSI and ISO; also known as alphachloralose. The IUPAC name is (R)-1,2-O-(2,2,2-trichloroethylidene)-α-D-glucofuranose, (I), in *C.A.* usage (I) *[15879-93-3]* formerly chloralose *[39598-39-5]* (formerly *[14798-36-8]*). It has been in use for many years in Europe on seed grain as a bird repellent.

Manufacture and properties. Produced by the reaction of trichloroacetaldehyde with glucose, alphachloralose is a crystalline powder, m.p. 187° $[α]_D^{22} + 19°$. Its solubility at 15° is 4·44 g/l water; soluble in diethyl ether, glacial acetic acid. Chloralose also exists in a beta-form, Registry No. *[16376-36-6]*, m.p. 227-230°, which is less soluble than the alpha-form in water, ethanol, diethyl ether. Alphachloralose will reduce Fehlings solution only after prolonged standing, and is hydrolysed to its components by acids.

Uses. Chloralose is a narcotic rendering birds easier to kill by other means; it is also used as a rodenticide in baits against mice, usually only by trained personnel. It retards metabolism and lowers body temperature to a fatal extent in small mammals. It is rapidly metabolised and hence non-cumulative.

Toxicology. The oral LD50 is: for rats 400 mg/kg; for mice 32 mg/kg; for birds 32-178 mg/kg.

Formulations. Bait (15 g a.i./kg grain) against birds; baits (≤ 40 g/kg) against mice. Baits should contain a warning or water-soluble dye.

Analysis. Product analysis is by hydrolysis and estimation of the trichloroacetaldehyde produced by adding pyridine and sodium hydroxide (K. C. Barrons & R. W. Hummer, *Agric. Chem.,* 1951, **6**(6), 48; G. I. Mills, *Anal. Chim. Acta.,* 1952, **7**, 70).

Chloramben

$$CO_2H$$

(chemical structure diagram)

ZR	BG	EG	CVQ		*chloramben*	$C_7H_5Cl_2NO_2$ (206·0)
ZR	BG	EG	CVQ	&ZH	*chloramben-ammonium*	$C_7H_8Cl_2N_2O_2$ (223·0)
ZR	BG	EG	CVO1		*chloramben-methyl*	$C_8H_7Cl_2NO_2$ (220·1)

Nomenclature and development. The common name chloramben is approved by BSI, ISO, ANSI and WSSA—exception India; WSSA formerly used name *amben*. The IUPAC and *C.A.* name is 3-amino-2,5-dichlorobenzoic acid *[133-90-4]*. The ammonium salt, Registry No. *[1076-46-6]* was introduced as a herbicide by Amchem Products, Inc. under the code number 'ACP M-629', trade mark 'Amiben' (formerly *'Amoben'*) and protected by USP 3014063; 3174842. Amchem Products, Inc. also introduced the methyl ester, chloramben-methyl, *C.A.* Registry No. *[7286-84-2]* under the code number 'Amchem 65-81B' and trade mark 'Vegiben'; a granular formulation *'Amchem 66-45'* is no longer available, nor is the corresponding amide *'Amiben 65-78'*.

Manufacture and properties. Produced by chlorinating benzoic acid, followed by nitration and reduction, it is a colourless crystalline solid, m.p. 200-201°, v.p. *c.* 7 x 10^{-3} mmHg at 100°. Its solubility at 25° is: 700 mg/l water; 172 g/kg ethanol. The technical product is a purplish-white powder, *c.* 90% pure, m.p. > 195°. It is stable to heat, to oxidation and to hydrolysis by acid or alkali but is decomposed by sodium hypochlorite solutions.

It forms water-soluble alkali metal and ammonium salts, there is no problem of precipitation with hard water. Chloramben-methyl is a colourless solid, m.p. 63-64°, solubility at 20° 120 mg/l water.

Uses. Chloramben is a selective pre-em. herbicide used for weed control in soyabeans, navy beans, groundnuts, sunflower, maize, sweet potatoes, seedling asparagus, squash, pumpkins and certain ornamentals at 2-4 kg a.e./ha. Chloramben-ammonium is rapidly leached in soil; esters are more strongly adsorbed and are leached to a lesser extent (S. R. McLane & M. D. Parkins, *Proc. Br. Weed Control Conf., 8th,* 1966, p.283). The *N*-glycoside was isolated from soyabeans, in amounts equivalent to chloramben applied, but little was recovered from barley, a susceptible crop. (S. R. Colby, *Science,* 1965, **150,** 619.)

Toxicology. The acute oral LD50 for male albino rats is 5620 mg/kg. The acute dermal LD50 for albino rats is > 3160 mg/kg; a single application of 3 mg caused a mild irritation which subsided within 24 h. In 2-y feeding trials rats receiving 10 000 mg/kg diet suffered no adverse effect.

Formulations. These include: chloramben-ammonium, 'Amiben', aqueous solution (240 g a.e./l) and 'Amiben Granular 10%', granules (100 g a.e./kg); chloramben-methyl, 'Vegiben', e.c. (240 g a.e./l).

Analysis. Product analysis is by i.r. spectroscopy or volumetric methods (AOAC Methods; *CIPAC Handbook,* in press; details of methods are available from Amchem Products, Inc.). See also: H. S. Segal & M. L. Sutherland, *Anal. Methods Pestic., Plant Growth Regul. Food Addit.,* 1967, **5,** 321; *Anal. Methods Pestic. Plant Growth Regul.,* 1972, **6,** 588.

Chlorbromuron

GR   BE   EMVN1&O1 $C_9H_{10}BrClN_2O_2$ (293·5)

Nomenclature and development. The common name chlorbromuron is approved by BSI, ISO, ANSI and WSSA. Its IUPAC name is 3-(4-bromo-3-chlorophenyl)-1-methoxy-1-methylurea (I), in *C.A.* usage *N'*-(4-bromo-3-chlorophenyl)-*N*-methoxy-*N*-methylurea *[13360-45-7]* formerly (I). Its herbicidal properties were described by D. H. Green *et al., Proc. Br. Weed Control Conf., 8th, 1966*, p. 363. It was introduced in 1961 by Ciba AG (now Ciba-Geigy AG) under the code number 'C 6313', trade mark 'Maloran' and protected by BP 965 313.

Manufacture and properties. Produced by the bromination of monolinuron (p.367), chlorbromuron forms tan-coloured crystals, m.p. 98°, v.p. 4 x 10^{-7} mmHg at 20°. Its solubility at 20° is 35 mg/l water; it is soluble in organic solvents.

Uses. Chlorbromuron is a pre- and post-em. herbicide, suitable for pre-em. use on carrots, soyabeans, sunflowers, peas and potatoes at 1·0-2·5 kg a.i./ha; for post-em. use on carrots and transplanted celery at 0·75-1·5 kg/ha. At these rates it persists in soil > 56 d.

Toxicology. The acute oral LD50 for rats is > 5000 mg/kg; the acute dermal LD50 for rats is > 2000 mg/kg. In 90-d feeding trials the 'no effect' level in male rats and dogs was > 316 mg/kg diet. It is slightly toxic to fish.

Formulation. 'Maloran': w.p. (500 g a.i./kg).

Analysis. Product analysis is by GLC with internal standard. Residues may be determined by hydrolysis to 4-bromo-3-chloroaniline, which is diazotised, coupled with *N*-(1-naphthyl)ethylenediamine and determined colorimetrically, or determined by GLC after suitable reaction. Particulars of these methods are available from Ciba-Geigy AG. See also: G. Voss, *Anal. Methods Pestic. Plant Growth Regul., 1973*, **7**, 569. Direct GLC with ECD may also be used: T. H. Byast *et al., Agric. Res. Counc. (G.B.) Weed Res. Organ., Tech. Rep.*, No. 15 (2nd Ed.), p.49.

Chlorbufam

GR   CMVOYI&1UUI $C_{11}H_{10}ClNO_2$ (223·7)

Nomenclature and development. The common name chlorbufam is approved by BSI and ISO—exception JMAF (BIPC). The IUPAC name is 1-methylprop-2-ynyl 3-chloro-phenylcarbamate, in *C.A.* usage 1-methyl-2-propynyl (3-chlorophenyl)carbamate *[1967-16-4]* formerly 1-methyl-2-propynyl *m*-chlorocarbanilate. Its herbicidal properties were first described by A. Fischer, *Z. Pflanzenkr. Pflanzenpathol. Pflanzenschutz,* 1960, **67,** 577. It was introduced in 1958 by BASF AG as a component of 'Alipur' under the name 'BiPC' and protected by DBP 1 034 912; 1 062 482.

Manufacture and properties. Produced by the reaction of 3-chlorophenyl isocyanate and 1-methylprop-2-yn-1-ol, it forms colourless crystals, m.p. 45-46°, v.p. 1·2 mmHg at 20°. Its solubility at 20° is: 540 mg/l water; 280 g/kg acetone; 95 g/kg ethanol; 286 g/kg methanol. It is unstable under acid and alkaline conditions; alcohols may cause transesterification.

Uses. Chlorbufam is a pre-em. herbicide generally used as a mixture with cycluron at 3-6 1 product/ha (for weed control in sugar beet, spinach and some vegetable crops) or with chloridazon at 3-6 kg product/ha (for use in onions, leeks and flower-bulb crops). The main weeds controlled are: *Sinapsis alba, Urtica circus, Raphanus raphanistrum, Stellaria media, Apera spica-venti, Matricaria, Polygonum, Atriplex* and *Poa* spp. Chlorbufam persists in sandy loam for *c.* 56 d.

Toxicology. The acute oral LD50 for rats is 2500 mg/kg; a slight temporary erythema was caused by 15-min application to the backs of white rabbits, significant after 20-h application.

Formulations. 'Alipur', an e.c. (100 g chlorbufam + 150 g cycluron/l); 'Alicep', a w.p. (200 g chlorbufam + 250 g chloridazon/kg).

Analysis. Acid hydrolysis produces 3-chloroaniline, estimated by titration (product analysis) or colorimetry (residue analysis).

Chlordane

$$
\begin{array}{c}
\text{CCl} \\
\text{ClC} \quad \text{CH} \!-\!\!\!-\text{CHCl} \\
\text{CCl}_2 \\
\text{ClC} \quad \text{CH} \quad \text{CHCl} \\
\text{CCl} \quad \text{CH}_2
\end{array}
$$

L   C555   A   IUTJ   AG   AG   BG   DG   EG   HG   IG   JG $C_{10}H_6Cl_8$ (409·8)

Nomenclature and development. The common name chlordane is approved by BSI and ISO; also known as ENT 9932. The IUPAC name is 1,2,4,5,6,7,8,8-octachloro-2,3,3a,4,7,7a-hexahydro-4,7-methanoindene formerly 1,2,4,5,6,7,8,8-octachloro-3a,4,7,7a-tetrahydro-4,7-methanoindane (I), in *C.A.* usage 1,2,4,5,6,7,8,8-octachloro-2,3,3a,4,7,7a-hexahydro-4,7-methano-1*H*-indene *[57-47-9]* formerly (I). Its insecticidal properties, first described by C. W. Kearns *et al.* (*J. Econ. Entomol.*, 1945, **38**, 661), were discovered independently by R. Riemschneider (*Chim. Ind. (Paris)*, 1950, **64**, 695) who used the code number 'M 410'. It was introduced in 1945 by Velsicol Chemical Corp. under the code number 'Velsicol 1068', trade mark 'Octachlor' and protected by BP 618432 (J. Hyman); USP 2598561 (Velsicol Chemical Corp.).

Manufacture and properties. It is produced by condensing hexachlorocyclopentadiene, made by chlorinating pentanes (USP 2509160) or cyclopentadiene (USP 2606910), with cyclopentadiene to produce 4,5,6,7,8,8-hexachloro-1,2,3a,4,7,7a-hexahydro-4,7-methanoindene, which is further chlorinated to chlordane. The technical product, *C.A.* Registry No. *[12789-03-6]*, is a viscous amber-coloured liquid, d^{25} 1·59-1·63, n_D^{25} 1·56-1·57, viscosity 75-120 x 10^{-6} m²/s at 54°; it is insoluble in water but soluble in most organic solvents including petroleum oils. The refined product has v.p. 1 x 10^{-5} mmHg at 25°.

Isomerism. Originally technical chlordane comprised 60-75% chlordane isomers, the remainder being related *endo*-compounds including heptachlor (see p.296) (R. Riemschneider, *World Rev. Pest Control*, 1963, **2** (4), 29; W. Benson *et al.*, *J. Agric. Food Chem.*, 1971, **19**, 857). The nomenclature at C(1) and C(2) of the stereoisomers of the octachloro compounds has been confused in the literature. The alpha- or *cis*-isomer (1α,2α,3aα,4β,7β,7aα) Registry No. *[5103-71-9]* (formerly *[22212-52-8]*) has m.p. 106-107°. The *trans*-isomer (1α,2β,3aα,4β,7β,7aα) *[5103-74-2]* usually known as gamma-occasionally as beta-chlordane, m.p. 104-105°, is less readily dehydrochlorinated than the *cis*-isomer. The term gamma-chlordane has also been applied to the 2,2,4,5,6,7,8,8-octachloro isomer *[5564-34-7]*, m.p. 131°, for which greater insecticidal activity and lower mammalian toxicity were claimed (K. H. Büchel & R. Fischer, *Z. Naturforsch. b*, 1966, **21**, 1112). Improvement in the manufacture resulted in a technical grade product containing 70% *cis*-, 25% *trans*-isomers and 1% heptachlor (W. P. Cochrane & R. Greenhalgh, *J. Assoc. Off. Anal. Chem.*, 1976, **59**, 696). This product, code number 'HCS-3260', has been called AG chlordane (W. Furness, *Proc. Br. Insectic. Fungic. Conf., 6th*, 1971, **2**, 541; A. I. Kovacs *et al., ibid., 7th*, 1973, **2**, 649).

Uses. Chlordane is a persistent, non-systemic stomach and contact insecticide; non-phytotoxic at insecticidal concentrations. It is mainly used against coleopterous pests, termites, wood-boring beetles, in ant baits and, in some countries, to reduce the earthworm population in lawns. Insecticidal activities of the components have been compared (S. J. Cristol, *Adv. Chem. Series*, 1950, **1**, 541; R. L. Metcalf, *Organic Insecticides*, p. 234).

continued

Toxicology. The acute oral LD50 for rats is 457-590 mg/kg. Rats fed 150 mg *trans*-chlordane/kg diet suffered no higher mortality than the controls but histopathological changes in the liver were apparent. The toxicology has been reviewed (*1967 Evaluations of Some Pesticide Residues in Food,* p.33). A high toxicity of the vapour to mice was due to traces of unreacted hexachlorocyclopentane, now reduced to insignificant amounts in technical chlordane.

Formulations. These include: e.c. (500 and 700 g a.i./l); kerosine solutions (20 and 200 g/l); dusts and granules (50 and 100 g/kg).

Analysis. Product analysis is by GLC or i.r. spectroscopy (M. Malina *et al., J. Assoc. Off. Anal. Chem.,* 1972, **55,** 972; M. Malina, *ibid.,* 1973, **56,** 591) or by total chlorine with colorimetric determination (*CIPAC Handbook,* 1970, **1,** 203; 1979, **1A,** in press; *Official Methods of Analysis: AOAC, 12th Ed.,* p.105). Residues may be determined by GLC with ECD (*ibid.,* p. 518). See also: *Pesticide Analytical Manual* (1975 Rev.); T. G. Bowery, *Anal. Methods Pestic., Plant Growth Regul. Food Addit.,* 1964, **2,** 49; *Anal. Methods Pestic. Plant Growth Regul.,* 1972, **6,** 315; H. K. Suzuki *et al., ibid.,* 1978, **10,** 45.

Chlordimeform

$$Cl-\langle\bigcirc\rangle-N = CHNMe_2$$
$$Me$$

GR Cl DNU1N1&1 $C_{10}H_{13}ClN_2$ (196·7)

Nomenclature and development. The common name chlordimeform is approved by BSI, ISO and ANSI—exceptions JMAF (chlorphenamidine) and New Zealand (chlorodimeform); also known as ENT 27335. Its IUPAC name is N^2-(4-chloro-o-tolyl)-N^1,N^1-dimethylformamidine (I), in *C.A.* usage N'-(4-chloro-2-methylphenyl)-N,N-dimethyl-methanimidamide *[6164-98-3]* formerly (I). It was introduced in 1966 by Schering AG under the code number 'Schering 36 268' and trade marks 'Fundal' and 'Spanone' and by Ciba AG (now Ciba-Geigy AG) under the code number 'C 8514' and trade mark 'Galecron'; protected by BP 1039930; DBP 1172081; USP 3378437. Its acaricidal properties were first described by V. Dittrich, *J. Econ. Entomol.,* 1966, **59,** 889.

Manufacture and properties. Produced by *(a)* Vilsmeier condensation of dimethyl-formamide with 4-chloro-o-toluidine and $POCl_3$ or $SOCl_2$ or $COCl_2$; *(b)* chlorination of N^1,N^1-dimethyl-N^2-o-tolylformamidine and liberating the free base with NaOH. Chlordimeform forms colourless crystals, m.p. 32°, b.p. 163-165°/14 mmHg, v.p. 3·6 x 10^{-4} mmHg at 20°, d^{30} 1·10. Its solubility at 20° is: 250 mg/l water; > 200 g/l in acetone, benzene, chloroform, ethyl acetate, hexane, methanol. The technical grade is > 96% pure.

It is hydrolysed in neutral and acidic media, first to 4-chloro-o-tolylformamide and then to 4-chloro-o-toluidine; it forms salts with acids, *e.g.* the hydrochloride (*C.A.* Registry No. *[19750-95-9]* also known as ENT 27567), m.p. 225-227° (decomp.), whose solubility is: > 500 g/l water; > 300 g/l methanol; 10-20 g/l chloroform; 1 g/l benzene or hexane. An aqueous solution of the hydrochloride (5 g/l; pH 3-4) is stable for several days at 20°.

Uses. Chlordimeform is an acaricide most effective against eggs and immature stages; normally used as an ovicide at 50 g a.i./100 l HV. It is also effective against the eggs and early instars of Lepidoptera, *e.g. Laspeyresia pomonella, Chilo suppressalis, Spodoptera littoralis, Trichoplusia ni, Heliothis* spp., at 0·2-1·0 kg/ha. Phytotoxic to some ornamental varieties.

Toxicology. The acute oral LD50 of the base is: for rats 340 mg/kg; for rabbits 625 mg/kg; of the hydrochloride for rats 355 mg/kg. The acute dermal LD50 of the hydrochloride for rabbits is > 4000 mg/kg, and irritation to rabbits is slight. In 2-y feeding trials the 'no effect' level in rats and dogs was 250 mg/kg diet. The LC50 (24 h) is: for trout 11·7 mg/l; for bluegill 1 mg/l. The LC50 (48 h) for Japanese killifish is 33 mg/l. It is non-toxic to honey bees.

Formulations. These include: w.s.p., 'Fundal 800' ('Fundal SP' in USA) and 'Galecron 80 SP' (800 g a.i./kg) as hydrochloride, 'Fundal 300' (300 g a.i./kg) as hydrochloride, 'Galecron 50 SP'; w.s.c., 'Galecron' (500 g a.i./l); e.c., 'Fundal 500 EC', 'Galecron 50 EC' (500 g a.i./l) as base; dust (20 g a.i./kg); granules (50 g a.i./kg); 'Ulvair 250'.

Analysis. Product analysis is by acidimetric titration or by GLC: G. Voss *et al., Anal. Methods Pestic. Plant Growth Regul.,* 1973, **7,** 211. Residues are determined by colorimetric or GLC analysis of the 4-chloro-o-toluidine recovered after hydrolysis: H. Geissbühler *et al., J. Agric. Food Chem.,* 1971, **19,** 365; or by GLC after conversion to 5-chloro-2-iodotoluene: L. Baunok & H. Geissbühler, *Bull. Environ. Contam. Toxicol.,* 1968, **3,** 7.

Chlorfenac

$$CH_2COH$$
$$\overset{O}{\underset{\|}{}}$$

QV1R BG CG FG   *chlorfenac* $C_8H_5Cl_3O_2$ (239·5)

OV1R BG CG FG &-NA-   *chlorfenac-sodium* $C_8H_4Cl_3NaO_2$ (261·5)

Nomenclature and development. The common name chlorfenac is approved by BSI and ISO—exceptions Canada and WSSA (fenac). The IUPAC name is (2,3,6-trichlorophenyl)acetic acid (I), in *C.A.* usage 2,3,6-trichlorobenzeneacetic acid *[85-34-7]* formerly (I). It was introduced as a herbicide by Amchem Products, Inc. in 1958 under the trade name 'Fenac'.

Manufacture and properties. Produced by the chlorination of toluene to give α,2,3,6-tetra-chlorotoluene which is converted to the nitrile and hydrolysed to the chlorfenac-sodium (BP 860310, to Hooker Chemical Corp.). Chlorfenac is a colourless solid, m.p. 156°, v.p. $8·5 \times 10^{-3}$ mmHg at 100°. Its solubility at 28° is 200 mg/l water; soluble in most organic solvents. It is stable and non-corrosive. It forms water-soluble salts with alkalies, *e.g.* chlorfenac-sodium, *C.A.* Registry No. *[2439-00-1]*.

Uses. Chlorfenac, usually applied as the sodium salt, is a herbicide absorbed primarily by roots; it resists leaching and is persistent in soil. It is used to control annual grasses, broad-leaved weeds, field bindweed, couch grass and other perennial weeds in industrial and other non-crop situations at rates up to 18 kg a.e./ha. It is also used for pre-em. weed control in sugarcane at 2·8-3·4 kg a.e./ha.

Toxicology. The acute oral LD50 for rats is 576-1780 mg/kg; the acute dermal LD50 for rabbits is 1440-3160 mg/kg. In 2-y feeding trials rats receiving 2000 mg/kg diet showed no ill-effect.

Formulation. Aqueous concentrate of chlorfenac-sodium (180 g a.e./l).

Analysis. Product analysis is by GLC. Residues may be determined by the methods given for chlorophenoxyalkanoic acids (G. Yip, *J. Assoc. Off. Anal. Chem.,* 1972, **45**, 367).

Chlorfenethol

$$Cl-\langle\ \rangle-\overset{\overset{\displaystyle OH}{|}}{\underset{\underset{\displaystyle Me}{|}}{C}}-\langle\ \rangle-Cl$$

GR DXQl&R DG $C_{14}H_{12}Cl_2O$ (267·2)

Nomenclature and development. The common name chlorfenethol is approved by BSI, ISO and ESA—exception JMAF (BCPE); also known as ENT 9624 and sometimes by the initials DMC or DCPC. The IUPAC name is 1,1-bis(4-chlorophenyl)ethanol formerly 1,1-di-(4-chlorophenyl)ethanol, in *C.A.* usage 4-chloro-α-(4-chlorophenyl)-α-methylbenzenemethanol *[80-06-8]* formerly 4,4'-dichloro-α-methylbenzhydrol. Its acaricidal properties were first described by O. Grummitt, *Science,* 1950, **111**, 361, and it was introduced in 1950 by the Sherwin-Williams Co. under the trade mark 'Dimite' and protected by USP 2 430 586.

Manufacture and properties. Produced by the reaction of 4,4'-dichlorobenzophenone with methylmagnesium bromide and treatment with water (O. Grummitt *et al., Anal. Chem.,* 152, **24**, 702), it forms colourless crystals, m.p. 69·5-70°. It is practically insoluble in water; soluble in most organic solvents, particularly polar ones. It is dehydrated on heating, forming 1,1-bis(4-chlorophenyl)ethylene, and is unstable to concentrated acids, but compatible with commonly-used pesticides. The dichlorobenzophenone used in its manufacture may contain the 2,4'- and 2,2'-dichloro isomers, which give equivalent amounts of the corresponding ethanols in the Grignard reaction.

Uses. Chlorfenethol is a non-systemic acaricide with a pronounced ovicidal activity, for use on fruit, vegetables and cotton at 50-75 g a.i./100 l. It is reported to act as a synergist for DDT against DDT-resistant insects (W. T. Summerford *et al., Science,* 1951, **114,** 6).

Toxicology. The acute oral LD50 for rats is 926-1391 mg/kg; rats tolerated 1000 mg/kg diet fed for 10 d (L. Peters, *Proc. Soc. Exp. Biol. Med.,* 1947, **72**, 304). The LC50 (48 h) for carp is 1·8 mg/l.

Formulations. These include 'Qikron W.P.' (Nippon Soda Co. Ltd), w.p. (500 g a.i./kg); 'Mitran W.P.', w.p. (250 g chlorfenethol + 250 g chlorfenson/kg).

Chlorfenprop-methyl

$$Cl - \langle\rangle - CH_2.CH.CO_2Me$$
$$\underset{Cl}{|}$$

GR DlYGVOl

$C_{10}H_{10}Cl_2O_2$ (233·1)

Nomenclature and development. The common name chlorfenprop-methyl is approved by BSI and ISO—exception USA. The IUPAC name is methyl 2-chloro-3-(4-chlorophenyl)propionate, in *C.A.* usage methyl α,4-dichlorobenzenepropanoate *[14437-17-3]* formerly methyl *p*,α-dichlorohydrocinnamate. It was introduced in 1968 by Bayer AG under the code number 'Bayer 70533', trade mark 'Bidisin' and protected by BP 1 077 194 and FP 1 476 247. Its use as a herbicide was described by L. Eue, *Z. Pflanzenkr. Pflanzenpathol. Pflanzenschutz,* 1968, Sonderheft IV, 211.

Manufacture and properties. Produced by the reaction of 4-chlorobenzenediazonium chloride with methyl acrylate, the pure compound is a colourless liquid, with a fennel-like odour, b.p. 110-113°/0·1 mmHg, v.p. 7 x 10^{-3} mmHg at 50°, n_D^{20} 1·532. Its solubility at 20° is 40 mg/l water; it is soluble in acetone, aromatic hydrocarbons, diethyl ether and fatty oils. The technical product is a light brown liquid, d_4^{20} 1·30.

Uses. Chlorfenprop-methyl is a specific herbicide for the control of wild oats and, being a contact herbicide, it should be used only after the emergence of wild oats which are most susceptible between the one-leaf stage and tillering. An application rate of 4 kg a.i. in 200-400 l water/ha is effective against *Avena fatua* but not against *A. sterilis.* This spray is well-tolerated by cereals (other than oats), by fodder crops, sugar beet and peas. See: W. Kampe, *Pflanzenschutz-Nachr. (Am. Ed.),* 1973, **26,** 299; H. J. Hewston, *ibid.,* p.317; H. Hack, *ibid.,* p.353; P. Zonderwijk, *ibid.,* 1974, **27,** 3.

Toxicology. The acute oral LD50 is: for rats *c.* 1190 mg/kg; for guinea-pigs and rabbits 500-1000 mg/kg; for cats > 1000 mg/kg; for dogs > 500 mg/kg; for chickens *c.* 1500 mg/kg. The acute dermal LD50 for rats is > 2000 mg/kg. The application of the compound to a rabbit's ear for 24 h caused reddening but an 8-h exposure caused no symptoms. In a 90-d feeding trial a level of 1000 mg/kg diet caused no ill-effect to rats. See: E. Loser & G. Kimmerle, *ibid.,* 1973, **26,** 368.

Formulations. An e.c. (500 and 800 g a.i./l).

Analysis. Product analysis is by GLC, details of methods are available from Bayer AG; for a residue method, see H. J. Jarczyk, *ibid.,* 1968, **21,** 360.

Chlorfenson

GR DSWOR DG $C_{12}H_8Cl_2O_3S$ (303·2)

Nomenclature and development. The common name chlorfenson is approved by BSI and ISO—exceptions: Argentine (ovatran), France (chlorofénizon), ANSI and Canada (ovex), USSR (ephirsulphonate) and JMAF (CPCBS); also known as ENT 16358. The IUPAC name is 4-chlorophenyl 4-chlorobenzenesulphonate, in *C.A.* usage 4-chloro-phenyl 4-chlorobenzenesulfonate *[80-33-1]* formerly *p*-chlorophenyl *p*-chlorobenzene-sulfonate. It was introduced in 1949 by The Dow Chemical Co. under the code number 'K 6451', trade mark 'Ovotran' and protected by USP 2528310. Its acaricidal activity was first reported by E. E. Kenaga & R. W. Hummer, *J. Econ. Entomol.,* 1949, **42**, 996.

Manufacture and properties. Produced by the condensation of 4-chlorobenzenesulphonyl chloride with 4-chlorophenol in the presence of alkali (R. H. Slaugh & E. C. Britton, *J. Am. Chem. Soc.,* 1950, **72**, 2808), it is a colourless crystalline solid, with a characteristic odour, m.p. 86·5°, of negligible v.p. at 25°. It is practically insoluble in water; moderately in ethanol, petroleum oils (20 g/kg 'Deobase'); readily in acetone, aromatic solvents. It is hydrolysed by alkali but is compatible with most pesticides. The technical grade is a colourless-tan flaky solid, m.p. *c.* 80°.

Uses. Chlorfenson is a non-systemic acaricide with long residual ovicidal activity but little insecticidal action. It is effective against mites of citrus and other fruit, vegetable crops and ornamentals at 17-32 g a.i./100 l. It is non-phytotoxic.

Toxicology. The acute oral LD50 for rats is *c.* 2000 mg/kg. It may cause skin irritation. In 130-d feeding trials no apparent effect was observed on rats receiving 300 mg/kg diet. The LC50 (48 h) for carp is 3·2 mg/l.

Formulations. These include: 'Sappiron' (Nippon Soda Co. Ltd), w.p. (500 g a.i./kg); 'Mitran WP' (Nippon Soda Co. Ltd), w.p. (250 g chlorfenson + 250 g chlorfenethol/kg).

Analysis. Product analysis is by hydrolysis with ethanolic potassium hydroxide and back titration (*CIPAC Handbook,* 1970, **1**, 213; *FAO Plant Prot. Bull.,* 1964, **12**(1), 3). Residues are determined by GLC or by alkaline hydrolysis to 4-chlorophenol, which forms a red derivative with 4-amino-2,3-dimethyl-1-phenyl-5-pyrazolone (4-aminoantipyrene) that is measured colorimetrically (A. H. Kutschinski & E. N. Luce, *Anal. Chem.,* 1952, **24**, 1188; F. A. Gunther & L. R. Jepson, *J. Econ. Entomol.,* 1954, **47**, 1027).

Chlorfenvinphos

$(EtO)_2\overset{\displaystyle O}{\overset{\|}{P}}.O$

(Z)

(E)

GR CG DYU1GOPO&O2&O2 $C_{12}H_{14}Cl_3O_4P$ (359·6)

Nomenclature and development. The common name chlorfenvinphos is approved by BSI, ISO and BPC—exceptions JMAF (CVP) and USA; also known as ENT 24969, OMS 1328. The IUPAC name is 2-chloro-1-(2,4-dichlorophenyl)vinyl diethyl phosphate (I), in *C.A.* usage 2-chloro-1-(2,4-dichlorophenyl)ethenyl diethyl phosphate formerly (I); Registry Nos. *[470-90-6]* (formerly *[2701-86-2]*), (Z + E)-isomers; *[17708-87-7]*, (Z)-isomer; *[18708-86-6]*, (E)-isomer. It was introduced in 1963 by the Shell Development Co. under the code number 'SD 7859', the trade mark 'Birlane' and protected by USP 2 956 073; 3 116 201, also by Ciba AG under the code number 'C 8949' and the trade mark *'Sapecron'*, and by Allied Chemical Corp. as 'GC 4072'—though no longer produced by these 2 companies or their successors. Its insecticidal properties were first described by W. F. Chamberlain *et. al., J. Econ. Entomol.,* 1962, **55,** 86.

Manufacture and properties. Produced by the reaction of triethyl phosphite with 2,2,2',4'-tetrachloroacetophenone, it is a mixture of the (Z)- and (E)-isomers, containing not less than 90%, the bulk of which is the (Z)-isomer, typical ratio (Z:E) 8·5:1. The technical material is an amber-coloured liquid, with a mild odour, m.p. —19 to —23°, b.p. 167-170°/0·5 mmHg, v.p. 2·6 x 10^{-3} mmHg at 80°, 1 x 10^{-4} mmHg at 45°, 7·5 x 10^{-6} mmHg (extrapolated) at 25°, 4 x 10^{-6} mmHg (extrapolated) at 20°, $d_{16·5}^{15·5}$ 1·36, n_D^{25} 1·5272. Its solubility at 23° is 145 mg/l water; miscible with acetone, ethanol, kerosine, propylene glycol, xylene. It is stable when stored in glass or polyethylene-lined containers. It is slowly hydrolysed by water, 50% decomposition occurs at 38° and pH 9·1 in > 400 h and at pH 1·1 in > 700 h. It may corrode iron, steel and brass on prolonged contact and the e.c. formulations are corrosive to tin plate.

Uses. Chlorfenvinphos is an insecticide effective for soil use to control root flies, root worms and cutworms at 2-4 kg. a.i./ha. As a foliage insecticide, it is recommended for the control of Colorado beetle on potato and scale insects on citrus at 200-400 g/ha and for stem borers on maize, sugar cane and rice at 550-2200 g/ha. It is non-phytotoxic at these concentrations; 50% loss from soil normally occurs in a few weeks. Metabolism and breakdown are reviewed by K. I. Beynon *et al., Residue Rev.,* 1973, **47,** 55.

Toxicology. The acute oral LD50 is: for rats 10-39 mg/kg; for mice 117-200 mg/kg; for rabbits 300-1000 mg/kg; for dogs > 12 000 mg/kg. The acute dermal LD50 is: for rats 31-108 mg/kg; for rabbits 417-4700 mg/kg. In 2-y feeding trials rats receiving 10 mg/kg diet showed no adverse effect on growth or food consumption. The LC50 (24 h) for harlequin fish is 0·36 mg/l.

Formulations. These include: e.c. (240 g a.i./l); w.p. (250 g/kg); dust (50 g/kg); granules (100 g/kg); liquid seed dressings.

Analysis. Product analysis is by i.r. spectrometry and GLC: particulars from the Shell International Chemical Co.; see also, *CIPAC Handbook,* 1979, **1A,** in press; FAO Specification (CP/66). Residues may be determined by GLC (K. I. Beynon *et al., J. Sci. Food Agric.,* 1968, **19,** 302).

Chlorflurecol-methyl

(Chlorflurenol-methyl)

L   B656   HHJ   EG   HVO1   HQ $\qquad$ $C_{15}H_{11}ClO_3$ (274·7)

Nomenclature and development. The common name chlorflurenol is approved by ISO—exceptions BSI, Canada and Denmark (chlorflurecol), Poland and USA—for 2-chloro-9-hydroxyfluorene-9-carboxylic acid (I), in *C.A.* usage 2-chloro-9-hydroxy-9*H*-fluorene-9-carboxylic acid *[2464-37-1]* formerly (I). The effects of derivatives of fluorene-9-carboxylic acids on plant growth were first described by G. Schneider, *Naturwissenschaften,* 1964, **51**, 416 and, since their activity differed from that of other growth-regulating compounds, G. Schneider *et al., Nature (London),* 1965, **208**, 1013, proposed the name morphactins. Among compounds of promise for growth retardation and suppression methyl 2-chloro-9-hydroxyfluorene-9-carboxylate, chlorflurecol-methyl, *C.A.* Registry No. *[2536-31-4]* was introduced in 1965 by E. Merck under code number 'IT 3456' and protected by BP 1 051 652; 1 051 653; 1 051 654.

Properties. Chlorflurecol-methyl forms colourless crystals, m.p. 152°. Its solubility at 20° is: 18 mg/l water; *c.* 1·6 g/l light petroleum (b.p. 50-70°); 2·4 g/l cyclohexane; 24 g/l propan-2-ol; 24 g/l carbon tetrachloride; 80 g/l ethanol; 150 g/l methanol; 260 g/l acetone. It is stable at room temperature, compatible with other growth regulators and with maleic hydrazide formulated as 'MH 30'.

Uses. Chlorflurecol-methyl is suggested for use as a general growth retardant at 2-4 kg a.i./ha and for soil application for weed suppression at 0·5-1·5 kg/ha; also in combination with other growth-active compounds.

Toxicology. The acute oral LD50 is: for rats > 12 800 mg/kg; for dogs > 6400 mg/kg. Non-toxic to honey bees.

Formulation. An e.c. (125 g a.i./kg).

Analysis. Product analysis is by u.v. spectroscopy or by chromotographic methods (W. P. Cochrane *et al., J. Assoc. Off. Anal. Chem.,* 1977, **60**, 728). Residues may be determined by the colorimetric measurement of 2-chlorofluoren-9-one 4-nitrophenylhydrazone, quantitatively formed from chlorflurecol-methyl under the chosen chemical conditions of the method. See also E. Amadori & W. Heupt, *Anal. Methods Pestic. Plant Growth Regul.,* 1978, **10**, 525.

Chloridazon

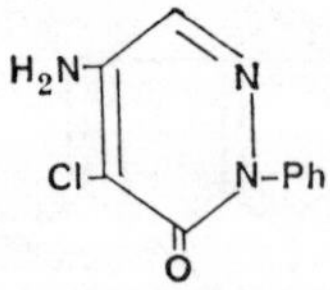

H_2N — N — N–Ph (structure) Cl — O

T6NNVJ BR& DG EZ

$C_{10}H_8ClN_3O$ (221·6)

Nomenclature and development. The common name chloridazon is proposed by ISO and has been be adopted by BSI and most of the members of ISO; the name pyrazon was formerly approved by many of these countries and is being retained by ANSI, Canada, Denmark, Poland and WSSA—exception JMAF (PCA). The IUPAC name is 5-amino-4-chloro-2-phenylpyridazin-3-one, in *C.A.* usage 5-amino-4-chloro-2-phenyl-3(2*H*)-pyridazinone *[1698-60-8]*. Its herbicidal properties were described by A. Fischer, *Weed Res.*, 1962, **2**, 177, and it was introduced in 1962 by BASF AG under the code number 'H 119', the trade mark 'Pyramin' and protected by DBP 1 105 232.

Manufacture and properties. Produced by the reaction of phenylhydrazine and 3,4-dichloro-2,5-dihydro-5-hydroxyfuran-2-one followed by treatment with ammonia, it is a pale yellowish solid, m.p. 205-206°, v.p. 7·4 x 10^{-2} mmHg at 40°. Its solubility at 20° is: 400 mg/l water; 28 g/kg acetone; 34 g/kg methanol; 0·7 g/kg benzene. It is stable, compatible with other pesticides and non-corrosive.

Uses. Pyrazon is a herbicide effective against broad-leaved weeds particularly for use on sugar beet and beet crops at 1·6-3·3 kg a.i./ha, applied pre-em. or after the late cotyledon stage. It persists in sandy loam for about 140 d and is decomposed to 5-amino-4-chloro-pyridazin-3-one which is non-phytotoxic.

Toxicology. The acute oral LD50 for rats is 2424 mg a.i. (as w.p.)/kg. It caused only a slight temporary erythema when applied for 20 h to the backs and ears of white rabbits. In 2-y feeding trials rats receiving 300 mg/kg diet suffered no detectable toxic effect.

Formulation. 'Pyramin', a w.p. (800 g tech./kg); 'Pyramin FL', an aqueous suspension (430 g/l).

Analysis. Product analysis is by i.r. spectroscopy. Residues may be determined by the colorimetric measurement of aniline produced on hydrolysis with ethanolic potassium hydroxide; particulars from BASF AG. Residues in soil may be determined by HPLC with u.v. detection (T. H. Byast *et al., Agric. Res. Counc. (G.B.) Weed Res. Organ., Tech. Rep.* No. 15 (2nd Ed.), p.65).

Chlormephos

$$\text{(EtO)}_2\overset{\displaystyle S}{\underset{\displaystyle \|}{P}}.\text{SCH}_2\text{Cl}$$

G1SPS&O2&O2 $C_5H_{12}ClO_2PS_2$ (234·7)

Nomenclature and development. The common name chlormephos is approved by BSI and ISO. The IUPAC and *C.A.* name is *S*-chloromethyl *O,O*-diethyl phosphorodithioate, Registry No. *[24934-91-6]*. It was introduced in 1968 by Murphy Chemical Ltd under the code number 'MC 2188' protected by BP 1 258 922; 817 360; 902 795. It has been developed in many countries under licence to Rhône-Poulenc Phytosanitaire with the trade mark 'Dotan'. It was first reported by D. G. Griffiths *et al., Ann. Appl. Biol.*, 1969, **64**, 21.

Manufacture and properties. Produced by the reaction between an alkali metal salt of *O,O*-diethyl hydrogen phosphorodithioate and bromochloromethane, it is a colourless liquid, b.p. 81-85°/0·1 mmHg, v.p. 5·7 x 10^{-2} mmHg at 30°, n_D 1·5244, d 1·260. Its solubility at 20° is 60 mg/l water; miscible with most organic solvents. The technical grade is *c.* 90-93% pure. It is stable to water but hydrolysed by dilute acid or alkali at 80°.

Uses. Chlormephos is a contact insecticide, effective when applied to soil for the control of wireworm, white grubs and millipedes at 2-4 kg a.i./ha for broadcast application and 0·3-0·4 kg a.i./ha for band treatment, in maize and sugar beet.

Toxicology. The acute oral LD50 for rats is 7 mg a.i./kg. The acute dermal LD50 for rats is 27 mg/kg. The acute dermal LD50 of the 5% granules is > 1600 mg/kg, according to the species tested. In 90-d feeding trials on rats the 'no effect' level was 0·39 mg/kg diet. The LC50 for fish is 1·5 mg/l.

Formulation. Granules (50 g a.i./kg).

Analysis. Product analysis is by GLC. Residue analysis is by GLC using a thermionic detector after extraction with dichloromethane and chromatographic clean-up. See also: V. P. Lynch, *Anal. Methods Pestic. Plant Growth Regul.*, 1978, **10**, 49.

Chlormequat chloride

$$Cl.CH_2CH_2N^+Me_3 \quad Cl^-$$

G2K1&1&1   *chlormequat* $C_5H_{13}ClN$ (122·6)

G2K1&1&1 &G   *chlormequat chloride* $C_5H_{13}Cl_2N$ (158·1)

Nomenclature and development. The common name chlormequat is approved by BSI and ISO for the 2-chloroethyltrimethylammonium ion (I), in *C.A.* usage 2-chloro-*N,N,N*-trimethylethanaminium ion *[7003-89-6]* formerly (I). It is mostly used as the chloride, *C.A.* Registry No. *[991-81-5]*, which is sometimes known by the trivial names chloro-choline chloride and CCC. The effects of chlormequat chloride on plant growth were first described by N. E. Tolbert, *Plant Physiol.*, 1960, **35**, 380; *J. Biol. Chem.*, 1960, **235**, 475, and it was introduced, in collaboration with the Michigan State University, by the American Cyanamid Co. in 1959 under the code number 'AC 38555', the trade mark 'Cycocel' and protected by BP 944807; FP 1264866; BelgP 593961.

Manufacture and properties. Produced by the reaction of 1,2-dichloroethane with trimethylamine, chlormequat chloride is a colourless crystalline solid, with a fish-like odour, and begins to decompose at 245°. Its solubility at 20° is > 1 kg/kg water; it is soluble in the lower alcohols; practically insoluble in diethyl ether, hydrocarbons. The technical product is 97-98% pure. The solid is extremely hygroscopic but its aqueous solutions are stable though corrosive to unprotected metals. It may be stored in glass, high-density plastic, rubber or epoxyresin-protected metal containers.

Uses. Chlormequat chloride is a plant growth regulator which influences the habit of certain plants by shortening and strengthening the stem, *e.g.* in wheat and poinsettias. It can also influence the developmental cycle resulting in increased flowering and harvest, *e.g.* in pears and tomatoes. An intensification of chlorophyll formation is often seen. The root system may also be increased, resulting in yield increases under dry conditions. Wheat is protected against damage by *Cercosporella herpotrichoides*. Chlormequat chloride is rapidly degraded in soil by enzyme activity and there is no influence on soil microflora or fauna.

Toxicology. The acute oral LD50 is: for male rats 670 mg/kg; for female mice 1020 mg/kg; for male guinea-pigs 620 mg/kg; for chickens 920 mg/kg. The acute dermal LD50 for male rabbits is 440 mg/kg. In 2-y feeding trials rats receiving 1000 mg/kg diet suffered no ill-effect. In rats 96% of the administered compound was excreted unchanged in the urine and faeces. The toxicity of chlormequat chloride is reduced by the addition of choline chloride (DBP 1215436) as several tests in laboratory animals showed.

Formulations. Aqueous solutions (118, 400, 500 and 725 g chlormequat chloride/1); dust (650 g/kg). In addition, a mixture (460 g chlormequat chloride + 320 g choline chloride/1) is sold as 'WR62', 'Cycocel 460' (American Cyanamid Corp. and BASF AG), 'BAS 06200W' and 'CCC Extra' (BASF AG).

Analysis. Product analysis is by potentiometric titration with silver nitrate and the method can be used to determine chlormequat chloride and choline chloride separately. Residue analysis is based on the colour intensity of the dipicrylamine salt at 415 nm: N. R. Pasarela & E. J. Orloski, *Anal. Methods Pestic. Plant Growth Regul.*, 1973, **7**, 523; *'Cycocel' Plant Growth Regulant Technical Manual 1966;* J. Jung & G. Henjes, *Landwirtsch. Forsch.*, 1964, **17**, 1.

Chloroacetic acid

$$Cl.CH_2.COOH$$

QV1G

chloroacetic acid $C_2H_3ClO_2$ (94·50)

OV1G   &-NA-

sodium chloroacetate $C_2H_2ClNaO_2$ (116·5)

Nomenclature and development. The IUPAC name chloroacetic acid (I) and the trivial name monochloroacetic acid are accepted in lieu of a common name by BSI and ISO, in *C.A.* usage (I), Registry No. *[79-11-8]*. Its uses as a herbicide were first described by A. E. Hitchcock *et al., Proc. Northeast. Weed Control Conf.,* 1951, p.105. Those of its sodium salt (*C.A.* Registry No. *[3926-62-3];* sometimes known as SMA or SMCA) were first reported by T. C. Breese & A. F. J. Wheeler, *Proc. Br. Weed Control Conf., 3rd,* 1956, p.759, and it was introduced in 1956 by ICI Ltd, Plant Protection Division, under the trade mark 'Monoxone' but is no longer produced by this firm.

Manufacture and properties. Produced by chlorinating acetic acid using iodine or sulphur as catalyst, it forms a deliquescent solid existing in 3 crystalline forms: alpha, m.p. 63°; beta, m.p. 55-56°; gamma, m.p. 50°. The liquid has b.p. 189°. The acid is very soluble in water; soluble in benzene, chloroform, ethanol, diethyl ether. The technical product has m.p. 61-63°. It is corrosive.

The sodium salt is a colourless crystalline solid, solubility at 20° 850 g/l water. The technical grade is *c.* 90% pure.

Uses. Sodium chloroacetate is a post-em. contact herbicide used to control a wide range of annual weeds at seedling stage in Brussels sprouts (20 kg a.i. in 225 l water/ha), kale (20-25 kg in 225-560 l/ha), leeks and onions (25 kg in 315-560 l/ha). It is also used in combination with atrazine for total weed control on industrial sites and other non-crop land.

Toxicology. The acute oral LD50 is: for rats 650 mg sodium chloroacetate/kg; for mice 165 mg/kg. It may cause irritation to the skin and eyes. The health of rats receiving 700 mg/kg diet for several months was not affected. At 18° the LC50 (24 h) for rainbow trout is 2000 mg/l, the LC50 (48 h) is 900 mg/l.

Formulations. These include: 'Herbon Somon' (Cropsafe), w.s.p. (900 g sodium chloroacetate/kg); 'A-Plus Granules' (Diamond Shamrock Agrochemicals Ltd), w.s.p. (atrazine + sodium chloroacetate).

Analysis. Product analysis is by chlorine content determined by alkaline hydrolysis and titration with silver nitrate, correcting for chloride ion and sodium dichloroacetate originally present.

Chlorobenzilate

GR   DXQR   DG&VO2
$C_{16}H_{14}Cl_2O_3$ (325·2)

Nomenclature and development. The common name chlorobenzilate is approved by BSI, ISO, ANSI and JMAF; also known as ENT 18596. The IUPAC name is ethyl 4,4'-dichlorobenzilate (I), in *C.A.* usage ethyl 4-chloro-α-(4-chlorophenyl)-α-hydroxy-benzeneacetate *[510-15-6]* formerly (I). Its acaricidal properties were first described by R. Gasser, *Experientia,* 1952, **8**, 65. It was introduced in 1952 by J. R. Geigy S.A. (now Ciba-Geigy AG) under the code number 'G 23992', trade marks 'Akar' and 'Folbex' and protected by SwissP 294599; BP 705037; USP 2745780.

Manufacture and properties. Produced by the reaction of 4,4'-dichlorobenzilic acid with diethyl sulphate, it is a colourless solid, m.p. 35-37°, b.p. 156-158°/0·07 mmHg, v.p. 6·8 x 10^{-6} mmHg at 20°. It is practically insoluble in water, but soluble in most organic solvents including petroleum oils.

Uses. Chlorobenzilate is a non-systemic acaricide with little insecticidal action. It is recommended for use against phytophagous mites on citrus, cotton, grapes, soyabeans, tea and vegetables at 30-60 g a.i./100 l or 1·0-1·5 kg/ha.

Toxicology. The acute oral LD50 for rats is 700-3100 mg/kg. In 2-y feeding trials the 'no effect' level was: in rats 40 mg/kg diet; in dogs 500 mg/kg diet. It is moderately toxic to fish.

Formulations. These include 'Akar 338', e.c. (250 g a.i./l); w.p. (250 g/kg); 'Akar 50', e.c. (500 g/l).

Analysis. Product analysis is: *(a)* by total chlorine content; *(b)* by saponification; *(c)* by acetylation of the hydroxy group; *(d)* by GLC with internal standard: E. Bartsch *et al., Residue Rev.,* 1971, **39**, 1; R. Suter *et al., Z. Anal. Chem.,* 1955 **147**, 173; A. Margot & K. Stammbach, *Anal. Methods Pestic., Plant Growth Regul. Food Addit.,* 1964, **2**, 65; *CIPAC Handbook,* 1979, **1A,** in press. Residue determination is: *(a)* by GLC with MCD, *Anal. Methods Pestic. Plant Growth Regul.,* 1972, **6**, 319; *(b)* by hydrolysis, nitration of the acid and reaction with sodium methoxide giving a red colour measured at 538 nm (H. J. Harris, *J. Agric. Food Chem.,* 1955, **3**, 939); *(c)* by hydrolysis and oxidation of the acid by chromium trioxide in glacial acetic acid (measuring the 4,4'-dichlorobenzophenone at 264 nm) or conversion to its 2,4-dinitrophenylhydrazone (measured at 519 nm); R. C. Blinn *et al., ibid.,* 1954, **2**, 1080; E. Bartsch *et al., loc. cit.*

Chloromethiuron

GR Cl DMYUS&N1&1   C$_{10}$H$_{13}$ClN$_2$S (228·7)

Nomenclature and development. The common name chloromethiuron is approved by BSI and ISO. The IUPAC name is 3-(4-chloro-*o*-tolyl)-1,1-dimethyl(thiourea), in *C.A.* usage *N′*-(4-chloro-2-methylphenyl)-*N*,*N*-dimethyl(thiourea) *[28217-97-2]*. It was synthesised and developed as an ixodicide by Ciba-Geigy AG under the code number 'CGA 13 444', trade mark 'Dipofene' and protected by BP 1 138 714; SwissP 541 282. Its acaricidal properties were first reported by M. Von Orelli *et al., Proc. World Vet. Congr., 20th,* 1975, p.659.

Properties. It is a crystalline solid, m.p. 175°, v.p. 8 x 10^{-9} mmHg at 20°. Its solubility at 20° is 50 mg/l water; slightly soluble in most organic solvents.

Uses. Chloromethiuron controls all tick species including strains resistant to other ixodicides. It can be used on cattle, sheep, horses and dogs in plunge dips at 1·8 g a.i./l.

Toxicology. The acute oral LD50 for rats is 2500 mg tech./kg; the acute dermal LD50 for rats is 2150 mg/kg.

Formulation. 'Dipofene 600 FW', flowable (600 g tech./kg).

Analysis. Details of methods are available from Ciba-Ceigy AG.

107

Chloroneb

1OR BG EG DO1 $C_8H_8Cl_2O_2$ (207·1)

Nomenclature and development. The common name chloroneb is approved by BSI, ISO and ANSI. The IUPAC and *C.A.* name is 1,4-dichloro-2,5-dimethoxybenzene, Registry No. *[2675-77-6]*. It was introduced in 1967 as an experimental fungicide by E. I. du Pont de Nemours & Co. (Inc.) under the code number 'Soil Fungicide 1823', trade mark 'Demosan' and protected by USP 3 265 564. Its activity was described by M. J. Fielding & R. C. Rhodes, *Proc. Cotton Dis. Counc.,* 1967, **27**, 56.

Manufacture and properties. Produced by the chlorination of 1,4-dimethoxybenzene, it is a colourless crystalline solid, with a musty odour, m.p. 133-135°, b.p. 268°, v.p. 3 x 10^{-3} mmHg at 25°. Its solubility at 25° is: 8 mg/l water; 115 g/kg acetone; 118 g/kg dimethylformamide; 133 g/kg dichloromethane; 89 g/kg xylene. It is stable at least up to 268° and in the presence of alkali or acid, but is subject to microbial decomposition under moist conditions in soil.

Uses. Chloroneb is a systemic fungicide which is taken up by the roots, concentrated in the roots and lower stem portions, rendering the plants fungistatic. It is highly fungistatic to *Rhizoctonia,* moderately so to *Pythium,* poorly to *Fusarium* and inactive against *Trichoderma* spp. It is used as a supplemental seed treatment, or as an in-furrow soil treatment at planting time, to control seedling diseases of cotton, beans and soyabeans; and as a supplemental seed treatment for sugar beet. Treatment rates are 2·44 g a.i./kg cotton seed, 1·63 g/kg for bean or soyabean seed, 2·44 kg sugar beet seed. Recommended rates for in-furrow treatment are: 90-135 g/ha (9030-10 540 m row) for cotton, and 68 g/ha for beans and soyabeans. It is also used for the control of snow mould *(Typhula)* at 13-20 g/100 m² and *Pythium* blight at 8·8 g/100 m² of turf grass.

Toxicology. The acute oral LD50 for rats is > 11 000 mg/kg; the approximate dermal lethal dose for rabbits is > 5000 mg/kg. A 50% aqueous suspension of the 65% w.p. caused no irritation to guinea-pigs and repeated applications did not result in skin sensitisation.

Formulations. These include: 'Demosan 65W', w.p. (650 g a.i./kg); 'Tersan SP' Turf fungicide (650 g/kg).

Analysis. Product analysis is by GLC (H. L. Pease & R. W. Reiser, *Anal. Methods Pestic. Plant Growth Regul.,* 1973, **7**, 657). Residue analysis is also by GLC (*idem. ibid.;* H. L. Pease, *J. Agric. Food Chem.,* 1967, **15**, 917).

Chlorophacinone

L56 BV DV CHJ CVYR&R DG $C_{23}H_{15}ClO_3$ (374·8)

Nomenclature and development. The common name chlorophacinone is approved by BSI, ISO and JMAF. The IUPAC name is 2-[2-(4-chlorophenyl)-2-phenylacetyl]indan-1,3-dione formerly 2-[2-(4-chlorophenyl)-2-phenylacetyl]indane-1,3-dione, in *C.A.* usage 2-[(4-chlorophenyl)phenylacetyl]-1*H*-indene-1,3(2*H*)-dione *[3691-35-8]* formerly 2-[(*p*-chlorophenyl)phenylacetyl]-1,3-indandione. It was introduced in 1961 by Lipha S.A. as a rodenticide under the code number 'LM 91' and trade marks 'Caid', 'Liphadione', 'Raviac' (Lipha S.A.), 'Drat' (May & Baker Ltd), 'Quick' (Rhône-Poulenc), 'Saviac' (Aulagne-Chimiotechnic). It is protected by USP 3 153 612; FP 1 269 638 (Lipha S.A.).

Manufacture and development. Bromination of phenylacetone gives 1-bromo-1-phenyl-acetone which is used to alkylate chlorobenzene and the resulting 1-(4-chlorophenyl)-1-phenylacetone condensed with dimethyl phthalate in the presence of sodium methoxide. Chlorophacinone is a yellow crystalline solid, m.p. 140°, with negligible v.p. at 20°. It is sparingly soluble in water but soluble in acetone, ethanol, ethyl acetate. It is stable, resistant to weathering and is non-corrosive; compatible with cereals, fruits, roots and other potential bait substrates.

Uses. Chlorophacinone is an anticoagulant rodenticide, a single dose of a 50 mg/kg bait killing *Rattus norvegicus* from the 5th d. It is normally incorporated as 50-250 mg/kg bait. It does not induce 'bait-shyness'. In mammals it uncouples oxidative phosphorylation in addition to its anticoagulant action.

Toxicology. Human volunteers tolerated a single dose of 20 mg a.i. with an uneventful recovery without treatment. A solution of 5 mg in 2 ml liquid paraffin applied to 100 cm² of a rabbit's shaved skin caused only a slight reduction of prothrombin rating. Administration of 15 daily doses of 2·25 mg to grey partridges produced no ill-effect.

Formulations. As 'Caid', 'Liphadione', solution (2.5 g a.i./l oil); 'Raviac', 'Quick', 'CX 14', prepared bait (50 mg/kg) on whole, cracked or milled grain.

Analysis. Chlorophacinone may be recovered from baits by extraction with ethyl acetate and chromatographic separation on alumina. It is oxidised by potassium permanganate to 4-chlorobenzophenone, which is estimated by GLC.

1-(4-Chlorophenoxy)-1-(imidazol-1-yl)-3,3-dimethylbutanone

T5N CNJ AYOR DG&VX1&1&1 $C_{15}H_{17}ClN_2O_2$ (292·5)

Nomenclature and development. The common name is not yet agreed. The IUPAC name is 1-(4-chlorophenoxy)-1-(imidazol-1-yl)-3,3-dimethylbutanone, in *C.A.* usage 1-(4-chloro-phenoxy)-1-(1*H*-imidazol-1-yl)-3,3-dimethyl-2-butanone *[38083-17-9]*. It was introduced in 1977 by Bayer AG under the code number 'BAY MEB 6401' and the trade mark 'Baysan'.

Properties. It is a colourless crystalline solid, m.p. 95·5°, v.p. 7·5 x 10^{-6} mmHg (extrapolated) at 50°. Its solubility at 20° is: 5·5 mg/l water; 100-200 g/kg propan-2-ol; 400-600 g/kg cyclohexanone.

Uses. It is a fungicide, effective against *Aspergillus, Penicillium, Candida* and *Paecilomyces* spp. on various household material, utensils and parts of buildings.

Toxicology. The acute oral LD50 for male rats is 400 mg/kg.

Formulation. An aerosol concentrate (5 g/l) combined with benzalkonium chloride (alkyl-benzyldimethylammonium chloride, 'Dimanin A').

Analysis. Product analysis is by GLC.

S-4-Chlorophenylthiomethyl O,O-dimethyl phosphorodithioate

$$(MeO)_2PS.S.CH_2.S—\langle \text{benzene ring} \rangle—Cl$$

GR   DS1SPS&O1&O1 $C_9H_{12}ClO_2PS_3$ (314·8)

Nomenclature and development. There is no common name; the compound is sometimes known as ENT 25599. The IUPAC and *C.A.* name is *S*-[[(4-chlorophenyl)thio]methyl] *O,O*-dimethyl phosphorodithioate *[953-17-3]*. It was introduced in 1958 by the Stauffer Chemical Co. under the code number 'R-1492' and trade marks 'Methyl Trithion' and 'Tri-Me', and is protected by USP 2 793 224. Its insecticidal properties were described by J. A. Harding, *J. Econ. Entomol.,* 1959, **52**, 1219.

Manufacture and properties. Produced by the reaction of chloromethyl 4-chlorophenyl sulphide with sodium *O,O*-dimethyl phosphorodithioate, it is a light yellow to amber liquid with a moderate mercaptan-like odour. The technical grade has f.p. *c.* —18°; v.p. 3 x 10⁻³ mmHg at 25°; d_{20}^{20} 1·34-1·35; n_D^{30} 1·6130. Its solubility at room temperature is *c.* 1 mg/l water; it is miscible with most organic solvents. The compound is moderately stable to heat and, because of its low solubility in water, resistant to hydrolysis. It is compatible with other common pesticides, non-corrosive to mild steel and may be stored in unlined steel drums.

Uses. The compound is a non-systemic acaricide similar in range of activity to its homologue, carbophenothion (p.85), but more effective against cotton boll weevil; generally used at 50-100 g a.i./100 l. It is non-phytotoxic.

Toxicology. The acute oral LD50 is: for male albino rats 157 mg/kg; for male albino mice 390 mg/kg. The acute dermal LD50 for albino rabbits is 2420 mg/kg.

Formulations. These include: e.c. (480 g a.i./l); dusts of various a.i. content.

Analysis. Particulars of the u.v. absorption method used for macro analysis and of a GLC method may be obtained from the Stauffer Chemical Co. For residue methods, see G. H. Batchelder *et. al., Anal. Methods Pestic., Plant Growth Regul. Food Addit.,* 1964, **2**, 313; *Anal. Methods Pestic. Plant Growth Regul.,* 1972, **6**, 443.

Chloropicrin

$$CCl_3.NO_2$$

WNXGGG $\qquad\qquad\qquad$ CCl_3NO_2 (164·4)

Nomenclature and development. The trivial name chloropicrin is accepted by BSI and ISO in lieu of a common name. The IUPAC and *C.A.* name is trichloronitromethane, Registry No. *[76-06-2]*. It use as an insecticide was protected in 1908 by BP 2387.

Manufacture and properties. Produced by the action of hypochlorites on nitromethane, it is a colourless liquid, m.p. —64°, b.p. 112·4°, v.p. 5·7 mmHg at 0°, 23·8 mmHg at 25°, n_D^{20} 1·595, d_4^{20} 1·656. Its solubility at 0° is 2·27 g/l water; miscible with acetone, benzene, carbon tetrachloride, diethyl ether, methanol. Chloropicrin is non-flammable and chemically rather inert. It is non-corrosive to copper, brass and bronze but attacks iron, zinc and other light metals; the formation of a protective coating permits storage in iron or galvanized containers.

Uses. Chloropicrin is an insecticide used for the fumigation of stored grain and of soil to control nematodes and other soil-dwelling pests; it is effective against soil fungi, except those forming sclerotia and is highly phytotoxic. Special regulations apply to its use in the UK.

Toxicology. It is lachrymatory and highly toxic, lethal on exposure for 30 min at 800 mg/m³, but the intense irritation of the mucous membranes serves to reduce hazards.

Formulations. It is used on its own, generally at 32-80 g/m³; also in other fumigants as a warning gas.

Analysis. Chloropicrin is determined by GLC. For determination in air: collect in propan-2-ol, oxidise and neutralise with HCl, react the liberated nitrite with sulphanilic acid, couple the diazonium salt with *N*-(1-naphthyl)ethylenediamine and measure at 540 nm: L. Feinsilver & F. W. Oberst, *Anal. Chem.,* 1953, **25,** 820.

Chloropropylate

GR DXQR DG&VOY1&1 $\qquad$ $C_{17}H_{16}Cl_2O_3$ (339·2)

Nomenclature and development. The common name chloropropylate is approved by BSI, ISO, ANSI, JMAF; also known as ENT 26999. The IUPAC name is isopropyl 4,4′-dichlorobenzilate (I), in *C.A.* usage 1-methylethyl 4-chloro-α-(4-chlorophenyl)-α-hydroxybenzeneacetate *[5836-10-2]* formerly (I). The acaricidal properties of esters of 4,4′-dichlorobenzilic acid were first described by R. Gasser, *Experientia*, 1952, **8**, 65, and of the isopropyl ester by F. Chabousson, *Phytiatr.-Phytopharm*, 1956, **5**, 203. It was introduced in 1964 by J. R. Geigy S.A. (now Ciba-Geigy AG) under the code number 'G 24163', trade mark 'Rospin' and protected by SwissP 294599; BP 705037; USP 2745780.

Manufacture and properties. Produced by the action of 4,4′-dichlorobenzilic acid with propan-2-ol, it is a colourless powder, m.p. 73°, b.p. 148-150°/0·5 mmHg, v.p. 1·8 x 10^{-7} mmHg at 20°. Its solubility at 20° is < 10 mg/l water, though it is soluble in most organic solvents.

Uses. Chloropropylate is a non-systemic contact acaricide suitable for use on fruit, nuts, tea, cotton, sugar beet, vegetables and ornamentals at 30-60 g a.i./100 l, applied to obtain full coverage of foliage. It is non-phytotoxic at these rates.

Toxicology. The acute oral LD50 for rats and mice is > 5000 mg/kg. In 2-y feeding trials the 'no effect' level was: in rats 40 mg/kg diet; in dogs 500 mg/kg diet. It is toxic to fish and slightly toxic to birds and honey bees.

Formulation. 'Rospin 25', e.c. (250 g/l).

Analysis. Product analysis is: *(a)* by total chlorine content; *(b)* by saponification; *(c)* by acetylation of the hydroxy group or *(d)* by GLC with internal standard: E. Bartsch *et al., Residue Rev.*, 1971, **39**, 1; R. Suter *et al., Z. Anal. Chem.*, 1955, **147**, 173; *CIPAC Handbook*, 1979, **1A**, in press.

113

Chlorothalonil

NCR BG CG DG FG ECN $C_8Cl_4N_2$ (265·9)

Nomenclature and development. The common name chlorothalonil is approved by BSI, ISO and ANSI—exception JMAF (TPN). The IUPAC name is tetrachloroisophthalonitrile (I), in *C.A.* usage 2,4,5,6-tetrachloro-1,3-benzenedicarbonitrile *[1897-45-6]* formerly (I). Its fungicidal properties were described by N. J. Turner *et al., Contrib. Boyce Thompson Inst.,* 1964, **22**, 303. It was introduced in 1963 by the Diamond Alkali Co. (now the Diamond Shamrock Corp.) under the trade marks, 'Bravo', 'Daconil 2787' and 'Exotherm Termil', and protected by USP 3 290 353; 3 331 735.

Manufacture and properties. Produced by the chlorination of isophthalonitrile, pure chlorothalonil forms colourless, odourless crystals, m.p. 250-251°, b.p. 350°, v.p. 0·01 mmHg at 40°, 9·2 mmHg at 170·4°. Its solubility at 25° is: 0·6 mg/kg water; 20 g/kg acetone, butanone, dimethyl sulphoxide; 30 g/kg cyclohexanone, dimethylformamide; < 10 g/kg kerosine; 80 g/kg xylene. The technical product is *c.* 98% pure and has a slight pungent odour. It is thermally stable under normal storage conditions, is stable to alkaline and acid aqueous solutions and to u.v. light; non-corrosive.

Uses. Chlorothalonil is effective against a broad range of plant pathogens attacking many vegetable and agronomic crops at rates of 0·63-2·52 kg a.i. ('Bravo 500 W-75')/ha; on turf at 0·46-1·8 g a.i. ('Daconil 2787 W-75')/m², on several ornamentals for control of *Botrytis* spp. and certain other diseases at 222-333 g a.i. ('Daconil 2787 W-75')/100 l; for sublimation in glasshouses at 2 g a.i./28 m³ using 'Exotherm Termil' to control *Botrytis* spp. on many ornamentals. Its algicidal properties were described by K. H. Goulding, *Proc. Br. Insectic. Fungic. Conf., 9th,* 1971, **2,** 621. Chlorothalonil ('Nopocide') is also used as a preservative in paints and adhesives.

Toxicology. The acute LD50 for albino rats is > 10000 mg/kg. The acute dermal LD50 for albino rabbits is > 10000 mg/kg. A single application of 100 mg to the eyes of rabbits caused marked eye irritation and corneal opacities in rabbits at 7 d. No toxic symptom was observed in 90-d feeding trials on rats at doses up to 2000 mg/kg diet. The acute oral LC50 is: for young mallard duck > 21 500 mg/l; for bobwhite quail 5200 mg/l. The estimated LC50 is: for rainbow trout 0·25 mg/l; for bluegill 0·39 mg/l; for channel catfish 0·43 mg/l.

Formulations. These include: 'Bravo W-75' and 'Daconil 2787 W-75 Fungicide', w.p. (750 g a.i./kg); 'Bravo 500' and 'Daconil Flowable', liquids (500 g/l); and 'Exotherm Termil', for sublimation in glasshouses.

Analysis. Products and residues are analysed by GLC; particulars from the Diamond Shamrock Corp. See also, D. L. Ballee *et al., Anal. Methods Pestic. Plant Growth Regul.,* 1976, **8**, 263.

Chloroxuron

Cl—⟨benzene ring⟩—O—⟨benzene ring⟩—NH.Ċ.NMe₂ (with O double-bonded to C)

GR DOR DMVNl&1 $C_{15}H_{15}ClN_2O_2$ (290·7)

Nomenclature and development. The common name chloroxuron is approved by BSI, ISO, ANSI, WSSA and JMAF—exceptions USSR (chloroxifenidim) and Germany. The IUPAC name is 3-[4-(4-chlorophenoxy)phenyl]-1,1-dimethylurea, in *C.A.* usage *N'*-[4-(4-chlorophenoxy)phenyl]-*N,N*-dimethylurea *[1982-47-4]* formerly 3-[*p*-(*p*-chlorophenoxy)phenyl]-1,1-dimethylurea. It was introduced in 1960 as a herbicide by Ciba AG (now Ciba-Geigy AG) under the code number 'C 1983', trade mark 'Tenoran' and protected by BP 913383; USP 3 119 682. Its activity was first described by the manufacturers, *Symp. New Herbic., 3rd,* 1961, p.88.

Manufacture and properties. Produced by the reaction of 4-(4-chlorophenoxy)phenylurea or 4-(4-chlorophenoxy)phenyl isocyanate with dimethylamine, it forms colourless crystals, m.p. 151-152°, v.p. 1·8 x 10^{-9} mmHg at 20°. Its solubility at 20° is 3·7 mg/l water; slightly soluble in benzene, ethanol; soluble in acetone, chloroform. Chloroxuron is compatible with other pesticides.

Uses. Chloroxuron is absorbed by roots and leaves and is recommended for the control of weeds in strawberries at 3·5-4·5 kg a.i./ha, or vegetables at 2·5-4·5 kg/ha pre-em. or after transplanting, or in soyabeans, ornamentals and conifers at similar rates. At these rates it dissipates to non-phytotoxic concentrations within 60 d.

Toxicology. The acute oral and acute dermal LD50 for rats is > 3000 mg/kg. Feeding rats daily at 10 mg/kg for 120 d or dogs 15 mg/kg for 90 d produced no ill-effect. Low fish toxicity was recorded in several species.

Formulation. A w.p. (500 g a.i./kg).

Analysis. Product analysis is by determination of the dimethylamine produced by hydrolysis: G. Voss *et al., Anal. Methods Pestic. Plant Growth Regul.,* 1973, **7**, 569. Residues may be determined by hydrolysis to 4-(4-chlorophenoxy)aniline which is diazotised and either coupled with *N*-(1-naphthyl)ethylenediamine and measured colorimetrically, or iodinated and measured by GLC with ECD: particulars from Ciba-Geigy AG; G. Voss *et al., loc. cit.* Alternatively, GLC with ECD may be used direct: T. H. Byast *et al., Agric. Res. Counc. (G.B.) Weed Res. Organ., Tech. Rep.,* No. 15 (2nd Ed.), p.49.

Chlorphonium chloride

GR	CG	D1P4&4&4		*chlorphonium ion*	$C_{19}H_{32}Cl_2P$	(362·3)
GR	CG	D1P4&4&4	&G	*chlorphonium chloride*	$C_{19}H_{32}Cl_3P$	(397·8)

Nomenclature and development. The common name chlorphonium is approved by BSI and ISO for the tributyl-(2,4-dichlorobenzyl)phosphonium ion (IUPAC) (I), in *C.A.* usage tributyl[(2,4-dichlorophenyl)methyl]phosphonium formerly (I). The chloride, *C.A.* Registry No. *[115-78-6],* was introduced by the Mobil Chemical Co. as a plant growth regulator, trade mark 'Phosfon', protected by USP 3 268 323.

Manufacture and properties. Produced by the reaction of 2,4-dichlorobenzyl chloride with tributylphosphine, chlorphonium chloride is a colourless crystalline solid, m.p. 114-120°, of negligible v.p. It is soluble in water, acetone, alcohols; insoluble in hexane, diethyl ether.

Uses. Chlorphonium chloride is used as a height retardant for lilies and chrysanthemums. It is also effective in rhododendrons. It is most effective as a soil treatment for potted plants.

Toxicology. The acute oral LD50 for rats is 178 mg tech./kg; the acute dermal LD50 for rabbits is 750 mg/kg. The technical grade and formulations are irritating to skin and eyes.

Formulations. 'Phosfon', liquid (100 g a.i./l); 'Phosfon' dusts (15 and 100 g/kg).

Analysis. Product analysis is by phosphorus and chlorine content, and by melting point.

Chlorphoxim

$$(EtO)_2\overset{\overset{S}{\|}}{P}.O.N=\overset{\overset{CN}{|}}{C}$$

NCYR   BG&UNOPS&O2&O2 $C_{12}H_{14}ClN_2O_3PS$ (332·7)

Nomenclature and development. The common name chlorphoxim is approved by BSI and ISO. The IUPAC name is *O,O*-diethyl 2-chloro-α-cyanobenzylideneamino-oxyphosphono-thioate formerly 2-chloro-α-(diethoxyphosphinothioyloxyimino)phenylacetonitrile, in *C.A.* usage 7-(2-chlorophenyl)-4-ethoxy-3,5-dioxa-6-aza-4-phosphaoct-6-ene-8-nitrile 4-sulfide *[14816-20-7]* formerly 2-chloro-α-[(diethoxyphosphinothioyloxy)imino]benzene-acetonitrile and *o*-chlorophenylglyoxylonitrile oxime *O,O*-diethyl phosphorothioate. Its insecticidal properties were first described by J. E. Hudson & W. O. Obudho, *Mosq. News,* 1972, **32**, 37. It was introduced in 1977 by Bayer AG under the code number 'BAY SRA 7747', trade mark 'Baythion C' and protected by DBP 1 238 902.

Properties. It is a colourless crystalline solid, m.p. 66·5° not distillable, v.p. < 7·5 x 10⁻⁶ mmHg at 20°. Its solubility at 20° is: 1·7 mg/kg water; 400-600 g/kg cyclohexanone; 400-600 g/kg toluene.

Uses. Chlorphoxim is an insecticide used to control mosquitoes and simulium flies.

Toxicology. The acute oral LD50 for rats is > 2500 mg/kg; the acute dermal LD50 for rats is > 500 mg/kg.

Formulations. A w.p. (500 g a.i./kg); ULV (200 g/l).

Analysis. Product analysis is by i.r. spectroscopy. Residues are determined by GLC: D. B. Leuck & M. C. Bowman, *J. Econ. Entomol.,* 1973, **66**, 798.

Chlorpropham

GR CMVOY1&1

$C_{10}H_{12}ClNO_2$ (213·7)

Nomenclature and development. The common name chlorpropham is approved by BSI, ISO, WSSA and JMAF—exception USSR (chlor-IFC); also known as ENT 18060 or the trivial terms CIPC and chloro-IPC. The IUPAC name is isopropyl 3-chlorophenyl-carbamate, in *C.A.* usage 1-methylethyl 3-chlorophenylcarbamate *[101-21-3]* formerly isopropyl *m*-chlorophenylcarbamate. It was introduced as a herbicide in 1951 (E. D. Witman & W. F. Newton, *Proc. Northeast. Weed Control Conf.*, 1951, p.45) and is protected by USP 2695225.

Manufacture and properties. Produced either by reacting 3-chloroaniline with isopropyl chloroformate, or propan-2-ol with 3-chlorophenyl isocyanate, it is a solid, m.p. 41·4°, d^{30} 1·180, n_D^{20} (supercooled) 1·5395. Its solubility at 25° is: 89 mg/l water; 100 g/kg kerosine; miscible with the lower alcohols, aromatic hydrocarbons and most organic solvents. The technical product is 98·5% pure, m.p. 38·5-40°. It is stable < 100° but is slowly hydrolysed in acidic and alkaline media.

Uses. Chlorpropham is a mitotic poison used as a pre-em. herbicide alone or, more frequently, in combination with other herbicides to increase the range of weeds controlled. Alone, it is used to control many germinating weeds and established chickweed in bulb crops, soft fruit and some vegetables; it is also used as an inhibitor of sprouting in ware potatoes (P. C. Martin & E. S. Schultz, *Am. Potato J.*, 1952, **29**, 268).

Toxicology. The oral LD50 for rats is 5000-7500 mg/kg. No toxic effect was observed when it was applied to shaved rabbits for 24 h. In 2-y feeding trials rats receiving 2000 mg/kg diet showed no toxic effect. Fish were not affected by concentrations of 5 mg/l.

Formulations. An e.c. (333 g a.i./l); granules (40, 50, 80 and 100 g/kg). Mixtures with diuron, endothal, fenuron, linuron, propham, pentanochlor are available.

Analysis. Product analysis is by GLC *(Anal. Methods Pestic. Plant Growth Regul.*, 1972, **6**, 612) or by hydrolysis measuring the carbon dioxide formed or titrating the liberated 3-chloroaniline with sodium nitrite (*CIPAC Handbook*, 1970, **1**, 223); FAO Specification (CP/73). Residues may be determined by GLC (*Anal. Methods Pestic. Plant Growth Regul., loc. cit.*; T. H. Byast *et al.*, *Agric. Res. Counc. (G.B.) Weed Res. Organ., Tech. Rep.*, No. 15 (2nd Ed.), p.17); by i.r. spectroscopy (C. E. Ferguson *et al.*, *J. Agric. Food Chem.*, 1963, **11**, 428); or by acid hydrolysis and colorimetric determination of the 3-chloroaniline formed (L. N. Gard & N. G. Rudd, *ibid.*, 1953, **1**, 630). See also: L. N. Gard & C. E. Ferguson, *Anal. Methods Pestic., Plant Growth Regul. Food Addit.*, 1964, **4**, 49.

Chlorpyrifos

T6NJ BOPS&O2&O2 CG EG FG $\qquad$ $C_9H_{11}Cl_3NO_3PS$ (350·6)

Nomenclature and development. The common name chlorpyrifos is approved by BSI, ISO, ANSI and BPC—exceptions France (chlorpyriphos-éthyl) and JMAF (chlorpyriphos); also known as ENT 27311. The IUPAC and *C.A.* name is *O,O*-diethyl *O*-(3,5,6-trichloro-2-pyridyl) phosphorothioate, Registry No. *[2921-88-2]*. It was introduced in 1965 by The Dow Chemical Co. under the code number 'DOWCO 179', the trade marks 'Dursban' and 'Lorsban' and protected by USP 3 244 586. Its insecticidal properties were first described by E. E. Kenaga *et al., J. Econ. Entomol.,* 1965, **58,** 1043.

Manufacture and properties. Produced by the reaction of 3,5,6-trichloropyridin-2-ol with *O,O*-diethyl phosphorochloridothioate, it forms colourless crystals, with a mild mercaptan odour, m.p. 42·5-43°, v.p. 1·87 x 10⁻⁵ mmHg at 25°. Its solubility at 35° is: 2 mg/l water; 790 g/kg octanes; 430 g/kg methanol; readily soluble in most other organic solvents. It is stable under normal storage conditions. The rate of hydrolysis in water increases with pH and temperature, the presence of copper and possibly other chelating metals. Under laboratory conditions, the time for 50% hydrolysis ranges from 1·5 d (water: 25°, pH 8) to 100 d (phosphate buffer: 15°, pH 7). It is compatible with non-alkaline pesticides but is corrosive to copper and brass.

Uses. Chlorpyrifos has a broad range of insecticidal activity and is effective by contact, ingestion and vapour action. It is not systemically active. Used for the control of mosquitoes (larvae and adults), flies, various soil and many foliar crop pests and household pests; also used for ectoparasite control on cattle and sheep. It is volatile enough to form insecticidal deposits on nearby untreated surfaces and is non-phytotoxic at insecticidal concentrations. It persists in soil for 60-120 d.

Toxicology. The acute oral LD50 is: for rats 135-163 mg/kg; for guinea-pigs 500 mg/kg; for chickens 32 mg/kg; for rabbits 1000-2000 mg/kg. The acute dermal LD50, in solutions, for rabbits is *c.* 2000 mg/kg. It is rapidly detoxified in the animal body. It is toxic to shrimps and to fish.

Formulations. These include: w.p. (250 g a.i./kg); e.c. (240 and 480 g/l); ULV (240 g/l); granules (10 and 100 g/kg).

Analysis. Product analysis is by u.v. spectroscopy. Residues may be determined by GLC. Details of both methods may be obtained from The Dow Chemical Co.

Chlorpyrifos-methyl

T6NJ BOPS&O1&O1 CG EG FG $C_7H_7Cl_3NO_3PS$ (322·5)

Nomenclature and development. The common name chlorpyrifos-methyl is approved by BSI, ISO and ANSI, in France (chlorpyriphos-méthyl); also known as ENT 27520. The IUPAC and *C.A.* name is *O,O*-dimethyl *O*-(3,5,6-trichloro-2-pyridyl) phosphorothioate, Registry No. *[5598-13-0]*. It was introduced in 1966 by The Dow Chemical Co. under the code number 'DOWCO 214', trade mark 'Reldan' and protected by USP 3 244 586.

Manufacture and properties. Produced by the reaction of 3,5,6-trichloropyridin-2-ol with *O,O*-dimethyl phosphorochloridothioate, it forms colourless crystals, with a slight mercaptan odour, m.p. 45·5-46·5°, v.p. 4·22 x 10^{-5} mmHg at 25°. Its solubility at 25° is 5 mg/l water; readily soluble in most organic solvents. It is stable under normal storage conditions. It is relatively stable in neutral media, but it is hydrolysed under both acidic (pH 4-6) and more readily under alkaline (pH 8-10) conditions.

Uses. Chlorpyrifos-methyl is an insecticide with a broad range of activity; effective by contact, ingestion and by vapour action but it is not systemic. It is used to control pests of stored grain, mosquitoes (adult), flies, aquatic larvae, household pests and various foliar crop pests. It is not persistent in soil.

Toxicology. The acute oral LD50 is: for rats 1630-2140 mg/kg; for guinea-pigs 2250 mg/kg; for rabbits 2000 mg/kg. Five chickens survived 7950 mg/kg fed by capsule. The acute percutaneous LD50 for rabbits is > 2000 mg/kg. It is readily metabolised in animals. It is relatively safe to most fish, though toxic to shrimps.

Formulation. An e.c. (240 g a.i./l).

Analysis. Product analysis is by u.v. spectroscopy. Residues may be determined by GLC. Details of these methods are available from The Dow Chemical Co.

Chlorthal-dimethyl

$$CO_2Me$$

(structure of tetrachloroterephthalate dimethyl ester)

$$CO_2Me$$

1OVR BG CG EG FG DVO1 $C_{10}H_6Cl_4O_4$ (332·0)

Nomenclature and development. The common name chlorthal is approved by BSI and ISO—exception USA—for 2,3,5,6-tetrachloroterephthalic acid, the dimethyl ester of which has the approved BSI and ISO name chlorthal-dimethyl—exceptions WSSA (DCPA) and JMAF (TCTP). The IUPAC name of the ester is dimethyl tetrachloro-terephthalate (I), in *C.A.* usage dimethyl 2,3,5,6-tetrachloro-1,4-benzenedicarboxylate *[1861-32-1]* formerly (I). Chlorthal-dimethyl was introduced in 1959 by the Diamond Shamrock Corp., formerly the Diamond Alkali Co., under the code number 'DAC 893' and the trade mark 'Dacthal'. Its use as a herbicide was first described by P. H. Schuldt *et al., Proc. Northeast. Weed Control Conf.,* 1960, p.42 and it is protected by USP 2 923 634; 3 052 712.

Manufacture and properties. The ester is produced by the chlorination of terephthaloyl chloride (N. Rabjohn, *J. Am. Chem. Soc.,* 1948, **70**, 3518) and reaction with methanol. It forms colourless crystals, m.p. 156°, v.p. < 0·5 mmHg at 40°. Its solubility at 25° is: < 0·5 mg/l water; 100 g/kg acetone; 250 g/kg benzene; 120 g/kg dioxane; 170 g/kg toluene; 140 g/kg xylene. It is stable in the pure state and in w.p. formulations and is non-corrosive.

Uses. Chlorthal-dimethyl is a pre-em. herbicide recommended for the control of annual weeds in many crops at 6-14 kg/ha. It is hydrolysed in soils with a 50% loss in 100 d in most soils.

Toxicology. The acute oral LD50 for rats is > 3000 mg/kg; the acute dermal LD50 for albino rabbits is > 10 000 mg/kg; a single application of 3 mg to the eyes of albino rabbits produced a mild irritation which subsided within 24 h. Rats and dogs fed at 10 000 mg/kg diet suffered no ill-effect. It is metabolised in animals to the monomethyl ester and to chlorthal, which are eliminated in the urine.

Formulations. 'Dacthal W-75', w.p. (750 g a.i./kg); 'Dimethyl-T', w.p. (750 g/kg); 'Dacthal G-5', 30/60 mesh granules (50 g/kg).

Analysis. Product analysis is by i.r. spectrometry or GLC; residues may be determined by colorimetric method or by coulometric GLC: details from Diamond Shamrock Corp. See also: P. H. Schuldt *et al., Contrib. Boyce Thompson Inst.,* 1961, **21**, 163; H. P. Burchfield & E. E. Storris, *Anal. Methods Pestic., Plant Growth Regul. Food Addit.,* 1964, **4**, 67; *Anal. Methods Pestic. Plant Growth Regul.,* 1972, **6**, 616; *Methods of Analysis—A.O.A.C.,* 1975, p.109.

Chlorthiamid

SUYZR BG FG $C_7H_5Cl_2NS$ (206·1)

Nomenclature and development. The common name chlorthiamid is approved by BSI and ISO—exception JMAF (DCNB). The IUPAC name is 2,6-dichloro(thiobenzamide) (I), in *C.A.* usage 2,6-dichlorobenzenecarbothiamide *[1918-13-4]* formerly (I). It was introduced in 1963 by Shell Research Ltd under the code number 'WL 5792', trade mark 'Prefix' and protected by BP 987 253. Its herbicidal properties were described by H. Sandford, *Proc. Br. Weed Control Conf., 7th,* 1964, p.208.

Properties. It is an almost colourless solid, m.p. 151-152°, v.p. 1 x 10⁻⁶ mmHg at 20°. Its solubility at 21° is 950 mg/l water; 50-100 g/kg aromatic and chlorinated hydrocarbons. It is stable at ≤ 90° and in acid solution but is converted into dichlobenil (p.166) in alkaline solution.

Uses. Chlorthiamid is toxic to germinating seeds. It is absorbed by the roots and, to some extent, by the leaves though there is little downward translocation. The phytotoxic symptoms resemble those of boron deficiency. For total weed control on non-crop areas it is recommended at 17-36 kg a.i./ha. The 7·5% granular formulation is preferred for selective weed control and is recommended at 6·6 kg a.i./ha in apples, 7·8 kg/ha in blackcurrants and gooseberries, 9·0-13·2 kg/ha in vines, 4·4 kg/ha in forest plantations and 6·6 kg/ha in certain ornamentals. It is also useful as a 'spot' treatment for the control of weeds such as docks and thistles. It should be applied in early spring before any vegetative growth takes place. Chlorthiamid is converted to dichlobenil in soil and the initial 50% loss from soil (chlorthiamid + dichlobenil) is *c.* 35 d under dry and 14 d under wet conditions. The fate in crops, soils and animals has been reviewed (K. I. Beynon & A. N. Wright, *Residue Rev.,* 1972, **43,** 23).

Toxicology. The acute oral LD50 is: for rats 757 mg/kg; for mice and chickens 500 mg/kg. The acute dermal LD50 for rats is > 1000 mg/kg. In 90-d feeding trials the 'no effect' level for rats was 100 mg/kg diet. The LC50 (24 h) for harlequin fish is 41 mg/l.

Formulations. These include: w.p. (500 g a.i./kg); granules (75 and 150 g/kg).

Analysis. Product analysis is by potentiometric titration with silver nitrate (*CIPAC Handbook,* 1979, **1A,** in press). Residues may be determined by GLC (K. I. Beynon *et al., J. Sci. Food Agric.,* 1966, **17,** 151; T. H. Byast *et al., Agric. Res. Counc. (G.B.) Weed Res. Organ., Tech. Rep.,* No. 15 (2nd Ed.), p.13). Details available from Shell International Chemical Co. Ltd.

Chlorthiophos

(I)   (II)   (III)

$C_{11}H_{15}Cl_2O_3PS_2$ (361·2)

(I)   2OPS&O2&OR   BG   EG   DS1

(II) 2OPS&O2&OR   CG   DG   FS1

(III) 2OPS&O2&OR   BG   DG   ES1

Nomenclature and development. The common name chlorthiophos is approved by BSI, ISO and ANSI for a reaction mixture; also known as ENT 27 635, OMS 1342. The IUPAC and *C.A.* name of the main component is *O*-2,5-dichloro-4-(methylthio)phenyl *O,O*-diethyl phosphorothioate (I) Registry No. *[21923-23-9]*, together with small quantities of *O*-4,5-dichloro-2-(methylthio)phenyl *O,O*-diethyl phosphorothioate (II) and *O*-2,4-dichloro-5-(methylthio)phenyl *O,O*-diethyl phosphorothioate (III). It was introduced in 1968 by C. H. Boehringer Sohn/Cela GmbH under the code number 'S 2957' and the protection of DBP 1298990; BP 1210826; USP 3600472. Sold under the trade mark 'Celathion' by Celamerck GmbH & Co. KG, its insecticidal properties were first described by H. Holtmann & E. Raddatz, *Proc. Br. Insectic. Fungic. Conf., 6th*, 1971, **2**, 485.

Manufacture and properties. Produced by the reaction of *O,O*-diethyl phosphoro-chloridothioate with a mixture of 2,5-dichloro-4-(methylthio)phenol, 2,4-dichloro-5-(methylthio)phenol and 4,5-dichloro-2-(methylthio)phenol it is a dark brown liquid, b.p. 155°/0·1 mmHg, which may contain crystals at low temperatures. It is practically insoluble in water, soluble in most organic solvents.

Uses. Chlorthiophos is a non-systemic stomach and contact insecticide, effective against Hemiptera, Diptera, Lepidoptera and with some acaricidal activity. Rates 25-75 g tech./100 l (300-1500 g/ha) depend on pests and the crop.

Toxicology. The acute oral LD50 is: for rats 8-11 mg tech./kg; for mice 91 mg/kg. The acute dermal LD50 for rabbits is 50-58 mg/kg. In 90-d feeding trials no toxic symptom, other than cholinesterase inhibition, was observed in rats receiving 32 mg/kg diet nor in dogs 7·3 mg/kg diet. Hens tolerated 2 mg/kg daily for 28 d without toxic symptom. Degradation in animal and plants is rapid; after 3-4 applications at field rates, residues (0·5-1·0 mg/kg) of chlorthiophos and its metabolites (the sulphoxide and sulphone) were found in apples 14-21 d, in cherries 7-14 d after the last application.

Formulations. These include: e.c. (500 g tech./l); w.p. (400 g/kg); granules (50 g/kg); dust (50 g/kg); ULV (500 g/kg).

Analysis. Product analysis is by bromometric titration; residue analysis by GLC after appropriate clean-up procedures.

Chlortoluron
(Chlorotoluron)

Me—⟨benzene ring⟩—NH.C.NMe$_2$ (with =O above the C)
Cl (attached to ring)

GR B1 EMVN1&1 $C_{10}H_{13}ClN_2O$ (212·7)

Nomenclature and development. The common name chlorotoluron is approved by ISO—exceptions BSI, France and New Zealand (chlortoluron). The IUPAC name is 3-(3-chloro-p-tolyl)-1,1-dimethylurea (I), in *C.A.* usage *N'*-(3-chloro-4-methylphenyl)-*N,N*-dimethylurea [15545-48-9] formerly (I). It was introduced in 1969 by Ciba AG (now Ciba-Geigy AG) under the code number 'C 2242', trade mark 'Dicuran' and protected by BelgP 728 267. Its herbicidal properties were first described by Y. L'Hermite *et al., C. R. Journ. Etud. Herbic. Conf., COLUMA, 5th,* 1969, **II**, 349.

Manufacture and properties. Produced by the action of 3-chloro-p-tolyl isocyanate with dimethylamine, it forms colourless crystals, m.p. 147-148°, v.p. 3·6 x 10^{-8} mmHg at 20°. Its solubility at 20° is 70 mg/l water; it is soluble in most organic solvents.

Uses. Chlortoluron is effective both as a residual soil-acting herbicide and as a contact foliar-acting spray against grass weeds and many broad-leaved weeds of cereal crops, especially against *Alopecurus myosuroides*. It is used on winter cereals immediately after sowing at 1·5-3·0 kg a.i./ha; for post-em., at the same rates when the cereal is at the 3-leaf stage to the end of tillering. It is combined with mecoprop ('Lumeton forte') to improve the control of *Gallium, Veronica* and *Papaver* spp.

Toxicology. The acute oral LD50 for rats is > 10000 mg/kg; the acute dermal LD50 for rats is > 2000 mg/kg. The 'no effect' level in 90-d feeding studies was estimated at 53 mg/kg daily in rats and 23 mg/kg daily in dogs. It has a low toxicity to birds and fish.

Formulations. 'Dicurane 80 WP', w.p. (800 g a.i./kg); 'Dicurane 500 FW', flowable (500 g/l); also in combination with mecoprop ('Lumeton forte' 60 WP).

Analysis. Product analysis is by solvent extraction, hydrolysis and titration of the dimethylamine. Residues may be determined by HPLC (T. H. Byast *et al., Agric. Res. Counc. (G.B.) Weed Res. Organ., Tech. Bull.,* No. 15 (2nd Ed.), p.49) or by alkaline hydrolysis, diazotisation of the 3-chloro-p-toluidine which is either coupled with *N*-(1-naphthyl)ethylenediamine and measured colorimetrically or determined by GLC after conversion to a volatile derivative: particulars are available from Ciba-Geigy AG.

Clofop-isobutyl

Cl—⟨ ⟩—O—⟨ ⟩—O.CHMe.CO.OBu^i

GR DOR DOY1&VO1Y1&1 $C_{19}H_{21}ClO_4$ (348·8)

Nomenclature and development. The common name clofop-isobutyl is approved by BSI and proposed by ISO. The IUPAC name is isobutyl ($\pm$)-2-[4-(4-chlorophenoxy)phenoxy]-propionate, in *C.A.* usage 2-methylpropyl ($\pm$)-2-[4-(4-chlorophenoxy)phenoxy]-propanoate *[51337-71-4]*. It was introduced in 1975 by Hoescht AG under the code number 'Hoe 22 870'. Its herbicidal properties were first described by F. Schwerdtle *et al., Mitt. Biol. Bundesanst. Land.-Forstwirtsch. Berlin-Dahlem.*, 1975, **165**, 171.

Properties. Pure clofop-isobutyl is a colourless crystalline solid, m.p. 39-40°. Its solubility at 22° is 180 mg/l water; highly soluble in acetone, xylene.

Uses. Clofop-isobutyl is a post-em. herbicide effective for control of annual grasses such as *Alopecurus myosuroides, Echinochloa crus-galli, Eleusine, Panicum* and *Setaria* spp. It is taken up through the leaves and roots of the weeds and is selective in nearly all cultivated dicotyledonous crops as well as in cereals.

Toxicology. The acute oral LD 50 for male rats is 1208 mg/kg. In 90-d feeding trials the 'no effect' level for dogs and male rats is 32 mg/kg diet.

Formulation. An e.c. (360 g a.i./l).

Analysis. Details of GLC methods are available from Hoechst AG.

Copper oxychloride

$$CuCl_2.3Cu(OH)_2$$

or

$$Cu_2Cl(OH)_3$$

.CU2.Q3.G $ClCu_2H_3O_3$ (213·6)

Nomenclature and development. The trivial name, copper oxychloride, is used for basic cupric chloride. The IUPAC name is dicopper chloride trihydroxide, in *C.A.* usage copper(II) chloride hydroxide *[1332-65-6]*. Preparations of copper oxychloride were introduced as fungicides in the early 1900's; trade marks of products subsequently used include 'Recop' (Sandoz), 'Cupravit', 'Fernacot' (ICI Ltd, Plant Protection Division), 'Pere-col' (ICI Ltd, Plant Protection Division).

Manufacture and properties. Produced by the action of air on scrap copper in a solution of copper dichloride plus sodium chloride, copper oxychloride is a green to bluish-green powder. The composition of the product varies with the conditions of manufacture but generally approaches that of the formula given above. Copper oxychloride is practically insoluble in cold or hot water but soluble, with formation of the corresponding cupric salts, in dilute acids. It is also soluble in ammonium hydroxide solutions after formation of a complex ion, but insoluble in organic solvents. It is strongly corrosive to iron and galvanised iron and should be packed other than in metal containers.

Uses. Copper oxychloride is used as a protectant fungicide for many disease-crop situations, including *Phytophthora infestans* on potatoes and *Pseudoperonospora humuli* on hops.

Toxicology. The acute oral LD50 for male rats is 1440 mg/kg. No toxicity to honey bees has been reported.

Formulation. A w.p., water-dispersible micro-granule (500 g copper/kg), or paste.

Analysis. The copper content and formulation characteristics may be determined: *CIPAC Handbook,* 1970, **1**, 226.

Coumachlor

(I)   (II)

T66 BOVJ DYR DG&1V1 EQ   *(I)* $C_{19}H_{15}ClO_4$ (342·8)

T B666 CO HVOT&&J DQ D1 FR DG   *(II)*

Nomenclature and development. The common name coumachlor is approved by BSI and ISO. The IUPAC name is 3-[1-(4-chlorophenyl)-3-oxobutyl]-4-hydroxycoumarin formerly 3-(α-acetonyl-4-chlorobenzyl)-4-hydroxycoumarin, in *C.A.* usage 3-[1-(4-chlorophenyl)-3-oxobutyl]-4-hydroxy-2*H*-1-benzopyran-2-one *[81-82-3]* formerly 3-(α-acetonyl-*p*-chlorobenzyl)-4-hydroxycoumarin. It was introduced in 1953 by J. R. Geigy S.A. (now Ciba-Geigy AG) as 'Geigy Rodenticide Exp. 332' and under trade marks 'Tomorin' and 'Ratilan'. Its use was first described by M. Reiff & R. Wiesmann, *Acta Trop.*, 1951, **8**, 97.

Manufacture and properties. Produced by the condensation of 4-hydroxycoumarin with 4-(4-chlorophenyl)but-3-en-2-one, it forms colourless crystals, m.p. 169-171°. It is practically insoluble in water, slightly soluble in benzene, diethyl ether; soluble in alcohols, acetone, chloroform. It forms water-soluble alkali metal salts. The technical product is a yellow crystalline powder, purity *c.* 70%, m.p. *c.* 103°; it is stable and persists in baits.

Uses. Coumachlor is an anticoagulant rodenticide. The acute oral LD50 for rats is 900-1200 mg/kg. The LD50 with repeated administration for 14-21 d to rats is 0·1-1·0 mg/kg daily.

Toxicology. Highly toxic to dogs and pigs.

Formulation. A bait (300 mg/kg) and tracking powder (10 g/kg).

Analysis. Product analysis is *(a)* by the Stepanow total chlorine method or *(b)* by GLC with internal standard after methylation of the hydroxy group; particulars from Ciba-Geigy AG. See also: K. Stammbach, *Anal. Methods Pestic., Plant Growth Regul. Food Addit.*, 1964, **3**, 185.

Coumatetralyl

T66 BOVJ EQ D- GL66&TJ $C_{19}H_{16}O_3$ (292·6)

Nomenclature and development. The common name coumatetralyl is approved by BSI and ISO—exception JMAF (coumarins). The IUPAC name is 4-hydroxy-3-(1,2,3,4-tetrahydro-1-naphthyl)coumarin (I), in *C.A.* usage 4-hydroxy-3-(1,2,3,4-tetrahydro-1-naphthalenyl)-2*H*-1-benzopyran-2-one *[5836-29-3]* formerly (I). It was introduced in 1956 by Bayer AG as a rodenticide, trade mark 'Racumin' and protected by DBP 1 079 382; USP 2 952 689. Its development is reviewed by G. Hermann & S. Hombrecher, *Pflanzenschutz-Nachr. (Am. Ed.)*, 1962, **15**, 89.

Manufacture and properties. Produced by the condensation of 4-hydroxycoumarin with 1,2,3,4-tetrahydro-1-naphthol, it is a yellowish-white crystalline powder, m.p. 172-176°. It is practically insoluble in water; slightly soluble in benzene, diethyl ether; soluble in acetone, ethanol. The enolic form has acidic properties and forms metal salts.

Uses. Coumatetralyl is an anticoagulant rodenticide which does not induce bait shyness. The sub-chronic LD50 (5 d) for rats is 0·3 mg/kg daily.

Toxicology. The LC50 (96 h) to fish is *c.* 1000 mg/l.

Formulations. 'Racumin 57' (7·5 g a.i./kg), used as a tracking powder or for bait (1 part 'Racumin 57' + 19 parts bait material); ready-prepared baits (375 mg/kg).

Analysis. Analysis is by u.v. spectroscopy: particulars from Bayer AG.

Crimidine

T6N CNJ BG DN1&1 F1 $C_7H_{10}ClN_3$ (171·6)

Nomenclature and development. The common name crimidine is approved by BSI and ISO. The IUPAC name is 2-chloro-4-dimethylamino-6-methylpyrimidine (I), in *C.A.* usage 2-chloro-*N,N*,6-trimethyl-4-pyrimidinamine *[535-89-7]* formerly (I). It was introduced in the early 1940's as a rodenticide by Bayer AG under the trade mark 'Castrix'. Its preparation is protected by USP 2 219 858 (to Winthrop).

Manufacture and properties. The condensation product of thiourea and ethyl 3-oxobutyrate is chlorinated and reacted with one equivalent of dimethylamine: *B.I.O.S. Final Report 1480* (1947). The technical product is a brownish wax containing *c.* 20% of the inactive 4-chloro-2-dimethylamino isomer. Pure crimidine has m.p. 87°, b.p. 140-147°/4 mmHg; it is practically insoluble in water, soluble in acetone, benzene, chloroform, diethyl ether and in dilute acids.

Uses. Crimidine is used as a rodenticide, the poisoned rats are non-toxic to predators.

Toxicology. The acute oral LD50 is: for rats 1·25 mg/kg; for rabbits 5 mg/kg. It is not a cumulative poison, but is rapidly metabolised. Birds are not harmed by the ingestion of a few grains of the formulation. Pyridoxine is an effective antidote in mice poisoned with crimidine (O. Karlog & E. Knudsen, *Nature (London)*, 1963, **200**, 790).

Formulation. 'Castrix-Giftkörner', impregnated wheat grains (1 g a.i./kg).

Analysis. Product analysis is by u.v. spectrometry at 267 nm: details from Bayer AG.

Crotoxyphos

$$(MeO)_2PO \quad \underset{Me}{\overset{O}{\parallel}} \quad \begin{array}{c} H \\ C = C \\ C - OCHPh \\ \parallel \quad \mid \\ O \quad Me \end{array}$$

1YR&OV1UY1&OPO&O1&O1 $C_{14}H_{19}O_6P$ (314·3)

Nomenclature and development. The common name crotoxyphos is approved by BSI, ISO and BPC (crotoxyfos); also known as ENT 24717. The IUPAC name is dimethyl (*E*)-1-methyl-2-(1-phenylethoxycarbonyl)vinyl phosphate also known as 1-methylbenzyl 3-(dimethoxyphosphinyloxy)isocrotonate, in *C.A.* usage 1-phenylethyl (*E*)-3-[(dimethoxy-phosphinyl)oxy]-2-butenoate *[7700-17-6]* formerly α-methylbenzyl (*E*)-3-hydroxy-crotonate ester with dimethyl phosphate. It was introduced in 1963 by the Shell Development Co. under the code number 'SD 4294', trade mark 'Ciodrin' and protected by USP 3068268; 3116201. Its insecticidal properties were described by C. P. Weidenback & R. L. Younger, *J. Econ. Entomol.,* 1962, **55**, 793.

Manufacture and properties. Produced by the reaction of trimethyl phosphite with 1-phenylethyl 2-chloro-3-oxobutyrate, the technical grade (80% pure) is a light straw-coloured liquid, with a mild ester odour, b.p. 135°/0·03 mmHg, v.p. 1·4 x 10⁻⁵ mmHg at 20°, d_{15}^{15} 1·2, n_D^{25} 1·5505. Its solubility at room temperature is 1 g/l water; slightly soluble in kerosine, saturated hydrocarbons; soluble in acetone, ethanol, propan-2-ol, xylene, chloroform and chlorinated hydrocarbons. It is stable when stored in glass, polyethylene or certain lined containers, stable in hydrocarbon solvents but is hydrolysed by water. In aqueous solution at 38°, 50% is decomposed in 35 h at pH 9, in 87 h at pH 1. It is slightly corrosive to mild steel, copper, lead, zinc and tin, but is non-corrosive to stainless steel 316, monel and aluminium 3003 and will not attack rigid PVC, fibre glass, reinforced polyester or the usual lacquers used for lining drums. It is incompatible with most mineral carriers, except synthetic silicas such as 'Hisil 233' and 'Colloidal Silica K320'. It is compatible with dichlorvos.

Uses. Crotoxyphos is an insecticide with a rapid action and moderately persistent residual effect on a wide range of external insect pests of livestock and is recommended for the control of flies, mites and ticks on cattle and pigs. Its fate in animals, soils and water has been reviewed (K. I. Beynon *et al., Residue Rev.,* 1973, **47**, 55); residues in milk and meat are negligible.

Toxicology. The acute oral LD50 is: for rats 52·3 mg/kg; for mice 90 mg/kg. The acute dermal LD50 for rabbits is 385 mg/kg. In 90-d feeding trials no effect on growth nor histopathological change was observed: for male rats receiving 900 mg/kg diet; for female rats receiving 300 mg/kg diet. The LC50 (48 h) for sheepshead minnow is > 1 mg/l.

Formulations. These include: e.c. (240 g a.i./l).

Analysis. Product analysis is by i.r. spectroscopy or by GLC. Residues may be determined by GLC with FPD. Details of these methods are available from Shell International Chemical Co. See also: P. E. Porter, *Anal. Methods Pestic., Plant Growth Regul. Food Addit.,* 1967, **5**, 243; *Anal. Methods Pestic. Plant Growth Regul.,* 1972, **6**, 325.

Crufomate

1X1&1&R   CG   DOPO&O1&M1 $C_{12}H_{19}ClNO_3P$ (291·7)

Nomenclature and development. The common name crufomate is approved by BSI, ISO and ANSI—exception Sweden; also known as ENT 25602-X. The IUPAC name is 4-*tert*-butyl-2-chlorophenyl methyl methylphosphoramidate (I), in *C.A.* usage 2-chloro-4-(1,1-dimethylethyl)phenyl methyl methylphosphoramidate *[299-86-5]* formerly (I). It was introduced in 1959 by The Dow Chemical Co. under code number 'DOWCO 132', trade mark 'Ruelene' and protected by BelgP 579 237. It was described by J. F. Landram and R. J. Shaver, *J. Parasitol.,* 1959, **45,** 55.

Manufacture and properties. Produced by treating phosphoryl chloride with, in turn, 4-*tert*-butyl-2-chlorophenol, ethanol and methylamine, it forms colourless crystals, m.p. 60°. The technical product is crystalline and on heating it decomposes before it boils. It is practically insoluble in water, light petroleum; readily soluble in acetone, acetonitrile, benzene, carbon tetrachloride. It is stable at pH ⩽ 7·0 but is incompatible with alkaline pesticides and unstable in strongly acidic media.

Uses. Crufomate is a systemic insecticide and anthelmintic used mainly for the treatment of cattle against warble flies, ectoparasites and helminths. It is not used in crop protection.

Toxicology. The acute oral LD50 is: for rats 770-950 mg/kg; for rabbits 400-600 mg/kg. Mild to moderate symptons of cholinesterase inhibition are seen in cattle treated orally with 100 mg/kg. It is not accumulated in the body fat and is non-hazardous to wildlife.

Formulations. Ready to use pour-on (70-320 g a.i./l).

Analysis. Product analysis is by i.r. spectroscopy, comparing with standards in carbon disulphide.

Cryolite

$$Na_3AlF_6$$

.NA3.AL-F6 AlF_6Na_3 (209.9)

Nomenclature and development. The mineralogical name cryolite is normally used; also known as ENT 24 984. The IUPAC name is aluminium trisodium fluoride, in *C.A.* usage cryolite *[15096-52-3]* (formerly *[1344-75-8]*); also known as sodium aluminofluoride, sodium fluoaluminate or sodium aluminium fluoride. It was first reported as an insecticide by S. Marcovitch & W. W. Stanley, *Tenn. Agric. Exp. Stn. Bull.,* No. 140, 1929.

Manufacture and properties. The naturally-occurring mineral is ≤ 98% pure. Aluminium trisodium fluoride is produced from aluminium fluoride, ammonium fluoride and sodium chloride (USP 1 475 155). The natural product consists of monoclinic crystals, d 2·95-3·00; the synthetic product is an amorphous powder. Both are practically insoluble in water [614 mg (synthetic product)/l] but are soluble in dilute alkali. It is incompatible with alkaline pesticides such as lime sulphur, Bordeaux mixture.

Uses. Cryolite is a stomach and contact insecticide, used generally as suspension (2 g/l) and it is non-phytotoxic at insecticidal concentrations.

Toxicology. It is of low acute toxicity to mammals; chronic intoxication is produced in some animal species by daily administration of 15-150 mg/kg (F. De Eds, *Medicine (Baltimore),* 1933, **12,** 1).

Analysis. Product analysis is by fluoride content: precipitating as lead chloride fluoride and determining chloride by titration with silver nitrate solution: AOAC Methods. Residues may be determined by titration with thorium tetranitrate after removal of fluoride by ashing with perchloric acid *(caution)* or by hydrochloric acid: AOAC Methods.

Cu_2O

.CU2.O Cu_2O (143·1)

Nomenclature and development. Known by the traditional chemical name cuprous oxide. The IUPAC name is dicopper oxide, in *C.A.* usage copper(I) oxide (Cu_2O) *[1317-39-1]*. It was introduced in 1932 by J. G. Horsfall, *N. Y. St. Agric. Exp. Stn. Bull.,* No. 615, 1932, as a seed protectant and was subsequently used for foliage protection under trade marks such as 'Perenox' (ICI Ltd, Plant Protection Division), 'Yellow Cuprocide' (Rohm & Hass Co.), 'Copper-Sandoz' (Sandoz) and 'Caocobre' (Sandoz).

Manufacture and properties. Produced by precipitation from copper salts by alkali in the presence of reducing agents or by the electrolytic oxidation of metallic copper, it is an amorphous powder ranging in colour from yellow to red. It is practically insoluble in water and organic solvents but is soluble in dilute mineral acids, to form copper(I) salt (or copper(II) salt plus metallic copper), and in aqueous solutions of ammonia and its salts. It is liable to oxidation to copper oxide and to conversion to a carbonate on exposure to moist air and is apt to corrode aluminium.

Uses. Cuprous oxide is a protective fungicide, used mainly for seed treatment and for foliage application against blight, downy mildews and rusts; is non-phytotoxic except to brassicas and 'copper-shy' varieties particularly under adverse weather conditions.

Toxicology. The acute oral LD50 for rats is 470 mg/kg. There have been instances of fatal copper poisoning in sheep grazing in orchards sprayed with cuprous oxide, but there is no significant history of ill-effect on birds or honey bees. Under conditions of moderate use and cultivation the hazard to earthworms, and hence to soil structure, is not considered significant.

Formulations. A w.p. or water dispersible micro-granule (500 g copper/kg); a stabiliser is necessary to delay oxidation and formation of carbonate.

Analysis. Total copper is determined by electrolytic or iodometric methods (AOAC Methods; *MAFF Tech. Bull.,* 1958, No. 1, p.16). Metallic copper in cuprous oxide may be determined (L. C. Hurd & A. R. Clark, *Ind. Eng. Chem. Anal. Ed.,* 1936, **8,** 380). CIPAC methods are available for total copper content of w.p. and dusts (*CIPAC Handbook,* 1970, **1,** 226). Copper residues are determined as for Bordeaux mixture (p.48).

Cyanazine

T6N CN ENJ BMX1&1&CN DM2 FG $C_9H_{13}ClN_6$ (240·7)

Nomenclature and development. The common name cyanazine is approved by BSI, ISO and WSSA—exception France. The IUPAC name is 2-(4-chloro-6-ethylamino-1,3,5-triazin-2-ylamino)-2-methylpropionitrile formerly 2-chloro-4-(1-cyano-1-methylethyl-amino)-6-ethylamino-1,3,5-triazine, in *C.A.* usage 2-[[4-chloro-6-(ethylamino)-1,3,5-triazin-2-yl]amino]-2-methylpropanenitrile *[21725-46-2]* formerly 2-[[4-chloro-6-(ethyl-amino)-s-triazin-2-yl]amino]-2-methylpropionitrile. It was described by W. J. Hughes *et al., Proc. North Cent. Weed Control Conf.,* 1966, p.27, and by T. Chapman *et al., Proc. Br. Weed Control Conf., 9th,* 1968, p.1018. It was first marketed in 1971 under the trade marks 'Bladex' (a w.p. for use on maize) and 'Fortrol' (a liquid suspension for use on peas) and protected by BP 1132306.

Properties. The technical product is a colourless crystalline solid, m.p. 166·5-167°, v.p. 1·6 x 10^{-9} mmHg at 20°, 1·0 x 10^{-8} mmHg at 30°. Its solubility at 25° is: 171 mg/l water; 15 g/l benzene; 210 g/l chloroform; 45 g/l ethanol; 15 g/l hexane. It is stable to heat and light and to hydrolysis in neutral and slightly acidic or basic media.

Uses. Cyanazine is a pre- and post-em. herbicide of short persistence. It is valuable for general weed control *(a)* applied pre-em. to the crop at 1-3 kg a.i./ha for maize, peas and broad beans; *(b)* applied post-em. at 0·25-0·4 kg/ha in combination with hormone weedkillers for the control of hard-to-kill broad-leaved weeds in wheat and barley applied during the early tillering stage. It also shows promise both alone and mixed with other herbicides in sugarcane, soyabean, cotton, sorghum and several other crops.

Laboratory tests show that the nitrile group is hydrolysed in plants and soil to the corresponding carboxylic acid. The chlorine atom may also be replaced by a hydroxy group and conjugates involving this group are sometimes formed. These degradation products have not been detected in crops following field use (K. I. Beynon *et al., Pestic. Sci.,* 1972, **3**, 293, 379, 389, 401).

Toxicology. The acute oral LD50 is: for rats 182 mg/kg; for mice 380 mg/kg; for white Leghorn chickens 750 mg/kg; for bobwhite quail 400-500 mg/kg. The acute percutaneous LD50 for rats is > 1200 mg/kg. In 2-y feeding trials the 'no effect' level was: for rats 12 mg/kg diet; for dogs 25 mg/kg diet. The LC50 for harlequin fish is 16 mg/l (24 h), 10 mg/l (48 h). Metabolism and excretion by mammals are rapid.

Formulations. These include: w.p. (500 g a.i./kg); a suspension concentrate (500 g/l); and a number of one-pack mixtures with other herbicides such as atrazine, linuron, MCPA, dichlorprop, mecoprop, 2,4-D.

Analysis. Product analysis is by i.r. spectrometry or by chromatographic methods (GLC, HPLC). Residues are determined by GLC using ECD. Details may be obtained from Shell International Chemical Co. Ltd.

Cyanofenphos

NCR   DOPS&R&O2

$C_{15}H_{14}NO_2PS$ (303·3)

Nomenclature and development. The common name cyanofenphos is approved by BSI and ISO—exception JMAF (CYP); also known as ENT 25832-a. The IUPAC and *C.A.* name is *O*-4-cyanophenyl *O*-ethyl phenylphosphonothioate, Registry No. *[13067-93-1]*. It was first described by Y. Nishizawa, *Bull. Agric. Chem. Soc. Japan,* 1960, **24,** 744, and was introduced in 1966 by Sumitomo Chemical Co. Ltd under the code number 'S-4087', trade mark 'Surecide' and protected by JapaneseP 410930; 410925; BP 929738.

Manufacture and properties. Produced by the condensation of 4-hydroxybenzonitrile with *O*-ethyl phosphonochloridothioate in the presence of an alkali-metal carbonate, it is a colourless crystalline solid, m.p. 83°, v.p. 1·32 x 10⁻⁵ mmHg at 25°. Its solubility at 30° is 0·6 mg/l water; moderately soluble in ketones, aromatic solvents. The technical grade is *c.* 92% pure.

Uses. Cyanofenphos is an insecticide effective against rice borer and gall midges, cotton bollworm in tropical areas and against lepidopterous larvae and other insect pests of vegetables in temperate zones.

Toxicology. The acute oral LD50 is: for rats 89 mg/kg; for male mice 43·7 mg/kg. The LC50 (48 h) for carp is 1·35 mg/l.

Formulations. An e.c. (250 g a.i./l); dust (15 g/kg).

Analysis. Product analysis is by u.v. spectroscopy after purification by TLC. Residues in milk and meat were determined by GLC using FPD: J. Miyamoto *et al., J. Pestic. Sci.,* 1977, **2,** 1.

$$\underset{\underset{\text{N.OMe}}{\overset{\|}{\text{}}}}{\text{EtNH.C.NH.C.C.CN}}\overset{\overset{\text{O}\quad\text{O}}{\|\quad\|}}{}$$

2MVMVYCN&UNO1

$C_7H_{10}N_4O_3$ (198.2)

Nomenclature and development. The common name is not yet agreed. The IUPAC name is 1-(2-cyano-2-methoxyiminoacetyl)-3-ethylurea, in *C.A.* usage 2-cyano-*N*-[(ethylamino)-carbonyl]-2-(methoxyimino)acetamide *[57966-95-7]*. Its fungicidal properties were described by J. M. Serres & G. A. Carraro, *Meded. Fac. Landbouwwet. Rijksuniv. Gent,* 1976, **41,** 645. It was introduced in 1974 by E. I. du Pont de Nemours & Co., Inc. under the code number 'DPX-3217', trade mark 'Curzate' and protected by USP 3957847; BP 1425622; 1425256.

Manufacture and properties. Produced by the reaction of 1-(2-cyano-2-hydroxyimino-acetyl)-3-ethylurea with dimethyl sulphate, it is a colourless crystalline solid, m.p. 160-161°, d^{25} 1·31, v.p. 6 x 10^{-7} mmHg (extrapolated) at 25°. Its solubility at 25° is: 1 g/kg water; 185 g/kg dimethylformamide; 105 g/kg acetone; 103 g/kg chloroform; 41 g/kg methanol; 2 g/kg benzene; < 1 g/kg hexane. It is stable under normal storage conditions and in neutral or slightly acidic media. Loss from soil is 50% in < 7 d.

Uses. It is a fungicide with local systemic action to control certain diseases during the incubation period. It is primarily effective on fungi belonging to the Peronosporales: *Phytophthora, Plasmopara* and *Peronospora* spp. It is used in combination with a preventive fungicide to improve residual activity. Typical rates in mixtures for control of grape downy mildew *(Plasmopara viticola)* and potato blight *(Phytophthora infestans)* are 100-120 g. a.i./ha. For example, grape downy mildew is usually controlled by treatment 3-6 d after infection and potato blight 1-3 d after inoculation.

Toxicology. The acute oral LD50 is: for rats 1196 mg tech./kg; for guinea-pigs 1096 mg/kg. The acute dermal LD50 for rabbits is > 3000 mg/kg. It is not a skin irritant or sensitiser; it is a very slight eye irritant. There was no evidence of cumulative toxicity from 10 doses of 200 mg/kg daily administered orally to rats over a 14-d period. The LC50 (8 d) is: for quail 2847 mg/kg diet; for mallard duck > 10000 mg/kg diet. The LC50 (96 h) is: for rainbow trout 18·7 mg/l; for sunfish 13·5 mg/l.

Analysis. Residue analysis is by GLC using a nitrogen-sensitive detector. Details are available from E. I. du Pont de Nemours & Co., Inc.

(S)-α-Cyano-3-phenoxybenzyl (1R)-*cis*-3-(2,2-dibromovinyl)-2,2-dimethylcyclopropanecarboxylate

L3TJ A1 A1 BVOYCN&R COR&& C1UYEE $C_{22}H_{19}Br_2NO_3$ (505·2)

Nomenclature and development. The common name decamethrin† has been proposed to BSI. The IUPAC name is (S)-α-cyano-3-phenoxybenzyl (1R)-*cis*-3-(2,2-dibromovinyl)-2,2-dimethylcyclopropanecarboxylate, in *C.A.* usage (1R[1α(S^*),3α])-cyano(3-phenoxyphenyl)methyl 3-(2,2-dibromoethenyl)-2,2-dimethylcyclopropanecarboxylate *[52918-63-5]*. This single isomer, first described by M. Elliott *et al., Nature (London),* 1974, **248**, 710, under the code number 'NRDC 161', has been introduced by Roussel Uclaf under the code number 'RU 22974' and the trade marks 'Decis' and 'K-othrin'.

Manufacture and properties. Esterification of (1R)-*cis*-3-(2,2-dibromovinyl)-2,2-dimethylcyclopropanecarboxylic acid with (RS)-α-hydroxy-3-phenoxyphenylacetonitrile yielded the mixed diastereoisomers, known by the code number 'NRDC 156', *C.A.* Registry No. *[52820-00-5]* (M. Elliott *et al., Pestic. Sci.,* 1978, **9**, 105). A solution of this in hexane deposited crystals of one isomer on cooling, recrystallisation yielded the pure single isomer 'NRDC 161'. The technical material now produced industrially by Roussel Uclaf contains > 98% 'NRDC 161' and is a colourless crystalline powder, m.p. 98-101°, [a]$_D$ + 61° (4% in benzene), v.p. 1·5 x 10^{-8} mmHg at 25°. Its solubility at 20° is < 2 µg/l water; soluble in acetone, dimethyl sulphoxide, dimethylformamide, benzene, xylene, cyclohexane, dioxane. 'NRDC 161', in contrast to the natural pyrethrins (p.459) is stable on exposure to air and sunlight.

Uses. It is a very potent insecticide, effective as a contact and stomach poison against a wide range of insects.. Its toxicity to normal *Musca domestica* is *c.* 1000 x, to *Anopheles stephensi* 220 x and to *Periplaneta americana* 30 x that of the pyrethrins (M. Elliott *et al., Annu. Rev. Entomol.,* 1978, **23**, 443). It controls numerous insect pests of field crops at 11 g/ha (J. Martel & R. Colas, *Beltwide Cotton Insect Conf.,* 1976; J. J. Herve *et al., Proc. 1977 Br. Crop. Prot. Conf., Pests. Dis.,* 1977, **2**, 613).

Toxicology. The acute oral LD50 for rats is: *c.* 135 mg a.i./kg; 134 mg/a.i. (as e.c.)/kg; 649 mg a.i. (as 10 g/l, ULV concentrate)/kg. The acute dermal LD50 for rabbits is > 2000 mg/kg.

Formulations. These include: e.c. (25 g a.i./l); ULV concentrate (4, 5, and 10 g/l); w.p. (25 and 50 g/kg); dusts (0·5 and 1·0 g/kg); granules (0·5 and 1·0 g/kg).

Analysis. Technical material and formulation analysis is by HPLC; i.r. spectroscopy or bromine determination may also be used. Residues in plant tissues may be determined by GLC with ECD after clean-up by liquid-liquid partition and column chromatography.

---

†The name decamethrin proved to be unacceptable and should not be used.

Cyanophos

$$(MeO)_2\overset{\overset{S}{\|}}{P}-O-\langle\text{C}_6\text{H}_4\rangle-CN$$

NCR DOPS&O1&O1 $C_9H_{10}NO_3PS$ (243·2)

Nomenclature and development. The common name cyanophos is approved by BSI and ISO—exception JMAF (CYAP); also known as ENT 25675. The IUPAC and *C.A.* name is *O*-4-cyanophenyl *O,O*-dimethyl phosphorothioate, Registry No. *[2636-26-2]*. It was introduced in 1960 by the Sumitomo Chemical Co. Ltd under the code number 'S-4084', trade mark 'Cyanox' and protected by JapaneseP 405852; 415199. Its insecticidal properties were first described by Y. Nishizawa, *Agric. Biol. Chem.,* 1961, **25,** 597.

Manufacture and properties. Produced by the reaction of 4-hydroxybenzonitrile with *O,O*-dimethyl phosphorochloridothioate, it is a clear amber liquid, m.p. 14-15°, n_D^{25} 1·5413, d_4^{25} 1·260. Its solubility at 30° is 46 mg/l water; miscible with most organic solvents. The technical grade is *c.* 93% pure. Stored under normal conditions it is stable for $\geqslant 2$ y and it is compatible with most other pesticides.

Uses. Cyanophos is an insecticide used at 25-50 g a.i./100 l for the control of lepidopterous pests on fruit, vegetables and ornamentals. It is also applicable for the control of household pests such as cockroaches and houseflies as well as mosquitoes.

Toxicology. The acute oral LD50 for rats is 610 mg/kg; the acute dermal LD50 for rats is 800 mg/kg. The LC50 (48 h) for carp is 5 mg/l.

Formulations. These include: e.c. (500 g a.i./l) and dust (30 g/kg) for agricultural use; liquid (10 g/l) and e.c. (50 g/l) for public health use.

Analysis. Product analysis is by GLC or by u.v. spectroscopy after purification by TLC. Residues may be determined by GLC using a flame thermionic detector. Details of methods are available from Sumitomo Chemical Co. Ltd.

138

Cycloate

$$\underset{\underset{Et}{|}}{EtS.\overset{\overset{O}{\parallel}}{C}.N}\diagdown\hexagon$$

L6TJ AN2&VS2 $C_{11}H_{21}NOS$ (215·4)

Nomenclature and development. The common name cycloate has been approved by BSI, WSSA and proposed by ISO—exception JMAF (hexylthiocarbam). The IUPAC name is *S*-ethyl *N*-cyclohexyl-*N*-ethyl(thiocarbamate) (I), in *C.A.* usage *S*-ethyl cyclohexylethyl-carbamothioate *[1134-23-2]* formerly (I). Cycloate was introduced, as a herbicide, by the Stauffer Chemical Co. under the code number 'R-2063', trade mark 'Ro-Neet' and the protection of USP 3175897. It was described by *Nat. Weed Comm. Can. Dep. Agric. Res. Rep.,* 1963, p.51.

Properties. It is a clear liquid, with an aromatic odour, b.p. 145°/10 mmHg, v.p. 6·2 x 10^{-3} mmHg at 25°, *d* 0·970, n_D^{30} 1·504. Its solubility at 22° is 85 mg/l water; miscible with most organic solvents including acetone, benzene, propan-2-ol, kerosine, methanol and xylene. It is stable and non-corrosive.

Uses. Cycloate is toxic to germinating seeds and is used for the control of annual broad-leaved weeds and grasses and of nut grass in sugar beet and spinach by pre-plant soil incorporation at 3-4 kg a.i./ha. It is non-persistent.

Toxicology. The acute oral LD50 for rats is 2000-3190 mg tech./kg; the acute dermal LD50 for rabbits is > 4640 mg/kg. In 90-d feeding trials no toxic symptom was noted in dogs receiving ≤ 240 mg/kg daily. The LC50 (96 h) for rainbow trout is 4·5 mg/l.

Formulations. An e.c. (720 g a.i./l); granules (100 g/kg).

Analysis. Product analysis is by GLC (*CIPAC Handbook,* 1979, **1A,** in press), particulars also available from the Stauffer Chemical Co. Residues may also be determined by hydrolysis, steam distillation of the amine so produced and colorimetric estimation of the cupric dithiocarbamate complex. See: J. R. Lane, *Anal. Methods Pestic., Plant Growth Regul. Food Addit.,* 1967, **5,** 491; W. A. Ja *et al., Anal. Methods Pestic. Plant Growth Regul.,* 1972, **6,** 686.

Cycloheximide

T6VMVTJ E1YQ- BL6VTJ D1 F1 $C_{15}H_{23}NO_4$ (281·4)

Nomenclature and development. The common name cycloheximide is approved by BSI, New Zealand and JMAF and proposed by ISO. The IUPAC name is 4-{(2R)-2-[(1S,3S,5S)-(3,5-dimethyl-2-oxocyclohexyl)]-2-hydroxyethyl}piperidine-2,6-dione   formerly 3-[2-(3,5-dimethyl-2-oxocyclohexyl)-2-hydroxyethyl]glutarimide (I), in *C.A.* usage [1S-[1α(S*), 3α,5β]]-4-[2-(3,5-dimethyl-2-oxocyclohexyl)-2-hydroxyethyl]-2,6-piperidinedione *[66-81-9]* formerly (I). It was isolated in 1946 by A. Whiffen *et al., J. Bacteriol.,* 1946, **52,** 610, as an antibiotic effective against certain fungi pathogenic to man; and identified by B. E. Leach *et al., J. Am. Chem. Soc.,* 1947, **69,** 474; J. H. Ford & B. E. Leach, *ibid.,* 1948, **70,** 1223, and by E. C. Kornfield *et al., ibid.,* 1949, **71,** 150. Its stereochemistry has been established (E. J. Eisenbraun *et al., ibid.,* 1958, **80,** 1261; F. Johnson *et al., ibid.,* 1965, **87,** 4612; 1966, **88,** 149). It was introduced for crop protection in 1948 by The Upjohn Co. under the trade mark 'Acti-dione'.

Manufacture and properties. Produced in culture by *Streptomyces griseus* and recovered as a by-product of streptomycin manufacture, cycloheximide forms colourless crystals, 115·5-117°, $[\alpha]_D^{25}$ —6·8° (20 g/l water). Its solubility at 2° is 21 g/l water; soluble in chloroform, propan-2-ol, other common organic solvents; sparingly soluble saturated hydrocarbons. Though stable in neutral or acid solution, it is rapidly decomposed by alkali at room temperature to form 2,4-dimethylcyclohexanone. Chlordane is reported to cause a rapid loss of activity.

Uses. Cycloheximide inhibits the growth, in culture, of many plant pathogenic fungi; its agricultural uses were reviewed (J. H. Ford *et al., Plant Dis. Rep.,* 1958, **42,** 680). Its growth regulating properties are used for promoting the abscission of fruit, *e.g.* oranges and olives.

Toxicology. The acute oral LD50 is: for rats 2 mg/kg; for mice 133 mg/kg; for guinea-pigs 65 mg/kg; for monkeys 60 mg/kg.

Formulations. These include: 'Acti-dione PM' (270 mg a.i./kg); 'Acti-dione TGF' (21 g/kg); 'Acti-Aid (42 g/l); 'Acti-dione', w.p. (1 g/kg); 'Acti-dione' thiram (7·5 g cycloheximide + 750 g thiram/kg); 'Acti-dione RZ' (13 g cycloheximide + 750 g quintozene/kg).

Analysis. Product analysis is by bioassay (G. C. Prescott *et al., J. Agric. Food Chem.,* 1956, **4,** 343; W. W. Kilgore, *Anal. Methods Pestic., Plant Growth Regul. Food Addit.,* 1964, 3, 1).

Cycluron

L8TJ AMVN1&1 $C_{11}H_{22}N_2O$ (198·3)

Nomenclature and development. The common name cycluron is approved by BSI, ISO and WSSA—exception JMAF (COMU); previously known as OMU. The IUPAC name is 3-cyclo-octyl-1,1-dimethylurea (I), in *C.A.* usage *N'*-cyclooctyl-*N,N*-dimethylurea *[2163-69-1]* formerly (I). Cycluron was introduced in 1958 by BASF AG as a component of their herbicide 'Alipur', protected by DBP 1027930; 1062482. The first report of its herbicidal activity was by A. Fischer, *Z. Pflanzenkr. Pflanzenpathol. Pflanzenschutz,* 1960, **67**, 577.

Manufacture and properties. Produced by the reaction of cyclo-octylamine with dimethylcarbamoyl chloride, it is a colourless crystalline solid, m.p. 138°. Its solubility at 20° is: 1·1 g/l water; 67 g/kg acetone; 55 g/kg benzene; 500 g/kg methanol. The technical product is *c.* 97% pure and has m.p. 134-138°. It is stable, compatible with other pesticides and non-corrosive.

Uses. Cycluron is used in 3:2 mixture with chlorbufam as a pre-em. herbicide at 3-6 l 'Alipur'/ha for the control of annual weeds in sugar beet and several vegetable crops. It is sensitive to soil conditions, particularly moisture.

Toxicology. The acute oral LD50 for rats is 2600 mg/kg. There was no local irritation to the backs or ears of white rabbits treated for 20 h with a 50% aqueous solution.

Formulation. 'Alipur', e.c. (150 g cycluron + 100 g chlorbufam/l).

Analysis. Macro analysis is by titration with perchloric acid in acetic anhydride; residues may be determined by GLC of the dimethylamine produced by hydrolysis.

Cyhexatin

L6TJ A-SN-Q- AL6TJ&- AL6TJ $C_{18}H_{34}OSn$ (385·2)

Nomenclature and development. The common name cyhexatin is approved by BSI, ISO and ANSI—exception JMAF (tricyclohexyltin hydroxide); also known as ENT 27395-X. The IUPAC name is tricyclohexyltin hydroxide, in *C.A.* usage tricyclohexylhydroxy-stannane *[13121-70-5]*. It was developed from a joint project of The Dow Chemical Co. and M & T Chemicals Inc. Its acaricidal properties were first described by W. E. Allison *et al., J. Econ. Entomol.,* 1968, **61,** 1254. It was introduced in 1968 by The Dow Chemical Co. under the code number 'DOWCO 213', trade mark 'Plictran' and protected by USP 3 264 177; 3 389 048.

Properties. The technical grade, purity 97%, is a colourless crystalline powder, m.p. 195-198°, of negligible v.p. at 25°. Its solubility at 25° is: < 1 mg/l water; 1·3 g/l acetone; 216 g/l chloroform; 37 g/l methanol; poorly soluble in most organic solvents. It is stable to 135° and in neutral and alkaline suspensions. It is degraded when exposed to u.v. light in thin layers.

Uses. Cyhexatin is an acaricide effective by contact against the motile stages of a wide range of phytophagous mites, the usual dosage being 20-30 g a.i./l. Cyhexatin is non-phytotoxic to deciduous fruit, vines, wind-break tree species, vegetables and most ornamentals grown in the open. Citrus (immature foliage and fruit in the early stages of development), as well as the seedlings and immature foliage of some glasshouse-grown ornamentals and vegetables, are susceptible to possible injury, usually in the form of localised spotting.

Toxicology. The acute oral LD50 is: for rats 540 mg tech./kg; for guinea-pigs 780 mg/kg; for chickens 654 mg/kg. The acute dermal LD50 for rabbits is > 2000 mg/kg. In 1-y feeding trials rats and dogs receiving 12 mg/kg diet showed indirect nutritional deficiencies; rats receiving 200 mg/kg diet for 112 d showed reduced weight gain because of the unpalatibility of the diet. The LC50 (8 d) is: for bobwhite quail 520 mg/kg diet; for mallard ducklings 3189 mg/kg. The LC50 (24 h) is: for large mouth bass 0·06 mg/l; for goldfish 0·55 mg/l. It is virtually non-hazardous to honey bees (LD50 32 µg/bee), most predacious mites and insects at the recommended rates of use.

Formulations. 'Plictran 50W', w.p. (500 g a.i./kg); 'Plictran 25W', w.p. (250 g/kg), available only in Europe, Middle East and Africa.

Analysis. Methods for determining organotin and total tin residues are available from The Dow Chemical Co. See also: M. E. Getzendaner & H. B. Corbin, *Anal. Methods Pestic. Plant Growth Regul.,* 1973, **7,** 417.

Cypermethrin

L3TJ A1 A1 BVOYCN&R COR&& ClUYGG $C_{22}H_{19}Cl_2NO_3$ (416·3)

Nomenclature and development. The common name cypermethrin is approved by BSI, New Zealand, BPC and proposed by ISO; also known as OMS 2002. The IUPAC name is *(RS)*-α-cyano-3-phenoxybenzyl *(1RS)-cis,trans*-3-(2,2-dichlorovinyl)-2,2-dimethyl-cyclopropanecarboxylate, in *C.A.* usage *(RS)*-cyano(3-phenoxyphenyl)methyl *(1RS)-cis,trans*-3-(2,2-dichloroethenyl)-2,2-dimethylcyclopropanecarboxylate *[52315-07-8]*. The ratio of *cis/trans* isomers, which may vary with the manufacturing process, should be stated. It was discovered by M. Elliott *et al., Pestic. Sci.,* 1975, **6**, 537 and given the number 'NRDC 149'. It has been developed by Shell International Chemical Co. Ltd under the code number 'WL 43 467', the trade marks 'Ripcord' (for agronomic use) and 'Barricade' (for veterinary use); and by ICI Ltd, Plant Protection Division under the code number 'PP 383' and the trade marks 'Cymbush', 'Imperator', 'Kafil Super' and 'CCN 52'. Its performance in field trials has been reported (M. H. Breese, *Pestic. Sci.,* 1977, **8**, 264; M. H. Breese & D. P. Highwood, *Proc. 1977 Br. Crop Prot. Conf.—Pests Dis.,* 1977, **2**, 641).

Manufacture and properties. Produced by the esterification of α-hydroxy-3-phenoxy-phenylacetonitrile with 3-(2,2-dichlorovinyl)-2,2-dimethylcyclopropanecarboxylic acid, the technical grade is a viscous yellowish-brown semi-solid mass, which is liquid at 60°, v.p. $3·8 \times 10^{-8}$ mmHg at 70° (for pure compound). Its solubility at 21° is 0·01-0·2 mg/l water; at 20° is: 103 g/l hexane, > 450 g/l acetone, cyclohexane, ethanol, xylene, chloroform. It is stable ≤ 220°; photochemical decomposition has been observed in laboratory tests, but field data indicate this does not adversely affect biological performance. It is more stable in acid than alkaline media, with optimum stability at pH 4.

Uses. Cypermethrin is a stomach and contact insecticide effective against a wide range of insect pests, particularly leaf and fruit-eating Lepidoptera and Coleoptera in cotton, fruit, vegetables, vines, tobacco and other crops at 25-150 g a.i./ha. If applied before infestations become well established, it will also give protection against Hemiptera in most crops, and soil surface sprays (100 g/ha) give good control of *Euxoa* spp. It has good residual activity on treated plants and no case of phytotoxicity has been reported. It also controls cattle ectoparasites, *e.g. Boophilus microplus,* at 150 mg/l bath, including strains resistant to organophosphorus compounds. Sheep scab (*Psoroptes ovis*) lice and ked are controlled by a single treatment at 10 mg/l. Good knock-down and residual control of biting flies in and around animal housing have been obtained following direct spray application to animals or structural surfaces. Its degradation in soil has been studied (T. R. Roberts & M. E. Standen, *Pestic. Sci.,* 1977, **8**, 305).

Toxicology. The acute oral LD50 is: for rats 303-4123 mg/kg (depending on the carrier and conditions used); for mice 138 mg/kg; for chickens > 2000 mg/kg. The acute dermal LD50 for rabbits is > 2400 mg/kg; it is a slight skin irritant, a mild eye irritant and can cause skin sensitisation. In 90-d feeding trials rats receiving 100 mg/kg diet showed no effect. The LC50 (96 h) for brown trout is 2·0-2·8 µg/l. It is relatively toxic to honey bees.

continued

143

Formulations. These include: e.c. (25-400 g a.i./l) for agronomic use; ULV (10-75 g/l); 'Barricade', e.c. (25-200 g/l) for veterinary and public health use. 'Ripcord', included in the e.c. formulations for agronomic use listed above, is also available in mixtures with other pesticides. Most commercial formulations contain a *cis/trans* isomer ratio of 40-50 : 60-50 *m/m*.

Analysis. Product analysis is by HPLC or, after column chromatography, by GLC with FID using an internal standard. Residues may be determined, after column chromatographic clean-up, by GLC with ECD. Details of methods are available from ICI Ltd, Plant Protection Division or Shell International Chemical Co. Ltd.

2,4-D

QV1OR BG DG
GR CG DO1VOY1&1
GR CG DO1VO2O4

2,4-D $C_8H_6Cl_2O_3$ (221·0)
2,4-D-isopropyl $C_{11}H_{12}Cl_2O_3$ (263·1)
2,4-D-(2-butoxyethyl) $C_{14}H_{18}Cl_2O_4$ (311·2)

Nomenclature and development. The common name 2,4-D is approved by BSI, ISO and WSSA—exception JMAF (2,4-PA). The IUPAC and *C.A.* name is (2,4-dichlorophenoxy)acetic acid, Registry No. *[94-75-7]*. The potent effects of its salts on plant growth were first described by P. W. Zimmerman & A. E. Hitchcock, *Contrib. Boyce Thompson Inst.*, 1942, **12**, 321. 2,4-D, various esters and salts have been marketed by many companies over the past 35 years; those mentioned below are currently the more important commercially.

Manufacture and properties. Produced by the condensation of sodium chloroacetate with 2,4-dichlorophenol, it is a colourless powder, with a slight phenolic odour, m.p. 140·5°, v.p. 0·4 mmHg at 160°. Its solubility at 25° is 620 mg/l water; soluble in aqueous alkali, alcohols, diethyl ether; insoluble in petroleum oils. It is corrosive.

It is a strong acid, pK_a 2·64, and forms water-soluble salts with alkali metals or amines. A sequestering agent is included in commercial formulations to prevent precipitation of the calcium or magnesium salts by hard water. The more important salts and their *C.A.* Registry Nos. are: sodium *[2702-72-9]* (solubility at room temperature 45 g/l water); diethanolamine *[5742-19-8]*; diethylamine *[20940-37-8]*; dimethylamine *[2008-39-1]*; ethanolamine *[3599-58-4]*; methylamine *[51173-63-8]*; triethanolamine *[2569-10-9]* (solubility at 30-32° 4·4 kg/kg water).

Esters frequently encountered are: 2,4-D-isopropyl, Registry No. *[94-11-1]*, colourless liquid, b.p. 130°/1 mmHg, crystallising in 2 forms, m.p. 5-10° and 20-25°, v.p. 1·05 x 10^{-2} mmHg at 25°, it is practically insoluble in water, soluble in alcohols, most oils; 2,4-D-(2-butoxyethyl), Registry No. *[1929-73-3]*; 2,4-D-butyl, Registry No. *[94-80-4]*; 2,4-D-iso-octyl, Registry No. *[25168-26-7]* (formerly *[1280-20-2]*).

Uses. 2,4-D, its salts and esters are systemic herbicides, widely used for weed control in cereals and other crops at 0·28-2·3 kg/ha, the highest rate persisting in soil *c.* 30 d.

Toxicology. The acute oral LD50 for rats is: 375 mg 2,4-D/kg; 666-805 mg 2,4-D-sodium/kg; 700 mg 2,4-D-isopropyl/kg. The toxicity and hazards to man, domestic animals and wildlife have been reviewed (J. M. Way, *Residue Rev.*, 1969, **26**, 37).

Formulations. Amine salts are usually marketed as solutions of declared a.e. content, methanol being included in the formulation to delay crystallisation on storage. Ester, such as the 2-butoxyethyl and iso-octyl are formulated as e.c. Mixtures with other herbicides are often used, including: aminotriazole + atrazine; 2,4-DB; dichlorprop; mecoprop; sodium chlorate; 2,4,5-T.

Analysis. Product analysis of 2,4-D, salts, esters and mixed combination products are by CIPAC methods (*CIPAC Handbook*, 1970, **1**, 241; 1979, **1A**, in press; *Herbicides 1977*, pp. 6-21). Residue analysis is by GLC of the methyl or butyl esters after esterification of the residue. (R. P. Marquardt *et al.*, *Anal. Methods Pestic., Plant Growth Regul. Food Addit.*, 1964, **4**, 95; D. J. Lisk, *ibid.*, 1967, **5**, 363; *Anal. Methods Pestic. Plant Growth Regul.*, 1972, **6**, 630; T. H. Byast *et al.*, *Agric. Res. Counc. (G.B.) Weed Res. Organ.*, *Tech. Rep.* No. 15 (2nd Ed.), p.31).

Dalapon

MeCCl$_2$.COOH

QVXGG1

OVXGG1 &-NA-

dalapon   C$_3$H$_4$Cl$_2$O$_2$ (143·0)

dalapon-sodium   C$_3$H$_3$Cl$_2$NaO$_2$ (165·0)

Nomenclature and development. The common name dalapon is approved by BSI, ANSI, WSSA and France but there is no ISO name; other common names used include, in Republic of South Africa (proprop), by JMAF (DPA); elsewhere the chemical name is used. The IUPAC name is 2,2-dichloropropionic acid (I), in *C.A.* usage 2,2-dichloropropanoic acid *[75-99-0]* formerly (I). The sodium salt, dalapon-sodium, Registry No. *[127-20-8]* was introduced as a herbicide in 1953 by The Dow Chemical Co. under the trade marks 'Dowpon' and 'Radapon' and protected by USP 2 642 354.

Manufacture and properties. Dalapon is produced by the chlorination of propionic acid and is a colourless liquid, b.p. 185-190°. Dalapon-sodium is a colourless to pale-tan hygroscopic powder which decomposes at 166·5° before melting. Its solubility at 25° is 502 g/l water; sparingly soluble in organic solvents other than alcohols. The calcium and magnesium salts are water-soluble. It is subject to hydrolysis, slight at 25° but comparatively rapid ⩾ 50°, so aqueous solutions should not be kept for any length of time. Alkali causes dehydrochlorination above 120°. Its solutions are corrosive to iron.

Uses. Dalapon-sodium is a selective contact herbicide, translocated from roots and foliage. It is used for the control of annual and perennial grasses at rates up to 23 kg a.e./ha on non-crop areas and at rates from 800 g a.e./ha on crops. It is readily broken down by soil micro-organisms (B. E. Day *et al., Soil Sci.*, 1963, **95**, 326).

Toxicology. The acute oral LD50 for rats is 7570-9330 mg/kg. The solid and concentrated solutions produced a non-permanent eye irritation. In 2-y feeding trials no effect was observed in rats receiving 15 mg/kg daily, there was a slight increase in kidney weight in rats at 50 mg/kg daily. The toxic effects of dalapon have been reviewed (E. E. Kenaga, *Residue Rev.*, 1974, **53**, 109).

Formulations. These include: 'Dowpon', w.s.p. (870 g a.e./kg) as sodium salt and numerous formulations of dalapon-sodium (usually of this strength); and mixture with other herbicides, *e.g.* MCPA, aminotriazole + diuron.

Analysis. Technical dalapon and dalapon-sodium are analysed by separating the complex formed with mercury dinitrate and copper dinitrate, decomposing this with potassium iodide and titrating the potassium hydroxide liberated (*CIPAC Handbook*, 1970, **1**, 274); GLC, after conversion to a suitable ester may also be used; specification (*Herbicides 1977*, p.45). Residues are determined by the GLC of a suitable ester, *e.g.* the butyl ester (T. H. Byast *et al., Agric. Res. Counc. (G.B.) Weed Res. Organ., Tech. Rep.*, No. 15 (2nd Ed.), p. 29) details are also available from The Dow Chemical Co. See also: G. N. Smith & E. H. Yonkers, *Anal. Methods Pestic., Plant Growth Regul. Food Addit.*, 1964, **4**, 79; *Anal. Methods Pestic. Plant Growth Regul.*, 1972, **6**, 621.

Daminozide

$$O \qquad O$$
$$Me_2N.NH.\overset{\|}{C}.CH_2.CH_2.\overset{\|}{C}.OH$$

QV2VMN1&1 $C_6H_{12}N_2O_3$ (160·2)

Nomenclature and development. The common name daminozide is approved by BSI, ANSI and ISO. The IUPAC name is *N*-dimethylaminosuccinamic acid, in *C.A.* usage butanedioic acid mono(2,2-dimethylhydrazide) *[1596-84-5]*, formerly succinic acid mono(2,2-dimethylhydrazide). It was introduced in 1962 by Uniroyal Inc. under the code number 'B-995' and the trade marks 'Alar' and 'B-Nine' protected by USP 3 240 799; 3 334 991. Its effects on plant growth were described by J. A. Riddell *et al.*, *Science*, 1962, **136**, 391.

Manufacture and properties. Produced by the reaction of succinic anhydride with 1,1-dimethylhydrazine, it forms colourless crystals, m.p. 154-156°. Its solubility at 25° is: 100 g/kg water; 25 g/kg acetone; 50 g/kg methanol; it is insoluble in simple hydrocarbons.

Uses. Daminozide is a plant growth regulator used in top fruit to improve the balance between vegetative growth and fruit production and to improve fruit quality. It is also used to modify the stem length and shape of ornamental plants.

Toxicology. The acute oral LD50 for rats is 8400 mg tech./kg; the acute dermal LD50 for rabbits is > 16 000 mg/kg. Rats and dogs fed for 2 y at levels up to 3000 mg (tech.)/kg diet showed no significant adverse responses. The LC50 for bobtail quail is > 4777 mg a.i. (as w.s.p.)/l, the highest concentration tested. The LC50 (96 h) is: for rainbow trout 306 mg a.i. (as w.s.p.)/l; for bluegill 552 mg a.i. (as w.s.p.)/l.

Formulations. These include: 'Alar-85', w.s.p. (850 g a.i./kg) including a wetting agent; 'B-Nine S.P.', s.p. (850 g/kg) for use on ornamentals; 'Kylar', s.p. (850 g/kg) for use on groundnuts.

Analysis. For product analysis extract with hot tetrahydrofuran, evaporate off the solvent and titrate with standard alkali. Residues may be determined by alkaline hydrolysis to liberate 1,1-dimethylhydrazine which is determined colorimetrically (V. P. Lynch, *J. Sci. Food Agric.*, 1969, **20**, 13; J. W. Dicks, *Pestic. Sci.*, 1971, **2**, 176). GLC of a silylated derivative may also be used (J. R. Lane, *Anal. Methods Pestic., Plant Growth Regul. Food Addit.*, 1967, **5**, 499). See also: *Anal. Methods Pestic. Plant Growth Regul.*, 1972, **6**, 697; V. P. Lynch, *ibid.*, 1976, **8**, 491.

Dazomet

T6NYS ENTJ A1 BUS E1 $C_5H_{10}N_2S_2$ (162·3)

Nomenclature and development. The common name dazomet is approved by BSI, ISO and WSSA—exception USSR (tiazon)—the WSSA originally used DMTT. The IUPAC name is tetrahydro-3,5-dimethyl-1,3,5-thiadiazine-2-thione, in *C.A.* usage tetrahydro-3,5-dimethyl-2*H*-1,3,5-thiadiazine-2-thione *[533-74-4]*. It was introduced in 1952 by the Stauffer Chemical Co. under code number 'N-521', by the Union Carbide Corp. under the code number 'Crag fungicide 974' and the trade mark 'Mylone', and by BASF AG under the trade mark 'Basamid'.

Manufacture and properties. Produced by the interaction of carbon disulphide, methylamine and formaldehyde (M. Delépine, *Bull. Soc. Chim. Fr.*, 1897, **15,** 891), it forms colourless crystals, m.p. 99·5° (decomp.). Its solubility at 20° is: 3 g/kg water; 15 g/kg ethanol; 173 g/kg acetone; 6 g/kg diethyl ether; 391 g/kg chloroform; 51 g/kg benzene; 0·4 g/kg cyclohexane. The technical product is 98-100% pure.

Dazomet is moderately stable but is sensitive to heat above 35°, and to moisture. Acid hydrolysis yields carbon disulphide but, in soil, dazomet breaks down to methyl(methylaminomethyl)dithiocarbamic acid which in turn yields methyl isothiocyanate. The dust formulation is not corrosive to tin-plate if kept dry.

Uses. Dazomet, by virtue of the release of methyl isothiocyanate, is a soil fumigant effective for the control of nematodes and soil fungi such as *Pythium, Rhizoctonia, Fusarium, Verticillium* spp. and *Colletotrichum atramentarium* when incorporated into glasshouse soils at rates of 370-395 kg a.i./ha. At these rates it is also effective against wireworms, millipedes and soil insects and suppresses the growth of many weeds. It is strongly phytotoxic and treated soil should not be planted until shown to be free of the compound and its decomposition products, generally within 90 d by the normal germination of cress seed sown on a sample of the treated soil: *Insecticide and Fungicide Handbook,* 5th Ed., p.253. The prill formulation is also recommended for outdoor use at 380-575 kg/ha depending on soil type; soil temperature should be at least 7° at a depth of 15 cm, the prills incorporated evenly to a depth of 20-22 cm and treated soil sealed with polythene or by use of a heavy flat roller to compact the surface. Cultivation is needed to aid release of the methyl isothiocyanate and the cress test should be applied, as above. Its behaviour in soil is summarised: N. Drescher & S. Otto, *Residue Rev.,* 1968, **23,** 49.

Toxicology. The acute oral LD50 is: for rats *c.* 640 mg/kg; for mice *c.* 280 mg/kg. The dust is irritant to the skin and eyes of rabbits.

Formulations. Dust (850 g a.i./kg); 'Basamid' prills (950 and 980 g/kg), see J. S. Taylor & J. L. Bedford, *Proc. Br. Insectic. Fungic. Conf., 7th,* 1973, **2,** 643.

Analysis. Product analysis is by acid hydrolysis, absorption in ethanolic KOH of the carbon disulphide produced and iodometric titration. See also, H. A. Stansbury, *Anal. Methods Pestic., Plant Growth Regul. Food Addit.,* 1964, **3,** 119.

2,4-DB

O.CH₂CH₂CH₂CO₂H

QV3OR   BG   DG

$C_{10}H_{10}Cl_2O_3$ (249·1)

Nomenclature and development. The common name 2,4-DB is approved by BSI and in New Zealand. The IUPAC name is 4-(2,4-dichlorophenoxy)butyric acid (I), in *C.A.* usage 4-(2,4-dichlorophenoxy)butanoic acid *[94-82-6]* formerly (I). It should be stated which ester or salt is present, *e.g.,* 2,4-DB-sodium *C.A.* Registry ‘No. *[10433-59-7],* 2,4-DB-potassium *[19480-40-1],* or methylammonium salt *[2758-42-1].* Though its properties as a growth regulator were described by M. E. Synerholm & P. W. Zimmerman, *Contib. Boyce Thompson Inst.,* 1947, **14,** 369, it was not introduced as a herbicide until 1957 by May & Baker Ltd, under the code number ‘MB 2878’ and trade mark ‘Embutox’; protected by CanadianP 570065.

Manufacture and properties. Produced by the reaction of 2,4-dichlorophenol and v-butyrolactone, it forms colourless crystals, m.p. 117-119°. Its solubility at 25° is 46 mg/l water but it is soluble in acetone, benzene, ethanol and diethyl ether. It has pK_a 4·8 and forms water-soluble alkali metal and amine salts but hard water will precipitate its calcium and magnesium salts. The acid, salts and esters are stable.

Uses. 2,4-DB is a translocatable herbicide of similar growth activity to 2,4-D, but is more selective since its activity is dependent on beta-oxidation to 2,4-D within the plant. It may be used on lucerne, undersown cereals and grassland at rates of 1·5-3 kg a.i./ha alone or in mixture with MCPA.

Toxicology. The acute oral LD50 of 2,4-DB for rats is 700 mg/kg; that of 2,4-DB-sodium is: for rats 1500 mg/kg, for mice *c.* 400 kg/mg.

Formulations. Typical formulations are ‘Embutox’, mixed sodium and potassium salts of 2,4-DB (300 g a.e./l); ‘Legumex DB’, containing 2,4-DB (240 g a.e.) plus MCPA (40 g a.e.)/l as alkali metal salts. Also numerous other mixtures.

Analysis. Product and residue analysis is by GLC after conversion to the methyl or butyl esters: W. H. Gutenmann, *Anal. Methods Pestic., Plant Growth Regul. Food Addit.,* 1967, **5,** 369; *Anal. Methods Pestic. Plant Growth Regul.,* 1972, **6,** 630, 636; C. E. McKone & R. J. Hance, *J. Chromatogr.,* 1972, **69,** 204; T. H. Byast *et al., Agric. Res. Counc. (G.B.) Weed Res. Organ., Tech. Rep.,* No. 15 (2nd Ed.), p.31.

DDT

(*pp'*-DDT)

GXGGYR   DG&R   DG $C_{14}H_9Cl_5$ (354·5)

Nomenclature and development. The common name DDT is approved by BSI, New Zealand, ESA and JMAF for a mixture of isomers the major component of which has the common name *pp'*-DDT, which is also known as ENT 1506, and the proportion of this isomer should be stated; other common names for the mixture were approved by BPC (dicophane), US Pharm. (chlorophenothane). The IUPAC name for *pp'*-DDT is 1,1,1-trichloro-2,2-bis(4-chlorophenyl)ethane formerly 1,1,1-trichloro-2,2-di-(4-chlorophenyl)ethane, in *C.A.* usage 1,1'-(2,2,2-trichloroethylidene)bis[4-chloro-benzene] *[50-29-3]* formerly 1,1,1-trichloro-2,2-bis(*p*-chlorophenyl)ethane; originally known as dichlorodiphenyltrichloroethane. It was introduced in 1942 by J. R. Geigy S.A. (now Ciba-Geigy AG) under the trade marks *'Gesarol', 'Guesarol'* and *'Neocid',* and protected by SwissP 226180; BP 547871 but is no longer produced by this company. Its insecticidal properties were first discovered by P. Müller (see: P. Langer *et al., Helv. Chim. Acta,* 1944, **27**, 892).

Manufacture and properties. DDT is produced by condensing trichloroacetaldehyde (1 mol) with chlorobenzene (2 mol) in the presence of sulphuric acid or chlorosulphonic acid. The technical product, a waxy solid of indefinite m.p., contains up to 30% of *op'*-DDT [1,1,1-trichloro-2-(2-chlorophenyl)-2-(4-chlorophenyl)ethane] which, being of insecticidal value, is not usually removed. *pp'*-DDT forms colourless crystals, m.p. 108·5°, v.p. 1·9 x 10^{-7} mmHg at 20°. It is practically insoluble in water; moderately soluble in hydroxylic or polar organic solvents, petroleum oils; readily soluble in most aromatic or chlorinated solvents. Technical DDT has similar solubility characteristics to *pp'*-DDT.

pp'-DDT is dehydrochlorinated at temperatures above its m.p. to the non-insecticidal 1,1-dichloro-2,2-bis(4-chlorophenyl)ethylene (sometimes known as DDE), a reaction catalysed by iron(III) or aluminium chlorides and by u.v. light and, in solution, by alkali or organic bases. It is generally stable to oxidation but should not be stored in iron containers or with alkaline compounds. Dehydrochlorination of DDT may occur > 50°. Transformation of residues during cooking and food processing has been discussed (T. E. Archer, *Residue Rev.,* 1976, **61**, 29).

Uses. DDT is a potent non-systemic stomach and contact insecticide of high persistence on solid surfaces but with little action on phytophagous mites. It is non-phytotoxic except to cucurbits and some varieties of barley. Its metabolism by microbial systems has been reviewed (R. E. Johnson, *ibid.,* p.1).

Toxicology. The acute oral LD50 for rats is 113-118 mg/kg; the acute dermal LD50 for female rats is 2510 mg/kg. Though stored in the body fat and excreted in the milk, 17 humans who ate 35 mg/man daily (*c.* 0·5 mg/kg daily) for 1·75 y suffered no ill-effect. The bioconcentration of DDT in the environment (A. Bevenue, *ibid.,* p.37) and effects on reproduction of higher animals (G. W.Ware, *ibid.,* 1975, **59**, 119) have been discussed.

continued

Formulations. These include: e.c., w.p., dusts and aerosol concentrates.

Analysis. Product analysis is normally by GLC which has largely replaced the classical methods based on chlorine content and, for *pp'*-DDT content, setting point (*WHO Tech. Rep. Ser.*, No. 24, 1951, p.43; *CIPAC Handbook*, 1970, **1**, 280; FAO Specification (CP/37); *Organochlorine Insecticides 1973*).

Conversion to TDE [1,1-dichloro-2,2-bis(4-chlorophenyl)ethane], which is no longer available commercially, also occurs *in vivo* and in the environment and must be considered when DDT-derived residues are considered. Residues are normally determined by GLC (D. M. Coulson *et al., J. Agric. Food Chem.*, 1960, **8**, 399; M. K. Baldwin *et al., Pestic. Sci.*, 1977, **8**, 110). Various colorimetric methods are now only of historical interest. See also: R. Miskus, *Anal. Methods Pestic., Plant Growth Regul. Food Addit.*, 1964, **2**, 97; *Anal. Methods Pestic. Plant Growth Regul.*, 1972, **6**, 340.

Dehydroacetic acid

T6OV DV CHJ CV1 F1 $C_8H_8O_4$ (168·1)

Nomenclature and development. The trivial name dehydroacetic acid is normally used; also known as DHA. The IUPAC name is 3-acetyl-6-methylpyran-2,4-dione, in *C.A.* usage 3-acetyl-6-methyl-2*H*-pyran-2,4(3*H*)-dione *[520-45-6]* also 3-acetyl-4-hydroxy-6-methyl-2*H*-pyran-2-one. Its fungicidal properties were described by P. A. Wolf, *Food Technol. (Chicago)*, 1950, **4**, 294. It was produced by The Dow Chemical Co. but they no longer market it.

Manufacture and properties. Produced by refluxing ethyl 3-oxobutyrate with sodium hydrogen carbonate, it is a colourless powder, m.p. 109-111° (subliming), b.p. 269°, v.p. 1·9 mmHg at 100°. Its solubility at 25° is: < 1 g/kg water; 220 g/kg acetone; 180 g/kg benzene; 50 g/kg methanol, diethyl ether; 30 g/kg ethanol, carbon tetrachloride; 7 g/kg heptane. It forms alkali metal salts, the sodium salt crystallising from water as a monohydrate. Its solubility at 25° is: 330 g/kg water; 480 g/kg propylene glycol; 140 g/kg methanol; sparingly soluble in non-hydroxylic organic solvents.

Uses. Dehydroacetic acid is used as a fungicide for the prevention of mould growth on fresh and dried fruit and vegetables and for the impregnation of food wraps.

Toxicology. The acute oral LD50 for rats is: 1000 mg dehydroacetic acid/kg; 1000 mg of the sodium salt/kg. It is non-irritating to the skin and there is no evidence of sensitisation (M. H. Seevers *et al., J. Pharmacol. Exp. Ther.*, 1950, **99**, 69, 84, 98). In 2-y feeding trials rats receiving 100 mg/kg daily showed no ill-effect, but in a 34-d trial rats receiving 300 mg/kg daily suffered weight loss (H. C. Spencer *et al., ibid.*, p. 57). No ill-effect was noted in humans receiving 10 mg/kg daily for 180 d.

Analysis. Residue analysis is by extraction with chloroform from acid solution, measuring the optical density at 307 nm.

Demephion

$$(MeO)_2P.O.CH_2CH_2.SMe \qquad (MeO)_2P.S.CH_2CH_2.SMe$$
$$\underset{S}{\|} \qquad\qquad\qquad\qquad \underset{O}{\|}$$

(I)   **(II)**

1S2OPS&O1&O1   *(I)* $C_5H_{13}O_3PS_2$ (216·2)

1S2SPO&O1&O1   *(II)*

Nomenclature and development. The common name demephion is approved by BSI and ISO—exception Germany—for a mixture of demephion-O (I) and demephion-S (II); the name methyl-demeton-methyl has also been used. Their IUPAC and *C.A.* names are *O,O*-dimethyl *O*-2-methylthioethyl phosphorothioate (I), Registry No. *[682-80-4]*, *O,O*-dimethyl *S*-2-methylthioethyl phosphorothioate (II), Registry No. *[2587-90-8]*; the mixed isomers have Registry No. *[8065-62-1]*. Its insecticidal properties were described by H. Rueppold, *Wiss. Z. Martin-Luther Univ. Halle-Wittenberg Math.-Naturwiss. Reihe,* 1955, **5**, 219. It was developed in 1954 by VEB Farbenfabrik Wolfen, protected by DDRP 7290; DBP 1014017. It was initially introduced into the UK by Atlas Products and Services Ltd under the trade name 'Atlasetox', and later as *'Cymetox'* by Cyanamid of Great Britain Ltd (who no longer market it), and 'Pyracide' by BASF United Kingdom Ltd. It is marketed in Eastern Europe as 'Tinox'.

Manufacture and properties. Produced by the stepwise esterification and transesterification of thiophosphoryl chloride, it is a straw-coloured liquid, a mixture of the thiono (I) and thiolo (II) isomers. (I) has b.p. 107° (decomp.)/0·1 mmHg, n_D^{25} 1·488, d_4^{20} 1·198; (II) has b.p. 65° (decomp.)/0·1 mmHg, n_D^{25} 1·508, d_4^{25} 1·218. At room temperature, the solubility of (I) in water is 300 mg/l, of (II) 3 g/l. The mixture is miscible with most aromatic solvents, chlorobenzene and ketones; immiscible with most aliphatic hydrocarbons. It is generally non-corrosive and compatible with most, except strongly alkaline, pesticides.

Uses. Demephion is a systemic insecticide and acaricide effective against sucking insects and non-phytotoxic to most crops at 170-510 g a.i./ha LV or 7·5-22·5 g/100 l HV. It is metabolised in plants to the sulphoxide, then slowly to the sulphone and to phosphates.

Toxicology. The acute oral LD50 of the 30% e.c. for rats is 0·046 ml/kg; the acute dermal LD50 for rats is 0·207 ml/kg. Rats fed 90 d at 5 mg a.i./kg diet suffered no ill-effect. It is harmful to fish and honey bees.

Formulations. 'Pyracide' e.c. (300 g a.i./l); 'Atlasetox'.

Analysis. Product analysis involves separating the 2 isomers by column chromatography; demephion-O is oxidised by perchloric acid and determined as phosphate; demephion-S is hydrolysed by alkali, the solution acidified and titrated with iodine. See also, demeton (p.154). Residue analysis is by isolation and oxidation to phosphate, which is measured colorimetrically by standard procedures.

Demeton

$$\underset{\substack{\|\\ \text{(EtO)}_2\text{P.O.CH}_2\text{CH}_2\text{.SEt}}}{\text{S}}$$

$$\underset{\substack{\|\\ \text{(EtO)}_2\text{P.S.CH}_2\text{CH}_2\text{.SEt}}}{\text{O}}$$

(I)   **(II)**

(I)   **(II)**
2S2OPS&O2&O2   2S2SPO&O2&O2 $C_8H_{19}O_3PS_2$ (258·3)

Nomenclature and development. The common name demeton is approved by BSI and ISO—exception USSR—for a reaction product comprising demeton-O and demeton-S; also known as ENT 17 295. Demeton-O **(I)** (in USSR, mercaptofos) has the IUPAC and *C.A.* name *O,O*-diethyl *O*-2-ethylthioethyl phosphorothioate, Registry No. *[298-03-3]*, formerly known as diethyl 2-ethylthioethyl phosphorothionate; demeton-S **(II)** (in USSR, mercaptofos teolovy) is *O,O*-diethyl *S*-2-ethylthioethyl phosphorothioate, Registry No. *[126-75-0]*, formerly diethyl 2-ethylthioethyl phosphorothiolate. The mixture of isomers has *C.A.* Registry No. *[8065-48-3]* (formerly *[8000-97-3]*). Its insecticidal properties were first described by G. Unterstenhöfer, *Meded. Landbouwwhogesch. Opzoekinsstn. Staat Gent,* 1952, **17**, 75 and it was introduced in 1951 by Farbenfabriken Bayer AG (now Bayer AG) under the code numbers 'Bayer 10 756' and 'E-1059', trade mark 'Systox' and protected by DBP 836 349; USP 2 571 989.

Manufacture and properties. Produced by the reaction of 2-(ethylthio)ethanol with *O,O*-diethyl phosphorochloridothioate, it is *c.* 65:35 mixture of **(I)** and **(II)**. Demeton-O **(I)** is a colourless oil, b.p. 123°/1 mmHg, d_4^{21} 1·119, n_D^{18} 1·4900, v.p. 2·84 x 10⁻⁴ mmHg at 20°; solubility at room temperature 60 mg/l water, soluble in most organic solvents. Demeton-S **(II)** is a colourless oil, b.p. 128°/1 mmHg, d_4^{21} 1·132, n_D^{18} 1·5000, v.p. 2·6 x 10⁻⁴ mmHg at 20°; solubility at room temperature 2 g/l water, soluble in most organic solvents.

The technical product is a light yellow oil with a pronounced mercaptan odour. It is hydrolysed by concentrated alkali, but is compatible with most non-alkaline pesticides except water-soluble mercury compounds. The mechanism of isomerisation of **(I)** to **(II)** was examined (T. R. Fukuto & R. L. Metcalf, *J. Am. Chem. Soc.,* 1954, **76**, 5103) and the self-alkylation of demeton described (D. F. Heath, *Nature (London),* 1957, **179**, 377).

Uses. Demeton is a systemic insecticide and acaricide with some fumigant action effective against sap-feeding insects and mites. Demeton-S rapidly penetrates into plants. The thioether sulphur of both isomers is oxidised, metabolically, to the sulphoxide and sulphone (R. B. March *et al., J. Econ. Entomol.,* 1955, **48**, 355). No marked phytotoxicity has been observed at field concentrations.

Toxicology. The acute oral LD50 is: for male rats 6-12 mg demeton/kg, for female rats 2·5-4·0 mg demeton/kg; for male rats 30 mg (I)/kg; 1·5 mg (II)/kg; 2·3 mg (II)-sulphoxide/kg; 1·9 mg (II)-sulphone/kg.

Formulations. These include e.c. of various a.i. content.

Analysis. Product analysis is by alkaline hydrolysis, determining either the acidity by back-titration (*CIPAC Handbook,* 1970, **1**, 302) or the thiol by extraction of a 2-(ethylthio)ethylthio-lead complex and titration with iodine *(ibid.)*. Residues may be determined by GLC (J. S. Thornton & C. A. Anderson, *J. Agric. Food Chem.,* 1968, **16**, 895; M. C. Bowman *et al., J. Assoc. Off. Anal. Chem.,* 1969, **52**, 157) or by oxidation to phosphoric acid which is measured by standard colorimetric procedures. See also: D. MacDougall, *Anal. Methods Pestic., Plant Growth Regul. Food Addit.,* 1964, **2**, 451; *Anal. Methods Pestic. Plant Growth Regul.,* 1972, **6**, 483.

Demeton-S-methyl

$$O$$
$$\|$$
$$(MeO)_2P.S.CH_2CH_2SEt$$

2S2SPO&O1&O1 $C_6H_{15}O_3PS_2$ (230·3)

Nomenclature and development. The common name demeton-S-methyl is approved by BSI, ISO and JMAF—exceptions USSR (methyl-mercaptofos teolovy) and USA. The IUPAC and *C.A.* name is *S*-2-ethylthioethyl *O,O*-dimethyl phosphorothioate, Registry No. *[919-86-8]*. It was introduced in 1957 by Farbenfabriken Bayer AG (now Bayer AG) under the code numbers 'Bayer 18436' and 'Bayer 25/154', trade mark 'Metasystox i'. It replaced demeton-methyl, a product (Registry No. *[8022-00-2]*) containing demeton-S-methyl and demeton-O-methyl (*O*-2-ethylthioethyl *O,O*-dimethyl phosphorothioate; Registry No. *[867-27-6]*), introduced in 1954 and protected by DBP 836 349; USP 2 571 989.

Manufacture and properties. Produced by the reaction of *O,O*-dimethyl hydrogen phosphorothioate with 2-chloroethyl ethyl sulphide, it is a pale yellow oil, b.p. 89°/0·15 mmHg, v.p. 3·6 x 10⁻⁴ mmHg at 20°, d_4^{20} 1·207, n_D^{20} 1·5065. Its solubility at room temperature is 3·3 g/l; soluble in most organic solvents.

Uses. Demeton-S-methyl is a systemic and contact insecticide and acaricide used on most agricultural and horticultural crops at 7·5-22 g a.i./100 l HV or 245 g/ha LV. Sprays are phytotoxic to some ornamentals especially certain chrysanthemum cultivars. It is metabolised in plants to the sulphoxide (oxydemeton-methyl, p.398) and sulphone (p.156) (T. R. Fukuto *et al., J. Econ. Entomol.,* 1955, **48**, 348; 1956, **49**, 147; 1957, **50**, 399).

Toxicology. The acute oral LD50 is: for rats 57-106 mg/kg; for male guinea-pigs 110 mg/kg. The acute dermal LD50 for male rats is 302 mg/kg.

Formulations. These include e.c. of various a.i. content. Product names include: 'Metasystox 55' (Bayer), 'Azotox' (Chafer), 'Campbell's DSM' (Campbell), 'Demetox' (ICI Ltd, Plant Protection Division), 'Duratox' (Shell).

Analysis. Product analysis is based on alkaline hydrolysis determining *(a)* the acidity by back-titration (*CIPAC Handbook,* 1970, **1**, 312) or *(b)* the thiol by extraction of a 2-(ethylthio)ethylthio-lead complex and titration with iodine *(ibid.)*. Residues may be determined by GLC (J. H. van der Merwe & W. B. Taylor, *Pflanzenschutz-Nachr. (Am. Ed.),* 1971, **24**, 259) or by oxidation to phosphate which is measured by standard colorimetric methods (E. Q. Laws & D. J. Webley, *Analyst (London),* 1959, **84**, 28; H. Tietz & H. Frehse, *Hoefchen-Briefe (Engl. Ed.),* 1960, **13**, 212). See also: D. MacDougall, *Anal. Methods Pestic., Plant Growth Regul. Food Addit.,* 1964, **2**, 295.

Demeton-S-methyl sulphone

(Demeton-S-methylsulphon)

$$(MeO)_2\overset{O}{\underset{}{P}}.S.CH_2CH_2.\overset{O}{\underset{O}{S}}.Et$$

WS2&2SPO&O1&O1 $\qquad\qquad\qquad\qquad\qquad$ $C_6H_{15}O_5PS_2$ (262·3)

Nomenclature and development. The common name demeton-S-methylsulphon is approved by ISO, spellings BSI (demeton-S-methyl sulphone) and France (déméton-S-méthylsulfone)—exception USA. The IUPAC name is *S*-2-ethylsulphonylethyl *O,O*-dimethyl phosphorothioate, in *C.A.* usage *S*-[2-(ethylsulfonyl)ethyl] *O,O*-dimethyl phosphorothioate *[17040-19-6]*. It was introduced in 1965 by Bayer AG after trials from 1954 under the code numbers 'Bayer 20 315', 'E 158', 'M 3/158' and the trade mark 'Metaisosystoxsulfon'.

Manufacture and properties. Produced by the oxidation of demeton-S-methyl with potassium permanganate (DBP 948 241), demeton-S-methyl sulphone is a colourless to yellowish crystalline solid, m.p. 60°, b.p. 120°/0·03 mmHg, v.p. 5 x 10^{-6} mmHg at 20°. It is readily soluble in alcohols; poorly in aromatic hydrocarbons. It is readily hydrolysed at pH > 7·0.

Uses. Demeton-S-methyl sulphone is a systemic insecticide effective against sap-feeding insects, sawflies and mites. Its range of action coincides with that of demeton-S-methyl (p.155) of which it is a metabolic product in plants and animals. It is normally used in combination with other insecticides, *e.g.* with azinphos-methyl to control a wide range of pests on pome fruit at (25 g azinphos-methyl + 7·5 g demeton-S-methyl sulphone/100 l HV).

Toxicology. The acute oral LD50 for rats is *c.* 37·5 mg a.i./kg; the acute dermal LD50 for rats is *c.* 500 g/kg; the acute intraperitoneal LD50 for rats is *c.* 20·8 mg/kg.

Formulations. Normally it is combined with other insecticides, *e.g.* 'Gusathion MS', w.p. (75 g demeton-S-methyl sulphone + 250 g azinphos-methyl/kg).

Analysis. Product analysis is by alkaline hydrolysis and back-titration, c.f. demeton-S-methyl (p.155). Residues are determined by GLC (M. C. Bowman *et al., J. Assoc. Off. Anal. Chem.,* 1969, **52**, 157; J. H. van der Merwe & W. B. Taylor, *Pflanzenschutz-Nachr. (Am. Ed.),* 1971, **24**, 259) or oxidation to phosphate which is measured by standard colorimetric methods: details of methods are available from Bayer AG.

Desmedipham

$$EtO.\overset{O}{\overset{||}{C}}.NH \longrightarrow O.\overset{O}{\overset{||}{C}}.NHPh$$

2OVMR   COVMR $C_{16}H_{16}N_2O_4$ (300·3)

Nomenclature and development. The common name desmedipham is approved by BSI, ISO, ANSI and WSSA. The IUPAC name is ethyl 3-phenylcarbamoyloxyphenyl-carbamate, also known as 3-ethoxycarbonylaminophenyl phenylcarbamate, in *C.A.* usage ethyl [3-[[(phenylamino)carbonyl]oxy]phenyl]carbamate *[13684-56-5]* formerly ethyl *m*-hydroxycarbanilate ester. It was introduced in 1971 by Schering AG under the code number 'Schering 38 107', trade mark 'Betanal AM' and protected by BP 1 127050. It was described by F. Arndt & G. Boroschewski, *Symp. New Herbic., 3rd,* 1969, p.141.

Manufacture and properties. Produced by the reaction of ethyl 3-hydroxyphenyl-carbamate and phenyl isocyanate, it forms colourless crystals, m.p. 120°, v.p. 3 x 10⁻⁹ mmHg at 25°. Its solubility at room temperature is: 7 mg/l water; 400 g/l acetone; 180 g/l methanol; 80 g/l chloroform; 1·6 g/l benzene. The technical product is *c.* 96% pure. Hydrolysis occurs under alkaline conditions.

Uses. Desmedipham is a post-em. herbicide used to control weeds, including *Amaranthus retroflexus,* in beet crops, in particular sugar beet, frequently sprayed in combination with phenmedipham. Application rates are 800-1000 g a.i./ha in 200-300 l water broadcast, or one-third of this amount for band spray. It acts through the leaves only, and does not depend on soil type and humidity under normal growing conditions. Due to its wide safety margin to the crop, spraying is merely timed according to the development stage of the weeds, with an optimum of weed control, when they have 2-4 true leaves.

Toxicology. The acute oral LD50 for rats is > 9600 mg a.i./kg or, for the formulated product 2100 mg/kg. The acute dermal LD50 for rats is 2000-10000 mg/kg. The LC50 (96 h) is: for rainbow trout 3·8 mg/l; for bluegill 13·4 mg/l.

Formulations. These include: 'Betanal AM' (in Canada and USA, 'Betanex'), e.c. (157 g a.i./l) for tank mix with phenmedipham ('Betanal'); 'Betanal AM11', e.c. (80 g desmedipham + 80 g phenmedipham/l).

Analysis. Product analysis is by quantitive TLC using a chromatogram spectrophotometer. Residue analysis is by alkaline hydrolysis and determination of the liberated aniline by bromination and GLC with ECD, or by diazotisation and coupling with *N*-(1-naphthyl)ethylenediamine for colorimetric measurement. Details of methods are available from Schering AG. See also: C. -H. Roder *et al., Anal. Methods Pestic. Plant Growth Regul.,* 1978, **10,** 293.

Desmetryne

(Desmetryn)

SMe

MeHN—⟨triazine ring⟩—NHPr^i

T6N CN ENJ BS1 DMY1&1 FM1 $C_8H_{15}N_5S$ (213·3)

Nomenclature and development. The common name desmetryn is approved by ISO and WSSA, spelling by BSI, France and JMAF (desmetryne)—exception Portugal. The IUPAC name is 2-isopropylamino-4-methylamino-6-methylthio-1,3,5-triazine, in *C.A.* usage *N*-methyl-*N'*-(1-methylethyl)-6-(methylthio)-1,3,5-triazine-2,4-diamine *[1014-69-3]*, formerly 2-(isopropylamino)-4-(methylamino)-6-(methylthio)-*s*-triazine. It was introduced in 1964 by J. R. Geigy S.A. (now Ciba-Geigy AG) under the code number 'G 34360', the trade mark 'Semeron' and protected by SwissP 337019; BP 814948. Its herbicidal properties were first described by C. Baker *et al., Weed Res.,* 1963, **3**, 109; J. G. Elliott & T. I. Cox, *Proc. Br. Weed Control Conf., 6th,* 1962, **2**, 759.

Manufacture and properties. Produced by treating trichloro-1,3,5-triazine with, in turn, one equivalent of isopropylamine, methylamine and methanethiol, it is a crystalline solid, m.p. 84-86°, v.p. 1·0 x 10^{-6} mmHg at 20°. Its solubility at 20° is 600 mg/l water; readily soluble in organic solvents. It is stable in neutral, slightly acidic or alkaline media.

Uses. Desmetryne is a selective post-em. herbicide, translocated from leaves and roots and of brief persistence in soil. It is effective for the control of *Chenopodium album* and other broad-leaved and grassy weeds in brassica crops. The recommended rate is *c.* 500 g a.i./ha.

Toxicology. The acute oral LD50 for rats is *c.* 1390 mg/kg. In 90-d feeding trials on rats at 100 mg/kg daily it caused only a slight decrease in body-weight gain. It has negligible toxicity to wildlife.

Formulation. 'Semeron 25', w.p. (250 g a.i./kg).

Analysis. Product analysis is by GLC with internal standard (*CIPAC Handbook,* 1979, **1A,** in press) or by titration with perchloric acid in acetic acid. Residues may be determined by GLC (C. A. Benfield & E. D. Chilwell, *Analyst (London),* 1964, **89,** 475; K. Ramsteiner *et al., J. Assoc. Off. Anal. Chem.,* 1974, **57,** 192), TLC (K. Stammbach *et al., Weed Res.,* 1964, **4,** 64) or u.v. spectrometry. Particulars of these methods are available from Ciba-Geigy AG.

2,4-DES-sodium

(Disul-sodium)

WSQO2OR BG DG

WSO&O2OR BG DG &-NA-

2,4-DES $C_8H_8Cl_2O_5S$ (303·1)

2,4-DES-sodium $C_8H_7Cl_2NaO_5S$ (325·1)

Nomenclature and development. The common name disul is approved by ISO—exceptions BSI, Australia and USSR (2,4-DES); WSSA (sesone)—for 2-(2,4-dichlorophenoxy)ethyl hydrogen sulphate (IUPAC), in *C.A.* usage 2-(2,4-dichlorophenoxy)ethyl hydrogen sulfate *[149-26-8]*. The herbicidal properties of sodium 2-(2,4-dichlorophenoxy)ethyl sulphate, 2,4-DES-sodium, Registry No. *[136-78-7]*, were first reported by L. J. King *et al., Contrib. Boyce Thompson Inst.,* 1950, **16**, 191. It was introduced in 1951 by the Union Carbide Corp. under the trade marks *'Crag Herbicide I', 'Crag Sesone',* and the protection of USP 2 573 769 but is no longer produced by them. It is available in England and Wales as a component of the herbicidal mixture 'Herbon Blue' (Cropsafe Ltd).

Manufacture and properties. Produced by the addition of oxirane to 2,4-dichlorophenol, treating the resulting alcohol with chlorosulphonic acid and the sodium salt prepared from the acid sulphate. 2,4-DES-sodium forms colourless crystals, m.p. 170° and of negligible v.p. at room temperature. Its solubility is 250 g/kg water, but it is insoluble in most organic solvents except methanol. The calcium salt is sufficiently soluble to avoid precipitation by hard water. 2,4-DES-sodium is hydrolysed by alkali to form 2-(2,4-dichlorophenoxy)ethanol and sodium hydrogen sulphate.

Uses. 2,4-DES-sodium is not itself phytotoxic, but is converted in moist soil to 2-(2,4-dichlorophenoxy)ethanol which is oxidised to 2,4-D. It is used as a mixture with simazine to control annual weeds pre-em. in maize, soft fruits, roses, certain established perennial crops and certain ornamental trees and shrubs.

Toxicology. The acute oral LD50 for rats is 730 mg/kg. In 2-y trials rats suffered no ill-effect at 2000 mg/kg diet. It is dangerous to fish.

Formulation. 'Herbon Blue', w.p. (2,4-DES-sodium + simazine).

Analysis. Complex with methylene blue chloride and measure the absorption at 630 nm, J. N. Hogsett & G. L. Funk, *Anal. Chem.,* 1954, **26**, 849.

Dialifos

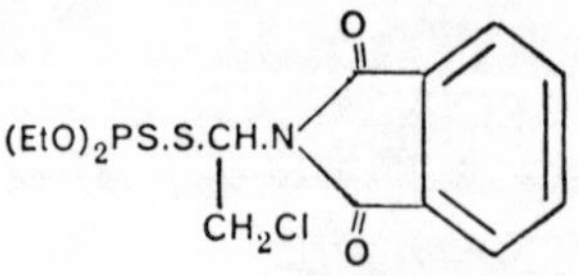

(EtO)$_2$PS.S.CH.N
        |
     CH$_2$Cl

T56 BVNVJ CY1GSPS&O2&O2   C$_{14}$H$_{17}$ClNO$_4$PS$_2$ (393·8)

Nomenclature and development. The common name dialifos is approved by BSI and ISO—exception ANSI and JMAF (dialifor); also known as ENT 27 320. The IUPAC name is *S*-2-chloro-1-phthalimidoethyl *O,O*-diethyl phosphorodithioate, in *C.A.* usage *S*-[2-chloro-1-(1,3-dihydro-1,3-dioxo-2*H*-isoindol-2-yl)ethyl] *O,O*-diethyl phosphoro-dithioate [*10311-84-9*] formerly *S*-(2-chloro-1-phthalimidoethyl) *O,O*-diethyl phosphoro-dithioate, also the *O,O*-diethyl phosphorodithioate *S*-ester with *N*-(2-chloro-1-mercaptoethyl)phthalimide. It was introduced in 1965 by Hercules Inc. under the code number 'Hercules 14503', the trade mark 'Torak' and protected by BP 1091738; USP 3 355 353. Its insecticidal properties were described by W. R. Corthan *et al., J. Econ. Entomol.,* 1967, **60,** 1151.

Manufacture and properties. Produced by the chlorination of *N*-vinylphthalimide followed by reaction with *O,O*-diethyl hydrogen phosphorodithioate, it is a colourless crystalline solid, m.p. 67-69°C. It is insoluble in water; slightly soluble in aliphatic hydrocarbons, alcohols; highly soluble in acetone, cyclohexanone, isophorone, xylene. The technical product and its formulations are stable > 2 y in normal storage conditions but it is readily hydrolysed by concentrated alkali. It is non-corrosive and compatible with most other pesticides.

Uses. Dialifos is a non-systemic insecticide and acaricide, effective in controlling many insects and mites common to apples, citrus, grapes, nut trees, potatoes and vegetables.

Toxicology. The acute oral LD50 is: for female rats 5 mg/kg; for male rats 43-53 mg/kg; for male mice 39 mg/kg; for female mice 65 mg/kg; for male dogs 97 mg/kg. The acute dermal LD50 for rabbits is 145 mg/kg.

Formulations. An e.c. (240 and 480 g tech./l).

Analysis. Product analysis is by u.v. spectrometry. Residues are determined by GLC with a thermionic detector. Details of methods are available from Hercules Inc.

Di-allate

(E)   (Z)

G1UYG1SVNY1&1&Y1&1 $C_{10}H_{17}Cl_2NOS$ (270·2)

Nomenclature and development. The common name di-allate is approved by BSI and ISO, WSSA (spelt diallate). The IUPAC name is S-2,3-dichloroallyl di-isopropylthio-carbamate (I), in *C.A.* usage S-(2,3-dichloro-2-propenyl) bis(1-methylethyl)carbamo-thioate *[2303-16-4]* formerly (I). This compound was introduced in 1960 by the Monsanto Co. under the code number 'CP 15 336', trade mark 'Avadex' and protected by USP 3 330 821. Herbicides employing this compound are protected by USP 3 330 643. Its herbicidal properties were first described by L. H. Hannah, *Proc. North Cent. Weed Control Conf.*, 1959, p.50.

Manufacture and properties. Produced by the condensation of di-isopropylamine, carbonyl sulphide and 1,2,3-trichloropropene, it is an amber-coloured liquid, b.p. 150°/9 mmHg, f.p. —8 to —15°. Its solubility at 25° is 14 mg/l water; miscible with ethanol and other organic solvents.

Isomerism. It exists as geometric isomers which have been separated, (F. H. A. Rummens, *Weed Sci.*, 1975, **23**, 7). The (Z)-isomer has m.p. < 0°, n_D^{40} 1·5124; (E)-isomer has m.p. 36°, n_D^{40} 1·5097. The (E)- is 65% less effective than the (Z)-isomer in reducing height of wild oats by 50%; a mixture (Z-, 42%) + (E-, 58%) is 2·6 times less effective than the (Z)-isomer (F. H. A. Rummens *et al., ibid.*, p.11).

Uses. Di-allate is a pre-drilling herbicide of particular value in the control of wild oats and blackgrass in brassicas, red beet and sugar beet. Being volatile, its immediate incorporation in the soil is necessary, usually at rates of 1·5 kg a.i./ha. Cultivated oats should not be sown in the year following application.

Toxicology. The acute oral LD50 is: for rats 395 mg/kg; for dogs 510 mg/kg. The acute dermal LD50 for rabbits is 2000-2500 mg/kg; the concentrate may cause irritation to the skin, eyes and mucous membranes. The 'no effect' level for rats and dogs is 125 mg daily.

Formulations. These include: e.c. (400 and 480 g a.i./l); granules (100 g/kg).

Analysis. Product analysis is by i.r. spectrometry at 7·8 µm against standards: particulars available from the Monsanto Co.

N,N-Diallyl-2,2-dichloroacetamide

$$Cl_2CH.\overset{\overset{\displaystyle O}{\|}}{C}.N(CH_2CH=CH_2)_2$$

GYGVN2U1&2U1 $C_8H_{11}Cl_2NO$ (208·1)

Nomenclature and development. There is no common name. The IUPAC name is *N,N*-diallyl-2,2-dichloroacetamide, in *C.A.* usage 2,2-dichloro-*N,N*-di-2-propenylacetamide *[37764-25-3]*. It was introduced in 1971 by the Stauffer Chemical Co. under the code number 'R-25788' and protected by DBP 2218097. Its use to enhance herbicidal selectivity was reported by F. Y. Chang *et al., Can. J. Plant Sci.,* 1972, **52,** 707.

Manufacture and properties. Produced by the reaction of dichloroacetyl chloride with diallylamine in chloroform, it is a clear viscous liquid, d_{20}^{20} 1·202, n_D^{30} 1·4990, v.p. 6 x 10^{-3} mmHg at 25°. Its solubility at 20° is *c.* 5 g/l water.

Uses. *N,N*-Diallyl-2,2-dichloroacetamide increases the tolerance of maize to thiocarbamate herbicides, including butylate, EPTC and vernolate, at rates of 140-700 g a.i./ha. The mixture with herbicide is applied when the soil is dry enough to permit incorporation in the top 7·5-10 cm and the maize is then planted without delay. The chemical structure/biological activity of analogues has been discussed (G. R. Stephenson, *J. Agric. Food Chem.,* 1978, **26,** 137). The metabolism in maize, and rats and the degradation in soil was reported (J. B. Miallis *et al., Chemistry and Mode of Action of Herbicide Antidotes,* p.109), the uptake by plant roots studied (R. A. Gray & G. K. Joo, *ibid.,* p.67) and the mode of action investigated (R. E. Wilkinson, *ibid.,* p.85).

Toxicology. The acute oral LD50 for female albino rats is 2000 mg/kg. The acute dermal LD50 for rabbits is > 4640 mg/kg; it is non-irritating in rabbit eye tests.

Formulations. These include: 'R-25788, 6E', e.c. (720 g/l); 'Eradicane 6E' (720 g EPTC + 60 g *N,N*-diallyl-2,2-dichloroacetamide/l); also mixtures with butylate and vernolate.

Diazinon

T6N CNJ BY1&1 DOPS&O2&O2 F1 $\qquad$ $C_{12}H_{21}N_2O_3PS$ (304·3)

Nomenclature and development. The common name diazinon is approved by BSI, ISO and ESA—exception BPC (dimpylate), USA; also known as ENT 19 507. The IUPAC name is *O,O*-diethyl *O*-2-isopropyl-6-methylpyrimidin-4-yl phosphorothioate, in *C.A.* usage *O,O*-diethyl *O*-[6-methyl-2-(1-methylethyl)-4-pyrimidinyl] phosphorothioate *[333-41-5]* formerly *O,O*-diethyl *O*-(2-isopropyl-6-methyl-4-pyrimidinyl) phosphorothioate. Its insecticidal properties were first described by R. Gasser, *Z. Naturforsch. B: Anorg. Chem. Org. Chem.*, 1953, **8**, 225, and it was introduced in 1952 by J. R. Geigy S.A. (now Ciba-Geigy AG) under the code number 'G 24480', trade marks 'Basudin', 'Diazitol', 'Neocidol' and 'Nucidol' and protected by BP 713 278; USP 2 754 243.

Manufacture and properties. Produced by condensing isobutyramidine with ethyl 3-oxobutyrate and treating the resulting pyrimidine with *O,O*-diethyl phosphoro-chloridothioate (H. Gysin, *Chimia,* 1954, **8**, 205, 221), it is a colourless liquid, b.p. 83-84°/2 x 10⁻⁴ mmHg, v.p. 1·4 x 10⁻⁴ mmHg at 20°, d^{20} 1·116-1·118, n^{20} 1·4978-1·4981. Its solubility at room temperature is 40 mg/l water; miscible with most organic solvents. It decomposes > 120° and is susceptible to oxidation; it is stable in alkaline media but is slowly hydrolysed by water and dilute acids. The presence of traces of water promotes hydrolysis on storage to the highly poisonous tetraethyl monothiopyrophosphate (A. Margot & H. Gysin, *Helv. Chim. Acta.,* 1957, **40**, 1562).

Uses. Diazinon is a non-systemic insecticide with some acaricidal action. Main applications are in rice, fruit trees, vineyards, sugarcane, corn, tobacco, potatoes and horticultural crops for a wide range of sucking and leaf-eating insects. Used also against flies in glasshouses, mushroom houses and against flies and ticks in veterinary practice. Decreases in residue levels in plants and animals have been discussed (E. Bartsch, *Residue Rev.,* 1974, **51**, 37).

Toxicology. The acute oral LD50 for rats is 300-850 mg/kg (values reported earlier were sometimes lower due to the formation of very toxic deterioration products). The acute dermal LD50 for rats is > 2150 mg/kg. In 2-y feeding trials the 'no effect' level was estimated as: for rats 0·1 mg/kg daily; for rhesus monkeys 0·05 mg/kg daily. Highly toxic to birds and honey bees; toxic to fish.

Formulations. Typical formulations for agricultural and horticultural use include: granules, 'Basudin 5' (50 g a.i./kg), 'Basudin 10' (100 g/kg); w.p., 'Basudin 40WP' (400 g/kg); seed dressing, 'Basudin 50SD'; e.c., 'Basudin 60 EC' (600 g/l), 'Diazitol Liquid'; 'Basudin Ulvair 500'; 'Basudin 20 Mushroom Aerosol'. For veterinary use: 'Neocidol 60'/'Nucidol 60', e.c. (600 g/l).

Analysis. Product analysis is by GLC with internal standard (CIPAC Handbook, 1979, **1A**, in press; *Anal. Methods Pestic. Plant Growth Regul.*, 1972, **6**, 345; D. O. Eberle, *J. Assoc. Off. Anal. Chem.,* 1974, **57**, 48). Residues may be determined by GLC with thermionic phosphorus detection or MCD (D. O. Eberle & D. Novak, *ibid.,* 1969, **52**, 1067) or by spectrometry after hydrolysis with hydrobromic acid (A. Margot & K. Stammbach, *Anal. Methods Pestic., Plant Growth Regul. Food Addit.,* 1964, **2**, 109). For review of analytical methods see D. O. Eberle, *Residue Rev.,* 1974, **51**, 1.

1,2-Dibromo-3-chloropropane

BrCH₂.CHBr.CH₂Cl

G1YE1E $C_3H_5Br_2Cl$ (236·3)

Nomenclature and development. The common name DBCP is approved by JMAF. The IUPAC and *C.A.* name is 1,2-dibromo-3-chloropropane, Registry No. *[96-12-8]*. It is the main component of a fumigant introduced in 1955 by The Dow Chemical Co. under the trade mark 'Fumazone' and by the Shell Development Co. under the code number 'OS1897' and trade mark 'Nemagon'. It was first described as a pesticide by C. W. McBeth & G. B. Bergeson, *Plant Dis. Rep.,* 1955, **39,** 223.

Manufacture and properties. Produced by the liquid phase addition of bromine to allyl chloride, it is an amber to dark brown liquid, with a mildly pungent odour, b.p. 196°, v.p. 0·8 mmHg at 21°, d_{20}^{20} 2·08, n_D^{25} 1·5518. Its solubility at room temperature is 1 g/kg water; miscible with hydrocarbon oils, acetone, propan-2-ol, methanol, 1,2-dichloropropane and 1,1,2-trichloroethane. Though stable in neutral and acidic media, it is converted by alkali to 2-bromoallyl alcohol. It is compatible with chlorinated hydrocarbon insecticides and with dry and liquid NPK fertilisers. It corrodes aluminium, magnesium and their alloys; if the water content is < 0·02% it is non-corrosive to steel and copper alloys.

Uses. 1,2-Dibromo-3-chloropropane is a soil fumigant and nematicide, effective against a wide range of nematodes, including root-knot nematodes, at 10-125 kg/ha. Soil temperatures at 15 cm depth should be 21-27° for best results. Many perennial plants tolerate high concentrations but others, including potatoes and tobacco, require a long aeration period before planting. It does not persist in soil to the extent that creates an accumulation problem.

Toxicology. The acute oral LD50 is: for rats 170-300 mg/kg; for mice 260-400 mg/kg. The acute dermal LD50 for rabbits is 1420 mg/kg. It is not markedly irritant to the eyes and mucous membranes. In 90-d feeding trials the lowest dose level causing a decrease in growth rate was: for female rats 150 mg/kg, for male rats 450 mg/kg. In the USA a standard of 1 µg/m³ in manufacturing plants has been set by the Occupational Safety and Health Administration. In lifetime studies in rats and mice it has been shown by the oral gavage to be carcinogenic (W. A. Olsen *et al., J. Nat. Cancer Inst.,* 1973, **51,** 1993). The LC50 (48 h) is: for bass 30-50 mg/l; for sunfish 50-125 mg/l.

Formulations. These include: 'Fumazone', e.c. (878 g a.i./kg); 'Nemagon', e.c. (870 g/kg and 145 g/l).

Analysis. Product analysis is by GLC (C. E. Castro, *Anal. Methods Pestic., Plant Growth Regul. Food Addit.,* 1964, **3,** 165). Residues may be determined by GLC with ECD (T. E. Archer *et al., J. Chromatogr.,* 1961, **6,** 457; H. Beckman & A. Bevenue, *J. Agric. Food Chem.,* 1963, **11,** 479). Details of methods are available from the Shell International Chemical Co. See also: *Anal. Methods Pestic. Plant Growth Regul.,* 1972, **6,** 714.

Dicamba

QVR BG EG FO1 $C_8H_6Cl_2O_3$ (221·0)

Nomenclature and development. The common name dicamba is approved by BSI, ISO, ANSI and WSSA—exceptions USSR (dianat) and JMAF (MDBA). The IUPAC name is 3,6-dichloro-*o*-anisic acid (I) also known as 3,6-dichloro-2-methoxybenzoic acid (II), in *C.A.* usage (II) *[1918-00-9]* formerly (I). It was introduced in 1965 by the Velsicol Chemical Corp. under the code number 'Velsicol 58-CS-11', the trade marks 'Banvel' and 'Mediben' and protected by USP 3013054.

Manufacture and properties. It is produced from 1,2,4-trichlorobenzene by hydrolysis in methanolic sodium hydroxide, treatment with carbon dioxide under pressure of the resulting 2,5-dichlorophenol and subsequent methylation by dimethyl sulphate. The resulting technical grade is a pale buff crystalline solid, purity 83-87%—the remainder consisting mainly of 3,5-dichloro-2-methoxybenzoic acid. Pure dicamba is a colourless solid, m.p. 114-116°, v.p. 3·75 x 10^{-3} mmHg at 100°. Its solubility at 25° is 4·5 g/l water; moderately soluble in xylene; readily soluble in ethanol and ketones. The solubility of the sodium salt is 380 g a.e./l water, of the dimethylammonium salt > 720 g a.e./l water. The acid is stable and resistant to oxidation and hydrolysis under normal conditions.

Uses. Dicamba is a translocatable post-em. herbicide for weed control in cereals (not undersown), it is usually formulated with one or more phenoxyalkanoic acids, or with herbicides of various other classes, so that the dosage of dicamba lies in the range 80-130 g a.e./ha. It is also used for the control of docks in established grassland and for the control of bracken. Its oil-soluble salt with (3-aminopropyl)octadec-9-enylamine (W. Furness & J. L. Forryan, *Proc. Br. Weed Control Conf., 8th,* 1966, p.405) and furfuryl ester are used for control of brush. It is rapidly degraded in soil and none was detected 300 d after an application of 6·7 kg/ha. It is metabolised in plants to the herbicidally inactive 2,5-dichloro-3-hydroxy-6-methoxybenzoic acid (W. H. Zick & T. R. Castro, *ibid.,* p.265).

Toxicology. The acute oral LD50 for rats is 2900 ± 800 mg/kg. There was no effect in 2-y feeding trials on rats receiving 500 mg/kg diet or dogs 50 mg/kg diet. The LC50 (96 h) is: for rainbow trout 28 mg/l; for bluegill 23 mg/l. The LC50 (48 h) for small carp is 465 mg dicamba-dimethylammonium/l.

Formulations. These include: 'Banvel-M' (20 g dicamba + 250 g MCPA/l) as potassium salts; 'Banvel-K' (20 g dicamba + 250 g 2,4-D/l) as potassium salts; 'Banvel-P' (27·5 g dicamba + 425 g mecoprop/l) as potassium salts. Dicamba is a component of numerous other formulations including ternary mixtures with benazolin and dichlorprop or mecoprop and MCPA.

Analysis. Product analysis is by i.r. spectroscopy (M. A. Malina, *Anal. Methods Pestic. Plant Growth Regul.,* 1973, **7**, 545; *CIPAC Handbook,* 1979, **1A**, in press). Residues in plants may be determined by GLC after esterification (M. Smith *et al., J. Assoc. Off. Agric. Chem.,* 1965, **48**, 1164; M. A. Malina, *loc. cit.;* H. K. Suzuki *et al., Anal. Methods Pestic. Plant Growth Regul.,* 1978, **10**, 305).

Dichlobenil

NCR BG FG $C_7H_3Cl_2N$ (172·0)

Nomenclature and development. The common name dichlobenil is approved by BSI, ISO, ANSI and WSSA—exception JMAF (DBN). The IUPAC and *C.A.* name is 2,6-dichloro-benzonitrile, Registry No. *[1194-65-6]*; also known as 2,6-dichlorophenyl cyanide. It was introduced in 1960 by Philips Duphar B.V. under the code number 'H 133', trade mark 'Casoron' and protected by DutchP 572 662; USP 3 027 248. Its herbicidal properties were first described by H. Koopman & J. Daams, *Nature (London)*, 1960, **186**, 89.

Manufacture and properties. Produced by oxidising 2,6-dichlorotoluene to 2,6-dichloro-benzaldehyde and conversion to the nitrile, it is a colourless to off-white crystalline solid, m.p. 145-146°, v.p. 5 x 10⁻⁴ mmHg at 25°. Its solubility at 20° is 18 mg/l water; slightly soluble in most organic solvents. The technical product, m.p. 139-145° is ≥ 94% pure. It is stable to heat and to acids but is hydrolysed by alkali to 2,6-dichlorobenzamide. It is non-corrosive and compatible with other herbicides.

Uses. Dichlobenil is a herbicide that inhibits actively-dividing meristems. It controls annual and perennial weeds in the seedling stages and many in more advanced stages. It is used as a selective weedkiller, pre- and post-em., *e.g.* post-em. control in fruit and other crops at 2·5-10 kg a.i./ha; in the control of aquatic weeds at 4·5-12 kg/ha; for total weed control ≤ 20 kg/ha. Its fate in crops, soils and animals has been reviewed (K. I. Beynon & A. N. Wright, *Residue Rev.*, 1972, **43**, 23; A. Verloop, *ibid.*, p.55).

Toxicology. The acute oral LD50 is: for rats 3160 mg/kg; for mice 2056 mg/kg; for guinea-pigs 150 mg/kg. The acute dermal LD50 for albino rabbits is 1350 mg/kg. Rats receiving 1000 mg/kg diet gained weight normally but showed changes in organ weight. The LC50 (48 h) is: for guppies > 18 mg/l; for water fleas 9·8 mg/l.

Formulations. These include: w.p., 'Casoron' (450 g a.i./kg); granules, 'Casoron G' (67·5 g/kg), 'Casoron GSR' (200 g/kg), 'Fydulan G' (67·5 g dichlobenil + 100 g dalapon/kg), 'Fydulex G' (75 g dichlobenil + 150 g dalapon/kg) and 'Fydusit G' (135 g dichlobenil + 15 g bromacil/kg).

Analysis. Product analysis is by GLC (*CIPAC Handbook,* in press) or spectrometry. Residues may be analysed by GLC (K. I. Beynon *et al., J. Sci. Food Agric.,* 1966, **17**, 151; T. H. Byast *et al., Agric. Res. Counc. (G.B.) Weed Res. Organ., Tech. Rep.,* No. 15 (2nd Ed.), p.13; A. van Rossum, *Anal. Methods Pestic. Plant Growth Regul.,* 1978, **10**, 311). Details are available from Philips-Duphar B.V.

Dichlofenthion

$$(EtO)_2 \overset{\overset{\displaystyle S}{\|}}{P} \cdot O \text{—}$$

GR   CG   DOPS&O2&O2

$C_{10}H_{13}Cl_2O_3PS$ (315·2)

Nomenclature and development. The common name dichlofenthion is approved by BSI, ISO, ANSI, BPC and the Society of Nematologists—exception JMAF (ECP); also known as ENT 17470. The IUPAC and *C.A.* name is *O*-2,4-dichlorophenyl *O,O*-diethyl phosphorothioate, Registry No. [97-17-6]. It was first reported as a nematicide by M. A. Manzelli, *Plant Dis. Rep.*, 1955, **39**, 400, and was introduced in 1956 by the Virginia-Carolina Chemical Corp. under the code number 'V-C 13 Nemacide' and protected by USP 2 761 806.

Manufacture and properties. Produced by the reaction of *O,O*-diethyl phosphoro-chloridothioate with 2,4-dichlorophenol in butanone in the presence of anhydrous sodium carbonate, it is a colourless liquid, b.p. 120-123°/0·2 mmHg, d_4^{20} 1·313, n_D^{25} 1·5318. Its solubility at 25° is 0·245 mg/l; miscible with most organic solvents including kerosine. The technical product has d_4^{20} 1·30-1·32, n_D^{25} 1·530-1·533 and is 95-97% pure. It is stable to hydrolysis except under strongly alkaline conditions. On heating at 175° for 8 h, 42% is converted into the *O,S*-diethyl isomer.

Uses. Dichlofenthion is a non-systemic nematicide and soil insecticide recommended for use in soil at 16·8-280 kg a.i./ha (M. A. Manzelli & V. H. Young, *ibid.*, 1957, **41**, 195). Rates against soil insects are at the low, and against nematodes at the high ends of this range.

Toxicology. The acute oral LD50 for male albino rats is 270 mg/kg; the acute dermal LD50 for rabbits is 6000 mg/kg. In 90-d feeding trials cholinesterase activity was not significantly reduced nor was there evidence of abnormal pathology or disturbance in rats receiving 0·75 mg/kg daily.

Formulations. These include 'Mobilawn': e.c. (250 and 750 g a.i./l); granules (100 g/kg).

Analysis. Product and residue analysis is by GLC.

Dichlofluanid

$$Me_2N.S.N.S.CCl_2F$$

GXGFSNR&SWN1&1

$C_9H_{11}Cl_2FN_2O_2S_2$ (333·2)

Nomenclature and development. The common name dichlofluanid is approved by BSI and ISO—exception USA. The IUPAC name is *N*-dichlorofluoromethylthio-*N'*,*N'*-dimethyl-*N*-phenylsulphamide formerly *N*-dichlorofluoromethanesulphenyl-*N'*,*N'*-dimethyl-*N*-phenylsulphamide, in *C.A.* usage 1,1-dichloro-*N*-[(dimethylamino)sulfonyl]-1-fluoro-*N*-phenylmethanesulfenamide *[1085-98-9]* formerly *N*-[(dichlorofluoromethyl)thio]-*N'*,*N'*-dimethyl-*N*-phenylsulfamide. Its fungicidal properties were first reported by H. W. K. Müller, *Erwerbsobstbau,* 1964, **6,** 67. It was introduced in 1965 by Bayer AG under the code numbers 'Bayer 47531', 'Kü 13-032-c', trade marks 'Euparen' and 'Elvaron' and protected by DAS 1 193 498. It is a member of the aryl(dichlorofluoromethylthio) compounds described by E. Kühle, *Angew. Chem.,* 1964, **76,** 807.

Manufacture and properties. Produced by the reaction of dichlorofluoromethanesulphenyl chloride with *N,N*-dimethyl-*N'*-phenylsulphamide, which is formed by reacting sulphuryl chloride with, in turn, dimethylamine and aniline, it is a colourless powder, m.p. 105·0-105·6°, v.p. 1×10^{-6} mmHg at 20°. It is practically insoluble in water; solubility: 15 g/l methanol; 70 g/l xylene. It is sensitive to light (T. Clark & D. A. M. Watkins, *Pestic. Sci.,* 1978, **9,** 225) but discoloration does not affect its biological activity. It is decomposed by strongly alkaline media and by polysulphides.

Uses. Dichlofluanid is a protective fungicide with a broad range of activity and used for the control of *Venturia* spp. on apples and pears, *Botrytis* spp. and downy mildews. It has some effect on powdery mildews and against red spider mites, but does not affect *Phytoseiulus persimilis,* the predator used in biological control of *Tetranychus urticae* in glasshouses. It is applied at 75-200 g a.i./100 l HV.

Toxicology. The acute oral LD50 for male rats is 500-2500 mg/kg, for female rats *c.* 525 mg/kg. The acute dermal LD50 (4 h) for male rats is 1000 mg/kg. In 2-y feeding trials on male and female rats, the 'no effect' level was 1500 mg/kg diet.

Formulations. A w.p. (500 g a.i./kg); dust (75 g/kg).

Analysis. Product analysis is by reaction with sodium methoxide and, after oxidation with hydrogen peroxide, titration of the chloride (*CIPAC Handbook,* in press). Residues may be determined by GLC (K. Vogeler & H. Niessen, *Pflanzenschutz-Nachr. (Am. Ed.),* 1967, **20,** 534). Details of these methods may be obtained from Bayer AG.

p-Dichlorobenzene

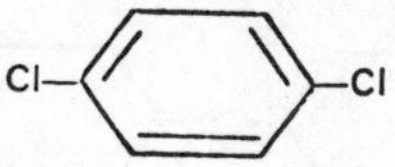

GR   DG

$C_6H_4Cl_2$ (147·0)

Nomenclature and development. The former chemical name p-dichlorobenzene (I) is accepted in lieu of a common name by BSI and ISO; also known as paracide, paradichlorobenzene and PDB. The IUPAC name is 1,4-dichlorobenzene (II), in *C.A.* usage (II) [*106-46-7*] formerly (I). It came into use as a fumigant *c.* 1913 (W. Moore, *J. Econ. Entomol.,* 1916, **9,** 71).

Manufacture and properties. Produced by the chlorination of benzene in the presence of suitable catalysts and separation of isomers by fractional distillation, it forms colourless crystals, with a strong odour, m.p. 53°, b.p. 173·4°, v.p. 1·0 mmHg at 25°, d_4^{20} 1·4581. Its solubility at 25° is 80 mg/l water; readily soluble in organic solvents. It is stable and non-corrosive.

Uses. p-Dichlorobenzene is an insecticidal fumigant which, being non-corrosive and non-staining, is used to prevent infestation by clothes moths. It is used for the control of mites in fungal cultures (E. W. Elliott, *Proc. Iowa Acad. Sci.,* 1948, **55,** 99) and for fumigating narcissus bulbs to control mites.

Toxicology. The acute oral LD50 is: for rats 500-5000 mg/kg; for mice 2950 mg/kg. It has been cleared of the suspicion that it may cause cataract in man (R. L. Hollingsworth *et al., AMA Arch. Ind. Health,* 1956, **14,** 138).

Formulation. It is generally used as pure crystals.

Analysis. Residues in soil may be determined by steam distillation into kerosine and noting the change in refractive index (R. D. Chisholm & L. Koblitsky, *J. Assoc. Off. Agric. Chem.,* 1943, **26,** 273).

1,1-Dichloro-2,2-bis(4-ethylphenyl)ethane

GYGYR D2&R D2 $C_{18}H_{20}Cl_2$ (307·3)

Nomenclature and development. The IUPAC name is 1,1-dichloro-2,2-bis(4-ethyl-phenyl)ethane formerly 1,1-dichloro-2,2-di-(4-ethylphenyl)ethane, in *C.A.* usage 1,1'-(2,2-dichloroethylidene)bis[4-ethylbenzene] *[72-56-0]* formerly 1,1-dichloro-2,2-bis[p-ethylphenyl]ethane. It was introduced in 1950 as an insecticide by Rohm & Haas Co. under the code number 'Q-137', the trade mark 'Perthane' and protected by USP 2 464 000; 2 881 111; 2 883 428.

Manufacture and properties. Produced by the condensation of dichloroacetaldehyde with ethylbenzene, it is a crystalline solid, m.p. 60-61°. The technical product is a wax, m.p. ≥ 40°; some decomposition occurs ≥ 52°. It is practically insoluble in water; soluble in most aromatic solvents and in dichloromethane.

Uses. It is a non-systemic insecticide with a useful specifity. It is recommended for the control of pear psylla, against leaf-hoppers and various larvae on vegetable crops at 1-16 kg a.i./ha. It is used domestically for the control of clothes moth and carpet beetles. It is non-phytotoxic and of moderate persistence in soil.

Toxicology. The acute oral LD50 is: for rats 8170 mg/kg; for mice 6600 mg/kg. In 2-y feeding trials rats receiving 500 mg/kg diet suffered no mortality or ill-effect on their blood.

Formulations. These include: 'Perthane E.C.', e.c. (450 g a.i./l); 'Perthane', a solution (750 g/l).

Analysis. Product analysis is by total chlorine using the Parr bomb (F. A. Gunther & R. C. Blinn, *Analysis of Insecticides and Acaricides,* 1955, p.370). Residues may be determined by extraction with chloroform, dehydrochlorination, reaction with sulphuric acid and colorimetry at 493 nm (J. R. W. Miles, *J. Agric. Food Chem.,* 1957, **5**, 249; C. F. Gordon *et al., ibid.,* 1962, **10**, 380). See also: C. F. Gordon, *Anal. Methods Pestic., Plant Growth Regul. Food Addit.,* 1964, **2**, 321; *Anal. Methods Pestic. Plant Growth Regul.,* 1972, **6**, 447.

2,3-Dichloro-*N*-(4-fluorophenyl)maleimide

T5VNVJ BR DF& DG EG $C_{10}H_4Cl_2FNO_2$ (260·1)

Nomenclature and development. The common name fluoromide is approved by JMAF. The IUPAC name is 2,3-dichloro-*N*-(4-fluorophenyl)maleimide, in *C.A.* usage 3,4-dichloro-1-(4-fluorophenyl)-1*H*-pyrrole-2,5-dione *[41205-21-4]*. It was introduced as a fungicide in 1970 by Mitsubishi Chemical Industries Ltd and Kumiai Chemical Industry Co. Ltd, under code number 'MK-23', the trade name 'Sparticide' and protected by JapaneseP 712681. Its activity has been reported (*Jpn. Pestic. Inf.*, 1978, No. 34, p.26).

Manufacture and properties. Produced by the reaction of 4-fluoroaniline on 2,3-dichloro-maleimide, it forms a pale yellow crystalline powder, m.p. 240·5-241·8°. Its solubility at 20° is: 5·9 mg/l water; 840 mg/kg methanol. It is stable in neutral or slightly acid media but alkali hydrolyses it to fungicidally inactive products.

Uses. It is used as a horticultural fungicide on many crops including apple and citrus at 2-5 kg a.i./ha.

Toxicology. The acute oral LD50 for rats or mice is > 15 000 mg a.i./kg. The acute dermal LD50 for mice is > 5000 mg/kg. In 2-y feeding trials the 'no effect' level was between 600 and 2000 mg/kg diet. LC50 (48 h) for carp is 5·6 mg/l.

Formulation. A w.p. (500 g a.i./kg).

Analysis. Product and residue analysis is by GLC.

Dichlorophen

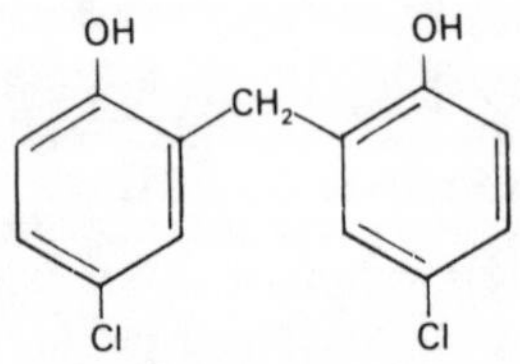

QR DG BIR BQ EG

$C_{13}H_{10}Cl_2O_2$ (269·1)

Nomenclature and development. The common name dichlorophen, is approved by BSI, ISO and BPC. The IUPAC name is 4,4'-dichloro-2,2'-methylenediphenol formerly bis(5-chloro-2-hydroxyphenyl)methane or 5,5'-dichloro-2,2'-dihydroxydiphenylmethane, in *C.A.* usage 2,2'-methylenebis(4-chlorophenol) *[97-23-4].* It was the best mildew preventive for use on cotton of a series of bisphenols listed by P. B. Marsh & M. L. Butler, *Ind. Eng. Chem.,* 1946, **38,** 701. It was introduced in 1946 by Sindar Corp. under the code number 'G4'.

Manufacture and properties. Produced by the addition of aqueous formaldehyde to 4-chlorophenol in methanolic sulphuric acid at —10 to 0° (USP 2334408) it forms colourless, odourless crystals, m.p. 177-178°, v.p. 10^{-10} mmHg at 25°, 10^{-4} mmHg at 100°. Its solubility at 25° is: 30 mg/l water; 800 g/l acetone; 530 g/l ethanol. The technical product is a light tan powder with a slight phenolic odour and m.p. $\geqslant 164°$. It is acidic (pK_1 7·6; pK_2 11·6) and readily soluble in aqueous alkali, forming salts.

Uses. Dichlorophen is a fungicide and bactericide recommended for the protection of textiles and materials including horticultural benches and equipment against moulds and algae. It is also used in the treatment of tapeworm infestation in man and animals and is the basis of a preparation against athlete's foot.

Toxicology. The acute oral LD50 is: for guinea-pigs 1250 mg/kg; for dogs 2000 mg/kg. In 90-d feeding trials rats receiving 2000 mg/kg diet showed no evidence of toxicity. Fifty patients treated with dichlorophen (6 g) on 2 successive days showed no ill-effect (H. Most, *J. Am. Med. Assoc.,* 1963, **874,** 185).

Formulation. Dichlorophen is used as such; as 'Panacide', an aqueous solution of the monosodium salt (400 g/l); or the anthelmintic, 'Antiphen'.

Analysis. Product analysis is by u.v. spectroscopy (J. R. Clements & S. H. Newburger, *J. Assoc. Off. Agric. Chem., 1954,* **37,** 190).

O-2,4-Dichlorophenyl *O*-ethyl phenylphosphonothioate

GR CG DOPS&R&O2

$C_{14}H_{13}Cl_2O_2PS$ (347·2)

Nomenclature and development. The common name EPBP is approved by JMAF. The IUPAC and *C.A.* name is *O*-2,4-dichlorophenyl *O*-ethyl phenylphosphonothioate, Registry No. *[3792-59-4]*. It was introduced in 1960 by Nissan Chemical Industries Ltd under the code number 'S-7', trade mark 'S-Seven' and protected by USP 3 318 764.

Manufacture and properties. Produced by the condensation of *O*-ethyl phenylphos-phonochloridothioate with sodium 2,4-dichlorophenoxide, it is a light brown oil, b.p. 206°/5 mmHg, v.p. 3·83 mmHg at 200°, n_D^{20} 1·5956, d_4^{24} 1·312. It is practically insoluble in water; soluble in most organic solvents. The technical grade is *c.* 90% pure. It is non-corrosive but incompatible with propanil.

Uses. It is a non-systemic contact and stomach insecticide used mainly for the control of soil-dwelling pests such as *Hylemya* spp., *Phyllotreta striolata, Agrotis fucosa, Rhizoglyphus echinopus,* etc., using the dust at 1·8–3·0 kg a.i./ha.

Toxicology. The acute oral LD50 and the acute subcutaneous LD50 for mice are 275 mg/kg and 784 mg/kg, respectively.

Formulation. Dust (30 g a.i./kg).

Analysis. Product analysis is by extraction with acetone and GLC using dibutyl sebacate as internal standard. Residues may be extracted by methanol and determined by GLC after clean-up by liquid-liquid partition.

2,4-Dichlorophenyl 3-methoxy-4-nitrophenyl ether

WNR BO1 DOR BG DG $C_{13}H_9Cl_2NO_4$ (314·1)

Nomenclature and development. The common name chlormethoxynil is approved by JMAF. The IUPAC name is 2,4-dichlorophenyl 3-methoxy-4-nitrophenyl ether, in *C.A.* usage 4-(2,4-dichlorophenoxy)-2-methoxy-1-nitrobenzene *[32861-85-1]*. It was introduced as a herbicide by Nihon Nohyaku Co. Ltd under the code number 'X-52' and protected by JapaneseP 600 441.

Properties. It forms yellow crystals, m.p. 113-114°. Its solubility at 15° is 0·3 mg/l water; soluble in organic solvents such as acetone, ethanol and benzene.

Uses. It is a herbicide used mainly for paddy field and upland rice at 1·5-2·5 kg a.i./ha.

Toxicology. The acute oral LD50 for mice is 33000 mg/kg. The LC50 (48 h) for carp is 237 mg/l.

Formulations. A w.p. (700 g a.i./kg); granules (70 g/kg).

Analysis. Product analysis is by GLC (F. Yamane & K. Tsuchiya, *Anal. Methods Pestic. Plant Growth Regul.*, 1978, **10**, 267).

3,6-Dichloropicolinic acid

T6NJ BVQ CG FG

$C_6H_3Cl_2NO_2$ (192·0)

Nomenclature and development. The name 3,6-dichloropicolinic acid (**I**) has been accepted in lieu of a common name by BSI, ANSI and ISO. The IUPAC name is (**I**) or 3,6-dichloropyridine-2-carboxylic acid, in *C.A.* usage 3,6-dichloro-2-pyridinecarboxylic acid *[1702-17-6]*. Its herbicidal properties were first described by T. Haagsma, *Down Earth,* 1975, **30** (4), 1 and it was introduced in 1975 by The Dow Chemical Co. under the code number 'DOWCO 290', the trade mark 'Lontrel' and protected by USP 3 317 549.

Manufacture and properties. Produced by the reduction of 3,4,5,6-tetrachloropyridine-2-carboxylic acid it forms colourless crystals, m.p. 151-152°, v.p. 1·2 x 10^{-5} mmHg at 25°. Its solubility at 20° is 1 g/l water; readily soluble in acetone, xylene, ethanol. It is acidic (pK_a 2·33) and forms salts. It is stable under normal storage conditions, but is corrosive to many metals.

Uses. 3,6-Dichloropicolinic acid is a foliar-applied growth-regulator herbicide, which is readily absorbed by leaves and roots and translocated throughout the plant. Effective against various problem broad-leaved weeds, such as many species of Leguminosae and Polygonaceae; Compositae are particularly susceptible. It is selective to graminaceous and brassica crops, sugar beet and flax. For weed control in cereals (not undersown) it is usually formulated with one or more phenoxyalkanoic acid or with herbicides of various other classes. A dosage of 50-90 g a.e./ha generally controls annual Compositae and some *Polygonum* spp. In sugar beet and oilseed rape it is applied either alone, or in combination with herbicides such as phenmedipham (sugar beet) and benazolin and propyzamide (oilseed rape).

It is not metabolised in plants. It undergoes microbial degradation in soils; incorporated at levels ranging from 0·25-1·0 mg/kg, 50% being lost by this route in *c.* 49 d under conditions favourable to microbial activity.

Toxicology. The acute oral LD50 for rats is > 4300-5000 mg/kg; the acute dermal LD50 for rabbits is > 2000 mg/kg. It is a severe eye irritant. In 90-d feeding trials there was no effect in rats receiving 150 mg/kg daily.

Formulations. These vary with crop and country. Examples include: for cereals 'Sel-oxone' (15·0 g 3,6-dichloropicolinic acid + 510 g mecoprop/l) as ethanolamine salts, in the UK; 'Seppic MMD' (17·5 g 3,6-dichloropicolinic acid + 450 g mecoprop + 100 g MCPA/l) as ethanolamine salts, in France; for sugar beet 'Lontrel S.F. 100' (100 g 3,6-dichloropicolinic acid/l) as ethanolamine salt, in France; on oilseed rape, 'Matrigon' (100 g 3,6-dichloropicolinic acid/l), as ethanolamine salt, in Sweden; 'Benzalox', w.p. (50 g 3,6-dichloropicolinic acid + 300 g benazolin/kg), in UK.

Analysis. Product analysis and residue determination is by GLC: details of methods are available from The Dow Chemical Co. Residue analysis is by GLC of suitable derivatives, *e.g.* the butyl ester (T. H. Byast *et al., Agric. Res. Counc. (G.B.) Weed Res. Organ., Tech. Rep.* No. 15 (2nd Ed.), p.31). See also A. J. Pik & G. W. Hodgson, *J. Assoc. Off. Anal. Chem.,* 1976, **59**, 264.

1,2-Dichloropropane

ClCH$_2$.CHCl.CH$_3$

GY1&1G   C$_3$H$_6$Cl$_2$ (113·0)

Nomenclature and development. The IUPAC and *C.A.* name 1,2-dichloropropane, *C.A.* Registry No. *[78-87-5]*, is accepted in lieu of a common name by BSI and ISO; also known as propylene dichloride and as ENT 15406. It was first examined as an insecticidal fumigant in 1925 by I. E. Neifert *et al., U.S. Dep. Agric. Bull.,* No. 1313, 1925.

Manufacture and properties. Produced by the chlorination of propene, it is a colourless liquid, b.p. 95·4°, f.p. —70°, v.p. 210 mmHg at 19·6°, d_{20}^{20} 1·1595, n_D^{25} 1·437. Its solubility at 20° is 2·7 g/kg water; soluble in ethanol, diethyl ether. It is flammable, flash point (Cleveland open cup) 21°.

Uses. 1,2-Dichloropropane is an insecticidal fumigant. It is a component of 'D-D' mixture and 'Vidden D'. It is far less toxic to nematodes (C. R. Youngson & C. I. A. Goring, *Plant Dis. Rep.,* 1970, **54,** 196; M. V. McKenry & I. J. Thomason, *Hilgardia,* 1974, **42,** 422) than (*Z*)- and (*E*)-1,3-dichloropropenes, but moves more freely in soil, from which the greatest loss is by volatilisation (T. R. Roberts & G. Stoydin, *Pestic. Sci.,* 1976, **7,** 325).

Toxicology. It is strongly narcotic but low concentrations cause an irritation of the respiratory tract. Exposure of guinea-pigs, rabbits and rats to the vapour has been studied (L. A. Heppel *et al., J. Ind. Hyg. Toxicol.,* 1946, **28,** 1).

Formulation. It is a component of 'D-D' and 'Vidden D' (see p.177).

Analysis. Product and residue analysis is by GLC.

Dichloropropane-dichloropropene mixture

Nomenclature and development. This reaction product, also known as ENT 8420, is often referred to by its trade mark 'D-D' (Shell Development Co.), *C.A.* Registry No. *[8003-19-8],* for the mixture (see under components), and was introduced by that company in 1942, its first use as a soil fumigant being described by W. Carter, *Science,* 1943, **97,** 383. It is also produced by The Dow Chemical Co. as 'Vidden D'.

Manufacture and properties. Produced by the high temperature chlorination of propene (H. P. A. Gross & G. Hearne, *Ind. Eng. Chem.,* 1939, **31,** 1530) it is a mixture of chlorinated hydrocarbons containing $\geqslant 50\%$ *m/m* (*E*)- and (*Z*)- 1,3-dichloropropenes (see p.178), *C.A.* Registry No. *[542-75-6],* the other main constituent being 1,2-dichloropropane (p.176), *C.A.* Registry No. *[78-87-5].* It has an organic chlorine content of $\geqslant 55.0\%$ *m/m.*

It is clear amber liquid, with a pungent odour, which flash distils over the range 59-115°, v.p. 35 mmHg at 20°, d_{20}^{20} 1·17-1·22. Its solubility at room temperature is *c.* 2 g/kg water; soluble in hydrocarbon and halogenated solvents, esters, ketones. The mixture is stable up to 500° but reacts with dilute organic bases, concentrated acids, halogens and some metal salts. It is corrosive to some metals—*e.g.,* aluminium.

Uses. Dichloropropane-dichloropropene is a pre-plant nematicide effective against soil nematodes including root knot, meadow, sting and dagger, spiral and sugar beet nematodes. The mixture is usually applied by injection into the soil or through tractor-drawn hollow tines, to a depth of 15-20 cm at 150-400 kg/ha (occasionally to a maximum of 1000 l/ha) depending on soil type and the following crop. The soil surface is sealed by rolling. Because the components are highly phytotoxic, a 7-d pre-planting interval should be allowed for every 75 l applied/ha. In wet or cold conditions (soil temperature < 15°) rather longer intervals may be required. It does not accumulate in the soil (M. V. McKenry & I. J. Thomason, *Hilgardia,* 1974, **42,** 393; T. R. Roberts & G. Stoydin, *Pestic. Sci.,* 1976, **7,** 325); and has no permanent harmful effect on soil micro-organisms. Tainting of potato tubers has been reported and attributed to impurities such as 2,2-dichloropropane and 1,2,3-trichloropropane (C. J. Shepherd, *Nature (London),* 1952, **170,** 1073).

Toxicology. The acute oral LD50 is: for rats 140 mg/kg; for mice 300 mg/kg. The dermal LD50 for rabbits is 2100 mg/kg. It is irritant to human skin.

Formulation. The mixture is used without formulation.

Analysis. Product and residue analysis is by GLC. Details can be obtained from Shell International Chemical Co. Ltd. See also T. R. Roberts & G. Stoydin, *loc cit.;* C. E. Castro, *Anal. Methods Pestic., Plant Growth Regul. Food Addit,* 1964, **3,** 151; *Anal. Methods Pestic. Plant Growth Regul.,* 1972, **6,** 710.

1,3-Dichloropropene

$$\underset{H}{\overset{Cl}{\diagdown}}C=C\underset{CH_2Cl}{\overset{H'}{\diagup}} \qquad \underset{H}{\overset{Cl}{\diagdown}}C=C\underset{H}{\overset{CH_2Cl}{\diagup}}$$

(E)   (Z)

G2U1G $C_3H_4Cl_2$ (111·0)

Nomenclature and development. The IUPAC name 1,3-dichloropropene (I) is accepted by BSI and ISO in lieu of a common name, in *C.A.* usage 1,3-dichloro-1-propene *[542-75-6]* formerly (I). Its properties as a soil fumigant were first described (*Down Earth,* 1956, **12** (2), 7) by The Dow Chemical Co. who introduced it in 1956 under the trade mark 'Telone'. It is also a component of 'D-D' and 'Vidden D'.

Properties. The technical product is a colourless to amber-coloured liquid, with a sweetish odour, f.p. < —50°, b.p. 104°, flash point 21°, d_4^{25} 1·217. Its solubility at 20° is 1 g/kg water; miscible with acetone, benzene, carbon tetrachloride, heptane, methanol.

It is a mixture of (E)-isomer, *C.A.* Registry No. *[10061-02-6],* b.p. 104·2°, d_4^{20} 1·224, n_D^{20} 1·4682; and (Z)-isomer *[10061-01-5],* b.p. 112°, d_4^{20} 1·217, n_D^{20} 1·4730.

Uses. Dichloropropene is a soil fumigant and nematicide, controlling root-knot nematodes at 187 l/ha at soil temperature of 5-27°, a double rate is needed on highly organic soils. It is non-persistent and is hydrolysed in soil to the corresponding 3-chloroallyl alcohols (C. E. Castro & N. O. Belser, *J. Agric. Food Chem.,* 1966, **14,** 69; M. V. McKenry & I. J. Thomason, *Hilgardia,* 1974, **42,** 393; T. R. Roberts & G. Stoydin, *Pestic. Sci.,* 1976, **7,** 325). The (Z)-isomer is more toxic than the (E)-isomer to nematodes (M. V. McKenry & I. J. Thomason, *loc. cit.;* C. R. Youngson & C. A. I. Goring, *Plant Dis. Rep.,* 1970, **54,** 196).

Toxicology. The acute oral LD50 for rats is 250-500 mg/kg; it is a skin vesicant very irritating and damaging to the eyes.

Formulations. These include: 'Telone' [42% (Z)- + 36·5% (E)-1,3-dichloropropene with 20·5% 1,2-dichloropropane and related compounds]; 'Telone II' [92% (Z)- + (E)-1,3-dichloropropene]. 1-Chloro-2,3-epoxyethane (epichlorohydrin) (1%) is added as a stabiliser in countries where this is permitted.

Analysis. Product analysis is by bromination, determining the excess of bromine; and determining the total chlorine and, from these figures, calculating the composition (F. A. Gunther & R. C. Blinn, *Analysis of Insecticides and Acaricides,* 1955, p.404). Residues may be determined by GLC (M. V. McKenry & I. J. Thomason, *loc. cit.;* T. R. Roberts & G. Stoydin, *loc. cit.*).

Dichlorprop

Me
O.CHCO₂H

QVY1&OR BG DG $C_9H_8Cl_2O_3$ (235·1)

Nomenclature and development. The common name dichlorprop is approved by BSI, ISO and WSSA—exception USSR (2,4-DP). The IUPAC name is (±)-2-(2,4-dichloro-phenoxy)propionic acid (I), in *C.A.* usage (±)-2-(2,4-dichlorophenoxy)propanoic acid *[120-36-5]* formerly (I). Although the growth-regulating properties of this compound were described in 1944 (P. W. Zimmerman & A. E. Hitchcock, *Proc. Am. Soc. Hortic. Sci.,* 1944, **45,** 353), it was not introduced commercially until 1961, by The Boots Co. Ltd under the code number 'RD 406' and trade mark 'Cornox RK'.

Manufacture and properties. It is produced either by the condensation of 2-chloro-propionic acid with 2,4-dichlorophenol or by the chlorination of 2-phenoxypropionic acid. Having an asymmetric carbon atom, it exists as 2 optically active forms of which only the (+)-form is herbicidally active. The commercial product contains equal amounts of the 2 forms. Dichlorprop is a colourless crystalline solid, m.p. 117·5-118·1°, of negligible v.p. at room temperature. Its solubility at 20° is 350 mg/l water; readily soluble in most organic solvents. The technical product, m.p. 114°, may have a slight phenolic odour. The acid is stable to heat and resistant to reduction, hydrolysis and atmospheric oxidation. Esters formed with the lower alcohols are volatile liquids. The solubility at 20° of its potassium, sodium and diethanolamine salts is 900, 660 and 750 g a.e./l water respectively. Salt formulations are compatible with similar formulations of other growth-active herbicides. The acid is corrosive to metals in the presence of water, but concentrated solutions (480 g a.e./l) do not corrode iron or tin plate if the pH is ≥8·6 and the temperature < 70°.

Uses. Dichlorprop is a post-em. translocated herbicide, particularly effective for the control of *Polygonum persicaria, P. lapathifolium* and *P. convolvulus.* It also controls *Galium aparine* and *Stellaria media* but is not consistently effective against *P. aviculare.* It is used at 2·7 kg a.e./ha and may be used alone or in combination with other herbicides such as 2,4-D, MCPA, mecoprop, ioxynil, bromoxynil, bentazone or 3,6-dichloropicolinic acid. It is also used, at much lower rates, to prevent premature drop of apples.

Toxicology. The acute oral LD50 is: for rats 800 mg/kg; for mice 400 mg/kg. The dermal LD50 for mice is 1400 mg/kg and it is not an irritant to eyes at 10 g/l or skin at 24 g/l. In 98-d feeding trials no toxic effect was observed in rats receiving 12·4 mg/kg daily, though slight liver hypertrophy occurred at 50 mg/kg daily.

Formulations. These include: aqueous solutions of alkali metal, particularly potassium, and amine salts; e.c. of esters; e.c. based on dried non-hygroscopic salts of dichlorprop.

Analysis. Product analysis is by GLC of the methyl ester or by titratable acid content (*CIPAC Handbook,* 1979, **1A,** in press). Residue analysis is by GLC (T. H. Byast *et al., Agric. Res. Counc. (G.B.) Weed Res. Organ., Tech. Rep.,* No. 15 (2nd Ed.), p.61).

179

Dichlorvos

$$\underset{\substack{\parallel \\ (MeO)_2P.O.CH=CCl_2}}{O}$$

GYGU1OPO&O1&O1 $C_4H_7Cl_2O_4P$ (221·0)

Nomenclature and development. The common name dichlorvos is approved by BSI, ISO, BPC and ESA—exception JMAF (DDVP) and USSR (DDVF); also known as ENT 20738 and sometimes by the trivial term DDVP. The IUPAC name is 2,2-dichlorovinyl dimethyl phosphate (I), in *C.A.* usage 2,2-dichloroethenyl dimethyl phosphate *[62-73-7]* formerly (I). The insecticidal properties of this organophosphate were first described in 1951 by Ciba AG in BP 775085 but it was given an incorrect structure in the patent; in 1955 it was reported as an insecticidal impurity of trichlorphon (p.531) (A. M. Martson *et al., J. Agric. Food Chem.*, 1955, **3**, 319). It was later introduced by Ciba AG (now Ciba-Geigy AG) under the trade marks 'Nogos', 'Nuvan' and protected by BP 775085; by the Shell Chemical Co. under the trade mark 'Vapona' and protected by USP 2956073; by Farbenfabriken Bayer AG under the code number 'Bayer 19149' and trade mark 'Dedevap'.

Manufacture and properties. Produced by condensing trimethyl phosphite with trichloroacetaldehyde, or by dehydrochlorination of trichlorphon (p.531) (W. F. Barthel *et al., J. Am. Chem. Soc.*, 1955, **77**, 2424; W. Lorenz *et al., ibid.*, p.2554), it is a colourless to amber liquid, with an aromatic odour, b.p. 35°/0·05 mmHg, v.p. 1·2 x 10^{-2} mmHg at 20°, d_4^{25} 1·415, n_D^{25} 1·4523. Its solubility at 20° is: *c.* 10 g/l water; 2-3 g/kg kerosine; miscible with most organic solvents and aerosol propellants. It is stable to heat but is hydrolysed by water, a saturated aqueous solution at room temperature is converted to dimethyl hydrogen phosphate and dichloroacetaldehyde at a rate of *c.* 3%/d, more rapidly in alkali. It is corrosive to iron and mild steel but non-corrosive to stainless steel, aluminium, nickel, 'Hastelloy 13', 'Teflon'.

Uses. Dichlorvos is a contact and stomach insecticide with fumigant and penetrant action. It is used as a household and public health fumigant, especially against diptera and mosquitoes; for the protection of stored products at 0·5-1·0 g a.i./100 m³; for crop protection against sucking and chewing insects at 300-1000 g/ha. It is non-phytotoxic (except to some chrysanthemum cultivars) and non-persistent. It is used as an anthelmintic for incorporation in animal feeds.

Toxicology. The acute oral LD50 for rats is 56-108 mg/kg; the acute dermal LD50 for rats is 75-210 mg/kg. In 90-d feeding trials rats receiving 1000 mg/kg diet showed no intoxication. The 4-h inhalation LC50 is: for mice 13·2 mg/m³; for rats 14·8 mg/m³. The LC50 (24 h) for bluegill is 1 mg/l. It is highly toxic to honey bees and toxic to birds.

Formulations. The water content and acidity of the technical product must be rigidly controlled to delay hydrolysis. Formulations include: 'Nogos 50EC', 'Nogos 100EC', 'Nuvan 50EC', 'Nuvan 100EC', e.c. (500 and 1000 g a.i./l); 'Nuvan 100SC', an oil soluble concentrate (1000 g/l); 'Vapona Pest Strip' (200 g/kg) solid solution in plastic; aerosol concentrates (4-10 g/l); granules (5 g/kg). The safety of resin strips is discussed by J. W. Gillet *et al., Residue Rev.*, 1972, **44**, 115, 161.

Analysis. Product analysis is by GLC or by i.r. spectroscopy (details of methods from Shell International Chemical Co. Ltd; *CIPAC Handbook*, 1979, **1A**, in press). Residues may be determined by GLC (details from Shell International Chemical Co. Ltd). See also P. E. Porter, *Anal. Methods Pestic., Plant Growth Regul. Food Addit.*, 1964, **2**, 561. For the sampling of atmospheres see, *Anal. Methods Pestic. Plant Growth Regul.*, 1972, **6**, 529.

Diclofop-methyl

GR CG DOR DOY1&VO1 $C_{16}H_{14}Cl_2O_4$ (341·2)

Nomenclature and development. The common name diclofop is approved by ANSI and WSSA for (±)-2-[4-(2,4-dichlorophenoxy)phenoxy]propionic acid (IUPAC), (±)-2-[4-(2,4-dichlorophenoxy)phenoxy]propanoic acid *(C.A.)*, Registry No. *[40843-25-2]*. The corresponding methyl ester, diclofop-methyl, Registry No. *[51338-27-3]*—a name approved directly by BSI and proposed by ISO—was introduced in 1975 by Hoescht AG under the code number 'Hoe 23 408'. Its herbicidal properties were first described by P. Langelüddeke *et al., Mitt. Biol. Bundesanst. Land.-Forstwirtsch. Berlin-Dahlem.*, 1975, **165,** 169.

Properties. Diclofop-methyl is a colourless crystalline solid, m.p. 39-41°. Its solubility at 22° is 0·3 mg/l water; highly soluble in acetone, xylene.

Uses. Diclofop-methyl is a post-em. herbicide effective for the control of annual grasses such as *Avena fatua, A. ludoviciana, Echinochloa crus-galli, Eleusine indica, Setaria faberii, S. lutescens, S. viridis, Panicum dichotomiflorum, Lolium multiflorum, Leptochloa* spp. and volunteer maize. It is effective after uptake by foliage and roots and is selective to wheat, barley, sugar beet, peas, field and dwarf beans, soyabeans, groundnuts, clover, lucerne, rape-seed, potatoes, tomatoes, cucumbers, lettuces, spinach, carrots, celery. Dosage: 0·7-1·1 kg a.i./ha. Studies on degradation in plants and soil (A. E. Smith, *J. Agric. Food Chem.,* 1977, **25,** 893; R. Martens, *Pestic. Sci.,* 1978, **9,** 127) show a rapid formation of diclofop, which is decomposed further. Diclofop-methyl and diclofop are both herbicidal but appear to act at different sites in *Avena* spp. and cereals (M. A. Shimabukuro *et al., Pestic. Biochem. Physiol.,* 1978, **8,** 199).

Toxicology. The acute oral LD50 for rats is 563-580 mg/kg. In 90-d feeding tests the 'no effect' level for male dogs was 80 mg/kg diet and in 2-y tests on rats 6·3 mg/kg diet.

Formulation. An e.c. (285 and 380 g a.i./l).

Analysis. Details of GLC methods are available from Hoescht AG.

Dicloran

ZR   BG   FG   DNW $C_6H_4Cl_2N_2O_2$ (207·0)

Nomenclature and development. The common name dicloran is approved by BSI and New Zealand; by JMAF (CNA); there is no ISO name. The IUPAC name is 2,6-dichloro-4-nitroaniline (I), in *C.A.* usage 2,6-dichloro-4-nitrobenzenamine *[99-30-9]* formerly (I). Its fungicidal properties were first described by N. G. Clark *et al., Chem. Ind. (London),* 1960, p.572; *J. Sci. Food Agric.,* 1961, **12,** 751. It was first introduced in 1959 by The Boots Co. Ltd under the code number 'RD 6584', trade mark 'Allisan' and protected by BP 845916. The trade mark 'Botran' is registered by the Upjohn Co.

Manufacture and properties. Produced by the chlorination of 4-nitroaniline, it is a yellow crystalline solid, m.p. 195°, v.p. 1·2 x 10^{-6} mmHg at 20°. It is practically insoluble in water; moderately soluble in polar organic solvents, *e.g.,* 34 g/kg acetone. The technical product is brownish-yellow and $\geqslant 90\%$ pure. It is stable to hydrolysis and to oxidation and is compatible with other pesticides.

Uses. Dicloran is essentially a protectant fungicide with little effect on spore germination, but which causes hyphal distortion and is effective against a wide range of fungal pathogens, particularly certain species of *Botrytis, Sclerotinia* and *Rhizopus.* It is used in a number of fruit, vegetable and ornamental crops and is basically non-phytotoxic but young lettuce plants, between stages of transplanting and the development of 2 true leaves, display some susceptibility to damage.

Toxicology. The acute oral LD50 for various species ranges from 1500 to 4000 mg/kg. In 2-y feeding trials rats receiving 1000 mg/kg diet suffered no ill-effect. It is rapidly metabolised in rats and is excreted in the urine as the sulphate conjugate of 3,5-dichloro-4-aminophenol. In plants it is metabolised to polar compounds.

Formulations. These include: w.p. (500 and 750 g a.i./kg); dusts (40-80 g/kg); smoke generator. Also 'Turbair Botryticide' ULV (dicloran + thiram).

Analysis. Product analysis is by i.r. spectroscopy (P. G. Marshall, *Analyst. (London),* 1960, **85,** 681). Residues may be determined by reduction to 2,6-dichloro-*p*-phenylenediamine which is oxidised in the presence of aniline to a dark blue dye and this measured colorimetrically at 655 nm (J. Roburn, *J. Sci. Food Agric.,* 1961, **12,** 766), by colorimetry at 464 nm of the yellow colour produced with concentrated alkali and acetone (W. W. Kilgore *et al., J. Agric. Food Chem.,* 1962, **10,** 399) or by GLC (K. W. Cheng & W. W. Kilgore, 'Botran' Symposium, Brook Lodge, November 1965). See also: W. W. Kilgore, *Anal. Methods Pestic., Plant Growth Regul. Food Addit.,* 1964, **3,** 61; *Anal. Methods Pestic. Plant Growth Regul.* 1972, **6,** 553.

Dicofol

GXGGXQR DG&R DG $C_{14}H_9Cl_5O$ (370·5)

Nomenclature and development. The common name dicofol is approved by BSI, ISO and ESA—exception JMAF (kelthane), Austria and Germany; also known as ENT 23 648. The IUPAC name is 2,2,2-trichloro-1,1-bis(4-chlorophenyl)ethanol formerly 2,2,2-trichloro-1,1-di-(4-chlorophenyl)ethanol, in *C.A.* usage 4-chloro-α-(4-chlorophenyl)-α-(trichloromethyl)benzenemethanol *[115-32-2]* formerly 4,4'-dichloro-α-(trichloromethyl)benzhydrol. Its acaricidal properties were first described by J. S. Barker & F. B. Maugham, *J. Econ. Entomol.,* 1956, **49,** 458, and it was introduced in 1955 by Rohm & Haas Co. under the code number 'FW-293', trade mark 'Kelthane' and protected by BP 778 209; USP 2 812 280; 2 812 362; 3 102 070; 3 194 730.

Manufacture and properties. Produced either by the chlorination of chlorfenethol (p.97) (S. Reuter & K. R. S. Ascher, *Experientia,* 1956, **12,** 316) or by the reaction of 1,1,1,2-tetrachloro-2,2-bis(4-chlorophenyl)ethane with silver acetate and hydrolysis of the resulting ester (E. D. Bergmann & A. Kaluszyner, *J. Org. Chem.,* 1958, **21,** 1306), it is a colourless solid, m.p. 78·5-79·5°. The technical product is a brown viscous oil, *c.* 80% pure, d^{25} 1·45. It is practically insoluble in water; soluble in most aliphatic and aromatic solvents. It is hydrolysed by alkali to 4,4'-dichlorobenzophenone and chloroform, but is compatible with all but highly alkaline pesticides. The w.p. formulations are sensitive to solvents and surfactants and these may affect acaricidal activity and phytotoxicity.

Uses. Dicofol is a non-systemic acaricide with little insecticidal activity, recommended for the control of mites on a wide range of crops at 0·56-4·5 kg a.i./ha. Although residues in soil decrease rapidly, traces may remain for ≥ 1 y.

Toxicology. The acute oral LD50 for rats is 668-842 mg/kg; the acute dermal LD50 for rabbits is 1870 mg/kg. In 1-y feeding trials dogs receiving 300 mg/kg showed no evidence of toxicity (R. Blackwell Smith *et al., Toxicol. Appl. Pharmacol.,* 1959, **1,** 119).

Formulations. These include: w.p. 'Kelthane AP' (185 g a.i./kg), 'Kelthane 35' (350 g/kg); e.c. 'Kelthane EC' (185 g/l), 'Kelthane MF' (420 g/l); dicofol 20 and dicofol emulsion (200 g/l); 'Kelthane Dust Base' (300 g/kg); 'Childion' (400 g dicofol + 125 g tetradifon/l).

Analysis. Product analysis is by GLC (AOAC Methods) or by determining the hydrolysable chloride (correcting for ionic chloride). Residues may be determined by GLC (D. B. Seba & C. E. Lane, *Bull. Environ. Contam. Toxicol.,* 1969, **4,** 297; AOAC Methods) or by hydrolysis estimating the chloroform produced (I. Rosenthal *et al., J. Agric. Food Chem.,* 1957, **5,** 514) or the 4,4'-dichlorobenzophenone at 264 nm (F. A. Gunther & R. C. Blinn, *ibid.,* p.517). See also: C. F. Gordon & R. J. Schuckert, *Anal. Methods Pestic., Plant Growth Regul. Food Addit.,* 1964, **2,** 263; *Anal. Methods Pestic. Plant Growth Regul.,* 1972, **6,** 415.

Dicrotophos

$$(MeO)_2P-O \quad \quad H \atop \quad C=C \quad CNMe_2$$

(E)   *(Z)*

1OPO&O1&OY1&U1VN1&1 $C_8H_{16}NO_5P$ (237·2)

Nomenclature and development. The common name dicrotophos is approved by BSI and ISO; also known as ENT 24482. The IUPAC name is dimethyl (*E*)-2-dimethylcarbamoyl-1-methylvinyl phosphate formerly 3-dimethoxyphosphinyloxy-*N,N*-dimethyliso-crotonamide, in *C.A.* usage 3-(dimethylamino)-1-methyl-3-oxo-1-propenyl dimethyl phosphate (*E*)-isomer *[141-66-2]* formerly dimethyl phosphate ester with 3-hydroxy-*N,N*-dimethylcrotonamide; the (*Z*)-isomer has the Registry No. *[18250-63-0]* and the mixed isomers *[3735-78-3]*. Dicrotophos was introduced in 1963 by Ciba AG (now Ciba-Geigy AG) under the code number 'C-709' and the trade marks 'Carbicron' and 'Ektafos'; and in 1965 by Shell Development Co. under the code number 'SD 3562', the trade mark 'Bidrin' and protected by USP 2956073; 3068268. Its insecticidal properties were first described by R. A. Corey, *J. Econ. Entomol.,* 1965, **58**, 112.

Manufacture and properties. Produced by reacting trimethyl phosphite with 2-chloro-*N,N*-dimethyl-3-oxobutyramide, it is an amber liquid, with a mild ester odour. The technical material, containing about 85% (*E*)-isomer, has b.p. 130°/0·1 mmHg, d_{15}^{15} 1·216, n_D^{23} 1·4680, v.p. 7 x 10^{-5} mmHg at 20°. It is miscible with water and organic solvents such as acetone, 4-hydroxy-4-methylpentan-2-one, propan-2-ol, ethanol; its solubility in diesel oil and kerosine is < 10 g/kg. It is stable when stored in glass or polythene containers up to 40°, but is decomposed after prolonged storage at 55°. The initial half-life of an aqueous solution at 38° is 50 d at pH 9·1, 100 d at pH 1·1. It is compatible with most other pesticides and is relatively non-corrosive to Monel, copper, nickel and aluminium, but somewhat corrosive to cast iron, mild steel, brass and stainless steel 304. It does not attack glass, polythene, stainless steel 316.

Uses. Dicrotophos, the (*E*)-isomer (which is more active than the (*Z*)-), is a systemic insecticide and acaricide of moderate persistence. It is effective against sucking, boring and chewing pests at dose rates of 300-600 g a.i./ha and is recommended for use on rice, cotton, coffee and other crops. It is non-phytotoxic except to certain varieties of fruit under some conditions. Its metabolism and breakdown have been reviewed (K. I. Beynon *et al., Residue Rev.,* 1973, **47**, 55).

Toxicology. The acute dermal LD50 for rats is 12·8-30 mg/kg. The acute dermal LD50 is: for rats 148-181 mg/kg; for rabbits 224 mg/kg. In 2-y feeding trials no toxicological effect was noted in rats receiving 1·0 mg/kg diet or dogs receiving 1·6 mg/kg diet. The 'no effect' level in a 3-generation reproduction study with rats was 2 mg/kg daily. It is not neurotoxic to hens. The LC50 (24 h) is: for mosquito fish 200 mg/l; for harlequin fish > 1000 mg/l. It is toxic to honey bees and birds.

Formulations. These include: w.s.c. 'Bidrin Technical' (850 g a.i./kg), 'Carbicron 500 SCW', 'Carbicron 1000 SCW'; e.c. (400 and 500 g/l); 'Carbicron 100 ULV', 'Carbicron 250 ULV' (for spraying undiluted); ULV solutions (240-500 g/l).

Analysis. Product analysis is by i.r. spectroscopy or GLC. Residues may be determined by GLC using phosphorus-sensitive detection (M. C. Bowman & M. Beroza, *J. Agric. Food Chem.,* 1967, **15**, 465; B. Y Giang & H. F. Beckman, *ibid.,* 1968, **16**, 899) or by clean-up, hydrolysis and measurement of the dimethylamine formed (R. T. Murphy, *ibid.,* 1965, **13**, 242). See also: P. E. Porter, *Anal. Methods Pestic., Plant Growth Regul. Food Addit.,* 1967, **5**, 213; *Anal. Methods Pestic. Plant Growth Regul.,* 1972, **6**, 287.

Dieldrin
(HEOD)

$C_{12}H_8Cl_6O$ (380·9)

T E3 D5 C555 A D- FO KUTJ AG AG BG JG KG LG

Nomenclature and development. The common name HEOD is approved by BSI and ISO—exceptions Canada, Denmark, USA and USSR (dieldrin). The name dieldrin is approved by BSI, ISO, BPC and JMAF for material containing ⩾85% HEOD—exceptions as above; also· known as ENT 16225. The IUPAC name is (1R,4S,5S,8R)-1,2,3,4,10,10-hexachloro-1,4,4a,5,6,7,8,8a-octahydro-6,7-epoxy-1,4:5,8-dimethanonaphthalene formerly 1,2,3,4,10,10-hexachloro-6,7-epoxy-1,4,4a,5,6,7,8,8a-octahydro-*exo*-1,4-*endo*-5,8-dimethanonaphthalene, in *C.A.* usage 3,4,5,6,9,9-hexachloro-1aα,2β,2aα,3β,6β,6aα,7β,-7aα-octahydro-2,7:3,6-dimethanonaphth[2,3-*b*]oxirene *[60-57-1]* formerly 1,2,3,4,10,10-hexachloro-6,7-epoxy-1,4,4a,5,6,7,8,8a-octahydro-*endo*-1,4-*exo*-5,8-dimethanonaphthalene. It was introduced in 1948 by J. Hyman & Co. under the code number 'Compound 497', trade mark 'Octalox' and protected by USP 2676547; and by the Shell International Chemical Co. Ltd. Its insecticidal properties were first described by C. W. Kearns *et al.*, *J. Econ. Entomol.*, 1949, **42**, 127.

Manufacture and properties. Produced by the oxidation of aldrin (p.6) with peracetic or perbenzoic acids (BP 692257), dieldrin forms colourless crystals, m.p. 175-176°, v.p. 3·1 x 10⁻⁶ mmHg at 20°, 5·4 x 10⁻⁶ mmHg at 25°, *d* 1·75. Its solubility at 25-29° is: 0·186 mg/l water; slightly soluble in petroleum oils; moderately soluble in acetone; soluble in aromatic solvents. The technical product, consisting of buff to light brown flakes, has a setting point ⩾95°. Dieldrin is stable to alkali, mild acids and to light; it gives no reaction with Grignard reagents and the epoxide ring is unusually stable though reacting with hydrogen bromide to give the bromohydrin. It is compatible with most other pesticides.

Uses. Dieldrin is a non-systemic and persistent insecticide of high contact and stomach activity to most insects. It is non-phytotoxic. It is formed from aldrin by metabolic oxidation in animals and by chemical oxidation in soils.

Toxicology. The acute oral LD50 for rats is 46 mg/kg; the acute dermal LD50 for rats is 10-102 mg/kg. The 'no toxic effect' level for rats and dogs is 1 mg/kg diet (A. I. T. Walker *et al.*, *Toxicol. Appl. Pharmacol.*, 1969, **15**, 345). The LC50 (96 h) is: for goldfish 0·037 mg/l, for bluegill 0·008 mg/l.

Formulations. These include: e.c. (180-200 g a.i./l); w.p. (500 and 750 g/kg); granules (20 and 50 g/kg), the addition of urea may be necessary for stability with certain carriers; oil solutions for ULV application (180-200 g/l); seed dressings, frequently incorporating fungicides.

Analysis. Product analysis is by GLC (details of methods are available from Shell International Chemical Co. Ltd) or by i.r. spectroscopy (*CIPAC Handbook*, 1970, **1**, 405; FAO Specification (CP/36), *Organochlorine Insecticides 1973*). Residues can be determined by GLC with ECD (K. Elgar *et al.*, *Analyst (London)*, 1966, **91**, 143). See also: P. E. Porter, *Anal. Methods Pestic., Plant Growth Regul. Food Addit.*, 1964, **2**, 143; *Anal. Methods Pestic. Plant Growth Regul.*, 1972, **6**, 268.

185

Dienochlor

$$C_{10}Cl_{10} \quad (474{\cdot}6)$$

L5 AHJ AG BG CG DG EG A- AL5 AHJ AG BG CG DG EG

Nomenclature and development. The common name dienochlor is approved by BSI and ISO; also known as ENT 25718. The IUPAC name is perchlorobi(cyclopenta-2,4-dienyl), in *C.A.* usage 1,1′,2,2′,3,3′,4,4′,5,5′,-decachlorobi-2,4-cyclopentadien-1-yl *[2227-17-0]*. It was introduced in 1960 by the Hooker Chemical Corp. under the trade mark 'Pentac' and protected by USP 2732409; 2934470. Its acaricidal properties were first described by W. W. Allen *et al., J. Econ. Entomol.,* 1964, **57,** 187.

Manufacture and properties. Produced by the catalytic reduction of hexachlorocyclopentadiene, it is a tan crystalline solid, m.p. 122-123°, v.p. 10^{-5} mmHg at 25°. It is insoluble in water; slightly soluble in hot ethanol, acetone, aliphatic hydrocarbons; moderately soluble in aromatic hydrocarbons. It is stable towards aqueous acids and bases though it loses activity when exposed to high temperature (6 h at 130° results in a 50% loss) or when exposed to direct sunlight or u.v. light. It is compatible with most other pesticides.

Uses. Dienochlor is a specific acaricide recommended for mite control on roses under glass at 25 g a.i./100 l using 2 applications with a 14-d interval; on other ornamentals one application is required except for severe attacks. It appears to act by an interference with oviposition for initial results require 3-5 d. It has no insecticidal activity and is non-phytotoxic.

Toxicology. The acute oral LD50 for male albino rats is > 3160 mg tech./kg; the dermal LD50 for albino rabbits is > 3160 mg/kg; a single application of 3 mg to rabbit eyes produced a slight irritation which subsided by the 6th d.

Formulation. 'Pentac WP', w.p. (500 g a.i./kg).

Analysis. Product analysis is by i.r. spectroscopy at 7·98 μm: details of methods are available from the Hooker Chemical Corp.

Diethatyl-ethyl

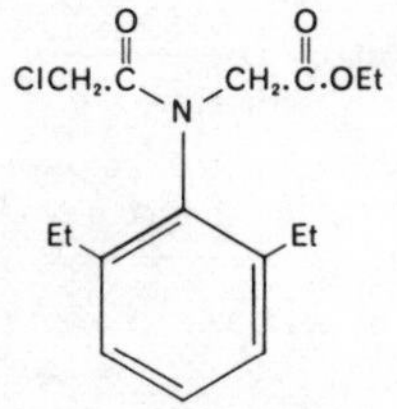

2OV1NV1GR   B2   F2

$C_{16}H_{22}ClNO_3$ (311·8)

Nomenclature and development. The common name diethatyl is approved by BSI and proposed by ISO and ANSI for *N*-chloroacetyl-*N*-(2,6-diethylphenyl)glycine (IUPAC and *C.A.*) *[38725-95-0]*. The ethyl ester, diethatyl-ethyl, Registry No. *[38727-55-8]* was introduced as a herbicide in 1974 by Hercules Inc. under the code number 'Hercules 22 234' and trade mark 'Antor'. Its activity was first described by S. K. Lehman, *Proc. North Cent. Weed Control Conf., 29th,* 1972. Patents are pending worldwide.

Manufacture and properties. Produced by the reaction of 2,6-diethylaniline with ethyl chloroacetate followed by reaction with chloroacetyl chloride, diethatyl-ethyl is a colourless crystalline solid, m.p. 49-50°. Its solubility at 25° is 105 mg/l water; soluble in common organic solvents (*e.g.* xylene, isophorone, chlorobenzene, cyclohexanone). It is hydrolysed under strongly acid or alkaline conditions. The technical product and formulations are stable under normal conditions for > 2 y.

Uses. Diethatyl-ethyl is used at 5-10 l/ha to control grasses (*Echinochloa, Digitaria, Panicum, Alopecurus* and *Setaria* spp.) in mixtures with broad-leaved-acting herbicides in crops such as sugar beet, red beet, winter wheat, soyabeans and spinach. It is also effective against the broad-leaved weed *Amaranthus* spp. It is applied pre-em., either soil-incorporated or to the soil surface.

Toxicology. The acute oral LD50 for albino rats is 2300-3700 mg/kg. The acute dermal LD50 for rabbits is 4000 mg/kg.

Formulation. An e.c. (480 g tech./l), crystallisation can occur below 0°.

Analysis. Product analysis is by GLC or HPLC. Residues are determined by GLC after suitable pre-treatment. Details of methods are available from Hercules Inc.

187

O,O-Diethyl O-5-phenylisoxazol-3-yl phosphorothioate

T5NOJ CR& EOPS&O2&O2 $C_{13}H_{16}NO_4PS$ (313·3)

Nomenclature and development. The common name isoxathion is approved by JMAF. The IUPAC and *C.A.* name is *O,O*-diethyl *O*-5-phenylisoxazol-3-yl phosphorothioate *[18854-01-8]*. It was introduced in 1972 by the Sankyo Co. Ltd under the code numbers 'E-48' and 'SI-6711', trade mark 'Karphos' and protected by JapaneseP 525850. Its insecticidal properties were described by N. Sampei *et al., Sankyo Kenkyusho Nempo*, 1970, **22**, 221.

Properties. It is a yellowish liquid, b.p. 160°/0·15 mmHg. It is almost insoluble in water; readily soluble in organic solvents. It is unstable to alkali and decomposes at high temperatures.

Uses. It is a contact insecticide with a wide range of action, controlling aphids and scale insects at 33-50 g a.i./100 l; also effective against borers, hoppers and gall midges of paddy rice and against caterpillars, beetles and mites on many crops.

Toxicology. The acute oral LD50 is: for rats 112 mg/kg; for mice 98·4 mg/kg. The acute dermal LD50 is: for rats > 450 mg/kg; for mice 193 mg/kg. In 90-d feeding trials rats and mice receiving 3·2 mg/kg diet showed no effect. The LC50 (48 h) for carp is 2·13 mg/l.

Formulations. An e.c. (500 g a.i./l); w.p. (400 g/kg); microgranule (30 g/kg); dusts (20 and 30 g/kg).

Analysis. Product and residue analysis is by TLC and GLC (T. Nakamura & K. Yamaoka, *Anal. Methods Pestic. Plant Growth Regul.*, 1978, **10**, 83).

Diethyltoluamide

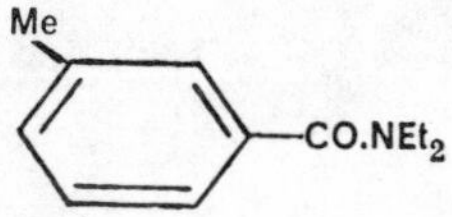

Me
CO.NEt₂

2N2&VR   Cl

$C_{12}H_{17}NO$ (191·3)

Nomenclature and development. The name diethyltoluamide is accepted by BSI in lieu of a common name, and as a common name by BPC—ESA use the name deet; also known as ENT 20218. The IUPAC name is *N,N*-diethyl-*m*-toluamide (I), in *C.A.* usage *N,N*-diethyl-3-methylbenzamide *[134-62-3]* formerly (I). It was developed as an insect repellent following the discovery by USDA of these properties in the related *N,N*-diethylbenzamide (S. I. Gertler, USP 2408389). It was introduced in 1955 by Hercules Inc. (who no longer manufacture or market it) under the trade mark 'Metadelphene' for the 95% *m*-isomer.

Manufacture and properties. Produced by the reaction of *m*-toluoyl chloride with diethylamine (E. Y. McCabe *et al., J. Org. Chem.,* 1954, **19**, 493), it is a colourless to amber liquid, b.p. 111°/1 mmHg, n_D^{25} 1·5206. It is practically insoluble in water; miscible with ethanol, propan-2-ol, propylene glycol, cottonseed oil. The technical product contains 85-95% *m*-isomer, d^{24} 0·996-0·998, viscosity 13·3 mPa s at 30°.

Uses. Diethyltoluamide is an insect repellent especially effective against mosquitoes (I. H. Gilbert *et al., J. Econ. Entomol.,* 1955, **48**, 741). The *o*- and *p*-isomers are highly repellent but less effective than the *m*-isomer.

Toxicology. The acute oral LD50 for male albino rats is *c.* 2000 mg/kg. In 200-d feeding trials rats receiving 10000 mg/kg diet suffered no ill-effect (A. M. Ambrose *et al., Toxicol. Appl. Pharmacol.,* 1959, **1**, 97). The undiluted compound may irritate mucous membranes, but daily applications to the face and arms caused only mild irritation.

Analysis. Product analysis is by i.r. spectroscopy or by GLC.

Difenacoum

T66 BOVJ EQ D- GL66&TJ IR DR $C_{31}H_{24}O_3$ (444·5)

Nomenclature and development. The common name difenacoum is approved by BSI and New Zealand and proposed by ISO. The IUPAC name is 3-(3-biphenyl-4-yl-1,2,3,4-tetrahydro-1-naphthyl)-4-hydroxycoumarin, in *C.A.* usage 3-[3-(1,1'-biphenyl)-4-yl-1,2,3,4-tetrahydro-1-naphthalenyl]-4-hydroxy-2*H*-1-benzopyran-2-one *[56073-07-5]*. It was introduced in 1974 by Sorex (London) Ltd under the trade mark 'Neosorexa'. Its rodenticidal properties were first described by M. Hadler., *J. Hyg.,* 1975, **74,** 441.

Manufacture and properties. Produced by the condensation of 4-hydroxycoumarin with 3-biphenyl-4-yl-1,2,3,4-tetrahydro-1-naphthol, it is an off-white powder, m.p. 215-219°. It is practically insoluble in water; slightly soluble in benzene, alcohols; soluble in acetone, chloroform, other chlorinated solvents. It does not readily form water-soluble alkali metal salts, but does form certain amine salts of limited solubility in water.

Uses. Difenacoum is an indirect anticoagulant rodenticide, more potent than earlier compounds and effective against rats and most mice resistant to other anticoagulants.

Toxicology. The acute oral LD50 is: for male rats 1·8 mg/kg; for pigs > 80 mg/kg. It is also of low toxicity to birds, cats and dogs.

Formulations. 'Neosorexa', concentrate (1 g a.i./kg); 'Neosorexa' ready-to-use bait (50 mg/kg); 'Ratak' (50 mg/kg).

Analysis. Product analysis is by u.v. spectroscopy: details from Sorex (London) Ltd. HPLC may also be used for product and residue analysis (D. E. Munday & A. F. Machin, *J. Chromatogr.,* 1977, **139,** 321).

Difenoxuron

MeO⟶⟨phenyl⟩⟶O⟶⟨phenyl⟩⟶NH.CO.NMe₂

1OR DOR DMVN1&1 $C_{16}H_{18}N_2O_3$ (286·3)

Nomenclature and development. The common name difenoxuron is approved by BSI and ISO. The IUPAC name is 3-[4-(4-methoxyphenoxy)phenyl]-1,1-dimethylurea, in *C.A.* usage *N'*-[4-(4-methoxyphenoxy)phenyl]-*N,N*-dimethylurea *[14214-32-5]* formerly 3-[*p*-(*p*-methoxyphenoxy)phenyl]-1,1-dimethylurea. It was introduced commercially in 1970 by Ciba-Geigy AG under the code number 'C 3470', trade mark 'Lironion' and protected by BP 913 383.

Manufacture and properties. Produced by the reaction of 4-(4-methoxyphenoxy)phenyl isocyanate with dimethylamine, it is a colourless crystalline solid, m.p. 138-139°, v.p. $9·3 \times 10^{-12}$ mmHg at 20°. Its solubility at 20° is 20 mg/l water; 10 g/kg propan-2-ol; 63 g/kg acetone; 156 g/kg dichloromethane.

Uses. Difenoxuron is a herbicide for selective use in onions at 2·5 kg a.i./ha at the 'whip' stage or later.

Toxicology. The acute oral LD50 is: for rats > 7750 mg/kg; for mice > 1000 mg/kg. The acute dermal LD50 for rats is > 2150 mg/kg. It is moderately toxic to fish.

Formulation. A w.p. (500 g a.i./kg).

Analysis. Details of methods are available from Ciba-Geigy AG.

Difenzoquat

T5KNJ A1 B1 CR& ER   *difenzoquat* $C_{17}H_{17}N_2$ (249·3)

T5KNJ A1 B1 CR& ER &WSO&O1

difenzoquat methyl sulphate $C_{18}H_{20}N_2O_4O_4S$ (360·4)

Nomenclature and development. The common name difenzoquat is approved by BSI, ANSI, New Zealand and WSSA, and proposed by ISO. The IUPAC name is 1,2-dimethyl-3,5-diphenylpyrazolium ion, in *C.A.* usage 1,2-dimethyl-3,5-diphenyl-1*H*-pyrazolium ion *[49866-87-7]*. The methyl sulphate, in *C.A.* methyl sulfate *[43222-48-6]*, was introduced by the American Cynamid Co. under the code number 'AC 84 777', the trade marks 'Avenge' and 'Finaven' and protected by BelgP 792801. Its herbicidal properties were described by T. R. O'Hare & C. B. Wingfield, *Proc. North Cent. Weed Control Conf.*, 1973, and by N. E. Shafer, *Proc. Br. Weed Control Conf., 12th*, 1974, **2**, 831.

Properties. The pure salt is a white solid, slightly hygroscopic, m.p. 155-157°. Its solubility is: at 25° 760 g methyl sulphate/l water, at 37° 780 g/l, at 56° 850 g/l; slightly soluble in alcohols, low molecular weight glycols; insoluble in petroleum derivatives. It is stable under normal storage conditions and is unaffected by light. Stability of the technical grade is excellent and in water is good at low pH but alkaline conditions cause precipitation from solution.

Uses. Difenzoquat methyl sulphate is a post-em. herbicide for control of wild oats, *Avena fatua*, *A. sterilis*, and *A. ludoviciana*, in wheat and barley; R. J. Winfield & J. J. B. Caldicott, *Pestic Sci.*, 1975, **6**, 297.

Toxicology. The acute oral LD50 for male albino rats is 470 g tech/kg; the acute dermal LD50 for male white rabbits is 3540 mg/kg. The LC50 (96 h) is: for bluegill 696 mg difenzoquat/l; for rainbow trout 694 mg/l.

Formulations. These include: aqueous solutions (354 g difenzoquat/kg; 400 g/l) and (227 g/kg; 250 g/l).

Analysis. Details of analytical methods are available from the American Cyanamid Co.

Diflubenzuron

GR DMVMVR BF FF $C_{14}H_9ClF_2N_2O_2$ (310·7)

Nomenclature and development. The common name diflubenzuron is approved by BSI, ANSI and New Zealand, and proposed by ISO; also known as ENT 29054, OMS 1804. The IUPAC name is 1-(4-chlorophenyl)-3-(2,6-difluorobenzoyl)urea, in *C.A.* usage *N*-[[(4-chlorophenyl)amino]carbonyl]-2,6-difluorobenzamide *[35367-38-5]*. It was introduced by Philips-Duphar B.V. under the code number 'PH 60-40', trade mark 'Dimilin' and protected by BP 1 324 293. Other designations include 'TH 6040' and 'PDD 6040-I'. The insecticidal properties of this class of compound were first described by J. J. Van Daalen *et al., Naturwissenschaften,* 1972, **59,** 312.

Properties. The technical product is colourless to yellowish-brown crystals, m.p. 210-230°; pure diflubenzuron has m.p. 230-232°. Its solubility at 20° is: 0·2 mg/l; almost insoluble in apolar solvents; moderate to good solubility in very polar solvents.

Activity. Diflubenzuron belongs to a new group of insecticides (K. Wellinga *et al., J. Agric. Food Chem.,* 1973, **21,** 348, 993) effective as a stomach poison, acting by interference with the deposition of insect chitin (R. Mulder & M. J. Gijswijt, *Pestic. Sci.,* 1973, **4,** 737). Hence all stages of insects that form new cuticles should be susceptible and this includes many insect species. A species-dependent ovicidal contact action has been demonstrated (*e.g. Spodoptera littoralis*) and prevention of egg eclosion after uptake by females (*e.g. Anthonomus grandis*); its biological activity has been reviewed (A. C. Grosscurt, *Pestic. Sci.,* 1978, **9,** 373). It has no systemic activity and does not penetrate plant tissue; hence sucking insects are, in general, unaffected, forming the basis of selectivity in favour of many insect predators and parasites.

Uses. Diflubenzuron is effective against: leaf-feeding larvae and leaf miners in forestry, top fruit, citrus, field crops, cotton, soyabeans and horticultural crops at 1·5-30 g a.i./100 l (pests controlled include: *Cydia pomonella, Psylla simulans* and the citrus rust mite); larvae of Sciaridae and Phoridae in mushroom crops at 1 g/m² casing at case mixing, or as a drench in 2·5 l water to the finished casing; mosquito larvae at 20-45 g/ha water surface; fly larvae as a surface application in animal housing at 0·5-1·0 g/m² surface. It is rapidly degraded in soil, 50% loss occurring in < 7 d; its fate in soil, plants and animals is reviewed (A. Verloop & C. D. Ferrell, *ACS Symp. Series,* 1977, No. 37, 237).

Toxicology. The acute oral LD50 for mice is > 4640 mg/kg; the intraperitoneal LD50 for mice is > 2150 mg/kg; the percutaneous LD50 for rabbits is > 2000 mg/kg. In 2-y feeding trials the 'no effect' level for rats was 40 mg/kg diet. No effect was observed in teratogenic, mutagenic and oncogenic studies. No toxic symptom was observed in mallard ducks and bobwhite quail at 4640 mg/kg diet. The LC50 (96 h) is: for bluegill 135 mg/kg; for rainbow trout 140 mg/l.

Formulation. Product and residue analysis is by HPLC or by GLC after conversion to a suitable derivative (B. Rabenort *et al., Anal. Methods Pestic. Plant Growth Regul.,* 1978, **10,** 57). A spectrophotometric method is available for the analysis of formulations.

Di-isopropyl 1,3-dithiolan-2-ylidenemalonate

T5SYSTJ BUYVOY1&1&VOY1&1 $C_{12}H_{18}O_4S_2$ (290·4)

Nomenclature and development. The common name isoprothiolane has been suggested by JMAF; there is no BSI or ISO name; also known as IPT. The IUPAC name is di-isopropyl 1,3-dithiolan-2-ylidenemalonate, in *C.A.* usage bis(1-methylethyl) 1,3-dithiolan-2-ylidenepropanedioate *[50512-35-1]*. It was introduced in 1975, after trials since 1968, by the Nihon Nohyaku Co. Ltd under the name 'IPT' and trade mark 'Fuji-one'. Its biological properties were described by F. Araki *et al., Proc. Br. Insectic. Fungic. Conf., 8th,* 1975, **2**, 715.

Properties. It forms colourless crystals, m.p. 50-54·5°, b.p. 167-169°/0·5 mmHg. Its solubility at 20° is 48 mg/l water.

Uses. It is a systemic fungicide effective against *Piricularia oryzae* of rice at 3·2-4·8 kg a.i./ha. The population of planthoppers on the treated rice was markedly decreased by an 'insectistatic' action *(idem. ibid.).*

Toxicology. The acute oral LD50 is: for male rats 1190 mg/kg; for male mice 1340 mg/kg. The LC50 (48 h) for carp is 6·7 mg/l.

Formulation. Granules (120 g a.i./kg), for application in paddy water.

Analysis. See T. Hattori & M. Kanauchi, *Anal. Methods Pestic. Plant Growth Regul.,* 1978, **10**, 229.

Dikegulac-sodium

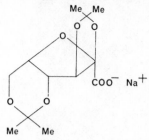

T B556 CO EO GO JO LOTJ

D1 D1 FVO K1 K1 &-NA-

$C_{12}H_{17}NaO_7$ (292·6)

Nomenclature and development. The common name dikegulac is approved by BSI and New Zealand and proposed by ISO for 2,3:4,6-di-*O*-isopropylidene-α-L-xylo-2-hexulo-furanosonic acid, in *C.A.* usage 2,3:4,6-bis-*O*-(1-methylethylidene)-α-L-xylo-2-hexulo-furanosonic acid [*18467-77-1*]. It was introduced in 1973 by F. Hoffmann-La Roche & Co. as a herbicide and plant growth regulator under the code number 'Ro 07-6145'. Its sodium salt, dikegulac-sodium, Registry No. [*52508-35-7*], was introduced in 1976 by Dr. R. Maag Ltd as a plant growth regulator under the trade mark 'Atrinal', and its properties were first described by P. Bocion *et al., Nature (London)*, 1975, **258**, 142.

Properties. Dikegulac-sodium is a colourless powder, m.p. > 300°, v.p. < 10^{-10} mmHg at room temperature. Its solubility at room temperature is: 590 g/kg water; 390 g/kg methanol; 230 g/kg ethanol; 63 g/kg chloroform; < 10 g/kg acetone, cyclohexanone, hexane. It is insensitive to light, non-corrosive and is stable, stored at room temperature in dry state, for at least 3 y.

Uses. Dikegulac-sodium is a systemic plant growth regulator which reduces apical dominance and increases the side-branching and flower-bud formation on ornamental plants (chemical pinching) and temporarily retards longitudinal growth on hedges and woody ornamentals. An additional side-branching effect on the lower, older parts of hedges is initiated and therefore improves the foliar coverage. The addition of a surfactant is necessary and hedges should not be treated until at least 4 y old. Rates of use range from 0·06-6·0 g a.i./l.

Toxicology. The acute oral LD50 of dikegulac-sodium is: for rats 18 000-31 000 mg/kg; for mice 19 500 mg/kg. The acute dermal LD50 for rats is > 2000 mg/kg; aqueous solutions (200 g/l) caused no skin irritation to guinea-pigs and (30 g/l) no eye irritation to rabbits. The acute inhalation LD50 for rats is > 0·4 mg/kg. In 90-d feeding trials no significant effect was observed: in rats receiving 2000 mg/kg daily; in dogs receiving 3000 mg/kg daily. The chronic oral LD50 in 5-d feeding trials was for Japanese quail, mallard ducks and chickens > 50 000 mg/kg diet. The LC50 (96 h) is: for bluegill > 10 000 mg/l; for goldfish, harlequins, rainbow trout and Japanese black carp > 5000 mg/l. The LD50 by oral and topical application to honey bees was much more than 0·1 mg/bee.

Formulation. 'Atrinal', soluble concentrate (200 g dikegulac-sodium/l). A special surfactant (37·5 g 2-[nonylphenoxy(polyethoxy)]ethanol/l) is used in conjunction with it.

Analysis. Product analysis is by quantitatively converting dikegulac-sodium to the ethyl ester which is determined by GLC using dibutyl phthalate as internal standard. For residue analysis: extract with water under alkaline conditions, partition the free acid (pH 1·8) into chloroform and esterify the ammonium salt with ethyl iodide; the ethyl ester is analysed by GLC after a column chromatographic clean-up.

Dimefox

$$O$$
$$(Me_2N)_2 \overset{\|}{P} . F$$

1N1&PO&FN1&1 $C_4H_{12}FN_2OP$ (154·1)

Nomenclature and development. The common name dimefox is approved by BSI and ISO; also known as ENT 19 109. The IUPAC and *C.A.* name is tetramethylphosphoro-diamidic fluoride, Registry No. *[115-26-4]*, also known as bis(dimethylamino)fluoro-phosphine oxide. It was introduced in 1949 by Pest Control Ltd as *'Pestox XIV'* (now obsolete), in 1953 by Murphy Chemical Ltd under the trade mark 'Terra Sytam' produced since 1968 by Wacker-Chemie GmbH. Its insecticidal properties were first described by H. Kükenthal & G. Schrader, *B.I.O.S. Final Report*, 1095, 1946 and its use on hops by G. A. Emery, *World Crops*, 1954, **6**(7). Its use is protected by BP 688 760; 741 662.

Manufacture and properties. Produced by the reaction of dimethylamine, phosphoryl chloride and potassium fluoride in toluene and a small amount of water, it is a colourless liquid, b.p. 67°/4 mmHg, v.p. 0·36 mmHg at 25°, d^{20} 1·115, n_D^{20} 1·4171. It is miscible with water and most organic solvents. It has a chloroform/water partition coefficient 15:1 in favour of chloroform. It is resistant to hydrolysis by alkali but is hydrolysed by acids, slowly oxidised by vigorous oxidising agents, rapidly by chlorine. It is compatible with other pesticides but the technical product slowly attacks metals. The technical material contains schradan (p.470) and tris(dimethylamino)phosphine oxide.

Uses. Dimefox is a systemic insecticide and acaricide and its main use is as a soil treatment for hop plants against aphids and red spider mites. It is effective at non-phytotoxic concentrations for 42-56 d.

Toxicology. The acute oral LD50 for rats is 1-2 mg/kg; the acute dermal LD50 for rats is 5 mg/kg. Vapour toxicity hazards are high.

Formulations. 'Terra Sytam' is a solution (500 g dimefox/l) containing also schradan and tris(dimethylamino)phosphine oxide (170 g/l); 'S 14/10' solution (100 g dimefox/l), ratio dimefox/schradan/tris(dimethylamino)phosphine oxide as 'Terra Sytam'.

Analysis. Product analysis for dimefox, schradan and tris(dimethylamino)phosphine oxide is preferably by GLC (details of method are available from Wacker-Chemie GmbH) or by selective acid hydrolysis at various temperatures and the determination of amine in the separated fractions *(CIPAC Handbook*, 1970, **1**, 329). Residue analysis for dimefox is by GLC with FID (Wacker-Chemie GmbH).

Dimethachlor

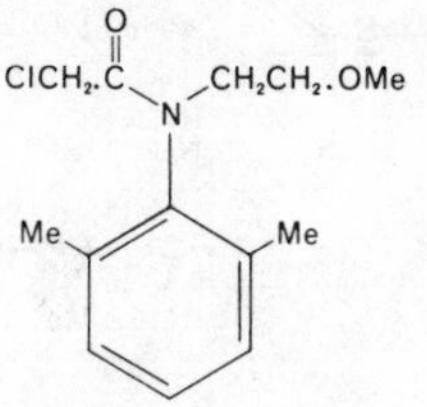

G1VN2O1&R B1 F1

$C_{13}H_{18}ClNO_2$ (255·7)

Nomenclature and development. The common name dimethachlor is approved by BSI and proposed by ISO. The IUPAC name is 2-chloro-*N*-(2-methoxyethyl)acet-2',6'-xylidide, in *C.A.* usage 2-chloro-*N*-(2,6-dimethylphenyl)-*N*-(2-methoxyethyl)acetamide *[50563-36-5]*. It was synthesised and developed as a herbicide by Ciba-Geigy Ltd under the code number 'CGA 17 020', trade mark 'Teridox' and protected by BP 1 422473. Its activity was first described by J. Cortier *et al., C. R. Journ. Etud. Herbic. Conf. COLUMA, 9th,* 1977, **1,** 99.

Properties. Pure dimethachlor is a colourless crystalline solid, m.p. 47°, v.p. $1·6 \times 10^{-5}$ mmHg at 20°. Its solubility at 20° is 2·1 g/l water; it is readily soluble in most organic solvents.

Uses. Dimethachlor is a selective herbicide for use on rape, effective against annual broad-leaved weeds and grasses *e.g. Alopecurus myosuroides, Apera spica-venti* and *Poa annua.* It is used immediately after sowing at 1·25-2·0 kg a.i./ha depending on the organic matter content of the soil.

Toxicology. The acute oral LD50 for rats is 1600 mg tech./kg; the acute dermal LD50 for rats is > 3170 mg/kg.

Formulation. 'Teridox 500 EC', e.c. (500 g a.i./l).

Analysis. Details of methods are available from Ciba-Geigy Ltd.

Dimethametryn

T6N CN ENJ BS1 DMY1&Y1&1 FM2 $C_{11}H_{19}N_2O$ (209·3)

Nomenclature and development. The common name dimethametryn is approved by BSI and ISO. The IUPAC name is 2-(1,2-dimethylpropylamino)-4-ethylamino-6-methylthio-1,3,5-triazine, in *C.A.* usage *N*-(1,2-dimethylpropyl)-*N'*-ethyl-6-(methylthio)-1,3,5-triazine-2,4-diamine *[22936-75-0]* formerly 2-[(1,2-dimethylpropyl)amino]-4-ethylamino-6-methylthio-*s*-triazine. It was introduced in 1969 by Ciba-Geigy AG as an experimental herbicide under the code number 'C 18898' and protected by SwissP 485 410; BP 1 191 585; USP 3 799 925. It was first described by D. H. Green & L. Ebner, *Proc. Br. Weed Control Conf., 11th,* 1972, p.822.

Properties. It is a solid, m.p. 65°, b.p. 151-153°/0·05 mmHg, v.p. 1·4 x 10^{-6} mmHg at 20°. Its solubility at 20° is 50 mg/l water; soluble in polar organic solvents.

Uses. Dimethametryn is a selective herbicide active against broad-leaved weeds in rice. Under the trade mark 'Avirosan' it is marketed in combination with piperophos for the control of both mono- and di-cotyledonous weeds. Rates of 1-2 kg total a.i./ha are used in transplanted rice and 2-3 kg/ha in seeded rice.

Toxicology. The acute oral LD50 for rats is 3000 mg/kg; the acute dermal LD50 for rats is > 2150 mg/kg. It is slightly toxic to fish.

Formulations. Mixtures of piperophos + dimethametryn (4:1); granules 'Avirosan 3G', 'Avirosan 5G'; e.c. 'Avirosan 500EC'.

Analysis. Product analysis is by GLC with internal standard. Residues may be determined by extraction with methanol, and measured, after liquid-liquid partition and column chromatographic clean-up, by GLC with thermionic detection. Particulars of these methods are available from Ciba-Geigy AG.

Dimethirimol

$$\text{Bu}$$

Me—⟨ring⟩—OH

N   N

NMe$_2$

T6N CNJ BN1&1 DQ E4 F1   C$_{11}$H$_{19}$N$_2$O (209·3)

Nomenclature and development. The common name dimethirimol is approved by BSI and ISO, in France (diméthyrimol). The IUPAC name is 5-butyl-2-dimethylamino-6-methyl-pyrimidin-4-ol, in *C.A.* usage 5-butyl-2-(dimethylamino)-6-methyl-4(1*H*)-pyrimidinone *[5221-53-4]* formerly 5-butyl-2-(dimethylamino)-6-methyl-4-pyrimidinol. It was introduced in 1968 by ICI Ltd, Plant Protection Division under the code number 'PP 675', and the trade mark 'Milcurb' as a systemic fungicide protected by BP 1 182 584. It was first described by R. S. Elias *et al., Nature (London),* 1968, **219.** 1160.

Manufacture and properties. Produced either by the condensation of ethyl 2-acetyl-hexanoate with 1,1-dimethylguanidinium sulphate in the presence of sodium ethoxide, or 5-butyl-6-methyl-2-methylthiopyrimidin-4-ol with dimethylammonium acetate, it forms colourless needles, m.p. 102°, v.p. 1·1 x 10^{-5} mmHg at 30°. Its solubility at 25° is: 1·2 g/l water; 45 g/l acetone; 1·2 kg/l chloroform; 65 g/l ethanol; 360 g/l xylene. It is stable to heat and to acidic and alkaline solutions, dissolving in aqueous solutions of strong acids to form salts. It is non-corrosive to metals, but acid formulations should not be stored in galvanised containers.

Uses. Dimethirimol is a fungicide, effective against the powdery mildews of cucurbits, chrysanthemum and cineraria. When applied to foliage it enters the plant and moves in the transpiration stream. It is relatively stable and loosely absorbed when applied to soil and, following root uptake, can protect the plant for $\geqslant$42 d. Soil applications of $\leqslant$ 1 g a.i./plant are safe on established plants but young plants with restricted root system should not be overdosed. It has no significant effect on insect pests, red spider mites or their predators.

Toxicology. The acute oral LD50 is: for rats 2350 mg/kg; for mice 800-1600 mg/kg; for guinea-pigs 500 mg/kg; for hens 4000 mg/kg; for honey bees 4000 mg/kg. Daily applications (14 d) of 500 mg/kg to the shaved skin of rabbits had no effect. In 2-y feeding trials the 'no effect' level was: for dogs 25 mg/kg; for rats 300 mg/kg diet. The LC50 at 16° for fingerling brown trout is: (96 h) 28 mg/l, (48 h) 33 mg/l, (24 h) 42 mg/l.

Formulations. Aqueous solutions of the hydrochloride: 'Milcurb' (125 g dimethirimol/l); in the UK only, 'PP 675' (12·5 g/l). The solution is acidic and should not be used in galvanised steel.

Analysis. Product analysis is by GLC (J. E. Bagness & W. G. Sharples, *Analyst (London),* 1974, **99,** 225) or by colorimetry at 303 nm. For residues extract with methanol and, after liquid-liquid partition clean-up, separate by TLC and either compare the spot with standards on the same plate viewed under u.v. light or elute with methanol and measure at 303 nm. Details are available from ICI Ltd, Plant Protection Division.

Dimethoate

$$\text{(MeO)}_2\text{P.S.CH}_2\text{.C.NHMe}$$
$$\overset{\text{S}}{\underset{\|}{}} \qquad \overset{\text{O}}{\underset{\|}{}}$$

1OPS&O1&S1VM1 $C_5H_{12}NO_3PS_2$ (229·2)

Nomenclature and development. The common name dimethoate is approved by BSI, ISO, ANSI and JMAF—exception USSR (fosfamid); also known as ENT 24650. The IUPAC name is *O,O*-dimethyl *S*-methylcarbamoylmethyl phosphorodithioate, in *C.A.* usage *O,O*-dimethyl *S*-[2-(methylamino)-2-oxoethyl] phosphorodithioate *[60-51-5]* formerly *O,O*-dimethyl phosphorodithioate *S*-ester with 2-mercapto-*N*-methylacetamide. It was first described by E. I. Hoegberg & J. T. Cassaday, *J. Am. Chem. Soc.*, 1951, **73**, 557, and was introduced in 1956 by American Cyanamid Co. under the code number 'E.I. 12880', trade marks 'Cygon' and 'Dimetate', and protected by USP 2494283; by Montedison S.p.A. (formerly Montecatini S.p.A.) under the code number 'L 395', trade marks 'Fostion MM', 'Rogor' and protected by BP 791824. Other trade marks are 'Roxion' (Cela), 'Perfekthion' (BASF). Its insecticidal properties were described in *Ital. Agric.*, 1955, **92**, 747.

Manufacture and properties. It is produced by: *(a)* the reaction of sodium *O,O*-dimethyl phosphorodithioate with 2-chloro-*N*-methylacetamide; *(b)* treating with methylamine the reaction product of sodium *O,O*-dimethyl phosphorodithioate and phenyl chloroacetate (Boehringer Sohn, 1958, DBP 1076662); *(c)* the reaction of methylamine with methyl (dimethoxyphosphinothioylthio)acetate. It forms colourless crystals, with a camphor-like odour, m.p. 51-52°, v.p. 8·5 x 10⁻⁶ mmHg at 25°, d_4^{50} 1·281, n_D^{65} 1·5334. Its solubility at 21° is 25 g/l water; soluble in most organic solvents except paraffins. It is stable in aqueous solution but is readily hydrolysed by aqueous alkali; heating converts it to the *O,S*-dimethyl phosphorodithioate isomer. It is incompatible with alkaline pesticides.

Uses. Dimethoate is a contact and systemic insecticide and acaricide effective against a broad range of insects and mites when applied at 0·3-0·7 kg a.i./ha on a wide range of crops. It is especially effective against houseflies and Diptera of medical importance. It is non-phytotoxic at recommended rates except to a few olive, citrus, fig and nut varieties.

Toxicology. The acute oral LD50 is: for rats 320-380 mg a.i. (as 94-96% tech.)/kg, 500-600 g a.i./kg; for male pheasants 15 mg/kg; for female duck 40 mg/kg; for sparrows 22 mg/kg; for blackbirds 26 mg/kg. No irritation developed after the application of 130 mg a.i. (as e.c.)/20 cm² shaved skin of rabbits. Tests on humans showed the highest tolerable dose 0·2 mg/kg daily; its mammalian toxicity is low enough for it to be considered for the systemic control of cattle grubs (F. W. Knapp *et al.*, *J. Econ. Entomol.*, 1960, **53**, 541). The LC50 (96 h) for mosquito fish is 40-60 mg/l. The LD50 for honey bees is 0·9 μg/bee.

Formulations. These include: e.c. (200, 400 and 500 g tech./l); 'Rogor AS' (300 g/l) for ULV application; w.p. (200 g/kg); granules (50 g/kg).

Analysis. Product analysis is by GLC with partition chromatography (*CIPAC Handbook*, 1979, **1A**, in press) or by TLC or an arsenometric method. Residues may be determined by GLC (W. Wagner & H. Frehse, *Pflanzenschutz-Nachr. (Am. Ed.)*, 1976, **29**, 54). Details of these methods may be obtained from Montedison S.p.A. See also: G. L. Sutherland, *Anal. Methods Pestic., Plant Growth Regul. Food Addit.*, 1964, **2**, 171; J. E. Boyd, *Anal. Methods Pestic. Plant Growth Regul.*, 1972, **6**, 357; P. de Pietri-Tonelli *et al.*, *Residue Rev.*, 1965, **11**, 60.

1,3-Di(methoxycarbonyl)prop-1-en-2-yl dimethyl phosphate

$$(MeO)_2\overset{\overset{O}{\|}}{P}.O.C=CH.\overset{\overset{O}{\|}}{C}.OMe$$
$$\overset{|}{C}H_2.\underset{\underset{O}{\|}}{C}.OMe$$

1OVlUYlVOl&OPO&Ol&Ol

$C_9H_{15}O_8P$ (282·2)

Nomenclature and development. There is no BSI or ISO common name; sometimes known as ENT 24833. The IUPAC name is 1,3-di(methoxycarbonyl)prop-1-en-2-yl dimethyl phosphate or dimethyl 3-(dimethoxyphosphinyloxy)pent-2-enedioate, in *C.A.* usage dimethyl 3-[(dimethoxyphosphinyl)oxy]-2-pentenedioate *[122-10-1]* formerly dimethyl 3-hydroxyglutaconate dimethyl phosphate. It was introduced in 1959 by Allied Chemical Corp. Agricultural Division (now Hopkins Agricultural Chemical Co.) under the code number 'GC-3707', trade mark 'Bomyl' and protected by USP 2 891 887.

Manufacture and properties. Produced by reacting dimethyl 3-oxopentanedioate with sulphuryl chloride and the resulting chloro compound with trimethyl phosphite, it is a yellow oil, b.p. 164°/2 mmHg, *d* 1·2. It is moderately soluble in water; soluble in acetone, ethanol, propylene glycol, xylene; practically insoluble in kerosine. It is hydrolysed by alkali with 50% loss in > 10 d at pH 5, > 4 d at pH 6, < 1 d at pH 9. It exists as 2 geometric isomers which, in contrast to other vinyl phosphates, have similar insecticidal and anticholinesterase properties (P. E. Newallis *et al.*, *J. Agric. Food Chem.*, 1967, **15**, 940).

Uses. It is a contact insecticide and acaricide, with some residual action, recommended for use: on cotton at 0·8-1·1 kg a.i./ha for the control of bollworms, boll weevil and aphids; for broadcast application at 140 g/ha against grasshoppers; as a sugar bait (5 g and 10 g a.i./l) in farm buildings, garbage dumps and picnic areas for control of flies, ants, cockroaches and other pests. It shows activity against snails, and is currently being evaluated against locusts.

Toxicology. The acute oral LD50 for albino rats is *c.* 32 mg/kg; the acute dermal LD50 for rabbits is 20-31 mg/kg.

Formulations. These include: w.p. (250 g a.i./kg); e.c. (120, 240 and 480 g/l); bait (10 and 22 g/kg).

Analysis. Product analysis is by GLC and i.r. spectroscopy. Details are available from Hopkins Agricultural Chemical Co.

Dimethylarsinic acid

$$O$$
$$\parallel$$
$$Me_2As.OH$$

Q-AS-O&1&1 $C_2H_7AsO_2$ (138·0)

Nomenclature and development. The IUPAC name dimethylarsinic acid (I) is proposed by BSI and ISO as an alternative to the trivial name cacodylic acid in lieu of a common name, WSSA have approved cacodylic acid as common name; in *C.A.* usage (I) *[75-60-5]* formerly hydroxydimethylarsine oxide. It was introduced as a herbicide in 1958 by Ansul Chemical Co. (which no longer exists) under the trade marks *'Ansar'*, replaced by 'Phytar', and protected by USP 3056668. Its herbicidal properties were described by R. G. Mowev & J. F. Cornman, *Proc. Northeast. Weed Control Conf.,* 1959, **13**, 62.

Manufacture and properties. Produced by the reduction of disodium methylarsonate by sulphur dioxide and methylation of the resulting disodium methylarsonite, it forms colourless crystals, m.p. 192-198°. Its solubility at 25° is: 2 kg/kg water; soluble in short-chain alcohols; insoluble in diethyl ether. The technical product is 65% pure, sodium chloride being one impurity. It is decomposed by powerful oxidising or reducing agents, but is compatible with most other herbicides. Its aqueous solution is mildly corrosive. The acid has pK_a 6·29 (D. Wauchope, *J. Agric. Food Chem.,* 1976, **24**, 717) and forms water-soluble sodium and potassium salts, both deliquescent, at *c.* pH 7.

Uses. Dimethylarsinic acid is used as a non-selective post-em. herbicide to control weeds in non-crop situations, for lawn renovation at 11-17 kg/ha, as a desiccant and defoliant for cotton at 1·1-1·7 kg/ha and for killing unwanted trees by injection. It is virtually inactivated on contact with soil.

Toxicology. The acute oral LD50 for rats is 1350 mg tech./kg; it is non-irritating to the skin and eyes of rabbits.

Formulations. These include: 'Phytar 560', w.s.c. of sodium salt (300 g a.e. + surfactant/l); 'Silvisar', liquid concentrate (720 g a.i./l).

Analysis. Methods are described by E. A. Dietz & L. O. Moore, *Anal. Methods Pestic. Plant Growth Regul.,* 1978, **10**, 385.

202

Dimethyl 4-(methylthio)phenyl phosphate

$$(MeO)_2\overset{\overset{O}{\|}}{P}.O-\!\!\!\!\langle\text{phenyl}\rangle\!\!\!\!-SMe$$

ISR   DOPO&O1&O1 $C_9H_{13}O_4PS$ (248·2)

Nomenclature and development. There is no BSI or ISO common name; sometimes known as ENT 25734. The IUPAC and *C.A.* name is dimethyl 4-(methylthio)phenyl phosphate, Registry No. *[3254-63-5]*. It was introduced in 1963 by the Allied Chemical Corp., Agricultural Division (now Hopkins Agricultural Chemical Co.), under the code number 'GC-6506' and protected by USP 3151022. Currently available under the code number 'HA-1200'.

Properties. It is a colourless liquid which decomposes at 269-284°, $d^{21.4}$ 1·273, n_D^{25} 1·5349. Its solubility at room temperature is: 98 mg/l water, 890 g/l acetone; 540 g/l dioxane; 580 g/l carbon tetrachloride; 860 g/l ethanol; > 1 kg/l xylene. It is hydrolysed by alkali (pH 9·5) at 37·5°.

Uses. It is a contact and systemic insecticide and acaricide for the control of aphids, mites and lepidopterous larvae at 0·3-1·1 kg a.i./ha. A soil drench at 1·1 kg/ha has given systemic control. Most tested crops were uninjured at rates giving effective insecticidal action; cotton seed germination was not affected by applications of 4·5 kg/ha but, at 2·2 kg/ha, bean seed germination and potato sprouting were reduced. It gives good control of boll weevil, pink bollworm and tobacco budworm. There is no evidence of crosslinked resistance in laboratory tests.

Toxicology. The acute oral LD50 for male albino rats is 6·5-7·5 mg/kg; the percutaneous LD50 for albino rabbits is 46-50 mg/kg. There was no significant difference in cholinesterase levels between control rats and those fed 0·35 mg/kg daily for 10 d. It is highly toxic to honey bees.

Formulations. A w.p. (250 g a.i./kg); e.c. (480 g/l); granules (100 g/kg).

Analysis. Residue analysis is by GLC. Details are available from Hopkins Agricultural Chemical Co.

Dimethyl phthalate

1OVR   BVO1 $C_{10}H_{10}O_4$ (194·2)

Nomenclature and development. The IUPAC name dimethyl phthalate (I) is accepted by BSI and ISO in lieu of a common name; also known as ENT 262 and DMP. The *C.A.* name is dimethyl 1,2-benzenedicarboxylate *[131-11-3]* formerly (I). It was introduced as an insect repellent during the 1939-45 war, having long been used as a plasticiser.

Manufacture and properties. Produced by the esterification of phthalic acid or anhydride with methanol, it is a colourless to pale yellow viscous liquid, b.p. 282-285°, 147·6°/10 mmHg, v.p. 0·01 mmHg at 20°, d_{20}^{20} 1·194, n_D^{20} 1·5168. Its solubility at room temperature is 4·3 g/kg water; soluble in petroleum oils; miscible with ethanol, diethyl ether and most organic solvents. It is stable though hydrolysed by alkali.

Uses. Dimethyl phthalate is an insect repellent used for personal protection from biting insects.

Toxicology. The acute oral LD50 for rats is 8200 mg/kg; the acute dermal LD50 (9 d) for rats is > 4800 mg/kg. It may cause smarting if applied to the eye or mucous membranes. In 2-y feeding trials there was no effect on the growth rate of rats receiving 20 000 mg/kg diet.

Formulations. Applied alone or incorporated into creams. It is often used in combination with 2-ethylhexane-1,3-diol (ethohexadiol) and butopyronoxyl in the ratio 6:2:2 (USP 2 356 801).

Analysis. Product analysis is either by alkaline hydrolysis and estimation of the resulting phthalic acid by standard methods, or by estimation of methoxy groups by standard methods.

Dimetilan

T5NNJ AVN1&1 COVN1&1 E1 $C_{10}H_{16}N_4O_3$ (240·3)

Nomenclature and development. The common name dimetilan is approved by BSI, New Zealand and ESA; also known as ENT 25922. The IUPAC name is 1-dimethylcarbamoyl-5-methylpyrazol-3-yl dimethylcarbamate, in *C.A.* usage 1-[(dimethylamino)carbonyl]-5-methyl-1*H*-pyrazol-3-yl dimethylcarbamate *[644-64-4]* formerly dimethylcarbamate ester of 3-hydroxy-*N,N*,5-trimethylpyrazole-1-carboxamide. It was described by H. Gysin, *Chimia,* 1954, **8**, 205, 221, and was introduced in 1962 by J. R. Geigy S.A. (now Ciba-Geigy AG) under the code numbers 'G 22870' and 'GS 13332', the trade mark 'Snip' and protected by SwissP 281946; 282655; BP 681376; USP 2681879.

Manufacture and properties. Produced by the reaction of 3-methylpyrazol-5-one with dimethylcarbamoyl chloride, it is a colourless solid, m.p.68-71°, b.p. 200-210°/13 mmHg, v.p. 1 x 10⁻⁴ mmHg at 20°. It is readily soluble in water, chloroform, dimethylformamide; soluble in ethanol, acetone, xylene and other organic solvents. The technical product (> 96% pure) is a light yellow to reddish-brown crystalline solid, m.p. 55-65°. It is hydrolysed by acids and by alkalies.

Uses. Dimetilan is a stomach poison used for the control of flies.

Toxicology. The acute oral LD50 is: for rats 64 mg/kg; for mice 60-65 mg/kg. The acute dermal LD50 for rats is > 2000 mg/kg. Livestock is more susceptible.

Formulation. A sugar bait.

Analysis. Product analysis is by GLC with internal standard. Residues may be determined by GLC with MCD or by TLC. Details of unpublished methods are available from Ciba-Geigy AG. See also: A. Margot & K. Stammbach, *Anal. Methods Pestic., Plant Growth Regul. Food Addit.,* 1964, **2**, 183; *Anal. Methods Pestic. Plant Growth Regul.,* 1972, **6**, 376.

Dinitramine

FXFFR BZ CNW ENW DN2&2

$C_{11}H_{13}F_3N_4O_4$ (322·2)

Nomenclature and development. The common name dinitramine is approved by BSI, ISO, ANSI and WSSA. The IUPAC name is N^1,N^1-diethyl-2,6-dinitro-4-trifluoromethyl-*m*-phenylenediamine, in *C.A.* usage N^3,N^3-diethyl-2,4-dinitro-6-(trifluoromethyl)-1,3-benzenediamine *[29091-05-2]* or N^1,N^1-diethyl-2,6-dinitro-4-(trifluoromethyl)-1,3-benzenediamine formerly N^1,N^1-diethyl-α,α,α-trifluoro-3,5-dinitrotoluene-2,4-diamine. It was introduced in 1971 by US Borax & Chemical Corp. under the code number 'USB 3584', trade mark 'Cobex' and protected by USP 3617252.

Manufacture and properties. Produced from 2,4-dichlorobenzotrifluoride by nitration, diethylamination and an ammonolysis, dinitramine is a yellow crystalline solid, m.p. 98-99°, v.p. 3·6 x 10⁻⁶ mmHg at 25° and decomposing above 200°. Its solubility at 25° is: 1·1 mg/l water; 120 g/kg ethanol; 640 g/kg acetone. The technical grade is > 83% pure. Both pure and technical dinitramine showed no significant decomposition after 2-y storage at ambient temperatures. No decomposition of a.i. or loss of emulsification properties were observed after storage of 'Cobex' at ambient temperatures for 5 y. It is subject to photodegradation. It is non-corrosive.

Uses. Dinitramine is a pre-em. herbicide which is incorporated in the soil against many annual grasses and broad-leaved weeds in cotton, soyabeans, sunflowers, beans, groundnuts, carrots, swedes, turnips, transplanted tomatoes, transplanted peppers and transplanted brassicas at 0·4-0·8 kg a.i./ha. It has little post-em. activity.

Toxicology. The acute oral LD50 for rats is 3000 mg a.i./kg; the acute dermal LD50 for rabbits is > 6800 mg/kg. In 90-d feeding trials no ill-effect was observed for rats and beagle dogs receiving 2000 mg/kg diet. The LC50 (96 h) is: for trout 6·6 mg/l; for bluegill 11 mg/l; for catfish 3·7 mg/l.

Formulation. An e.c. (250 g/kg; 240 g/l).

Analysis. Product analysis is by GLC, details from US Borax & Chemical Corp. Residues may be analysed by GLC with ECD after a liquid-liquid partition and column chromatographic clean-up. (H. C. Newsom & E. M. Mitchell, *J. Agric. Food Chem.*, 1972, **20**, 1222; H. C. Newsom, *Anal. Methods Pestic. Plant Growth Regul.*, 1976, **8**, 359).

Dinobuton

2Y1&R   CNW   ENW   BOVOY1&1 $C_{14}H_{18}N_2O_7$ (326·3)

Nomenclature and development. The common name dinobuton is approved by BSI and ISO; also known as ENT 27244. The IUPAC name is 2-*sec*-butyl-4,6-dinitrophenyl isopropyl carbonate (I), in *C.A.* usage 1-methylethyl 2-(1-methylpropyl)-4,6-dinitrophenyl carbonate *[973-21-7]* formerly (I). It was introduced in 1963 by Murphy Chemical Ltd under the code number 'MC 1053' and the trade marks *'Acrex'* and *'Sytasol'* and protected by BP 1019451. It was first described by M. Pianka & C. B. F. Smith, *Chem. Ind. (London),* 1965, p.1216. It is now manufactured by KenoGard AB under the trade mark 'Acrex'.

Manufacture and properties. Produced by the condensation of isopropyl chloroformate with an alkali metal salt of dinoseb (p.209), it forms pale yellow crystals, m.p. 61-62°, with negligible v.p. at room temperature. It is practically insoluble in water; soluble in aliphatic hydrocarbons, ethanol, fatty oils; highly soluble in lower aliphatic ketones, aromatic hydrocarbons. The technical product is 97% pure, m.p. 58-60°. It is non-corrosive but, to prevent hydrolysis, alkaline components must be avoided in its formulation. It is compatible with non-alkaline pesticides, but is deactivated by carbaryl.

Uses. Dinobuton is a non-systemic acaricide and a fungicide active against powdery mildews; it is recommended for glasshouse and field use against red spider mites and powdery mildews of vegetables, apples and cotton at 50 g a.i./100 l.

Toxicology. The acute oral LD50 is: for mice 2540 mg/kg; for rats 140 mg/kg; for hens 150 mg/kg. The acute dermal LD50 for rats is > 5000 mg/kg. The maximum 'no effect' level for dogs is 4·5 mg/kg daily, for rats 3-6 mg/kg daily. It acts as a metabolic stimulant, high doses causing loss of body weight. Its persistence in soils is brief.

Formulation. An e.c. (300 g a.i./l).

Analysis. Product and residue analysis is by chromatography on an alumina column and colorimetric measurement of the resulting dinoseb at 378 nm (*CIPAC Handbook,* in press; V. P. Lynch, *Anal. Methods Pestic. Plant Growth Regul.,* 1976, **8**, 275). Details are available from KenoGard AB.

Dinocap

(i)

(ii)

$C_{18}H_{24}N_2O_6$ (364·3)

6Y1&R	CNW	ENW	DOV1U2		6Y1&R	CNW	ENW	BOV1U2	$n = 0$
5Y2&R	CNW	ENW	DOV1U2		5Y2&R	CNW	ENW	BOV1U2	$n = 1$
4Y3&R	CNW	ENW	DOV1U2		4Y3&R	CNW	ENW	BOV1U2	$n = 2$

Nomenclature and development. The common name dinocap is approved by BSI and ISO for an isomeric reaction mixture—exception JMAF (DPC); also known as ENT 24727. The substance was originally thought to be 2-(1-methylheptyl)-4,6-dinitrophenyl crotonate (IUPAC) (I; ii, $n = 0$), in *C.A.* usage 2-(1-methylheptyl)-4,6-dinitrophenyl (*E*)-2-butenoate *[131-72-6]* formerly (I). Dinocap, *C.A.* Registry No. *[34300-45-3]* is a mixture of 2,4-dinitro-6-octylphenyl crotonates (ii) and 2,6-dinitro-4-octylphenyl crotonates (i), octyl here being a mixture of the 1-methylheptyl-, 1-ethylhexyl- and 1-propylpentyl- isomers; it normally contains 4-5 parts of (ii) and 2 parts of (i). It was introduced in 1946 by Rohm & Haas Co. under the code number 'CR-1693', trade mark 'Karathane' (formerly *'Arathane'*) and protected by USP 2526660; 2810767. The trade mark 'Crotothane' is registered by May & Baker Ltd.

Manufacture and properties. Produced by the nitration of the condensation product of phenol and octan-1-ol and esterification with crotonyl chloride, it is a dark brown liquid, b.p. 138-140°/0·05 mmHg, d 1·10. It is insoluble in water; soluble in most organic solvents.

Isomers. The nitrated phenols used for esterification with crotonyl chloride (A. H. M. Kirby & L. D. Hunter, *Nature (London)*, 1965, **208**, 189) and those obtained by hydrolysing commercial dinocap (D. R. Clifford *et al.*, *Chem. Ind. (London)*, 1965, p.1654) were shown to be a mixture of three 2,4-dinitro-6-octylphenols (*cf.* ii) and three 2,6-dinitro-4-octylphenols (*cf.* i) in which the side chains have $n = 0$, 1 and 2. The latter phenols are more effective fungicides than the former phenols (A. H. M. Kirby *et al.*, *Ann. Appl. Biol.*, 1966, **57**, 211; R. J. W. Bird *et al.*, *ibid*, p.223); the reverse is the case for acaricidal action (Kirby *et al.*, *loc. cit.*).

Uses. Dinocap is a non-systemic acaricide and contact fungicide, recommended for the control of *Panonychus ulmi* on apples and of powdery mildews on various fruits, grape vines and ornamentals at 70-1120 g a.i./ha.

Toxicology. The acute oral LD50 for rats is 980-1190 mg/kg (F. S. Larson *et al.*, *Arch. Int. Pharmacodyn. Ther.*, 1959, **119**, 31). In 1-y feeding trials dogs receiving 50 mg/kg diet suffered no loss in weight; but cataracts were produced in white Pekin ducks at this dose level.

Formulations. These include: w.p. (250 g a.i./kg); e.c. (500 g/l).

Analysis. Product analysis is by the Kjeldahl method (U.S. Dep. Agric., Agric. Res. Div., Method 580·0). Residues may be analysed by GLC (*Anal. Methods Pestic. Plant Growth Regul.*, 1972, **6**, 568) or by colorimetry at 442-444 nm after clean-up by steam distillation (I. Rosenthal *et al.*, *J. Agric. Food Chem.*, 1957, **5**, 914) or liquid-liquid partition (W. W. Kilgore & K. W. Cheng, *ibid.*, 1963, **11**, 477; Kilgore, *Anal. Methods Pestic., Plant Growth Regul. Food Addit.*, 1963, **3**, 107).

Dinoseb

WNR BQ ENW CY2&1 $C_{10}H_{12}N_2O_5$ (240·2)

Nomenclature and development. The common name dinoseb is approved by BSI, ISO, ANSI and WSSA—exception JMAF (DNBP); also known as ENT 1122. The IUPAC name is 2-*sec*-butyl-4,6-dinitrophenol (I), in *C.A.* usage 2-(1-methylpropyl)-4,6-dinitrophenol [88-85-7] formerly (I). It was first described as a herbicide by A. S. Crafts, *Science,* 1945, **101,** 417, and was developed by The Dow Chemical Co. under the code number 'DN 289', trade mark 'Premerge' and protected by USP 2192197.

Manufacture and properties. Produced by the nitration of 2-*sec*-butylphenol, obtained by the reaction of but-2-ene with phenol and separation of the 4-alkyl isomer, it is an orange-brown liquid 95-98% pure, m.p. 30-40°. Pure dinoseb has m.p. 38-42°. Its solubility at room temperature is *c.* 100 mg tech./l water; soluble in petroleum oils and most organic solvents. It is an acidic phenol, pK_a 4·62, and forms salts, some of which are water-soluble, with inorganic and organic bases. It is corrosive to mild steel in the presence of water.

Uses. Dinoseb is a contact herbicide used, as the ammonium or an amine salt, for post-em. annual weed control on peas, seedling lucerne, cereals and undersown cereals at ≤ 2 kg a.i./ha. For pre-em. control of annual weeds in peas, beans and potatoes and pre-harvest desiccation of potatoes and leguminous seed crops it is recommended at 2·5 kg (as e.c.)/ha.

Toxicology. The acute oral LD50 for rats is 58 mg/kg; the acute dermal LD50 for rabbits is 80-200 mg/kg. In 180-d feeding trials rats receiving 100 mg/kg diet suffered no ill-effect.

Formulations. Aqueous concentrates of its amine salts; solutions in oil.

Analysis. Product analysis is by a column chromatographic separation and colorimetry in alkali at 375 nm (C. E. Bowman & L. Westenberg, *Chem. Weekbl.,* 1956, **52,** 827; *CIPAC Handbook,* 1970, **1,** 337). Residues are determined by GLC of the methyl ether after chromatographic clean-up and treatment with diazomethane (J. R. Lane, *Anal. Methods Pestic., Plant Growth Regul. Food Addit.,* 1967, **5,** 385; *Anal. Methods Pestic. Plant Growth Regul.,* 1972, **6,** 639).

Dinoseb acetate

2Y1&R   CNW   ENW   BOV1 $C_{12}H_{14}N_2O_6$ (282·3)

Nomenclature and development. The common name dinoseb is approved by BSI, ISO, ANSI and WSSA—exception JMAF (DNBP). Its acetate ester, common name dinoseb acetate approved by BSI and ISO—exception JMAF (DNBPA)—has the IUPAC name 2-*sec*-butyl-4,6-dinitrophenyl acetate (I), in *C.A.* usage 2-(1-methylpropyl)-4,6-dinitro-phenyl acetate *[2813-95-8]* formerly (I). It was introduced in 1958 by Hoechst AG under the code number 'Hoe 2904', trade marks 'Aretit' and 'Ivosit' and protected by DBP 1 088 757; USP 3 130037; BP 909 372. Its herbicidal properties were first described by K. Haertel, *Meded. Landbouwhogesch. Opzoekingsstn. Staat Gent,* 1960, **25,** 1422.

Manufacture and properties. Produced by the acetylation of dinoseb (p.209), it is a brown oil, with an aromatic vinegar-like odour, m.p. 26-27°, v.p. 6×10^{-4} mmHg at 20°. The technical grade is a viscous brown oil, *c.* 94% pure. Its solubility at 20° is 1·6 g tech./l water; soluble in aromatic solvents. The ester is slowly hydrolysed in the presence of water and is sensitive to acids or alkali. The technical product is slightly corrosive.

Uses. Dinoseb acetate is a post-em. herbicide recommended against annual broad-leaved weeds in cereals, maize, peas, beans, potatoes and lucerne at 1·5-2·5 kg a.i./ha. It is also used in combination with monolinuron pre-em. in dwarf beans and potatoes.

Toxicology. The acute oral LD50 for rats is 60-65 mg/kg. The application of 200 mg w.p./kg 5 times to the shaved skin of rabbits caused no irritation. In 90-d feeding trials there was no symptom in rats receiving 50 mg/kg diet or dogs 10 mg/kg diet.

Formulations. These include: w.p. (376 g a.i./kg); e.c. (492 g/l); w.p., mixtures with monolinuron (in various ratios).

Analysis. Product analysis is by extraction with heptane, hydrolysis with ethanolic sodium hydroxide, the dinoseb formed being steam distilled and measured in dilute sodium hydroxide at 420 nm.

Dinoterb

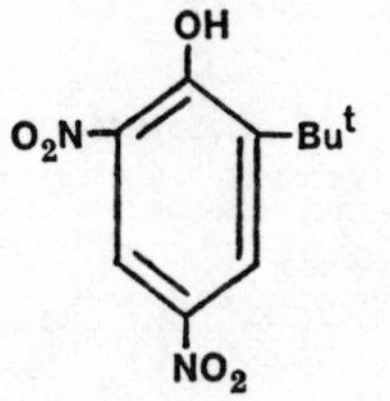

WNR BQ ENW CX1&1&1 $C_{10}H_{12}N_2O_5$ (240·2)

Nomenclature and development. The common name dinoterb is approved by BSI and ISO. The IUPAC name is 2-*tert*-butyl-4,6-dinitrophenol (I), in *C.A.* usage 2-(1,1-dimethyl-ethyl)-4,6-dinitrophenol *[1420-07-1]* formerly (I). Its herbicidal properties were first described by P. Poignant & P. Crisinel, *C. R. Journ. Etud. Herbic. Conf. COLUMNA, 4th,* 1967, p.196. It was introduced by Pépro (now a subsidiary of Rhône Poulenc Phytosanitaire) as the ammonium salt, *C.A.* Registry No. *[6365-83-9]* (FP 1475686; BP 1126658) or in combination with mecoprop (FP 1532332; USP 3565601) or isoproturon.

Manufacture and properties. Produced by the nitration of 2-*tert*-butylphenol, it is a yellow solid, m.p. 125·5-126·5°. It is practically insoluble in water but dissolves in aqueous alkali forming salts; its solubility is *c.* 200 g/kg ethyl acetate, cyclohexanone, dimethyl sulphoxide; *c.* 100 g/kg alcohols, glycols, aliphatic hydrocarbons.

Uses. Dinoterb is a contact herbicide used as the ammonium salt for post-em. control of annual weeds in cereals and lucerne, and pre-em. in peas and beans.

Toxicology. The acute oral LD50 for mice is *c.* 25 mg/kg; the acute dermal LD50 for guinea-pigs is 150 mg/kg.

Formulations. These include: 'Herbogil Super A', an aqueous paste of dinoterb-ammonium (500 g a.e./l); 'DM68', in combination with mecoprop; 'Tolkan A' and 'Tolkan V', in combination with isoproturon.

Analysis. Product analysis is by GLC (*CIPAC Handbook,* in press).

211

Dioxacarb

$$O$$
$$O.C.NHMe$$

T5O   COTJ   BR   BOVM1 $C_{11}H_{13}NO_4$ (223·2)

Nomenclature and development. The common name dioxacarb is approved by BSI, ISO, ANSI and JMAF; also known as ENT 27389. The IUPAC name is 2-(1,3-dioxolan-2-yl)-phenyl methylcarbamate (I), in *C.A.* usage (I) *[6988-21-2]* formerly *o*-1,3-dioxolan-2-ylphenyl methylcarbamate. It was introduced in 1968 by Ciba AG (now Ciba-Geigy AG) under the code number 'C 8353', trade marks 'Famid' and 'Elocron' and protected by BP 1122633. Its insecticidal properties were described by F. Bachmann & J. B. Legge, *J. Sci. Food Agric., Suppl.,* 1968, p.39; E. F. Nikles, *J. Agric. Food Chem.,* 1969, **17**, 939.

Manufacture and properties. Produced by the condensation of salicylaldehyde with ethylene glycol and reaction of the product with methyl isocyanate, it forms colourless crystals, m.p. 114-115°, v.p. 3×10^{-7} mmHg at 20°. Its solubility at 20° is: 6 g/l water; 280 g/l acetone; 235 g/l cyclohexanone; 550 g/l dimethylformamide; 80 g/l ethanol; 180 mg/l cyclohexane; 345 g/l dichloromethane; 9 g/l xylene. It is compatible with most pesticides and is non-corrosive. In aqueous media at 20°, 50% loss occurs in: 40 min at pH 3; 3 d at pH 5; 60 d at pH 7; 20 h at pH 9; 2 h at pH 10. It is rapidly decomposed in soil and is unsuitable for use against soil insects.

Uses. Dioxacarb is a contact and stomach insecticide used against cockroaches (including those resistant to organochlorine and organophosphorus insecticides) and against a wide range of household and stored products pests, for wall application at 0·5-2·0 g a.i./m². It is effective against a wide range of sucking and chewing foliage pests including aphids resistant to organophosphorus compounds, Colorado beetle, plant and leaf hoppers of rice, capsids; recommended against sucking pests at 250-500 g/ha, against chewing insects 500-750 g/ha. It has a rapid 'knockdown' action and persists on foliage 5-7 d, on wall surfaces *c.* 0·5 y depending on porosity.

Toxicology. The acute oral LD50 for rats is 60-80 mg tech./kg. The acute dermal LD50 is: for rats *c.* 3000 mg/kg; for rabbits 1950 mg/kg. In 90-d feeding trials no ill-effect was observed for rats receiving 10 mg/kg diet or dogs 2 mg/kg diet. It is relatively non-toxic to birds, fish and wildlife, but is toxic to honey bees.

Formulations. These include: 'Elocron 50 WP', w.p. (500 g/kg); 'Elocron 400 SCW', w.s.c. (400 g/l); 'Ulvair 250'.

Analysis. Product analysis is by titration of the hydrochloric acid produced on treatment with hydroxylamine hydrochloride (A. Becker & W. Zeizler, *Z. Anal. Chem.,* 1971, **257**, 125). Residues may be determined by GLC after conversion to a volatile derivative with 1-fluoro-2,4-dinitrobenzene, or hydrolysis to salicylaldehyde which is separated and determined by polarography: details of these methods are available from Ciba-Geigy AG.

Dioxathion

T60 DOTJ BSPS&O2&O2 CSPS&O2&O2 $\qquad$ $C_{12}H_{26}O_6P_2S_4$ (456·5)

Nomenclature and development. The common name dioxathion is approved by BSI, ISO and ANSI—exceptions Turkey and USSR (delnav), JMAF (dioxane phosphate) and Italy; also known as ENT 22879. The IUPAC name is S,S'-(1,4-dioxane-2,3-diyl) O,O,O',O'-tetraethyl di(phosphorodithioate) (I) formerly 1,4-dioxan-2,3-diyl S,S-di-(O,O-diethyl phosphorodithioate), in *C.A.* usage (I) *[78-34-2]* formerly S,S-*p*-dioxane-2,3-diyl O,O,O',O'-tetraethyl di(phosphorodithioate). It was introduced in 1954 by Hercules Inc. under the code number 'Hercules AC258', the trade mark 'Delnav' and protected by USP 2725328; 2815350. Its insecticidal properties were first described by W. R. Diveley *et al., J. Am. Chem. Soc.,* 1959, **81**, 139.

Manufacture and properties. Produced by the condensation of 2,3-dichloro-1,4-dioxane with O,O-diethyl hydrogen phosphorodithioate, the technical product is a brown liquid, d_4^{26} 1·257, n_D^{20} 1·5409, viscosity 0·117 Pa s at 25°. It consists of 68-75% dioxathion as a mixture of about 24% *cis-*, 48% *trans-* isomers and 30% related compounds. It is insoluble in water; solubility 10 g/kg hexane, kerosine; soluble in most other organic solvents. It is stable in water but is hydrolysed by alkali or on heating. It is unstable to iron or tin surfaces or when mixed with certain carriers.

Uses. Dioxathion is a non-systemic insecticide and acaricide especially useful for the treatment of livestock to control external pests including ticks. It is recommended against phytophagous mites on fruit (including citrus), nuts and ornamentals. The *cis-*isomer is somewhat more toxic to flies and rats than the *trans-* isomer (B. W. Arthur & J. E. Casida, *J. Econ. Entomol.,* 1959, **52**, 20).

Toxicology. The acute oral LD50 is: for male albino rats 43 mg/kg; for females 23 mg/kg. The acute dermal LD50 is: for albino male rats 235 mg/kg; for females 63 mg/kg. Cholinesterase inhibition was observed in female rats receiving 10 mg/kg diet, and in dogs receiving 10 doses of 0·8 mg/kg fed over 14 d, though there was no other sign of toxicity in the latter case. Relatively harmless to honey bees.

Formulations. These include e.c. (480 and 960 g tech./l); mixtures with other insecticides, for control of livestock pests.

Analysis. Product analysis is by i.r. spectrometry: details from Hercules Inc. Residues may be determined by conversion to 2,3-dichloro-1,4-dioxane which is hydrolysed and measured colorimetrically as glyoxal 2,4-dinitrophenylosazone (C. L. Dunn, *J. Agric. Food Chem.,* 1958, **6**, 203).

Diphacinone

L56　BV　DV　CHJ　CVYR&R　　　　　　　　　　　　$C_{23}H_{16}O_3$　(340·4)

Nomenclature and development. The common name diphacinone is approved by BSI, ISO and ANSI—exceptions Turkey (diphacin), BPC (diphenadione), Italy. The IUPAC name is 2-(diphenylacetyl)indan-1,3-dione (I) formerly 2-(diphenylacetyl)indane-1,3-dione, in *C.A.* usage 2-(diphenylacetyl)-1*H*-indene-1,3(2*H*)-dione *[82-66-6]* formerly (I). It has been in use as an anticoagulant rodenticide since the early 1950's, trade marks including 'Diphacin' and 'Ramik' (Velsicol Chemical Corp.). Its activity was reported by J. T. Correll *et al., Proc. Soc. Exp. Biol. Med.,* 1952, **80**, 139.

Manufacture and properties. Produced by the condensation of dimethyl phthalate with 1,1-diphenylacetone in the presence of sodium ethoxide (USP 2672483, Upjohn Co.), it forms yellow crystals, m.p. 145°. It is slightly soluble in water; soluble in acetone. It forms alkali metal salts, the sodium salt being sparingly soluble in water. Diphacinone is stable to hydrolysis and towards mild oxidising agents. It is non-corrosive.

Uses. Diphacinone is an anticoagulant rodenticide with an acute oral LD50 for rats *c.* 3 mg/kg; the chronic LD50 for albino rats is 0·1 mg/kg daily.

Toxicology. The acute oral LD50 is: for dogs 3·0-7·5 mg/kg; for cats 14·7 mg/kg; for pigs 150 mg/kg.

Formulations. Concentrates (1 g a.i./kg) for the preparation of baits include: 'Diphacin 110' on corn starch, for the preparation of dry baits using 1 part product + 19 parts good quality feedingstuff; 'Diphacin 120' (1·06 g diphacinone-sodium + 999 g sugar), for dilution at 20 g product/l. Optimum concentration of a.i. in the bait is 50-125 mg/kg depending on the pest species. Three bait forms, 'Ramik Brown', 'Ramik Red' and 'Ramik Green', flavoured as apple, meat and fish, respectively, are extruded pellet baits (50 mg a.i./kg).

Analysis. Product analysis is by u.v. or i.r. spectrometry. Soil residues may be determined by oxidation with chromic acid, the benzophenone formed being measured by GLC or HPLC.

Diphenamid

$$O$$
$$\|$$
$$Ph_2CH.C.NMe_2$$

1N1&VYR&R $C_{16}H_{17}NO$ (239·3)

Nomenclature and development. The common name diphenamid is approved by BSI, ISO, ANSI, WSSA and JMAF—exception Germany. The IUPAC name is *N,N*-dimethyl-2,2-diphenylacetamide (I), in *C.A.* usage *N,N*-dimethyl-α-phenylbenzeneacetamide *[957-51-7]* formerly (I). It was introduced as a herbicide in 1960 by Eli Lilly & Co. under the code number 'L- 34 314' and trade mark 'Dymid' and by The Upjohn Co. under the trade name 'Enide'. Its herbicidal properties were first described by E. F. Alder *et al., Proc. North Cent. Weed Control Conf.,* 1960, p.55, and it is protected by USP 3 120434.

Manufacture and properties. Produced by the reaction of diphenylmethane with dimethyl-carbamoyl chloride in the presence of sodium hydride, it forms colourless crystals, m.p. 134·5-135·5°, v.p. negligible at room temperature. Its solubility at 27° is 260 mg/l water; moderately soluble in polar organic solvents. The technical product is an off-white crystalline solid, m.p. 132-134°, and is stable at room temperature for 7 d in acid (pH 3) or alkali (pH 9) and non-corrosive.

Uses. Diphenamid is a selective pre-em. herbicide used to control most grass weeds and many broad-leaved weeds in a wide range of crops including: cotton, soyabeans, potatoes, brassicas, fruit trees, soft fruit, many vegetables, tobacco and ornamentals. It is generally used at 4-6 kg a.i./ha. It is metabolised in plants and animals by *N*-demethylation.

Toxicology. The acute oral LD50 for rats is 1050 mg/kg; the acute dermal LD50 for rats is > 225 mg/kg. In 2-y feeding trials dogs and rats receiving 2000 mg/kg diet suffered no unusual effect on their physiology or fertility.

Formulations. These include: 'Enide 90W', w.p. (900 g a.i./kg); 'Enide 50W', w.p. (500 g/kg); 'Enide granular' (200 g/kg).

Analysis. Residues may be determined by GLC with FID: G. A. Boyack *et al., J. Agric. Food Chem.,* 1966, **14,** 312; J. B. Tepe *et al., Anal. Methods Pestic., Plant Growth Regul. Food Addit.,* 1967, **5,** 375; *Anal. Methods Pestic. Plant Growth Regul.,* 1972, **6,** 637.

Dipropetryn

T6N CN ENJ BS2 DMY1&1 FMY1&1 $C_{11}H_{21}N_5S$ (255·4)

Nomenclature and development. The common name dipropetryn is approved by BSI, ISO, ANSI and WSSA. The IUPAC name is 2-ethylthio-4,6-bis(isopropylamino)-1,3,5-triazine, formerly 2-ethylthio-4,6-di(isopropylamino)-1,3,5-triazine, in *C.A.* usage 6-(ethylthio)-*N,N'*-bis(1-methylethyl)-1,3,5-triazine-2,4-diamine *[4147-51-7]* formerly 2-ethylthio-4,6-bis(isopropylamino)-*s*-triazine. It was developed by J. R. Geigy S.A. under the code number 'GS 16068' and introduced in 1971 by Ciba-Geigy AG under the trade mark 'Cotofor' and, in the USA, 'Sancap'.

Manufacture and properties. Produced by reacting propazine (p.448) with thiourea and subsequently diethyl sulphate, it is a colourless solid, m.p. 104-106°, v.p. 7·3 x 10⁻⁷ mmHg at 20°. Its solubility at 20° is 16 mg/l water; soluble in organic solvents.

Uses. Dipropetryn is a pre-em. herbicide used for weed control in cotton and melons at 1·25-3·5 kg a.i./ha.

Toxicology. The acute oral LD50 for rats is 3900-4200 mg/kg; the acute dermal LD50 for rabbits is > 10000 mg/kg. In 160-d feeding trials the 'no effect' level for dogs was 400 mg/kg diet. It is slightly toxic to birds and moderately toxic to fish.

Formulations. A w.p. (800 g/kg); 500 FW, flowable (500 g/l).

Analysis. Product analysis is by GLC with internal standard or by titration with perchloric acid in acetic acid: details are available from Ciba-Geigy AG. Residues may be determined by the methods given under ametryne (p.25).

Dipropyl isocinchomeronate

$$PrO_2C \diagdown \text{(pyridine ring)} \diagup CO_2Pr$$

T6NJ BVO3 EVO3 $C_{13}H_{17}NO_4$ (251·3)

Nomenclature and development. The trivial name dipropyl isocinchomeronate is normally used; also known as ENT 17595. The IUPAC name is dipropyl pyridine-2,5-dicarboxylate, in *C.A.* usage dipropyl 2,5-pyridinedicarboxylate *[136-45-8]*. Its insect repellent properties were discovered in 1951 at the University of Illinois (L. D. Goodhue & R. E. Stansbury, *J. Econ. Entomol.,* 1953, **46,** 982) and it was introduced commercially by the McLaughlin Gormley Co. under the name 'MGK Repellent 326' and protected by USP 2 757 120; 3 067 091.

Manufacture and properties. Produced by the oxidation of 5-ethyl-2-methylpyridine and esterification with propan-1-ol, it is an amber liquid, with a mild aromatic odour, b.p. 150°/1 mmHg, n_D^{25} 1·4979, d^{20} 1·082. The technical grade has d^{20} 1·106-1·130. It is practically insoluble in water; miscible with ethanol, propan-2-ol, methanol, kerosine. It is stable, except to sunlight, and non-corrosive. It is hydrolysed by alkali and is incompatible with alkaline pesticides or with high concentrations of dichlorvos.

Uses. Dipropyl isocinchomeronate is a fly repellent effective against *Musca domestica, Tabatius* and *Chrysops* spp., especially *M. vetustissima* of Australia. It lacks persistence at high humidity, a defect overcome by the addition of 7-hydroxy-4-methylcoumarin (USP 2 824 824).

Toxicology. The acute oral LD50 for rats is 5230-7230 mg/kg; the acute dermal LD50 for rats is 9400 mg/kg.

Formulations. Oil sprays or in ethanol or other carriers for personal use (2-50 g a.i./l). Other repellents may be included.

Analysis. Product and residue analysis is by GLC or by hydrolysis to isocinchomeronic acid and measuring the absorption of the hydrochloride at 227, 247 and 273 nm: details are available from McLaughlin Gormley King Co.

Diquat

2A⁻

T	B666	GK	JK&T&J		

diquat cation $C_{12}H_{12}N_2$ (184·2)

T	B666	GK	JK&T&J	&E	&E

diquat dibromide $C_{12}H_{12}Br_2N_2$ (344·0)

Nomenclature and development. The common name diquat is approved by BSI, ISO, ANSI, WSSA and JMAF—exceptions Germany (deiquat), USSR (region). The IUPAC name is 1,1'-ethylene-2,2'-bipyridyldiylium ion, in *C.A.* usage 6,7-dihydrodipyrido[1,2-a:2',1'-c]pyrazinediium ion, Registry Nos.: free ion *[2764-72-9]*, dibromide *[85-00-7]*, dibromide monohydrate *[6385-62-2]*. The dibromide was introduced in 1957 by ICI Ltd, Plant Protection Division under the code number 'FB/2' and the trade marks 'Reglone', 'Weedol' and 'Pathclear'; also 'Aquacide' (Chipman Chemical Co. Ltd). Its herbicidal properties were first described by R. C. Brian *et al.*, *Nature (London)*, 1958, **181**, 446 and are protected by BP 785 732.

Manufacture and properties. Diquat dibromide is produced by the reaction of 2,2'-bipyridyl, produced by the oxidative coupling of pyridine over heated Raney nickel, and 1,2-dibromoethane in water, yielding a technical product > 95% pure from which the solid salt is not isolated. The dibromide forms a monohydrate, colourless to yellow crystals, decomposing > 300°. Its solubility at 20° is 700 g/l water; slightly soluble in alcohols and hydroxylic solvents; practically insoluble in non-polar organic solvents. It is non-volatile and has no measurable vapour pressure. It is stable in acid and neutral solution but unstable under alkaline conditions. It may be inactivated by inert clays and by anionic surfactants. The unformulated compound is corrosive to most common metals, especially aluminium and zinc, but the formulated product contains corrosion inhibitors enabling aqueous solutions to be applied through spray machinery.

Uses. Diquat is a contact desiccant and herbicide with some systemic properties. It is rapidly absorbed by green plant tissues and is inactivated upon contact with soil. Its uses include potato haulm destruction (420-840 g diquat/ha); seed crop desiccation (392-784 g/ha); aquatic weed control (588-2950 g/ha). Where grasses predominate, a diquat/paraquat formulation is used.

Toxicology. The acute oral LD50 is: for rats 231 mg/kg; for mice 125 mg/kg; for dogs 100-200 mg/kg; for cows 30 mg/kg; for hens 200-400 mg/kg. The only pathological symptom associated with long-term feeding trials is the occurrence of bilateral cataracts. In 2-y studies on rats the 'no effect' level was 25 mg/kg diet; cataracts appeared after 124 d at 35 mg/kg diet. In 4-y studies on dogs the 'no effect' level was 50 mg/kg diet. Diquat is absorbed through human skin only after prolonged exposure; shorter exposure can cause irritation and a delay in the healing of cuts and wounds. It is an irritant to eyes and can cause temporary damage to nails and nose bleeding if inhaled.

Formulations. These include: aqueous concentrates (140 and 200 g diquat/l) as dibromide; water-soluble granules (25 g diquat + 25 g paraquat/kg) as dibromides.

continued

Analysis. Diquat may be identified by the green colour given by the free radical formed with alkaline sodium dithionite. Product analysis is by u.v. spectroscopy in acetate buffer at 310 nm (*CIPAC Handbook,* 1970, **1,** 342); impurities may be measured by GLC (*ibid.,* 1979, **1A,** in press; *Herbicides 1977,* p.48). Residues may be determined by extraction on a cation exchange column and, after reduction, measurement of the extinction at 379 nm and other wavelengths (A. Calderbank *et al., Analyst (London),* 1961, **86,** 569; A. Calderbank & S. H. Yuen, *ibid.,* 1965, **90,** 95; T. H. Byast *et al., Agric. Res. Counc. (G.B.) Weed Res. Organ. Tech. Rep.,* No. 15 (2nd Ed.), p.35) or by polarography (T. H. Byast, *et al., loc. cit.*). Details are available from ICI Ltd, Plant Protection Division. See also: J. E. Pack, *Anal. Methods Pestic., Plant Growth Regul. Food Addit.,* 1967, **5,** 397; J. B. Leary, *Anal. Methods Pestic. Plant Growth Regul.,* 1978, **10,** 321.

Disodium Octaborate

approximate composition

$$Na_2B_8O_{13}.4H_2O$$

.NA2.B8-O13.QH4 $B_8H_8Na_2O_{17}$ (412·5)

Nomenclature and development. The trivial name disodium octaborate is used, in *C.A.* usage disodium borate ($Na_2B_8O_{13}$) *[12008-41-2]* (for anhydrous salt). It was introduced as a non-selective herbicide by the US Borax & Chemical Corp. under the trade mark 'Polybor' and protected by USP 2998 310.

Manufacture and properties. Produced by spray-drying a solution containing disodium tetraborate and boric acid in the correct proportions, it is not a true compound, but a convenient homogeneous form exploiting the enhancement of water solubility when borax and boric acid are mixed. It is an amorphous colourless powder, composition *c.* $Na_2B_8O_{13}.4H_2O$, m.p. *c.* 195°. Its solubility at 20° is 95 g/l water. It is stable, non-flammable and non-corrosive.

Uses. Disodium octaborate is non-selective herbicide used on crop-free ground at 490-2400 kg/ha; it is generally used in combination with sodium chlorate. It is also used as a stump treatment as a solution (100 g/l) in conifer plantations as a protection against infection by *Fomes annosus.*

Toxicology. The acute oral LD50 for guinea-pigs is 5300 mg/kg.

Analysis. Product analysis is by standard methods such as the mannitol NaOH titration (AOAC Methods). Residues may be determined spectrophotometrically (C. M. Austin & J. S. McHargue, *J. Assoc. Off. Agric. Chem.,* 1948, **31**, 427).

Disulfoton

$$S$$
$$\|$$
$$(EtO)_2P.S.CH_2CH_2.SEt$$

2S2SPS&O2&O2 $C_8H_{19}O_2PS_3$ (274·4)

Nomenclature and development. The common name disulfoton is approved by BSI and ISO—exceptions USSR (M-74) and JMAF (ethylthiodemeton); also known as ENT 23347, thiodemeton. The IUPAC and *C.A.* name is *O,O*-diethyl *S*-2-ethylthioethyl phosphorodithioate *[298-04-4]*. It was introduced in 1956 by Bayer AG under the code numbers 'Bayer 19639' and 'S 276' and the trade marks 'Disyston' ('Di-Syston' in USA), 'Dithiosystox' and protected by DBP 917668; 947369; USP 2759010; and by Sandoz AG under the trade marks 'Frumin AL' (1964) and 'Solvirex' (1966).

Manufacture and properties. Produced by the reaction of *O,O*-diethyl hydrogen phosphorodithioate with 2-chloroethyl ethyl sulphide, it is a colourless oil, with a characteristic odour, b.p. 62°/0·01 mmHg, v.p. 1·8 x 10^{-4} mmHg at 20°, d_4^{20} 1·144, n_D^{20} 1·5496. The technical product is a dark yellowish oil. Its solubility at room temperature is 25 mg/l water; readily soluble in most organic solvents. It is relatively stable to hydrolysis < pH 8·0.

Uses. Disulfoton is a systemic insecticide and acaricide used mainly for treatment of seed and as granules for soil application to protect seedlings from insect attack. It is metabolised in plants to the sulphoxide and sulphone and the corresponding derivatives of the phosphorothioate, demeton-S (R. L. Metcalf *et al., J. Econ. Entomol.,* 1957, **50**, 338). Its fate in soils is discussed by I. Takase, *Pflanzenschutz-Nachr. (Am. Ed.),* 1972, **25**, 43.

Toxicology. The acute oral LD50 for rats is 2·6-8·6 mg/kg; the acute dermal LD50 for male rats is *c.* 20 mg/kg. In 90-d feeding trials the 'no effect' level for rats was 1 mg/kg diet. It is probably metabolised in animals to demeton-S (T. J. Bombinski & K. P. DuBois, *AMA Arch. Ind. Health,* 1958, **17**, 192).

Formulations. These include: 'Disyston' and 'Frumin AL', seed dressing powder (500 g a.i./kg); 'Disyston' and 'Solvirex', granules (50-100 g/kg) on various mineral carriers or carbon.

Analysis. Product analysis is: by GLC (details available from Sandoz AG); by alkaline hydrolysis and estimation of the thiol with 5,5'-dithiobis(2-nitrobenzoic acid) at 617 nm (T. D. Talbott *et al., J. Agric. Food Chem.,* 1976, **24**, 155); by hydrolysis with dilute ethanolic sodium hydroxide and back titration; by hydrolysis with dilute alkali containing lead acetate; separation of the 2-(ethylthio)ethylthio-lead complex which is titrated with iodine (W. Pilz, *Z. Anal. Chem.,* 1958, **164**, 241); or by phosphorus determination after paper chromatography (Sandoz AG). Residues may be determined by GLC (see demeton, p.154) (M. C. Bowman & M. Beroza, *J. Assoc. Off. Anal. Chem.,* 1969, **52**, 1231), or phosphorus determination after separation by paper chromatographic separation. See also: D. MacDougall & T. E. Archer, *Anal. Methods Pestic., Plant Growth Regul. Food Addit.,* 1964, **2**, 187; *Anal. Methods Pestic. Plant Growth Regul.,* 1972, **6**, 377.

Ditalimfos

$$(EtO)_2 \overset{\overset{S}{\|}}{P}-N$$

T56 BVNVJ CPS&O2&O2 $C_{12}H_{14}NO_4PS$ (299·3)

Nomenclature and development. The common name ditalimfos is approved by BSI, ANSI, New Zealand and proposed by ISO. The IUPAC name is O,O-diethyl phthalimidophos-phonothioate (I), in *C.A.* usage O,O-diethyl (1,3-dihydro-1,3-dioxo-2*H*-isoindol-2-yl)-phosphonothioate *[5131-24-8]* formerly (I). Its fungicidal properties were first described by H. Tolkmith, *Nature (London),* 1966, **211,** 522 and it was introduced in 1966 by The Dow Chemical Co. under the code number 'DOWCO 199', trade mark 'Plondrel' and protected by BelgP 661 891; BP 1 034 493.

Manufacture and properties. Produced by the reaction between O,O-diethyl phosphoro-chloridothioate with the potassium salt of phthalimide, it forms colourless plates, with a mild thiophosphate odour, m.p. 83-84°, v.p. $1·45 \times 10^{-6}$ mmHg at 25°. Its solubility at room temperature is *c.* 133 mg/l water; soluble in hexane, cyclohexane, ethanol; very soluble in benzene, carbon tetrachloride, ethyl acetate, xylene. It is stable to u.v. light but its stability is reduced at pH > 8 or at temperatures above its m.p. There is no indication during storage of the thiono-thiolo isomerisation which is characteristic of some phosphorothioate esters.

Uses. Ditalimfos is a non-systemic foliar fungicide with protectant and curative activity. It is used to control powdery mildew on ornamentals (primarily roses) and vegetables (cucurbits) under glass, as well as under field conditions, at 30-50 g a.i./100 l; on apples (25-50 g/100 l) and cereals (500-550 g/ha). It is also used against *Venturia inaequalis* on apples at 37·5-100 g/100 l. It is liable to 'russet' certain apple cultivars, particularly Golden Delicious. It is compatible with most pesticides with the exception of alkaline materials.

Toxicology. The acute oral LD50 is: for rats 4930-5660 mg/kg; for male guinea-pigs 5660 mg/kg; for chickens 4500 mg/kg. The acute dermal LD50 for rabbits is > 2000 mg/kg. In 102-d feeding trials rats receiving 51 mg/kg daily showed no sign of toxicity. It is a primary skin irritant and may produce an allergic response in some sensitive individuals. Hazards to wildlife are low; the LD50 for honey bees is > 100 µg/bee.

Formulations. A w.p. (500 g a.i./kg); e.c. (200 g/l).

Analysis. Product analysis is by u.v. spectroscopy. Residues may be determined by GLC. Details of both methods are available from The Dow Chemical Co.

Dithianon

T C666 BV DS GS IVJ ECN FCN $C_{14}H_4N_2O_2S_2$ (296·3)

Nomenclature and development. The common name dithianon is approved by BSI, ISO and JMAF—exception Italy. The IUPAC name is 5,10-dihydro-5,10-dioxonaphtho-[2,3-b]-1,4-dithiin-2,3-dicarbonitrile (I) formerly 2,3-dicyano-1,4-dithia-anthraquinone, in *C.A.* usage (I) *[3347-22-6]* formerly 5,10-dihydro-5,10-dioxonaphtho[2,3-b]-p-dithiin-2,3-dicarbonitrile. It was introduced by E. Merck in 1962 under the code numbers 'IT931', 'MV119A', and trade mark 'Delan'; protected by BP 857 383. Its fungicidal properties were described by J. Berker *et al.*, *Proc. Br. Insectic. Fungic. Conf., 2nd,* 1963, p.351.

Manufacture and properties. Produced by the reaction of disodium 2,3-dimercapto-maleonitrile (from carbon disulphide and sodium cyanide) with 1,4-naphthoquinone, it forms brown crystals, m.p. 225°. It is practically insoluble in water but soluble in dioxane, chlorobenzene, chloroform. It is decomposed under alkaline conditions and is incompatible with petroleum oil sprays.

Uses. Dithianon is a protectant fungicide effective against many foliar diseases of pome and stone fruit but not against powdery mildews. Pathogens controlled include *Venturia* spp. (apple, pear and cherry), *Microthyriella rubi* (apple), *Coccomyces hiemalis* and *Clasterosporium carpophilum* (cherry), *Monilia* spp., *Taphrina deformans* and *Tranzschelia discolor* (peach and apricot), *Plasmopara viticola* (grape), *Pseudoperonospora humuli* (hop), *Elsinoe fawcetti* and *Phomopsis citri* (citrus), *Mycosphaerella fragrariae* and *Diplocarpon earliana* (strawberry), *Colletotrichum coffeanum* (coffee), *Cronartium ribicola, Drepanopeziza ribis* and *Plasmopara ribicola* (currant). The average rate is 50 g a.i./100 l. It is non-phytotoxic at fungicidal concentrations.

Toxicology. The acute oral LD50 is: for rats 638 mg/kg; for dogs > 900 mg/kg; for guinea-pigs 110 mg/kg. It is of low dermal toxicity. Rats fed for 120 d at 20 mg/kg daily suffered neither toxic nor carcinogenic effects during the succeeding 1·7 y. Non-toxic to honey bees.

Formulation. A w.p. (750 g a.i./kg); dispersion in water (250 g/l).

Analysis. Product analysis after chromatographic clean-up is by nitrogen estimation or reaction with 2,4-dinitrophenylhydrazine measuring the absorbance at 550 nm in alkaline solution (S. H. Yuen, *Analyst (London),* 1969, **94,** 1095). Residues may be determined by a colorimetric measurement (H. Sieper & H. Pies, *Z. Anal. Chem.,* 1968, **242,** 234). See also: E. Amodori & W. Heupt, *Anal. Methods Pestic. Plant Growth Regul.,* 1978, **10,** 181.

Diuron

GR BG DMVN1&1 $C_9H_{10}Cl_2N_2O$ (233·1)

Nomenclature and development. The common name diuron is approved by BSI, ISO, ANSI and WSSA—exception USSR (dichlorfenidim), JMAF (DCMU) and Sweden. The IUPAC name is 3-(3,4-dichlorophenyl)-1,1-dimethylurea (I), in *C.A.* usage N'-(3,4-dichlorophenyl)-*N,N*-dimethylurea *[330-54-1]* formerly (I). Its herbicidal properties were first reported by H. C. Bucha & C. W. Todd, *Science,* 1951, **114,** 493, and it was introduced in 1954 by the E. I. du Pont de Nemours & Co. (Inc.) under the trade mark 'Karmex' and protected by USP 2655445; BP 691403; 692589.

Manufacture and properties. Produced by the reaction of 3,4-dichlorophenyl isocyanate with dimethylamine, it is a colourless solid, m.p. 158-159°, v.p. $3·1 \times 10^{-6}$ mmHg at 50°. Its solubility is: at 25° 42 mg/l water; at 27° 53 g/kg acetone; sparingly soluble in hydrocarbons. It is stable to oxidation and moisture and its rate of hydrolysis is negligible at ordinary temperatures and under normal conditions, but is increased under acid or alkaline conditions or at elevated temperatures. It decomposes at 189° and is non-corrosive.

Uses. Diuron is a herbicide that inhibits photosynthesis and is used for general weed control on non-crop areas at 10-30 kg a.i./ha; for subsequent annual maintenance 5-10 kg/ha will prevent re-infestation by seedlings. It is also used selectively on crops such as sugarcane, citrus, pineapple, cotton, asparagus, temperate tree and bush fruits at 0·6-4·8 kg/ha. Phytotoxic residues in soil disappear within 1 season at these lower rates. It is degraded in soil by demethylation.

Toxicology. The acute oral LD50 for rats is 3400 mg/kg. The concentrated material may cause irritation to the eyes and mucous membranes but a 50% water paste was not irritating to the intact skin of guinea-pigs. In 2-y feeding trials the 'no effect' level was: for rats 250 mg/kg diet; for dogs 125 mg/kg diet (H. C. Hodge *et al., Food Cosmet. Toxicol.,* 1967, **5,** 513).

Formulations. These include: 'Karmex' weed killer, w.p. (800 g a.i./kg); 'Krovar' I, w.p. (400 g diuron + 400 g bromacil/kg); 'Krovar' II, w.p. (270 g diuron + 530 g bromacil/kg).

Analysis. Product analysis is by hydrolysis and titration of the amine liberated (*CIPAC Handbook,* 1979, **1A,** in press; details available from E. I. du Pont de Nemours & Co., Inc.). Residues may be determined by GLC (T. H. Byast *et al., Agric. Res. Counc. (G.B.) Weed Res. Organ., Tech. Rep.,* No. 15 (2nd Ed.), p.49) or by hydrolysis, isolation and diazotisation of the 3,4-dichloroaniline formed, and coupling with *N*-(1-naphthyl)ethylenediamine, the extinction being measured at 560 nm (H. L. Pease, *J. Agric. Food Chem.,* 1961, **10,** 279; R. L. Dalton & H. L. Pease, *J. Assoc. Off. Agric. Chem.,* 1962, **45,** 377). See also: *Anal. Methods Pestic. Plant Growth Regul.,* 1972, **6,** 664.

DNOC

WNR BQ Cl ENW $C_7H_6N_2O_5$ (198·1)

Nomenclature and development. The common name DNOC is approved by BSI, ISO, WSSA and JMAF; also known as ENT 154 and originally as DNC. The IUPAC name is 4,6-dinitro-*o*-cresol (I), in *C.A.* usage 2-methyl-4,6-dinitrophenol *[534-52-1]* formerly (I). It was first used as an insecticide in 1892 by Fr Bayer & Co. under the trade mark 'Antinonnin' and protected by BP 3301; it was introduced as a herbicide in 1932 by G. Truffaut et Cie under the trade mark 'Sinox' and protected by BP 425 295.

Manufacture and properties. Produced by the sulphonation of *o*-cresol, followed by controlled nitration, it forms yellowish crystals, m.p. 86°, v.p. 1·05 x 10⁻⁴ mmHg at 25°. Its solubility at 15° is 130 mg/l water; soluble in most organic solvents. The ammonium, sodium, potassium and calcium salts are water-soluble. The technical grade is 95-98% pure, m.p. 83-85°. It is explosive and is usually moistened with up to 10% water to reduce the hazard, though it is corrosive to mild steel in the presence of moisture.

Uses. DNOC is a non-systemic stomach poison and contact insecticide, ovicidal to the eggs of certain insects and spider mites. It is strongly phytotoxic and its use as an insecticide is limited to dormant sprays especially for fruit trees or on waste ground—*e.g.* against locusts. It is used as a contact herbicide for the control of broad-leaved weeds in cereals at ≤ 10 kg/ha and, in e.c. formulations, for the pre-harvest desiccation of potatoes and leguminous seed crops.

Toxicology. The acute oral LD50 is: for rats 25-40 mg/kg; for sheep 200 mg DNOC-sodium/kg. Rats tolerated 100 mg/kg diet. DNOC acts as a cumulative poison in man though there is little evidence of accumulation in laboratory animals (D. G. Harvey *et al., Br. Med. J.,* 1951, **2**, 13; E. King & D. G. Harvey, *Biochem. J.,* 1953, **53**, 185, 196).

Formulations. Pastes of the sodium or amine salts, ammonium sulphate being added as 'activator'; solutions in petroleum oil.

Analysis. Product analysis is by recovering the DNOC by methods which avoid reduction, and estimating by iodometric titration (R. L. Wain, *Ann. Appl. Biol.,* 1942, **29**, 301) or by titration with titanium trichloride in an inert atmosphere (W. Fischer, *Z. Anal. Chem.,* 1938, **112**, 91). See also: *MAFF Tech. Bull.,* No. 1, 1958; *CIPAC Handbook,* 1970, **1**, 348. Residues may be determined by colorimetry in alkaline solution (A. W. Avens *et al., J. Econ. Entomol.,* 1948, **41**, 432; W. H. Parker, *Analyst (London),* 1949, **74**, 646).

Dodemorph

T6N DOTJ C1 E1 A- AL-12-TJ   *dodemorph* $C_{18}H_{35}NO$ (281·5)

T6N DOTJ C1 E1 A- AL-12-TJ &QV1   *dodemorph acetate* $C_{20}H_{39}NO_3$ (341·5)

Nomenclature and development. The common name dodemorph is approved by BSI and ISO. The IUPAC and *C.A.* name is 4-cyclododecyl-2,6-dimethylmorpholine *[1593-77-7]*, though up to 1973 *C.A.* used the erroneous name 4-cyclododecyl-3,5-dimethylmorpholine *[35842-13-2]*. The acetate salt, dodemorph acetate—with *C.A.* usage correct, Registry No. *[31717-87-0]*—was introduced in 1967 by BASF AG under the code number 'BAS 238F' and protected by DBP 1 198 125. The fungicidal properties of morpholines with large alkyl groups attached to the nitrogen were first described by K. H. König *et al., Angew. Chem.,* 1965, **77**, 327; those of dodemorph by J. Kradel & E. -H. Pommer, *Proc. Br. Insectic. Fungic. Conf., 4th,* 1967, 170.

Properties. Dodemorph acetate is a yellow liquid, with a characteristic odour, *d c.* 0·93. It is miscible with water, stable in unopened containers for ≥ 1 y and is compatible with other pesticides. The decomposition products are flammable. Dodemorph is practically insoluble in water.

Uses. Dodemorph is an eradicant fungicide which is absorbed by foliage and roots to give a systemic protective action. The 40% concentrate is effective against powdery mildews when applied, at 14-d intervals, at 100-180 g a.i./100 l on roses outdoors, 100-150 g/100 l under glass, and on cucumbers at 50 g/100 l. It is phytotoxic to cinerarias and begonias.

Toxicology. The acute oral LD50 for rats is 1800 mg a.i. (as 4·5 ml 40% formulation)/kg; it causes severe skin and eye irritation to rabbits. The LC50 for guppies is *c.* 40 mg/l. It is harmless to honey bees.

Formulations. These include: a concentrate 'Meltatox' ('BASF F238') (400 g a.i./l); 'Badilin Rosenfluid' ('BAS 31 002F') (280 g dodemorph + 58 g dodine/l).

Analysis. Product analysis is by potentiometric titration with perchloric acid in glacial acetic acid. Residues in crops may be analysed by colorimetry of a complex formed with methyl orange.

Dodine

$$\left[\begin{array}{c} NH_2 \\ \| \\ C_{12}H_{25}NH.C.NH_2 \end{array} \right]^{+} \quad \left[\begin{array}{c} O \\ \| \\ CH_3.C.O \end{array} \right]^{-}$$

MUYZM12

MUYZM12 &QV1

free base $C_{13}H_{29}N_3$ (227·4)

dodine $C_{15}H_{33}N_3O_2$ (287·4)

Nomenclature and development. The common name dodine is approved by BSI, ISO and ANSI—exceptions France (doguadine), USSR (tsitrex) and JMAF (guanidine) for 1-dodecylguanidinium acetate (IUPAC); prior to 1969 BSI applied the name dodine to the free base. The *C.A.* name is dodecylguanidine monoacetate *[2439-10-3].* Its fungicidal properties were first described by B. Cation, *Plant Dis. Rep.,* 1957, **41,** 1029, and it was introduced in 1956 by American Cyanamid Co. under the code number 'AC 5223', the trade marks 'Cyprex', 'Melprex' and protected by USP 2 867 562.

Manufacture and properties. Produced by the reaction of a 1-dodecyl halide with sodium cyanamide and treatment of the product with ammonia, its acetate forms colourless crystals, m.p. 136°. It is soluble in water, ethanol; insoluble in most organic solvents. It is stable under moderately alkaline or acid conditions but the free base, Registry No. *[112-65-2],* is liberated by concentrated alkali. Dodine is compatible with anionic surfactants and with hard water.

Uses. Dodine is a protective fungicide recommended for the control of a number of major fungal diseases of fruit, nut and vegetable crops and on certain ornamentals and shade trees. Foliar application is recommended, especially against *Venturia* spp. (on apple and pear) and cherry leaf spot, against which dodine has some eradicant action. The usual rate is 30-80 g a.i./100 l.

Toxicology. The acute oral LD50 for male rats is *c.* 1000 mg/kg; the acute dermal LD50 for rabbits is > 1500 mg/kg. In 2-y feeding trials rats receiving 800 mg/kg diet suffered a slight retardation of growth with no effect on reproduction or lactation. Non-toxic to honey bees at normal rates of usage.

Formulations. These include: w.p. (650 and 800 g a.i./kg); liquid (200 and 250 g/l); dust (750 g/kg).

Analysis. Product analysis is by titration in non-aqueous media (*CIPAC Handbook,* in press). Residues may be determined by colorimetric measurement of the complex formed between dodine and bromocresol purple (W. A. Stellar *et al., J. Agric. Food Chem.,* 1960, **8,** 460; N. R. Pasarela, *J. Assoc. Off. Agric. Chem.,* 1964, **47,** 300; G. L. Sutherland, *Anal. Methods Pestic., Plant Growth Regul. Food Addit.,* 1964, **3,** 41).

Drazoxolon

T5NOVYJ   DUNMR   BG&   El $C_{10}H_8ClN_3O_2$ (237·6)

Nomenclature and development. The common name drazoxolon is approved by BSI and ISO. The IUPAC name is 4-(2-chlorophenylhydrazono)-3-methyl-5-isoxazolone, in *C.A.* usage 3-methyl-4,5-isoxazoledione 4-[(2-chlorophenyl)hydrazone] *[5707-69-7]*. It was synthesised in 1960 by ICI Ltd, Plant Protection Division and introduced as a fungicide under the code number 'PP 781' and protected by BP 999097. Described by W. H. Read, *World Rev. Pest Control*, 1966, **5**, 45; M. J. Geoghegan, *Proc. Br. Insectic. Fungic. Conf., 4th*, 1967, **1**, 451.

Manufacture and properties. Produced by coupling 2-chlorobenzenediazonium chloride with ethyl 3-oxobutyrate followed by reaction with hydroxylamine, it is a yellow crystalline solid, with a faint odour, m.p. 167°, v.p. 4 x 10⁻⁶ mmHg at 30°. Its solubility is: almost insoluble in water, aliphatic hydrocarbons; 40 g/kg aromatic hydrocarbons; *c.* 100 g/kg chloroform; 10 g/kg ethanol; 50 g/kg ketones. It is insoluble in and stable to acids; it dissolves in aqueous alkali forming salts. It is non-corrosive to the usual materials used for packaging and spray appliances but, if long storage is anticipated, polythene liners should be used in spray drums. Aqueous formulations should not be stored in metal containers. It is incompatible with lime sulphur and should not be used with dodine on apples because of phytotoxicity.

Uses. Drazoxolon is effective in the control of powdery mildews on blackcurrants, roses and other crops and for the control of certain other foliar diseases; also as a seed treatment for the control of *Pythium* and *Fusarium* spp. on such crops as peas, beans, maize, grass and as a soil treatment to control damping-off in seedling ornamentals.

Toxicology. The acute oral LD50 is: for rats 126 mg/kg; for mice 129 mg/kg; for rabbits 100-120 mg/kg; for guinea-pigs 12·5-25 mg/kg; for dogs 17 mg/kg; for cats 50-100 mg/kg; for hens *c.* 100 mg/kg. It has no marked irritant effect on the skin and eyes, but some sensitisation may occur and prolonged contact should be avoided. It causes some lung irritation and vapour concentrations > 0·5 mg/m³ must be avoided. In 90-d feeding trials there was no significant abnormality in rats receiving 30 mg/kg diet nor in dogs receiving 2 mg/kg daily.

Formulations. These include: 'Mil-Col', an aqueous suspension (300 g a.i./l); 'SAIsan', a col (300 g/l); 'Ganocide', a grease formulation (100 g/kg) for the control of *Ganoderma* spp. on rubber.

Analysis. Product analysis is by treatment with aqueous sodium hydroxide to form the sodium salt, which is separated from insoluble material, the solution acidified, extracted with chloroform, the colour of which is measured at 400 nm. Residues are determined in the same way after clean-up of the chloroform extract. Details of methods are available from ICI Ltd, Plant Protection Division. See also: S. H. Yuen, *Anal. Methods Pestic. Plant Growth Regul.*, 1973, **7**, 665.

Edifenphos

$$\begin{array}{c} O \\ \parallel \\ EtO.P(SPh)_2 \end{array}$$

2OPO&SR&SR $C_{14}H_{15}O_2PS_2$ (310·4)

Nomenclature and development. The common name edifenphos is approved by BSI and ISO—exception JMAF (EDDP). The IUPAC and *C.A.* name is *O*-ethyl *S,S*-diphenyl phosphorodithioate, Registry No. *[17109-49-8]*. It was introduced in 1968 by Bayer AG under the code number 'Bayer 78 418', trade mark 'Hinosan' and protected by BelgP 686 048; DAS 1 493 736. It was first described by H. Scheinpflug & H. F. Jung, *Pflanzenschutz-Nachr. (Am. Ed.)*, 1968, **21**, 79.

Manufacture and properties. Produced by the reaction of ethyl phosphorodichloridoate with thiophenol, it is a clear yellow to light-brown liquid, with an odour characteristic of thiophenol, b.p. 154°/0·01 mmHg, d_4^{20} *c.* 1·23, n_D^{22} 1·61. It is practically insoluble in water; soluble in acetone, xylene. At 25° 50% loss occurs in 19 d at pH 7, in 2 d at pH 9.

Uses. Edifenphos is a fungicide with specific action against *Piricularia oryzae* on rice at 30-50 g a.i./l water using 800-1200 l/ha; 1 or 2 applications to wet paddy in the nursery, 2 or 3 applications after transplanting or in fields of broadcast rice before tillering has ceased. It is also effective against *Pellicularia sasakii* and ear blight and is well tolerated by rice varieties at effective fungicidal rates. It should not be used within 10 d before or after an application of propanil.

Toxicology. The acute oral LD50 is: for male rats 340 mg tech. (in ethanol + propylene glycol)/kg; for female rats 150 mg/kg; for mice 218 mg/kg; for guinea-pigs and rabbits 350-400 mg/kg; for hens *c.* 750 mg tech. (in 'Lutrol')/kg. A 24-h application of the technical product to a rabbit's ear caused no injury. The LC50 for mirror carp is: 1·3 mg a.i./l (using e.c. of 500 g a.i./l) and 0·9 mg a.i./l (using dust of 20 g a.i./kg)/l.

Formulations. These include: e.c. (300, 400 and 500 g a.i./l); dusts (15, 20 and 25 g/kg).

Analysis. Details of methods for product and residue analysis are available from Bayer AG. Residues in rice may be extracted into acetone and estimated, after liquid-liquid partition clean-up, by GLC with thermionic detection (K. Vogeler, *ibid.*, p.317).

Eglinazine-ethyl

| T6N | CN | ENJ | BM2 | DM1VQ | FG | | *eglinazine* | $C_7H_{10}ClN_5O_2$ | (237·6) |
| T6N | CN | ENJ | BM2 | DM1VO2 | FG | | *eglinazine-ethyl* | $C_9H_{14}ClN_5O_2$ | (259·7) |

Nomenclature and development. The common name eglinazine is approved by BSI and proposed by ISO. The IUPAC name is *N*-(4-chloro-6-ethylamino-1,3,5-triazin-2-yl)glycine (I), in *C.A.* usage (I) *[68228-19-3]* formerly *N*-[4-chloro-6-(ethylamino)-*s*-triazin-2-yl]glycine. The ethyl ester, Registry No. *[6616-80-4]*, was introduced in 1972 by Nitrokémia Ipartelepek as a herbicide under the code number 'MG-06'.

Manufacture and properties. Produced by the reaction of trichloro-1,3,5-triazine with 1 molecular proportion of ethyl glycinate hydrochloride, followed by 1 molecular proportion of ethylamine in the presence of an acid binding agent, it forms colourless crystals, m.p. 228-230°. Its solubility at 25° is 300 mg/l water; soluble in organic solvents. It is stable in neutral, weakly acidic or weakly basic media at room temperature, but on heating it is hydrolysed by acid or alkali to give the corresponding herbicidally inactive triazin-2-ol. It is compatible with other pesticides when used at normal rates and is non-corrosive. It is rapidly decomposed in soil, 50% loss occurring in 12-18 d.

Uses. Eglinazine-ethyl is a pre-em. herbicide used in cereals at 3 kg a.i./ha, and is especially effective against *Apera spica-venti* and *Matricaria inodora* though its herbicidal range is narrow.

Toxicology. The acute oral LD50 is: for rats and mice > 10000 mg/kg; for rabbits 3000 mg/kg; for guinea-pigs > 3375 mg/kg. The acute intraperitoneal LD50 for rats and mice is > 7100 mg/kg.

Formulation. A w.p. (500 g a.i./kg).

Analysis. Product analysis is by GLC. Residues may be determined by GLC with TID.

Endosulfan

$$C_9H_6Cl_6O_3S \quad (406\cdot9)$$

T C755 A EOSO KUTJ AG AG BG FO JG KG LG

Nomenclature and development. The common name endosulfan is approved by BSI, ISO and ANSI—exceptions Iran and USSR (thiodan), JMAF (benzoepin) and Italy; also known as ENT 23979, OMS 570. The IUPAC name is 1,4,5,6,7,7-hexachloro-8,9,10-trinorborn-5-en-2,3-ylenedimethyl sulphite formerly 6,7,8,9,10,10-hexachloro-1,5,5a,6,9,9a-hexahydro-6,9-methano-2,4,3-benzo[e]dioxathiepin 3-oxide, in *C.A.* usage 6,7,8,9,10,10-hexachloro-1,5,5a,6,9,9a-hexahydro-6,9-methano-2,4,3-benzodioxathiepin 3-oxide *[115-29-7]* formerly 1,4,5,6,7,7-hexachloro-5-norbornene-2,3-dimethanol cyclic sulfite. Its insecticidal properties were first described by W. Findenbrink, *Nachrichtenbl. Dtsch. Pflanzenschutzdienstes (Brawnschweig), 1956, **8**, 183. It was introduced in 1956 by Hoechst AG under the code number 'Hoe 2671', trade mark 'Thiodan' and protected by DBP 1015797; USP 2799685; BP 810602. In the USA it was developed by the Agricultural Division of the FMC Corp. as 'FMC 5462', trade mark 'Thiodan'. Other trade marks include 'Cyclodan', 'Beosit', 'Malix', 'Thimul' and 'Thifor'.

Manufacture and properties. Produced by reacting thionyl chloride with the hydrolysed Diels-Alder adduct of hexachlorocyclopentadiene and (Z)-but-2-enylene diacetate, the technical grade is a brownish crystalline solid, m.p. 70-100°, v.p. 1×10^{-5} mmHg at 25°, and has an odour of sulphur dioxide. It is practically insoluble in water; moderately soluble in most organic solvents. It is stable to sunlight, compatible with non-alkaline pesticides, but subject to slow hydrolysis to the diol and sulphur dioxide.

Isomers. Endosulfan is a mixture of 2 stereoisomers: alpha-endosulfan, endosulfan (I), *C.A.* Registry No. *[959-98-8]* (formerly *[33213-66-0]*), stereochemistry 3α,5aβ,6α,9α,9aβ, m.p. 108-110°, provides 70% of the technical grade; beta-endosulfan, endosulfan (II), Registry No. *[33213-65-9]* (formerly *[891-86-1]* and *[19670-15-6]*), stereochemistry 3α,5aα,6β,9β,9aα, m.p. 208-210°, accounts for 30% technical endosulfan. Earlier reports on the stereochemistry of these isomers gave conflicting reports (*e.g.* W. Reimschneider, *World Rev. Pest Control,* 1963, **2**(4), 19).

Uses. Endosulfan is a broad spectrum non-systemic contact and stomach insecticide effective against numerous insects and certain mites attacking crops. The 2 stereoisomers have comparable LD50 values for *Musca domestica* (D. A. Lindquist & P. A. Dahm, *J. Econ. Entomol.,* 1957, **50**, 483).

Toxicology. The acute oral LD50 is: for rats 80-110 mg tech. (in oil)/kg, 76 mg alpha-isomer/kg, 240 mg beta-isomer/kg; for dogs 76·7 mg tech./kg; for mallard duck 200-750 mg/kg; for ring-necked pheasant 620-1000 mg/kg. The acute dermal LD50 for rabbits is 359 mg (in oil)/kg. In feeding trials rats receiving 30 mg/kg diet for 2 y and dogs 30 mg/kg diet for 1 y showed no ill-effect. It is highly toxic to fish but, in practical use, should be harmless to wildlife and honey bees.

continued

Endosulfan

Toxicology—*continued*

It is metabolised, in plants and mammals, to the corresponding sulphate which is toxicologically similar to endosulfan (the sulphite). For most fruits and vegetables 50% of the residue is lost in 3-7 d. There is no accumulation in milk, fat or muscle and it is excreted as conjugates of the diol and as other highly polar metabolites depending on the species (H. Maier-Bode, *Residue Rev.,* 1968, **22,** 1).

Formulations. These include: e.c. (161, 357 and 480 g a.i./l); w.p. (164, 329, 470 g/kg); dusts (30-47 g/kg); granules (10, 30, 40 and 50 g/kg); ULV (242, 497 and 604 g/l). Combinations are available with other pesticides, *e.g.* dimethoate and parathion-methyl at various concentrations.

Analysis. Product analysis is by i.r. spectrometry (*CIPAC Handbook,* 1970, **1,** 360) or by GLC (G. Zweig & T. E. Archer, *J. Agric. Food Chem.,* 1960, **8,** 190, 403; FAO Specification (CP/49); *Organochlorine Insecticides 1973*). Residues may be determined by GLC with MCD (D. M. Coulson *et al., J. Agric. Food Chem.,* 1960, **8,** 399; J. Burke & P. A. Mills, *J. Assoc. Off. Agric. Chem.,* 1963, **46,** 177) or with ECD (L. R. Mitchell, *J. Assoc. Off. Anal. Chem.,* 1976, **59,** 209). See also: J. R. Graham *et al., Anal. Methods Pestic., Plant Growth Regul. Food Addit.,* 1964, **2,** 507; *Anal. Methods Pestic. Plant Growth Regul.,* 1972, **6,** 488, 511.

Endothal

(Endothal-sodium)

H₂C structures (chemical diagrams)

T55 A AOTJ CVQ DVQ   *endothal* $C_8H_{10}O_5$ (186·2)

T55 A AOTJ CVO DVO &-NA- 2   *endothal-sodium* $C_8H_8Na_2O_5$ (230·2)

Nomenclature and development. The common name endothal is approved by BSI and New Zealand, spelling by ANSI, Canada and WSSA (endothall), a salt being indicated by the appropriate suffix *e.g.* endothal-sodium; the ISO common name endothal-sodium is defined for the disodium salt—exception Italy. The IUPAC and *C.A.* name is 7-oxabicyclo-[2.2.1]heptane-2,3-dicarboxylic acid, Registry No. *[145-73-3]*, also known as 3,6-epoxy-cyclohexane-1,2-dicarboxylic acid. Its herbicidal properties were first described by N. Tischler *et al., Proc. Northeast. Weed Control Conf.*, 1951, p.51. It was introduced in 1954 by Sharples Chemical Corp. (now merged with Pennwalt Corp.) and protected by USP 2 550 494; 2 576 080; 2 576 081.

Manufacture and properties. Produced by reduction of the Diels-Alder condensation product of furan and maleic anhydride, the *exo-cis* (in *C.A.* nomenclature *endo-cis*) hydrate, is a colourless solid, m.p. 144°. Its solubility at 20° is: 100 g/kg water; 70 g/kg acetone; 100 mg/kg benzene; 75 g/kg dioxane; 280 g/kg methanol. It is stable to light but is converted to the anhydride at 90°. It is non-flammable and non-corrosive to metals. Of the 3 isomers of endothal, the *exo-cis*-isomer has, in general, the greatest biological activity.

Uses. Salts of endothal are recommended for the pre- and post-em. control of weeds in sugar beet, red beet and spinach at 2-6 kg a.i./ha, sometimes used in combination with propham. The combination is also used as a desiccant on lucerne and on potato haulm, for the defoliation of cotton and to control aquatic weeds and algae.

Toxicology. The acute oral LD50 for rats is: 51 mg endothal/kg; 182-197 mg endothal-sodium (for 19·2% formulation)/kg. The latter causes light to moderate skin irritation. In 2-y feeding trials rats receiving 1000 mg/kg diet suffered no ill-effect.

Formulations. These include: 'Des-I-Cate', endothal mono(ethyldimethylammonium) salt (5·2 g a.e. + ammonium sulphate/kg); 'Aquathol Plus', endothal-potassium + fenoprop-potassium (157 g a.e. + 22·2 g a.e./kg).

Analysis. Residues may be determined by GLC with MCD, details of methods are available from Pennwalt Corp. See also: R. Carlson *et al., Anal. Methods Pestic. Plant Growth Regul.*, 1978, **10**, 327.

Endrin

$$C_{12}H_8Cl_6O \quad (380.9)$$

T E3 D5 C555 A D- FO KUTJ AG AG BG JG KG LG

Nomenclature and development. The common name endrin is approved by BSI, ISO, ESA and JMAF—exception Republic of South Africa (nendrin) and India; also known as ENT 17251. The IUPAC name is (1*R*,4*S*,5*R*,8*S*)-1,2,3,4,10,10-hexachloro-1,4,4a,5,6,7,8,8a-octahydro-6,7-epoxy-1,4:5,8-dimethanonaphthalene formerly 1,2,3,4,10,10-hexachloro-6,7-epoxy-1,4,4a,5,6,7,8,8a-octahydro-*exo*-1,4-*exo*-5,8-dimethanonaphthalene, in *C.A.* usage 3,4,5,6,9,9-hexachloro-1aα,2β,2aβ,3α,6α,6aβ,7β,7aα-octahydro-2,7:3,6-dimethano-napth[2,3-*b*]oxirene *[72-20-8]* formerly 1,2,3,4,10,10-hexachloro-6,7-epoxy-1,4,4a,5,6,7,8,8a-octahydro-*endo*-1,4-*endo*-5,8-dimethanonaphthalene. It was introduced in 1951 by J. Hyman & Co. under the code number 'Experimental Insecticide 269' and protected by USP 2676132; subsequently it has been developed by Shell International Chemical Co.

Manufacture and properties. It is produced by condensing vinyl chloride with hexachloro-cyclopentadiene, dehydrochlorinating the adduct and subsequent reaction with cyclopentadiene to form isodrin (BSI common name), which is epoxidised by peracetic or perbenzoic acid. Endrin is a colourless crystalline solid, m.p. 226-230° (decomp.), v.p. 2×10^{-7} mmHg at 25°. It is practically insoluble in water; sparingly soluble in alcohols and petroleum hydrocarbons; moderately soluble in acetone, benzene. The technical product is a light tan powder ≥ 92% pure. It is stable to alkali and acids, but concentrated acids or heating > 200° cause a rearrangement to a less insecticidal derivative. It is compatible with other pesticides.

Uses. Endrin is a non-systemic and persistent insecticide used mainly on field crops and in particular on cotton. It is non-phytotoxic at insecticidal concentrations.

Toxicology. The acute oral LD50 for rats is 7·5-17·5 mg/kg; the acute dermal LD50 for female rats is 15 mg/kg. In 2-y feeding trials rats receiving 1 mg/kg diet showed no ill-effect. It is highly toxic to fish. The fate of endrin given in the diet to lactating cows and laying hens is reported by M. K. Baldwin *et al., Pestic. Sci.,* 1976, **7,** 575.

Formulations. These include: e.c. (192 and 200 g a.i./l), stabilising agents to counter dehydrochlorination are not necessary; w.p. (500 g/kg); granules (10-50 g/kg); dusts (10 and 20 g/kg). Dusts on certain carriers require the addition of up to 5% hexamethylenetetramine (hexamine, BPC) to prevent loss of insecticidal properties by rearrangement.

Analysis. Product analysis is by GLC (details available from Shell International Chemical Co.) or by i.r. spectroscopy (*CIPAC Handbook,* 1970, **1,** 378; FAO Specification (CP/46); *Organochlorine Insecticides 1973).* Residues may be determined by GLC with ECD (K. Elgar *et al., Analyst (London),* 1966, **91,** 143). See also: L. C. Terriere, *Anal. Methods Pestic., Plant Growth Regul. Food Addit.,* 1964, **2,** 209; *Anal. Methods Pestic. Plant Growth Regul.,* 1972, **6,** 393.

EPN

$$\underset{\underset{\text{OEt}}{|}}{\overset{\overset{\text{S}}{\|}}{\text{Ph.P.O}}} - \text{\Large\textcircled{}} - \text{NO}_2$$

WNR DOPS&R&O2 $C_{14}H_{14}NO_4PS$ (323·3)

Nomenclature and development. The common name EPN is aproved by ESA and JMAF, there is no BSI or ISO name; also known as ENT 17 298. The IUPAC name is *O*-ethyl *O*-4-nitrophenyl phenylphosphonothioate (I), in *C.A.* usage (I) *[2104-64-5]* formerly *O*-ethyl *O-p*-nitrophenyl phenylphosphonothioate. It was introduced in 1949 by E. I. du Pont de Nemours & Co. (Inc.) and protected by USP 2 503 390 and subsequently manufactured by Nissan Chemical Industries Ltd.

Manufacture and properties. Produced by condensation of phenylphosphonothioic dichloride with ethanol in the presence of pyridine in benzene at room temperature, followed by reaction with sodium 4-nitrophenoxide in boiling chlorobenzene, it is a light yellow crystalline powder, m.p. 36°, v.p. 3 x 10^{-4} mmHg at 100°. It is practically insoluble in water; soluble in most organic solvents. The technical product is a dark amber-coloured liquid, d^{25} 1·27, n_D^{30} 1·5978. It is stable in neutral and acidic media, but is hydrolysed by alkali and so incompatible with alkaline pesticides. On heating in a sealed tube it is converted to the *S*-ethyl isomer (R. L. Metcalf & R. B. March, *J. Econ. Entomol.*, 1953, **46**, 288).

Uses. EPN is a non-systemic insecticide and acaricide with contact and stomach action. It is effective at 0·5-1·0 kg a.i./ha against a wide range of lepidopterous larvae, especially bollworms (*Heliothis* spp. and *Pectinophora gossypiella*) and leaf worm of cotton, rice stem borers and other leaf-eating larvae on fruit and vegetables. It is non-phytotoxic except on some apple varieties.

Toxicology. The acute oral LD50 is: for male rats 33-42 mg/kg; for female rats 14 mg/kg; for mice 50-100 mg/kg; the lethal range for dogs is 20-45 mg/kg. The acute dermal LD50 for rats is 110-230 mg/kg.

Formulations. These include: e.c. (450 and 480 g a.i./l); granules; also emulsifiable combinations with other insecticides such as parathion-methyl.

Analysis. Product analysis is by GLC or by reduction and potentiometric titration, details available from E. I. du Pont de Nemours & Co. (Inc.). Residues may be determined by GLC (H. L. Pease & J. J. Kirkland, *J. Agric. Food Chem.*, 1967, **15**, 187).

Epoxyethane

(Ethylene oxide)

T3OTJ

C_2H_4O (44·05)

Nomenclature and development. The trivial name epoxyethane is accepted by BSI in lieu of common name, and similarly ethylene oxide (I) by ISO, ESA and JMAF. The IUPAC and *C.A.* name is oxirane *[75-21-8]* formerly (I). It was introduced in 1928 as an insecticidal fumigant by R. T. Cotton & R. C. Roark, *Ind. Eng. Chem.*, 1928, **20**, 805.

Manufacture and properties. Produced by the reaction of chlorine and ethylene in the presence of water to form 2-chloroethanol which is treated with calcium hydroxide, or by the oxidation of ethylene in the presence of a silver catalyst, it is a mobile colourless liquid, b.p. 10·7°, m.p. —111°, v.p. 1095 mmHg at 20°, d_4^7 0·8, n_D^8 1·3599. It is miscible with water and most organic solvents. Although highly reactive and capable of many additive reactions, epoxyethane is fairly stable to water, slowly yielding ethylene glycol. It is flammable in air at concentrations > 55 g/m³ and is relatively non-corrosive.

Uses. Epoxyethane is an insecticidal fumigant used mainly for the vault fumigation of stored food products (H. D. Young & R. L. Busbey, *U.S. Dep. Agric., Bureau Entomol. Plant Quarantine*, 1935). There is some evidence that it has fungicidal and bacterial properties (C. R. Phillips & S. Kayser, *Am. J. Hyg.*, 1949, **50**, 270) and a lethal action on soil microflora (F. H. Dalton & C. Hurwitz, *Soil Sci.*, 1948, **66**, 233).

Toxicology. The intolerable irritation it causes to the eyes and nose serves as a warning of its presence, lessening the risk of accidental poisoning; its toxicology has been reviewed (R. L. Hollingsworth *et al.*, *AMA Arch. Ind. Health*, 1956, **13**, 217).

Formulation. It is usually marketed as a mixture of 1 part with 7-12 parts of carbon dioxide, the latter to reduce flammability.

Analysis. Residues may be determined by the estimation of the hydroxide ion produced by its reaction with magnesium chloride solution (O. F. Lubatti, *J. Soc. Chem. Ind.*, 1932, **51**, 361T; El Khishen, *J. Sci. Food Agric.*, 1950, **1**, 71).

EPTC

$$\text{Pr}_2\text{N.C.SEt}$$
$$\overset{\text{O}}{\overset{\|}{}}$$

3N3&VS2 $C_9H_{19}NOS$ (189·3)

Nomenclature and development. The common name EPTC is approved by BSI, ISO, WSSA and JMAF. The IUPAC name is *S*-ethyl dipropylthiocarbamate (I), in *C.A.* usage *S*-ethyl dipropylcarbamothioate *[759-94-4]* formerly (I). It was introduced in 1954 as a herbicide by the Stauffer Chemical Co. under code number 'R-1608', trade mark 'Eptam' and protected by USP 2913327. Its properties were described by J. Antognini *et al., Proc. Northeast. Weed Control Conf.,* 1957, p.2.

Manufacture and properties. Produced either by the reaction of ethanethiol with dipropylcarbamoyl chloride, which is formed from dipropylamine and carbonyl chloride, or by the reaction of dipropylamine with *S*-ethyl chlorothioformate, it is a liquid, with aromatic odour, b.p. 127°/20 mmHg, v.p. 3·4 x 10⁻² mmHg at 35°, d^{30} 0·9546, n_D^{30} 1·4750. Its solubility at 20° is 365 mg/l water; miscible with benzene, propan-2-ol, methanol, toluene, xylene. It is stable and non-corrosive.

Uses. EPTC kills germinating seeds and inhibits bud development from underground portions of some perennial weeds. It is recommended for pre-planting use at 3-6 kg a.i./ha on a wide range of crops. Incorporation into the soil mechanically or by irrigation is necessary to avoid loss by volatilisation. EPTC is particularly useful for the control of *Agropyron repens* and perennial *Cyperus* spp.

Toxicology. The acute oral LD50 for male albino rats is 1652 mg/kg; the acute dermal LD50 for male albino rabbits is *c.* 10000 mg/kg. Rats fed 326 mg/kg for 21 d showed no symptom other than excitability and loss of weight. The sub-chronic LC50 (7 d) for bobwhite quail was 20000 mg/kg diet. The LC50 (48 h) for killifish is > 10 mg/l; the LC50 (96 h) is: for bluegill 27 mg/l; for rainbow trout 19 mg/l.

Formulations. These include: e.c. (720 and 840 g a.i./kg); granules (23 and 100 g/kg).

Analysis. Product analysis is by GLC (*CIPAC Handbook,* in press). Residues may be determined by GLC (R. E. Hughes & V. H. Freed, *J. Agric. Food Chem.,* 1961, **91**, 381; T. H. Byast *et al., Agric. Res. Counc. (G.B.) Weed Res. Organ., Tech. Rep.,* No. 15 (2nd Ed.), p.17). See also: G. G. Patchett *et al., Anal. Methods Pestic., Plant Growth Regul. Food Addit.,* 1964, **4**, 117; W. Y. Ja *et al., Anal. Methods Pestic. Plant Growth Regul.,* 1972, **6**, 644.

Etacelasil

$$(MeOCH_2CH_2O)_3SiCH_2CH_2Cl$$

1O2O-SI-2GO2O1&O2O1

$C_{11}H_{25}ClO_6Si$ (316·9)

Nomenclature and development. The common name etacelasil is approved by BSI and proposed by ISO. The IUPAC name is 2-chloroethyltris(2-methoxyethoxy)silane, in *C.A.* usage 6-(2-chloroethyl)-6-(2-methoxyethoxy)-2,5,7,10-tetraoxa-6-silaundecane *[37894-46-5]*. It was synthesised and developed as a plant growth regulator by Ciba-Geigy AG under the code number 'CGA 13 586', trade mark 'Alsol' and protected by BP 1 371 804. Its use as an abscission agent was described by J. Rufener & D. Pietà, *Riv. Ortoflorofruttic. Ital.*, 1974, **4**, 274.

Properties. It is a colourless liquid, b.p. 85°/1 x 10⁻³ mmHg, v.p. 2 x 10⁻⁴ mmHg at 20°. Its solubility at 20° is 25 ml/l water; it is soluble in most organic solvents.

Uses. Etacelasil is an abscission agent for olives. It acts by releasing ethylene. Depending on the olive variety, a spray (1-2 g a.i./l) is recommended with harvest 6-10 d after application.

Toxicology. The acute oral LD50 for rats is 2066 mg a.i./kg; the acute dermal LD50 for rats is > 3100 mg/kg. It has negligible toxicity to wildlife.

Formulations. 'Alsol' 200 SCW and 'Alsol' 800 SCW, soluble concentrate (200 and 800 g a.i./l).

Analysis. Details of methods are available from Ciba-Geigy Ltd.

238

Etem

(I)   (II)

T55 ANYSS FN EU&TJ BUS   (I) $C_4H_4N_2S_3$ (176·3)

T7MYSYMTJ BUS DUS   (II) $C_4H_6N_2S_3$ (178·3)

Nomenclature and development. The common name etem is approved by BSI (but see below). The IUPAC name is 5,6-dihydroimidazo[2,1-c][1,2,4]-dithiazole-3-thione (I), in *C.A.* usage 5,6-dihydro-3*H*-imidazo[2,1-c]-1,2,4-dithiazole-3-thione [33813-20-6]. Compound (I) was shown in 1971 to be one of the oxidation products of nabam (p.370) (C. W. Pluijgers *et al., Tetrahedron Lett.,* 1971, p.1371; W. R. Benson *et al., J. Assoc. Off. Anal. Chem.,* 1972, **55**, 44; M. Alvarez *et al., Tetrahedron Lett.,* 1973, p.939).

Earlier, the structure (II), hexahydro-1,3,6-thiadiazepine-2,7-dithione, also known as ethylenethiuram monosulphide (ETM—as by JMAF), in *C.A.* usage tetrahydro-1,3,6-thiadiazepine-2,7-dithione [5782-83-3], had been given to this oxidation product of nabam (G. D. Thorn & R. A. Ludwig, *Can. J. Chem.,* 1954, **32**, 872; R. A. Ludwig & G. D. Thorn, *Plant Dis. Rep.,* 1953, **37**, 127; R. A. Ludwig *et al., Can. J. Bot.,* 1954, **32**, 48; G. D. Thorn, *Can. J. Chem.,* 1960, **38**, 2349, 2358). Until 1974 the BSI common name etem referred to structure (II), a code number UCP/21 has been used and the trade mark 'Vegita' (Kumiai Chemical Industry Co. Ltd and Hokko Chemical Industry Co. Ltd).

Manufacture and properties. When produced by the aerial or deliberate chemical oxidation of ethylenebis(dithiocarbamates), etem is contaminated by sulphur. The pure compound is a yellow crystalline powder, m.p. 121-124° (from chloroform). Structure (I) has been confirmed by an independent synthesis (R. J. S. Beer & A. Naylor, *Tetrahedron Lett.,* 1973, p.2989). Etem has also been identified as one of the oxidation products of other ethylenebis(dithiocarbamates), for example maneb, and zineb (R. A. Ludwig *et al., Can. J. Bot.,* 1955, **33**, 42).

Analysis. Residue analysis is by GLC (W. H. Newsome, *J. Agric. Food Chem.,* 1975, **23**, 348) or by polarography (R. Engst & W. Schnaak, *Z. Lebensm.-Unters. -Forsch.,* 1970, **144**, 81).

Ethalfluralin

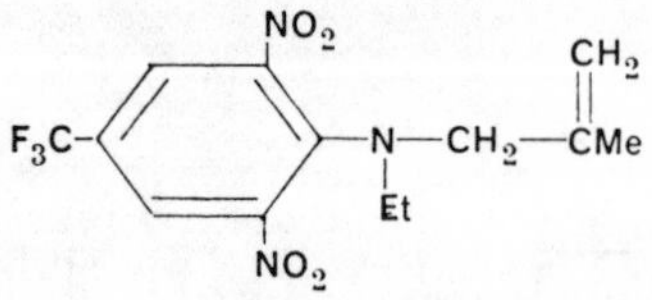

F_3C— (benzene ring with NO₂ at top, NO₂ at bottom) —N—CH₂—CMe, with CH₂ above CMe, and Et below N

FXFFR   CNW   ENW   DN2&1Y1&U1 $C_{13}H_{14}F_3N_3O_4$ (333·3)

Nomenclature and development. The common name ethalfluralin is approved by BSI, ANSI and WSSA, and proposed by ISO. The IUPAC name is *N*-ethyl-α,α,α-trifluoro-*N*-(2-methylallyl)-2,6-dinitro-*p*-toluidine (I) formerly *N*-ethyl-*N*-(2-methylallyl)-2,6-dinitro-4-trifluoromethylaniline, in *C.A.* usage, *N*-ethyl-*N*-(2-methyl-2-propenyl)-2,6-dinitro-4-(trifluoromethyl)benzenamine *[55283-68-6]*, formerly (I). It was introduced in 1974 by Eli Lilly & Co. under the code number 'EL-161', trade mark 'Sonalan' or 'Sonalen'. Its herbicidal properties were described by G. Skylakakis *et al., Proc. Br. Weed Control Conf., 12th,* 1974, **2**, 795.

Manufacture and properties. Produced by the nitration of 4-chloro-α,α,α-trifluoro-toluene and reaction with *N*-ethyl(2-methylallyl)amine, it is a yellow-orange crystalline solid, m.p. 55-56°, v.p. 8·2 x 10^{-5} mmHg at 25°. Its solubility in water at pH7 and 25° is 0·2 mg/l but it is soluble in most organic solvents.

Uses. Ethalfluralin is a pre-planting herbicide which controls susceptible weed species as their seeds germinate. Established weeds are tolerant. When soil-incorporated at 1·0-1·25 kg a.i./ha, it has a residual action on numerous broad-leaved and annual grass weeds in cotton, dry beans and soyabeans. A mixture with atrazine is recommended for pre- and post-em. surface application in maize.

Toxicology. The acute oral LD50 is: for rats and mice > 10000 g/kg; for cats, dogs, bobwhite quail and mallard ducks > 200 mg/kg. The dermal LD50 for rabbits is > 2000 mg/kg and there were only slight irritant effects to skin and eyes. In 90-d feeding trials the 'no effect' level was: in rats 1100 mg/kg diet; in dogs 27·5 mg/kg daily. The LC50 is: for bluegill 0·012 mg/l; for rainbow trout 0·0075 mg/l; for goldfish 0·1 mg/l.

Formulations. These include the e.c. (333 g a.i./l); 'Maizor', w.p. (240 g ethalfluralin + 300 g atrazine/kg).

Analysis. Product analysis is by GLC using FID or by spectrophotometry. Residues are determined by GLC using ECD. Details available from Elanco Products. See also, E. W. Day, *Anal. Methods Pestic. Plant Growth Regul.,* 1978, **10**, 341.

Ethephon

$$ClCH_2CH_2\overset{\overset{\displaystyle O}{\|}}{P}(OH)_2$$

QPQO&2G $C_2H_6ClO_3P$ (144·5)

Nomenclature and development. The common name ethephon is approved by ANSI, chlorethephon by New Zealand; there is no BSI or ISO name. The IUPAC and *C.A.* name is 2-chloroethylphosphonic acid, Registry No. *[16672-87-0]*. Its plant-growth regulating properties were discovered in 1965 by Amchem Products, Inc. and it was introduced under the trade names 'Ethrel' and 'Florel' (Amchem Products, Inc.) and 'Cepha' (GAF Corp.), and protected by USP 3879188; 3896163; 3897486.

Manufacture and properties. Produced by the rearrangement of tris(2-chloroethyl) phosphite to bis(2-chloroethyl) 2-chloroethylphosphonate and controlled hydrolysis, it is a colourless solid, m.p. 74-75° (decomp.). It is very soluble in water, lower alcohols and glycols; sparingly soluble in non-polar organic solvents. It is non-flammable but is corrosive having pH 0·8. It is stable at pH ≤ 3 but decomposes at higher pH values.

Uses. Ethephon is used to accelerate the pre-harvest ripening of fruit and vegetables including: apples at 24-48 g a.i./100 l (fenoprop or 1-naphthylacetic acid being included in the spray to prevent premature fruit drop); blackcurrants at 18-36 g/100 l; tomatoes at 48 g/100 l to stimulate end-of-season ripening of the fruit; cherries, coffee, citrus, tobacco; also as a post-harvest treatment of lemons; to shorten stems of forced daffodils; to stimulate branching and basal bud initiation for geraniums and roses. It acts by releasing ethylene within the plant or fruit. Plant response depends upon climatic conditions and so rates recommended may vary with area and country.

Toxicology. The acute oral LD50 for rats is *c.* 4000 mg/kg. It is a dermal and eye irritant. The LC50 (8 d) for mallard duck was > 10000 mg/kg diet. In 2-y feeding trials rats receiving ≤ 12500 mg/kg diet showed no ill-effect except at top dose levels towards the end of the trial. The LC50 (96 h) is: for bluegill 300 mg/l; for rainbow trout 350 g/l.

Formulations. These include: 'Ethrel Plant Regulator' (in the USA), liquid concentrate (240 g a.i./l); 'Ethrel C' (in the UK, ICI Ltd, Plant Protection Division), liquid concentrate (480 g/l).

Analysis. Product analysis is by measuring the ethylene produced on treatment with concentrated sodium hydroxide solution. Residues may be determined by reaction with diazomethane to form the dimethyl ester which is measured by GLC with FID or FPD.

241

Ethidimuron

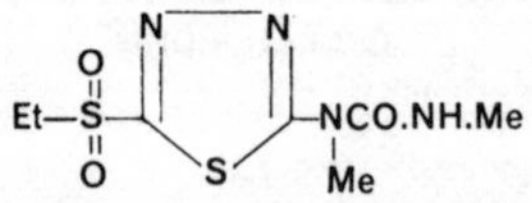

Et—S
||
O
(structure: 1,3,4-thiadiazole ring with N—N at top, S at bottom, Et—SO₂ on left, —NCO.NH.Me and Me substituent on right)

$$Et{-}\underset{O}{\overset{O}{S}}{-}\quad N{=}N \quad {-}NCO.NH.Me$$

T5NN DSJ CSW2 EN1&VM1 $C_7H_{12}N_4O_3S_2$ (264·2)

Nomenclature and development. The common name ethidimuron is approved by BSI and proposed by ISO. The IUPAC name is 1-(5-ethylsulphonyl-1,3,4-thiadiazol-2-yl)-1,3-dimethylurea, in *C.A.* usage *N*-(5-ethylsulfonyl-1,3,4-thiadiazol-2-yl)-*N*,*N'*-dimethylurea *[30043-49-3]*. It was introduced in 1975 by Bayer AG under the code number 'BAY MET 1486', trade mark 'Ustilan' and protected by BelgP 743615. Its herbicidal properties were described by L. Eue *et al., C. R. Journ. Etud. Herbic. Conf. COLUMA, 4th,* 1973, **1,** 14.

Properties. It forms colourless crystals, m.p. 156°. Its solubility is very low in water; 10-20 mg/kg cyclohexanone, dichloromethane.

Uses. Ethidimuron is a herbicide for total weed destruction on non-crop areas at 7 kg a.i./ha and for use in sugarcane plantations at 0·7-1·4 kg/ha.

Toxicology. The acute oral LD50 is: for rats > 5000 mg/kg; for mice > 2500 mg/kg; for Japanese quail 300-400 mg/kg; for canaries 1000 mg/kg. The acute dermal LD50 for rats is > 1000 mg/kg. In 90-d feeding trials the 'no effect' level for rats was > 1000 mg/kg diet. The LC50 for golden orfe is > 1000 mg/l. It is not harmful to honey bees.

Formulations. These include: w.p. (700 g a.i./kg); combinations of various composition with dichlorprop and aminotriazole as w.p. or granules.

Analysis. Product analysis is by i.r. spectroscopy.

Ethiofencarb

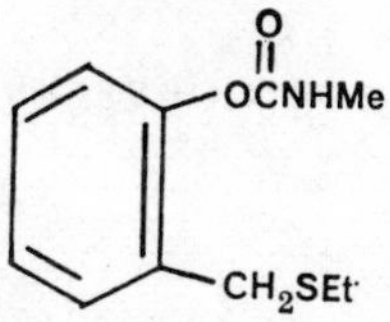

2S1R   BOVM1

$C_{11}H_{15}NO_2S$ (225·3)

Nomenclature and development. The common name ethiofencarb is approved by BSI and ISO, in France (éthiophencarbe). The IUPAC name is 2-(ethylthiomethyl)phenyl methylcarbamate (I), in *C.A.* usage (I) *[29973-13-5]* formerly α-(ethylthio)-*o*-tolyl methylcarbamate. It was introduced by Bayer AG under the code number 'HOX 1901', trade mark 'Croneton' and protected by DOS 1910588; BelgP 746649. Its insecticidal properties were described by J. Hammann & H. Hoffmann, *Pflanzenschutz-Nachr. (Am. Ed.)*, 1974, **27**, 267.

Manufacture and properties. Produced by the reaction of 2-(ethylthiomethyl)phenol with methyl isocyanate, it is a yellow oil, m.p. 33·4°, v.p. 1×10^{-4} mmHg at 30°, d_4^{20} 1·147. Its solubility at 20° is: 1·82 g/kg water; > 600 g/kg dichloromethane, propan-1-ol, toluene.

Uses. Ethiofencarb is a soil- and foliar-applied systemic insecticide with a specific effect against aphids at *c.* 50 g a.i./100 l (R. Homeyer, *ibid.*, 1976, **29**, 254). Its breakdown in soil and water have been studied (G. Dräger, *ibid.*, 1977, **30**, 18).

Toxicology. The acute oral LD50 is: for male rats 411-499 mg/kg; for mice 224-256 mg/kg; for Japanese quail 155 mg/kg; for canaries *c.* 100 mg/kg; for hens *c.* 1000 mg/kg. The acute dermal LD50 for male rats is > 1150 mg/kg. In 90-d feeding trials rats receiving 500 mg/kg diet showed no symptom. The LC50 (96 h) is: for carp 10-20 mg/l; for golden orfe 8-10 mg/l; for goldfish 20-40 mg/l.

Formulations. 'Croneton', e.c. (500 g a.i./l); granules (100 g/kg).

Analysis. Residues may be determined by GLC (G. Dräger, *ibid.*, 1974, **27**, 144).

243

Ethion

$$S \quad\quad S$$
$$(EtO)_2\overset{\parallel}{P}.S.CH_2.S.\overset{\parallel}{P}(OEt)_2$$

2OPS&O2&S1SPS&O2&O2 $C_9H_{22}O_4P_2S_4$ (384·5)

Nomenclature and development. The common name ethion is approved by BSI, ISO, ANSI and JMAF—exceptions France (diéthion), India and the Republic of South Africa (diethion), Italy, Portugal and Turkey; also known as ENT 24 105. The IUPAC and *C.A.* name is *O,O,O',O'*-tetraethyl *S,S'*-methylene di(phosphorodithioate) *[563-12-2]* formerly *S,S'*-methylene bis(*O,O*-diethyl phosphorodithioate). First reported in *Chem. Eng. News,* 1957, **35,** 87, it was introduced in 1956 by FMC Corp. (Agricultural Chemical Division) under the code number 'FMC 1240' and protected by BP 872 221; USP 2 873 228.

Manufacture and properties. Produced by the reaction of dibromomethane with *O,O*-diethyl hydrogen phosphorodithioate in ethanol at controlled pH, it is a colourless to amber-coloured liquid, f.p. —12° to —15°, v.p. 1·5 x 10^{-6} mmHg, d_4^{20} 1·22, n_D^{20} 1·5490. It is sparingly soluble in water; soluble in most organic solvents including kerosine, petroleum oils. The technical grade has d_4^{20} 1·215-1·230, n_D^{20} 1·530-1·542. It is non-corrosive, is slowly oxidised in air and is hydrolysed by acids and alkalies.

Uses. Ethion is a non-systemic insecticide and acaricide used on apples, particularly with petroleum oils on dormant trees, to kill eggs and scales. It is non-phytotoxic except to the apple varieties maturing at the same time as or before Early McIntosh. Other uses include the control of *Boophilus* spp. on cattle and of citrus mite.

Toxicology. The acute oral LD50 is: for rats 208 mg a.i./kg; for female rats 24·4 mg tech./kg. The acute dermal LD50 for rabbits is 915 mg/kg. In 28-d feeding trials female albino rats receiving 300 mg/kg diet survived without any effect on growth rate though there was evidence of cholinesterase inhibition > 10 mg/kg diet.

Formulations. These include: w.p. (250 g a.i./kg); e.c. (400 and 960 g/l); dust (40 g/kg); granules (50, 80 and 100 g/kg); various oils and combinations with other pesticides.

Analysis. Product analysis is by i.r. spectroscopy (F. A. Gunther *et al., J. Agric. Food Chem.,* 1962, **10,** 224), or by hydrolysis to *O,O*-diethyl hydrogen phosphorodithioate, conversion to the copper salt and colorimetry at 418 nm (J. R. Graham & E. F. Orwell, *ibid.,* 1963, **11,** 67). Residues may be determined by GLC (*Pestic. Anal. Manual 2,* Sections 120, 123; details of other published methods are available from FMC Corp.). See also: J. R. Graham, *Anal. Methods Pestic., Plant Growth Regul. Food Addit.,* 1964, **2,** 233; *Anal. Methods Pestic. Plant Growth Regul.,* 1972, **6,** 396.

Ethirimol

T6N CNJ BM2 DQ E4 F1 $C_{11}H_{19}N_3O$ (209·3)

Nomenclature and development. The common name ethirimol is approved by BSI and ISO, in France (éthyrimol). The IUPAC name is 5-butyl-2-ethylamino-6-methyl-pyrimidin-4-ol, in *C.A.* usage 5-butyl-2-(ethylamino)-6-methyl-4(1*H*)-pyrimidinone *[23947-60-6]* formerly 5-butyl-2-(ethylamino)-6-methyl-4-pyrimidinol. It was introduced in 1968 by ICI Ltd, Plant Protection Division under the code number 'PP 149' and protected by BP 1 182 584. Its fungicidal properties were first described by R. M. Bebbington *et al.*, *Chem. Ind. (London)*, 1969, p.1512.

Manufacture and properties. Produced by condensing 1-ethylguanidine with ethyl 2-acetylhexanoate, it is a colourless crystalline solid, m.p. 159-160° (with a phase change at *c.* 140°), v.p. 2×10^{-6} mmHg at 25°. Its solubility at room temperature is 200 mg/l water; it is sparingly soluble in acetone; slightly soluble in 4-hydroxy-4-methyl-pentan-2-one, ethanol; soluble in chloroform, trichloroethylene and aqueous solutions of strong acids and bases. It is stable to heat and in alkaline and acid solutions. It is non-corrosive to metals, but acidic solutions should not be stored in galvanised steel.

Uses. Ethirimol is a systemic fungicide effective against powdery mildews of cereals and other field crops. It is effective when applied as a seed dressing or to the soil in the root zone; the plant roots then continue to absorb it giving protection to foliage throughout critical growth stages. When used as a foliage spray it penetrates the leaf and moves towards the leaf margins. Ethirimol has no significant insecticidal properties and no effect on soil fauna.

Toxicology. The acute oral LD50 is: for female rats 6340 mg/kg; for male rabbits 1000-2000 mg/kg; for female cats > 1000 mg/kg; for female guinea-pigs 500-1000 mg/kg; for hens 4000 mg/kg. The application of 50 mg/l to the eyes of rabbits for 1 d caused only slight irritation; it is without effect on the skin of rabbits. In 2-y feeding trials the 'no effect' level was: for rats 200 mg/kg diet; for dogs 30 mg/kg daily. The LC50 (96 h) for brown trout fingerlings is 20 mg/l. It is harmless to honey bees.

Formulations. These include: 'Milstem', an aqueous suspension (col) (500 g a.i./l) used undiluted as a seed dressing; 'Milgo', col (250 g/l) used as a spray; 'Milcap' (70 g ethirimol + 360 g captafol/l); 'Milcurb' Super (250 g/l).

Analysis. Product analysis is by GLC (J. E. Bagness & W. G. Sharples, *Analyst (London)*, 1974, **99**, 225), or by acidifying, extracting the neutral buffered solution with chloroform, evaporating and dissolving the residue in methanol and colorimetry at 297 nm. Residue analysis is by extraction with methanol, clean-up and TLC, detecting under u.v., eluting corresponding area with methanol and measurement at 297 nm. Details are available from ICI Ltd, Plant Protection Division. See also: M. J. Edwards, *Anal. Methods Pestic. Plant Growth Regul.*, 1976, **8**, 285.

Ethofumesate

T56 BOT&J CO2 D1 D1 GOSW1 $C_{13}H_{18}O_5S$ (286·3)

Nomenclature and development. The common name ethofumesate is approved by BSI, ISO and WSSA. The IUPAC name is (±)-2-ethoxy-2,3-dihydro-3,3-dimethylbenzo-furan-5-yl methanesulphonate, in *C.A.* usage (±)-2-ethoxy-2,3-dihydro-3,3-dimethyl-benzofuran-5-yl methanesulfonate *[26225-79-6]*. It was introduced in 1974 by Fisons Ltd Agrochemical Division, under the code number 'NC 8438', trade mark 'Nortron' and protected by BP 1271659. Its herbicidal properties were first described by R. K. Pfeiffer, *Symp. New Herbic., 3rd,* 1969, p.1.

Manufacture and properties. It is produced by reaction of *p*-benzoquinone with the enamine from morpholine and isobutyraldehyde to give 2,3-dihydro-3,3,-dimethyl-2-morpholinobenzofuran-5-ol which is mesylated and the product converted to ethofumesate. It is a colourless crystalline solid, m.p. 70-72°, v.p. 6·45 x 10^{-7} mmHg at 25°. Its solubility at 25° is: 110 mg/l water; 100 g/kg ethanol; 400 g/kg acetone, chloroform, dioxane, benzene; 4 g/kg hexane. It is stable to hydrolysis in water at pH 7; acid hydrolysis gives 2,3-dihydro-2-hydroxy-3,3-dimethylbenzofuran-5-yl methananesulphonate and ethanol; alkaline hydrolysis gives 2-ethoxy-2,3-dihydro-3,3-dimethylbenzofuran-5-ol.

Uses. Ethofumesate is a selective herbicide. Sugar beet shows a very high tolerance, pre-sowing incorporated or pre-em. at 1·0-3·0 kg a.i./ha. Mixtures with other residual herbicides are normally recommended, *e.g.* post-em. at 0·5-1·0 kg/ha in mixtures with phenmedipham. It controls many important grass and broad-leaved weeds with a good persistence of activity in the soil, high tolerance is shown by other beet crops, onions, sunflowers and tobacco. Selectivity also occurs among grasses, *e.g.* the control of *Hordeum* spp. in pastures in New Zealand, and of annual weed grasses in ryegrass. It is degraded biologically in soil, the rate depending on climate, soil type and microbial activity. The time for 50% loss ranges from < 35 d under moist, warm conditions, to > 98 d under dry, cold conditions.

Toxicology. The acute oral LD50 for rats is > 6400 mg/kg; the acute dermal LD50 for rats is > 1440 mg/kg. In 2-y feeding trials the 'no effect' level in rats was ⩾ 1000 mg/kg. diet. Toxicity to birds and honey bees is very low and to fish moderate.

Formulation. An e.c. (200 g a.i./l).

Analysis. Residues of ethofumesate and its metabolites may be extracted from sugar beet with dichloromethane + methanol (organo-soluble components) and with aqueous methanol (water-soluble components). The former, and the aglycones liberated by acid hydrolysis of the latter, are cleaned-up by liquid-liquid partition and column chromatography, and determined by GLC with FPD. See R. J. Whiteoak *et al., Anal. Methods Pestic. Plant Growth Regul.,* 1978, **10**, 353. Residues in soil are determined by GLC with FPD (T. H. Byast *et al., Agric. Res. Counc. (G.B.) Weed Res. Organ., Tech Rep.,* No. 15 (2nd Ed.), p.27).

Ethohexadiol

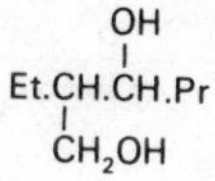

OH
|
Et.CH.CH.Pr
|
CH$_2$OH

QY3&Y2&1Q

C$_8$H$_{18}$O$_2$ (146·2)

Nomenclature and development. The common name ethohexadiol is approved by the US Pharm., and ethyl hexanediol by ESA; also known as ENT 375 or ethyl hexylene glycol. The IUPAC name is 2-ethylhexane-1,3-diol, in *C.A.* usage 2-ethyl-1,3-hexanediol *[94-96-2]*. It was first described as an insect repellent by P. Granett & H. L. Haynes, *J. Econ. Entomol.*, 1945, **38**, 671 and it was introduced under the code number 'Rutgers 612' or 'Rutgers 6-12' and protected by USP 2 407 205.

Manufacture and properties. Produced by the hydrogenation of 2-ethyl-3-hydroxyhexanal or by the condensation of butyraldehyde with aluminium magnesium ethoxide and the hydrolysis of the ester so formed (M. S. Kulpinski & F. F. Nord, *J. Org. Chem.*, 1943, **8**, 256), it is a colourless liquid, b.p. 244°, f.p. < —40°, v.p. < 0·1 mmHg at 20°, d_{20}^{20} 0·9422, n_D^{20} 1·4511. Its solubility at 20° is 6 g/l water; miscible with chloroform, ethanol, diethyl ether. The technical product, b.p. 240-250°, d_{20}^{20} 0·939-0·943, has a faint odour of witch hazel. It is stable under normal storage conditions. Both hydroxyl groups can be esterfied, the secondary group with difficulty. It is without chemical or solvent action on clothing and most plastics.

Uses. Ethohexadiol is an insect repellent effective against most biting insects; its properties have been reviewed (W. V. King, *J. Econ. Entomol.*, 1951, **44**, 339; B. V. Travis & C. N. Smith, *ibid.*, p.428).

Toxicology. The acute oral LD50 for rabbits is 2600 mg/kg; the acute dermal LD50 (90 d) for rabbits is 2000 mg/kg (J. H. Draize *et al.*, *J. Pharmacol. Exp. Ther.*, 1944, **82**, 377). Its widespread use confirms its lack of hazard and irritant properties in man.

Formulation. A mixture with dimethyl phthalate and butopyronoxyl (2:6:2) USP 2 356 801.

Analysis. Product analysis is by reaction with acetic anhydride in pyridine and titration of the acetic acid liberated in the acetylation.

Ethoprophos

$$\text{EtO.P(SPr)}_2$$
$$\overset{\overset{\displaystyle O}{\parallel}}{}$$

3SPO&S3&O2 $C_8H_{19}O_2PS_2$ (242·3)

Nomenclature and development. The common name ethoprophos is approved by BSI and ISO—exception ANSI and Society of Nematologists (in USA) (ethoprop). The IUPAC and *C.A.* name is *O*-ethyl *S,S*-dipropyl phosphorodithioate *[13194-48-4]*. It was introduced as a nematicide and soil insecticide by Mobil Chemical Co. under the code number 'VC9-104' and the trade marks 'Prophos' and 'Mocap' and protected by USP 3 112 244; 3 268 393. In the USA it was first registered in 1967 for use on tobacco and reported by S. J. Locascio, *Proc. Fla. St. Hortic. Soc.,* 1966, **79,** 170.

Manufacture and properties. Produced by the reaction of ethyl phosphorodichloridate and propane-1-thiol in the presence of aqueous sodium hydroxide, it is a clear pale yellow liquid, b.p. 86-91°/0·2 mmHg, v.p. 3·5 x 10⁻⁴ mmHg at 26°, d_4^{20} 1·094. Its solubility is 750 mg/l water; very soluble in most organic solvents. It is very stable in water up to 100° but is rapidly hydrolysed in alkaline media at 25°.

Uses. Ethoprophos is a non-systemic, non-fumigant nematicide and soil insecticide used at rates of 1·6-6·6 kg a.i./ha for most crops.

Toxicology. The acute oral LD50 for albino rats is 62 mg/kg; the acute dermal LD50 for albino rabbits is 26 mg/kg, the granular formulation being much less toxic. In 90-d feeding trials rats and dogs receiving 100 mg/kg diet showed depression of cholinesterase levels but no other effect on pathology or histology.

Formulations. 'Mocap', granules (100 g a.i./kg); e.c. (700 g/l).

Analysis. Product analysis is by GLC and comparison with an internal standard. Residues may be determined by GLC with MCD (sulphur) or FPD (sulphur and phosphorus).

Ethylene dibromide

Br.CH$_2$.CH$_2$.Br

E2E

C$_2$H$_4$Br$_2$ (187·9)

Nomenclature and development. The traditional name ethylene dibromide is accepted in lieu of a common name by BSI and ISO—exception JMAF (EDB); also known as ENT 15349. The IUPAC and *C.A.* name is 1,2-dibromoethane *[106-93-4]*. Its use as a fumigant was first reported by I. E. Neifert *et al., U.S. Dep. Agric. Bull.,* No. 1313, 1925. It was introduced in 1946 by The Dow Chemical Co. under the trade mark 'Bromofume' and protected by USP 2448265; 2473984.

Manufacture and properties. Produced by the bromination of ethylene, it is a colourless liquid, b.p. 131·5°, m.p. 9·3°, v.p. 11·0 mmHg at 25°, d_{25}^{25} 2·172, n_D^{20} 1·5379. Its solubility at 30° is 4·3 g/kg water; soluble in ethanol, diethyl ether and most organic solvents. It is stable and non-flammable.

Uses. Ethylene dibromide is an insecticidal fumigant used against pests of stored products (J. Aman *et al., Ann. Appl. Biol.,* 1946, **33,** 389); for the treatment of fruit and vegetables (J. W. Balock, *Science,* 1951, **114,** 122); for the spot treatment of flour mills; for soil treatment against certain insects and nematodes (M. W. Stone, *U.S. Dep. Agric. Bull.,* E-786, 1969). Planting must be delayed until 8 d after soil treatment because of its phytotoxicity.

Toxicology. The acute oral LD50 for male rats is 146 mg/kg; dermal applications, if confined, will cause severe burning. Rats tolerated 7-h exposures, 5 d/week, for 0·5 y at rates < 210 mg/m³.

Formulations. 'Dowfume 85' (1·44 kg a.i./l) in an inert solvent, for soil application; for mill use it is marketed in solution in carbon tetrachloride.

Analysis. Product analysis is by GLC (*CIPAC Handbook,* in press; AOAC Methods). Residue analysis is by distillation of the sample with water and toluene using a Dean-Stark trap, and analysis of the dried organic layer by GLC (R. Bielorai & E. Alumot, *J. Agric. Food Chem.,* 1966, **14,** 622); or by GLC after extraction with aqueous acetone (S. G. Heuser & K. A. Scudamore, *J. Sci. Food Agric.,* 1969, **20,** 566; *Pestic Sci.,* 1970, **1,** 240). Atmospheres can be sampled by passing through ethanolamine and determining bromide by Volhard titration (W. B. Sinclair & P. R. Crandell, *J. Econ. Entomol.,* 1952, **45,** 80). See also: C. E. Castro, *Anal. Methods Pestic., Plant Growth Regul. Food Addit.,* 1964, **3,** 155; *Anal. Methods Pestic. Plant Growth Regul.,* 1972, **6,** 711.

Ethylene dichloride

$$Cl.CH_2.CH_2.Cl$$

G2G $\qquad\qquad\qquad\qquad\qquad\qquad\qquad\qquad\qquad\qquad\qquad$ $C_2H_4Cl_2$ (98·96)

Nomenclature and development. The traditional name ethylene dichloride is accepted in lieu of a common name by BSI and ISO—exception JMAF (EDC). The IUPAC and *C.A.* name is 1,2-dichloroethane *[107-06-2]*. It was introduced in 1927 as a component of insecticidal fumigants (R. T. Cotton & R. C. Roark, *J. Econ. Entomol.,* 1927, **20**, 636).

Manufacture and properties. Produced by the chlorination of ethylene, it is a colourless liquid, with a chloroform-like odour, b.p. 83·5°, m.p. —36°, v.p. 78 mmHg at 20°, d_4^{20} 1·2569, n_D^{20} 1·4443. Its solubility at room temperature is 4·3 g/l water; soluble in most organic solvents. It is flammable, flash point (Abel-Penchy) 12-15°, lower and upper limits of flammability in air 275 and 700 g/m³. It is stable, resistant to oxidation and non-corrosive.

Uses. Ethylene dichloride is an insecticidal fumigant used mainly for the fumigation of stored products (W. A. Gersdorff, *U.S. Dep. Agric. Misc. Publ.,* No. 117, 1932) using 130-225 g/m³.

Toxicology. The acute oral LD50 is: for rats 670-890 mg/kg; for mice 870-950 mg/kg; for rabbits 860-970 mg/kg.

Formulation. Mixture with carbon tetrachloride in a ratio of 3:1 to reduce the fire hazard.

Analysis. Product and residue analysis is by GLC (S. G. Heuser & K. A. Scudamore, *J. Sci. Food Agric.,* 1969, **20**, 566; *Analyst (London),* 1974, **99**, 570).

N-(2-Ethylhexyl)-8,9,10-trinorborn-5-ene-2,3-dicarboximide

```
         CH
        /|
   HC  | CH—CO
   ||   |        \
        CH₂        N—CH₂.CHEt.Bu
   ||   |        /
   HC  | CH—CO
        \|
         CH
```

T   C555   A   DVNV   IUTJ   E1Y4&2 $C_{17}H_{25}NO_2$ (275·4)

Nomenclature and development. The trivial name *N*-octylbicycloheptenedicarboximide has been used; also known as ENT 8184. The IUPAC name is *N*-(2-ethylhexyl)-8,9,10-trinorborn-5-ene-2,3-dicarboximide also known as *N*-(2-ethylhexyl)bicyclo[2.2.1]hept-5-ene-2,3-dicarboximide, in *C.A.* usage 2-(2-ethylhexyl)-3a,4,7,7a-tetrahydro-4,7-methano-1*H*-isoindole-1,3(2*H*)-dione *[113-48-4]*. Its use as a synergist was described by R. H. Nelson *et al., Soap Chem. Spec.,* 1949, **25**(1), 120. It was introduced in 1944 by Van Dyke Co. under the name 'Van Dyke 264' and in 1949 by McLaughlin Gormley King Co. under the name 'MGK 264', formerly *'Octacide 264'* and protected by USP 2 476 512.

Manufacture and properties. Produced by heating 2-ethylhexylamine with the condensation product of cyclopentadiene and maleic anhydride, it is a liquid f.p. < —20°, n_D^{25} 1·4982-1·4988. The technical grade, d^{20} 1·040-1·060, is *c.* 98% pure, the main impurity being 2-ethylhexanol. It is practically insoluble in water; miscible with most organic solvents including petroleum oils and fluorinated hydrocarbons. Chemically it is very stable, non-corrosive, compatible with most pesticides and a solvent for DDT and HCH.

Uses. Its main use is an insecticidal synergist for pyrethroids. Slight phytotoxicity is shown at concentrations > 10 g/kg.

Toxicology. The acute oral LD50 for rats is 3640 mg/kg; the acute dermal LD50 for rabbits is 470 mg/kg. In 2-y feeding trials no toxic effect was observed in rats receiving 1000 mg/kg diet or pigs 300 mg/kg diet. In a 3-generation teratogenic study in rats there was no systemic or gross teratologic effect at rates ≤ 36·4 mg/kg daily, but at 364 mg/kg daily there was some effect on female fertility, litter size and growth ratio.

Formulations. It is incorporated, at 5-20 g/kg, with pyrethroids in aerosol concentrates, oil concentrates and dusts.

Analysis. Details of GLC methods are available from McLaughlin Gormley King Co.

S-Ethylsulphinylmethyl *O,O*-di-isopropyl phosphorodithioate

$$\underset{\text{(Me}_2\text{CHO)}_2\text{P.S.CH}_2\text{.S.Et}}{\overset{\displaystyle \underset{\|}{\overset{S}{}} \qquad \overset{O}{\underset{\|}{}}}{}}$$

OS2&1SPS&OY1&1&OY1&1 $C_9H_{21}O_3PS_3$ (304·4)

Nomenclature and development. The common name IPSP is approved by JMAF. The IUPAC name is *S*-ethylsulphinylmethyl *O,O*-di-isopropyl phosphorodithioate, in *C.A.* usage *S*-[(ethylsulfinyl)methyl] *O,O*-bis(1-methylethyl) phosphorodithioate *[5827-05-4]* formerly *S*-ethylsulfinylmethyl *O,O*-diisopropyl phosphorodithioate. It was introduced in 1963 by Hokko Chemical Industry Co. under the code number 'PSP-204', trade mark 'Aphidan' and protected by JapaneseP 531 126; USP 3 408 426; BP 1 068 628.

Manufacture and properties. Produced by the condensation of *O,O*-di-isopropyl hydrogen phosphorodithioate, formaldehyde and ethanethiol, followed by oxidation of the thioether, it is a colourless liquid, v.p. 1·5 x 10^{-5} mmHg at 27°, n_D^{20} 1·5260, d_4^{20} 1·1696. Its solubility at 15° is 1·5 g/l water; miscible with most organic solvents except hexane. The technical grade is *c.* 90% pure. It is moderately stable but reducing substances must be avoided. It is non-corrosive.

Uses. It is a systemic insecticide effective for aphid control on potatoes and vegetables by soil treatment; granules applied at 150-250 g a.i./ha or 50-100 mg/plant give effective control for *c.* 40 d.

Toxicology. The acute oral LD50 for mice is 84·5 mg/kg.

Formulations. Granules (50 g a.i./kg).

Analysis. Product analysis is by TLC followed by colorimetric determination; details are available from Hokko Chemical Industry Co. Ltd.

S-(2-Ethylsulphinyl-1-methylethyl) *O,O*-dimethyl phosphorothioate

$$\text{(MeO)}_2\overset{\overset{\text{O}}{\|}}{\text{P}}.\text{S.}\overset{\overset{\text{Me}}{|}}{\text{CH}}\text{CH}_2.\overset{\overset{\text{O}}{\|}}{\text{S}}.\text{Et}$$

OS2&1Y1&SPO&O1&O1 $\qquad\qquad\qquad\qquad$ C$_7$H$_{17}$O$_4$PS$_2$ (260·3)

Nomenclature and development. The common name ESP is approved by JMAF, there is no BSI or ISO name; also known as ENT 25647. The IUPAC name is *S*-(2-ethylsulphinyl-1-methylethyl) *O,O*-dimethyl phosphorothioate, in *C.A.* usage *S*-[2-(ethylsulfinyl)-1-methylethyl] *O,O*-dimethyl phosphorothioate *[2674-91-1]*. It was introduced in 1960 by Bayer AG under the code numbers 'Bayer 23655', 'S 410', trade marks 'Metasystox S' and 'Estox' and protected by DBP 1 035 958; USP 2 952 700.

Manufacture and properties. Produced by the oxidation of *S*-2-ethylthio-1-methylethyl *O,O*-dimethyl phosphorothioate (I) with hydrogen peroxide at pH 4, it is a yellow oil, almost odourless, b.p. 115°/0·02 mmHg, v.p. 4·7 x 10^{-6} mmHg at 20°, d_4^{20} 1·257, n_D^{25} 1·5149. It is soluble in water, chlorinated hydrocarbons, alcohols, ketones; sparingly soluble in light petroleum. It is oxidised by potassium permanganate to the corresponding sulphone. Both it and the sulphone are more resistant to hydrolysis under acid conditions than the sulphide (I) which is too unstable for practical use. It is more resistant than the sulphone to alkaline hydrolysis.

Uses. It is a systemic and contact insecticide and acaricide effective against sap-feeding insects and mites at 25 g a.i./100 l.

Toxicology. The acute oral LD50 for rats is 105 mg/kg; the intraperitoneal LD50 for guinea-pigs is 100 mg/kg; the acute dermal LD50 for male rats is 800 mg/kg. The growth of rats fed at 10 mg/kg daily for 50 d was unaffected.

Formulation. An e.c. (500 g a.i./l).

Analysis. Product analysis is by reduction with titanium disulphate and back titration of the excess with iron trichloride: particulars are available from Bayer AG. Residues may be determined by oxidation to phosphate which is measured by standard colorimetric methods (H. Tietz & H. Frehse, *Hoefchen-Briefe (Engl. Ed.)*, 1960, **13**, 212).

Etridiazole

EtO—⟨thiadiazole ring⟩—CCl₃

T5NS DNJ CO2 EXGGG

$C_5H_5Cl_3N_2OS$ (247·5)

Nomenclature and development. The common name etridiazole is approved by BSI and New Zealand and proposed by ISO—exception JMAF (echlomezol). The IUPAC and *C.A.* name is 5-ethoxy-3-(trichloromethyl)-1,2,4-thiadiazole *[2593-15-9]*. It was introduced in 1969 by Olin Chemicals as a soil fungicide and a component of seed dressings under the code number 'OM 2424', trade mark 'Terrazole' and protected by USP 3 260 588; 3 260 725.

Properties. It is a pale yellow liquid, with a persistent odour, f.p. 20°, v.p. 1 x 10⁻⁴ mmHg at room temperature. Its solubility at 25° is 50 mg/l water; soluble in acetone, carbon tetrachloride, ethanol, 'Skellysolve'. The technical product is a reddish-brown liquid 95-97% pure, the main impurities being 5-chloro-3-(trichloromethyl)-1,2,4-thiadiazole and trichloroacetamide. It is stable and suffered no loss of biological activity during 3 y storage under normal conditions.

Uses. Etridiazole is a fungicide recommended for the control of *Pythium* and *Phytophthora* spp. and diseases of turf, vegetables, fruit, cotton, groundnuts and ornamentals at 450-900 g a.i. in 2500 l water/ha. It is also used, in combination with quintozene, as a soil fungicide and seed treatment for the control of pre- and post-em. cotton seedling diseases caused by *Rhizoctonia, Pythium, Phytophthora* and *Fusarium* spp.

Toxicology. The acute oral LD50 for mice is 2000 mg/kg.

Formulations. These include: w.p. 'Terrazole' (Olin), 'Aaterra' (Aagrunol B.V.) (350 g a.i./kg); 'Truban' (Mallinckrodt) (300 g/kg); e.c. 'Terrazole 25% Emulsifiable' (Olin), 'Koban' (Mallinckrodt) (250 g/l); 'Terrazole 44% Emulsifiable' (Olin) (440 g/l); 'Pansoil EC' (Sankyo) (400 g/l); dusts 'Pansoil' 4% dust (Sankyo) (40 g/kg); seed treatment 'Terra-Coat L21' (Olin); 'Terrachlor Super X' (Olin); 'Terra-Coat L 205' (Olin) liquid seed treatment (58 g etridiazole + 232 g quintozene/l); 'Ban-rot' (Mallinckrodt), w.p. (150 g etridiazole + 250 g thiophanate-methyl/kg). Various other combinations are available for use as soil fungicides.

Etrimfos

$$(MeO)_2\overset{\displaystyle S}{\underset{\displaystyle \|}{P}}O \quad\text{—}\quad OEt$$

T6N   CNJ   B2   DOPS&O1&O1   FO2 $C_{10}H_{17}N_2O_4PS$ (292·3)

Nomenclature and development. The common name etrimfos is approved by BSI, ANSI and New Zealand and proposed by ISO. The IUPAC name is *O*-6-ethoxy-2-ethyl-pyrimidin-4-yl *O,O*-dimethyl phosphorothioate, in *C.A.* usage *O*-(6-ethoxy-2-ethyl-4-pyrimidinyl) *O,O*-dimethyl phosphorothioate *[38260-54-7]*. It was introduced in 1972 by Sandoz Ltd under the code number 'SAN 197 I', trade mark 'Ekamet' and protected by DOS 2 209 554. Its insecticidal properties were described by H. J. Knutti & F. W. Reisser, *Proc. Br. Insectic. Fungic. Conf., 8th,* 1975, **2**, 675.

Manufacture and properties. Produced by the reaction of the sodium salt of 6-ethoxy-2-ethylpyrimidin-4-ol with *O,O*-dimethyl phosphorochloridothioate in an inert solvent, it is a colourless oil, v.p. 4·9 x 10^{-5} mmHg at 20°, n_D^{20} 1·5068. Its solubility at 24° is 40 mg/l water; soluble in acetone, diethyl ether, ethanol, kerosine, xylene. Measured in aqueous buffered solution at 25° containing 50 mg/l, 50% loss occurred in 16 d at pH 6 and 14 d at pH 9.

Uses. Etrimfos is a broad-range non-systemic insecticide effective against species of Lepidoptera, Coleoptera, Diptera and, to a variable extent, Hemiptera, at 0·25-0·75 kg a.i./ha. It is used mainly on fruit trees, citrus, olives, grapes, potatoes, vegetables, paddy maize and lucerne. Granular applications against Pyralidae in paddy rice require 1·0-1·5 kg a.i./ha. It has a moderate residual activity lasting 7-14 d.

Toxicology. The acute oral LD50 for male rats is 1800 mg/kg. The dermal LD50 is: for male rats > 2000 mg/kg; for male rabbits > 500 mg/kg. In 180-d feeding trials the 'no effect' level for dogs was 12 mg/kg diet; in 90-d trials for rats it was 9 mg/kg diet. The LC50 (96 h) for carp is 13·3 mg/l. It is toxic to honey bees by contact.

Formulations. These include: 'Ekamet', e.c. (520 g a.i./l = 500 g/kg); granules (50 g/kg).

Analysis. Product analysis is by GLC; or by paper chromatography, followed by combustion and subsequent determination of phosphate by standard methods. Residues may be determined by GLC. Details of methods may be obtained from Sandoz AG.

255

Fenaminosulf

$$Me_2N-\text{⟨benzene ring⟩}-N{=}N{-}\overset{\displaystyle O}{\underset{\displaystyle O}{S}}{-}O^-\quad Na^+$$

WSO&NUNR   DN1&1   &-NA- $C_8H_{10}N_3NaO_3S$ (251·2)

Nomenclature and development. The common name fenaminosulf is approved by BSI and ISO, in France (phénaminosulf)—exception JMAF (DAPA). The IUPAC name is sodium 4-dimethylaminobenzenediazosulphonate, in *C.A.* usage sodium [4-(dimethyl-amino)phenyl]diazenesulfonate *[140-56-7]* formerly sodium *p*-(dimethylamino)benzene-diazosulfonate. It was introduced in 1955 under the code number 'Bayer 22555', trade marks 'Bayer 5072' and 'Lesan' (formerly *'Dexon'*) and protected by DAS 1 028 828. Its development as a fungicide is described by E. Urbschat, *Angew. Chem.,* 1960, **72,** 981.

Manufacture and properties. Produced by the reaction of diazotised 4-dimethylamino-aniline with sodium sulphite, it is a yellowish-brown powder. Its solubility at 25° is 20-30 g/l water; soluble in dimethylformamide; insoluble in diethyl ether, benzene, petroleum oils. Its aqueous solution is intense orange and is sensitive to light but is stabilised by sodium sulphite; it is stable in alkaline media.

Uses. Fenaminosulf is a seed and soil fungicide for the control of *Pythium, Aphanomyces* and *Phytophthora* spp. in soil at 45-110 g a.i./100 kg seed (depending on the crop).

Toxicology. The acute oral LD50 is: for rats 60 mg a.i./kg; for guinea-pigs 150 mg/kg.

Formulation. A w.p. (700 g a.i./kg).

Analysis. Product analysis is by polarography; particulars from Bayer AG. Residues may be determined by dialysis into 1% sodium sulphite with 5 drops of toluene as preservative; coupling with resorcinol in alkaline solution, extracting the colour with benzene and spectroscopic measurement at 450 nm (C. A. Anderson & J. M. Adams, *J. Agric. Food Chem.,* 1963, **11,** 474). See also: D. MacDougall, *Anal. Methods Pestic., Plant Growth Regul. Food Addit.,* 1964, **3,** 49.

Fenamiphos

1Y1&MPO&O2&OR Cl DS1
$C_{13}H_{22}NO_3PS$ (303·4)

Nomenclature and development. The common name fenamiphos is approved by BSI and ISO, in France (phénamiphos). The IUPAC name is ethyl 4-methylthio-*m*-tolyl isopropylphosphoramidate (I) also known as ethyl 3-methyl-4-(methylthio)phenyl isopropylphosphoramidate, in *C.A.* usage ethyl 3-methyl-4-(methylthio)phenyl 1-methylethylphosphoramidate *[22224-92-6]* formerly (I). It was described by G. Schrader *et al.,* (DBP 1 121 882; USP 2 978 479) and was introduced in 1969 by the Chemagro Division of the Baychem Corp. as a nematicide under the code number 'Bay 68 138' and trade mark 'Nemacur', formerly *'Nemacur P'*. Its nematicidal properties were first described by R. O'Bannon & A. L. Taylor, *Plant Dis. Rep.,* 1969, **51**, 995; B. Homeyer, *Pflanzenschutz-Nachr. (Am. Ed.),* 1971, **24**, 48.

Manufacture and properties. Produced by the reaction of 4-methylthio-*m*-cresol with ethyl isopropylamidophosphorochloridate, the latter produced by successive reactions of ethanol and isopropylamine on phosphoryl chloride, it is a colourless solid, m.p. 49°, v.p. 1 x 10^{-6} mmHg at 30°. Its solubility at room temperature is *c.* 700 mg/l water.

Uses. Fenamiphos is a systemic nematicide, active against ecto- and endo-parasitic, free-living, cyst-forming and root-knot nematodes and is recommended for broadcast application (with or without soil incorporation) at 5-20 kg a.i./ha; for band application at 7-40 g/100 m of row in 30-45 cm wide bands; for bare root dip at 100-400 mg/l for 5-30 min (W. M. Zeck, *ibid.,* p.114; A. Vilarbedo, *ibid.,* p.153; W. M. Burnett & I. M. Inglis, *ibid.,* p.170). Its metabolism has been studied (T. B. Waggoner & A. M. Khasawinah, *Residue Rev.,* 1974, **53**, 79).

Toxicology. The acute oral LD50 is: for rats 15·3-19·4 mg/kg; for dogs 10 mg/kg; for male guinea-pigs 75-100 mg/kg; for hens 12 mg/kg. The acute dermal LD50 for male rats is *c.* 500 mg/kg. In 2-y feeding trials rats receiving 3 mg/kg diet showed no symptom of poisoning (E. Löser & G. Kimmerle, *Pflanzenschutz-Nachr. (Am. Ed.),* 1971, **24**, 69).

Formulations. An e.c. (400 g a.i./l); granules (50 and 100 g/kg).

Analysis. Product and residue analysis is by GLC. Particulars of methods are available from Bayer AG. See also: J. S. Thornton, *J. Agric. Food Chem.,* 1971, **19**, 890.

Fenarimol

T6N CNJ EXQR BG&R DG$\qquad\qquad\qquad$ $C_{17}H_{12}Cl_2N_2O$ (331·2)

Nomenclature and development. The common name fenarimol is approved by BSI, ANSI and New Zealand, and proposed by ISO. The IUPAC name is 2,4'-dichloro-α-(pyrimidin-5-yl)benzhydryl alcohol formerly 2,4'-dichloro-α-(pyrimidin-5-yl)diphenyl-methanol, in *C.A.* usage α-(2-chlorophenyl)-α-(4-chlorophenyl)-5-pyrimidinemethanol *[60168-88-9]*. It was introduced in 1975 by Eli Lilly & Co. under the code number 'EL-222', the trade marks 'Bloc', 'Rimidin', 'Rubigan' and protected by BP 1 218 623. It was first reported by I. F. Brown *et al., Proc. Am. Phytopathol. Soc.,* 1975, **2**, 31.

Manufacture and properties. Produced by condensing pyrimidin-5-yllithium with 2,4'-dichlorobenzophenone, the latter being prepared by the Friedel-Crafts reaction of 2-chlorobenzoyl chloride with chlorobenzene, fenarimol is a colourless crystalline solid, m.p. 117-119°, v.p. 1 x 10^{-7} mmHg at 25°. Its solubility at 25° is 13·7 mg/l water at pH 7; it is soluble in acetone, acetonitrile, benzene, chloroform, methanol. It breaks down rapidly in sunlight to form many minor degradation products.

Uses. Fenarimol is a fungicide with a protective, curative and eradicative activity, recommended for use on apples (against *Podosphaera leucotricha* and *Venturia inaequalis*), on grapes (against *Uncinula necator*), on cucurbits (against *Erysiphe cichoracearum* and *Sphaerotheca fuliginea*) and on roses (against *S. pannosa*) as well as on other crops at rates of 10-90 mg/l, in sprays applied every 7-21 d.

Toxicology. The acute oral LD50 is: for adult mice > 4000 mg/kg; for rats 2500 mg/kg. Doses of fenarimol > 200 mg/kg were not acutely toxic to beagle dogs, mallard ducks and bobwhite quail. It induced no skin irritation and only slight eye irritation in rabbits. No teratogenic effects were observed after rabbits were fed ≤ 35 mg/kg daily. In 90-d feeding trials there was no effect on: rats receiving 50 mg/kg diet; mice 365 mg/kg diet; dogs 800 mg/kg diet. The LC50 (96 h) for bluegill is 0·91 mg/l.

Formulations. These include e.c. (40 and 120 g a.i./l) and w.p. (60 g/kg).

Analysis. Formulation analysis is by GLC with FID following dissolution in or extraction with chloroform or methanol. Residues in soil and plant tissue may be determined by GLC with ECD following suitable clean-up. Detailed analytical procedures and other information are available from Eli Lilly & Co.

Fenbutatin oxide

$$\left(\begin{array}{c} \text{Me} \\ | \\ \text{Ph}\text{--}\text{C}\text{--}\text{CH}_2. \\ | \\ \text{Me} \end{array}\right)_3 \text{Sn O Sn} \left(\begin{array}{c} \text{Me} \\ | \\ .\text{CH}_2\text{--}\text{C}\text{--}\text{Ph} \\ | \\ \text{Me} \end{array}\right)_3$$

$C_{60}H_{78}OSn_2$ (1053)

1X1&R&1-SN-1X1&1&R&1X1&1&R&O-SN-1X1&1&R&1X1&1&R&1X1&1&R

Nomenclature and development. The common name fenbutatin oxide is approved by BSI and New Zealand and proposed by ISO; also known as ENT 27738. The IUPAC name is bis[tris(2-methyl-2-phenylpropyl)tin] oxide, in *C.A.* usage hexakis(2-methyl-2-phenyl-propyl)distannoxane *[13356-08-6]*. It was introduced by Shell Development Co. in the USA under the code number 'SD 14114' and the trade mark 'Vendex' (USA only), 'Torque' (in other countries) and protected by USP 3657451.

Properties. The technical material is a colourless crystalline powder, m.p. 138-139°. It is insoluble in water; its solubility at 23° is: 6 g/l acetone; 380 g/l dichloromethane; 140 g/l benzene. It is thermally and photochemically stable. Water causes conversion of fenbutatin oxide to tris(2-methyl-2-phenylpropyl)tin hydroxide which is reconverted to the parent compound slowly at room temperature and rapidly at 98°.

Uses. Fenbutatin oxide gives effective and long-lasting control, at spray concentrations of 18-30 g a.i./100 l, of the mobile stages of a wide range of phytophagous mites on top fruit, citrus, vines, vegetables and ornamentals, including glasshouse crops. No phytotoxicity has been seen on these crops at concentrations twice those recommended. Effects on predacious mites and beneficial insects are minimal.

Toxicology. The acute oral LD50 is: for rats 2630 mg tech./kg; for mice 1450 mg/kg; for dogs > 1500 mg/kg. The acute percutaneous LD50 for rabbits is > 2000 mg/kg. In feeding trials ≤ 2-y duration there was no effect of toxicological significance in rats receiving 100 mg/kg diet or dogs 30 mg/kg daily. In an 8-d feeding study on bobwhite quail the LC50 was 5065 mg/kg diet. The LC50 (48 h) for rainbow trout is 0.27 mg a.i. (as w.p.)/l. Fenbutatin oxide is relatively non-toxic to honey bees with an oral LD50 > 100 µg/bee.

Formulations. A w.p. (500 g a.i./kg); suspension concentrate (550 g/l).

Analysis. Product analysis is by non-aqueous titration. Residues may be determined by GLC. Details of the methods are available from Shell International Chemical Co. Ltd. See also: *Anal. Methods Pestic. Plant Growth Regul.*, 1978, **10**, 139.

Fenchlorphos

$$(MeO)_2\overset{\overset{S}{\|}}{P}\text{-}O\text{-}$$

GR BG DG   EOPS&O1&O1 $C_8H_8Cl_3O_3PS$ (321·5)

Nomenclature and development. The common name fenchlorphos is approved by BSI and ISO—exception Canada and ANSI (ronnel); also known as ENT 23 284. The IUPAC and *C.A.* name is *O,O*-dimethyl *O*-2,4,5-trichlorophenyl phosphorothioate *[299-84-3]*. It was introduced in 1954 by The Dow Chemical Co., under the code numbers 'Dow ET-57' (drug grade), 'Dow ET-14' (tech. grade); the trade marks 'Nankor' and 'Trolene' for the drug grade, 'Korlan' for the tech. grade; and protected by USP 2 599 516; BP 699 064.

Manufacture and properties. Produced by the reaction of *O*-2,4,5-trichlorophenyl phosphorodichloridothioate, made from sodium 2,4,5-trichlorophenoxide and thiophosphoryl chloride with sodium methoxide in methanol, it is a colourless crystalline powder, softening at 35-37° finally melting at 40-42°, v.p. 8×10^{-4} mmHg at 25°. Its solubility at room temperature is 40 mg/l water; readily soluble in most organic solvents including refined kerosine. It is stable at temperatures $\leqslant 60°$ and in neutral and acidic media though hydrolysed by dilute alkali to the demethyl compound (F. W. Plapp & J. E. Casida, *J. Agric. Food Chem.*, 1958, **6**, 662). It is incompatible with alkaline pesticides.

Uses. Fenchlorphos is a systemic insecticide used as a residual spray for the control of flies and other household pests. It is also used as a spray against ectoparasites of cattle, sheep and pigs; as a pour-on and backrubber application for lice and hornfly control on cattle; as a dip against itch-mite (*Psoregates* spp.) and other ectoparasites of sheep. Also used as a shampoo for the control of ectoparasites on dogs; in feed and mineral licks against cattle ectoparasites and in various preparations (oils and aerosol concentrates) as a wound dressing against blowfly in livestock and for spot applications for the control of ticks on cattle.

Toxicology. The acute oral LD50 for rats is 1740 mg/kg; the acute dermal LD50 for rats is 2000 mg/kg. Rats tolerated 15 mg/kg daily for 105 d (D. D. McCollister *et al., ibid.*, 1959, **7**, 689). It is non-hazardous to wildlife.

Formulations. These include: e.c. (240 and 480 g a.i./l).

Analysis. Product analysis is by i.r. spectroscopy (R. P. Marquardt, *Anal. Methods Pestic., Plant Growth Regul. Food Addit.*, 1964, **2**, 427), details are available from The Dow Chemical Co. Residues in bovine fat may be determined by GLC (*Anal. Methods Pestic. Plant Growth Regul.*, 1972, **6**, 473) or by acid hydrolysis, the 2,4,5-trichlorophenol formed is steam distilled and estimated by colorimetry after reaction with 4-amino-1,5-dimethyl-2-phenylpyrazol-3-one (4-aminoantipyrine) (S. Gottlieb & P. B. March, *Ind. Eng. Chem., Anal. Ed.*, 1946, **18**, 16).

Fenfuram

T5OJ B1 CVMR $C_{12}H_{11}NO_2$ (201·2)

Nomenclature and development. The common name fenfuram is approved by BSI and New Zealand, and proposed by ISO. The IUPAC name is 2-methyl-3-furanilide, in *C.A.* usage 2-methyl-*N*-phenyl-3-furancarboxamide *[24691-80-3]*. It was developed in 1974 as a fungicide by Shell Research Ltd under the code number 'WL 22361' and protected by BP 1 215 066. It is being manufactured and developed commercially by KenoGard AB.

Properties. The technical product is a cream-coloured crystalline solid, m.p. 109-110°, v.p. 1·1 x 10^{-3} mmHg at 70° (1·5 x 10^{-7} mmHg at 20° by extrapolation). Its solubility at 20° is 0·1 g/l water; 300 g/l acetone; 340 g/l cyclohexanone; 145 g/l methanol. It is thermally and photochemically stable, is fairly stable at neutral pH, but is hydrolysed under strongly alkaline and strongly acidic conditions.

Uses. Fenfuram is a fungicide highly active as a seed dressing for use against the smuts and bunts of temperate cereals. At 0·3-1·5 g a.i./kg seed, it controls *Tilletia* and *Ustilago* spp. including those within the seed, such as *U. nuda*.

Toxicology. The acute oral LD50 for rats is 12,900 mg a.i./kg. In 90-d feeding trials the 'no effect' level was: for dogs 300 mg/kg diet; for rats 17 mg/kg daily. The Ames test for mutagenicity was negative.

Formulations. Seed dressings are available as liquids or powders either containing fenfuram alone or in combination with other fungicides *e.g.* guazatine.

Analysis. Product analysis is by GLC with FID. Residues in barley and wheat may be determined by GLC with a nitrogen-specific detector. Further information may be obtained from KenoGard AB.

261

Fenitrothion

WNR B1 DOPS&O1&O1 $C_9H_{12}NO_5PS$ (277·2)

Nomenclature and development. The common name fenitrothion is approved by BSI, ISO and BPC—exception JMAF (MEP); also known as ENT 25715, OMS 45. The IUPAC name is *O,O*-dimethyl *O*-4-nitro-*m*-tolyl phosphorothioate (I), in *C.A.* usage *O,O*-dimethyl *O*-(3-methyl-4-nitrophenyl) phosphorothioate *[122-14-5]* formerly (I). It was described by Y. Nishizawa *et al., Bull. Agric. Chem. Soc. Jpn.,* 1960, **24**, 744; *Agric. Biol. Chem.,* 1961, **25**, 605. It was introduced in 1959 by Sumitomo Chemical Co. and independently by Bayer AG and later by American Cyanamid Co. Code numbers include: 'Bayer 41831', 'S 5660', 'S-1102A' (Bayer AG); 'AC 47300' (American Cyanamid Co.). Trade marks include: 'Folithion' (Bayer AG), 'Sumithion' (Sumitomo Chemical Co.); 'Accothion', 'Cytel' and 'Cyfen' (American Cyanamid Co.). Protected by BelgP 594 669 (Sumitomo Chemical Co.); BelgP 596 091 (Bayer AG).

Manufacture and properties. Produced by the reaction of *O,O*-dimethyl phosphoro-chloridothioate with an alkali metal 3-methyl-4-nitrophenoxide, it is a brownish-yellow liquid, b.p. 140-145° (decomp.)/0·1 mmHg, v.p. 6 x 10^{-6} mmHg at 20°, d_4^{25} 1·3227, n_D^{25} 1·5528. It is practically insoluble in water; soluble in most organic solvents, but sparingly soluble in aliphatic hydrocarbons. It is hydrolysed by alkali, in 10mM sodium hydroxide 50% loss occurs in 4·5 h at 30°. It is readily converted to the *O,S*-dimethyl isomer on attempted distillation.

Uses. Fenitrothion is a contact insecticide, particularly effective against rice stem borers, and with some effect against red spider mites but of low ovicidal activity. Its properties have been summarised, *Residue Rev.,* 1976, **60.**

Toxicology. The acute oral LD50 is: for rats 250-500 mg tech./kg; for mice 870 mg/kg (G. Schrader, *Angew. Chem.* 1961, **73**, 331). The acute dermal LD50 for mice is > 3000 mg/kg. The LC50 (48 h) for carp is 4·4 mg/l (Y. Hishiuchi & V. Hashimoto, *Bochu-Kaguku,* 1967, **32**, 5). Its relatively low mammalian toxicity had been described earlier (J. Drabek & J. Pelikan, *Chem. Prum.,* 1956, **6**, 293).

Formulations. These include: e.c. (500 g/l and 1000 g a.i./l); w.p. (400 g and 500 g/kg); granules (30 and 50 g/kg); dust (50 g/kg).

Analysis. Product analysis is by the Averell & Norris method (AOAC Methods; *CIPAC Handbook,* 1979, **1A**, in press; *FAO Specification* (CP/62)). Residues may be determined by GLC (E. Möllhoff, *Pflanzenschutz-Nachr. (Am. Ed.),* 1968, **21**, 331) or by hydrolysis to 4-nitro-*m*-cresol and a comparison of the absorbance in acid and alkaline solution at 400 nm.

Fenoprop

QVY1&OR BG DG EG   *fenoprop* $C_9H_7Cl_3O_3$ (269·5)

QVY1&OR BG DG EG &Q2N2Q2Q   *triethanolamine salt* $C_{15}H_{22}Cl_3NO_6$ (418·7)

Nomenclature and development. The common name fenoprop is approved by BSI and ISO—exceptions ANSI and WSSA (silvex); France (as alternative to ISO name), USSR and JMAF (2,4,5-TP). The IUPAC name is (±)-2-(2,4,5-trichlorophenoxy)propionic acid (I), in *C.A.* usage (±)-2-(2,4,5-trichlorophenoxy)propanoic acid *[93-72-1]* formerly (I). The plant growth regulating activity of its salts was first reported by M. E. Synerholm & P. W. Zimmerman, *Contrib. Boyce Thompson Inst.*, 1945, **14**, 91. An ester formulation of low volatility was introduced in 1953 by The Dow Chemical Co. as a herbicide under the trade mark 'Kuron'; the triethanolamine salt, Registry No. *[17369-89-0]*, was introduced by Amchem Products, Inc. as a plant growth regulator under the trade mark 'Fruitone T'.

Manufacture and properties. Produced by the reaction of 2,4,5-trichlorophenol with sodium 2-chloropropionate it is a colourless powder, m.p. 179-181°. It solubility at 25° is: 140 mg/l water; 180 g/kg acetone; 134 g/kg methanol; 98 g/kg diethyl ether; 860 mg/kg heptane. Its lower alkyl esters are slightly volatile but its 2-butoxy-1-methylethyl esters (propylene glycol butyl ether esters) are practically non-volatile. Formulations and their dilutions are non-corrosive to spray equipment. Fenoprop is acidic, pK_a 2·84, forming water-soluble alkali metal and amine salts. They are soluble in lower alcohols, acetone; insoluble in most non-polar organic solvents, aromatic or chlorinated hydrocarbons.

Uses. Fenoprop is a hormone-type weedkiller, absorbed by leaves and stems and translocated. It is recommended for brush control at 2-4 kg a.e./100 l; for the control of submergent and emergent aquatic weeds; for broad-leaved weed control in certain crops 0·75-1·5 kg a.e./ha; at lower rates in mixture with mecoprop for the control of a wide range of annual weeds in cereals. The triethanolamine salt is used as a spray (1·4 g a.e./100 l) 7-14 d before harvest to reduce the pre-harvest drop of apples.

Toxicology. The acute oral LD50 for rats is 650 mg a.i./kg; 500-1000 mg mixed butyl or propylene glycol butyl ether esters/kg; *c.* 3940 mg a.e. (as triethanolamine salt)/kg. The acute dermal LD50 for rabbits is > 3200 mg a.e. (as triethanolamine salt)/kg. Fenoprop, its triethanolamine salt and undiluted esters are eye irritants. The LC50 (8 d) for mallard duck and quail is > 12800 mg a.e. (as triethanolamine salt)/kg diet.

Formulations. These include, as herbicides: 'Kuron', mixed propylene glycol ether esters (480 g a.e./l); mixtures with other herbicides, *e.g.* with mecoprop 'SB CMPP + 2,4,5-TP' (Stokes Bomford). As plant growth regulator: 'Fruitone T' (Amchem Products Inc.), triethanolamine salt (34 and 68 g a.e./l).

Analysis. Product analysis is by total chlorine content (AOAC Methods). Residues may be determined by GLC with ECD. See also: R. P. Marquardt, *Anal. Methods Pestic., Plant Growth Regul. Food Addit.*, 1964, **4**, 211; *Anal. Methods Pestic. Plant Growth Regul.*, 1972, **6**, 688.

Fensulfothion

$$\underset{(EtO)_2P.O-}{\overset{\overset{S}{\|}}{}} \hspace{-0.3cm} \left\langle \hspace{-0.2cm} \bigcirc \hspace{-0.2cm} \right\rangle \hspace{-0.3cm} \underset{S=O}{\overset{Me}{|}}$$

OS1&R   DOPS&O2&O2 $C_{11}H_{17}O_4PS_2$ (308·3)

Nomenclature and development. The common name fensulfothion is approved by BSI, ISO, ESA and Society of Nematologists (in USA); also known as ENT 24945. The IUPAC name is *O,O*-diethyl *O*-4-methylsulphinylphenyl phosphorothioate, in *C.A.* usage *O,O*-diethyl *O*-4-(methylsulfinyl)phenyl phosphorothioate *[115-90-2]* formerly *O,O*-diethyl *O-p*-(methylsulfinyl)phenyl phosphorothioate. It was developed by G. Schrader & E. Schegk and introduced in 1957 by Bayer AG under the code numbers 'Bayer 25 141', 'S 767', trade marks 'Dasanit' and 'Terracur P' and protected by DBP 1 101 406; USP 3 042 703.

Manufacture and properties. Produced by the condensation of *O,O*-diethyl phosphoro-chloridothioate with either 4-methylsulphinylphenol or with 4-(methylthio)phenol and oxidation of the resultant ester by hydrogen peroxide, it is a yellow oil, b.p. 138-141°/0·01 mmHg, d_4^{20} 1·202, n_D^{25} 1·540. Its solubility at 25° is 1·54 g/l water; soluble in most organic solvents. It is readily oxidised to the sulphone and apparently isomerises readily to the *O,S*-diethyl isomer (E. Benjamini *et al., J. Econ. Entomol.,* 1959, **52**, 94, 99).

Uses. Fensulfothion is an insecticide and nematicide active against free-living, cyst-forming and root-knot nematodes; it is recommended for soil treatment, has long persistence and some systemic activity (B. Homeyer, *Pflanzenschutz-Nachr. (Am. Ed.),* 1971, **24**, 367; W. Kolbe, *ibid.,* p.431; R. Dern, *ibid.,* p.476; H. F. Jung & K. Iwaya, *ibid.,* p.486; C. K. Proude, *ibid.,* p.503).

Toxicology. The acute oral LD50 for male rats is 4·7-10·5 mg/kg; the acute dermal LD50 for female rats is 3·5 mg (in xylene)/kg, for male rats 30 mg/kg. In 490-d feeding trials rats receiving 1 mg/kg diet showed no ill-effect (G. Löser & G. Kimmerle, *ibid.,* p.407).

Formulations. These include: soluble concentrate (600 g a.i./l); w.p. (250 g/kg); dust (100 g/kg); granules (25, 50 and 100 g/kg) for foliar and soil application.

Analysis. Product analysis is by the reaction of the sulphoxide group with an excess of dititanium trisulphate and back titration with iron trichloride: details from Bayer AG. Residues may be determined by GLC (details of procedures may be obtained from Bayer AG) or by extraction, clean-up and conversion to phosphate which is measured by standard colorimetric methods. See also: C. A. Anderson, *Anal. Methods Pestic. Plant Growth Regul.,* 1973, **7**, 253.

Fenthion

$$\text{(MeO)}_2\text{P.O}-\overset{\underset{\displaystyle \|}{\text{S}}}{}-\left\langle\right\rangle-\text{SMe}$$

Me

1SR   B1   DOPS&O1&O1 $C_{10}H_{15}O_3PS_2$ (278·3)

Nomenclature and development. The common name fenthion is approved by BSI, ISO and BPC—exception JMAF (MPP); also known as ENT 25 540, OMS 2. The IUPAC name is *O,O*-dimethyl *O*-4-methylthio-*m*-tolyl phosphorothioate (I), in *C.A.* usage *O,O*-dimethyl *O*-[3-methyl-4-(methylthio)phenyl] phosphorothioate *[55-38-9]* formerly (I). It was developed by G. Schrader & E. Schegk (see G. Schrader, *Hoefchen-Briefe (Engl. Ed.)*, 1960, **13**, 1) and was introduced in 1957 by Bayer AG under the code numbers 'Bayer 29 493', 'S 1752', trade marks 'Baycid', 'Baytex', 'Entex', 'Lebaycid', 'Mercaptophos', 'Queletox', 'Tiguvon' and protected by DBP 1 116 656; USP 3 042 703.

Manufacture and properties. Produced by the condensation of 4-(methylthio)-*m*-cresol and *O,O*-dimethyl phosphorochloridothioate, it is a colourless liquid, b.p. 87°/0·01 mmHg, v.p. 3×10^{-5} mmHg at 20°, d_4^{20} 1·250, n_D^{20} 1·5698. Its solubility at room temperature is 54-56 mg/l water, readily soluble in most organic solvents and in glyceride oils. The technical product is a brown oil 95-98% pure, with a weak garlic odour. It is stable $\leqslant$ 210°, to light and to alkali.

Uses. Fenthion is a contact and stomach insecticide with a useful penetrant and persistent action. It is effective, for example, against fruit flies, leaf hoppers and cereal bugs. Oxidation to the sulphoxide and sulphone, both highly insecticidal, proceeds in plants (H. Niessen *et al., Pflanzenschutz-Nachr. (Am. Ed.)*, 1962, **15**, 125). It is also used against weaver birds *(Quelea quelea)* in the tropics.

Toxicology. The acute oral LD50 is: for male rats 190-315 mg/kg; for female rats 245-615 mg/kg. The acute dermal LD50 for rats is 330-500 mg/kg; it is more toxic to dogs and poultry. In 1-y feeding trials dogs receiving 50 mg/kg diet showed no loss of weight or food consumption. The acute oral LD50 of the sulphoxide for rats is 125 mg/kg, of the sulphone 125 mg/kg.

Formulations. These included: 'Baytex', w.p. (400 g/kg), fogging concentrate (600 g/l), e.c. (500 and 1000 g/l) and granular (20 g/kg) for household use; 'Lebaycid', w.p. (250 and 400 g/kg), e.c. (500 and 1000 g/l), ULV (1000 g/l) and dust (30 g/kg) for use against crop pests; 'Queletox' for use against weaver birds; 'Tiguvon' various formulations against animal parasites.

Analysis. Product analysis is by u.v. spectrometry at 252 nm (F. B. Ibrahim & J. C. Cavagnol, *J. Agric. Food Chem.*, 1966, **44**, 369; *CIPAC Handbook,* in press) or by paper chromatography (details available from Bayer AG). Residues may be determined by GLC (R. J. Anderson *et al., J. Agric. Food Chem.*, 1966, **14**, 619; M. C. Bowman & M. Beroza, *ibid.*, 1968, **16**, 399; *idem., J. Assoc. Off. Anal. Chem.*, 1969, **52**, 1231) or by extraction, clean-up and oxidation to phosphate which is measured by standard colorimetric methods (H. Frehse *et al., Pflanzenschutz-Nachr. (Am. Ed.)*, 1962, **15**, 152, 164; H. Maier-Bode, *Anz. Schaedlingskde.*, 1962, **35**, 49). See also D. MacDougall, *Anal. Methods Pestic., Plant Growth Regul. Food Addit.*, 1964, **2**, 43; *Anal. Methods Pestic. Plant Growth Regul.*, 1972, **6**, 301.

Fentin acetate

$$\text{Ph}_3\text{SnO.C.Me} \atop \overset{\text{O}}{\overset{\|}{}}$$

1VO-SN-R&R&R $C_{20}H_{18}O_2Sn$ (409·0)

Nomenclature and development. The common name fentin acetate is approved by BSI; the name fentin (the anion being stated) is approved by ISO—exception (for the acetate) USSR (asetat fenolovo), Republic of South Africa and USA (chemical name); also known as ENT 25 208. The IUPAC name is triphenyltin acetate, in *C.A.* usage (acetyloxy)triphenylstannane *[900-95-8]* formerly acetoxytriphenylstannane. The fungicidal properties of organotin compounds were first explored by G. J. M. van der Kerk & J. G. A. Luijten, *J. Appl. Chem.,* 1954, **4**, 314; 1956, **6**, 56. Fentin acetate was introduced in 1954 by Hoechst AG under the code numbers 'VP 1940' and 'Hoe 2824', trade mark 'Brestan' and protected by DBP 950 970.

Manufacture and properties. Produced by the reaction of sodium acetate with fentin chloride, obtained by the action of tin tetrachloride on tetraphenyltin, it forms colourless crystals, m.p. 118-122°, v.p. $1·33 \times 10^{-6}$ mmHg at 30°. Its solubility at 20° is 28 mg/l water (*cf* R. D. Barnes *et al., Pestic. Sci.,* 1973, **4**, 305); poorly soluble in most organic solvents. The technical product is 90-95% pure, m.p. 120-125°. It is stable when dry, but is relatively easily decomposed when exposed to air or light, finally forming insoluble tin compounds. It is compatible with common pesticides but should not be mixed with oil emulsions.

Uses. Fentin acetate is a non-systemic fungicide effective against much the same range of fungi as the copper fungicides but at about one-tenth the dosage. It is recommended for the control of *Phytophthora infestans* on potatoes, *Ramularia* spp. of sugar beet and celery at 160-260 g a.i./ha; though phytotoxic to vine, hop, fruit, ornamental and glasshouse plants, it is safe on potatoes, beet, celery, cocoa and rice and may be used for algal control on paddy rice. It has an 'antifeeding' effect on caterpillars (J. Meisner & K. R. S. Ascher, *Z. Pflanzenkr. Pflanzenpathol. Pflanzenschutz,* 1965, **72**, 458; K. R. S. Ascher & J. Meisner, *Z. Pflanzenkr. Pflanzenschutz.,* 1969, **76**, 564).

Toxicology. The acute oral LD50 for rats is 125 mg/kg; the acute dermal LD50 for rats is 500 mg/kg though it is irritant to mucous membranes. In 2-y feeding trials dogs and guinea-pigs receiving 10 mg/kg diet showed no symptom. It is harmless to honey bees.

Formulations. A w.p. (190 and 540 g a.i./kg); also as mixtures with maneb ('Brestan').

Analysis. Product analysis is by extraction with carbon tetrachloride, hydrolysis of the extract to fentin hydroxide which is measured by non-aqueous titration (*CIPAC Handbook,* 1979, **1A**, in press). Residues may be determined by extraction with dichloromethane, partitioning with a solution of ethylenediaminetetra-acetic acid and measured either by colorimetry with dithizone (G. A. Lloyd *et al., J. Sci. Food Agric.,* 1962, **13**, 353) or by polarography of the tetrabromide (S. Gorbach & R. Bock, *Z. Anal. Chem.,* 1958, **163**, 419).

Fentin hydroxide

Ph$_3$SnOH

Q-SN-R&R&R   C$_{18}$H$_{16}$OSn   (367·0)

Nomenclature and development. The common name fentin hydroxide is approved by BSI: the name fentin (the anion being stated) is approved by ISO—exception (for the hydroxide) USSR (gidrookie fenolovo) and the Republic of South Africa and USA (chemical name); also known as ENT 28009. The IUPAC name is triphenyltin hydroxide, in *C.A.* usage hydroxytriphenylstannane *[76-87-9]*. The fungicidal properties of organotin compounds were first explored by G. J. M. van der Kerk & J. G. A. Luijten, *J. Appl. Chem.,* 1954, **4,** 314; 1956, **6,** 56, and fentin hydroxide was introduced by Philips-Duphar B. V., under the trade mark 'Du-Ter'.

Manufacture and properties. Produced by hydrolysis of fentin chloride obtained by the reaction of tin tetrachloride with phenylmagnesium chloride, it is a colourless powder, m.p. 116-120°. The technical product is ⩾94% pure, m.p. 116-120°. It is practically insoluble in water; moderately soluble in most organic solvents. It is stable at room temperature but dehydration may occur at elevated temperatures. The formulated w.p. is compatible with w.p. formulations of most other pesticides, but should not be mixed with liquid formulations because of possible phytotoxicity.

Uses. Fentin hydroxide is a non-systemic fungicide effective against many of the fungi susceptible to copper fungicides. It is recommended for the control of early and late blights of potato at 250-350 g a.i./ha, of leaf spot on sugar beet at 350-400 g/ha, of blast diseases of rice at 220-450 g/ha, of coffee berry disease at 1 kg/ha, of brown spot disease of tobacco at 200-350 g/ha.

Toxicology. The acute oral LD50 is: for male rats 108 mg/kg; for female mice 209 mg/kg; for female guinea-pigs 27·1 mg/kg. The LC50 (8 d) for bobwhite quail is 38·5 mg (as 475 g/kg w.p.)/kg diet.

Formulations. These include: 'Du-Ter', w.p. (190 g a.i./kg); 'Du-Ter Extra', w.p. (475 g/kg). These formulations are slightly hygroscopic and should not be mixed with liquid formulations.

Analysis. Product analysis is by potentiometric titration after purification with alkaline alumina (*CIPAC Handbook,* 1979, **1A,** in press). Residues may be determined by extraction and conversion to tetravalent tin which is determined colorimetrically with pyrocatechol violet. See also fentin acetate (p.266).

Fenuron

$$O$$
$$\|$$
$$Ph.NH.C.NMe_2$$

1N1&VMR $C_9H_{12}N_2O$ (164·2)

Nomenclature and development. The common name fenuron is approved by BSI, ISO, ANSI and WSSA—exception USSR (fenidim), Sweden and Portugal; it has also been called PDU. The IUPAC name is 1,1-dimethyl-3-phenylurea (I), in *C.A.* usage *N,N*-dimethyl-*N'*-phenylurea *[101-42-8]* formerly (I). Its herbicidal action was first described by H. C. Bucha & C. W. Todd, *Science,* 1951, **114,** 493, and it was introduced in 1957 by E. I. du Pont de Nemours & Co. (Inc.) under the trade mark *'Dybar'* and protected by USP 2655447; BP 691403; 692589, but is no longer produced by this company.

Manufacture and properties. Produced by the reaction of phenyl isocyanate with dimethylamine, it is a colourless crystalline solid, m.p. 133-134°, d_{20}^{20} 1·08, v.p. $1·6 \times 10^{-4}$ mmHg at 60°. Its solubility at 25° is 3·85 g/l water; sparingly soluble in hydrocarbons. It is stable to oxidation but is subject to microbial decompostition in soil.

Uses. Fenuron is a herbicide that inhibits photosynthesis and is absorbed through the roots. It is recommended for the control of woody plants by basal application at 2-30 kg/ha and, in mixtures with chlorpropham, for the control of many germinating weeds and established chickweed pre-em. in many vegetable crops, pre-em. or post-planting in onions and leeks, and in soft fruit, top fruit and nursery trees.

Toxicology. The acute oral LD50 is 6400 mg/kg; a 33% aqueous paste was practically non-irritating to the intact skin of guinea-pigs. In 90-d feeding trials no apparent effect was observed in rats receiving 500 mg/kg diet.

Formulations. Combinations with other herbicides, *e.g.* chlorpropham, propham.

Analysis. Product analysis is based on a hydrolytic method (W. K. Lowen & H. M. Baker, *Anal. Chem.,* 1952, **24,** 1476). Residues may be determined by colorimetry (R. L. Dalton & H. L. Pease, *J. Assoc. Off. Agric. Chem.,* 1962, **45,** 377).

Fenuron-TCA

$$\text{Ph.NHC.NHMe}_2 \quad \text{Cl}_3\text{C.C.O}^-$$

1N1&VMR &QVXGGG

$C_{11}H_{13}Cl_3N_2O_3$ (327·6)

Nomenclature and development. The common name fenuron-TCA is approved by WSSA for fenuron trichloroacetate. The IUPAC name is 1,1-dimethyl-3-phenyluronium trichloroacetate, in *C.A.* usage trichloroacetic acid compound with *N,N*-dimethyl-*N'*-phenylurea (1:1) *[4482-55-7]* formerly trichloroacetic acid compound with 1,1-dimethyl-3-phenylurea (1:1). This salt of fenuron was introduced in 1956 by the Allied Chemical Corp. Agricultural Division (now Hopkins Agricultural Chemical Co.) under the code number 'GC-2603', trade mark 'Urab' and protected by USP 2 782 112; 2 801 911.

Manufacture and properties. Produced by the interaction of fenuron (p.268) and trichloroacetic acid, it is a colourless crystalline solid, m.p. 65-68°. Its solubility at room temperature is: 4·8 g/l water; 666 g/kg 1,2-dichloroethane; 567 g/kg trichloroethylene; poorly soluble in petroleum oils.

Uses. Fenuron-TCA combines the herbicidal actions of fenuron and TCA (p.494) and it is recommended for total weed control on non-crop areas and for the control of woody plants.

Toxicology. The acute oral LD50 for albino rats is 4000-5700 mg/kg.

Formulations. These include: an oil-miscible concentrate (360 g/l); granules (20 and 220 g/kg).

Analysis. Product analysis is by total nitrogen determination by the Kjeldahl method, or total chlorine content using the Parr bomb. Details are available from Hopkins Agricultural Chemical Co.

Fenvalerate

1Y1&YR DG&VOYCN&R COR $C_{25}H_{22}ClNO_3$ (419·9)

Nomenclature and development. The common name fenvalerate is approved by BSI and proposed by ISO; also known as OMS 2000. The IUPAC name is (*RS*)-α-cyano-3-phenoxybenzyl (*RS*)-2-(4-chlorophenyl)-3-methylbutyrate, in *C.A.* usage (*RS*)-cyano(3-phenoxyphenyl)methyl (*RS*)-4-chloro-α-(1-methylethyl)benzeneacetate *[51630-58-1]*. It was introduced by the Sumitomo Chemical Co. Ltd in 1972 under the code number 'S-5602', the trade mark 'Sumicidin' and protected by BP 1439615; USP 4062968. It is also being developed in some countries by Shell International Chemical Co. Ltd under the code number 'WL 43 775' and trade mark 'Belmark' (in USA, 'Pydrin'). It was first reported by N.Ohno *et al.*, *Agric. Biol. Chem.,* 1974, **38**, 881; *Pestic. Sci.,* 1976, **7**, 247, and the results of field trials by M. D. Mowlam *et al.*, *Proc. 1977 Br. Crop Prot. Conf.,—Pests Dis.,* 1977, **2**, 649.

Manufacture and properties. Produced by the esterification of 2-(4-chlorophenyl)-3-methylbutyric acid with α-hydroxy-3-phenoxyphenylacetonitrile, it is a viscous yellow liquid, b.p. 300°/37 mmHg, v.p. 2·3 x 10⁻⁷ mmHg (extrapolated) at 20°, d_{25}^{25} 1·175. Its solubility at 20° is: < 1 mg tech./l water; 77 g/l hexane; > 450 g/l acetone, cyclohexanone, ethanol, xylene, chloroform. It is highly stable thermally. Photochemical degradation has been observed in laboratory studies but field data indicate that this does not adversely affect biological performance. It is more stable in acid than alkaline media with optimum stability at *c.* pH 4.

Uses. Fenvalerate is a highly active contact insecticide effective against a broad range of pests, including strains resistant to organochlorine, organophosphorus and carbamate insecticides. It gives control of leaf- and fruit-eating insects on a range of crops, including cotton, fruit, vines and vegetables at 25-250 g a.i./ha. *Myzus persicae* is controlled by sprays of 2-5 g a.i./100 l and cutworms by soil-surface sprays at 50-70 g/ha. Its stability in sunlight and resistance to rain wash-off ensure good residual activity on treated plants. It is also used in public health and animal husbandry; 100 mg/m² wall can provide 60-d residual control of flies in animal housing.

Toxicology. The acute oral LD50 is: for rats 300-630 mg tech./kg; for female mice 100-200 mg/kg, for male mice 200-300 mg/kg; for domestic fowl 1600 mg/kg. Moderate skin and eye irritation was observed in rabbits. In 2-y feeding trials no effect was observed for rats receiving 250 mg/kg diet. The LC50 (96 h) for rainbow trout is 3·6 µg/l; the LC50 (48 h) for carp is < 0·1 mg/l.

Formulations. These include: 'Belmark', e.c. (25-300 g a.i./l); ULV concentrates (50-75 g/l). ULV and e.c. mixtures ('Belmark') with other insecticides are also available.

Analysis. Product analysis is by GLC or HPLC. Residues may be determined by GLC with ECD (Y. W. Lee, *J. Assoc. Off. Anal. Chem.,* 1978, **61**, 869). Details of methods are available from Shell International Chemical Co. Ltd or from Sumitomo Chemical Co. Ltd.

Ferbam

$$\left[\begin{array}{c} \overset{\displaystyle S}{\underset{\displaystyle \|}{}} \\ Me_2NC.S^- \end{array} \right]_3 \quad Fe^{3+}$$

SUYS&N1&1   3   &-FE- $C_9H_{18}FeN_3S_6$ (416·5)

Nomenclature and development. The common name ferbam is approved by BSI, ISO and JMAF—exception Germany; also known as ENT 14 689. The IUPAC name is iron tris(dimethyldithiocarbamate) formerly ferric dimethyldithiocarbamate, in *C.A.* usage (OC-6-11)-tris(dimethylcarbamodithioato-*S,S'*)iron *[14484-64-1]* formerly tris(dimethyl-dithiocarbamato)iron. It was introduced *c.* 1940 by E. I. du Pont de Nemours & Co. (Inc.) (who no longer manufacture or market it) under the trade mark 'Fermate' and protected by USP 1 972 961.

Manufacture and properties. Produced by the precipitation from sodium dimethyldithio-carbamate with soluble iron(III) salts, it is a black powder, decomp. > 180°, v.p. negligible at room temperature. Its solubility at room temperature is 130 mg/l water; soluble in organic solvents of high dielectric constant such as acetonitrile, chloroform, pyridine. Its chemical properties are suggestive of co-ordination, but it tends to decompose on prolonged storage or exposure to moisture and heat. It is compatible with other pesticides, but not with copper or mercury compounds or alkaline pesticides.

Uses. Ferbam is used mainly for the protection of foliage against fungal pathogens including *Taphrina deformans* of peaches at 175 g/100 l. It is generally non-phytotoxic but the black spray residue may be objectionable.

Toxicology. The acute oral LD50 for rats is > 4000 mg/kg. In 2-y feeding trials the 'no effect' level was: for rats 250 mg/kg diet; for dogs 5 mg/kg daily. It is not stored in the body tissues (H. C. Hodge *et al., J. Pharmacol. Exp. Ther.,* 1956, **118,** 174).

Formulation. A w.p. (760 g a.i./kg).

Analysis. Product analysis is by dissolving in chloroform and either colorimetry at 410 nm (*CIPAC Handbook,* 1970, **1,** 397) or acid hydrolysis of the extract, the liberated carbon disulphide being absorbed in alkali and the dithiocarbonate estimated by titration with iodine (D. G. Clarke *et al., Anal. Chem.,* 1951, **23,** 1842). Residues may be determined by acid hydrolysis and reaction of the liberated carbon disulphide with a copper salt plus an amine and suitable colorimetry (H. L. Pease, *J. Assoc. Off. Anal., Chem.,* 1957, **40,** 1113; D. Dickenson, *Analyst (London),* 1946, **71,** 327; W. K. Lowen, *Anal. Chem.,* 1951, **23,** 1846).

Flamprop-isopropyl (racemate)

GR   BF   ENVR&Y1&VOY1&1 $C_{19}H_{19}ClFNO_3$ (363·8)

Nomenclature and development. The common name flamprop-isopropyl has been approved by BSI and New Zealand and proposed by ISO; the common name flamprop is approved by WSSA for the parent acid (*C.A.* Registry No. *[58667-63-3]*). The IUPAC name of the racemic ester is isopropyl *N*-benzoyl-*N*-(3-chloro-4-fluorophenyl)-DL-alaninate formerly known as isopropyl (±)-2-[*N*-(3-chloro-4-fluorophenyl)benzamido]propionate or isopropyl (±)-2-(*N*-benzoyl-3-chloro-4-fluoroanilino)propionate, in *C.A.* usage 1-methylethyl *N*-benzoyl-*N*-(3-chloro-4-fluorophenyl)-DL-alaninate *[52756-22-6]*. It was introduced in 1972 by Shell Research Ltd under the code number 'WL 29 762', trade mark 'Barnon' and protected by BP 1 164 160; 1 289 283; BPA 46223/72. Its herbicidal properties were first described by A. Mouillac *et al., C. R. Journ. Etud. Herbic. COLUMA, 7th,* 1973, **2**, 363; A. P. Warley *et al., Proc. Br. Weed Control Conf., 12th,* 1974, **3**, 883.

Properties. The technical material is an almost colourless crystalline powder, m.p. 56-57°. Its solubility at 20° is *c.* 18 mg/l water; readily soluble in most organic solvents. It is photochemically stable and very stable to hydrolysis at $2 < pH < 8$; it is rapidly hydrolysed to propan-2-ol and the parent acid at $pH > 8$.

Uses. Flamprop-isopropyl has given consistently $> 90\%$ control of all *Avena* spp. when applied at 1·0 kg a.i./ha to barley between the end of tillering and the appearance of the first node. At least twice the recommended dose, applied within these crop growth stages, is tolerated by barley. Crop competition is necessary to ensure the effective control of wild oat (B. Jeffcoat & W. N. Harries, *Pestic. Sci.,* 1975, **6**, 283).

Toxicology. The acute oral LD50 is: for rats > 3000 mg/kg; for mice 2554 mg/kg. The acute percutaneous LD50 for rats is > 600 mg a.i. (as 200 g/l e.c.)/kg. In 90-d feeding trials no changes of toxicological significance were seen in rats receiving 500 mg/kg diet. It is of moderate to low toxicity to fish.

Formulation. An e.c. (200 and 250 g a.i./l).

Analysis. Product analysis is by GLC or i.r. spectroscopy. Residues may be determined by GLC. Details of the methods are available from Shell International Chemical Co. Ltd.

Flamprop-isopropyl ((R)-(-)-enantiomorph)

GR   BF   ENVR&Y1&VOY1&1

$C_{19}H_{19}ClFNO_3$ (363·8)

Nomenclature and development. The common name flamprop-isopropyl has been accepted by BSI and proposed by ISO for racemic flamprop-isopropyl (p.272); suitable qualification of this common name to represent the (R)-enantiomorph is still under discussion within ISO. The IUPAC name is isopropyl N-benzoyl-N-(3-chloro-4-fluoro-phenyl)-L-alaninate formerly isopropyl (R)-(-)-N-benzoyl-N-(3-chloro-4-fluorophenyl)-2-aminopropionate, in *C.A.* usage 1-methylethyl N-benzoyl-N-(3-chloro-4-fluorophenyl)-L-alaninate [57973-67-8]. It was introduced by Shell Research Ltd under the code number 'WL 43425' and protected by BP 1437711; internationally the trade mark 'Suffix BW' has been applied though trade marks may vary with the country. Its herbicidal properties were described by R. M. Scott *et al., Proc. 1976 Br. Crop Prot. Conf.—Weeds,* 1976, **2**, 723 and its development has been discussed (D. Jordan, *Span,* 1977, **20**, 21).

Properties. Pure flamprop-isopropyl (R)-isomer is a colourless crystalline solid, m.p. 72·5-74·5°, v.p. 2·38 x 10^{-7} mmHg at 20°. The technical material is an off-white crystalline powder, m.p. 70-71°. Its solubility at 20° is *c.* 10 mg/l water; > 400 g/l acetone; 677 g/l cyclohexanone; 147 g/l ethanol; 6 g/l hexane. It is stable to heat, to light and to hydrolysis at 2 < pH < 8. Hydrolysis to propan-2-ol and the parent acid (*C.A.* Registry No. [57353-42-1]) occurs at pH > 8.

Uses. The (R)-isomer has given consistently > 90% control of *Avena* spp. (wild oats) at 600 g a.i./ha to barley between mid-tillering and the formation of the first node and to wheat between mid-tillering and the formation of the second node. Best results are achieved when the growing conditions are good and the crop is growing vigorously. When used within the above timing, the compound is selective at twice the recommended dose. A useful degree of suppression of blackgrass *(Alopecurus myosuroides)* can often be obtained in addition to wild oat control.

Toxicology. The acute oral LD50 is: for rats and mice > 4000 mg/kg; for domestic fowl > 2000 mg/kg. The acute percutaneous LD50 for rats is > 1600 mg/kg. In 90-d feeding trials no toxicological effect was observed in rats receiving 50 mg/kg diet or dogs 30 mg/kg diet. The LC50 (96 h) for rainbow trout is 3·3 mg/l.

Formulation. An e.c. (200 g a.i./l).

Analysis. Product analysis is by optical rotation and GLC. Residues can be determined by GLC with ECD. Details of the methods are available from Shell International Chemical Co. Ltd.

Flamprop-methyl (racemate)

Ph.C(=O)—N(—CH(Me).C.OMe(=O))— [structure: N-benzoyl-N-(3-chloro-4-fluorophenyl) group]

GR   BF   ENVR&Y1&VO1 $C_{17}H_{15}ClFNO_3$ (335·8)

Nomenclature and development. The common name flamprop-methyl has been approved by BSI and New Zealand and proposed by ISO; the common name flamprop is approved by WSSA for the parent acid (*C.A.* Registry No. *[58667-63-3]*). The IUPAC name for the racemic ester is methyl *N*-benzoyl-*N*-(3-chloro-4-fluorophenyl)-DL-alaninate formerly methyl (±)-*N*-benzoyl-*N*-(3-chloro-4-fluorophenyl)-2-aminopropionate, in *C.A.* usage methyl *N*-benzoyl-*N*-(3-chloro-4-fluorophenyl)-DL-alaninate *[52756-25-9]*. It was introduced by Shell Research Ltd under the code number 'WL 29 761', the trade mark 'Mataven' and protected by BP 1437711. Its herbicidal properties were first described by E. Haddock *et al., Proc. Br. Weed Control Conf., 12th,* 1974, **1**, 9 and the reasons given for the choice of this member of a related series of compounds (B. Jeffcoat *et al., Pestic. Sci.,* 1977, **8**, 1).

Properties. Pure flamprop-methyl is a colourless crystalline solid, m.p. 84-86°, v.p. 3·5 x 10^{-7} mmHg at 20°. The technical material is an off-white crystalline powder, m.p. 81-82°. Its solubility at 22° is *c.* 35 mg/l water; at 20° > 500 g/l acetone, 414 g/l cyclohexanone, 135 g/l ethanol, 7 g/l hexane. It is stable to heat, to light and to hydrolysis at 2 < pH < 7. Hydrolysis to methanol and the parent acid occurs at pH > 7.

Uses. Flamprop-methyl has given consistently > 90% control of *Avena fatua, A. ludoviciana* and other species of wild oats at 525 g a.i./ha to the majority of both winter- and spring-sown wheats from mid-tillering to before the formation of the second node. Best results are achieved when the crop is growing vigorously and is not subject to check or stress factors. Tests on more than 60 varieties of spring and winter wheat have indicated that, when applied as recommended, it is selective at 3-4 times the recommended dose. A useful degree of suppression of blackgrass (*Alopecurus myosuroides*) can often be obtained in addition to wild oat control.

Toxicology. The acute oral LD50 is: for rats 1210 mg/kg; for mice 720 mg/kg; for domestic fowl > 1000 mg/kg. The acute percutaneous LD50 for rats is > 294 mg a.i. (as e.c.)/kg. In 2-y feeding trials no toxicological effect was noted for rats receiving 2·5 mg/kg diet or dogs 10 mg/kg diet. The LC50 (96 h) for rainbow trout is 4·7 mg/l.

Formulation. An e.c. (105 g a.i./l).

Analysis. Product analysis is by i.r. spectroscopy or by GLC. Residues may be determined by GLC with ECD. Details of methods are available from Shell International Chemical Co. Ltd.

Fluometuron

FXFFR   CMVN1&1

$C_{10}H_{11}F_3N_2O$ (232·2)

Nomenclature and development. The common name fluometuron is approved by BSI, ISO, ANSI and WSSA. The IUPAC name is 1,1-dimethyl-3-(α,α,α-trifluoro-*m*-tolyl)urea (I) formerly 1,1-dimethyl-3-(3-trifluoromethylphenyl)urea, in *C.A.* usage *N,N*-dimethyl-*N'*-[3-(trifluoromethyl)phenyl]urea *[2164-17-2]* formerly (I). Its herbicidal properties were described by C. J. Counselman *et al., Proc. South. Weed Conf., 17th,* 1964, p.189. It was introduced in 1960 by Ciba AG (now Ciba-Geigy AG) under the code number 'C 2059', trade mark 'Cotoran' and protected by BP 914 779; USP 3 134 665.

Manufacture and properties. Produced by the reaction of dimethylamine with α,α,α-trifluoro-*m*-tolyl isocyanate, it forms colourless crystals, m.p. 163-164·5°, v.p. 5 x 10⁻⁷ mmHg at 20°. Its solubility at 20° is 105 mg/l water; soluble in most organic solvents.

Uses. Fluometuron is a herbicide mainly absorbed through the roots; it has weak foliar activity, and is especially suitable for the control of broad-leaved and grass weeds in cotton at 1·0-1·5 kg a.i./ha. It is of intermediate persistence, 50% loss occurring in 60-75 d according to soil conditions.

Toxicology. The acute oral LD50 is: for rats 6416- > 8000 mg/kg; for dogs > 10000 mg/kg. The acute dermal LD50 for rats is > 2000 mg/kg. In 1-y feeding trials no effect was observed for rats receiving 10 mg/kg daily or dogs 15 mg/kg daily. It has negligible toxicity to birds and wildlife.

Formulations. A w.p. (500 g a.i./kg and 800 g/kg); flowable (500 g/l); 'Cotoran Multi' 50 WP, in combination with metolachlor.

Analysis. Product analysis is by GLC (*CIPAC Handbook,* in press). Residues may be determined by alkaline hydrolysis to α,α,α-trifluoro-*m*-toluidine which is purified, diazotised, converted to α,α,α-trifluoro-3-iodotoluene and this measured by GLC with ECD. Details of these methods are available from Ciba-Geigy AG. See also G. Voss *et al., Anal. Methods Pestic. Plant Growth Regul.,* 1973, **7**, 569.

Fluoroacetamide

$$\overset{\text{O}}{\underset{\text{FCH}_2.\overset{\|}{\text{C}}.\text{NH}_2}{}}$$

ZV1F C_2H_4FNO (77·06)

Nomenclature and development. The name fluoroacetamide is accepted in lieu of a common name by BSI and ISO. The IUPAC and *C.A.* name is 2-fluoroacetamide, Registry No. *[640-19-7]*. Although it has insecticidal properties, it is considered too toxic to mammals for commercial use as an insecticide. Its use as a rodenticide, as 'compound 1081' was suggested by C. Chapman & M. A. Phillips, *J. Sci. Food Agric.,* 1955, **6**, 231.

Properties. It is a colourless crystalline powder, m.p. 108°. It is very soluble in water; soluble in acetone; moderately soluble in ethanol; sparingly soluble in aliphatic and aromatic hydrocarbons.

Uses. Fluoroacetamide is a rodenticide and an internal mammalian poison subject to many restrictions in use. Normally it is used as a bait (20 g a.i./kg) in areas to which the public have no access such as sewers and locked warehouses.

Toxicology. The acute oral LD50 is: for *Rattus norvegicus c.* 13 mg/kg (C. Chapman & M. A. Phillips, *loc. cit.*; E. W. Bentley & J. H. Greaves, *J. Hyg.,* 1960, **58**, 125; E. W. Bentley *et al., ibid.,* 1961, **59**, 413).

Formulation. Bait, dyed cereal base (30 g a.i./kg) which is mixed with water for use.

Analysis. Product analysis is by reaction with sodium and precipitation as lead chloride fluoride (*cf WHO Manual,* (2nd Ed.), SRT/5).

Fluorodifen

WNR   DOR   BNW   DXFFF $\qquad$ $C_{13}H_7F_3N_2O_5$ (328·2)

Nomenclature and development. The common name fluorodifen is approved by BSI, ISO, ANSI, WSSA and JMAF. The IUPAC name is 4-nitrophenyl α,α,α-trifluoro-2-nitro-*p*-tolyl ether formerly 4-nitrophenyl 2-nitro-4-trifluoromethylphenyl ether, in *C.A.* usage 2-nitro-1-(4-nitrophenoxy)-4-trifluoromethylbenzene *[15457-05-3]* formerly *p*-nitrophenyl α,α,α-trifluoro-2-nitro-*p*-tolyl ether. It was introduced in 1968 by Ciba AG (now Ciba-Geigy AG) under the code number 'C 6989', trade mark 'Preforan', and protected by BP 1 033 163. Its herbicidal properties were first described by L. Ebner *et al., Proc. Br. Weed Control Conf., 9th,* 1968, p.1026.

Manufacture and properties. Produced by the reaction of sodium 4-nitrophenoxide with 4-chloro-α,α,α-trifluoro-3-nitrotoluene, it forms yellowish-brown crystals, m.p. 94°, v.p. 7 x 10⁻⁸ mmHg at 20°. It solubility at 20° is 2 mg/l water; soluble in organic solvents.

Uses. Fluorodifen is a pre- and post-em. herbicide acting by contact. It is recommended for pre-em. use on rice, beans and soyabeans. It is not recommended for surface seeded rice pre-em., but it is safe on drilled rice. When used on crops other than rice, herbicidal action persists for 56-84 d in dry soil. The dosage range is 3-4 kg a.i./ha.

Toxicology. The acute oral LD50 for rats is 9000 mg/kg; the acute dermal LD50 for rats is > 3000 mg/kg. The toxic response to fish varies with the species from moderate to high.

Formulation. An e.c. (300 g a.i./l).

Analysis. Product analysis is by GLC with internal standard. Residues are determined by GLC using ECD. Details of these methods are available from Ciba-Geigy AG.

Fluotrimazole

T5NN DNJ AXR&R&R CXFFF $\qquad$ $C_{22}H_{16}F_3N_3$ (379·4)

Nomenclature and development. The common name fluotrimazole is approved by BSI, New Zealand and proposed by ISO. The IUPAC name is 1-(3-trifluoromethyltrityl)-1,2,4-triazole, in *C.A.* usage 1-[diphenyl[3-(trifluoromethyl)phenyl]methyl]-1H-1,2,4-triazole *[57381-79-0]*. It was introduced in 1973 by Bayer AG under the code number 'BUE 0620', trade mark 'Persulon' and protected by DBP 1 795 249; USP 3 682 950. Its fungicidal properties were first described by F. Grewe & K. H. Büchel, *Mitt. Biol. Bundesanst. Land-Forstwirtsch. Berlin-Dahlem,* 1973, **151,** 208.

Properties. It is a colourless crystalline solid, m.p. 132°. Its solubility at 20° is: 1·5 µg/l water; 400 g/kg dichloromethane; 200 g/kg cyclohexanone; 100 g/kg toluene; 50 g/kg propan-2-ol. It is stable in 0·1M sodium hydroxide, but undergoes *c.* 40% degradation in 0·2M sulphuric acid in 24 h.

Uses. Fluotrimazole is a fungicide with specific action against powdery mildews on melons, cucumbers, peaches, grapes and barley.

Toxicology. The acute oral LD50 is: for rats > 5000 mg/kg; for canaries > 1000 mg/kg. The acute dermal LD50 for rats is > 1000 mg/kg. In 90-d feeding trials the 'no effect' level was: for rats 800 mg/kg diet; for dogs > 5000 mg/kg diet. The LC50 (96 h) for the golden orfe is > 100 mg/l. It is not harmful to honey bees.

Formulations. These include: w.p. (500 g a.i./kg); e.c. (125 g/l).

Analysis. Product analysis is by i.r. spectroscopy. Residues may be determined by GLC with FID (W. Sprecht, *Pflanzenschutz-Nachr. (Am. Ed.),* 1977, **30,** 55).

Flurecol-butyl

(Flurenol-butyl)

L B656 HHJ HVO4 HQ $\qquad$ $C_{18}H_{18}O_3$ (282·3)

Nomenclature and development. The common name flurenol is approved by ISO— exceptions BSI, Canada and Denmark (flurecol), and USA—for (IUPAC) 9-hydroxy-fluorene-9-carboxylic acid (I), in *C.A.* usage 9-hydroxy-9*H*-fluorene-9-carboxylic acid *[467-69-6]* formerly (I). The effects on plant growth of derivatives of fluorene-9-carboxylic acid were first described by G. Schneider *Naturwissenschaften*, 1964, **51**, 416, who, because their activity differed from other plant-growth regulators, proposed that they be called morphactins (G. Schneider *et al., Nature (London),* 1965, **208**, 1013). The butyl ester of flurecol, flurecol-butyl, *C.A.* Registry No. *[2314-09-2]*, was introduced in 1964 by E. Merck under the code number 'IT 3233'; protected by BP 1 051 652; 1 051 653.

Manufacture and properties. Produced from 9-hydroxyfluorene-9-carboxylic acid, by rearrangement of 9,10-phenanthraquinone, and esterified with butan-1-ol, flurecol-butyl forms colourless crystals, m.p. 71°. Its solubility at 20° is: 36·5 mg/l water; *c.* 7 g/l light petroleum (b.p. 50-70°); 35 g/l cyclohexane; 250 g/l propan-2-ol; 550 g/l carbon tetrachloride; 700 g/l ethanol; 950 g/l benzene; 1·45 kg/l acetone; 1·5 kg/l methanol. It is stable at room temperature, compatible with other herbicides and non-corrosive.

Uses. Flurecol-butyl is absorbed via leaves and roots causing a general inhibition of plant growth, but is mainly used in conjunction with phenoxyalkanoic acid herbicides, whose action it potentiates. The combination is used for weed control in cereals at rates of 2-4 l 'Aniten'/ha, which persists in the soil for about 6 weeks.

Toxicology. The acute oral LD50 for rats and dogs is > 5000 mg/kg. It is non-toxic to honey bees and of low toxicity to fish.

Formulation. An e.c. (125 g flurecol-butyl/kg); 'Aniten', e.c. (100 g flurecol-butyl + 400 g MCPA-isooctyl/kg).

Analysis. Product analysis is by quantitative saponification or u.v. spectrometry. Residues may be determined by the colorimetric measurement of fluoren-9-one 4-nitrophenylhydrazone, formed quantitatively from flurecol-butyl under the chosen chemical conditions of the analytical method.

Fluridone

T6N DVJ A1 CR CXFFF& ER $C_{19}H_{14}F_3NO$ (329·3)

Nomenclature and development. The common name fluridone is approved by BSI, ANSI, New Zealand and WSSA, and proposed by ISO. The IUPAC name is 1-methyl-3-phenyl-5-(α,α,α-trifluoro-*m*-tolyl)-4-pyridone, in *C.A.* usage 1-methyl-3-phenyl-5-[3-(trifluoromethyl)phenyl]-4(1*H*)-pyridinone *[59756-60-4]*. It was introduced in 1976 by Eli Lilly & Co. under the code number 'EL-171' and protected by BP 1 521 092. Its herbicidal properties were first described by T. W. Waldrep & H. M. Taylor, *J. Agric. Food Chem.*, 1976, **24**, 1250.

Manufacture and properties. Produced by the condensation of (α,α,α-trifluro-*m*-tolyl)-acetonitrile with ethyl phenylacetate followed by acidification, and cyclisation of the resulting intermediate with methylamine and methyl formate, it is an off-white crystalline solid, m.p. 151-154°, v.p. 1 x 10^{-7} mmHg at 25°. Its solubility is *c.* 12 mg/l water at pH 7; moderately soluble in acetone, chloroform, ethanol, ethyl acetate; less soluble in diethyl ether, benzene; almost insoluble in hexane. It is moderately susceptible to decomposition by u.v. light in aqueous solution and is stable to hydrolysis at pH 3-9.

Uses. Fluoridone is a pre-em. selective herbicide for use particularly on cotton, controlling annual grass and broad-leaved weeds and certain perennial species. It is also recommended for testing as an aquatic herbicide. Its mode of action is by inhibition of carotenoid biosynthesis; selectivity between cotton (tolerant) and maize, soyabean and rice (susceptible) being due to retention of fluridone by cotton roots whilst it is translocated to the foliage of the other species (D. F. Berard *et al.*, *Weed Sci.*, 1978, **26**, 296).

Toxicology. The acute oral LD50 is: for rats and mice > 10000 mg/kg; for dogs > 500 mg/kg; for cats > 250 mg/kg. In 90-d feeding trials no treatment-related effect was noted in rats receiving 1400 mg/kg diet.

Formulations. A w.p. (500 g a.i./kg); flowable (500 g/l); pellets (50 g/kg).

Analysis. Product analysis is by GLC with FID following extraction with chloroform or chloroform + methanol. Residues in soil and plant tissue may be determined by GLC with nitrogen-specific detection following suitable clean-up, or ECD after clean-up and formation of a suitable derivative (S. D. West, *J. Agric. Food Chem.*, 1978, **26**, 644). Details of analytical procedures and other information may be obtained from Eli Lilly & Co.

Folpet

T56 BVNVJ CSXGGG   C$_9$H$_4$Cl$_3$NO$_2$S (296·6)

Nomenclature and development. The common name folpet is approved by BSI, ANSI, New Zealand and JMAF—exception France (folpel). The IUPAC name is *N*-(trichloro-methylthio)phthalimide (I) formerly *N*-(trichloromethanesulphenyl)phthalimide, in *C.A.* usage 2-[(trichloromethyl)thio]-1*H*-isoindole-1,3(2*H*)-dione *[133-07-3]* formerly (I). Its fungicidal properties were first described by A. R. Kittleson, *Science,* 1952, **115,** 84. It was introduced in 1949 by the Standard Oil Development Co. and later by the Chevron Chemical Co. under the trade mark 'Phaltan' and protected by USP 2 553 770; 2 553 771; 2 553 776.

Manufacture and properties. Produced by the reaction of trichloromethanesulphenyl chloride with sodium phthalimide, it forms colourless crystals, m.p. 177°, v.p. < 10^{-5} mmHg at 20°. Its solubility at room temperature is 1 mg/l water; slightly soluble in organic solvents. It is non-corrosive though its decomposition products are corrosive. It is slowly hydrolysed by water and rapidly by concentrated alkali.

Uses. Folpet is a protective fungicide used mainly for foliage application against *Alternaria, Botrytis, Rhizoctonia, Pythium* and *Venturia* spp., leaf spot, powdery and downy mildews. Crops on which it is used include: apples, citrus, soft fruit, grapes, cucumbers, melons, onions, lettuces, tomatoes and flowering ornamentals. Rates are 140-560 g a.i./100 l depending on crop and pathogen. It is non-phytotoxic except to D'Anjou pears and it can produce russeting of sensitive apple varieties (Golden Delicious, Red Delicious and Stayman Winesap) if applied early in cropping.

Toxicology. Rats survived oral administration of 10 000 mg/kg; the acute dermal LD50 for albino rabbits is > 22 600 mg/kg. It can cause irritation of the mucuous membranes; contact with eyes and skin, or inhalation of dust or spray mists can result in local irritation. In 1·4-y feeding trials no ill-effect, histopathological change or significant difference in tumour incidence was noted in albino rats receiving 10 000 tech. mg/kg diet, no adverse effect was noted for dogs receiving 1500 mg/kg 5 d/week. No significant effect on reproductive performance was observed over 3 generations in rats at 1000 mg/kg diet; no teratogenic effect was noted in 2 species of monkeys, hamsters and rats. Treatment of male mice by Bateman's dominant lethal procedure did not induce dominant lethal mutations as measured by increase in early post-implantation deaths.

Formulations. 'Ortho Phaltan 50W', w.p. (500 g a.i./kg); w.p. (750 g/kg); Murphy's rose fungicide (folpet + dinocap).

Analysis. Product analysis is by i.r. spectroscopy, HPLC (A. A. Carlstrom, *J. Assoc. Off. Anal. Chem.,* 1978, **60,** 1157) or by total chlorine content after alkaline hydrolysis. Residues may be determined by reaction with resorcinol (A. R. Kittleson, *Anal. Chem.,* 1952, **24,** 1173). See also: J. N. Ospenson, *Anal. Methods Pestic., Plant Growth Regul. Food Addit.,* 1964, **3,** 137; *Anal. Methods Pestic. Plant Growth Regul.,* 1972, **6,** 546.

Fonofos

(S)-isomer (R)-isomer

2OPS&2&SR $C_{10}H_{15}OPS_2$ (246·3)

Nomenclature and development. The common name fonofos is approved by BSI and ISO. The IUPAC and *C.A.* name is (±)-*O*-ethyl *S*-phenyl ethylphosphonodithioate, Registry No. *[944-22-9]*. Its insecticidal properties were first described by J. J. Menn & K. Szabo, *J. Econ. Entomol.,* 1965, **58,** 734, and it was introduced in 1967 by the Stauffer Chemical Co. under the code number 'N-2790', trade mark 'Dyfonate' and protected by USP 2988474.

Properties. It is a pale yellow liquid, with a pungent mercaptan-like odour, b.p. 130°/0·1 mmHg, v.p. 2·1 x 10^{-4} mmHg at 25°, d_{25}^{25} 1·16, n_D^{30} 1·5883. It is practically insoluble in water; miscible with organic solvents such as kerosine, 4-methylpentan-2-one, xylene. It is stable under normal conditions.

Having 4 different groups attached to the phosphorus atom, fonofos exists as chiral isomers and these have been isolated (R. Allahyari *et al., J. Agric. Food Chem.,* 1977, **25,** 471; P. W. Lee *et al., Pestic. Biochem. Physiol.,* 1978, **8,** 146, 158). The optical rotations of the chiral isomers of fonofos are reversed on solution in cyclohexane, methanol, carbon tetrachloride.

Uses. Fonofos is an insecticide suitable for the control of soil insects such as rootworms, symphylids, wireworms and crickets at 1·0-1·5 kg a.i./ha. It has caused some injury to seeds when placed in their proximity and persists in soil for *c.* 56 d. A combined granule with disulfoton is used for brassicas to extend the activity to control aphids: C. Sinclair & T. J. Purnell, *Proc. 1977 Br. Crop Prot. Conf. Pests. Dis.,* 1977, **2,** 589.

Fonofos (*R*)-isomer is more toxic to insects and mice and a more potent inhibitor of cholinesterase than the (*S*)-isomer. The analogous *O*-ethyl *S*-phenyl ethylphosphonothioate behaves similarly, its (*S*)-isomer (which corresponds sterically to fonofos (*R*)-isomer) having much greater biological activity than the (*R*)-isomer.

Toxicology. The acute oral LD50 of the racemate for male albino rats is 7·94-17·5 mg/kg. The acute dermal LD50 is: for rabbits *c.* 150 mg/kg; for guinea-pigs 278 mg/kg.

Formulations. 'Dyfonate', granules (50 and 100 g a.i./kg). 'Doubledown', granules (40 g fonofos + 60 g disulfoton/kg).

Analysis. Product and residue analysis is by GLC with ECD: J. E. Barney *et al., Anal. Methods Pestic. Plant Growth Regul.,* 1973, **7,** 269. Particulars of methods are available from the Stauffer Chemical Co.

Formaldehyde

HCHO

VHH

CH_2O (30.03)

Nomenclature and development. The IUPAC and *C.A.* name formaldehyde, Registry No. *[50-00-0],* is accepted in lieu of a common name by BSI, ISO and JMAF, also known as methanal. The common name formalin applies to the aqueous solution containing 400 g a.i./l. It was introduced as a disinfectant by Loew in 1888 and was first used as a seed disinfectant by Geuther in 1896.

Manufacture and properties. Produced by the catalytic oxidation of methanol vapour, it is a flammable colourless gas, with a pungent irritating odour, b.p. —19·5°, m.p. —92°, d_4^{20} 0·815. It is very soluble in water, ethanol, diethyl ether. It is very reactive chemically and a powerful reducing agent.

Formalin is a colourless liquid, with a pungent odour, d_{25}^{25} 1·081-1·085, n_D^{20} 1·3746. It is miscible with water, ethanol, acetone. It is chemically reactive, the formaldehyde polymerising on standing and is a powerful reducing agent.

Uses. Formaldehyde is a powerful bactericide and fungicide, used as a soil fumigant and, at one time, for seed treatment though this use is limited by phytotoxicity. It is also used to fumigate glasshouse structures after cropping.

Toxicology. The acute oral LD50 for rats is 800 mg formalin/kg. Formaldehyde vapour is very irritating to the mucous membranes and toxic to animals, including man.

Formulations. An aqueous solution to which methanol is added to delay polymerisation; most pharmacopoeias require a content of 370-410 g formaldehyde/kg.

Analysis. Product analysis is by oxidation by hydrogen peroxide to form formic acid which is determined by titration (AOAC Methods; *MAFF Tech. Bull.,* No. 1, 1958, p.36; *CIPAC Handbook,* in press).

Formetanate

N:CH.NMe₂ (structure diagram)

O.CO.NHMe

1N1&1UNR COVM1 $C_{11}H_{15}N_3O_2$ (221·3)

Nomenclature and development. The common name formetanate is approved by BSI, ISO and ANSI. The IUPAC name is 3-dimethylaminomethyleneaminophenyl methylcarbamate, in *C.A.* usage *N,N*-dimethyl-*N'*-[3-[(methylamino)carbonyl]oxy]phenyl]methanimidamide *[22259-30-9]* formerly *N'*-(*m*-hydroxyphenyl)-*N,N*-dimethylformamidine methylcarbamate ester. It was introduced in 1967 by Schering AG as a combined formulation 'Fundal forte 750' with chlordimeform (p.95). In 1970 its hydrochloride (also known as ENT 27566), *C.A.* Registry No. *[23422-53-9]*, was introduced under the trade mark 'Dicarzol'; formetanate had the code number 'Schering 36056' and is protected by DBP 1169194; BP 987381; USP 3336186. It was first described by W. R. Steinhausen, *Z. Angew. Zool.,* 1968, **55,** 107.

Manufacture and properties. Produced by the reaction of 3-dimethylaminomethyleneaminophenol with methyl isocyanate, it is a yellowish crystalline solid, m.p. 102-103°, and non-volatile. Its solubility at room temperature is: < 1 g/l water; *c.* 100 g/l acetone, chloroform; > 200 g/l methanol. In aqueous methanol 50% decomposition occurs in 130 h at pH 4, in 17 h at pH 7, in 100 min at pH 9. Because the base is of low stability, it is formulated as the hydrochloride, a white powder decomposing at 200-202°, solubility: > 500 g/l water; *c.* 250 g/l methanol; < 1 g/l acetone, chloroform, hexane. The hydrochloride is very slowly hydrolysed at pH < 4 and unchanged when stored for 7 d at 60°.

Uses. Formetanate is an acaricide effective against the motile stages of fruit spider mites at 420 g a.i./ha (LV) or 19 g/100 l (HV); against glasshouse spider mite on roses, one spray at 25 g/100 l, on chrysanthemums 2 sprays at 47·5 g/100 l are recommended. It is effective as an insecticide against several plant bugs, beet fly and thrips.

Toxicology. The acute oral LD50 of the hydrochloride is: for rats 21 mg/kg; for mice 18 mg/kg; for beagle dogs 19·1 mg/kg; for chickens 21·5 mg/kg. The acute dermal LD50 is: for rats > 5600 mg/kg; for rabbits > 10200 mg/kg. The sub-acute LC50 (10 d) was: for pheasants and bobwhite quail > 4640 mg/kg diet; for ducks 6810 mg/kg diet. In 2-y feeding trials dogs and rats receiving 200 mg/kg diet showed no significant abnormality. The LC50 (96 h) is: for rainbow trout 2·8 mg/l, for bluegill 20 mg/l, for black bullhead 75 mg/l.

Formulations. These include: in France, 'Dicarzol 200', w.s.p. (250 g a.i./kg) as hydrochloride; in North America, 'Carzol SP', w.s.p. (920 g/kg) as hydrochloride; in some countries as a combined formulation with chlordimeform.

Analysis. Product analysis is by titration of the hydrochloride with alkali. Residues may be determined by separating the a.i. by TLC, followed by hydrolysis to 3-aminophenol which is either diazotised, coupled with *N*-(1-naphthyl)ethylenediamine and the colour measured at 555 nm, or brominated and estimated by GLC with ECD (N. A. Jenny & K. Kossman, *Anal. Methods Plant Growth Regul.,* 1973, **7,** 279).

Formothion

$$(MeO)_2\overset{\displaystyle S}{\underset{\displaystyle \parallel}{P}}.S.CH_2\overset{\displaystyle O}{\underset{\displaystyle \parallel}{C}}.\underset{\displaystyle Me}{N}.\overset{\displaystyle O}{\underset{\displaystyle \parallel}{C}}.H$$

VHN1&V1SPS&O1&O1 $C_6H_{12}NO_4PS_2$ (257·3)

Nomenclature and development. The common name formothion is approved by BSI, ISO and JMAF—exception Germany; also known as ENT 27257, OMS 698. The IUPAC name is S-(N-formyl-N-methylcarbamoylmethyl) O,O-dimethyl phosphorodithioate, in C.A. usage S-[2-(formylmethylamino)-2-oxoethyl] O,O-dimethyl phosphorodithioate [2540-82-1] formerly O,O-dimethyl phosphorodithioate S-ester with N-formyl-2-mercapto-N-methylacetamide. It was introduced in 1959 by Sandoz AG under the code numbers 'J-38', 'SAN 6913I', trade marks 'Anthio', 'Aflix' and protected by USP 3 176 035; 3 178 337. Its biological properties were first described by C. Klotzsche, *Mitt. Geb. Lebensmittelunters. Hyg.*, 1961, **52**, 341.

Manufacture and properties. Produced by the reaction of 2-chloro-N-formyl-N-methyl-acetamide with ammonium O,O-dimethyl phosphorodithioate, it is a yellow viscous oil or crystalline mass, which cannot be distilled without decomposition, m.p. 25-26°, v.p. 6×10^{-6} mmHg at 20°, d_4^{20} 1·361; n_D^{20} 1·5541. Its solubility at 24° is 2·6 g/l water; miscible with alcohols, chloroform, diethyl ether, acetone, xylene. It is stable in non-polar solvents, but is hydrolysed by alkali and incompatible with alkaline pesticides.

Uses. Formothion is a contact and systemic insecticide, effective against a wide range of sucking insects, thrips, aphids, jassids, psyllids, scales, bugs, whiteflies, fruit flies, as well as against some chewing insects, *Cydia pomonella*, Epilachna beetles and dipterous mining larvae on a variety of crops, fruit trees, citrus and other tropical fruits, olives, grapes, hops, cotton, tobacco, field crops, vegetables and ornamentals. It is used at 113-175 g a.i./ha. In plants it is metabolised to dimethoate, omethoate, (dimethoxy-phosphinthioylthio)acetic acid and bis(dimethylthiophosphoryl) disulphide: in animals to (dimethoxyphosphinthioylthio)acetic acid and to polar metabolites, In loamy soil 50% loss occurs in 14 d.

Toxicology. The acute oral LD50 for albino rats is 365-500 mg/kg; the acute dermal LD50 for male albino rats is > 1000 mg/kg. In 2-y feeding trials rats and dogs receiving 80 mg/kg diet showed no ill-effect. Hazards to wild life are low.

Formulations. 'Anthio': e.c. (245 g and 337 g a.i./l); ULV (360 g/l); 'Anthiomix' and 'Sandothion': e.c. (210 g formothion + 210 g fenitrothion/l).

Analysis. Product analysis is by GLC and paper chromatography (*CIPAC Handbook*, 1979, **1A**, in press). Residues may be determined by GLC, by TLC or by oxidation to phosphate which is determined by standard methods. Details of methods may be obtained from Sandoz AG. See also M. Wisson *et al.*, *Anal. Methods Pestic. Plant Growth Regul.*, 1976, **8**, 123.

Fosamine-ammonium

$$\begin{array}{c}\quad\ \ \overset{O}{\underset{}{\parallel}}\ \overset{O}{\underset{}{\parallel}} \\ EtO-\underset{\underset{O^-}{\mid}}{P}-C-NH_2 \qquad NH_4{}^+ \end{array}$$

ZVPQO&O2   *fosamine* $C_3H_8NO_4P$ (153·1)

ZVPQO&O2 &ZH   *fosamine-ammonium* $C_3H_{11}N_2O_4P$ (170·1)

Nomenclature and development. The common name fosamine is approved by BSI and ANSI and proposed by ISO for ethyl hydrogen carbamoylphosphonate (IUPAC), in *C.A.* usage ethyl hydrogen (aminocarbonyl)phosphonate *[59682-52-9]*. Ammonium ethyl carbamoylphosphonate, fosamine-ammonium, (*C.A.* Registry No. *[25954-13-6]*) was introduced in 1974 by E. I. du Pont de Nemours & Co. (Inc.) under the code number 'DXP 1108', trade mark 'Krenite' and protected by USP 3 627 507; 3 846 512. Its herbicidal properties were first described by O. C. Zoebisch *et al., Proc. Northeast. Weed Sci. Soc.,* 1974, **28**, 347.

Manufacture and properties. Produced by the reaction of triethyl phosphite with methyl chloroformate and ammonia, fosamine-ammonium is a colourless crystalline solid, m.p. 175°, *d* 1·33, v.p. 4×10^{-6} mmHg at 25°. Its solubility at 25° is: 1·79 kg/kg water; 300 mg/kg acetone; 400 mg/kg benzene; 1·4 g/kg dimethylformamide; 12 g/kg ethanol; 158 g/kg methanol. Aqueous formulations and spray tank solutions are stable, but dilute solutions (50 g/l) are subject to decomposition under acid conditions. It is rapidly decomposed in soil.

Uses. Fosamine-ammonium is a contact herbicide effective against woody plants at rates of 6·8-13·5 kg a.i./ha, applied during the 60-d period before autumn coloration. The addition of non-ionic surfactant is recommended. It may be used on non-crops areas including land adjacent to water supplies. It is also used for control of *Convolvulus arvensis* and *Pteridium aquilinum* and for selective use in forestry (conifer release).

Toxicology. The acute oral LD50 of the formulated product is: for rats 24 000 mg/kg; for guinea-pigs 7380 mg/kg; for mallard duck and bobwhite quail > 10 000 mg/kg. The acute dermal LD50 of the product for rabbits is > 4000 mg/kg. Administration of 0·1 ml product to the rabbit eye caused no irritation and it is not a skin irritant nor a sensitiser. Rats fed for 90 d with 1000 mg tech./kg diet suffered no ill-effect. The LC50 (96 h) of the product is: for bluegill 670 mg/l; for rainbow trout and fathead minnow > 1000 mg/l.

Formulation. 'Krenite' brush control agent, a water-soluble liquid (415 g fosamine-ammonium/kg).

Analysis. Residues in plant tissues and in soil can be determined by conversion to a silyl derivative and subsequent GLC: details are available from E. I. du Pont de Nemours & Co. (Inc.).

Fosthietan

$(EtO)_2\overset{\overset{O}{\|}}{P}-N=$

T4SYS   DHJ   BUNPO&O2&O2 $C_6H_{12}NO_3PS_2$ (241·3)

Nomenclature and development. The common name fosthietan is approved by BSI and proposed by ISO. The IUPAC and *C.A.* name is diethyl 1,3-dithietan-2-ylidenephosphoramidate, Registry No. *[21548-32-3]*, also known as 2-(diethoxyphosphinylimino)-1,3-dithietane or cyclic methylene diethoxy-phosphinyl dithioimidocarbonate. It was introduced in 1972 by American Cyanamid Co. under the code number 'AC 64 475' and protected by USP 3476837. Its biological properties were described by W. K. Whitney & J. L. Aston, *Proc. Br. Insectic. Fungic. Conf., 8th,* 1975, **2,** 625. 'Nem-a-tak' is the registered trade mark in the USA; 'Acconem' and 'Geofos' are proposed trade marks elsewhere.

Manufacture and properties. Produced by reacting diethyl phosphorochloridate with sodium thiocyanate, sodium hydrosulphide and dibromomethane, the technical product is a yellow liquid, with a mercaptan-like odour, d^{25} 1·3, v.p. 6·5 x 10^{-6} mmHg at 25°. Its solubility at 25° is 50 g/kg water; soluble in acetone, chloroform, methanol, toluene.

Uses. Fosthietan is a broad-range nematicide and insecticide, effective in soil applications of *c.* 1-5 kg a.i./ha. Although there are some phytotoxicity problems on dicotyledonous crops at the higher rates, it appears that they can be resolved by the use of correct dosage rates and appropriate application methods and timing.

Toxicology. The acute oral LD50 for rats is 5·7 mg tech./kg. The dermal LD50 (24 h) for rabbits is 54 mg a.i./kg; 3124 mg (5% granule)/kg. Nonmutagenic to bacteria. In soil 50% chemical degradation occurs in *c.* 10-42 d, depending on various environmental factors.

Formulations. These include: granules (30-150 g a.i./kg); w.s.c. (250 g/l).

Analysis. Details of methods for residue analysis are available from American Cyanamid Co.

287

Fuberidazole

T56 BM DNJ C- BT5OJ $C_{11}H_8N_2O$ (184·2)

Nomenclature and development. The common name fuberidazole is approved by BSI and ISO. The IUPAC name is 2-(2-furyl)benzimidazole (I), in *C.A.* usage 2-(2-furanyl)-1*H*-benzimidazole *[3878-19-1]* formerly (I). It was introduced in 1966 by Bayer AG as a fungicide after tests begun in 1957 under the code numbers 'Bayer 33172', 'W VII/117', trade mark 'Voronit' and protected by DAS 1209799. Its fungicidal properties were described by G. Schuhmann, *Nachrichtenbl. Dtsch. Pflanzenschutzdienstes (Braunschweig),* 1969, **20**, 1.

Manufacture and properties. Produced by the reaction of *o*-phenylenediamine and 2-furaldehyde with copper diacetate in an oxidising medium (R. Weidenhagen, *Ber. Dtsch. Chem. Ges.,* 1936, **69**, 2271), it forms a fine crystalline powder, m.p. 286° (decomp.). Its solubility at room temperature is: 78 mg/kg water; *c.* 50 g/kg propan-2-ol; *c.* 10 g/kg dichloromethane, toluene, light petroleum; 1·6 g/kg 0·1M hydrochloric acid; 2·3 g/kg 0·1M sodium hydroxide. It is unstable to light.

Uses. Fuberidazole is a fungicide used for the treatment of seed against diseases caused by *Fusarium* spp., particularly *F. nivale* on rye and *F. culmorum* on peas.

Toxicology. The acute oral LD50 for rats is *c.* 1100 mg/kg; the acute dermal LD50 (7 d) for rats is *c.* 1000 mg/kg; the acute intraperitoneal LD50 for rats is *c.* 100 mg/kg. In 90-d feeding trials rats receiving 1500 mg/kg diet showed no ill-effect.

Formulations. Seed dressings alone or in combination with: anthraquinone; hexachlorobenzene, in countries where this is allowed ('Voronit'); quintozene ('Voronit spezial'); sodium dimethyldithiocarbamate ('Neo-Voronit').

Analysis. Product analysis is by u.v. spectroscopy measuring at 321 nm in neutral methanol: details are available from Bayer AG. Residue analysis is by GLC (W. Sprecht, *Pflanzenschutz-Nachr. (Am. Ed.),* 1977, **30**, 55).

Furalaxyl

T5OJ BVNR B1 F1&Y1&VO1 $C_{17}H_{19}NO_4$ (301·3)

Nomenclature and development. The common name furalaxyl is approved by BSI and proposed by ISO. The IUPAC name is methyl N-(2-furoyl)-N-(2,6-xylyl)-DL-alaninate, in *C.A.* usage methyl N-(2,6-dimethylphenyl)-N-(2-furanylcarbonyl)-DL-alaninate *[57646-30-7]*. It was synthesised and developed as a fungicide by Ciba-Geigy AG under the code number 'CGA 38140', trade mark 'Fongarid' and protected by BP 1 488 810; 1 498 199. Its properties were first described by F. J. Schwinn *et al., Meded. Fac. Landbouwwet. Rijksuniv. Gent,* 1977, **42**, 1181; J. M. Smith *et al., Proc. 1977 Br. Crop Prot. Conf. Pests. Dis.,* 1977, **2**, 633.

Properties. It is a crystalline solid with dimorphic forms, m.p. 70° and 84°, v.p. 5·3 x 10^{-7} mmHg at 20°. Its solubility is 230 mg/l water; it is soluble in most organic solvents.

Uses. Furalaxyl is a residual fungicide with systemic properties suitable for preventative and also curative control of diseases caused by air- and soil-borne Oomycetes. It is mainly used against *Pythium* spp. and *Phytophthora* spp. attacking ornamentals. Soil drenches as well as foliar applications are effective.

Toxicology. The acute oral LD50 for rats is 940 mg/kg; the acute dermal LD50 for rats is > 3100 mg/kg. It is of negligible toxicity to wildlife.

Formulation. 'Fongarid 50 WP', w.p. (500 g a.i./kg).

Analysis. Details of methods are available from Ciba-Geigy AG.

Gamma-HCH

(Gamma-BHC)

L6TJ AG BG CG DG EG FG $C_6H_6Cl_6$ (290·8)

Nomenclature and development. The common names gamma-HCH and gamma-BHC are approved by ISO; the former is more commonly used in Europe and approved by BSI— exceptions: New Zealand and JMAF (gamma-BHC), ESA, EPA and BPC (gamma benzene hexachloride), Sweden (gamma-hexaklor), USSR (lindane), USA; also known as ENT 7796. Apart from USSR, the common name lindane is approved by ISO members for grades of HCH/BHC in which the gamma-isomer content is > 99%. The IUPAC and *C.A.* name is 1α,2α,3β,4α,5α,6β-hexachlorocyclohexane, Registry No. *[58-89-9]*. Its insecticidal properties were discovered in the early 1940's by ICI Ltd and reported by R. E. Slade, *Chem. Ind. (London)*, 1945, p.314. It was introduced by ICI, Plant Protection Ltd (now ICI Ltd, Plant Protection Division) under the trade mark *'Gammexane'*; there are numerous other trade marks held by other companies.

Manufacture and properties. Produced by the selective crystallisation of crude HCH (p.295), *e.g.* as in USP 2 502 258, it forms colourless crystals, m.p. 112·9°, v.p. 9·4 x 10⁻⁶ mmHg at 20°. Its solubility at room temperature is 10 mg/l water; slightly soluble in petroleum oils; soluble in acetone, aromatic and chlorinated hydrocarbons.

Lindane has m.p. > 112°. It is stable to light, heat, air and concentrated acids but is dehydrochlorinated by alkali.

Uses. Gamma-HCH is the main insecticidal component of HCH. It acts as a stomach poison, by contact and has some fumigant action. It is effective against a wide range of soil-dwelling and phytophagous insects, those hazardous to public health, other pests and some animal ectoparasites. Except for foliar application to cucurbits or hydrangeas, it is non-phytotoxic at insecticidal concentrations. It shows less 'tainting' side effects than HCH but should not be used on crops intended for canning. It is used as a foliar spray, as soil applications, as a seed treatment and in baits for rodent control.

Toxicology. The acute oral LD50 for rats is 88-91 mg/kg; the acute dermal LD50 for rats is 900-1000 mg/kg. Rats receiving 800 mg/kg diet for long periods suffered no ill-effect (O. G. Fitzhugh *et al., J. Pharmacol. Exp. Ther.*, 1950, **100**, 59). Regulations and side-effects of lindane have been summarised (E. Ullmann, *Lindane,* 1972, Supplements, 1974, 1976).

Formulations. These include: w.p., e.c., dusts, smokes and baits; frequently combined with fungicides or other insecticides, especially when intended for seed treatment.

Analysis. Product analysis is by a cryoscopic method (*CIPAC Handbook,* 1970, **1**, 71; FAO Specification (CP/34); *Organochlorine Insecticides 1973*). Residues may be determined by GLC (J. Burke *et al., J. Assoc. Off. Agric. Chem.*, 1964, **47**, 326, 845).

Gibberellic acid

$C_{19}H_{22}O_6$ (346·4)

T C5 C6556/C-F/JP C- 3ACJ P CX EY JXOV OUTJ BVQ EU1 FQ M1 NQ

Nomenclature and development. The trivial name gibberellic acid is accepted by BSI and proposed by ISO in lieu of a common name; also known as gibberellin A_3, or (ambiguously) as GA_3. The IUPAC name is ($3S,3aS,4S,4aS,6S,8aR,8bR,11S$)-6,11-dihydroxy-3-methyl-12-methylene-2-oxo-4a,6-ethano-3,8b-prop-1-enoperhydroindeno-[1,2-*b*]furan-4-carboxylic acid, in *C.A.* usage 2β,4aα,7-trihydroxy-1β-methyl-8-methyl-ene-4aα,4bβ-gibb-3-ene-1α,10β-dicarboxylic acid 1→4a lactone *[77-06-5]*. It was discovered by E. Kurosawa, *Trans. Nat. Hist. Soc. (Formosa)* 1926, **16,** 213, who called it gibberellin A, but it attracted little attention outside Japan until ICI, Plant Protection Ltd began further investigations in 1955. The ICI compound had biological properties similar to the Japanese one but, although chemically related, it was distinct from gibberellin A and was called gibberellic acid and assigned the trade marks 'Berelex', 'Activol' and 'Activol' GA. It and other members of the gibberellin group (52 are known) occur naturally in a wide variety of plant species. The establishment of its chemical structure and stereochemistry have been reviewed (J. F. Grove, *Q. Rev. Chem. Soc.,* 1961, **15,** 56; B. E. Cross *et al., Adv. Chem. Series,* 1961, No. 28, 13; J. McMillan & N. Takahashi, *Nature (London),* 1968, **217,** 170, 590.

Manufacture and properties. Produced by fermentation using the fungus *Gibberella fujikuroi,* it is a crystalline solid, m.p. 223-225° (decomp.). Its solubility is 5 g/l water; very soluble in alcohols; soluble in ethyl acetate, butyl acetate; slightly soluble in diethyl ether; insoluble in chloroform. It is an acid pK_a 4·0, forming water-soluble potassium (50 g potassium gibberellate/l water), sodium and ammonium salts; most other metal salts show a moderate solubility in water. Dry gibberellic acid is stable at room temperature, but is slowly hydrolysed in aqueous or aqueous-alcoholic solutions and decomposed by heat. It undergoes a rearrangement in alkaline solution, in hot acid solution it forms gibberellenic acid and later allogibberic acid—all these derivatives show little, if any, biological activity.

Uses. Gibberellic acid is a plant growth regulator its uses including: malting of barley; improved seedless and seed grape production; delayed harvesting of citrus; improved fruit set of pears and clementines; advanced cropping of artichokes and rhubarb. See also: J. N. Turner, *Outlook Agric.,* 1972, **7,** 14; A. Lang, *Annu. Rev. Plant Physiol.,* 1970, **21,** 537; J. McMillan, *Recent Adv. Phytochem.,* 1974, **7,** 1.

Toxicology. The acute oral LD50 for rats and mice is > 15 000 mg/kg. It is non-irritant to skin and eyes. In 90-d feeding trials for rats and dogs the 'no effect' level was > 1000 mg/kg diet.

Formulations. These include: an effervescent water-soluble tablet (1 g a.i.); w.s.p. (100 g/kg); unformulated crystalline a.i.

Analysis. See: V. W. Winkler, *Anal. Methods Pestic. Plant Growth Regul.,* 1978, **10,** 545.

Glyphosate

$$\underset{\text{HOC.CH}_2\text{.NH.CH}_2\text{P(OH)}_2}{\overset{\displaystyle O \qquad\qquad O}{\overset{\displaystyle \|\qquad\qquad\quad \|}{}}}$$

QV1M1PQQO   *glyphosate* $C_3H_8NO_5P$ (169·1)

QV1M1PQQO &ZY1&1   *glyphosate mono(isopropylamine) salt* $C_6H_{17}N_2O_5P$ (228·2)

Nomenclature and development. The common name glyphosate is approved by BSI, ISO, ANSI and WSSA. The IUPAC and *C.A.* name is *N*-(phosphonomethyl)glycine, Registry No. *[1071-83-6]*. It was introduced in 1971 by Monsanto Co. under the code number 'MON-0573' and protected by USP 3 799 758. Its herbicidal properties were first described by D. D. Baird *et al., Proc. North Cent. Weed Control Conf.,* 1971, **26,** 64. The principal formulation, trade mark 'Roundup', is the mono(isopropylamine) salt, Registry No. *[38641-94-0]*.

Manufacture and properties. It is a non-volatile white solid, m.p. *c.* 230° (decomp.), bulk density 0·5 g/cm³. Its solubility at 25° is 12 g/l water; insoluble in common organic solvents. Glyphosate-isopropylammonium is completely water soluble.

Uses. Glyphosate is a non-selective, non-residual post-em. herbicide; it is very effective on deep-rooted perennial species, and annual and biennial species of grasses, sedges and broad-leaved weeds. Excellent control of most species has been obtained at rates of 0·7-5·6 kg a.i./ha with annual species requiring the lower rates and some perennial species requiring > 2·9 kg/ha. In addition, better control of most weeds is obtained if applications are made at the later stages of plant maturity.

Toxicology. The acute oral LD50 for rats is 4320 mg/kg; the acute dermal LD50 for rabbits is > 7940 mg.

Formulation. 'Roundup'; a water-based solution (480 g glyphosate-isopropylammonium/l = 360 g a.e./l).

Analysis. Residues may be determined by extraction, liquid-liquid partition and ion-exchange chromatographic clean-up, conversion to the *N*-trifluoroacetyl derivative of the trimethyl ester and measurement by GLC with phosphorus-specific FPD, or by conversion to the *N*-nitroso derivative which is measured polarographically (T. H. Byast *et al., Agric. Res. Counc. (G.B.) Weed Res. Organ., Tech. Rep.,* No. 15 (2nd Ed.), p.63).

Glyphosine

$$\text{HO}\overset{\overset{\displaystyle O}{\|}}{\text{C}}\text{.CH}_2\text{.N[CH}_2\overset{\overset{\displaystyle O}{\|}}{\text{P}}\text{(OH)}_2]_2$$

QPQO&1N1VQ1PQQO $\qquad\qquad\qquad\qquad\qquad$ C$_4$H$_{11}$NO$_8$P$_2$ (263·1)

Nomenclature and development. The common name glyphosine is approved by BSI, ISO and ANSI. The IUPAC name is *N,N*-bis(phosphonomethyl)glycine (I) formerly *N,N*-di-(phosphonomethyl)glycine, in *C.A.* usage (I) *[2439-99-8]*. It was introduced in 1969 by Monsanto Co. under the code number 'CP-41 845', trade mark 'Polaris' and protected by USP 3 556 762. Its effects on carbohydrate deposition were described by C. A. Porter & L. E. Ahlrichs, *Hawaii Sugar Tech. Rep., 1971,* 1972, **30,** 71.

Properties. It is a white solid of negligible volatility at room temperature. Its solubility at 20° is 248 g/l water.

Uses. Glyphosine is a plant growth regulator used to hasten ripening of sugarcane and to increase the sucrose content. Effective rates range from 2·4-4·5 kg a.i./ha.

Toxicology. The acute oral LD50 for rats is 3925 mg/kg. It is a moderate skin irritant and is classified as a severe eye irritant.

Formulation. 'Polaris plant growth regulator', w.s.p. (850 g a.i./kg).

Analysis. Residues may be determined by extraction, ion-exchange chromatographic clean-up, conversion to the pentamethyl ester and measurement by GLC using a phosphorus-specific FPD.

Guazatine

$$\underset{\underset{H_2N.C.NH.(CH_2)_8.NH.(CH_2)_8.NH.C.NH_2}{\|}}{NH} \qquad \underset{\underset{}{\|}}{NH}$$

MUYZM8M8MYZUM $C_{18}H_{41}N_7$ (355·6)

Nomenclature and development. The common name guazatine is approved by BSI and ISO; BSI used the common name *guanoctine* from 1970-1972. The IUPAC name is bis(8-guanidino-octyl)amine formerly di-(8-guanidino-octyl)amine, in *C.A.* usage *N,N‴*-(iminodi-8,1-octanediyl)bisguanidine *[13516-27-3]* formerly 1,1′-[(iminobis(octamethylene)diguanidine]. The triacetate, *C.A.* Registry No. *[39202-40-9],* was introduced in 1968 by Evans Medical Ltd under the code number 'EM 379', 'MC 25' (Murphy Chemical Ltd), the trade mark 'Panoctine' (KenoGard AB) and protected by BP 1 114 155. The fungicidal properties of its salts were described by W. S. Catling *et al., Congr. Plant Pathol., 1st,* 1968, p.27 (Abstr.).

Manufacture and properties. Produced by the amidination of technical bis(8-aminooctyl)amine, the base is not isolated. Its salts are white crystalline solids, *e.g.* the triacetate, m.p. 140°, which is readily soluble in water at room temperature but is insoluble in organic solvents. It is stable to air and water with good compatibility with other pesticides.

Uses. Guazatine salts are fungicides, especially effective as: cereal seed dressings at 0·6-0·8 g a.i./kg seed, being repellent to hens, pigeons, pheasants and ravens; potato and sugarcane dips; and for the control of *Piricualaria oryzae* in rice by foliar application.

Toxicology. The acute oral LD50 is: for rats 230-260 mg guazatine triacetate/kg; for mice *c.* 300 mg/kg; for Japanese quail 263 mg/kg; for hens 125 mg/kg. The acute dermal LD50 for albino rabbits is *c.* 1100 mg/kg. In 90-d feeding trials for rats the 'no effect' level was 800 mg/kg diet.

Formulations. These include: liquids and powders; seed dressings in combination with other fungicides, *e.g.* fenfuram.

Analysis. Product analysis is by potentiometric titration with perchloric acid in acetic anhydride [assaying as the triacetate of di-guanidated bis(8-aminooctyl)amine]. Residues may be determined by hydrolysis to the triamine, followed by GC-MS determination. Particulars from KenoGard AB.

HCH

(BHC)

L6TJ   AG   BG   CG   DG   EG   FG $C_6H_6Cl_6$ (290·8)

Nomenclature and development. The common names HCH and BHC are approved by ISO; the former is more commonly used in Europe and is approved by BSI—exceptions include: New Zealand and JMAF (BHC), ESA, EPA and BPC (benzene hexachloride), Sweden (hexaklor), USSR (hexacloran) and USA; also known as ENT 8601. The names cover the mixed stereoisomers known in IUPAC and *C.A.* usage as 1,2,3,4,5,6-hexachlorocyclohexane, Registry No. *[608-73-1]*. Its insecticidal properties were discovered in the early 1940's by ICI Ltd. It was reported by R. E. Slade, *Chem. Ind. (London)*, 1945, p.134; and by A. Dupire & M. Racourt, *C. R. Hebd. Seances Acad. Agric. Fr.*, 1942, **20**, 470. The insecticidal activity is due mainly to the gamma-isomer, see gamma-HCH (p.290), which has largely replaced the mixed isomers for commercial use.

Manufacture and properties. Produced by the chlorination of benzene under u.v. light, it is an off-white to brown powder, with a persistent musty odour. Being a mixture, HCH has no precise physical properties. The various stereoisomers are as follows:

Isomer	Chlorine atom orientation	Physical properties	ENT No.	C.A. Registry No.
alpha	1α,2α,3β,4α,5β,6β	m.p. 159-160°, v.p. 0·06 mmHg at 40°	9232	*[319-84-6]*
beta	1α,2β,3α,4β,5α,6β	m.p. 312°, v.p. 0·17 mmHg at 40°		*[319-85-7]*
gamma	1α,2α,3β,4α,5α,6β	see gamma-HCH (p.290)	7796	*[58-89-9]*
delta	1α,2α,3α,4β,5α,6β	m.p. 138-139°	9234	*[319-86-8]*
epsilon	1α,2α,3α,4β,5β,6β	m.p. 218·5-219·3°		*[6108-10-7]*
eta	1α,2α,3α,4α,5β,6β			
theta	1α,2α,3α,4β,5α,6β			
zeta	1α,2α,3α,4α,5α,6α			

The isomers are sparingly soluble in water; soluble in aromatic hydrocarbons, chloroform. HCH is stable to light, air, heat, concentrated acids; apart from beta-HCH, the isomers are dehydrochlorinated by alkali at room temperature.

Uses. HCH is a persistent stomach poison and contact insecticide with fumigant action, its activity being due to the gamma-isomer. It is non-phytotoxic, except to cucurbits, at insecticidal concentrations, but may cause root deformation and polyploidy at higher concentrations. It taints certain crops seriously, especially blackcurrants, carrots and potatoes. It is also used in baits as a rodenticide.

Toxicology. The toxicities to test animals of the various isomers differ and various values are quoted in the literature. Typical figures for the acute oral LD50 for rats include: 500 mg alpha-isomer/kg; 6000 mg beta-isomer/kg; 1000 mg delta-isomer/kg. The beta-, unlike the alpha- and delta-isomers, has high chronic and cumulative toxicity.

Formulations. These include: w.p. and e.c. of various a.i. content. The concentration of gamma-isomer should be stated.

Analysis. Product analysis is by column chromatography, total chlorine content and polarography (*CIPAC Handbook*, 1970, **1**, 32; FAO Specification (CP/44); *Organochlorine Insecticides 1973*). Residues may be determined by GLC (D. B. Harper & R. J. McAnally, *Pestic. Sci.*, 1977, **8**, 35).

Heptachlor

L C555 A DU IUTJ AG AG BG FG HG IG JG $C_{10}H_5Cl_7$ (373·3)

Nomenclature and development. The common name heptachlor is approved by BSI and ISO; also known as ENT 15 152. The IUPAC name is 1,4,5,6,7,8,8-heptachloro-3a,4,7,7a-tetrahydro-4,7-methanoindene (I), in *C.A.* usage 1,4,5,6,7,8,8-heptachloro-3a,4,7,7a-tetrahydro-4,7-methano-1*H*-indene *[76-44-8]* formerly (I). It was first isolated from technical chlordane (p.93) and was introduced in 1948 by Velsicol Chemical Corp., under the code numbers 'E3314', 'Velsicol 104'.

Manufacture and properties. Produced by the chlorination of the adduct of hexachloro-cyclopentadiene and cyclopentadiene, using sulphonyl chloride in the presence of benzoyl peroxide (BP 618 432) or chlorine in the dark and the presence of fuller's earth (USP 2 576 666), it is a colourless crystalline solid, with a mild camphor-like odour, m.p. 95-96°, v.p. 3×10^{-4} mmHg at 25°. Its solubility at 25-29° is: 56 µg/l water; 45 g/l ethanol; 189 g/l kerosine. The technical product contains *c.* 72% heptachlor and 28% of related compounds and is a soft wax, m.p. 46-74°, d^9 1·57-1·59, viscosity 0·05-0·75 Pa s at 90°. It is stable to light, moisture, air and moderate heat and is not readily dehydrochlorinated (W. M. Rogoff & R. L. Metcalf, *J. Econ. Entomol.*, 1951, **44**, 910). It is susceptible to epoxidation and is compatible with most pesticides and fertilisers.

Uses. Heptachlor is a non-systemic stomach and contact insecticide with some fumigant action.

Toxicology. The acute oral LD50 for rats is 100-162 mg/kg; the acute dermal LD50 for rats is 195-250 mg/kg. Epoxidation is an important metabolic reaction first observed in dogs and rats (B. Davidow & J. L. Radomski, *J. Pharmacol. Exp. Ther.*, 1953, **107**, 259, 266) the 2,3-epoxide being highly persistent and biologically active.

Formulations. These include: e.c., w.p., dusts and granules of various a.i. content.

Analysis. Product analysis is by GLC (*Official Methods of Analysis AOAC*, 12th Ed., p.112) or by reaction with silver nitrate when the chlorine atom at $C_{(1)}$ forms silver chloride (*ibid.; CIPAC Handbook*, 1970, **1**, 420); FAO Specification (CP/47); *Organochlorine Insecticides 1973*). Residues may be determined by GLC (*AOAC Methods, loc. cit.* p.518; D. M. Coulson & L. A. Cavanagh, *Anal. Chem.*, 1960, **32**, 1245; D. M. Coulson *et al., J. Agric. Food Chem.*, 1960, **8**, 399). See also: T. G Bowery, *Anal. Methods Pestic., Plant Growth Regul. Food Addit.*, 1964, **2**, 245; *Anal. Methods Pestic. Plant Growth Regul.*, 1972, **6**, 404; H. K. Suzuki *et al., ibid.*, 1978, **10**, 73.

Heptenophos

L45 BU EUTJ BOPO&O1&O1 CG $C_9H_{12}ClO_4P$ (250·6)

Nomenclature and development. The common name heptenophos is approved by BSI and ISO. The IUPAC and *C.A.* name is 7-chlorobicyclo[3.2.0]-hepta-2,6-dien-6-yl dimethyl phosphate, Registry No. *[23560-59-0].* It was introduced in 1970 by Hoechst AG under the code number 'Hoe 2982', and the trade marks 'Hostaquick' (for crop protection use) and 'Ragadan' (for veterinary use). It was first reported (under an erroneous structure in *C.A.*, Registry No. *[34783-40-9]*) in *Biul. Inst. Ochr. Rosl.,* 1971, No. 50, p.109. Its activity has been outlined against crop pests (R. T. Hewson, *Proc. Br. Insectic. Fungic. Conf., 8th,* 1975, **2,** 697) and veterinary pests (W. Bonin, *ibid.,* p.705).

Manufacture and properties. Produced by the Perkow reaction between trimethyl phosphite and 7,7-dichlorobicyclo[3.2.0]hepta-2-en-6-one, it is a pale amber liquid, b.p. 64°/0·075 mmHg, v.p. 7·5 x 10⁻⁴ mmHg at 20°, d_4^{20} 1·294. Its solubility at 20° is 2·2 g tech./l water; miscible with most organic solvents.

Uses. Heptenophos is an insecticide with quick initial action and short residual effect. It penetrates plant tissue and is rapidly translocated, controlling sucking insects and certain Diptera. It is also effective against ectoparasites (lice, fleas, mites and ticks) of cattle, dogs, sheep, pigs and pets.

Toxicology. The acute oral LD50 is: for rats 96-121 mg/kg; for dogs > 500 mg/kg. The acute dermal LD50 for rats is *c.* 2900 mg/kg. In 2-y feeding trials the 'no effect' level was: for dogs 12 mg/kg diet; for rats 15 mg/kg diet.

Formulations. 'Hostaquick', e.c. (565 g a.i./l) for crop use; 'Ragadan', e.c. (250 g/l) and w.p. (400 g/kg) for veterinary use.

Analysis. Details of GLC methods are available from Hoechst AG.

Hexachlorobenzene

GR BG CG DG EG FG C_6Cl_6 (284·8)

Nomenclature and development. The IUPAC and *C.A.* name hexachlorobenzene, Registry No. *[118-74-1]* is accepted by BSI and ISO in lieu of a common name. It was introduced for seed treatment in 1945 and reported by H. Yersin *et al., C.R. Seances Acad. Agric. Fr.,* 1945, **31,** 24.

Manufacture and properties. Produced by the chlorination of benzene in the presence of suitable catalysts, it forms colourless crystals, m.p. 226°, v.p. 1·09 x 10^{-5} mmHg at 20°, d^{23} 2·044. It is practically insoluble in water and in cold ethanol; soluble in hot benzene. The technical product has m.p. ⩾ 220°.

Uses. Hexachlorobenzene is a selective fungicide suitable for the control of *Tilletia caries* on wheat (M. Lansade, *Parasitica,* 1949, **5,** 1). It was found effective against dwarf bunt, attributed to a fumigant action on the bunt spore (W. N. Siang & C. S. Holton, *Plant Dis. Rep.,* 1953, **37,** 63).

Toxicology. The acute oral LD50 for rats is 10000 mg/kg; guinea-pigs tolerated doses > 3000 mg/kg. It may cause a slight irritation to the skin. Its use is banned in some countries.

Formulations. Dust (100 g a.i./kg); also in combination with other seed protectants, *e.g.* fuberidazole ('Voronit'),

Analysis. Residues may be determined by GLC (T. Stijve, *Mitt. Geb. Lebensmittelunters. Hyg.,* 1971, **62,** 406).

Hexazinone

T6NVNVNJ A1 FN1&1 C- AL6TJ $\qquad$ $C_{12}H_{20}N_4O_2$ (252·3)

Nomenclature and development. The common name hexazinone has been approved by BSI and New Zealand and proposed by ISO and ANSI. The IUPAC name is 3-cyclohexyl-6-dimethylamino-1-methyl-1,3,5-triazine-2,4-dione, in *C.A.* usage 3-cyclohexyl-6-(dimethylamino)-1-methyl-1,3,5-triazine-2,4(1*H*,3*H*)-dione *[51235-04-2]*. It was introduced in 1976 by E. I. du Pont de Nemours & Co. (Inc.) under the code number 'DPX 3674', trade mark 'Velpar' and protected by USP 3902887. Its herbicidal activity was described by T. J. Hernandez *et al., Proc. North Cent. Weed Control Conf.*, 1974, p.138; *Proc. South. Weed Sci. Soc., 28th*, 1975, p.247.

Properties. It is a colourless crystalline solid, m.p. 115-117°, v.p. 6·4 x 10^{-5} mmHg at 86° (2 x 10^{-7} mmHg at 25°, extrapolated), *d* 1·25. Its solubility at 25° is: 33 g/kg water; 790 g/kg acetone; 940 g/kg benzene; 3·88 kg/kg chloroform; 836 g/kg dimethylformamide; 3 g/kg hexane; 2·65 kg/kg methanol; 386 g/kg toluene. It is stable in aqueous solution at pH 5-9 at temperatures up to 37°. Subject to microbial decomposition in soil.

Uses. Hexazinone is a post-em. contact herbicide effective against many annual and biennial weeds and, except for Johnson grass, most perennial weeds at 6-12 kg a.i./ha. At 2-5 kg a.i./ha it provides short term control. It is used on non-crop areas, but not on sites adjacent to deciduous trees or other desirable plants. The addition of a non-ionic surfactant is recommended.

Toxicology. The acute oral LD50 is: for rats 1690 mg tech./kg; for guinea-pigs 860 mg/kg. The acute dermal LD50 for rabbits is > 5278 mg/kg. Administration of 0·1 ml to rabbit eye caused reversible conjunctival irritation and it is classified as an eye irritant. The LC50 (8 d) for mallard duck and bobwhite quail is > 5000 mg/kg diet. The LC50 (96 h) is: for bluegill 370-420 mg/l; for fathead minnow 274 mg/l.

Formulation. A w.s.p. (900 g a.i./kg).

Analysis. Residues may be determined by HPLC (T. H. Byast *et al., Agric. Res. Counc. (G.B.) Weed Res. Organ., Tech. Rep.*, No. 15 (2nd Ed.), p.40).

Hydrogen cyanide

HCN

NCH   CHN   (27·03)

Nomenclature and development. The IUPAC name hydrogen cyanide is accepted by ISO in lieu of a common name. The *C.A.* name is hydrocyanic acid, Registry No. *[74-90-8]*. It was first used an an insecticidal fumigant in 1886 by D. W. Coquillett (cited by L. O. Howard *U.S. Dep. Agric. Yearbook,* 1899, p.150).

Manufacture and properties. Produced by the catalytic oxidation of ammonia-methane mixtures (L. Andrussow, *Angew. Chem.,* 1935, **48,** 593), it is a colourless liquid, with an odour of almonds, b.p. 26°, d_4^{20} 0·699, m.p. —14°. It is soluble in water, ethanol, diethyl ether.

Uses. Hydrogen cyanide is an insecticide used for the fumigation of flour mills, stored grain and ships. It is phytotoxic though usually not at fumigant concentrations, *e.g.* as used in glasshouses in some countries, but must not be used on crops bearing copper residues or that have been treated for copper deficiency.

Toxicology. It is very toxic to mammals and exposure for 30 min to concentrations of 360 mg/m³ are fatal to man.

Formulation. It is usually packed in metal containers, with or without an added irritant 'warning gas'; anhydrous oxalic acid (2 g/l) may be added to prevent polymerisation. Also absorbed on porous material such as Kieselguhr or cardboard discs. It may be generated at the site of use by the action of mineral acids on sodium cyanide (p.478) or of moisture on calcium cyanide (p.75).

Analysis. It may be determined by dissolving in water and titration with aqueous silver nitrate.

2-Hydroxyethyl octyl sulphide

Me(CH$_2$)$_7$.S.CH$_2$CH$_2$OH

Q2S8   C$_{10}$H$_{22}$OS   (190·3)

Nomenclature and development. Normally known by the trivial name 2-hydroxyethyl octyl sulphide, the IUPAC and *C.A.* name is 2-(octylthio)ethanol *[3547-33-9]*. Its insect-repelling properties were discovered in 1955 by the Phillips Petroleum Co. (L. D. Goodhue, *J. Econ. Entomol.,* 1960, **53,** 805). It was introduced as an insect repellent under the code number 'Phillips R-874' by that company and by the McLaughlin Gormley King Co. under the name 'MGK Repellent R-874'; protected by USP 2 863 799.

Manufacture and properties. Produced by the reaction of octane-1-thiol with ethylene oxide, it is a pale amber liquid, with a mild mercaptan-like odour, b.p. 98°/0·1 mmHg, m.p. 0°, d_4^{20} 0·925-0·935, n_D^{20} 1·470-1·478. It is slightly soluble in water; miscible with most organic solvents including refined kerosine, though a co-solvent such as propan-2-ol is required with the latter to maintain solution at low temperatures. It is stable and non-corrosive and is compatible with chlordane, diazinon and fenchlorphos.

Uses. Its chief use is as a cockroach repellent at 1 g a.i./m². Another important use is as a repellent against flies, ants and crawling insects in pressurised 'patio' foggers. It is not normally applied to plants though lucerne, maize and melons were not damaged at 700-1000 g/ha.

Toxicology. The acute LD50 for rats is 8530 mg/kg; the acute dermal LD50 for albino rabbits is 13 590 mg/kg. The application of 0·05 mg to the cornea of albino rats produced corneal necrosis in 2 of the 5 eyes tested but the corneas healed without opacities. In 90-d feeding trials rats receiving 20 000 mg/kg diet showed no effect on the blood, urine, organs or tissues examined.

Formulations. An e.c. or oil solution (10-50 g a.i./l).

Analysis. Details of a GLC method are available from McLaughlin Gormley King Co.

Imazalil

T5N CNJ A1YO2U1&R BG DG   *imazalil* $C_{14}H_{14}Cl_2N_2O$ (297·2)

T5N CNJ A1YO2U1&R BG DG 2 &WSQQ

          imazalil sulphate $C_{28}H_{30}Cl_4N_4O_6S$ (692·4)

T5N CNJ A1YO2U1&R BG DG &WNQ

          imazalil nitrate $C_{14}H_{15}Cl_2N_3O_4$ (360·2)

Nomenclature and development. The common name imazalil is approved by BSI and New Zealand, and proposed by ISO. The IUPAC name is 1-(β-allyloxy-2,4-dichloro-phenethyl)imidazole, in *C.A.* usage 1-[2-(2,4-dichlorophenyl)-2-(2-propenyloxy)ethyl]-1*H*-imidazole *[35554-44-0]*. It was introduced in 1972 by Janssen Pharmaceutica as an experimental fungicide under the code numbers 'R 23979' (base), 'R 27180' (sulphate) Registry No. *[58594-72-2]*, 'R 18531' (nitrate) *[33586-66-2]*. Its fungicidal properties were first described by E. Laville, *Fruits*, 1973, **28**, 545.

Properties. Imazalil is a slightly yellowish to brownish oil, d^{23} 1·2429, n_D^{20} 1·5643, v.p. 7 x 10^{-8} mmHg at 20°. It is slightly soluble in water; freely soluble in organic solvents. It is chemically stable at room temperature in the absence of light, and thermally stable to *c.* 285°. Imazalil sulphate is an almost colourless to beige-coloured powder, freely soluble in water, alcohols and slightly soluble in apolar organic solvents.

Uses. Imazalil is a systemic fungicide effective against a wide range of fungi affecting fruit, vegetables and ornamentals. Due to its high specific activity against *Helminthosporium, Fusarium* and *Septoria* spp. it is recommended as a seed treatment for the control of cereal diseases. Storage decay on citrus, bananas and other fruit can be controlled by post-harvest spray or dip in water or wax emulsions. Imazalil is highly effective against benzimidazole-resistant strains of *Penicillium* spp. Typical use rates are: for seed dressing 4-5 g a.i./100 kg seed; for vegetables and ornamentals 5-30 g/100 l and for post-harvest treatment 2-4 g/t fruit.

Toxicology. The acute oral LD50 for rats is 320 mg/kg; the acute dermal LD50 for rats is 4200-4880 mg/kg.

Formulations. These include: imazalil, e.c. (200, 500 or 700 g a.i./l); imazalil sulphate, w.s.p. (750 g base/kg). It is formulated as liquid and powder seed dressings also in combination with other fungicides such as: guazatine, fenfuram, carboxin, thiabendazole, thiophanate-methyl, carbendazim.

Analysis. Product analysis and formulation content is by GLC with an internal standard. For residue analysis, extraction and clean-up precedes the determination of the base by GLC with ECD or FID. HPLC with u.v. detection at 202 nm is also possible: J. Wynants, *Proc. Int. Soc. Citriculture, Orlando,* 1977, **3**, H-38.

4-(Indol-3-yl)butyric acid

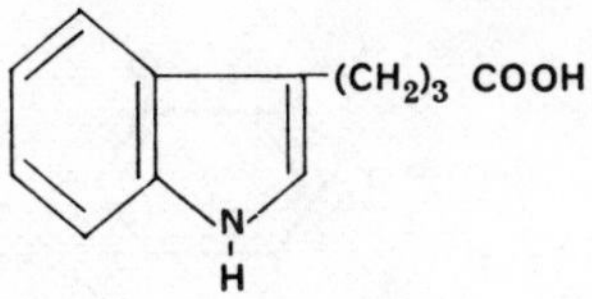

T56   BMJ   D3VQ

$C_{12}H_{13}NO_2$ (203·2)

Nomenclature and development. The IUPAC name 4-(indol-3-yl)butyric acid is accepted by BSI and proposed by ISO in lieu of a common name, in *C.A.* usage 1*H*-indole-3-butanoic acid *[133-32-4]* formerly indole-3-butyric acid; sometimes known as IBA. It was first prepared by R. W. Jackson & R. F. Manske, *J. Am. Chem. Soc.*, 1930, **52**, 5029. The ability of the acid and its simple esters to stimulate root formation in cuttings was reported by P. W. Zimmerman & F. Wilcoxon, *Contrib. Boyce Thompson Inst.*, 1935, **7**, 209; P. W. Zimmerman & A. E. Hitchcock, *ibid.*, p.439. The preparation of the acid is protected by USP 3 051 723 (Union Carbide Corp.) and it is a component of the 'Seradix' products (May & Baker Ltd), 'Rootone' (Amchem Products, Inc.).

Manufacture and properties. Produced by the interaction of indole, γ-butyrolactone and sodium hydroxide followed by acidification, it forms colourless or slightly yellow crystals, m.p. 123-125°. It is practically insoluble in water; soluble in acetone, diethyl ether, ethanol. It is non-flammable and non-corrosive.

Uses. 4-(Indol-3-yl)butyric acid is used as an aid to rooting cuttings (J. van Overbeek, *Encylopedia of Plant Physiology*, 1961, **14**, 1145).

Toxicology. The intraperitoneal LD50 for mice is 100 mg/kg (H. H. Anderson *et al., Proc. Soc. Exp. Biol. Med.*, 1936, **34**, 138).

Formulations. These include the following powders: 'Seradix No. 1' (1 g a.i./kg) for soft wood, 'Seradix No. 2' (3 g/kg) for semi-hardwoods, 'Seradix No. 3' (8 g/kg) for hardwood; 'Rootone', in combination with 1-naphthylacetic acid).

Analysis. GLC methods are used after conversion to a suitable volatile derivative.

Iodofenphos

(Jodfenphos)

IR BG EG DOPS&O1&O1 $C_8H_8Cl_2IO_3PS$ (413·0)

Nomenclature and development. The common name jodfenphos is approved by ISO—exceptions BSI, Canada, France, New Zealand, BPC and ESA (iodofenphos); also known as OMS 1211. The IUPAC and *C.A.* name is *O*-2,5-dichloro-4-iodophenyl *O,O*-dimethyl phosphorothioate, Registry No. *[18181-70-9]*. It was introduced in 1966 by Ciba AG (now Ciba-Geigy AG) under the code number 'C 9491', trade mark 'Nuvanol N' and protected by BP 1 057 609. Its insecticidal properties were described by B. C. Haddow & T. G. Marks, *Proc. Br. Insectic. Fungic. Conf., 5th,* 1969, **2,** 531.

Manufacture and properties. Produced by the reaction of *O,O*-dimethyl phosphoro-chloridothioate with sodium 2,5-dichloro-4-iodophenoxide it forms colourless crystals, m.p. 76°, v.p. 8 x 10⁻⁷ mmHg at 20°. Its solubility at 20° is: < 2 mg/l water; 480 g/l acetone; 610 g/l benzene; 860 g/l dichloromethane.

Uses. Iodofenphos is a non-systemic contact and stomach insecticide and acaricide used for the protection of stored products and in public hygiene. Wall applications of 1-2 g a.i./m² persist for 90 d; a weekly dose of 0·3 g/m² controls flies on refuse tips and it is used for sheep dips at 0·5 g/l.

Toxicology. The acute oral LD50 is: for rats 2100 mg/kg; for dogs > 3000 mg/kg. The acute dermal LD50 for rats is > 2000 mg/kg. In 90-d feeding trials the 'no effect' level was: for rats 5 mg/kg diet; for dogs 15 mg/kg diet. It is of low toxicity to birds but toxic to honey bees and highly toxic to fish.

Formulations. These include: 'Nuvanol N', w.p. (500 g a.i./kg), e.c. (200 g/l), dip concentrate (200 g/l), powder (50 g/kg); 'Elocril 50 WP', w.p. (500 g/kg). Products previously sold under the trade mark *'Alfacron'* are no longer available.

Analysis. Product analysis is by GLC with internal standard. Residues are determined by GLC using ECD or FPD (N. Burkhard & G. Voss, *Pestic. Sci.,* 1972, **3,** 183). Details of these methods are also available from Ciba-Geigy AG.

Ioxynil

QR BI FI DCN

C₇H₃I₂NO (370·9)

$C_7H_3I_2NO$ (370·9)

Nomenclature and development. The common name ioxynil is approved by BSI, ISO and WSSA—exception JMAF (who apply it to the octanoate, p.306). The IUPAC name is 4-hydroxy-3,5-di-iodobenzonitrile (I), also known as 4-hydroxy-3,5-di-iodophenyl cyanide, in *C.A.* usage (I) *[1689-83-4]*. Its herbicidal properties were reported independently by R. L. Wain, *Nature (London)*, 1963, **200**, 28, by K. Carpenter & B. J. Heywood, *ibid.*, p.28 and by R. D. Hart *et al.*, *Proc. Br. Weed Control Conf., 7th,* 1964, p.3. It was introduced in 1963 by Amchem Products, Inc. under the code number 'ACP 63-303' and trade mark 'Certrol'; by May & Baker Ltd under the code number 'MB 8873' and trade mark 'Actril'.

Manufacture and properties. Produced by the iodination of 4-hydroxybenzaldehyde, conversion to the oxime, reaction with acetic anhydride to give 4-cyano-2,6-di-iodophenyl acetate and hydrolysis to ioxynil (*cf.* K. Auwers & J. Reis, *Ber. Dtsch. Chem. Ges.,* 1896, **29**, 2355). It forms a colourless solid, m.p. 209°, practically non-volatile at room temperature though it sublimes at *c.* 140° at 0·1 mmHg. Its solubility at 25° is: 50 mg/l water; 70 g/l acetone; 20 g/l methanol; 340 g/l tetrahydrofuran. It is stable, non-corrosive and compatible with other pesticides. The technical product is a cream powder, with a faint phenolic odour but sternutatory, *c.* 95% pure, m.p. 200°.

It is an acid, pK_a 3·96, forming salts (solubilities at 20-25°: sodium, Registry No. *[2961-62-8],* 140 g/l water; lithium 220 g/l; potassium 107 g/l. Its esters with long-chain alkanoic acids, *e.g.* octanoic acid, are soluble in petroleum oils.

Uses. Ioxynil is a contact herbicide with some systemic activity, developed mainly for the control of broad-leaved weeds in cereals at rates of 275-825 g a.i./ha; it has little or no residual activity.

Toxicology. The acute oral LD50 is: for rats 110 mg ioxynil/kg, 112 mg ioxynil-sodium/kg, 305 mg ioxynil-lithium/kg; for dogs 140-280 mg ioxynil/kg; for pheasants 75 mg ioxynil/kg, 35 mg ioxynil-sodium/kg; for hens 200 mg ioxynil/kg, 120 mg ioxynil-sodium/kg. In 30-d feeding trials rats receiving 111 mg/kg diet suffered no ill-effect; growth rate was depressed at 333 mg/kg diet. The LC50 (24 h) for fish is 3·6 mg/l.

Formulations. These include aqueous concentrates of the alkali metal salts, water-soluble amine salts, oil-soluble amine salts. Typical are: 'Actril C', a mixture of potassium salts of ioxynil and mecoprop (total 300 g a.i./l); 'Certrol PA', a mixture of the potassium salts of ioxynil, dichlorprop and MCPA (total 450 g a.i./l); 'Oxytril' mixtures with bromoxynil.

Analysis. Product analysis is by total iodine using the Parr bomb, or by titration of the phenol. Residues may be determined by GLC of the methyl ether (T. H. Byast *et al., Agric. Res. Counc. (G.B.) Weed Res. Organ., Tech. Rep.,* No. 15 (2nd Ed.), p.13) or by i.r. spectroscopy of ioxynil. See also: H. S. Segal & M. L. Sutherland, *Anal. Methods Pestic., Plant Growth Regul. Food Addit.,* 1967, **5**, 423; *Anal. Methods Pestic. Plant Growth Regul.,* 1972, **6**, 654.

Ioxynil octanoate

CN

I — — I

O.C.(CH₂)₆Me
‖
O

NCR CI EI DOV7 $C_{15}H_{17}I_2NO_2$ (497·1)

Nomenclature and development. The common name ioxynil is approved by BSI, ISO and WSSA for 4-hydroxy-3,5-di-iodobenzonitrile, the introduction of which as a herbicide is described on p.305. Ioxynil octanoate (referred to by JMAF as ioxynil), has the IUPAC name 4-cyano-2,6-di-iodophenyl octanoate (I), in *C.A.* usage (I) *[3861-47-0]* formerly 4-hydroxy-3,5-di-iodobenzonitrile octanoate ester. It was introduced in 1963 by May & Baker Ltd under the code numbers 'MB 11641', '15 830 RP', trade marks 'Actril' and 'Totril' and protected by BP 1 067 033.

Manufacture and properties. Produced by the esterification of ioxynil with octanoyl chloride, it is a cream wax, m.p. 59-60°, of low volatility. The technical material has a slight ester odour. It is practically insoluble in water; its solubility at 25° is: 100 g/l acetone; 90 g/l methanol; 500 g/l xylene. It is stable in storage, does not react with most other pesticides and is only slightly corrosive. It is readily hydrolysed by dilute alkali.

Uses. Ioxynil octanoate is a contact herbicide with a limited systemic action, used for the post-em. control of annual broad-leaved weeds in cereals, sugarcane, onions, leeks and newly-sown turf at 275-770 g a.i./ha. It is rapidly broken down in soil by micro-organisms and chemical processes with 50% loss in *c.* 10 d. In plants, hydrolysis leads to ioxynil and conversion of the nitrile group to amide and then to the free acid with some de-iodination.

Toxicology. The acute oral LD50 is: for rats 390 mg/kg; for mice 2300 mg/kg. In 90-d feeding trials rats receiving 4 mg/kg daily and dogs 4·5 mg/kg daily suffered no ill-effect.

Formulations. These include the following e.c. products: 'Totril' (250 g ioxynil/l); 'Actril D', a mixture of ioxynil octanoate and 2,4-D-iso-octyl (total 350 g/l); 'Oxytril CM', the octanoates of ioxynil and bromoxynil (total 400 g/l); 'Oxytril P', dichlorprop-iso-octyl and the octanoates of ioxynil and bromoxynil (total 550 g/l).

Analysis. Product analysis is by iodine determination. Residues may be determined by GLC or by i.r. spectroscopy using the intense band at 4·5 μm due to aromatic nitrile.

306

Iprodione

T5NVNV EHJ AVMY1&1 CR CG EG $C_{13}H_{13}Cl_2N_3O_3$ (330·2)

Nomenclature and development. The common name iprodione has been approved by BSI and France and proposed by ISO. The IUPAC name is 3-(3,5-dichlorophenyl)-*N*-isopropyl-2,4-dioxoimidazolidine-1-carboxamide, in *C.A.* usage 3-(3,5-dichlorophenyl)-*N*-(1-methylethyl)-2,4-dioxo-1-imidazolinecarboxamide *[36734-19-7]*. It was introduced in 1971 by Rhône-Poulenc Phytosanitaire as a fungicide under the code numbers '26 019 RP', 'ROP 500F', 'NRC 910', 'LFA 2043' and 'FA 2071', the trade mark 'Rovral' and protected by BP 1 312 536; USP 3 755 350; FP 2 120 222. Its fungicidal properties were first described by L. Lacroix *et al., Phytiatr. Phytopharm.*, 1974, **23**, 165.

Manufacture and properties. Produced by the reaction of isopropyl isocyanate and 3-(3,5-dichlorophenyl)imidazolidine-2,4-dione which is prepared by the cyclisation of 5-(3,5-dichlorophenyl)hydantoic acid, obtained from glycine and 3,5-dichlorophenyl isocyanate, it forms colourless crystals, m.p. *c.*136°, v.p. $< 10^{-6}$ mmHg at 20°. Its solubility at 20° is: 13 mg/l water; 25 g/l ethanol, methanol; 300 g/l acetone, acetophenone, anisole; 500 g/l dichloromethane, dimethylformamide, 1-methyl-2-pyrrolidone. It is stable under normal conditions and non-corrosive.

Uses. Iprodione is a fungicide which is particularly effective against *Botrytis, Monilia* and *Sclerotium* spp. It is also active against other fungi including *Alternaria, Helminthosporium, Rhizoctonia, Corticium, Typhula* and *Fusarium* spp. It is used mainly on vines, deciduous tree fruits, soft fruit, vegetables, ornamentals and cereals, at 0·5-1·0 kg a.i./ha, and on turf at 0·3-1·2 g/m². The dose level for seed treatment varies with the crop.

Toxicology. The acute oral LD50 is: for rats 3500 mg/kg; for mice 4000 mg/kg; for bobwhite quail 930 mg/kg; for mallard duck 10 400 mg/kg. No toxic effect was observed when it was applied dermally at: for rats 2500 mg/kg; for rabbits 1000 mg/kg. In 1·5-y feeding trials rats receiving 1000 mg/kg diet showed no toxic effect; neither did dogs receiving 2400 mg/kg daily. It is practically non-toxic to honey bees.

Formulation. 'Rovral', w.p. (500 g a.i./kg).

Analysis. Product and residue analysis is by GLC (*Anal. Methods Pestic. Plant Growth Regul.*, in press).

307

Isocarbamid

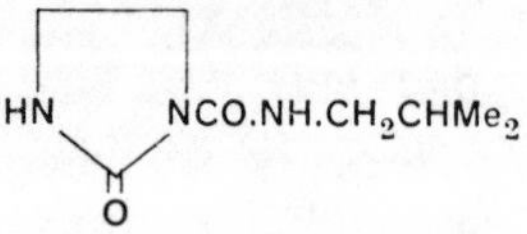

HN⎯⎯⎯NCO.NH.CH$_2$CHMe$_2$

O

T5MVNTJ   CVM1Y1&1   C$_8$H$_{15}$N$_3$O$_2$ (185·2)

Nomenclature and development. The common name isocarbamid is approved by BSI and proposed by ISO. The IUPAC name is *N*-isobutyl-2-oxoimidazolidine-1-carboxamide, in *C.A.* usage *N*-(2-methylpropyl)-2-oxo-1-imidazolidinecarboxamide *[30979-48-7]*. It was introduced in 1973 by Bayer AG under the code number 'MNF 0166', trade mark 'Merpelan AZ' (only as a 2-component mixture) and protected by DAS 1 795 117; USP 3 875 180.

Properties. It is a colourless crystalline solid, m.p. 95-96°, v.p. < 0·1 mmHg at 50°. Its solubility at 20° is 1·3 g/l water; 130 g/kg cyclohexanone; 281 g/kg dichloromethane. It is stable in alkaline and acid media.

Uses. Isocarbamid is a selective herbicide for use in sugar beet and fodder beet at rates of 3-4 kg 'Merpelan AZ'/ha for pre-em. application.

Toxicology. The acute oral LD50 is: for male rats > 2500 mg/kg; for female dogs > 500 mg/kg. The acute dermal LD50 for male rats is > 500 mg/kg. In 90-d feeding trials the 'no effect' level was: for rats 800 mg/kg diet; for dogs > 5000 mg/kg diet. Its toxicity to birds and fish is low; it is not harmful to honey bees.

Formulations. 'Merpelan AZ' ('Terratop' in UK), w.p. (630 g isocarbamid + 130 g lenacil/kg).

Analysis. Residues may be determined by GLC (H. J. Jarczyk, *Pflanzenschutz-Nachr., (Am. Ed.)*, 1974, **27**, 130).

Isofenphos

1Y1&OVR BOPS&O2&MY1&1 $C_{15}H_{24}NO_4PS$ (345·4)

Nomenclature and development. The common name isofenphos is approved by BSI and ISO, in France (isophenphos)—exceptions Canada and USA. The IUPAC name is *O*-ethyl *O*-2-isopropoxycarbonylphenyl isopropylphosphoramidothioate, in *C.A.* usage 1-methylethyl 2-[[ethoxy[(1-methylethyl)amino]phosphinothioyl]oxy]benzoate *[25311-71-1]*. It was introduced in 1974 by Bayer AG under the code number 'BAY SRA 12 869', trade mark 'Oftanol' and protected by DBP 1 668 047. Early references to its insecticidal properties include B. Homeyer, *Meded. Fac. Landbouwwet., Rijksuniv. Gent,* 1974, **39**, 789; *Mitt. Biol. Bundesanst. Land-Forstwirtsch. Berlin-Dahlem,* 1975, **165**, 208.

Properties. It is a colourless oil, v.p. 4 x 10^{-6} mmHg at 20°, d_4^{20} 1·13. Its solubility at 20° is 23·8 mg/kg water; > 600 g/kg dichloromethane, cyclohexanone.

Uses. Isofenphos is a contact and stomach insecticide and is translocated from roots to a limited extent in plants. It is effective against soil-dwelling insects when broadcast at 5 kg a.i./ha, against leaf-eating pests at 50-100 g/100 l, and is used on maize, vegetables and rape.

Toxicology. The acute oral LD50 is: for rats 28-38·7 mg/kg; for mice 91·3-127 mg/kg. The LC50 for Japanese quail is 5-12·5 mg/kg diet. The LC50 (96 h) for goldfish is 2 mg/l.

Formulations. These include: e.c. (500 g a.i./l); w.p. (400 g/kg); seed dressing; granules.

Analysis. Product analysis is by GLC. Residues may be determined by GLC with FID (K. Wagner, *Pflanzenschutz-Nachr. (Am. Ed.),* 1976, **29**, 67).

Isomethiozin

T6NN DNVJ CS1 DNU1Y1&1 FX1&1&1 $C_{12}H_{20}N_4OS$ (268·4)

Nomenclature and development. The common name isomethiozin is approved by BSI and New Zealand and proposed by ISO. The IUPAC name is 6-*tert*-butyl-4-isobutylideneamino-3-methylthio-1,2,4-triazin-5-one, in *C.A.* usage 6-(1,1-dimethylethyl)-4-[(2-methyl-propylidene)amino]-3-(methylthio)-1,2,4-triazin-5(4*H*)-one *[57052-04-7]*. It was introduced in 1975 by Bayer AG under the code number 'BAY DIC 1577', trade mark 'Tantizon' and protected by BP 1 182 801. Early references to its herbicidal properties include H. Hack, *Mitt. Biol. Landfortwirtsch Berlin-Dahlem*, 1975, **165**, 179; *idem, Pflanzenschutz-Nachr. (Am. Ed.)*, 1975, **28**, 241.

Properties. It is a colourless crystalline solid, m.p. 159·3°, v.p. 3·5 x 10^{-7} mmHg at 20°. Its solubility at 20° is: 10 mg/kg water; 103 g/kg cyclohexanone; 152 g/kg dichloromethane.

Uses. Isomethiozin is a post-em. herbicide, recommended for use in areas where there is a heavy grass weed infestation. It is suitable for spring application to winter wheat and, in combination with dichlorprop, to spring cereals at 3·0-4·0 kg/ha (*idem, ibid.;* W. Kolbe *ibid.,* p.257).

Toxicology. The acute oral LD50 is: for rats > 10000 mg/kg; for mice > 2500 mg/kg. In 90-d feeding trials the 'no effect' level was: for rats 100 mg/kg diet; for dogs 500 mg/kg diet. The LC50 (96 h) is: for goldfish 10-12 mg/l; for carp 10-20 mg/l. Its toxicity to birds is low and it is not harmful to honey bees.

Formulations. A w.p. (700 g a.i./kg); various combinations with dichlorprop.

Analysis. Product analysis is by i.r. spectroscopy. Residues may be determined by GLC with FID (H. J. Jarczyk, *ibid.,* p.287).

Isoprocarb

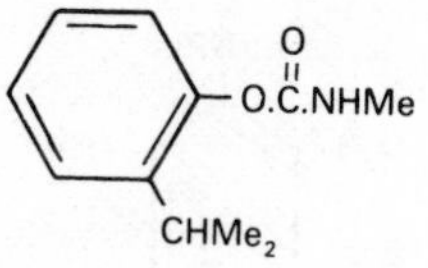

1Y1&R   BOVM1

$C_{11}H_{15}NO_2$ (193·2)

Nomenclature and development. The common name isoprocarb is approved by BSI and ISO—exception JMAF (MIPC) and USA; also known as ENT 25670. The IUPAC name is 2-isopropylphenyl methylcarbamate or *o*-cumenyl methylcarbamate (I), in *C.A.* usage 2-(1-methylethyl)phenyl methylcarbamate *[2631-40-5]* formerly (I). It was introduced in 1970 by Bayer AG under the code number 'Bayer 105 807' and trade mark 'Etrofolan', and by Mitsubishi Chemical Industries Ltd under the trade mark 'Mipcin'. Its use is mentioned by H. F. Jung & K. Iwaya, *Pflanzenschutz-Nachr. (Am. Ed.)*, 1972, **25**, 85.

Properties. The technical product is a colourless crystalline solid, m.p. 88-93°. It is insoluble in water; readily soluble in acetone, methanol. It is compatible with most conventional pesticides, but should not be combined with products having an alkaline reaction nor used within 10 d prior to or after application of propanil.

Uses. Isoprocarb is a contact insecticide of low mammalian toxicity effective against pests of rice at rates of 0·5-1·0 kg a.i./ha or 1·0-1·5 kg/ha for granular application to the water surface. Also for the control of aphids, leafhoppers, capsid bugs and other pests of deciduous fruit and other crops. It has a moderately long residual activity.

Toxicology. The acute oral LD50 is: for rats 403-485 mg/kg; for mice 487-512 mg/kg; for guinea-pigs and rabbits *c.* 500 mg/kg. The acute dermal LD50 for male rats is > 500 mg/kg and it is non-irritant to the eyes and ears of rabbits. The LC50 (48 h) for carp is 4·2 mg/l. It is harmful to honey bees.

Formulations. These include: e.c. (200 g a.i./l); w.p. (500 g/kg); 15% thermal fog; granules (50 g/kg); dusts.

Analysis. Product analysis is by u.v. spectroscopy; details of methods are available from Bayer AG.

Isopropalin

3N3&R BNW FNW DY1&1 $C_{15}H_{23}N_3O_4$ (309·4)

Nomenclature and development. The common name isopropalin is approved by BSI, ISO, ANSI and WSSA. The IUPAC name is 4-isopropyl-2,6-dinitro-*N,N*-dipropylaniline, in *C.A.* usage 4-(1-methylethyl)-2,6-dinitro-*N,N*-dipropylbenzenamine *[33820-53-0]* formerly 2,6-dinitro-*N,N*-dipropylcumidine. It was introduced in 1969 by Eli Lilly & Co. under the code number 'EL-179', trade mark 'Paarlan' and protected by USP 3 257 190. Its herbicidal properties were first described by L. R. Guse, *Proc. North Cent. Weed Control Conf.*, 1969, p.44; G. J. Shoop, *ibid.*, p.19.

Manufacture and properties. Produced by the reaction of 1-chloro-4-isopropyl-2,6-dinitrobenzene with dipropylamine, it is a red-orange liquid readily soluble in organic solvents. At 25°, its solubility is 0·1 mg/l water. It is stable though susceptible to decomposition by u.v. irradiation.

Uses. Isopropalin is a pre-plant incorporated herbicide at 1-2 kg a.i./ha for the control of broad-leaved and grass weeds in direct seeded peppers and tomatoes and in transplanted tobacco.

Toxicology. The acute oral LD50 of the technical material for mice and rats is > 5000 mg/kg. For rabbits, chickens, mallard ducks and dogs, the highest dose administered, 2000 mg/kg, produced no deaths. There were no deaths when bobwhite and Japanese quail were given a single oral dose of 1000 mg/kg. In 90-d tests the 'no effect' level in rats and dogs was > 250 mg/kg diet. Only very slight skin irritation to rabbits was observed with dermal exposure at 2000 mg/kg. The LC50 (96 h) is for: fathead minnow > 0·1 mg/l; goldfish > 0·15 mg/l.

Formulation. An e.c. (720 g a.i./l).

Analysis. Identification is by GLC; particulars from Eli Lilly & Co.; see also W. S. Johnson & O. D. Decherer, *Anal. Methods Pestic. Plant Growth Regul.*, 1976, **8**, 369.

312

S-2-Isopropylthioethyl *O,O*-dimethyl phosphorodithioate

$$\underset{\underset{(MeO)_2P.S.CH_2CH_2S.CHMe_2}{\overset{\|}{}}}{\overset{S}{}}$$

1Y1&S2SPS&O1&O1 $C_7H_{17}O_2PS_3$ (260·4)

Nomenclature and development. The common name isothioate is approved by JMAF. The IUPAC name is *S*-2-isopropylthioethyl *O,O*-dimethyl phosphorodithioate (I), in *C.A.* usage *O,O*-dimethyl *S*-[2-[(1-methylethyl)thio]ethyl] phosphorodithioate *[36614-38-7]* formerly (I). It was introduced by Nihon Nohyaku Co. Ltd under the trade mark 'Hosdon' and protected by JapaneseP 624714.

Properties. It is a light yellowish-brown liquid, with an aromatic odour, v.p. 2·2 x 10^{-3} mmHg at 20°. Its solubility at 25° is 97 mg/l water.

Uses. It is a systemic insecticide also active in the vapour phase. It is effective against aphids when used as a seed dressing or when applied to foliage at 1·0-1·5 kg/ha.

Toxicology. The acute oral LD50 is: for rats 150-170 mg/kg; for mice 50-80 mg/kg. The acute dermal LD50 for male mice is 240 mg/kg.

Formulations. Seed dressing (330 g a.i./l); granules (50 g/kg).

Analysis. Product analysis is by GLC (T. Nakagawa & M. Kanauchi, *Anal. Methods Pestic. Plant Growth Regul.*, 1978, **10**, 75).

Isoproturon

$$\text{Me}_2\text{CH} - \!\!\!\!\!\!\!\!\!\!\!\!\!\!\!\!\!\! - \text{NH.}\overset{\overset{\displaystyle O}{\|}}{\text{C}}.\text{NMe}_2$$

1Y1&R   DMVN1&1 $C_{12}H_{18}N_2O$ (206·3)

Nomenclature and development. The common name isoproturon is approved by BSI and New Zealand, and proposed by ISO. The IUPAC name is 3-(4-isopropylphenyl)-1,1-dimethylurea, in *C.A.* usage *N,N*-dimethyl-*N'*-[4-(1-methylethyl)phenyl]urea *[34123-59-6]* formerly 3-*p*-cumenyl-1,1-dimethylurea. It was introduced as a herbicide by Hoechst AG under the code number 'Hoe 16410' and trade mark 'Arelon', by Ciba-Geigy AG under the code number 'CGA 18731' and trade mark 'Graminon' and by Rhône-Poulenc Phytosanitaire.

Manufacture and properties. Produced by the reaction of 4-isopropylphenyl isocyanate with dimethylamine, it is a colourless powder, m.p. 155-156°, v.p. 2·5 x 10^{-8} mmHg at 20°. Its solubility at 20° is 55 mg/l water; soluble in most organic solvents. It is stable to light, acids and alkali.

Uses. Isoproturon controls annual grasses, *e.g., Alopecurus myosuroides, Apera spica-venti, Avena fatua* and *Poa annua,* and broad-leaved weeds in wheat, barley and rye at 1·0-1·5 kg a.i./ha.

Toxicology. The acute oral LD50 for rats is > 4640 mg/kg; the acute dermal LD50 for rats is > 3170 mg/kg. In 90-d feeding trials the 'no effect' level was: for rats 400 mg/kg diet; for dogs 50 mg/kg diet.

Formulations. These include: w.p. (500, 700, 750 and 800 g a.i./kg); 500 FW, flowable (500 g/l); also mixture with mecoprop (270 g isoproturon + 430 g mecoprop/kg).

Analysis. Product analysis is by titration. Residues may be determined by GLC. Details of these methods are available from Hoechst AG or Ciba-Geigy AG.

Kasugamycin

kasugamycin hydrochloride $C_{14}H_{25}N_3O_9,HCl,H_2O$ (433·8)

T6OTJ B1 CMYVQUM EZ FO- AL6TJ BQ CQ DQ EQ FQ &GH &QH

Nomenclature and development. The common name kasugamycin is approved by JMAF for an antibiotic formed by *Streptomyces kasugaenis* n. sp. discovered by H. Umezawa *et al., J. Antibiot. (Tokyo)*, 1965, **18**, 101 and whose structure was elucidated by Y. Suhara *et al., Tetrahedron Lett.*, 1966, p.1239. Its IUPAC name is [5-amino-2-methyl-6-(2,3,4,5,6-pentahydroxycyclohexyloxy)tetrahydropyran-3-yl]amino-α-iminoacetic acid, in *C.A.* usage D-3-*O*-[2-amino-4-[(1-carboxyiminomethyl)amino]-2,3,4,6-tetradeoxy-α-D-arabino-hexopyranosyl]-D-chiro-inositol *[6980-18-3]* formerly 3-*O*-[2-amino-4-[(1-carboxyformidoyl)amino]-2,3,4,6-tetradeoxy-α-D-arabino-hexopyranosyl]inositol. It was introduced in 1963 by the Institute of Microbial Chemistry and later by Hokko Chemical Industry Co. Ltd, under the trade mark 'Kasumin' and protected by BelgP 657659; BP 1094566.

Manufacture and properties. Produced by the fermentation of *S. kasugaensis* and extracted from the culture medium by strongly acidic ion-exchange resins, the hydrochloride, *C.A.* Registry No. *[19408-46-9]*, forms colourless crystals, decomposing at 236-239°. Its solubility at 25° is 125 g/l water; insoluble in most organic solvents. The acidic solution is stable when stored for 70 d at 50°.

Uses. Kasugamycin is fungitoxic to *Piricularia oryzae* of rice, giving effective control at 17-22 g a.i. (as 2% product) in 1700-2800 l water/ha; also effective against *Pseudomonas phaseolicola* of bean. It is non-phytotoxic to rice but slight injury has been observed on soyabeans.

Toxicology. The acute oral LD50 is: for rats 22 000 mg/kg; for mice 20 900 mg/kg. It is non-irritant. In 90-d trials rats receiving 1000 mg/kg daily showed no abnormality. In 10-d trials it had no effect on guppies or goldfish at 10 000 mg/l.

Formulations. These include: powder (2 g a.i./kg); w.p. (20 g/kg); liquid (20 g/l).

Analysis. Product analysis is by cup assay with *Pseudomonas fluorescens* NIHJ B-254 or *Piricularia oryzae* P-2 (*J. Antibiot. (Tokyo)*, 1968, **21**, 49).

Lead arsenate

PbHAsO$_4$

. PB . . AS-O3-Q   AsHO$_4$Pb   (347·1)

Nomenclature and development. The traditional name lead arsenate is accepted by ISO in lieu of a common name; also known as gypsine. The IUPAC name is lead hydrogen arsenate, in *C.A.* usage lead(II) arsenate (1:1) salt *[7784-40-9]*. It was first used as an insecticide by Moulton (cited by C. H. Fernald, *Hatch. Agric. Exp. Stn. Bull.* No. 24, 1894) against gypsy moth.

Manufacture and properties. Produced by the reaction of lead oxide (PbO) with an aqueous dispersion of diarsenic pentoxide, which is formed by oxidation of diarsenic trioxide, it is a colourless powder, d_{15}^{15} 5·786-5·930. It is practically insoluble in water. It is stable to light and is not decomposed by carbon dioxide, but is decomposed by alkalies. A number of 'basic' lead arsenates have been described (R. H. Robinson & H. V. Tartar, *Ore. Agric. Exp. Stn. Bull.*, No. 128, 1915), but these are of doubtful existence under ordinary conditions.

Uses. Lead arsenate is a non-systemic stomach insecticide with little contact action. It is used to control caterpillars of top fruit and does not harm predators of the fruit tree red spider mite. It is non-phytotoxic though the addition of components causing the production of water-soluble arsenic compounds may produce leaf damage.

Toxicology. The acute oral LD50 is: for rats 100 mg/kg; for chickens 450 mg/kg.

Formulations. A paste (*c.* 284 g PbO, 140 g As$_2$O$_5$/kg; water soluble As$_2$O$_5$ $\leqslant$ 2·5 g/kg); dispersible powder (*c.* 620 g PbO, 320 g As$_2$O$_5$/kg; water soluble As$_2$O$_5$ $\leqslant$ 2·5 g/kg).

Analysis. Product analysis is by gravimetric methods for lead after conversion to lead sulphate (*MAFF Tech. Bull.*, No. 1, p.38) or lead chromate (AOAC Methods); total arsenic being determined by conversion to arsenic trichloride (AOAC Methods). See also: *CIPAC Handbook*, 1979, **1A,** in press. Residues may be determined by ashing, separating the lead as the dithizone complex or as sulphide and colorimetric measurement (AOAC Methods), and arsenic by the Gutzeit method (AOAC Methods).

Lenacil

T56 FMVNVT&J H- AL6TJ $C_{13}H_{18}N_2O_2$ (234·3)

Nomenclature and development. The common name lenacil is approved by BSI, ISO, ANSI, WSSA and JMAF—exception Germany. The IUPAC name is 3-cyclohexyl-6,7-dihydro-1*H*-cyclopentapyrimidine-2,4-dione, in *C.A.* usage 3-cyclohexyl-6,7-dihydro-1*H*-cyclopentapyrimidine-2,4(*3H,5H*)-dione *[2164-08-1]*; also known as 3-cyclohexyl-5,6-trimethyleneuracil. The herbicidal activity of certain uracils was first reported by H. C. Bucha *et al., Science,* 1962, **137,** 537 and lenacil was described by G. W. Cussans, *Proc. Br. Weed Control Conf., 7th,* 1964, p.671. It was introduced in 1966 by E. I. du Pont de Nemours & Co. (Inc.) under the code number 'Du Pont 634', the trade mark 'Venzar' and protected by USP 3 235 360.

Manufacture and properties. Produced by condensing cyclohexylurea with ethyl 2-oxocyclopentanecarboxylate using phosphoric acid, heating the intermediate with sodium methoxide and subsequent acidification, it is a colourless crystalline solid, m.p. 315·6-316·8°. Its solubility at 25° is: 6 mg/l water; < 10 g/kg in most organic solvents; soluble in pyridine. It is stable and non-corrosive.

Uses. Lenacil is a herbicide absorbed by the roots that inhibits photosynthesis. It is recommended for: pre-plant incorporation or pre-em. treatment of sugar, fodder and red beet at 0·6-1·2 kg a.i./ha, shallow incorporation into the soil improves efficacy under dry conditions and allows reduction of the rates; for pre-em. treatment of spinach, strawberries and various ornamentals at 1·0-2·0 kg/ha, and of flax at 0·4-1·0 kg/ha. No adverse effects occur on crops such as wheat, oats, barley or potatoes following sugar beet in the normal crop rotation.

Toxicology. The acute oral LD50 for rats is > 11 000 mg/kg. A 10% suspension of the w.p. formulation applied to the intact and abraded skin of guinea-pigs proved only mildly irritating and there was no sensitisation.

Formulations. 'Venzar', w.p. (800 g a.i./kg); also mixtures with other sugar beet herbicides, *e.g.* cycloate or phenmedipham.

Analysis. Residues may be determined by GLC (H. L. Pease, *J. Sci. Food Agric.,* 1966, **17,** 121) or by HPLC (T. H. Byast *et al., Agric. Res. Counc. (G.B.) Weed Res. Organ., Tech. Bull.,* No. 15 (2nd Ed.), p.49).

Leptophos

GR   DG   BE   EOPS&R&O1 $C_{13}H_{10}BrCl_2O_2PS$ (412·1)

Nomenclature and development. The common name leptophos is approved by BSI, ISO and ANSI exception JMAF (MBCP). The IUPAC and *C.A.* name is *O*-4-bromo-2,5-dichlorophenyl *O*-methyl phenylphosphonothioate, Registry No. *[21609-90-5]*. It was introduced in 1969 by Veliscol Chemical Corp. under the code number 'VCS-506', trade marks 'Phosvel' and 'Abar' and protected by USP 3 459 836. Its insecticidal properties were described by A. K. Azab, *Proc. Br. Insectic. Fungic. Conf., 5th,* 1969, **2,** 550.

Manufacture and properties. Produced by the reaction of phenylphosphonothioic dichloride with methanol and trimethylamine in toluene at room temperature, followed by reaction with potassium 4-bromo-2,5-dichlorophenoxide in boiling toluene, it is a colourless amorphous solid, m.p. 70·2-70·6°, d^{25} 1·53. Its solubility at 25° is: 2·4 mg/l water; 470 g/l acetone; 1·3 kg/l benzene; 142 g/l cyclohexane; 59 g/l heptane; 24 g/l propan-2-ol. It is stable to acids under long exposure at normal temperature, but is slowly hydrolysed under strongly alkaline conditions.

Uses. Leptophos is a non-systemic insecticide. The e.c. is particularly effective against Lepidoptera, *e.g. Spodoptera littoralis* on cotton at 1·5 kg/ha; on vegetables and fruit at 100 g/100 l. Granules at 1·5 kg a.i./ha are effective for the control of *Ostrinia nubilalis* on maize. There is evidence that it has a positive temperature coefficient against Lepidoptera and may therefore be more useful in warmer climates. It is moderately persistent.

Toxicology. The acute oral LD50 for male albino rats is *c.* 50 mg/kg; the acute dermal LD50 for albino rabbits is > 800mg/kg. In 90-d feeding trials no ill-effect was observed in rats receiving 10 mg/kg diet or dogs 30 mg/kg diet. The LC50 (96 h) is: for rainbow trout 0·01 mg/l; for goldfish 0·13 mg/l. The acute oral LD50 for rats of the phosphonate analogue is 105-133 mg/kg.

Formulations. These include: e.c. (360 g a.i./l); w.p. (450 g/kg); dust (30 g/kg); granules (50 g/kg).

Analysis. Product analysis is by i.r. spectroscopy at 9·62 μm or by GLC. Residues are determined by GLC with FPD or TID. Details of the methods may be obtained from Veliscol Chemical Corp.

Lime sulphur

$$CaS_x$$

$$CaS_x$$

Nomenclature and development. Lime sulphur is the trivial name for an aqueous solution of calcium polysulphides, in *C.A.* usage calcium sulfide (CaS_x) *[1344-81-6]*; previously known as eau Grison after Grison who used it in 1852 as a fungicide; in 1886, F. Dusey showed it was effective against San José scale.

Manufacture and properties. Produced by dissolving sulphur in aqueous suspensions of calcium hydroxide, preferably under pressure and in the absence of air, it is a deep orange malodorous liquid, $d^{15.6} \geqslant 1.28$. It is an aqueous solution of calcium polysulphides and, in small amount, calcium thiosulphate. It is slightly alkaline due to hydrolysis, which is accelerated by dilution. It is decomposed by carbon dioxide, by acids and by the soluble salts of those metals that form insoluble sulphides; leading to the precipitation of sulphur and the formation of hydrogen sulphide or insoluble metal sulphides. Lime sulphur is of limited compatability with other pesticides.

Uses. Lime sulphur is a direct fungicide—*e.g.* on powdery mildews, an action correlated with the content of polysulphide sulphur (E. Salmon *et al., J. Agric. Sci.,* 1916, **7**, 473). The sulphur residue left by decomposition functions as a protective fungicide. Its toxicity to scales is attributed to its softening action on the scale wax (G. D. Schafer, *Mich. Agric. Exp. Stn. Tech. Bull.,* No. 11, 1911; No. 21, 1915). It is phytotoxic, especially to 'sulphur-shy' varieties.

Formulations. It is used without formulation, usually at dilutions of *c.* 10 ml/l.

Analysis. Product analysis for content of total sulphide sulphur (S_x), sulphide sulphur (S) and thiosulphate is by AOAC Methods. See also: W. Goodwin & H. Martin, *J. Agric. Sci.,* 1925, **15**, 96; *MAFF Tech. Bull.* No. 1, 1958, p.41; *CIPAC Handbook,* 1979, **1A**, in press; FAO Specification (CP/58).

Linuron

GR BG DMVN1&O1 $C_9H_{10}Cl_2N_2O_2$ (249·1)

Nomenclature and development. The common name linuron is approved by BSI, ISO, ANSI, WSSA and JMAF. The IUPAC name is 3-(3,4-dichlorophenyl)-1-methoxy-1-methylurea (I), in *C.A.* usage *N'*-(3,4-dichlorophenyl)-*N*-methoxy-*N*-methylurea *[330-55-2]* formerly (I). It was introduced in 1960 as a herbicide by E. I. du Pont de Nemours & Co. (Inc.) under the code number 'Du Pont Herbicide 326' and trade mark 'Lorox' and by Hoechst AG under the code number 'Hoe 2810' and trade mark 'Afalon'. Its properties were described by K. Härtel, *Meded. Landbouwhogesch. Opzoekingsstn. Staat Gent,* 1962, **27**, 1275. It is protected by USP 2960534; 3079244; DBP 1028986; BP 852422.

Manufacture and properties. Produced by the reaction of 3,4-dichlorophenyl isocyanate with *O,N*-dimethylhydroxylamine, it is a colourless crystalline solid, m.p. 93-94°, v.p. 1·5 x 10⁻⁵ mmHg at 24°. Its solubility at 25° is 75 mg/l water; slightly soluble in aliphatic hydrocarbons; moderately soluble in ethanol and common aromatic solvents; soluble in acetone. It is stable at its m.p. and in solution, but is slowly decomposed in acids and bases and in moist soil. It is non-corrosive.

Uses. Linuron is a pre- and post-em. selective herbicide acting by inhibition of photosynthesis. It is recommended for pre-em. use in soyabeans at 0·5-2·5 kg a.i./ha; in cotton (lay-by) at 0·5-1·0 kg/ha, in potatoes at 0·5-2·5 kg/ha; in maize, beans and peas at 0·75-1·0 kg/ha; in asparagus at 1·0-1·5 kg/ha; for pre- or post-em. use in carrots and winter wheat at 0·5-1·0 kg/ha. At these rates, phytotoxic concentrations generally disappear from soil in 120 d (H. Börner, *Z. Pflanzenkr. Pflanzenpathol. Pflanzenschutz,* 1965, **72**, 516).

Toxicology. The acute oral LD50 is: for rats 4000 mg/kg; for dogs *c.* 500 mg/kg. The application of a 10% aqueous suspension caused mild to moderate irritation to the intact skin of guinea-pigs but no sensitisation. In 2-y feeding trials rats and dogs receiving 125 mg/kg diet showed no ill-effect (H. C. Hodge *et al., Food Cosmet. Toxicol.,* 1968, **6**, 171).

Formulations. A w.p. (475 g a.i./kg); liquid formulations and mixtures with other herbicides, *e.g.* lenacil, monolinuron or metoxuron, are also available.

Analysis. Product analysis is by hydrolysis to 2,4-dichloroaniline, which is measured colorimetrically by diazotisation and coupling to form a suitable dye, or by non-aqueous titration with perchloric acid (*CIPAC Handbook,* 1979, **1A**, in press). Residue analysis is by GLC or HPLC (T. H. Byast *et al., Agric. Res. Counc. (G.B.) Weed Res. Organ., Tech. Bull.,* No. 15 (2nd Ed.), p.49). See also: H. L. Pease *et al., Anal. Methods Pestic., Plant Growth Regul. Food Addit.,* 1967, **5**, 433; *Anal. Methods Pestic. Plant Growth Regul.,* 1972, **6**, 659.

Malathion

$$\underset{\text{CH}_2\text{CO}_2\text{Et}}{\overset{\displaystyle\overset{\text{S}}{\underset{\|}{}}}{(\text{MeO})_2\overset{|}{\text{P}}.\text{S.CH.CO}_2\text{Et}}}$$

2OV1YVO2&SPS&O1&O1 $C_{10}H_{19}O_6PS_2$ (330·3)

Nomenclature and development. The common name malathion is approved by BSI, ISO, ESA, BPC and JMAF—exceptions Australia and New Zealand (maldison), Republic of South Africa (mercaptothion), USSR (carbofos) and Germany; also known as ENT 17 034. The IUPAC name is S-1,2-bis(ethoxycarbonyl)ethyl O,O-dimethyl phosphorodithioate, also known as diethyl (dimethoxyphosphinothioylthio)succinate, in C.A. usage diethyl (dimethoxyphosphinothioyl)thiobutanedioate [121-75-5] formerly diethyl mercaptosuccinate S-ester with O,O-dimethyl phosphorodithioate. It was introduced in 1950 by American Cyanamid Co. under the code number 'EI 4049' and protected by USP 2 578 652. 'Cythion' is an American Cyanamid Co. trade mark for premium grade deodorised malathion; other trade marks include 'Malathion', 'Malathiozol' and 'Malathiozoo'.

Manufacture and properties. Produced by the addition of O,O-dimethyl hydrogen phosphorodithioate to diethyl maleate, hydroquinone being present to suppress the polymerisation of the latter, it is a clear amber liquid, m.p. 2·85°, b.p. 156-157°/0·7 mmHg, v.p. 4×10^{-5} mmHg at 30°, d_4^{25} 1·23, n_D^{25} 1·4985. Its solubility at room temperature is 145 mg/l water; miscible with most organic solvents; of limited solubility in petroleum oils (350 g light petroleum/l malathion). The technical product is c. 95% pure.

Uses. Malathion is a non-systemic insecticide and acaricide of low mammalian toxicity and brief to moderate persistence. It is generally non-phytotoxic, but may damage cucumber, string bean and squash under glasshouse conditions. In addition to a wide range of agricultural and horticultural uses, it is used for the control of mosquitoes, flies, household insects, animal ectoparasites, and human head and body lice. Metabolism is by hydrolysis of the carboxylate or phosphorodithioate esters and/or oxidation to the phosphorothioate (sometimes known as maloxon); methods for their separation have been reported (N. W. H. Houx et al., Pestic. Sci., 1979, **10**, in press).

Toxicology. The acute oral LD50 for rats is 2800 mg/kg; the acute dermal LD50 (24 h) for rabbits is 4100 mg/kg. In 1·75-y feeding trials rats receiving 100 mg tech./kg diet showed normal weight gain. It is highly toxic to honey bees.

Formulations. These include: e.c. (25-1000 g a.i./l); w.p. (250 and 500 g/kg); dusts (usually 40 g/kg); ULV concentrates (920 g/l).

Analysis. Product analysis is by GLC (CIPAC Handbook, 1979, **1A**, in press). Residues on a wide range of crops may be determined by GLC: details are available from American Cyanamid Co. See also: G. L. Sutherland, Anal. Methods. Pestic., Plant Growth Regul. Food Addit., 1964, **2**, 283; J. E. Boyd, Anal. Methods Pestic. Plant Growth Regul., 1972, **6**, 418.

Maleic hydrazide

T6NMVJ FQ

$C_4H_4N_2O_2$ (112·1)

Nomenclature and development. The traditional name maleic hydrazide is accepted in lieu of a common name by BSI, ISO and JMAF—exception WSSA (MH). The IUPAC name is 6-hydroxy-3(2*H*)-pyridazinone also known as 1,2-dihydropyridazine-3,6-dione, in *C.A.* usage 1,2-dihydro-3,6-pyridazinedione *[123-33-1]* *or* 6-hydroxy-3(2*H*)-pyrid-azinone *[10071-13-3]*. Its plant-growth regulating activity was first described by D. L. Schoene & O. L. Hoffmann, *Science,* 1949, **109,** 588. It was introduced in 1948 by U.S. Rubber Co. under the code number 'MH-30' and protected by USP 2 575 954; 2 614 916; 2 614 917; 2 805 926.

Manufacture and properties. Produced by the condensation of maleic anhydride and hydrazine, it is a colourless crystalline powder, m.p. 296-298°. Its solubility at 25° is: 6 g/kg water; 1 g/kg ethanol; 24 g/kg dimethylformamide. The technical product is $\geqslant 97\%$ pure, m.p. $\geqslant 292°$, d^{25} 1·60. Maleic hydrazide is a monobasic acid (D. M. Miller & R. W. White, *Can., J. Chem.,* 1956, **34,** 510) forming salts of which the alkali metal and amine salts are water-soluble, but the calcium salt is precipitated by hard water. The solubility of the diethanolamine salt at 25° is 700 g/kg water. Maleic hydrazide is stable to hydrolysis, but is decomposed by concentrated acids with the release of nitrogen.

Uses. Maleic hydrazide is translocated in plants, inhibiting cell division but not extension. It is used for the retardation of growth of grass, hedges and trees, for the inhibition of sprouting of potatoes and onions and for the prevention of sucker development in tobacco, at 2-5 kg a.i./ha. Inclusion of 2,4-D improves the control of broad-leaved weeds.

Toxicology. The acute oral LD50 for rats is 6950 mg sodium salt/kg, 2340 mg diethanolamine salt/kg. It is non-irritant. In 2-y feeding trials rats receiving 50 000 mg sodium salt/kg diet showed no toxic effect. It is non-carcinogenic (J. M. Barnes *et al., Nature (London),* 1957, **180,** 62).

Formulations. These include: 'MH-30' and 'MH-36' (Bayer) aqueous solution of the diethanolamine salt (360 g a.i./l); a formulation of the potassium salt and a combination with 2,4-D are also available.

Analysis. Product analysis is by differential titration: details of method are available from Uniroyal Inc. Residues may be determined by reduction and hydrolysis to split off hydrazine which is separated by distillation and estimated colorimetrically (P. R. Wood, *Anal. Chem.,* 1953, **25,** 1879; J. R. Lane, *J. Assoc. Off. Agric. Chem.,* 1963, **46,** 261; *Anal. Methods Pestic., Plant Growth Regul. Food Addit.,* 1964, **4,** 147).

Malonoben

NCYCN&U1R   DQ   CX1&1&1   EX1&1&1 $C_{18}H_{22}N_2O$ (258·4)

Nomenclature and development. The common name malonoben is approved by BSI and ANSI and proposed by ISO; also known as ENT 27190. The IUPAC name is 2-(3,5-di-*tert*-butyl-4-hydroxybenzylidene)malononitrile (I), in *C.A.* usage [[3,5-bis(1,1-dimethylethyl)-4-hydroxyphenyl]methylene]propanedinitrile *[10537-47-0]* formerly (I). It was introduced by Gulf Oil Chemicals Co. in 1972 as an experimental acaricide and insecticide under the code numbers 'GCP-5126' and 'S-15126'. Its activity was first reported by H. Fukashi *et al., Agric. Biol. Chem.,* 1971, **35**, 2003; the results of field trials have been reported (A. van der Eijk *et al., Proc. 1977 Br. Crop Prot. Conf. Pests Dis.,* 1977, **2**, 581).

Manufacture and properties. Produced by the condensation of malononitrile with 3,5-di-*tert*-butyl-4-hydroxybenzaldehyde, it is a crystalline solid, m.p. 140-141°. It is a powerful uncoupler *in vitro* of respiratory-chain phosphorylation (S. Muraoka & H. Terada, *Biochim. Biophys. Acta,* 1972, **275**, 271).

Uses. Malonoben is a contact and stomach poison effective against motile stages of phytophagous mites, including strains resistant to organophosphorus and carbamate acaricides. It controls *Tetranychus, Panonychus, Aculus, Phyllocoptruta, Brevipalpus* and *Eotetranychus* spp. in pome and stone fruits, citrus, grapes, nut trees, ornamentals, cotton, seed lucerne and beans at 30-60 g a.i./100 l. It is also effective against *Typhlocyba pomaria, Archips argyrospilus* and *Scirtothrips citri* at the same rates and is relatively safe to *Amblyseius fallacis*, a predatory mite in apple orchards.

Toxicology. The acute oral LD50 for rats is 87 mg tech./kg; the acute dermal LD50 for rabbits is 2000 mg/kg. It is relatively non-toxic to honey bees.

Formulations. They include: 'GCP-5126-2EC', e.c. (240 g a.i./l); 'GCP-5126-4L', flowable liquid (480 g/l); 'GCP-5126-50W', w.p. (500 g/kg).

Mancozeb

$$\left[\begin{array}{c} \overset{\text{S}}{\underset{\|}{}} \qquad \overset{\text{S}}{\underset{\|}{}} \\ -\text{S.C.NH.CH}_2\text{CH}_2\text{.NH.C.S.Mn}- \end{array} \right]_x \quad (\text{Zn})_y$$

Nomenclature and development. The common name mancozeb is approved by BSI and ISO—exception JMAF (manzeb). It is a complex of a zinc salt and maneb and contains *c.* 20% manganese *m/m* and *c.* 2·5% zinc *m/m*, the anion present should be stated. The product is indexed in *C.A.* under [[1,2-ethanediylbis[carbamodithioato]](2⁻)]manganese mixture with [[1,2-ethanediylbis[carbamodithioato]](2⁻)]zinc *[8018-01-7]* (formerly *[8065-67-5]*), *i.e.* maneb (mixture with zineb), formerly [ethylenebis[dithiocarbamato]]-manganese *[12427-38-2]* mixture with [ethylenebis[dithiocarbamato]]zinc. It was first described in *Fungic. Nematic. Tests,* 1961, **17,** and was introduced in 1961 by Rohm & Haas Co. under the trade mark 'Dithane M-45' and protected by BP 996 264; USP 3 379 610; 2 974 156. Evidence indicates that mancozeb does not include inorganic anions as part of its structure; it is thought the polymer depicted above gives a better representation than a cyclic structure (Rohm & Haas Co.).

Properties. It is a greyish-yellow powder, which decomposes without melting, flash point (Tag open cup) 137·8°. It is practically insoluble in water and in most organic solvents. It is stable under normal storage conditions. It is decomposed at high temperatures by moisture, and by acid conditions, but is compatible with all commonly used pesticides. It may cause greasing with some emulsifiable concentrates.

Uses. Mancozeb is a protective fungicide generally used at 1·4-1·9 kg a.i./ha and effective against a wide range of foliage diseases including *Phytophthora infestans* on potatoes and *Fulvia fulva* on tomatoes. It is also used, in a combined formulation with zineb, against a wide range of foliage fungal diseases including *Venturia inaequalis* of pome fruit and various rust diseases.

Toxicology. The acute oral LD50 for rats is > 8000 mg/kg. It may cause skin irritation on repeated exposure.

Formulations. These include: 'Dithane M-45', 'Dithane 945' w.p. (⩾ 800 g mancozeb/kg); 'Karamate' (mancozeb + zineb).

Analysis. Product analysis is by decomposition with acid and measuring the carbon disulphide liberated, either by GLC or by conversion to dithiocarbonate with alkali and titration with iodine (*CIPAC Handbook,* 1979, **1A,** in press). Residues may be determined by digestion with acid to liberate carbon disulphide which is estimated by GLC or colorimetrically (C. F. Gordon *et al., J. Assoc. Off. Anal. Chem.,* 1967, **50,** 1102).

Maneb

$$\left[\ -S.\overset{\overset{S}{\|}}{C}.NH.CH_2CH_2.NH.\overset{\overset{S}{\|}}{C}.S\ -Mn-\ \right]_x$$

SUYS&M2MYS&US &-MN- $\qquad\qquad$ $(C_4H_6MnN_2S_4)x$ $\quad$ $(265\cdot3)x$

Nomenclature and development. The common name maneb is approved by BSI, ISO and JMAF; also known as ENT 14875, and MEB. The IUPAC name is polymeric manganese ethylenebis(dithiocarbamate), in *C.A.* usage manganese 1,2-ethanediylbis(carbamo-dithioate) complex *[12427-38-2]* formerly [1,2-ethanediylbis[carbamodithioato](2-)]-manganese and [ethylenebis[dithiocarbamato]]manganese. It was introduced in 1950 by E. I. du Pont de Nemours & Co. (Inc.) under the trade mark 'Manzate' and by Rohm & Haas Co. under the trade mark 'Dithane M-22' and is protected by USP 2 504 404; 2 710 822.

Manufacture and properties. Produced by the reaction of a water-soluble ethylenebis(dithiocarbamate) with manganese sulphate or dichloride, it is a yellow crystalline solid, which decomposes before melting, d 1·92. It is slightly soluble in water; insoluble in most organic solvents. The technical product is a light-coloured solid. It is stable under normal storage conditions, but decomposes more or less rapidly on exposure to moisture or to acids. In the presence of moisture, decomposition proceeds with the formation of etem (p.239) (R. A. Ludwig *et al., Can. J. Bot.,* 1955, **33**, 42; C. W. Pluijgers *et al., Tetrahedron Lett.,* 1971, p.1371).

Uses. Maneb is a protective fungicide effective against many diseases of foliage, particularly the blights of potato and tomato, at spray concentrations of 150-200 g/100 l.

Toxicology. The acute oral LD50 for rats is 6750 mg/kg. In feeding trials no ill-effect was observed: for rats receiving 250 mg/kg diet for 2 y; for dogs receiving 80 mg/kg diet for 1 y.

Formulation. A w.p. (800 g a.i./kg).

Analysis. Product analysis is by decomposition with acid and measurement of the liberated carbon disulphide, either by GLC or by titration with iodine of the dithiocarbonate formed in the presence of alkali (W. K. Lowen, *J. Assoc. Off. Agric. Chem.,* 1953, **36**, 484; M. Levitsky & W. K. Lowen, *ibid.,* 1954, **37**, 555; *CIPAC Handbook,* 1970, **1**, 463; 1979, **1A**, in press). Note: aged samples may give low results because etem cannot be determined by acid decomposition. Residues may be determined by reaction to form carbon disulphide which is measured either by GLC or by standard colorimetric methods (H. L. Pease, *J. Assoc. Off. Agric. Chem.,* 1957, **40**, 1113; G. E. Keppel, *J. Assoc. Off. Anal. Chem.,* 1969, **52**, 162).

MCPA

QV1OR DG B1

$C_9H_9ClO_3$ (200·6)

Nomenclature and development. The common name MCPA is approved by BSI, ISO and WSSA—exceptions USSR (metaxon) and JMAF (MCP). The IUPAC name is (4-chloro-*o*-tolyloxy)acetic acid formerly (4-chloro-2-methylphenoxy)acetic acid (I), in *C.A.* usage (I) *[94-74-6]* formerly [(4-chloro-*o*-tolyl)oxy]acetic acid. The discovery of its plant-growth regulating activity is described by R. E. Slade, *Nature (London),* 1945, **155,** 498 and it was introduced in 1945 by ICI Ltd. There are many trade marks for its formulations including: 'Agroxone' (ICI Ltd, Plant Protection Division), 'Agritox' (May & Baker Ltd), 'Cornox M' (The Boots Co. Ltd).

Manufacture and properties. Produced by the condensation of 4-chloro-*o*-cresol with sodium chloroacetate or chloroacetic acid in alkaline solution, pure MCPA is a colourless crystalline solid, m.p. 118-119°. Its solubility at room temperature is 825 mg/l water. As the chlorination of *o*-cresol gives a proportion of the isomeric 6-chloro-*o*-cresol, the crude MCPA contains the inactive 6-chloro-*o*-tolyloxy-isomer, nowadays limited in amount. The crude product is 85-95% pure, m.p. 100-115°.

It is an acid, pK_a 3·07, forming water-soluble alkali metal and amine salts, though precipitation may occur with hard water. Solutions of the alkali metal salts are alkaline in reaction and will corrode aluminium and zinc. Oil-soluble esters may be prepared.

Uses. MCPA is a systemic hormone-type selective herbicide, readily absorbed by leaves and roots. It is used for the control of annual and perennial weeds in cereals, grassland and turf at 0·28-2·25 kg a.i./ha.

Toxicology. The acute oral LD50 is: for rats 700 mg/kg; for mice 550 mg/kg. In feeding trials cows tolerated 30 mg/kg daily for 21 d without detectable symptoms (S. Dalgaard-Mikkelsen *et al., Nord. Veterinaermed.,* 1959, **11,** 469); rats receiving 100 mg/kg diet for 210 d were unaffected apart from slight kidney enlargement. Safe for fish at 10 mg/l. The presence of unreacted chlorophenols may cause taint.

Formulations. These include: aqueous concentrates of the salts (200-500 g/l); e.c. of esters (200-500 g/l); w.s.p. MCPA-sodium (750-800 g a.e./l); widely used in mixture with other herbicides.

Analysis. Product analysis is by i.r. spectrosocopy (*CIPAC Handbook,* 1970, **1,** 483), by liquid/liquid chromatography (*ibid.,* p.477) or by GLC (*CIPAC Handbook,* in press; FAO Specification (CP/48); *Herbicides 1977,* pp. 22-29). For mixtures with dicamba see *CIPAC Handbook,* 1979, **1A,** in press. Residues can be determined by GLC after conversion to the methyl ester (J. Lest, *Anal. Methods Pestic., Plant Growth Regul. Food Addit.,* 1967, **5,** 439; *Anal. Methods Pestic. Plant Growth Regul.,* 1972, **6,** 663).

326

MCPB

QV3OR DG B1 $C_{11}H_{13}ClO_3$ (228·7)

Nomenclature and development. The common name MCPB is approved by BSI, ISO, WSSA and JMAF—exceptions France (2,4-MCPB) and USSR (2M-4Kh-M). The IUPAC name is 4-(4-chloro-*o*-tolyloxy)butyric acid (I) formerly 4-(4-chloro-2-methylphenoxy)butyric acid, in *C.A.* usage 4-(4-chloro-2-methylphenoxy)butanoic acid *[94-81-5]* formerly (I). Its herbicidal activity was described by R. L. Wain & F. Wightman, *Proc. Roy. Soc.,* 1955, **142B**, 525, and it was introduced in 1954 by May & Baker Ltd under the code number 'MB 3046', the trade mark 'Tropotox' and protected by BP 758 980.

Manufacture and properties. Produced by the reaction of sodium 4-chloro-*o*-cresolate with γ-butyrolactone, the technical product, *c.* 92% pure, has m.p. 99-100°. Pure MCPB is a colourless crystalline solid, m.p. 100°. Its solubility at room temperature is: 44 mg/l water; 150 g/l ethanol; > 200 g/l acetone. It is an acid, pK_a 4·84, forming water-soluble alkali metal and amine salts, though precipitation occurs with hard water.

Uses. MCPB owes its properties as a selective herbicide to the ability of susceptible plants to translocate it (R. C. Kirkwood *et al., Pestic. Sci.,* 1972, **3**, 307) and to oxidise it to MCPA. It is used for the control of susceptible weeds in undersown cereals, peas, and established grassland at 1·7-3·4 kg a.e./ha.

Toxicology. The acute oral LD50 for rats is 680 mg a.i./kg, 690 mg MCPB-sodium/kg.

Formulations. 'Tropotox' is an aqueous solution of MCPB-sodium, *C.A.* Registry No. *[6062-26-6]*, (400 g a.e./l); mixtures with other herbicides, *e.g.* MCPA (as sodium/ potassium salts) or with benazolin are also available.

Analysis. Product analysis is by the determination of extractable acids (*CIPAC Handbook,* in press). Residues may be determined by i.r. spectroscopy (H. A. Glastonbury & M. D. Stevenson, *J. Sci. Food Agric.,* 1959, **10**, 379). See also; A. Guardiglii, *Anal. Methods Pestic. Plant Growth Regul.,* 1976, **8**, 397.

Mecarbam

$$\underset{\text{Me}}{(EtO)_2\overset{\overset{\displaystyle S}{\|}}{P}.S.CH_2.\overset{\overset{\displaystyle O}{\|}}{C}.N.\overset{\overset{\displaystyle O}{\|}}{C}.OEt}$$

2OVN1&V1SPS&O2&O2 $\qquad\qquad\qquad\qquad\qquad$ $C_{10}H_{20}NO_5PS_2$ (329·4)

Nomenclature and development. The common name mecarbam is approved by BSI, ISO and JMAF—exceptions France and USSR. The IUPAC name is *S*-(*N*-ethoxycarbonyl-*N*-methylcarbamoylmethyl) *O,O*-diethyl phosphorodithioate, in *C.A.* usage ethyl 6-ethoxy-2-methyl-3-oxo-7-oxa-5-thia-2-aza-6-phosphanonanoate 6-sulfide *[2595-54-2]* formerly ethyl [[(diethoxyphosphinothioyl)thio]acetyl]methylcarbamate, or ethyl (mercaptoacetyl)-methylcarbamate *S*-ester with *O,O*-diethyl phosphorodithioate. It was introduced in 1958 by Murphy Chemical Ltd. It was first described by M. Pianka, *Chem. Ind. (London),* 1961, p.324, and is protected by BP 867780. Its code numbers were 'P474' and 'MC474', trade marks include 'Murfotox' (Murphy Chemical Ltd), 'Pestan' (Takeda Chemical Industries) and formerly *'Afos'*.

Manufacture and properties. Produced by the reaction of *O,O*-diethyl hydrogen phosphorodithioate with ethyl (chloroacetyl)methylcarbamate in the presence of an acid-binding agent, it is a light brown to pale yellow oil, b.p. 144°/0·02 mmHg, of negligible v.p. at room temperature, d_{20}^{20} 1·223, n_D^{20} 1·5138. Its solubility at room temperature is: < 1 g/l water; < 50 g/kg in aliphatic hydrocarbons; miscible with alcohols, aromatic hydrocarbons, ketones, esters. The technical product is ≥ 85% pure, d_{20}^{20} 1·223. Though subject to hydrolysis, it is compatible with all but highly alkaline pesticides. The technical product slowly attacks metals.

Uses. Mecarbam is an insecticide and acaricide with slight systemic properties. It is used for the control of scale insects and other Hemiptera, olive fly and other fruit flies at 60 g a.i./100 l; against leaf hoppers and stem flies of rice as a dust (15 g/kg); and against root fly larvae of cabbage, onions, carrots and celery. At recommended rates of use it persists in soil for 28-42 d.

Toxicology. The acute oral LD50 is: for rats 36-53 mg/kg; for mice 106 mg/kg. The acute dermal LD50 for rats is > 1220 mg/kg. In 0·5-y feeding trials rats receiving 1·6 mg/kg daily suffered no ill-effect, but at 4·56 mg/kg daily there was a slight depression of growth rate.

Formulations. These include: 'Murfotox', e.c. (500, 680 and 900 g a.i./l; w.p. (250 g/kg); dust (15 and 40 g/kg); 'Murfotox Oil' (50 g/l) in petroleum oil; 'Pestan', e.c. (400 g/l).

Analysis. Product analysis is by GLC with FID. Residues may be determined, after liquid extraction and column chromatography, by GLC with ECD. Details of both methods are available from Murphy Chemical Ltd. See also: V. P. Lynch, *Anal. Methods Pestic. Plant Growth Regul.,* 1976, **8,** 135.

Mecoprop

Me
Me
Cl——O.CHCO$_2$H

QVY1&OR DG B1   C$_{10}$H$_{11}$ClO$_3$ (214·6)

Nomenclature and development. The common name mecoprop is approved by BSI, ISO and WSSA —exceptions Denmark (mechlorprop) and JMAF (MCPP); the initials CMPP have also been used. The IUPAC name is ($\pm$)-2-(4-chloro-o-tolyloxy)propionic acid (I) formerly ($\pm$)-2-(4-chloro-2-methylphenoxy)propionic acid, in *C.A.* usage 2-(4-chloro-2-methylphenoxy)propanoic acid *[93-65-2]* formerly (I). Its plant growth activity was first reported by C. H. Fawcett *et al., Ann. Appl. Biol.,* 1953, **40,** 232, and its use as a herbicide by G. B. Lush & E. L. Leafe, *Proc. Br. Weed Control Conf., 3rd,* 1956, pp.625, 633. It was introduced in 1957 by The Boots Co. Ltd under the code number 'RD 4593', trade mark 'Iso-Cornox' and protected by BP 820180; 822973; 825875.

Manufacture and properties. Produced by the condensation of 2-chloropropionic acid with 4-chloro-o-cresol giving the racemate (in which doses are normally expressed) of which only the (+)-form is active as a herbicide. Mecoprop forms colourless crystals, m.p. 94-95°. Its solubility at 20° is 620 mg/l water; readily soluble in most organic solvents. Many of its salts are highly soluble in water, *e.g.*: at 15°, 460 g mecoprop-sodium/l water; at 0°, 795 g mecoprop-potassium/l water; at 20° 580 g diethanolamine salt/l water. The technical product, which may have a slight phenolic odour, has m.p. $\geqslant$ 90°. It is stable to heat and resistant to reduction, hydrolysis and atmospheric oxidation. It is corrosive to metals in the presence of moisture, but solutions of the potassium salt do not corrode brass, iron or mild steel provided that the pH is $\geqslant$ 8·6 and the temperature < 80°.

Uses. Mecoprop is a systemic post-em. herbicide, recommended for the control of cleavers, chickweed and other weeds in cereals at 1·5-2·7 kg/ha; it is also used in combination with other herbicides such as 2,4-D, MCPA, dicamba, dichlorprop, ioxynil, 2,3,6-TBA and 3,6-dichloropicolinic acid.

Toxicology. The acute oral LD50 is: for rats 930 mg/kg; for mice 650 mg/kg. In 21-d feeding trials rats receiving 65 mg/kg daily showed no ill-effect; those receiving 100 mg/kg diet for 210 d suffered only a slight enlargement of the kidneys.

Formulations. These include: aqueous solutions of alkali metal, particularly potassium, or amine salts; e.c. of esters; e.c. based on dried non-hygroscopic mecoprop salts; concentrates range up to 650 g a.e./l. Also mixtures of: mecoprop + 2,4-D (as amine salts), mecoprop + fenoprop (as alkali metal salts).

Analysis. Product analysis is by GLC of the methyl ester (details from The Boots Co. Ltd; L. A. Haddock & L. G. Phillips, *Analyst. (London),* 1959, **84,** 94) or by titration of extractable acids (*CIPAC Handbook,* 1979, **1A,** in press).

Mefluidide

FXFFSWMR B1 D1 EMV1 $C_{11}H_{13}F_3N_2O_3S$ (310·3)

Nomenclature and development. The common name mefluidide is approved by BSI, ANSI, New Zealand and WSSA, and proposed by ISO. The IUPAC name is 5'-(trifluoro-methanesulphonamido)acet-2',4'-xylidide, in *C.A.* usage *N*-[2,4-dimethyl-5-[[(trifluoro-methyl)sulfonyl]amino]phenyl]acetamide *[53780-34-0]*. It was introduced and described as a plant growth regulator by 3M Company at the North Cent. Weed Control Conf., under the code number 'MBR 12 325', trade mark 'Embark' and protected by USP 3 894 078.

Manufacture and properties. Produced by the action of either trifluoromethanesulphonyl fluoride or trifluoromethanesulphonic anhydride on 5'-aminoacet-2',4'-xylidide, which is prepared by acetylation of 2,4-xylidine and subsequent nitration and reduction, it is a colourless crystalline solid, m.p. 183-185°, v.p. $< 10^{-4}$ mmHg at 25°; pK_a 4·6. Its solubility at 23° is: 180 mg/l water; 310 mg/l benzene; 2·1 g/l dichloromethane; 17 g/l octan-1-ol; 310 g/l methanol; 350 g/l acetone. It is stable at elevated temperatures, but the acetamido moiety is hydrolysed on refluxing mefluidide in acidic or alkaline solutions. It is degraded in aqueous solutions exposed to u.v. radiation; aqueous solutions and suspensions are mildly corrosive to metals on prolonged exposure.

Uses. Mefluidide is a plant growth regulator and herbicide used: for the suppression of growth and seed production of turf grasses, trees and woody ornamentals; as an agent for enhancing sucrose content of sugarcane; for the control of growth and seed production of weeds in various crops, particularly rhizomatous *Sorghum halepense* in soyabeans. The optimum rates for these uses range from 0·3 to 1·1 kg a.i./ha.

Toxicology. The acute oral LD50 is: for mice > 1920 mg tech./kg; for rats > 4000 mg/kg; for mallard ducks > 4640 mg/kg. Emesis occurs with dogs at 500 mg/kg. The acute dermal LD50 for rabbits is > 4000 mg/kg; it is mildly irritating to the rabbit eye. No teratogenic effect was observed in rabbits and no mutagenic effect was observed in *Salmonella typhimurium*. The LC50 (5 d) for mallard duck and bobwhite quail observed on the 8th d was > 10000 mg/kg diet. The LC50 (96 h) for rainbow trout and bluegill is > 100 mg/l.

Formulations. 'Embark' Plant Growth Regulator/Herbicide 2-S, solution (240 g a.i./l).

Analysis. Analysis is by GLC following the methylation with diazomethane; details may be obtained from Agrichemicals Project, 3M Company.

Menazon

$$(MeO)_2P.S.CH_2\text{---}$$

S (double bond on P)

[structure: dimethyl phosphorodithioate linked via S-CH₂ to a 1,3,5-triazine ring bearing two NH₂ groups]

T6N CN ENJ BZ DZ F1SPS&O1&O1 $C_6H_{12}N_5O_2PS_2$ (281·3)

Nomenclature and development. The common name menazon is approved by BSI, ISO, ESA and JMAF—exceptions France (azidithion), Germany and Italy; also known as ENT 25 760. The IUPAC name is S-(4,6-diamino-1,3,5-triazin-2-yl)methyl O,O-dimethyl phosphorodithioate (I), in $C.A.$ usage (I) *[78-57-9]* formerly S-[(4,6-diamino-s-triazin-2-yl)methyl] O,O-dimethyl phosphorodithioate. It was described by A. Calderbank *et al., Chem. Ind. (London),* 1961, p.630, and was introduced in 1961 by ICI Ltd, Plant Protection Division under the code number 'PP 175' and subsequently under the trade marks 'Sayfos', 'Saphizon', 'Saphicol' and protected by BP 899 701.

Manufacture and properties. Produced by the reaction of 2,4-diamino-6-chloromethyl-1,3,5-triazine with sodium O,O-dimethyl phosphorodithioate or of biguanide with ethyl (dimethoxyphosphinothioylthio)acetate, it forms colourless crystals, with little or no odour, which decompose with partial melting at 160-162°, v.p. 1 x 10⁻⁶ mmHg at 25°. Its solubility at 20° is 0·24 g/l water; at room temperature: 200 g/kg 2-ethoxyethanol, 250 g/kg 2-methoxyethanol, 150 g/kg tetrahydrofurfuryl alcohol, 100 g/kg ethylene glycol. The technical product is 90-95% pure and has a mercaptan-like odour. The solid is stable ≤ 50°, but is unstable under acid (< pH 4) or alkaline (> pH 8) conditions. It forms a crystalline hydrochloride. It is compatible with all but strongly alkaline pesticides, but may be decomposed by the reactive surfaces of some 'inert' fillers.

Uses. Menazon is a systemic aphicide used mainly as a tuber and seed treatment for the protection of seedlings; for foliage application; as a soil drench and a root dip. It is non-phytotoxic.

Toxicology. The acute oral LD50 for female rats is 1950 mg/kg; rabbits treated on the shaved back for 24 h with 500-800 mg/kg suffered no local or systemic toxic effect. In 90-d feeding trials the 'no effect' level for rats was 30 mg/kg diet; in a 2-y feeding trial rats receiving 4000, 1000 and 250 mg/kg diet showed no significant effect other than on cholinesterase. The LC50 (48 h) for harlequin fish is 220 mg/l. It is toxic to honey bees and should not be applied at flowering.

Formulations. These include: w.p. (700 g a.i./kg) and a liquid suspension ('Col') (400 g/l).

Analysis. Product analysis is by colorimetry at 430 nm of the vanadomolybdate complex after oxidation to phosphoric acid. Residues may be determined by standard colorimetric methods for phosphate after oxidation, clean-up being by ion exchange chromatography and separation by paper chromatography (A. Calderbank & J. B. Turner, *Analyst. (London),* 1962, **87,** 273). Details are available from ICI Ltd, Plant Protection Division. See also: A. Calderbank, *Anal. Methods Pestic. Plant Growth Regul.,* 1973, **7,** 317.

Mephosfolan

$$(EtO)_2\overset{\overset{O}{\|}}{P}.N{=}\!\!\!\begin{array}{c} S \\ \\ S \end{array}\!\!\!\rangle\!\!-\!\! Me$$

T5SYSTJ BUNPO&O2&O2 D1 $C_8H_{16}NO_3PS_2$ (269·3)

Nomenclature and development. The common name mephosfolan is approved by BSI and ISO, in France (méphospholan); also known as ENT 25991. The IUPAC name is diethyl (4-methyl-1,3-dithiolan-2-ylidene)phosphoramidate (I) also known as 2-(diethoxy-phosphinylimino)-4-methyl-1,3-dithiolane, in *C.A.* usage (I) *[950-10-7]* formerly *P,P*-diethyl cyclic propylene ester of phosphonodithioimidocarbonic acid. It was introduced in 1963 by the American Cyanamid Co. under the code number 'EI 47470', trade mark 'Cytrolane' and protected by BP 974138; FP 1327386.

Manufacture and properties. Produced by the reaction of 2-imino-4-methyl-1,3-dithiolane with diethyl phosphorochloridate, it is a yellow to amber liquid, b.p. 120°/1 x 10⁻³ mmHg, n_D^{26} 1·539. Its solubility at 25° is 57 g/kg water; soluble in acetone, ethanol, benzene, 1,2-dichloroethane. It is stable in water under neutral conditions, but is hydrolysed by acid or alkali (pH < 2 or > 9).

Uses. Mephosfolan is a contact and stomach insecticide with systemic activity following root or foliar absorption and is used for the control of *Spodoptera* spp., stem borers, bollworms, whiteflies, mites and aphids on major crops such as cotton, maize, vegetables, fruit, and on field crops.

Toxicology. The acute oral LD50 is: for rats 3·9-8·9 mg tech./kg; for albino mice 11·3 mg/kg. The acute dermal LD50 for male albino rabbits is: 28·7 mg/kg (24 h); 185 mg/kg (4 h); 340-> 5000 mg granules/kg (24 h). In 90-d feeding trials male albino rats receiving ≤ 15 mg/kg diet no significant effect was observed on weight gain, but there was a reduction in erythrocyte and brain cholinesterase activity.

Formulations. These include: e.c. (250 and 750 g a.i./l); granules (20-100 g/kg).

Analysis. Product analysis is by u.v. spectroscopy. Residues may be determined by GLC (N. R. Pasarela & E. J. Orloski, *Anal. Methods Pestic. Plant Growth Regul.,* 1973, **7**, 231). See also: R. C. Blinn & J. E. Boyd, *J. Assoc. Off. Agric. Chem.,* 1964, **47**, 1106.

Mercuric chloride

HgCl₂

.HG..G2   Cl₂Hg   (271·5)

Nomenclature and development. The traditional chemical name mercuric chloride is accepted by ISO in lieu of a common name. The IUPAC name is mercury dichloride, in *C.A.* usage mercury chloride (HgCl₂) *[7487-94-7];* it has also been called corrosive sublimate. It was first used for crop protection in 1891 by H. L. Bolley, *N.D. Agric. Exp. Stn. Bull.,* No. 4, 1891. Owing to its high toxicity to mammals it has largely been replaced by mercurous chloride or other compounds.

Manufacture and properties. Produced by the direct combination of chlorine and mercury or by heating a mixture of mercury sulphate and sodium chloride and collecting the sublimate, it is a colourless crystalline powder, m.p. 277°, sublimes *c.* 300°, v.p. 1·4 x 10⁻⁴ mmHg at 35°, *d* 5·32. Its solubility at 20° is 69 g/l water, at 100° 613 g/l water; soluble in ethanol, diethyl ether, pyridine. It is unstable to alkalies which precipitate mercury chloride oxide; it is readily reduced chemically or by sunlight to dimercury dichloride and to metallic mercury.

Uses. Mercuric chloride is a general poison. Being strongly phytotoxic, its use in crop protection is limited to soil application. It was, at one time, popular for the control of potato scab, club root of brassicas and as an insecticide against various root maggots of crucifers.

Toxicology. The acute oral LD50 for rats is 1-5 mg/kg.

Formulations. These include: w.p. (mixture with mercurous chloride); dust (mixture with mercurous chloride and malchite green).

Analysis. Product analysis is by AOAC Methods; see also *CIPAC Handbook,* in press.

Mercuric oxide

HgO

. HG . . O   HgO   (216·6)

Nomenclature and development. The traditional chemical name mercuric oxide is accepted by ISO in lieu of a common name. The IUPAC name is mercury oxide, in *C.A.* usage mercury oxide (HgO) *[21908-53-2]* formerly *[1344-45-2]*). The fungicidal action of mercury compounds has long been known. It is available under the trade mark 'Santar' (Sandoz AG).

Manufacture and properties. Produced by heating mercury in oxygen, it is a red powder (yellow when finely divided) which darkens *c.* 400° and decomposes into its constituent elements at *c.* 500°. It is insoluble in water, but dissolves in acids forming the corresponding mercury salt. It is decomposed by light.

Uses. Mercuric oxide is used as a paint against apple canker and as a protective seal on bark injuries and pruning cuts on fruit trees and ornamental shrubs and trees.

Toxicology. It is extremely poisonous orally to all animals.

Formulation. 'Santar', paint.

Analysis. Product analysis for mercury content is by standard methods (*CIPAC Handbook,* in press).

Mercurous chloride

Hg_2Cl_2

.HG2.G2 Cl_2Hg_2 (472·1)

Nomenclature and development. The traditional name mercurous chloride is accepted by ISO in lieu of a common name; also known as calomel. The IUPAC name is dimercury dichloride, in *C.A.* usage mercury chloride (Hg_2Cl_2) *[7546-30-7]*. Although long used as an insecticide, it first came into general use in 1929 as a substitute for mercuric chloride against root maggots of crucifers following the work of H. Glasgow, *J. Econ. Entomol.,* 1929, **22,** 335.

Manufacture and properties. It is produced by the direct chlorination of mercury or by heating a mixture of dimercury sulphate and sodium chloride, the distillate being washed free from mercury dichloride. It is a colourless powder, which sublimes at 400-500°, *d* 7·15. Its solubility at 18° is 2 mg/l water; soluble in cold dilute acids, ethanol, most organic solvents. In the presence of water, it slowly dissociates to form mercury and mercury dichloride, a reaction accelerated by alkali.

Uses. Mercurous chloride is limited to soil applications in crop protection use because of its phytotoxicity. It is recommended to control root maggots; club root of brassicas; white rot of onions; as a fungicide and moss-killer on turf.

Toxicology. The acute oral LD50 for rats is 210 mg/kg.

Formulations. Dust (40 g a.i./kg) on a non-alkaline carrier; also with lawn sand and ferrous sulphate; or with mercuric chloride (dust and w.p.).

Analysis. Product analysis is either by reacting with standard iodine solution and titrating the excess with sodium thiosulphate (AOAC Methods) or by oxidation with sulphuric acid and potassium nitrate titrating the mercuric ions with potassium thiocyanate (*CIPAC Handbook,* 1970, **1,** 514). Residues may be determined colorimetrically using the complex formed with dithizone.

Metalaxyl

1OVY1&NV1O1&R   B1   F1 $C_{15}H_{21}NO_4$ (279·3)

Nomenclature and development. The common name metalaxyl is approved by BSI and proposed by ISO. The IUPAC and *C.A.* name is methyl *N*-(2-methoxyacetyl)-*N*-(2,6-xylyl)-DL-alaninate, Registry No. *[57837-19-1]*. It was synthesised by Ciba-Geigy AG and developed under the code number 'GGA 48 988', trade mark 'Ridomil' and protected by BP 1 500 581. Its fungicidal properties were first described by F. Schwinn *et al., Mitt. Biol. Bundesanst. Land.-Forstwirtsch. Berlin-Dahlem,* 1977, **178,** 145; P. A. Urech *et al., Proc. 1977 Br. Crop Prot. Conf. Pests Dis.,* 1977, **2,** 623.

Properties. It is a crystalline solid, m.p. 71-72°, v.p. 2·2 x 10⁻⁶ mmHg at 20°. Its solubility at 20° is 7·1 g/l; soluble in most organic solvents.

Uses. It is a residual fungicide with systemic properties suitable for preventative and curative control of diseases caused by air- and soil-borne Oomycetes. It is recommended for use in potatoes at 200-300 g a.i./ha against *Phytophthora infestans,* in grape vines at 20-30 g/100 l against *Plasmopara viticola,* in tobacco against *Perenospora tabacina* and in hops against *Pseudoperonospora humuli* using 20-30 g metalaxyl + 40-60 g folpet/100 l or 20-30 g metalaxyl + copper fungicides at 50-70 g copper/100 l, and in lettuces against *Bremia lactucae* at 20 g/100 l HV or a soil treatment at 50 g/m³ compost.

Toxicology. The acute oral and dermal LD50 for rats are 669 mg/kg and > 3100 mg/kg, respectively. It is of negligible toxicity to wildlife.

Formulations. 'Ridomil 50 WP', w.p. (500 g a.i./kg); 'Ridomil 25 WP', w.p. (250 g/kg).

Analysis. Details of analytical methods are available from Ciba-Geigy AG.

Metaldehyde

```
Me—CH—O—CH.Me
    |         |
    O         O
    |         |
Me—CH—O—CH.Me
```

T8O CO EO GOTJ B1 D1 F1 H1 $C_8H_{16}O_4$ (176·2)

Nomenclature and development. The trivial name metaldehyde is accepted by BSI and ISO in lieu of a common name; also known as metacetaldehyde. The IUPAC name is 2,4,6,8-tetramethyl-1,3,5,7-tetraoxacyclo-octane (I), in *C.A.* usage (I) *[108-62-3]* or acetaldehyde homopolymer *[9002-91-9]*. Its slug-killing properties were first reported by G. W. Thomas, *Gard. Chron.,* 1936, **100,** 453.

Manufacture and properties. Produced by the polymerisation of acetaldehyde in ethanol in the presence of acid, it forms colourless crystals, m.p. (in sealed tube) 246°, it readily sublimes at 110-120°. Its solubility at 17° is 200 mg/l water; soluble in benzene, chloroform, sparingly soluble in ethanol, diethyl ether. It is subject to depolymerisation, hence a need for care in storage for which soldered tinplate is unsuitable. It is flammable. It does not give the tests for an aldehyde group.

Uses. Metaldehyde is a molluscicide, especially effective against slugs, death following immobilisation and exposure to low r.h. Neither acetaldehyde nor paraldehyde (a trimer) share the biological action.

Toxicology. The acute oral LD50 for dogs is 600-1000 mg/kg.

Formulations. Slug baits (25-40 g a.i./kg) in a protein-rich milling offal such as bran.

Analysis. Product analysis is either by hydrolysis, distilling the acetaldehyde into aqueous sodium hydrogen sulphite, treating the adduct with sodium hydrogen carbonate and estimating the liberated sulphite by titration with iodine (*CIPAC Handbook,* 1970, **1,** 532) or by heating with hydroxylamine hydrochloride and titrating the liberated acid with alkali (*MAFF Tech. Bull.,* No. 1, 1958, p.58).

Metamitron

T6NN DNVJ C1 DZ FR $C_{10}H_{10}N_4O$ (202·2)

Nomenclature and development. The common name metamitron is approved by BSI and New Zealand, and proposed by ISO. The IUPAC name is 4-amino-4,5-dihydro-3-methyl-6-phenyl-1,2,4-triazin-5-one, in *C.A.* usage 4-amino-3-methyl-6-phenyl-1,2,4-triazin-5(4*H*)-one *[41394-05-2]*. It was introduced in 1975 by Bayer AG under the code number 'BAY DRW 1139', trade mark 'Goltix' and protected by USP 3847914; BelgP 799854; BP 1368416. Its herbicidal properties were described by H. Hack and by R. R. Schmidt *et al.,* at the 3rd Int. Meeting of Selective Weed Control in Beet Crops, Paris, 1975.

Properties. It is a crystalline solid, m.p. 166·6°, v.p. 1 x 10⁻⁴ mmHg at 20-70°. Its solubility at 20° is: 1·8 g/l water; 10-50 g/kg cyclohexanone, dichloromethane. It is stable in acid media but unstable under strongly alkaline conditions.

Uses. Metamitron is a herbicide with a high selectivity to sugar and fodder beets in which crops it is used to control broad-leaved and grass weeds. It is generally recommended at 3·5-5 kg a.i./ha.

Toxicology. The acute oral LD50 is: for male rats 3343 mg/kg; for female rats 1832 mg/kg; for mice 1450-1463 mg/kg. The acute dermal LD50 for rats is > 1000 mg/kg. In 90-d feeding trials the 'no effect' level was: for rats 460 mg/kg diet; for dogs 500 mg/kg diet. The LC50 (96 h) for goldfish is > 100 mg/l; the LC50 for canaries is > 1000 mg/kg daily. It is not harmful to honey bees.

Formulation. A w.p. (700 g a.i./kg).

Analysis. Product analysis is by i.r. spectroscopy.

Methabenzthiazuron

T56 BN DSJ CN1&VM1 $C_{10}H_{11}N_3OS$ (221·3)

Nomenclature and development. The common name methabenzthiazuron is approved by BSI and ISO—exception USA. The IUPAC name is 1-(benzothiazol-2-yl)-1,3-dimethylurea, in *C.A.* usage *N*-2-benzothiazolyl-*N,N'*-dimethylurea *[18691-97-9]* formerly 1-(2-benzothiazolyl)-1,3-dimethylurea. It was introduced in 1968 by Bayer AG as an experimental herbicide under the code number 'Bayer 74283', trade mark 'Tribunil' and protected by BP 1085430. It was described by H. Hack, *Pflanzenschutz-Nachr. (Am. Ed.)*, 1969, **22**, 341.

Manufacture and properties. Produced by the reaction of 2-methylaminobenzothiazole with methyl isocyanate, it is a colourless crystalline solid, m.p. 119-120°, v.p. *c.* 1 x 10⁻⁶ mmHg at 20°. Its solubility at 20°: is 59 mg/l water; 116 g/l acetone; *c.* 100 g/l dimethylformamide; 65·9 g/l methanol.

Uses. Methabenzthiazuron is a selective herbicide recommended for the control of blackgrass, meadow grass and certain other annual weeds pre-em. or early post-em. in winter cereals at 2·0-3·5 kg a.i./ha, in spring wheat and oats at 1·75-2·0 kg/ha, in broad beans and peas at 2·0-2·8 kg/ha. Safety to winter barley varieties and timing of applications have been reviewed (W. Kolbe, *ibid.*, 1974, **27**, 90; H. Hack, *ibid.*, p.167). It is not recommended on crops under-sown with clover. The soil should be moist at the time of treatment for methabenzthiazuron is absorbed mainly through the roots; weeds die within 14-20 d after treatment but root-propagated weeds survive. It is broken down in the soil and has no after-effects on subsequent crops.

Toxicology. The acute oral LD50 is: for male rats > 2500 mg/kg; for rabbits, cats, dogs and female mice > 1000 mg/kg; for guinea-pigs > 2500 mg/kg. The acute dermal LD50 for female rats is > 500 mg/kg. There was no symptom of skin injury to the forearms of human volunteers. In 90-d feeding trials rats receiving 150 mg/kg diet showed no ill-effect.

Formulations. These include: w.p. (700 g a.i./kg); also in combination with dichlorprop or other phenoxyalkanoic acids.

Analysis. Product analysis is by TLC, measuring the a.i. by u.v. spectrometry at 267 nm. Residues may be determined, after column chromatographic clean-up, by GLC with FID (H. J. Jarczyk, *ibid.*, 1972, **25**, 21; T. H. Byast *et al., Agric. Res. Counc. (G.B.) Weed Res. Organ., Tech. Bull.*, No. 15 (2nd Ed.), p.49) or by HPLC (T. H. Byast *et al., ibid.*).

Methamidophos

1SPZO&O1 $C_2H_8NO_2PS$ (141·1)

Nomenclature and development. The common name methamidophos is approved by BSI, ISO and ANSI—exception Canada; also known as ENT 27 396. The IUPAC and *C.A.* name is *O,S*-dimethyl phosphoramidothioate, Registry No. *[10265-92-6]*. It was introduced in 1969 by the Chevron Chemical Co. under the code number 'Ortho 9006' and trade mark 'Monitor', and by Bayer AG under the code numbers 'Bayer 71 628' and 'SRA 5172', and trade mark 'Tamaron'. Its insecticidal uses are protected by USP 3 309 266; DBP 1 210 835.

Manufacture and properties. Produced by the isomerisation of *O,O*-dimethyl phosphoramidothioate, it is a low melting solid, off-white in the technical grade with a pungent odour. The pure compound has m.p. 44·5°, v.p. 3 x 10⁻⁴ mmHg at 30°, n_D^{40} 1·5092, $d^{44.5}$ 1·31. It is readily soluble at room temperature in water, ethanol; 20-25 g/l chloroform, dichloromethane, diethyl ether; < 10 g/l kerosine; < 100 g/l benzene, xylene. It is stable at ambient temperature but at pH 9 and 37°, 50% is decomposed in 120 h; and at pH 2 and 40° in 140 h. It is not distillable. The technical compound and concentrates are slightly corrosive to mild steel and copper-containing alloys; concentrates should not be stored in mild steel containers. It is compatible with most pesticides including oil sprays, but mixing with alkaline sprays such as lime sulphur or Bordeaux mixture should be avoided.

Uses. Methamidophos is an insecticide and acaricide effective against a broad range of insect pests on many crops (I. Hammann, *Pflanzenschutz-Nachr. (Am. Ed.)*, 1970, **23**, 140). It is systemic in action when applied to the base or trunk of deciduous trees; defoliation has occurred when applied as a foliage spray to deciduous fruit. It is especially useful on brassica crops, cotton, head lettuce and potatoes. At 0·5-1·0 kg/ha, its contact effectiveness persists for 7-21 d.

Toxicology. The acute oral LD50 is: for rats and mice 30 mg/kg; for guinea-pigs 30-50 mg/kg; for rabbits 10-30 mg/kg; for hens 25 mg/kg; for bobwhite quail 57·5 mg/kg. The acute dermal LD50 for male rats is 50-110 mg/kg. In 2-y feeding trials: dogs receiving 0·75 mg/kg daily showed no significant abnormalities; rats receiving 10 mg/kg diet showed no effect. The LC50 (96 h) is: for trout 51 mg/l; for guppies 46 mg/l.

Formulations. These include: 'Monitor Spray Concentrate', water miscible conc. (717 g/l); 'Tamaron', w.s.c. (600 g/l).

Analysis. Product analysis is by i.r. spectroscopy or by GLC. Residues, after liquid-liquid partition clean-up, may be measured by GLC with FID (E. Möllhoff, *ibid.*, 1971, **24**, 252). Details of these methods are obtainable from Chevron Chemical Co. and from Bayer AG. See also: J. B. Leary, *Anal. Methods Pestic. Plant Growth Regul.*, 1973, **7**, 339.

Metham-sodium

(Metam-sodium)

$$\underset{\text{MeNH.C.S}^-}{\overset{\overset{\displaystyle S}{\|}}{}}\quad Na^+$$

SUYS&M1 &-NA- $C_2H_4NNaS_2$ (129·2)

Nomenclature and development. The common name metham is approved by BSI, spelling in Canada, France and New Zealand (metam) for methyldithiocarbamic acid (I), in *C.A.* usage methylcarbamodithioic acid *[144-54-7]* formerly (I); ISO defined the common name metam-sodium for the corresponding sodium salt, *C.A.* Registry No. *[137-42-8]*—exceptions, those listed above, USSR (karbation, for sodium salt), WSSA (metham, for the sodium salt), JMAF (carbam, for both the ammonium and sodium salts). The fungicidal properties of metham-sodium were first described by H. L. Klöpping, Thesis, University of Utrecht, 1951; A. J. Overman & D. S. Burgis, *Proc. Fla. St. Hortic. Soc.,* 1956, **69,** 250. It was introduced in 1955 by Stauffer Chemical Co. under the code number 'N-869' and trade mark 'Vapam'; other trade marks include *'VPM'* (formerly by E. I. du Pont de Nemours & Co. (Inc.)) and 'Sistan', E. I. du Pont de Nemours & Co. (Inc.). It is protected by BP 789690; USP 2766554; 2791605.

Manufacture and properties. Produced by the reaction of methylamine with carbon disulphide in the presence of sodium hydroxide, metham-sodium forms a colourless crystalline dihydrate, *C.A.* Registry No. *[6734-80-1]*. Its solubility at 20° is 722 g dihydrate/l water; moderately soluble in ethanol; practically insoluble in most organic solvents. It is stable in concentrated aqueous solution, but unstable to dilution, decomposition being promoted by soil, acids and heavy metal salts. It is corrosive to brass, copper and zinc.

Uses. Metham-sodium is a soil fungicide, nematicide and herbicide with a fumigant action, applied at *c.* 1100 l 32·7% solution in 2000- 12000 l water/ha. Its activity is due to decomposition to methyl isothiocyanate (p.356). It is phytotoxic and the planting in treated soil must be delayed until decomposition and aeration are complete as shown by the normal germination of cress seed sown on a sample of treated soil (P. M. Smith & C. R. Worthing, *Insecticide and Fungicide Handbook,* 5th Ed., p.253). Under moist conditions this occurs within 14 d.

Toxicology. The acute oral LD50 is: for male albino rats 820 mg/kg; for albino mice 285 mg/kg. The acute dermal LD50 for rabbits is 800 mg/kg. It is an irritant to the eyes, skin and mucous membranes; any injury should be treated as a burn.

Formulations. Solutions of metham-sodium are available under various trade marks.

Analysis. Product analysis is by acid hydrolysis to liberate carbon disulphide, which is converted to dithiocarbonate and measured by titration with iodine (*CIPAC Handbook,* 1970, **1,** 537), or by measurement of the yellow colour produced on reaction with dilute aqueous copper dichloride (R. A. Gray & H. G. Streim, *Phytopathology,* 1962, **52,** 734). Residues in soil may be determined by conversion to 1-methyl(thiourea) by treatment with ammonia. See also: R. A. Gray, *Anal. Methods Pestic., Plant Growth Regul. Food Addit.,* 1964, **3,** 177; *Anal. Methods Pestic. Plant Growth Regul.,* 1972, **6,** 717.

Methazole

T5NOVNVJ   AR   CG   DG&   D1 $C_9H_6Cl_2N_2O_3$ (261·1)

Nomenclature and development. The common name methazole is approved by BSI, ANSI and WSSA. The IUPAC and *C.A.* name is 2-(3,4-dichlorophenyl)-4-methyl-1,2,4-oxadiazolidine-2,5-dione, Registry No. *[20354-26-1]*. It was introduced in 1968 by Velsicol Chemical Corp. under the code number 'VCS-438'. Its general herbicidal properties were described by W. Furness, *Proc. Congr. Int. Prot. Plant., 7th,* Paris, 1970, p.314; and, with special reference to weed control in onions (*idem, C. R. Journ. Etud. Herbic., Conf. COLUMA, 6th,* 1971, p.449). Trade marks include: 'Probe' (Velsicol Chemical Corp.); 'Paxilon' (Fisons Ltd).

Manufacture and properties. Produced (USP 3 437 664) by the cyclisation, with a haloformate, of 1-(3,4-dichlorophenyl)-1-hydroxy-3-methylurea, which is formed by the reaction of *N*-(3,4-dichlorophenyl)hydroxylamine with methyl isocyanate, the technical product is a light tan solid, $\geqslant 95\%$ pure, m.p. 123-124°. Its solubility at 25° is: 1·5 mg/l water; 55 g/l xylene; more soluble in other organic solvents.

Uses. Methazole is a selective herbicide recommended for the control of certain grasses and many broad-leaved weeds when applied pre-em. in cotton, garlic and potatoes at $\leqslant 6$ kg a.i./ha; as a directed spray on to soil or emerged weeds in established vines, tea, stone-fruits, citrus and nuts at < 10 kg/ha; as a directed spray in cotton 15-cm tall at $\leqslant 6$ kg/ha. It is also used for weed control in onions which have passed the 2-leaf stage at $\leqslant 2·5$ kg/ha and in winter dormant lucerne more than 1-y-old at $\leqslant 2·25$ kg/ha. It is preferably applied to moist soil.

Toxicology. The acute oral LD50 for albino rats is 1350 mg tech./kg; the acute dermal LD50 for albino rabbits is $\geqslant 900$ mg a.i. (as w.p.)/kg. It is mildly irritating to the skin and moderately to the eyes. In 90-d feeding trials rats and dogs receiving 50 mg/kg diet suffered no significant gross or histopathological changes. The LC50 (96 h) for goldfish and rainbow trout is 3 mg/l.

Formulations. A w.p. (750 g a.i./kg); granules (50 g/kg).

Analysis. Product analysis is by i.r. spectroscopy at 13·25 μm. Residues may be determined by HPLC or by GLC with ECD (D. M. Whiteacre *et al., Anal. Methods Pestic. Plant Growth Regul.,* 1978, **10**, 367).

Methidathion

$$\text{(MeO)}_2\text{P.S.CH}_2\text{—N} \cdots\cdots \text{(thiadiazolinone ring with OMe)}$$

T5NNVSJ B1SPS&O1&O1 EO1 $C_6H_{11}N_2O_4PS_3$ (302·3)

Nomenclature and development. The common name methidathion is approved by BSI, ISO and ANSI—exception JMAF (DMTP); also known as ENT 27193. The IUPAC name is S-2,3-dihydro-5-methoxy-2-oxo-1,3,4-thiadiazol-3-ylmethyl O,O-dimethyl phosphorodithioate, in *C.A.* usage S-[(5-methoxy-2-oxo-1,3,4-thiadiazol-3(2*H*)-yl)methyl] O,O-dimethyl phosphorodithioate *[950-37-8]* formerly O,O-dimethyl phosphorodithioate S-ester with 4-(mercaptomethyl)-2-methoxy-Δ²-1,3,4-thiadiazolin-5-one. It was introduced in 1966 by J. R. Geigy S.A. (now Ciba-Geigy AG) under the code number 'GS 13005', the trade marks 'Supracide' and 'Ultracide Ciba-Geigy' and protected by SwissP 392521; 395637; BP1008451; USP 3230230; 3240668; FP 1335755. Its insecticidal properties were described by H. Grob *et al., Proc. Br. Insectic. Fungic. Conf., 3rd,* 1965, p.451.

Manufacture and properties. Produced by the treatment of 5-methoxy-1,3,4-thiadiazol-2-one with formaldehyde, subsequent conversion to a 3-halogenomethyl derivative by thionyl chloride or phosphorous tribromide and reaction with O,O-dimethyl hydrogen phosphorodithioate, it forms colourless crystals, m.p. 39-40°, v.p. 1 x 10⁻⁶ mmHg at 20°. Its solubility at 25° is 240 mg/l water; readily soluble in most organic solvents. It is hydrolysed by alkali.

Uses. Methidathion is a non-systemic insecticide with some acaricidal activity. It controls a wide range of sucking and leaf-eating insects with a specific use against scale insects. The application rate for sucking insects is 30-60 g a.i./100 l on fruit, 250-800 g/ha on field crops. Foliar penetration enables it to be used against leafrollers. It is non-phytotoxic to all plants tested. It is rapidly metabolised by plants and animals and excreted by the latter (H. O. Esser & P. W. Müller, *Experientia,* 1966, **22,** 36; H. O. Esser *et al., Helv. Chim. Acta,* 1968, **51,** 513).

Toxicology. The acute oral LD50 is: for rats 25-54 mg/kg; for mice 25-70 mg/kg. The acute dermal LD50 for rats is 1546-1663 mg/kg. In 2-y feeding trials the 'no effect' level was: for rats 0·2 mg/kg daily; for rhesus monkeys 0·25 mg/kg daily. It is toxic to birds and honey bees; highly toxic to fish.

Formulations. These include: 'Supracide 40'/'Ultracide 40 Ciba-Geigy', e.c. (400 g a.i./l) and w.p. (400 g/kg); 'Supracide 20'/'Ultracide 20 Ciba-Geigy', e.c. (200 g/l) and w.p. (200 g/kg); 'Supracide'/'Ultracide' Ulvair 250; 'Supracide'/'Ultracide' combi 40 EC, e.c. (150 g methidathion + 250 g DDT/l).

Analysis. Product analysis is by GLC or by titration with perchloric acid in the presence of mercury diacetate. Residues may be determined by TLC and GLC (D. O. Eberle & W. D. Hörmann, *J. Assoc. Off. Anal. Chem.,* 1971, **54,** 150; D. O. Eberle *et al., J. Agric. Food Chem.,* 1967, **15,** 213; G. Dupuis *et al., J. Econ. Entomol.,* 1971, **64,** 588). Details of unpublished methods are available from Ciba-Geigy AG. See also: D. O. Eberle & R. Suter, *Anal. Methods Pestic. Plant Growth Regul.,* 1976, **8,** 141.

Methiocarb

(Mercaptodimethur)

1SR B1 F1 DOVM1 $C_{11}H_{15}NO_2S$ (225·3)

Nomenclature and development. The common name methiocarb is approved by BSI, New Zealand, the Republic of South Africa, Turkey and ESA, the common name mercapto-dimethur has been approved by ISO—exceptions the countries listed above that use methiocarb, and Canada; also known as ENT 25726, OMS 93. The IUPAC name is 4-methylthio-3,5-xylyl methylcarbamate (I) also known as 3,5-dimethyl-4-(methylthio)-phenyl methylcarbamate (II), in *C.A.* usage (II) *[2032-65-7]* formerly (I). It was introduced in 1962 by Bayer AG under the code numbers 'Bayer 37344', 'H321', the trade marks 'Mesurol' and 'Draza' and protected by FP 1275658; DBP 1162352. Its insecticidal properties were first described by G. Unterstenhöfer, *Pflanzenschutz-Nachr. (Am. Ed.),* 1962, **15,** 181.

Manufacture and properties. Produced by the reaction of 4-methylthio-3,5-xylenol with methyl isocyanate, it is a colourless crystalline powder, m.p. 117-118°, with negligible v.p. at room temperature. It is practically insoluble in water; soluble in most organic solvents. It is hydrolysed by alkali.

Uses. Methiocarb is a non-systemic insecticide and acaricide with a broad range of action and good residual activity, used at 50-100 g a.i./100 l. It is also a powerful molluscicide used as pellets at 200 g a.i./ha; and bird repellent when used as a seed dressing (G. Hermann & W. Kolbe, *ibid.,* 1971, **24,** 279).

Toxicology. The acute oral LD50 is: for male rats 100 mg/kg; for guinea-pigs 40 mg/kg. The acute dermal LD50 for male rats is 350-400 mg/kg. In 1·67-y feeding trials no symptom was noted in rats receiving 100 mg/kg diet.

Formulations. These include: 'Mesurol', w.p. (500 and 750 g a.i./kg); dust (30 g/kg); pellets for slug control (40 g/kg).

Analysis. Product analysis is by u.v. spectroscopy. Residues may be determined by u.v. spectroscopy after hydrolysis to give 4-methylthio-3,5-xylenol, measuring the difference between alkaline and acid spectra at 266 nm in aqueous methanol (H. Niessen & H. Frehse, *ibid.,* 1963, **16,** 205). Residues of the parent compound and some of its metabolites, including the corresponding sulphoxide and sulphone, may be determined by GLC (M. C. Bowman & M. Beroza, *J. Assoc. Off. Anal. Chem.,* 1969, **52,** 1054).

Methomyl

MeS,   O
             ‖
    C = N—OCNHMe
Me

1SY1&UNOVM1 $C_5H_{10}N_2O_2S$ (162·2)

Nomenclature and development. The common name methomyl is approved by BSI, ISO, ANSI and JMAF. The IUPAC name is *S*-methyl *N*-(methylcarbamoyloxy)thioacetimidate, in *C.A.* usage methyl *N*-[[(methylamino)carbonyl]oxy]ethanimidothioate *[16752-77-5]* formerly methyl *N*-[(methylcarbamoyl)oxy]thioacetimidate. It was introduced in 1966 as an experimental insecticide and nematicide by E. I. du Pont de Nemours & Co. (Inc.) under the code number 'Du Pont 1179', trade mark 'Lannate' and protected by USP 3576834; 3639633. It was first described in Europe by G. A. Roodhans & N. B. Joy, *Meded. Rijksfac. Landbouwwet. Gent.*, 1968, **33**, 833.

Manufacture and properties. Produced by the reaction of methyl *N*-hydroxythioacetimidate with methyl isocyanate, it is a colourless crystalline solid, with a slight sulphurous odour, m.p. 78-79°, v.p. 5×10^{-5} mmHg at 25.°. Its solubility at 25° is: 58 g/l water; 730 g/kg acetone; 420 g/kg ethanol; 220 g/kg propan-2-ol; 1 kg/kg methanol; 30 g/kg toluene. The aqueous solution is non-corrosive. It is stable as a solid, in aqueous solution under normal conditions but is subject to decomposition in moist soil.

Uses. Methomyl is used as a foliar treatment for control of many insects including aphids, armyworms, cabbage looper, tobacco budworm, tomato fruitworm, cotton leaf perforator, cotton bollworm. Following soil treatment it is taken up by roots and translocated, controlling certain foliage insects, but this method is generally less effective than foliar treatment.

Toxicology. The acute oral LD50 for white rats is: 17-24 mg a.i./kg; 31 mg a.i. (as 240 g/l liquid)/kg; 47 mg a.i. (as w.p.)/kg. The acute dermal LD50 for rabbits is: > 5000 mg a.i./kg; 4080 mg a.i. (as 240 g/l liquid)/kg. The sub-acute LC50 (8 d) was: for Pekin duck 1890 mg/kg diet; for bobwhite quail 3680 mg/kg diet. The LC50 (96 h) is: for rainbow trout 3·4 mg/l; for bluegill 0·87 mg/l; for goldfish > 0·1 mg/l. No adverse effects were noted when field-caged bobwhite quail and albino rabbits (with food and water exposed) were oversprayed 6 times (at 5-d intervals) with 1 kg a.i. (as w.s.p.) in 280 l water/ha. Relatively non-toxic to honey bees once the spray has dried.

Formulations. These include: 'Lannate' insecticide, w.s.p. (900 g a.i./kg); 'Lannate' 25 w.p. insecticide, w.p. (250 g/kg); 'Lannate L' insecticide, water-soluble liquids (200, 240 and 360 g/kg).

Analysis. Residues are determined by HPLC (J. E. Thean *et al., J. Assoc. Off. Anal. Chem.*, 1978, **61**, 15) or by GLC (H. L. Pease & J. J. Kirkland, *J. Agric. Food Chem.*, 1968, **16**, 554; I. H. Williams, *Pestic. Sci.*,1972, **3**, 179). Use of FPD instead of MCD gives improved results (details available from E. I. du Pont de Nemours & Co. (Inc.)). See also: R. E. Leitch & H. L. Pease, *Anal. Methods Pestic. Plant Growth Regul.*, 1973, **7**, 331.

Methoprotryne

T6N CN ENJ BS1 DMY1&1 FM3O1

$C_{11}H_{21}N_5OS$ (271·4)

Nomenclature and development. The common name methoprotryne is approved by BSI and ISO, in France (metoprotryne). The IUPAC name is 2-isopropylamino-4-(3-methoxy-propylamino)-6-methylthio-1,3,5-triazine, in *C.A.* usage *N*-(3-methoxypropyl)-*N'*-(1-methylethyl)-6-(methylthio)-1,3,5-triazine-2,4-diamine *[841-06-5]* formerly 2-(isopropyl-amino)-4-[(3-methoxypropyl)amino]-6-(methylthio)-*s*-triazine. It was introduced in 1965 by J. R. Geigy S.A. (now Ciba-Geigy AG) under the code number 'G 36 393', trade mark 'Gesaran' and protected by SwissP 394 704; BP 927 348; USP 3 326 914. Its herbicidal properties were first reported by A. Gast *et al.*, *Z. Pflanzenkr. Pflanzenpathol. Pflanzenschutz*, 1965, **72**, 325.

Manufacture and properties. Produced either by reacting 2-chloro-4-isopropylamino-6-(3-methoxypropylamino)-1,3,5-triazine with methanethiol in the presence of an equivalent of sodium hydroxide, or by reacting 2-isopropylamino-4-mercapto-6-(3-methoxy-propylamino)-1,3,5-triazine with a methylating agent in the presence of sodium hydroxide, it is a crystalline solid, m.p. 68-70°, v.p. 2·85 x 10^{-7} mmHg at 20°. Its solubility at 20° is 320 mg/l water; soluble in most organic solvents.

Uses. Methoprotryne is a herbicide used in combination with simazine post-em. for the control of weed grasses in winter-sown cereals at 2·5-4·0 kg 'Gesaran 2079'/ha.

Toxicology. The acute oral LD50 is: for rats > 5000 mg/kg; for mice 2400 mg/kg. In 90-d feeding trials rats receiving 60 mg/kg daily suffered no ill-effect; 300 mg/kg daily had marginal effects. Its toxicity to fish is low.

Formulation. 'Gesaran 2079', w.p. (225 g methoprotryne + 50 g simazine/kg).

Analysis. Product analysis is by GLC with internal standard (*CIPAC Handbook*, 1979, **1A**, in press) or by titration with perchloric acid in acetic acid (H. P. Bosshardt *et al.*, *J. Assoc. Off. Anal. Chem.*, 1971, **54**, 749). Residues may be determined by GLC, TLC or by u.v. spectroscopy (*cf.* ametryne). Details of these methods are available from Ciba-Geigy AG.

2-Methoxy-4*H*-benzo-1,3,2-dioxaphosphorin 2-sulphide

T66 BOPO EHJ CS CO1 $C_8H_9O_3PS$ (216·2)

Nomenclature and development. The common name salithion is approved by JMAF. The IUPAC name is 2-methoxy-4*H*-benzo-1,3,2-dioxaphosphorin 2-sulphide, in *C.A.* usage 2-methoxy-4*H*-benzo-1,3,2-dioxaphosphorin 2-sulfide *[3811-49-2]* formerly cyclic *O,O*-(methylene-*o*-phenylene)phosphorothioate *O*-methyl ester. It was introduced in 1962 by Sumitomo Chemical Co. Ltd under the trade mark 'Salithion' and protected by JapaneseP 467645; 446747; BP 987378; FP 1360130. Its properties were described by M. Eto & Y. Oshima, *Agric. Biol. Chem.*, 1962, **26**, 452.

Manufacture and properties. Produced by the reaction of salicyl alcohol with *O*-methyl phosphorodichloridothioate, it is a colourless crystalline solid, m.p. 55-56°, v.p. $4·7 \times 10^{-3}$ mmHg at 25°. Its solubility at 30° is 58 mg/l water; soluble in acetone, benzene, ethanol, diethyl ether; moderately soluble in cyclohexanone, 4-methylpentan-2-one, toluene and xylene. It is stable in weakly acidic or alkaline media.

Uses. It is an insecticide effective at 25 g a.i./100 l against a wide range of pests of fruit, rice, vegetable and fibre crops, particularly against rice gall midge and parathion-resistant cotton bollworm.

Toxicology. The acute oral LD50 is: for rats 180 mg/kg; for mice 91·3 mg/kg. The LC50 (48 h) is: for carp 3·55 mg/l; for goldfish 2·8 mg/l.

Formulations. These include: e.c. (250 g a.i./l); w.p. (250 g/kg); granules (50 and 100 g/kg).

Analysis. Product analysis is by u.v. spectroscopy at 505 nm after purification by TLC. Residues may be determined by GLC with FPD; details are available from Sumitomo Chemical Co. Ltd. See also: M. Eto & J. Miyamoto, *Anal. Methods Pestic. Plant Growth Regul.*, 1973, **7**, 431.

Methoxychlor

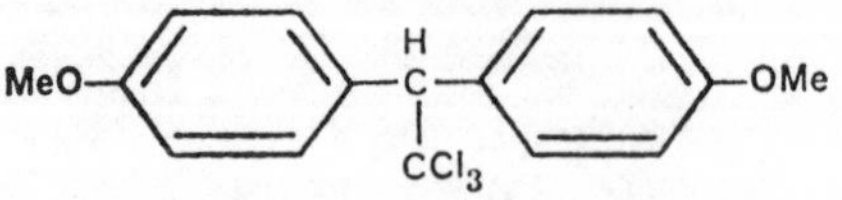

MeO—⟨benzene ring⟩—CH(CCl₃)—⟨benzene ring⟩—OMe

The structure shows two methoxy-substituted benzene rings connected by a central carbon bearing H and CCl₃ group.

GXGGYR DO1&R DO1 $C_{16}H_{15}Cl_3O_2$ (345·7)

Nomenclature and development. The common name methoxychlor is approved by BSI and ISO; also known as ENT 1716, DMTD. The IUPAC name is 1,1,1-trichloro-2,2-bis(4-methoxyphenyl)ethane formerly 1,1,1-trichloro-2,2-di(4-methoxyphenyl)ethane, in *C.A.* usage 1,1'-(2,2,2-trichloroethylidene)bis[4-methoxybenzene] *[72-43-5]* formerly 1,1,1-trichloro-2,2-bis(*p*-methoxyphenyl)ethane. Its insecticidal properties were first reported by P. Läuger *et al., Helv. Chim. Acta,* 1944, **27**, 892. It was introduced *c.* 1945 by J. R. Geigy AG (now Ciba-Geigy AG, who no longer manufacture or market it) and protected by SwissP 226 180; BP 547 871; and by E. I. du Pont de Nemours & Co. (Inc.) under the trade mark 'Marlate' and protected by USP 2 420 928; 2 477 655; BP 624 561.

Manufacture and properties. Produced by condensing trichloroacetaldehyde with anisole, the technical product, setting point 77°, is a grey flaky powder, d^{25} 1·41, and contains *c.* 88% of the *pp'*-isomer, the bulk of the remainder being the *op'*-isomer (*c.f.* DDT, p. 150). It is practically insoluble in water; moderately soluble in ethanol, petroleum oils; readily soluble in most aromatic solvents. The pure *pp'*-isomer forms colourless crystals, m.p. 89°. It is resistant to heat and oxidation. It is susceptible to dehydrochlorination by heavy metal catalysts but less readily by alcoholic alkali.

Uses. Methoxychlor is a non-systemic contact and stomach insecticide with little aphicidal or acaricidal activity. Its range of activity coincides with that of DDT. It shows little or no tendency to be stored in the body fat or to be excreted in milk. It is therefore recommended for fly control in dairy barns and on lactating dairy cows (F. Kunze *et al., Fed. Proc.,* 1950, **9**, 293).

Toxicology. The acute oral LD50 for rats is 6000 mg/kg. In 2-y feeding trials rats receiving 200 mg/kg diet suffered no ill-effect, those receiving 1600 mg/kg diet showed a reduction in growth (H. C. Hodge *et al., J. Pharmacol. Exp. Ther.,* 1950, **99**, 140; 1952, **104**, 60).

Formulations. These include: w.p. (600 g a.i./kg); e.c. (240 g/l); dusts; aerosol concentrates.

Analysis. Product analysis is by total chlorine content after separation of the a.i. by TLC (H. L. Pease, *J. Assoc. Off. Anal. Chem.,* 1976, **58**, 40). Residues may be determined by GLC (J. Solomon & W. L. Lockhart, *ibid.,* 1978, **60**, 690). See also: W. K. Lowen *et al., Anal. Methods Pestic., Plant Growth Regul. Food Addit.* 1964, **2**, 303; *Anal. Methods Pestic. Plant Growth Regul.* 1972, **6**, 441.

4-Methoxy-3,3'-dimethylbenzophenone

1OR B1 DVR C1 $C_{16}H_{16}O_2$ (240·3)

Nomenclature and development. The IUPAC name is 4-methoxy-3,3'-dimethylbenzo-phenone (I), in *C.A.* usage (4-methoxy-3-methylphenyl)(3-methylphenyl)methanone *[41295-28-7]* formerly (I). It was introduced in 1971 by Nippon Kayaku Co. Ltd under the code number 'NK-049', trade mark 'Kayametone' and protected by JapaneseP 51-5446; BP 1 355 926; USP 3 873 304. Its herbicidal properties were first described in the *Proc. 4th Asian-Pacific Weed Sci. Soc. Conf., N.Z.,* 1973, p.215.

Manufacture and properties. Produced by the Friedel-Crafts reaction of *m*-toluoyl chloride with 2-methylanisole, it forms fine colourless crystals, m.p. 62·0-62·5°. Its solubility at 20° is 2 mg/l water; soluble in most organic solvents. It is stable under acid and alkaline conditions but is slowly decomposed by sunlight.

Uses. It is a selective pre-em. herbicide effective against annual grasses and broad-leaved weeds in paddy rice and vegetable crops at 3-5 kg a.i./ha; also used in rice pre-em. in combination with bensulide at 30-50 kg granules/ha. It produces chlorosis and inhibits photosynthesis. It is biodegradable, leaving no residue problems.

Toxicology. The acute oral LD50 for rats and mice is > 4000 mg/kg. In 90-d feeding trials the 'no effect' level was: for rats 1500 mg/kg diet; for mice 1000 mg/kg diet. The LC50 (48 h) is: for carp 3·2 mg/l; for goldfish 10 mg/l.

Formulations. These include: 'Kayametone', w.p. (500 g a.i./kg) for use on vegetables and paddy rice; 'Kayaphenone', granules (80 g 4-methoxy-3,3'-dimethylbenzophenone + 30 g bensulide/kg).

Analysis. Product and residue analysis is by GLC.

2-Methoxyethylmercury chloride

MeO(CH$_2$)$_2$.HgCl

G-HG-2O1   C$_3$H$_7$ClHgO   (295·1)

Nomenclature and development. The IUPAC name 2-methoxyethylmercury chloride is accepted by BSI, New Zealand and the Republic of South Africa in lieu of a common name. Its *C.A.* name is chloro(2-methoxyethyl)mercury *[123-88-6]*. It was introduced *c.* 1928 by I. G. Farbenindustrie AG (W. Bonrath, *Nachr. Schaedlingsbekaempfung I. G. Farbenindustrie (Leverkusen)*, 1935, **10**, 23) under the trade marks 'Agallol', 'Aretan', 'Ceresan Universal Nassbeize' (Bayer AG).

Manufacture and properties. Produced by the action of ethylene on a solution of mercury diacetate in methanol giving 2-methoxyethylmercury acetate (W. Schoeller *et al., Ber. Dtsch. Chem. Ges.*, 1913, **46**, 2864) from which the chloride is produced by precipitation from a solution of a soluble chloride, it is a colourless crystalline powder, m.p. 65°, v.p. 1 x 10^{-3} mmHg at 35° (G. F. Phillips *et al., J. Sci. Food Agric.*, 1959, **10**, 604). Its solubility at room temperature is 50 g/l water; readily soluble in acetone, ethanol. 2-Methoxyethylmercury salts, though stable to alkali, are decomposed by hydrochloric acid to ethylene, methanol and the mercury(II) salt. (J. Chatt, *Chem. Rev.*, 1951, **48**, 7).

Uses. 2-Methoxyethylmercury chloride is effective against the major seed-borne diseases of cereals and as a dip for seed potato against *Phoma solanicola* f. *foveata.*

Toxicology. The acute oral LD50 is: for rats 30 mg/kg; for mice 47 mg/kg.

Formulations. These include: 'Ceresan Universal Feuchtbeize' (20 g mercury/kg); 'Ceresan Universal Nassbeize' (25 and 36 g mercury/kg) for use as a slurry seed dressing; 'Agallol' (30 g mercury/kg) for potato dips; 'Aretan 6' (60 g mercury/kg) for dips.

Analysis. Product analysis is by digestion with sulphuric and nitric acids, dilution and decomposition of nitrous acid by urea, and titration with potassium iodide. Particulars may be obtained from Bayer AG. See also: *CIPAC Handbook*, 1970, **1**, 503.

2-Methoxyethylmercury silicate

MeOCH₂CH₂Hg.silicate

Nomenclature and development. The trivial name 2-methoxyethylmercury silicate is accepted in lieu of a common name by New Zealand and the Republic of South Africa. The compound is an organomercury silicate but there is some doubt about whether it is an orthosilicate or a metasilicate and therefore about its true molecular composition. It was introduced in 1935 by I. G. Farbenindustrie AG under the trade mark 'Ceresan Universal Trockenbeize' (Bayer AG).

Manufacture and properties. Produced by precipitation from solutions of a soluble silicate and 2-methoxyethylmercury acetate, which is obtained by the action of ethylene on a solution of mercury diacetate in methanol (W. Stoeller *et al., Ber. Dtsch.Chem. Ges.,* 1913, **46**, 2864), it is a colourless crystalline powder, v.p. $3 \cdot 3 \times 10^{-3}$ mmHg at 35°. It is practically insoluble in water. 2-Methoxyethylmercury salts, though stable to alkali, are decomposed by halogen acids to form ethylene, methanol and a mercury(II) salt (J. Chatt, *Chem. Rev.* 1951, **48**, 7).

Uses. 2-Methoxyethylmercury silicate is used in seed treatments against various seed-borne diseases of cereals.

Toxicology. The acute oral LD50 for rats is 1140 mg (formulated product)/kg.

Formulations. These include: 'Ceresan Universal Trockenbeize' (17·5 g Hg/kg); mixtures with hexachlorobenzene and anthraquinone ('Ceresan-Morkit'), with hexachlorobenzene, gamma-HCH and anthraquinone ('Ceresan Gamma M') and with hexachlorobenzene.

Analysis. Product analysis is by digestion with sulphuric and nitric or perchloric acids, decomposition of nitrous acid by urea and measurement of mercury by titration with potassium iodide: details from Bayer AG. See also: *CIPAC Handbook,* 1970, **1**, 503.

Methylarsonic acid

$$\text{MeAs(OH)}_2 \quad (\text{with } =O)$$

Q-AS-QO&1	*methylarsonic acid* CH_5AsO_3	(140·0)
Q-AS-O&O&1 &-NA-	*MSMA* CH_4AsNaO_3	(162·0)
O-AS-O&O&1 &-NA-2	*DSMA* $CH_3AsNa_2O_3$	(183·9)

Nomenclature and development. The common name MAA is approved by WSSA. The IUPAC name is methylarsonic acid (I), in *C.A.* usage (I) *[124-58-3]* formerly methanearsonic acid. Several of its salts have been used since 1956 as herbicides. The common name MSMA is approved by WSSA for sodium hydrogen methylarsonate, in *C.A.* usage monosodium methylarsonate *[2163-80-6]* formerly monosodium methane-arsonate; WSSA and JMAF approve the name DSMA for the corresponding disodium salt Registry No. *[144-21-8]*; WSSA has also approved the names CMA for calcium bis(hydrogen methylarsonate) Registry No. *[5902-95-4]* and MAMA for ammonium hydrogen methylarsonate. MSMA and DSMA were introduced by Ansul Chemical Co. (which no longer exists) under the trade mark 'Ansar' their use being protected by USP 2 678 265; they are now produced by Diamond Shamrock Corp., MSMA having the trade marks 'Daconate' and 'Bueno'. CMA has the trade mark 'Calcar'.

Manufacture and properties. Methylarsonic acid is a strong dibasic acid, m.p. 161°; freely soluble in water, soluble in ethanol.

MSMA is produced by the reaction of diarsenic trioxide with methyl chloride and aqueous sodium hydroxide or by the methylation of sodium arsenite by dimethyl sulphate (USP 2 889 347) and forms a colourless crystalline hydrate, MSMA 1·5 H_2O, m.p. 113-116°. Its solubility at 20° is 1·4 kg (as anhydrous salt)/kg water; soluble in methanol; insoluble in most organic solvents.

DSMA is produced by the reaction of methyl chloride on sodium arsenite, and is a colourless crystalline solid, m.p., 132-139°, which slowly decomposes at elevated temperatures. The anhydrous salt absorbs water to form a stable hexahydrate. Its solubility at 20° is 279 g (as anhydrous salt)/kg water; soluble in methanol; practically insoluble in organic solvents.

MSMA and DSMA are stable to hydrolysis, MSMA being formed from DSMA at pH 6-7. They are decomposed by strong oxidising and reducing agents, are non-flammable and non-corrosive to iron, rubber and most plastics.

Uses. MSMA and DSMA are selective pre-em. contact herbicides with some systemic properties. MSMA is recommended for the control of grass weeds: in cotton at 2·24 kg MSMA in 150 l water/ha; in sugarcane at 3·3 kg in 200 l/ha; as directed sprays under tree crops almost to run-off at 460 g/100 l; and on non-crop areas at 0·55-1·1 kg/100 l. The addition of a surfactant is advisable for unformulated products; air temperatures should be > 21° for maximum effectiveness. DSMA is recommended for control of grass weeds in cotton at 2·5 kg in 375 l/ha; for turf treatment; and for weed control on uncropped land.

Toxicology. The acute oral LD50 for young albino rats is 900 mg MSMA/kg, 1800 mg DSMA/kg. Applications of methylarsonic acid are only mildly irritant to the skin of rabbits. In feeding trials dogs receiving 100 mg acid/kg diet showed no significant effect on weight or health. The LC50 (48 h) for bluegill is > 1000 mg MSMA or DSMA/l, but the addition of surfactants increases the toxicity.

continued

Methylarsonic acid—*continued*

Formulations. These include for MSMA: 'Ansar 529', 'Daconate' and 'Bueno', w.s.c. (480 g/l); 'Ansar 529 HC', 'Daconate 6' and 'Bueno 6' (720 g/l); 'Arsonate Liquid' (791 g/l); 'Super Arsonate' (959 g/l). For DSMA: 'Ansar 8100', w.s.p. (810 g/kg); 'Diamond Shamrock DSMA Liquid' (270 g/l).

Analysis. Product analysis is by acid-base titration; total arsenic is determined by a wet oxidation procedure (G. R. Robertson, *J. Am. Chem. Soc.,* 1921, **43,** 152). See also: E. A. Dietz & L. O. Moore, *Anal. Methods Pestic. Plant Growth Regul.,* 1978, **10,** 385.

Methyl bromide

CH_3Br

E1 CH_3Br (94·94)

Nomenclature and development. The traditional chemical name methyl bromide is accepted in lieu of a common name by BSI and ISO. The IUPAC and *C.A.* name is bromomethane, Registry No. *[74-83-9].* Its insecticidal activity was first reported in 1932 by Goupil, *Rev. Pathol. Veg. Entomol. Agric. Fr.,* 1932, **19**, 169.

Manufacture and properties. Produced by the action of hydrobromic acid on methanol, it is a colourless gas, b.p. 4·5°, forming a colourless liquid, with a chloroform-like odour, f.p. —93°, $d°$ 1·73. The specific heat of the liquid at 0° is 0·50 J/g. Its solubility at 25° is 13·4 g/kg water and it forms a crystalline hydrate with ice water; soluble in most organic solvents. It is stable, non-corrosive and non-flammable.

Uses. Methyl bromide has high insecticidal and some acaricidal properties and is used for space fumigation and for the fumigation of plants and plant products in stores, mills and ships. It is a soil fumigant used for the control of nematodes, fungi and weeds and to sterilise mushroom houses and compost. Residues of bromide ion and methyl bromide in soils fumigated with the latter and their effects are reviewed (G. A. Maw & R. J. Kempton, *Soils Fert.,* 1973, **36**, 41).

Toxicology. It is highly toxic to man, the threshold limit value is 65 mg/m³, above which concentration respirators must be worn: see *'Fumigations Using Methyl Bromide, Guidance Note',* 1976; *'Fumigation of Small Enclosures with Methyl Bromide',* 1972; *'Fumigation of Soil and Compost under Gas-proof Sheeting with Methyl Bromide',* 1973; *'Fumigation with Methyl Bromide under Gas-proof Sheets'* (3rd Ed.), 1974. In many countries its use is restricted to trained personnel.

Formulation. It is packed in glass ampoules (up to 50 ml), in metal cans and cylinders for direct use. Chloropicrin is sometimes added, up to 2%, as a warning agent.

Analysis. Methyl bromide may be detected by the halide lamp, a non-specific test given by all volatile halides; by commercially available detector tubes (H. K. Heseltine, *Pest. Tech.,* 1959, **1**, 253). Mixtures with chloropicrin, *e.g.* 'Dowfume MC-2' are analysed by i.r. spectroscopy, depending on the measurement of absorbance at 10·45 μm for methyl bromide and 6·2 μm for chloropicrin: details are available from The Dow Chemical Co. For its estimation in air a commercial thermal conductivity meter may be used (H. K. Heseltine *et al., Chem. Ind. (London),* 1958, p.1287); or GLC (H. J. Kolbezen & F. J. Abu-El Haj, *Pestic. Sci.,* 1972, **3**, 73). It reacts with soil or organic material to form bromide ion. Residues of methyl bromide and bromide ion may be distinguished (S. G. Heuser & K. A. Scudamore, *ibid.,* 1970, **1**, 244).

1,1'-Methylenedi(thiosemicarbazide)

$$\underset{\underset{H_2N.C.NH.NH.CH_2.NH.NH.C.NH_2}{\parallel}}{S} \qquad \underset{\parallel}{S}$$

SUYZMM1MMYZUS $C_3H_{10}N_6S_2$ (194·3)

Nomenclature and development. The IUPAC name is 1,1'-methylenedi(thiosemicarbazide), in *C.A.* usage 2,2'-methylenebis(hydrazinecarbothioamide) *[39603-48-0]*. It was introduced in 1971 by Nippon Kayaku Co. Ltd under the code number 'NK-15 561', and trade mark 'Kayanex'. Its rodenticidal properties were described in JapaneseP 50-20 126; BP 1 351 710; USP 3 826 841.

Manufacture and properties. Produced by the reaction of thiosemicarbazide with formaldehyde, it is a colourless crystalline powder, which decomposes at 171-174°. It is practically insoluble in water and in common organic solvents; soluble in dimethyl sulphoxide. It is slowly decomposed in water, more rapidly under acid or alkaline conditions.

Uses. It is a quick-acting rodenticide effective against *Rattus norvegicus, R. rattus* and the Japanese field vole, *Microtus montebelli*. Vitamin B_6 is an effective antidote.

Toxicology. The acute oral LD50 is: for mice 30·4-50 mg/kg; for guinea-pigs 32-36 mg/kg; for chickens 120-150 mg; for cats *c.* 150 mg/kg; for dogs > 1500 mg/kg. It induces vomiting in cats and dogs. In 90-d feeding trials the 'no effect' level was: for mice 100 mg/kg diet; for rats 50 mg/kg diet.

Formulations. Baits: (5, 10 and 20 g a.i./kg).

Analysis. Product analysis is by reaction with hot methanolic acetic acid to give thiosemicarbazide which is measured colorimetrically at 352 nm after conversion to 4-nitrobenzaldehyde thiosemicarbazone.

Methyl isothiocyanate

$$Me—N=C=S$$

SCN1 C_2H_3NS (73·11)

Nomenclature and development. The IUPAC name methyl isothiocyanate (I) is accepted in lieu of a common name by BSI and ISO, in *C.A.* usage isothiocyanatomethane *[556-61-6]* formerly (I). It was introduced in 1959 by Schering AG as a nematicide under the trade mark 'Trapex' and protected by USP 3 113 908.

Manufacture and properties. Produced by the reaction of sodium methyldithiocarbamate with ethyl chloroformate, it has a pungent horse-radish-like odour, and forms colourless crystals, m.p. 35°, b.p. 119°, v.p. 20·7 mmHg at 20°, d^{37} 1·069. Its solubility at 20° is 7·6 g/l water; readily soluble in acetone, benzene, cyclohexanone, ethanol, methanol, dichloromethane, light petroleum. The technical product is > 94% pure, b.p. 117-119°.

Uses. Methyl isothiocyanate is a soil fumigant used at a rate of 17·5-25 g/m² for the control of soil fungi, insects and nematodes and against weed seeds. It is phytotoxic and planting must be delayed until decomposition is complete, about 21 d at soil temperatures of 12-18°, 56 d at 0-6°; the cress germination test (P. M. Smith & C. R. Worthing, *Insecticide and Fungicide Handbook*, 5th Ed., p.253) should be used before planting to check the absence of residues.

Toxicology. The acute oral LD50 is: for male rats 175 mg/kg; for male mice 90 mg/kg. The acute dermal LD50 is: for male rats 2780 mg/kg; for male mice 1820 mg/kg. Prolonged skin contact causes irritation. In 0·5-y feeding trials rats tolerated 30 mg/kg daily.

Formulations. These include: 'Trapex', e.c. (175 g a.i./l), flash point (closed) *c.* 20°; 'Di-Trapex' ('Vorlex' in North America) (235 g/l in dichloropropane-dichloropropene mixture) do not store in aluminium containers; 'Di-Trapex CP' as 'Di-Trapex', with added chloropicrin (trichloronitromethane). These preparations are moderately corrosive and appliances should be cleaned with mineral oil and not water.

Analysis. Product and residue analysis is by GLC; details are available from Schering AG. See also: M. Ottnad *et al., Anal. Methods Pestic. Plant Growth Regul.*, 1978, **10**, 563.

5-Methylisoxazol-3-ol

Me—⟨isoxazole ring⟩—OH
N
O

T5NOJ C1 EQ $C_4H_5NO_2$ (99·15)

Nomenclature and development. The common name hydroxyisoxazole is approved by JMAF; also known as hymexazol. The IUPAC name is 5-methylisoxazol-3-ol, in *C.A.* usage 5-methyl-3-isoxazolol *[10004-44-1]*. It was introduced in 1970 by Sankyo Co. Ltd under the code numbers 'F-319' and 'SF-6505', trade mark 'Tachigaren' and protected by JapaneseP 518 249; 532 202. Its fungicidal properties were described by I. Iwai & N. Nakarura, *Chem. Pharm. Bull.*, 1966, **14**, 1277. Its chemistry, biological properties and toxicology have been comprehensively reviewed (K. Tomita *et al., Sankyo Kenkyusho Nempo,* 1973, **25**, 1).

Manufacture and properties. Produced by the reaction of hydroxylamine and alkali with ethyl 3-chlorobut-2-enoate, which is prepared from the action of phosphorus penta-chloride on ethyl 3-oxobutyrate, it forms colourless crystals, m.p. 86°. Its solubility at 25° is 85 g/l water; readily soluble in organic solvents. The technical grade is *c.* 98% pure. It is stable to acid and alkali and is non-corrosive.

Uses. It is a soil fungicide and a plant growth promoter. It is effective against soil-borne diseases caused by *Fusarium, Aphanomyces, Pythium* and *Corticium* spp. and others when applied as a soil drench at 30-60 g a.i./100 l to paddy seedlings, carnations, sugar beet, forest tree seedlings and other crops. It is also used as a seed dressing for sugar beet at 5-10 g/kg seed.

Toxicology. The acute oral LD50 is: for rats 3909-4678 mg/kg; for mice 1968-2148 mg/kg; for chickens > 1000 mg/kg. The acute dermal LD50 for rats is > 10000 mg/kg; for mice > 2000 mg/kg. In 90-d feeding trials the 'no effect' level for rats and mice was 2500-5000 mg/kg diet. The LC50 (48 h) for carp and Japanese killifish is > 40 mg/l.

Formulations. These include: liquid (30 g a.i./l); dust (40 g/kg); seed dressing (700 g/kg).

Analysis. Product analysis is by GLC. Residues may be determined by GLC after conversion to *O,O*-diethyl *O*-5-methylisoxazol-3-yl phosphorothioate (T. Nakamura *et al., Anal. Methods Pestic. Plant Growth Regul.,* 1978, **10**, 215).

357

4-(Methylthio)phenyl dipropyl phosphate

$$(PrO)_2P-O-\!\!\!\bigcirc\!\!\!-SMe$$
$$\overset{\|}{O}$$

3OPO&O3&OR DS1 $C_{13}H_{21}O_4PS$ (304·3)

Nomenclature and development. The common name propaphos is approved by JMAF. The IUPAC and *C.A.* name is 4-(methylthio)phenyl dipropyl phosphate, Registry No. *[7292-16-2]*. It was introduced in 1967 by Nippon Kayaku Co. Ltd under the code number 'NK-1158', trade mark 'Kayaphos' and protected by Japanese P 482 500; 462 729. Its insecticidal properties were first described in *Jpn. Pestic. Inf.*, 1970, No. 4, p.7.

Manufacture and properties. Produced by the reaction of 4-(methylthio)phenol with dipropyl phosphorochloridate in the presence of sodium hydroxide, it is a colourless liquid, b.p. 175-177°/0·85 mmHg. Its solubility at 25° is 125 mg/l; soluble in most organic solvents. It is stable in neutral and acid media but is slowly decomposed in alkaline media.

Uses. It is a systemic insecticide effective in the control of the rice pests *Laodelphax striatella, Nephotettix cincticeps, Chilo suppressalis, Oulema oryzae* in paddy rice at 600-800 g a.i./ha. It is effective against strains resistant to carbamate and other organophosphorus insecticides.

Toxicology. The acute oral LD50 is: for rats 70 mg/kg; for mice 90 mg/kg; for rabbits 82·5 mg/kg. The acute dermal LD50 for mice is 156 mg/kg. In 90-d feeding trials the 'no effect' level was: for rats 100 mg/kg diet; for mice 5 mg/kg diet. The LC50 (48 h) for carp is 4·8 mg/l.

Formulations. These include: e.c. (500 g a.i./l); dust (20 g/kg); granules (50 g/kg).

Analysis. Product and residue analysis is by GLC.

Metobromuron

ER   DMVN1&O1 $C_9H_{11}BrN_2O_2$ (259·1)

Nomenclature and development. The common name metobromuron is approved by BSI, ISO, ANSI and WSSA. The IUPAC name is 3-(4-bromophenyl)-1-methoxy-1-methylurea, in *C.A.* usage *N'*-(4-bromophenyl)-*N*-methoxy-*N*-methylurea *[3060-89-7]* formerly 3-(*p*-bromophenyl)-1-methoxy-1-methylurea. Its herbicidal properties were described by J. Schuler & L. Ebner, *Proc. Br. Weed Control Conf., 7th,* 1964, p.450. It was introduced in 1963 by Ciba AG (now Ciba-Geigy AG) under the code number 'Ciba 3126', trade mark 'Patoran' and protected by BP 965 313; USP 3 223 721.

Manufacture and properties. Produced either by the bromination of 1-methoxy-1-methyl-3-phenylurea or by the reaction of 4-bromophenyl isocyanate with *O,N*-dimethyl-hydroxylamine, it forms colourless crystals, m.p. 95·5-96°, v.p. 3 x 10^{-6} mmHg at 20°. Its solubility at 20° is 330 mg/l water; soluble in most organic solvents.

Uses. Metobromuron is a herbicide absorbed by roots and leaves and is recommended for pre-em. use in potatoes, beans, soyabeans at 1·5-2·0 kg a.i./ha, on tobacco and tomatoes prior to transplanting at 1·5-2·5 kg/ha. At these concentrations it is degraded to non-phytotoxic levels within 90 d.

Toxicology. The acute oral LD50 for rats is 2000-3000 mg/kg; the acute dermal LD50 for rats is > 3000 mg/kg. In 2-y feeding trials rats receiving 250 mg/kg diet and dogs 100 mg/kg diet were unaffected. It is of low toxicity to wildlife.

Formulations. These include: w.p. (500 g a.i./kg); also in combination with metolachlor ('Galex', 'Tobacron') or with terbutryne ('Igrater').

Analysis. Product analysis is by GLC with internal standard. Residues may be determined by alkaline hydrolysis, steam distillation and extraction of the 4-bromoaniline which is diazotised and coupled with *N*-(1-naphthyl)ethylenediamine prior to colorimetry; alternatively the 4-bromoaniline is diazotised, iodinated and measured by GLC with ECD. Particulars of these methods are available from Ciba-Geigy AG. See also: G. Voss *et al., Anal Methods Pestic. Plant Growth Regul.,* 1973, **7**, 569.

2R   Cl   BNV1GY1&1O1 $C_{15}H_{22}ClNO_2$ (283·8)

Nomenclature and development. The common name metachlor is approved by BSI, ANSI, New Zealand and WSSA, and proposed by ISO. The IUPAC name is 2-chloro-6'-ethyl-*N*-(2-methoxy-1-methylethyl)acet-*o*-toluidide also known as α-chloro-2'-ethyl-6'-methyl-*N*-(1-methyl-2-methoxyethyl)acetanilide, in *C.A.* usage 2-chloro-*N*-(2-ethyl-6-methylphenyl)-*N*-(2-methoxy-1-methylethyl)acetamide *[51218-45-2]*. It was introduced in 1974 by Ciba-Geigy AG under the code number 'CGA 24 705', trade mark 'Dual' and protected by DOS 2 320 340. Its herbicidal properties were described by H. R. Gerber *et al., Proc. Br. Weed Control Conf., 12th,* 1974, **2,** 787.

Properties. It is a colourless liquid, b.p. 100°/1 x 10^{-3} mmHg, v.p. 1·3 x 10^{-5} mmHg at 20°. Its solubility at 20° is 530 mg/l water; soluble in most organic solvents.

Uses. Metolachlor is a germination inhibitor active mainly on grasses at 1·0-2·5 kg a.i./ha, rates depending on soil and climatic conditions. It is selective in maize, soyabeans, groundnuts and other crops and used, in combination with atrazine, for weed control in maize and sugarcane. It is rapidly metabolised in plants.

Toxicology. The acute oral LD50 for rats is 2780 mg/kg; the acute dermal LD50 for rats is > 3170 mg/kg; it causes slight skin irritation and no eye irritation to rabbits. No toxic effect was observed in sub-chronic feeding tests on rats and dogs. It is toxic to honey bees and slightly to moderately toxic to fish.

Formulations. These include: 'Dual 500 EC', 'Dual 720 EC', e.c. (500 and 720 g a.i./l); also used in combination with atrazine ('Primagram' and 'Primextra'), with prometryne ('Codal'), with fluometuron ('Coturan multi'), with metobromuron ('Galex', Tobacron').

Analysis. Details of analytical methods are available on request from Ciba-Geigy AG.

Metoxuron

1OR BG DMVN1&1 $C_{10}H_{13}ClN_2O_2$ (228·7)

Nomenclature and development. The common name metoxuron is approved by BSI and ISO. The IUPAC name is 3-(3-chloro-4-methoxyphenyl)-1,1-dimethylurea (I), in *C.A.* usage *N'*-(3-chloro-4-methoxyphenyl)-*N,N*-dimethylurea *[19937-59-8]* formerly (I). It was introduced in 1968 by Sandoz AG under the code numbers 'SAN 6915H', trade mark 'Dosanex', and in 1971 under code number 'SAN 7102H', trade mark 'Purivel' and protected by BP 1 165 160; FP 1 497 868. Its herbicidal properties were first described by W. Berg, *Z. Pflanzenkr. Pflanzenpathol. Pflanzenschutz*, 1968, **75**, 233.

Manufacture and properties. Produced by the reaction of 3-chloro-4-methoxyphenyl isocyanate with dimethylamine, it is a colourless crystalline powder, m.p. 126-127°, v.p. 3·2 x 10⁻⁸ mmHg at 20°. Its solubility at 24° is 678 mg/l water; soluble in acetone, cyclohexanone, hot ethanol; moderately soluble in benzene, cold ethanol; practically insoluble in light petroleum. It is stable upon storage and in water. The technical grade is *c.* 97% pure.

Uses. Metoxuron is a selective herbicide for use in cereals and carrots, particularly against blackgrass, silky bent grass, wild oats, ryegrass, canary grass, and most annual broad-leaved weeds. It is applied to winter and some spring wheats, winter barley and winter rye at early post-em. at 2·4-3·2 kg a.i./ha or at late post-em. at 3·2-4·0 kg/ha. On irrigated wheat in light to medium soil types good results against canary grass and wild oats are obtained with 1·2-1·6 kg/ha. Most winter wheats, winter barley and winter rye, and some spring wheat varieties show a high tolerance but some damage has occurred on a few varieties. On carrots it is used pre- and post-em. at 2·4-4·3 kg/ha, depending on the soil type. So far as is known, all varieties are tolerant. Other uses are as a defoliant for hemp and flax at 3·2 kg/ha; as a potato haulm killer at 2·0 kg/ha. It inhibits the Hill reaction. A 50% loss occurs in 10-30 d according to soil type. In plants, the main metabolic reactions are *N*-demethylation and hydrolysis of the urea moiety.

Toxicology. The acute oral LD50 for rats is 3200 mg/kg; the acute dermal LD50 for albino rats is > 2000 mg/kg. In 90-d feeding trials dogs receiving 2500 mg/kg diet showed no toxic symptom. Chicks receiving 1250 mg/kg diet for 42 d showed no significant abnormality. Moderately toxic to fish. Safe to honey bees.

Formulations. These include: 'Dosanex' and 'Sulerex', w.p. (800 g a.i./kg); 'Dosanex FL' and 'Dosaflo', flowable (500 g/l); 'Dosanex G' and 'Dosamet', granules (400 g/kg); 'Certosan', flowable (285 g metoxuron + 240 g DNOC/l); 'Dosamix', w.p. (720 g metoxuron + 80 g simazine/kg); 'Dosater', flowable (335 g metoxuron + 195 g mecoprop/l); 'Riflex', granules (140 g metoxuron + 60 g tri-allate/kg): 'Purivel' and 'Piruvel', w.p. (800 g/kg) are special formulations for pre-harvest defoliation of hemp and potatoes.

Analysis. Product analysis is by modification of the method of W. K. Lowen & H. N. Baker, *Anal. Chem.*, 1952, **24**, 1475, or by TLC and subsequent u.v. spectroscopy of the eluted compound (*CIPAC Handbook*, 1979, **1A**, in press). Residues are extracted with acetonitrile, cleaned up by column chromatography and solvent partition and determined by TLC or by a colorimetric method. Details of these methods are available from Sandoz AG. See also: M. Wisson, *Anal. Methods Pestic. Plant Growth Regul.*, 1976, **8**, 417.

Metribuzin

T6NN DNVJ CS1 DZ FX1&1&1 $C_8H_{14}N_4OS$ (214·3)

Nomenclature and development. The common name metribuzin is approved by BSI, ISO and WSSA. The IUPAC name is 4-amino-6-*tert*-butyl-4,5-dihydro-3-methylthio-1,2,4-triazin-5-one, in *C.A.* usage 4-amino-6-(1,1-dimethylethyl)-3-(methylthio)-1,2,4-triazin-5(4*H*)-one *[21087-64-9]* formerly 4-amino-6-*tert*-butyl-3-(methylthio)-*as*-triazin-5(4*H*)-one. It was introduced in 1971 by Bayer AG under the code numbers 'Bayer 94337', Bayer 6159H', 'Bayer 6443H' and 'DIC 1468', trade mark 'Sencor' ('Sencorex' in Great Britain; 'Sencoral' in France) and protected by BelgP 697083, and by E. I. du Pont de Nemours & Co. (Inc.) under the trade name 'Lexone' protected by USP 3905801. Particulars of its mode of action were first reported by W. Draber *et al., Naturwissenschaften,* 1968, **55,** 446. Its properties and use have been reviewed (L. Eue, *Pflanzenschutz-Nachr. (Am. Ed.),* 1972, **25,** 175).

Properties. It is a colourless crystalline solid, m.p. 125°, v.p. 2 x 10^{-4} mmHg at 60°, < 10^{-5} mmHg at 20°. Its solubility at 20° is: 1·2 g/l water; 450 g/l methanol; 130 g/l toluene. The technical product is white to yellowish and 90-95% pure.

Uses. Metribuzin is a herbicide for use in soyabeans at 500-700 g a.i./ha (L. R. Hawf & T. B. Waggoner, *ibid.,* 1973, **26,** 35); also in potatoes, tomatoes, lucerne, asparagus, sugar beet and other crops (A. D. Cohick, *ibid.,* p.23; H. Hack & H. Lembrich, *ibid.,* 1972, **25,** 361; W. Kolbe & K. Zimmer, *ibid.,* p.210; W. Kampe, *ibid.,* p.283).

Toxicology. The acute oral LD50 is: for rats 2200-2345 mg/kg; for mice 698-711 mg/kg. In 2-y feeding trials no ill-effect was noted for rats or dogs receiving 100 mg/kg diet (E. Löser & G. Kimmerle, *ibid.,* p.186).

Formulation. A w.p. (350, 500 and 700 g a.i./kg).

Analysis. Product analysis is by i.r. spectroscopy (J. W. Betker *et al., J. Assoc. Off. Anal. Chem.,* 1976, **59,** 278). Residues may be determined by GLC (T. H. Byast *et al., Agric. Res. Counc. (G.B.) Weed Res. Organ., Tech. Rep.,* No. 15 (2nd Ed.), pp. 40, 67). Details of both methods are available from Bayer AG. See also: C. A. Anderson, *Anal. Methods Pestic. Plant Growth Regul.,* 1976, **8,** 453.

Mevinphos

(E)   *(Z)*

1OV1UY1&OPO&O1&O1 $C_7H_{13}O_6P$ (224·1)

Nomenclature and development. The common name mevinphos is approved by BSI and ISO—exception USSR; also known as ENT 22 374. The IUPAC name is 2-methoxy-carbonyl-1-methylvinyl dimethyl phosphate, also known as methyl 3-(dimethoxy-phosphinyloxy)but-2-enoate, in *C.A.* usage methyl 3-[(dimethoxyphosphinyl)oxy]-2-butenoate *[7786-34-7]* formerly methyl 3-hydroxycrotonate dimethyl phosphate ester. Its insecticidal properties were first described by R. A. Corey *et al., J. Econ. Entomol.,* 1953, **46,** 386, and it was introduced in 1953 by Shell Chemical Co. under the code number 'OS-2046', trade mark 'Phosdrin' and protected by USP 2 685 552. 'Phosdrin' is 100% insecticidally active and contains ⩾60% *m/m* of the (*E*)-isomer, *C.A.* Registry No. *[26718-65-0]* (formerly *[298-01-1]*); *c.* 20% *m/m* of the (*Z*)-isomer, Registry No. *[338-45-4],* is likely to be present.

Manufacture and properties. Produced by the reaction of trimethyl phosphite with methyl 2-chloro-3-oxobutyrate, the technical product is a pale yellow to orange liquid, b.p. 99-103°/0·03 mmHg, d_4^{20} 1·25, n_D^{25} 1·4493. The (*E*)-isomer has m.p. 21°, d^{20} 1·2345, n_D^{20} 1·4452; the (*Z*)-isomer, m.p. 6·9, d^{20} 1·245, n_D^{20} 1·4524. The technical product is miscible with: water, alcohols, ketones, chlorinated hydrocarbons, aromatic hydrocarbons; only slightly soluble in aliphatic hydrocarbons. It is stable when stored at ordinary temperature but is hydrolysed in aqueous solution, 50% loss occurring in 1·4 h at pH 11, 3 d at pH 9, 35 d at pH 7, and 120 d at pH 6. It is compatible with alkaline pesticides and fertilisers. It is corrosive to cast iron, mild and some stainless steels, and brass; relatively non-corrosive to monel, copper, nickel and aluminium; non-corrosive to glass and many plastics but passes slowly through thin films of polyethylene.

Uses. Mevinphos is a contact and systemic insecticide and acaricide with short residual activity. The technical product (60% (*E*)-isomer) is effective against sap-feeding insects at 125-250 g/ha, mites and beetles at 200-300 g/ha, caterpillars at 250-500 g/ha. It is non-phytotoxic. Its metabolism and degradation have been reviewed (K. I. Beynon *et al., Residue Rev.,* 1973, **47,** 55).

Toxicology. The acute oral LD50 is: for rats 3-12 mg tech./kg; for mice 7-18 mg/kg. The acute percutaneous LD50 is: for rabbits 16-34 mg/kg; for rats 1-90 mg/kg. In 2-y feeding trials no effect on general health and no significant pathological or clinical chemical effect was observed for rats receiving 4 mg/kg diet and dogs 5 mg/kg diet. The LC50 (24 h) for mosquito fish is 0·8 mg/l.

Formulations. Being water-soluble, formulation is unnecessary but e.c. (100, 240, 480 and 500 g tech./l) are available, also w.s.c.

Analysis. Product analysis is by GLC with FID (*CIPAC Handbook,* in press). Residues may be determined using GLC with FPD. Details can be obtained from Shell International Chemical Co. Ltd. See also: P. E. Porter *et al., Anal. Methods Pestic., Plant Growth Regul. Food Addit.,* 1964, **2,** 351; *Anal. Methods Pestic. Plant Growth Regul.,* 1972, **6,** 450.

Molinate

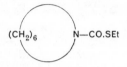

T7NTJ AVS2 $C_9H_{17}NOS$ (187·3)

Nomenclature and development. The common name molinate is approved by BSI, ISO and JMAF—exception Germany. The IUPAC name is *S*-ethyl *N,N*-hexamethylenethiocarbamate, also known as *S*-ethyl perhydroazepine-1-thiocarboxylate, in *C.A.* usage *S*-ethyl hexahydro-1*H*-azepine-1-carbothioate *[2212-67-1]*. It is one of a series of thiolcarbamates introduced in 1954 by Stauffer Chemical Co. and has the code number 'R-4572', trade mark 'Ordram' and is protected by USP 3 198 786.

Manufacture and properties. Produced by the reaction of *S*-ethyl chlorothioformate with perhydroazepine, it is a clear liquid, with an aromatic odour, b.p. 202°/10 mmHg, v.p. 5·6 x 10⁻³ mmHg at 25°, d_{20}^{20} 1·065, n_D^{30} 1·5124. Its solubility at 21° is: 800 mg/l water; soluble in acetone, benzene, propan-2-ol, methanol, xylene. It is stable to hydrolysis and is non-corrosive.

Uses. Molinate is toxic to germinating broad-leaved and grassy weeds and is particularly useful for the control of *Echinochloa* spp. in rice at 2-4 kg a.i./ha. It is applied either before planting to water-seeded or shallow soil-seeded rice or post-flood, post-em. on other types of rice culture. It is rapidly taken up by the plant roots.

Toxicology. The acute oral LD50 for male rats is 720 mg/kg; the acute dermal LD50 for rabbits is > 10000 mg/kg. It is rapidly metabolised by rats, about 50% to carbon dioxide, 25% excreted in the urine and 7-20% in the faeces in 3 d. The LC50 (96 h) is: for rainbow trout 1·3 mg/l; for goldfish 30 mg/l. At recommended rates it has had no detectable effects on fish in ditches draining water from treated rice fields in California.

Formulations. These include: e.c. (720 and 960 g a.i./l); granules (50 and 100 g/kg).

Analysis. Product and residue analysis is by GLC (details from Stauffer Chemical Co.; *CIPAC Handbook,* in press). Residues may also be determined colorimetrically after hydrolysis, steam distillation of the amine and reaction to form the copper dithiocarbamate complex. See also: G. R. Patchett & G. H. Batchelder, *Anal. Methods Pestic., Plant Growth Regul. Food Addit.,* 1967, **5,** 469; G. R. Patchett *et al., Anal. Methods Pestic. Plant Growth Regul.,* 1972, **6,** 668.

Monalide

Cl—⟨ ⟩—NH.CO.CMe₂.Pr

GR   DMVX3&1&1 $C_{13}H_{18}ClNO$ (239·7)

Nomenclature and development. The common name monalide is approved by BSI and ISO. The IUPAC name is 4'-chloro-2,2-dimethylvaleranilide (I), formerly 4'-chloro-α,α-dimethylvaleranilide or *N*-(4-chlorophenyl)-2,2-dimethylvaleramide, in *C.A.* usage *N*-(4-chlorophenyl)-2,2-dimethylpentanamide *[7287-36-7]* formerly (I). It was introduced in 1963 by Schering AG under the code number 'Schering 35 830', the trade mark 'Potablan' and protected by BP 971 819. Its herbicidal properties were described by F. Arndt, *Z. Pflanzenkr. Pflanzenpathol. Pflanzenschutz,* 1965, Sonderheft III, p.277.

Manufacture and properties. Produced by the reaction of 2,2-dimethylvaleric acid with thionyl chloride and 4-chloroaniline, it is a colourless crystalline solid, m.p. 87-88°. Its solubility at 23° is: 22·8 mg/l water; < 10 g/l light petroleum; *c.* 100 g/l xylene; *c.* 500 g/l cyclohexanone. It is stable to hydrolysis and to high temperatures.

Uses. Monalide is a post-em. herbicide, absorbed by leaves and roots and effective for the weeding of umbelliferous crops at 4 kg a.i. in 400-800 l water/ha. Because of its short residual action, monalide is not recommended for use on tropical crops.

Toxicology. The acute oral LD50 for rats is > 4000 mg/kg; the acute dermal LD50 for rats and rabbits is > 800 g a.i. (as e.c.)/kg. Rats fed 5 times a week for 28 d with doses of 150 mg/kg showed no significant toxic symptom; those fed 900 mg/kg suffered a slight alopecia and the female, but not the male, rats showed a dilation of the suprarenal gland and liver.

Formulations. 'Potablan', e.c. (200 g a.i./l); 'Potablan S', e.c. (200 g monalide + 50 g linuron/l).

Analysis. Residue analysis is by TLC or GLC: details may be obtained from Schering AG.

Monocrotophos

(E)   (Z)

1OPO&O1&OY1&U1VM1 $C_7H_{14}NO_5P$ (223·2)

Nomenclature and development. The common name monocrotophos is approved by BSI and ISO; also known as ENT 27129. The IUPAC name is dimethyl (E)-1-methyl-2-methylcarbamoylvinyl phosphate formerly 3-(dimethoxyphosphinyloxy)-N-methyliso-crotonamide, in C.A. usage dimethyl (E)-1-methyl-3-(methylamino)-3-oxo-1-propenyl phosphate [6923-22-4] (formerly [919-44-8]) formerly dimethyl phosphate ester with (E)-3-hydroxy-N-methylcrotonamide; the (Z)-isomer has the Registry No. [919-44-8] and the mixed isomers [2157-98-4]. It was introduced in 1965 by Ciba AG (now Ciba-Geigy AG) under the code number 'C 1414' and the trade mark 'Nuvacron' and by Shell Development Co. under the code number 'SD 9129' and trade mark 'Azodrin'.

Properties. It is a crystalline solid, with a mild ester odour, m.p. 54-55°, v.p. 7 x 10^{-6} mmHg at 20°. The technical product is a reddish-brown semi-solid mass, m.p. 25-30°. It is miscible with water; soluble in acetone, ethanol; sparingly soluble in xylene; almost insoluble in diesel oils, kerosine. It is unstable in low mol. wt alcohols and glycols; stable in ketones and higher mol. wt alcohols and glycols. Its rate of hydrolysis in water in the pH range 1-7 is almost independent of pH (in solution containing 2 mg monocrotophos/l water, 50% loss occurs in c. 22 d at 38°); the rate of hydrolysis increases rapidly at pH > 7. It may be corrosive to black iron, drum steel, stainless steel 304 and brass, but does not attack glass, aluminium or stainless steel 316. It is incompatible with alkaline pesticides.

Uses. Monocrotophos is a fast-acting insecticide with both systemic and contact action, used against a wide range of pests including mites, sucking insects, leaf-eating beetles, bollworms and other larvae on a variety of crops. Dosage rates against mites and sucking insects are 250-500 g/ha, against lepidopterous larvae 500-1000 g/ha. It persists for 7-14 d. It has caused phytotoxicity in cool conditions to Golden Delicious and Ananas Reinette apples, to cherries, peaches and to sorghum varieties related to Red Swazi. Its metabolism and breakdown have been reviewed (K. I. Beynon et al., Residue Rev., 1973, **47**, 55).

Toxicology. The acute oral LD50 for rats is 14-23 mg/kg; the acute percutaneous LD50 for rabbits is 336 mg/kg. In 2-y feeding trials no toxicological effect was observed for rats receiving 1·0 mg/kg diet or dogs 1·6 mg/kg diet. The LC50 (24 h) is: for rainbow trout 12 mg/l; for bluegill 23 mg/l. It is highly toxic to birds and honey bees.

Formulations. These include: concentrates (200, 400 and 600 g a.i./l); 200 SCW and 400 SCW, w.s.c. (200 and 400 g/l); 250 ULV, ULV concentrate (250 g/l); granules (50 g/kg); 'Nuvaron combi', e.c. mixtures with DDT (various proportions).

Analysis. Product analysis is by i.r. spectrometry or by GLC: details can be obtained from Shell International Chemical Co. Ltd. Residues are determined by GLC using phosphorus-sensitive detectors (M. C. Bowman & M. J. Beroza, J. Agric. Food Chem., 1967, **15**, 465; B. Y. Giang & H. F. Beckman, ibid., 1968, **16**, 899). See also: P. E. Porter, Anal. Methods Pestic., Plant Growth Regul. Food Addit., 1967, **5**, 193; Anal. Methods Pestic. Plant Growth Regul., 1972, **6**, 287.

Monolinuron

$$Cl-\langle \text{ring} \rangle-NH.CO.NMe.OMe$$

GR DMVN1&O1 $C_9H_{11}ClN_2O_2$ (214·7)

Nomenclature and development. The common name monolinuron is approved by BSI, ISO and WSSA. The IUPAC name is 3-(4-chlorophenyl)-1-methoxy-1-methylurea (I), in *C.A.* usage *N'*-(4-chlorophenyl)-*N*-methoxy-*N*-methylurea *[1746-81-2]* formerly (I). Its herbicidal properties were first described by K. Härtel, *Meded. Landbouwhogesch. Opzoekingsstn. Staat Gent,* 1962, **27,** 1275, and it was introduced in 1958 by Hoechst AG under the code number 'Hoe 2747', trade mark 'Aresin', and protected by DBP 1 028 986; USP 2 960 534; 3 079 244; BP 852 422.

Manufacture and properties. Produced by the reaction of 4-chlorophenyl isocyanate with hydroxylamine followed by reaction with dimethyl sulphate, it forms colourless crystals, m.p. 79-80°, v.p. 1·5 x 10^{-4} mmHg at 22°. Its solubility at 25° is 735 mg tech./l water; soluble in acetone, dioxane, ethanol, xylene. It is stable at its m.p. and in solution but slowly decomposes in acids and bases, and in moist soil. It is non-corrosive.

Uses. Monolinuron is absorbed by leaves and roots and is used as a pre-em. herbicide effective against annual grasses and broad-leaved weeds. It is recommended for pre-em. use on maize and beans at 0·5-1·0 kg a.i./ha, on potatoes at 1·0-1·5 kg/ha, on asparagus at 1·0-2·0 kg/ha, in vineyards at 2·0-3·0 kg/ha; also on blackcurrants and ornamental shrubs before bud burst at 1·0-2·0 kg/ha. It is degraded in soil (H. Börner, *Z. Pflanzenkr. Pflanzenpathol. Pflanzenschutz,* 1965, **72,** 516).

Toxicology. The acute oral LD50 for rats is 2250 mg/kg. In 2-y feeding trials the 'no effect' level for rats was 250 mg/kg diet.

Formulations. A w.p. (475 g a.i./kg); mixtures with dinoseb acetate or with linuron, and a colloidal suspension with paraquat are also available.

Analysis. For analysis see H. L. Pease *et al., Anal. Methods Pestic., Plant Growth Regul. Food Addit.,* 1967, **5,** 443. Residues may be determined by the colorimetric method of R. L. Dalton & H. L. Pease, *J. Assoc. Off. Agric. Chem.,* 1962, **45,** 377.

Monuron

$$Cl-\!\!\bigcirc\!\!-NH.\overset{\overset{\displaystyle O}{\|}}{C}.NMe_2$$

GR   DMVN1&1 $C_9H_{11}ClN_2O$ (198·7)

Nomenclature and development. The common name monuron is approved by BSI, ISO, ANSI and WSSA—exceptions USSR (chlorfenidim) and Portugal; previously known as CMU. The IUPAC name is 3-(4-chlorophenyl)-1,1-dimethylurea, in *C.A.* usage *N'*-(4-chlorophenyl)-*N*,*N*-dimethylurea *[150-68-5]* formerly 3-(*p*-chlorophenyl)-1,1-dimethylurea. Its herbicidal properties were described by H. C. Bucha & C. W. Todd, *Science,* 1951, **114,** 493 and it was introduced in 1952 by E. I. du Pont de Nemours & Co. (Inc.), who no longer manufacture or market it, under the trade mark *'Telvar'* and protected by USP 2 655 445; BP 691 403; 692 589.

Manufacture and properties. Produced by the reaction of 4-chlorophenyl isocyanate with dimethylamine, it is a colourless crystalline solid, m.p. 174-175°, v.p. 5 x 10^{-7} mmHg at 25°, d_{20}^{20} 1·27. Its solubility at 25° is 230 mg/l water; at 27° is 52 g/kg acetone; sparingly soluble in petroleum oils and in polar organic solvents. It is stable to moisture and oxidation at room temperature but is decomposed at 185-200°. The rate of hydrolysis at room temperature and pH 7 is negligible, but is increased at elevated temperatures and under acidic or alkaline conditions. It is subject to slow decomposition in moist soils, is non-corrosive and non-flammable.

Uses. Monuron is an inhibitor of photosynthesis and is absorbed via the roots. It is recommended for total weed control of non-crop areas at 10-30 kg a.i./ha; for subsequent annual maintenance, treatment at 5-10kg/ha will suffice.

Toxicology. The acute oral LD50 for rats is 3600 mg/kg; application to the intact or abraded skin of guinea-pigs produced no irritation or sensitisation. The 'no effect' level for rats and dogs is 250-500 mg/kg diet (M. C. Hodge *et al., AMA. Arch. Ind. Health,* 1958, **17,** 45).

Formulation. A w.p. (800 g a.i./kg).

Analysis. Product analysis is by hydrolysis and titration of the liberated amine (details from E. I. du Pont de Nemours & Co. (Inc.); *CIPAC Handbook,* 1979, **1A,** in press). Residues may be determined by GLC (J. J. Kirkland, *Anal. Chem.,* 1964, **34,** 428; T. H. Byast *et al., Agric. Res. Counc. (G.B.) Weed Res. Organ., Tech. Rep.,* No. 15 (2nd Ed.), p.49) or by HPLC (T. H. Byast *et al., loc. cit.*). See also: *Anal. Methods Pestic. Plant Growth Regul.,* 1972, **6,** 664.

Cl—⟨benzene ring⟩—NH.C.NHMe$_2$ CCl$_3$CO$_2^-$ (with O double bond and + over C)

GR DMVN1&1 &QVXGGG   C$_{11}$H$_{12}$Cl$_4$N$_2$O$_3$ (326·0)

Nomenclature and development. The common name monuron-TCA is approved by WSSA for monuron trichloroacetate. The IUPAC name is 3-(4-chlorophenyl)-1,1-dimethyluronium trichloroacetate, in *C.A.* usage trichloroacetic acid compound with *N'*-(4-chlorophenyl)-*N,N*-dimethylurea (1:1) *[140-41-0]* formerly trichloroacetic acid compound with 3-(*p*-chlorophenyl)-1,1-dimethylurea (1:1). The salt of monuron (p.368) was introduced as a herbicide in 1956 by Allied Chemical Corp. Agricultural Division (now Hopkins Agricultural Chemical Co.) under the code number 'GC-2996', trade mark 'Urox', and protected by USP 2 782 112; 2 801 911.

Manufacture and properties. Produced by the reaction of monuron and trichloroacetic acid, it is a crystalline solid, m.p. 78-81°. Its solubility at room temperature is: 918 mg/l water; 177 g/kg methanol; 91 g/kg xylene; 400 g/kg 1,2-dichloroethane. It is acidic in reaction and is incompatible with alkaline materials.

Uses. Monuron-TCA is a general herbicide effective in the total weed control of uncropped areas at rates of 10-15 kg/ha.

Toxicology. The acute oral LD50 for rats is 2300-3700 mg (in corn oil)/kg. It is an irritant to the skin and mucous membranes.

Formulations. These include: 'Urox' weedkiller: granules (55, 110 and 220 g/kg); oil-miscible concentrate; oil/water miscible concentrate; oil/water miscible concentrate + 2,4-D.

Analysis. Product analysis is by i.r. spectroscopy and/or total chlorine and nitrogen determination. Details are available from Hopkins Agricultural Chemical Co.

Nabam

$$Na^+ \ {}^-S.\overset{\overset{\displaystyle S}{\|}}{C}.NHCH_2CH_2NH.\overset{\overset{\displaystyle S}{\|}}{C}.S^- \ Na^+$$

SUYS&M2MYS&US &-NA- 2 $\qquad\qquad\qquad\qquad$ $C_4H_6N_2Na_2S_4$ (256·3)

Nomenclature and development. The common name nabam is approved by BSI and ISO. The IUPAC name is disodium ethylenebis(dithiocarbamate) (I), in *C.A.* usage disodium 1,2-ethanediylbis(carbamodithioate) *[142-59-6]* formerly (I). Its fungicidal activity was first described by A. E. Dimond *et al., Phytopathology,* 1943, **33**, 1095, and it was introduced by E. I. du Pont de Nemours & Co. (Inc.) under the trade mark *'Parzate'* and by Rohm & Haas Co. under the trade mark *'Dithane D-14'* and protected by USP 2 317 765—though neither company now manufacture it.

Manufacture and properties. Produced by the interaction of ethylenediamine and carbon disulphide in the presence of sodium hydroxide, it forms colourless crystals of the hexahydrate. Its solubility at room temperature is *c.* 200 g (as anydrous salt)/l water, forming a yellow solution. On aeration, aqueous solutions deposit yellow mixtures of which the main fungicidal components are sulphur and etem (p.239) (C. W. Pluijgers *et al., Tetrahedron Lett.,* 1971, p.1371).

Uses. Nabam has been superseded, as a protective fungicide, by zineb though the latter may be prepared as a tank mix from solutions of nabam and zinc sulphate. Nabam is too phytotoxic for general use on foliage; soil applications were reported to have a systemic action on *Phytophthora fragariae* (E. M. Stoddard, *Phytopathology,* 1951, **41**, 858).

Toxicology. The acute oral LD50 for rats is 395 mg/kg. A goitrogenic effect was noted in rats receiving 1000-2500 mg/kg diet for 10 d (R. B. Smith *et al., J. Pharmacol Exp. Ther.,* 1953, **109**, 159).

Formulation. 'Cambell's X-Spor' (aqueous solution).

Analysis. Product analysis is by heating with dilute sulphuric acid, the carbon disulphide liberated converted by alkali to dithiocarbonate which is measured by titration with iodine (*CIPAC Handbook,* 1970, **1**, 539).

Naled

$$O$$
$$\|$$
$$(MeO)_2P.O.CHBr.CBrCl_2$$

GXGEYEOPO&O1&O1 $C_4H_7Br_2Cl_2O_4P$ (380·8)

Nomenclature and development. The common name naled is approved by BSI, ISO, ANSI and ESA—exception the Republic of South Africa (bromchlophos), Denmark (dibrom), JMAF (BRP), Germany and Sweden; also known as ENT 24 988. The IUPAC and *C.A.* name is 1,2-dibromo-2,2-dichloroethyl dimethyl phosphate, Registry No. *[300-76-5]*. It was introduced in 1956 by Chevron Chemical Co. under the code number 'RE-4355', the trade mark 'Dibrom' and protected by BP 855 157; USP 2 971 882. Its biological properties were first reported by J. M. Grayson & B. D. Perkins, *Pest Control,* 1960, **28**(6), 9.

Manufacture and properties. Produced by the action of bromine, for 10-11 h at 0-30° on dichlorvos (p.180) in carbon tetrachloride, the technical product is a yellow liquid, with a slightly pungent odour, b.p. 110°/0·5 mmHg, v.p. 2 x 10^{-3} mmHg at 20°, d_{20}^{20} 1·97, n_D^{28} 1·5108 and is *c.* 93% pure. It is practically insoluble in water; slightly soluble in aliphatic solvents; readily soluble in aromatic solvents. Pure naled has m.p. 26°. It is stable under anhydrous conditions, but is rapidly hydrolysed in water, 90-100% in 48 h at room temperature and by alkali. It is degraded by sunlight. It is stable in brown glass containers but, in the presence of metals and reducing agents, rapidly loses bromine and reverts to dichlorvos.

Uses. Naled is a fast-acting non-systemic contact and stomach insecticide and acaricide with some fumigant action. It is recommended for use against adult mosquitoes and flies, on many crop plants and also in glasshouses and mushroom houses at *c.* 8·1 g a.i./100 m³.

Toxicology. The acute oral LD50 for rats is 430 mg/kg; the acute dermal LD50 for rabbits is 1100 mg/kg. In feeding trials no toxic effect was observed on rats receiving 100 mg tech. (99% pure)/kg diet for 84 d, nor on albino rats receiving 100 mg tech. (91% pure)/kg diet for 2 y. The LC50 (24 h) is: for goldfish 2-4 mg/l; for crabs 0·33 mg/l; there was no mortality to mosquito fish or tadpoles when it was applied at 560 g/ha.

Formulations. An e.c. (960 g a.i./l); dust (40 g/kg).

Analysis. Product, formulation and residue analysis is by GLC; details available from Chevron Chemical Co. See also: D. E. Pack *et al., Anal. Methods Pestic., Plant Growth Regul. Food Addit.,* 1964, **2,** 125; J. B. Leary, *Anal. Methods Pestic. Plant Growth Regul.,* 1972, **6,** 350.

Naphthalene

L66J $C_{10}H_8$ (129·2)

Nomenclature and development. The IUPAC and *C.A.* name naphthalene, Registry No. *[91-20-3]*, is accepted by BSI and ISO, in France (naphtalène), in lieu of a common name. It has been in long use as a household fumigant against clothes moths.

Manufacture and properties. Produced from intermediate distillates of coal tar, it is separated by centrifugation and purified by sublimation. It forms colourless flaky crystals, m.p. 80°, b.p. 218°, v.p. 4·92 x 10^{-2} mmHg at 20°, d_4^{15} 1·517. Its solubility at room temperature is: 30 mg/l water; 77 g/l ethanol, methanol; 285 g/l benzene, toluene; 500 g/l chloroform, carbon tetrachloride; very soluble in diethyl ether, 1,2-dichloromethane. It is flammable, flash point 79° (open cup), 88° (closed cup) but otherwise stable.

Uses. Naphthalene is an insecticidal fumigant of restricted usefulness and somewhat low potency; of doubtful value as a moth repellent (W. S. Abbott & S. C. Billings, *J. Econ. Entomol.* 1935, **28**, 493). Of some use as a soil fumigant but rapidly decomposed by soil organisms.

Toxicology. The acute oral LD50 for rats is 2200 mg/kg.

Formulation. Only pure grades, free from dust should be used for fumigation; heat only in specifically-designed lamps.

Analysis. Product analysis is by distillation from mixtures in the presence of ethanol and precipitation as the picrate (D. S. Binnington & W. F. Giddes, *Ind. Eng. Chem. Anal. Ed.,* 1934, **6**, 461; W. L. Millar, *J. Assoc. Off. Agric. Chem.,* 1934, **17**, 308). For estimation in air, collect in ethanol and measure the colour given with modified Denigès reagent (M. C. Robbins, *AMA Arch. Ind. Hyg.,* 1951, **4**, 85).

Naphthalic anhydride

T666   1A   M   CVOVJ $C_{12}H_6O_3$ (198·2)

Nomenclature and development. The trivial name is naphthalic anhydride. The IUPAC name is naphthalene-1,8-dicarboxylic anhydride, in *C.A.* usage 1*H*,3*H*-naphtho[1,8-*cd*]-pyran-1,3-dione *[81-84-5]*. It was introduced in 1972 by Gulf Oil Corp. under the trade mark 'Protect' and protected by USP 3 131 509; 3 564 768. The activity of naphthalic anhydride in increasing the selectivity of thiocarbamate herbicides to maize was first reported in 1969 (O. L. Hoffman, *Weed Sci. Soc. Am., Abstr.,* 1969, No. 12; G. A. Wicks & C. R. Fenster, *Weed Sci.,* 1971, **19,** 565).

Properties. It is a light tan crystalline solid, m.p. 270-274°. It is relatively insoluble in water and most non-polar solvents; solubility 13·9 g/l dimethylformamide. It is stable under normal storage conditions, is non-corrosive and non-hygroscopic. The technical product is ≥ 98% pure.

Uses. Naphthalic anhydride is used as a seed coating, at 5 g/kg, prior to planting and protects the crop from damage by EPTC, vernolate and related herbicides. Germination of maize seed planted 2 y after treatment was not affected. It shows promise in enhancing the selectivity of several herbicides between wild and cultivated rice (C. Parker & M. L. Dean, *Pestic. Sci.,* 1976, **7,** 403). No metabolism occurs in maize and no residues are found after the seedling stage. See also: G. R. Stephenson & F. Y. Chang, *Chemistry and Mode of Action of Herbicide Antidotes,* p.35.

Toxicology. The acute oral LD50 is: for rats 12 300 mg/kg; for mallard duck > 6810 mg/kg; for bobwhite quail 4100 mg/kg. The acute dermal LD50 for rabbits is > 2025 mg/kg; it is moderately irritating to the eyes of albino rabbits. Inhalation of the dust in air by rats for 4 h at a concentration of 820 mg/m³ revealed no adverse toxic effect. In 90-d feeding trials no toxic symptom was noted for rats and dogs receiving 500 mg/kg diet. The LC50 (96 h) is: for rainbow trout 6·0 mg/l; for bluegill 4·9 mg/l.

Formulation. Naphthalic anhydride is used directly as the 98% technical product, 'Protect'.

Analysis. Residues in crops may be determined, after clean-up on an anion-exchange resin, by GLC with ECD (J. R. Riden, *Anal. Methods Pestic. Plant Growth Regul.,* 1976, **8,** 483).

Naphthenic acids

Nomenclature and development. Naphthenic acids is the traditional name for a group of carboxylic acids derived from crude petroleum oils and includes alkylcyclopentane- and alkylcyclohexane-carboxylic acids containing 9-10 carbon atoms. The copper, *C.A.* Registry No. *[1338-02-9]*, and zinc salts have long been used to protect wood from fungal attack. They are marketed under various trade marks, including 'Cuprinol'.

Manufacture and properties. Produced by fractionation and partition of the crude petroleum fraction, the acids are reacted with soluble copper or zinc salts. Copper naphthenate is a deep greenish viscous oil, v.p. $< 1 \times 10^{-3}$ mmHg at 100°; the zinc salt is similar but almost colourless. They are practically insoluble in water; moderately soluble in petroleum oils; soluble in most organic solvents.

Uses. Copper and zinc naphthenates are generally toxic to fungi, on account of both the metal and acid content. They are extremely phytotoxic, their main use being for the preservation of wooden seed boxes, benches etc.

Formulations. 'Cuprinol green', e.c. (copper naphthenate in oil); 'Cuprinol clear', e.c. (zinc naphthenate in oil).

Analysis. Product analysis is by breaking the emulsion with alkali and sodium chloride, partitioning between dilute sulphuric acid and diethyl ether and either estimating the copper content electrolytically (*MAFF Tech. Bull.,* No. 1, 1958, p.16), or by reaction with potassium iodide and titration of the liberated iodine (V. G. Hiatt, *J. Assoc. Off. Agric. Chem.,* 1964, **47,** 253).

1-Naphthylacetamide

$$CH_2 . \overset{\displaystyle O}{\overset{\displaystyle \|}{C}} . NH_2$$

L66J   B1VZ $C_{12}H_{11}NO$ (185·2)

Nomenclature and development. The IUPAC name 1-naphthylacetamide is proposed by BSI and ISO in lieu of a common name, in *C.A.* usage 1-naphthaleneacetamide *[86-86-2]*; also known as α-naphthaleneacetamide and NAD. It was introduced as a thinning agent for apples and pears by Amchem Products, Inc. under the trade mark 'Amid-Thin'.

Manufacture and properties. Produced by the controlled hydrolysis of 1-naphthyl-acetonitrile, obtained by chloromethylation of naphthalene followed by reaction with sodium cyanide, it forms colourless crystals, m.p. 184°. It is sparingly soluble in water; soluble in acetone, ethanol, propan-2-ol; insoluble in kerosine. It is stable under normal storage conditions and non-flammable.

Uses. 1-Naphthylacetamide is used for thinning many apple and pear varieties at 2·5-5·0 g a.i./100 l, within a few days after petal fall. It induces formation of an abscission zone in the peduncle.

Toxicology. The acute oral LD50 for rats is *c.* 6400 mg/kg; the acute dermal LD50 for rabbits is > 5000 mg/kg. It is not a skin or eye irritant.

Formulation. 'Amid-Thin' W, w.p. (84 g a.i./kg).

Analysis. Product and residue analysis is by GLC.

1-Naphthylacetic acid

$$CH_2.\overset{\overset{\displaystyle O}{\|}}{C}.OH$$

L66J	B1VQ	*1-naphthylacetic acid*	$C_{12}H_{10}O_2$ (186·2)
L66J	B1VO2	*ethyl 1-naphthylacetate*	$C_{14}H_{14}O_2$ (214·3)

Nomenclature and development. The IUPAC name 1-naphthylacetic acid or 1-naphthalene-acetic acid (I) is accepted by BSI and ISO in lieu of a common name; also known as NAA. In *C.A.* usage (I) *[86-87-3]* formerly α-naphthaleneacetic acid. It was introduced as a plant growth regulator by Amchem Products, Inc. under the trade marks 'NAA-800', 'Fruitone-N' and 'Rootone' and by ICI Ltd, Plant Protection Division under the trade mark 'Phyomone'. Its plant growth regulating activity was reported by F. E. Gardiner *et al.*, *Science*, 1939, **90**, 208.

Manufacture and properties. Produced by the hydrolysis of 1-naphthylacetonitrile, obtained by chloromethylation of naphthalene and subsequent reaction with sodium cyanide, it is a colourless crystalline powder, m.p. 134-135°. Its solubility at 20° is 420 mg/l water; at 26°: 10·6 g/l carbon tetrachloride, 55 g/l xylene; very soluble in acetone, ethanol, propan-2-ol. The technical grade has m.p. 125-128°. It is stable and non-flammable. It forms water-soluble salts with bases.

Esterification with ethanol gives the ethyl ester, *C.A.* Registry No. *[2122-70-5]*, a colourless liquid, b.p. 158-160°/3 mmHg, d^{25} 1·106. It is practically insoluble in water; miscible with most organic solvents.

Uses. 1-Naphthylacetic acid is a plant growth regulator used to prevent pre-harvest drop of apples, pears and mangoes at 0·5-1·0 g a.i./100 l applied at first sign of sound mature fruit drop. It is also used for thinning apples and pears at 0·5-5·0 g/100 l after the calyx period or after petal fall. It is used to stimulate rooting of cuttings, sometimes in combination with 4-(indol-3-yl)butyric acid.

Ethyl 1-naphthylacetate is used as a bark paint to reduce water-shoot production in pruned apple and pear trees and regrowth of sprouts of some ornamental trees.

Toxicology. The acute oral LD50 is: for rats *c.* 1000-5900 mg acid/kg, *c.* 3580 mg ethyl ester/kg; for mice 670 mg a.e. (as sodium salt)/kg. The acute dermal LD50 for rabbits is > 5000 mg acid/kg, > 5000 mg ethyl ester/kg. Slight to moderate irritation of the rabbit skin has been reported on prolonged contact. The chronic LC50 (8 d) for mallard duck and bobwhite quail was > 10000 mg/kg diet.

Formulations. These include: 'Phyomone', w.p. (182 g acid/kg); 'Fruitone-N', w.s.c. (35 g sodium salt/l); w.s.c. of potassium salt; 'Boots Hormone Rooting Powder' (1-naphthylacetic acid + 4-(indol-3-yl)butyric acid + thiram); 'Tree-Hold' (Amchem Products, Inc.), paint (ethyl 1-naphthylacetate + phenylmercury acetate).

Analysis. Product analysis is by titration with alkali after extraction of the acid into diethyl ether, or by GLC of the ethyl ester. Residues on apples may be determined by u.v. spectrometry at 283 nm or by colorimetry at 370 nm after nitration (C. A. Bache *et al., J. Agric. Food Chem.*, 1962, **10**, 365). Details are available from ICI Ltd, Plant Protection Division.

2-Naphthyloxyacetic acid

L66J   CO1VQ $C_{12}H_{10}O_3$ (202·2)

Nomenclature and development. The IUPAC name 2-naphthyloxyacetic acid is accepted by BSI and ISO in lieu of a common name; also known (erroneously) as 2-naphthoxy-acetic acid or β-naphthoxyacetic acid, in *C.A.* usage 2-naphthalenyloxyacetic acid *[120-23-0]* formerly (I). Its effect of increasing fruit setting was reported by S. C. Bausor, *Am. J. Bot.,* 1939, **26,** 415. It was introduced in 1946 by Synchemicals Ltd under the trade mark 'Betapal'.

Manufacture and properties. Produced by the condensation of 2-naphthol with chloroacetic acid in an alkaline medium, the technical product forms green crystals. The pure acid is a colourless crystalline solid m.p. 156°. It is almost insoluble in water at room temperature; soluble in ethanol, diethyl ether. It forms water-soluble alkali metal and amine salts.

Uses. 2-Naphthyloxyacetic acid is used as a fruit setting spray on tomatoes, holly, pineapples, grapes and strawberries. The normal rate is 40-60 mg a.e./l.

Toxicology. The acute oral LD50 for rats is 600 mg/kg.

Formulation. 'Betapal', solution of the triethanolamine salt (16 g a.e./l) with added surfactant.

Analysis. Residues may be determined by HPLC (T. E. Archer & J. D. Stokes, *J. Agric. Food Chem.,* 1978, **26,** 452).

Napropamide

O.CHMe.CO.NEt$_2$

L66J   BOYl&VN2&2 $C_{17}H_{21}NO_2$ (271·4)

Nomenclature and development. The common name napropamide is approved by BSI, ISO and WSSA. The IUPAC name is *N,N*-diethyl-2-(1-naphthyloxy)propionamide (I), in *C.A.* usage *N,N*-diethyl-2-(1-naphthalenyloxy)propanamide *[15299-99-7]* formerly (I). It was introduced in 1971 by the Stauffer Chemical Co. under the code number 'R-7465', trade mark 'Devrinol' and protected by USP 3480671. Its herbicidal properties were described by B. J. Van den Brink *et al., Symp. New Herbic., 3rd,* 1969, p.35.

Manufacture and properties. Produced by the reaction of (*RS*)-2-(1-naphthyloxy)propionic acid with diethylamine, it is a brown solid, m.p. 74·8-75·5°. Its solubility at 20° is 73 mg/l water; very soluble in acetone, ethanol, xylene.

Uses. It is a herbicide effective against annual grasses and certain annual broad-leaved weeds and of promise for the weeding of asparagus, brassica crops, potatoes, lima beans, oil seed rape, groundnuts, peppers, sunflowers, tobacco, tomatoes, tree fruits and vines. It is also effective against perennial grass weeds. For soil treatments, use 2-4 kg a.i./ha incorporated into the soil within 2 d of application; for pre-em. surface treatment, use 3-6 kg/ha applied to the weed-free soil surface and follow with irrigation if no rain falls within 2 d. The (*R*)-(—)-isomer (Registry No. *[41643-35-0]*) is 8 times as toxic to 3 weed species as the (*S*)-(+)-isomer (Registry No. *[41643-36-1]*) (J. H. Chan *et al., J. Agric. Food Chem.,* 1975, **23**, 1008).

Toxicology. The acute oral LD50 for rats is > 5000 mg tech./kg; the acute dermal LD50 for rabbits is > 4640 mg/kg. It is not a primary eye nor skin irritant.

Formulations. These include: w.p. (500 g a.i./kg); e.c. (240 g/l).

Analysis. Product and residue analysis is by GLC: details are available from the Stauffer Chemical Co. See also: G. G. Patchett *et al., Anal. Methods Pestic. Plant Growth Regul.,* 1976, **8**, 347.

Naptalam

L66J BMVR BVQ $C_{18}H_{13}NO_3$ (291·3)

Nomenclature and development. The common name naptalam is approved by BSI, ISO and WSSA—exceptions Turkey (alanap) and JMAF (NPA). The IUPAC name is *N*-1-naphthylphthalamic acid (I), in *C.A.* usage 2-[(1-naphthalenylamino)carbonyl]benzoic acid *[132-55-1]* formerly (I). The plant growth activity of *N*-arylphthalamic acids was reported by O. L. Hoffman & A. E. Smith, *Science,* 1949, **109**, 588. Naptalam-sodium was introduced as a herbicide in 1950 by Uniroyal Inc. under the trade mark 'Alanap', and protected by USP 2 556 664; 2 556 665.

Manufacture and properties. Produced by the reaction of phthalic acid with 1-naphthylamine in benzene, xylene or kerosine at room temperature, it is a crystalline solid, m.p. 185°. Its solubility at room temperature is 200 mg/l water; slightly soluble in acetone, benzene, ethanol. The solubility of its sodium salt, *C.A.* Registry No. *[132-67-2]*, at room temperature is 300 g/kg water. Naptalam is hydrolysed in solutions of pH > 9·5 and is unstable at elevated temperatures, tending to form *N*-(1-naphthyl)phthalimide. It is non-corrosive and non-explosive.

Uses. Naptalam inhibits seed germination and is recommended as a pre-em. herbicide for use in cucurbits, soyabeans, potatoes and groundnuts usually at 2·0-5·5 kg/ha. It persists in soil for 21-56 d.

Toxicology. The acute oral LD50 for rats is 8200 mg a.i./kg, 1770 mg naptalam-sodium/kg. Inhalation may cause slight irritation. In 90-d feeding trials no effect was observed on rats or beagle dogs receiving 1000 mg/kg diet.

Formulations. These include: 'Alanap', naptalam-sodium (240 g a.i./l); also in combination with chlorpropham or dinoseb.

Analysis. Product analysis is by nitrogen determination using the Kjeldahl method. Residues may be determined by hydrolysis to give 1-naphthylamine which is purified and coupled with diazotised sulphanilic acid and the colour measured at 534 nm (A. E. Smith & G. M. Stone, *Anal. Methods Pestic., Plant Growth Regul. Food Addit.,* 1964, **4**, 1).

Neburon

GR BG DMVN4&1 $C_{12}H_{16}Cl_2N_2O$ (275·2)

Nomenclature and development. The common name neburon is approved by BSI, ISO, ANSI and WSSA—exception the Republic of South Africa (neburea). The IUPAC name is 1-butyl-3-(3,4-dichlorophenyl)-1-methylurea (I), in *C.A.* usage *N*-butyl-*N'*-(3,4-dichlorophenyl)-*N*-methylurea *[555-37-3]* formerly (I). Its herbicidal action was first described by H. C. Bucha & C. W. Todd, *Science,* 1951, **144**, 493 and it was introduced in 1957 by E. I. du Pont de Nemours & Co. (Inc.) under the trade mark 'Kloben' and protected by USP 2 655 444; 2 655 445.

Manufacture and properties. Produced by reacting 3,4-dichlorophenyl isocyanate with *N*-methylbutylamine in dioxane at 20-30°, it is a colourless crystalline solid, m.p. 102-103°. Its solubility at 24° is 4·8 mg/l; sparingly soluble in common hydrocarbon solvents. It is stable to oxidation and moisture under normal storage conditions.

Uses. Neburon inhibits photosynthesis and is absorbed through the roots. It is recommended for the pre-em. control of annual weeds and grasses in wheat, lucerne, strawberries and nursery plantings of certain woody ornamentals at 2-3 kg a.i./ha.

Toxicology. The acute oral LD50 for rats is > 11 000 mg/kg. A suspension (150 g a.i./l dimethyl phthalate) applied to the shaved backs of guinea-pigs produced only mild irritation and no sensitisation.

Formulation. 'Kloben' weedkiller, w.p. (600 g a.i./kg).

Analysis. Residues may be determined by GLC with ECD (T. H. Byast et al., *Agric. Res. Counc. (G.B.) Weed Res. Organ., Tech. Rep.,* No. 15 (2nd Ed.), p.49). See also: *Anal. Methods Pestic., Plant Growth Regul. Food Addit.,* 1964, **4,** 157; *Anal. Methods Pestic. Plant Growth Regul.,* 1972, **6,** 664.

Niclosamide

WNR	CG	DMVR	BQ	EG		niclosamide	$C_{13}H_8Cl_2N_2O_4$	(327·1)
WNR	CG	DMVR	BQ	EG	&Z2Q	ethanolamine salt	$C_{15}H_{15}Cl_2N_3O_5$	(388·2)

Nomenclature and development. The common name niclosamide is approved by BSI, ISO and BPC—exception Germany (niclosamid, for veterinary use; clonitralid, for the ethanolamine salt in crop protection use). The IUPAC name is 2′,5-dichloro-4′-nitro-salicylanilide (I), formerly 5-chloro-N-(2-chloro-4-nitrophenyl)salicylamide, in *C.A.* usage 5-chloro-N-(2-chloro-4-nitrophenyl)-2-hydroxybenzamide [50-65-7] formerly (I). Its molluscicidal properties were first described by R. Gönnert & E. Schraufstätter, *Proc. Int. Conf. Trop. Med. Malar.,* 1958, **2**, 5, and its chemical development is discussed by R. Gönnert *et al., Z. Naturforsch. Teil B.,* 1961, **16**, 95. The ethanolamine salt, *C.A.* Registry No. [1420-04-8], was introduced in 1959 by Bayer AG under the code numbers 'Bayer 25 648', 'Bayer 73' and 'SR 73'. The ready-for-use formulations have the trade marks 'Bayluscid' or 'Bayluscide' and are protected by USP 3 079 297; 3 113 067.

Manufacture and properties. Produced by the condensation of 5-chlorosalicylic acid with 2-chloro-4-nitroaniline, it is an almost colourless solid, m.p. 230°. Its solubility at room temperature is 5-8 mg/l water.

The ethanolamine salt is a yellow solid, m.p. 216°. Its solubility at room temperature is 180-280 mg/l water. It is stable to heat and is hydrolysed only by concentrated acid or alkali. Its biological activity is not affected by hard water.

Uses. Niclosamide is a powerful molluscicide giving a total kill of *Australorbis glabratus* at 0·3 mg/l. It is non-phytotoxic at field concentrations. As 'Yomesan' it is used in the treatment of tapeworm infestation.

Toxicology. Rats survived oral doses of 5000 mg ethanolamine salt/kg; cats vomited an oral dose of 500 mg/kg but tolerated 250 mg/kg without ill-effect. The acute intraperitoneal LD50 for rats is 250 mg/kg. Rats and rabbits survived oral doses of 5000 mg niclosamide/kg. It is toxic to fish and zooplankton.

Formulations. These include: e.c. (250 g a.i./l); w.p. (700 g/kg).

Analysis. Product analysis is by reduction with dititanium trisulphate and back titration with iron trichloride. Residues in water may be estimated: by extracting from acidified water with pentyl acetate, treating with methanolic sodium hydroxide and measuring the yellow colour at 385 nm (organic matter interferes); by adding a buffered solution of Safranin TH extra Conc. (Hoescht AG) and extracting with pentyl acetate, measuring the red colour (R. Strufe, *Pflanzenschutz-Nachr. (Am. Ed.),* 1962, **15**, 42); or by titration: details from Bayer AG.

Nicotine

T6NJ C- BT5NTJ A1 $C_{10}H_{14}N_2$ (162·2)

Nomenclature and development. The trivial name nicotine is accepted in lieu of a common name by BSI, ISO and JMAF. The IUPAC name is (S)-3-(1-methylpyrrolidin-2-yl)-pyridine, in *C.A.* usage (S)-3-(1-methyl-2-pyrrolidinyl)pyridine *[54-11-5]* formerly nicotine. Extracts of tobacco have long been used against sucking insects, but are now replaced by technical nicotine (95-98% alkaloids) or by nicotine sulphate (40% alkaloids).

Manufacture and properties. Produced from waste tobacco, *Nicotiana tabacum,* or from *N. rustica* either by steam distillation in the presence of alkali, or by extraction with trichlorethylene in the presence of alkali and extraction from the solvent by dilute sulphuric acid, it is a colourless liquid, b.p. 247°, m.p. —80°, v.p. 4·25 x 10^{-2} mmHg at 25°, d_4^{20} 1·009, $n_D^{22.4}$ 1·5239, $[\alpha]_D^{20}$ —161·55°. It is miscible with water below 60°, forming a hydrate, and above 210°; miscible with ethanol, diethyl ether; readily soluble in most organic solvents. It darkens slowly and becomes viscous on exposure to air. It is a base, pK_1 6·16, pK_2 10·96 forming salts with acids. The predominant component of the crude alkaloid is (-)-nicotine, small amounts of related alkaloids may be present.

Uses. Nicotine is a non-persistent, non-systemic contact insecticide with some ovicidal properties (H. Shaw & W. Steer, *J. Pomol.,* 1938, **16,** 364). It is used as a fumigant in closed spaces, *e.g.* in glasshouses.

Toxicology. The acute oral LD50 for rats is 50-60 mg/kg; the acute dermal LD50 by single application for rabbits is 50 mg/kg. It is toxic to man by inhalation and by dermal contact.

Formulations. Nicotine is marketed as the technical alkaloid (950 g a.i./l) or as nicotine sulphate (400 g/kg); the addition of soap or alkali is required to dilutions of the latter to liberate the nicotine; also as dusts 30-50 g/kg. For fumigation, nicotine is applied to a heated metal surface or nicotine 'shreds' are burnt.

Analysis. Product analysis is by steam distillation and precipitation as silicotungstate (AOAC Methods; *MAFF Tech. Bull.* No. 1, 1958, p.59; *CIPAC Handbook,* 1970, **1,** 543). For nicotine sulphate, see *ibid.,* 1979, **1A,** in press. Residues may be determined by GLC, after clean-up by liquid-liquid partition (R. J. Martin, *J. Assoc. Off. Anal. Chem.,* 1967, **50,** 939).

Nitralin

WSl&R   CNW   ENW   DN3&3 $C_{13}H_{19}N_3O_6S$ (345·4)

Nomenclature and development. The common name nitralin is approved by BSI, ISO, WSSA and JMAF—exception Canada. The IUPAC name is 4-methylsulphonyl-2,6-dinitro-*N,N*-dipropylaniline, in *C.A.* usage 4-(methylsulfonyl)-2,6-dinitro-*N,N*-dipropyl-benzenamine *[4726-14-1]* formerly 4-(methylsulfonyl)-2,6-dinitro-*N,N*-dipropylaniline. Its herbicidal properties were first described by J. B. Regan *et al., Proc. Northeast. Weed Control Conf.,* 1966, p.36 and by R. H. Schieferstein & W. J. Hughes, *Proc. Br. Weed Control Conf., 8th,* 1966, p.377. It was introduced in 1966 by Shell Research Ltd, under the code number 'SD 11831', trade mark 'Planavin' and protected by BelgP 672 199.

Properties. It crystallises in orange prisms, m.p. 151-152°, v.p. 9·3 x 10^{-9} mmHg at 20°, 3·3 x 10^{-8} mmHg at 30°. Its solubility at 22° is: 0·6 mg/l water; 360 g/l acetone; 330 g/l dimethyl sulphoxide; 250 g/l 2-nitropropane; sparingly soluble in common hydrocarbon and aromatic solvents and alcohols. Contact of the solid with concentrated bases or excessive heat should be avoided.

Uses. Nitralin is a selective herbicide acting by disruption of primary cell wall formation during cell division. Pre-em. weed control of annual grasses and many broad-leaved weeds is successful in cotton, soyabeans, groundnuts, colza and transplanted crops such as tobacco, tomatoes, brassicas using 0·5-0·75 kg a.i./ha on light soils, 1·0-1·5 kg/ha on heavier soils (≤ 5% o.m.). It is not used in highly organic or peat soils. Shallow mechanical incorporation into the upper 2-4 cm of soil is recommended. It is relatively immobile in soil and 50% loss occurs in *c.* 30-50 d.

Toxicology. The acute oral LD50 for rats and mice is > 2000 mg/kg; the acute percutaneous LD50 for rabbits is > 2000 mg/kg. In 2-y feeding trials the 'no effect' level for rats and dogs was 2000 mg/kg diet. The LC50 (96 h) for fish is 46-68 mg/l.

Formulations. A w.p. (750 g a.i./kg); WDL 4, water-dispersible liquid (425 g/l).

Analysis. Product analysis is by i.r. spectroscopy or by GLC. Residues may be determined by GLC with ECD: details of methods are available from Shell International Chemical Co. Ltd. See also: *Anal. Methods. Pestic. Plant Growth Regul.,* 1973, **7,** 625.

Nitrapyrin

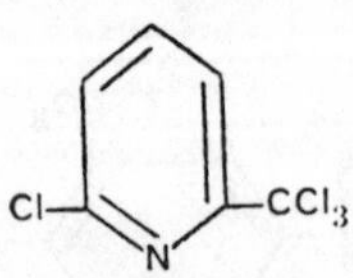

T6NJ BXGGG FG

$C_6H_3Cl_4N$ (230·9)

Nomenclature and development. The common name nitrapyrin is approved by BSI, ISO and ANSI. The IUPAC and *C.A.* name is 2-chloro-6-trichloromethylpyridine, Registry No. *[1929-82-4]*. It was introduced in 1962 by The Dow Chemical Co. as a soil bactericide under the code number 'DOWCO 163', trade mark 'N-Serve' and protected by USP 3 135 594; BP 961 109.

Properties. It is a colourless crystalline solid, m.p. 62-63°, v.p. 2·8 x 10^{-3} mmHg at 23°. It is practically insoluble in water; soluble in anhydrous ammonia, most organic solvents including acetone, ethanol, xylene.

Uses. Nitrapyrin is a nitrogen stabiliser because of its highly selective action as a soil bactericide against *Nitrosomonas* spp., the micro-organisms that oxidise ammonium ions in soil. It is hydrolysed in soil to 6-chloropicolinic acid which is absorbed by plants and is the principal metabolite (C. T. Redeman *et al., J. Agric. Food Chem.*, 1964, **12**, 207; 1965, **13**, 518).

Toxicology. The acute oral LD50 is: for rats 1072-1231 mg/kg; for chickens 235 mg/kg. In 90-d feeding trials the 'no effect' level was: for rats 300 mg/kg diet; for dogs 600 mg/kg diet.

Formulations. These include: technical grade (≥ 90% pure); concentrates, emulsifiable or non-emulsifiable (240 g a.i./l).

Analysis. Residues of 6-chloropicolinic acid, the main degradation product, may be determined by GLC of the methyl ester (D. J. Jensen, *ibid.,* 1971, **19**, 897; C. T. Redeman, *Bull. Environ. Contam. Toxicol.,* 1967, **2**, 289).

Nitrilacarb

$$\text{Me} \qquad \text{O}$$
$$| \qquad\qquad ||$$
$$\text{NC.CH}_2\text{CH}_2\text{.C.CH=N.O.C.NHMe}$$
$$|$$
$$\text{Me}$$

NC2X1&1&1UNOVM1   *nitrilacarb* $C_9H_{15}N_3O_2$ (197·2)

NC2X1&1&1UNOVM1   & . ZN . . G2

nitrilacarb zinc chloride complex $C_9H_{15}Cl_2N_3O_2Zn$ (333·5)

Nomenclature and development. The common name nitrilacarb is approved by BSI and ANSI and proposed by ISO. The IUPAC name is 4,4-dimethyl-5-(methylcarbamoyloxy-imino)pentanenitrile, in *C.A.* usage 4,4-dimethyl-5-[[[(methylamino)carbonyl]oxy]-imino]pentanenitrile *[29672-19-3]*. It was introduced in 1973 by American Cyanamid Co. under the code number 'CL 72 613', normally used as the 1:1 complex with zinc dichloride, code number 'AC 85 258', *C.A.* Registry No. *[61332-32-9]* or *[58270-08-9]*, proposed trade mark 'Accotril' and protected by USP 3 681 505; 3 621 049. Its insecticidal properties were described by W. K. Whitney & J. L. Aston, *Proc. Br. Insectic. Fungic. Conf., 8th,* 1975, **2,** 633.

Manufacture and properties. Produced by reacting methyl isocyanate with 5-hydroxy-imino-4,4-dimethylpentanenitrile to form the carbamate, which is then complexed with zinc dichloride, the technical grade is an almost colourless powder, m.p. 120-125°, ≏ 0·5 g/cm³. The complex is soluble in water, acetone, acetonitrile, alcohols; slightly soluble in chloroform; practically insoluble in benzene, diethyl ether, hexane, toluene, xylene. It is very hygroscopic and must be kept in tightly closed containers when not in use. Stored in the original containers ≤ 25°, the product is stable for ≥ 1·5 y.

Uses. Nitrilacarb is effective against a wide range of phytophagous mites and aphids as well as several other important pests, including whiteflies, thrips, leafhoppers and *Leptinotarsa decemlineata,* and is capable of translaminar activity. Rates of use are 25-75 g nitrilacarb/100 l; *L. decemlineata* is controlled by 0·5 kg/ha.

Toxicology. The acute oral LD50 of the zinc dichloride complex of technical grade nitrilacarb is: for male rats 9 mg/kg; for mice 6-18 mg/kg. The acute dermal LD50 (and duration of exposure) for rabbits is: 857 mg tech./kg (24 h), > 5000 mg tech./kg (4 h); > 500 mg a.i. (as w.p.)/kg (24 h), > 1250 mg a.i. (as w.p.)/kg (4 h). Aqueous solutions (10 g nitrilacarb/l) were non-irritating to rabbits' eyes and skin, but undiluted technical zinc dichloride complex is corrosive to eyes and irritating to skin. Sub-acute feeding studies on rats and dogs showed that nitrilacarb is tolerated at much higher levels than predicted from the acute oral LD50 values.

Formulation. 'AC 85 258 25-WP', w.p. (250 g nitrilacarb/kg) as zinc dichloride complex.

Analysis. Residues may be determined by GLC: details are available from American Cyanamid Co.

Nitrofen

WNR DOR BG DG $\qquad$ $C_{12}H_7Cl_2NO_3$ (284·1)

Nomenclature and development. The common name nitrofen is approved by BSI, ISO and WSSA—exceptions JMAF (NIP), Canada and Germany. The IUPAC name is 2,4-dichlorophenyl 4-nitrophenyl ether, in *C.A.* usage 2,4-dichloro-1-(4-nitrophenoxy)-benzene [1836-75-5] formerly 2,4-dichlorophenyl *p*-nitrophenyl ether. It was introduced *c.* 1964 as a herbicide by Rohm & Haas Co. under the code number 'FW-925', the trade marks 'Tok E-25' and 'Tokkorn' and protected by BP 974475; 3080225.

Manufacture and properties. Produced by the condensation of 1-chloro-4-nitrobenzene with an alkali 2,4-dichlorophenoxide, it is a crystalline solid, m.p. 70-71°, v.p. 8 x 10^{-6} mmHg at 40°. Its solubility at 22° is 0·7-1·2 mg/l water.

Uses. Nitrofen is a selective herbicide, toxic to a number of broad-leaved and grassy weeds. It is effective when left as a thin layer on the soil surface; activity is rapidly lost on incorporation in soil. It is used for weed control in cereals pre-em. at 2 kg a.i./ha. Many vegetable crops tolerate applications pre-em. $\leqslant$ 3 kg/ha and post-em. $\leqslant$ 1 kg/ha.

Toxicology. The acute oral LD50 is: for rats 630-755 mg a.i. (as e.c.)/kg; for rabbits 300-510 mg a.i. (as e.c.)/kg. Neither the a.i. nor the formulations caused irritation to rabbit skin or visible toxic effects.

Formulations. These include: 'Tok E-25', 'Tokkorn', e.c. (250 g a.i./kg; 240 g/l); 'Tok WP-50', w.p. (500 g/kg).

Analysis. Product analysis is by GLC. Residues may be determined by GLC after extraction with dichloromethane and chromatographic clean-up. See also: I. L Adler & B. M. James, *Anal. Methods Pestic. Plant Growth Regul.*, 1978, **10**, 403.

Nitrothal-isopropyl

1Y1&OVR   CNW   EVOY1&1 $C_{14}H_{17}NO_6$ (295·3)

Nomenclature and development. The common name nitrothal-isopropyl is approved by BSI and New Zealand and proposed by ISO. The IUPAC name is di-isopropyl 5-nitro-isophthalate, in *C.A.* usage bis(1-methylethyl) 5-nitro-1,3-benzenedicarboxylate *[10552-74-6]*. It was introduced in 1973, after trials since 1969, by BASF AG under the code number 'BAS 30 000F'. Its fungicidal properties were reported by E. L. Frick, *Rep. East Malling Res. Stn., 1972*, 1973, p.152 and by W. H. Phillips *et al., Proc. Br. Insectic. Fungic. Conf., 7th, 1973,* **2,** 673.

Manufacture and properties. Produced by the esterification of 5-nitroisophthalic acid with propan-2-ol, it is a yellow crystalline solid, m.p. 65°. It is practically insoluble in water; very soluble in most organic solvents.

Uses. Nitrothal-isopropyl is a non-systemic fungicide effective against powdery mildews. Sprays of 50 g a.i./100 l, at intervals of up to 14 d have controlled apple mildew without effect on red spider mite. It is used in combinations with other fungicides.

Toxicology. The acute oral LD50 is: for rats 6400 mg 50% w.p./kg; for rabbits > 10 000 mg/kg. In 90-d feeding trials the 'no effect' level was: for rats 500 mg/kg diet; for dogs 20 000 mg/kg diet.

Formulations. These include: 'Kumulan' ('BAS 38 501F') (167 g nitrothal-isopropyl + 533 g sulphur/kg); 'Pallinal' ('BAS 37 900F'), *C.A.* Registry No. *[55257-78-8]*, (125 g nitrothal-isopropyl + 600 g zineb-polyethylenethiuram disulphide mixed precipitation/kg).

Analysis. Product analysis is by GLC after separation by column chromatography. Residues may be determined by extraction with light petroleum, clean-up and measurement by GLC with ECD. Details of methods are available from BASF AG.

Norbormide

$$C_{33}H_{25}N_3O_3 \quad (511 \cdot 6)$$

T C555 A AY DVMV IUTJ AUYR&- BT6NJ& IXQR&- BT6NJ

Nomenclature and development. The common name norbormide is approved by BSI, ISO and ANSI—exception France (nobormide); also known as ENT 51 762. The IUPAC name is 5-(α-hydroxy-α-2-pyridylbenzyl)-7-(α-2-pyridylbenzylidene)bicyclo[2.2.1]hept-5-ene-2,3-dicarboximide, in *C.A.* usage 3a,4,7,7a-tetrahydro-5-(hydroxyphenyl-2-pyridyl-methyl)-7-(phenyl-2-pyridylmethylene)-4,7-methano-1*H*-isoindole-1,3(2*H*)-dione *[991-42-4]* formerly 5-(α-hydroxy-α-2-pyridylbenzyl)-7-(α-2-pyridylbenzylidene)-5-norbornene-2,3-dicarboximide. Its rodenticidal properties were first described by A. P. Roszkowski *et al., Science,* 1964, **144,** 412, and it was introduced in 1964 by the McNeil Laboratories Inc. under the code number 'McN-1025', trade marks 'Shoxin', 'Raticate' and protected by BP 1 059 405.

Manufacture and properties. Produced by condensing cyclopentadiene with 2-benzoyl-pyridine and subsequent reaction with maleimide (R. J. Mohrbacher *et al., J. Org. Chem.,* 1966, **31,** 2141), it is a colourless to off-white crystalline powder, m.p. > 160°. Its solubility at room temperature is 60 mg/l water; at 30°: 14 mg/l ethanol; > 150 mg/l chloroform; 1 mg/l diethyl ether; 29 mg/l 0·1 M hydrochloric acid. It is a mixture of stereo-isomers, stable at room temperature when dry, and to boiling water; hydrolysed by alkali. It is non-corrosive.

Uses. Norbormide is a selective rodenticide lethal to all members of the genus *Rattus* that have been tested; the acute oral LD50 for *R. rattus* is 52 mg/kg, for *R. norvegicus* 11·5 mg/kg, for *R. hawaiiensis c.* 10 mg/kg. It is relatively non-lethal to other rodents.

Toxicology. The acute oral LD50 is: for mice 2250 mg/kg; for hamsters 140 mg/kg; for prairie dogs > 1000 mg/kg; a single oral dose of 1000 mg/kg had no effect on cats, dogs, monkeys, chickens, turkeys; nor was it lethal to any of 40 other animal species tested (A. P. Roszkowski, *J. Pharmacol. Exp. Ther.,* 1965, **149,** 288). Three adult male volunteers received 15 mg/kg daily for 3 d without ill-effect.

Formulation. 'Raticate', concentrate (5-10 g a.i./kg) for bait preparation in cereal offals, fish meal, etc.; usually with a warning dye.

Analysis. Baits may be analysed by extraction with aqueous buffer (pH 6) and chloroform, the latter is washed with 0·1 M hydrochloric acid, the norbormide content of which is measured by u.v. spectroscopy (C. A. Janick *et al., J. Pharm. Sci.,* 1966, **55,** 1077).

Norflurazon

T6NNVJ BR CXFFF& DG EM1 $\qquad$ $C_{12}H_9ClF_3N_3O$ (303·7)

Nomenclature and development. The common name norflurazon is approved by BSI, ISO, ANSI and WSSA. The IUPAC name is 4-chloro-5-methylamino-2-(α,α,α-trifluoro-*m*-tolyl)pyridazin-3-one formerly 4-chloro-5-methylamino-2-(3-trifluoromethylphenyl)-pyridazin-3-one, in *C.A.* usage 4-chloro-5-(methylamino)-2-[3-(trifluoromethyl)phenyl]-3(2*H*)-pyridazinone *[27314-13-2]* formerly 4-chloro-5-methylamino-2-(α,α,α-trifluoro-*m*-tolyl)-3(2*H*)-pyridazinone. It was introduced in 1971 by Sandoz AG under the code numbers 'H 52 143' and 'H 9789', trade marks 'Zorial', 'Evital', 'Solicam' and protected by USP 3 644 355.

Manufacture and properties. Produced by the reaction of α,α,α-trifluoro-*m*-tolyl-hydrazine with (*Z*)-2,3-dichloro-3-formylacrylic acid followed by reaction with methylamine, it is a colourless crystalline solid, m.p. 174-180°, v.p. 2 x 10^{-8} mmHg at 20°. Its solubility at 23° is 28 mg/l water; moderately soluble in acetone, hot ethanol; sparingly soluble in hydrocarbons. It is stable under alkaline and acid conditions, is non-corrosive but is susceptible to light. The technical grade is *c.* 97% pure.

Uses. Norflurazon inhibits photosynthesis by reduction of carotenoid biosynthesis. It is a selective herbicide used in cotton, stone fruits, pome fruits, nuts and cranberries. It is effective against many annual broad-leaved weeds at 1-4 kg a.i./ha; and also suppresses perennial grass and sedge species, *e.g. Cynodon, Sorghum, Agropyron* and *Cyperus*. In soil 50% loss occurs in 21-28 d according to soil type and method of application. In plants a main metabolic reaction is *N*-demethylation.

Toxicology. The acute oral LD50 is: for rats > 8000 mg/kg; for bobwhite quail and mallard duck > 1250 mg/kg. The acute dermal LD50 for rabbits is > 20000 mg/kg. In 90-d feeding trials the 'no effect' level was: for rats 50 mg/kg daily; for dogs 12·5 mg/kg daily. The LC50 for catfish and goldfish is > 200 mg/l.

Formulations. These include: 'Zorial', w.p. (800 g a.i./kg); 'Evital', granules (50 g/kg); 'Solicam', w.p. (800 g/kg).

Analysis. Product analysis is by TLC, followed by u.v. spectroscopy of the eluted a.i. Residues may be determined by GLC with ECD: details from Sandoz AG. See also: S. S. Brady *et al., Anal. Methods Pestic. Plant Growth Regul.,* 1978, **10**, 415.

Nuarimol

T6N CNJ EXQR BG&R DF $\qquad$ $C_{17}H_{12}ClFN_2O$ (314·7)

Nomenclature and development. The common name nuarimol is approved by BSI, ANSI and New Zealand, and proposed by ISO. The IUPAC name is 2-chloro-4'-fluoro-α-(pyrimidin-5-yl)benzhydryl alcohol, in *C.A.* usage α-(2-chlorophenyl)-α-(4-fluorophenyl)-5-pyrimidinemethanol *[63284-71-9]*. It was introduced in 1975 by Eli Lilly & Co. under the code number 'EL-228', trade marks 'Trimidal' and 'Triminol' and protected by BP 1 218 623.

Manufacture and properties. Produced by the condensation of pyrimidin-5-yllithium with 2-chloro-4'-fluorobenzophenone, the latter being prepared by the Friedel-Crafts reaction of 2-chlorobenzoyl chloride with fluorobenzene, it is a colourless crystalline solid, m.p. 126-127°, v.p. $< 2 \times 10^{-8}$ mmHg at 25°. Its solubility at 25° is 26 mg/l water at pH7; soluble in acetone, acetonitrile, benzene, chloroform, methanol; slightly soluble in hexane. The material breaks down rapidly in sunlight to form a large number of minor degradation products.

Uses. Nuarimol is a systemic fungicide with activity against a wide range of plant pathogenic fungi. It is used as a foliar spray for control of *Erysiphe graminis* on barley at 40 g a.i./ha and as a seed dressing on wheat and barley for control of *E. graminis, Leptosphaeria nodorum, Pyrenophora graminea, Fusarium* and *Ustilago* spp., at rates of 100-200 mg/kg seed.

Toxicology. The acute oral LD50 is: for rats 1250-2000 mg/kg; for mice 2500-3000 mg/kg. Oral doses of 500 mg/kg had no effect on beagle dogs. In 90-d feeding trials it produced no effect on rats receiving 50 mg/kg diet or mice 225 mg/kg diet. No observable effects were produced in bluegill exposed to 1·1 mg/l in a continuous flow-through system for 7 d.

Formulation. An e.c. (90 g a.i./l).

Analysis. Technical product and formulation analysis is by GLC with FID following dissolution in or extraction with chloroform or methanol. Residues in plant tissue and soils are measured by GLC with ECD following suitable clean-up of sample extracts. Detailed analytical procedures and information on nuarimol may be obtained from Eli Lilly & Co.

1,4,4a,5a,6,9,9a,9b-Octahydrodibenzofuran-4a-carbaldehyde

HCO

T B656 HO DU KUTJ GVH $C_{13}H_{16}O_2$ (204·3)

Nomenclature and development. The IUPAC name is 1,4,4a,5a,6,9,9a,9b-octahydrodi-benzofuran-4a-carbaldehyde, in *C.A.* usage 1,5a,6,9,9a,9b-hexahydro-4a(4*H*)-dibenzo-furancarboxaldehyde *[126-15-8]*; also known as ENT 17596 and 2,3:4,5-bis(2-butylene)-tetrahydro-2-furaldehyde. It was introduced in 1949 by Phillips Petroleum Co. and its repellent action on cockroaches was first described by L. D. Goodhue & C. Linnaid, *J. Econ. Entomol.*, 1952, **45**, 133. It is marketed as 'MGK Repellent 11' by McLaughlin Gormley King Co.

Manufacture and properties. Produced by the condensation of buta-1,3-diene with 2-furaldehyde (USP 2 683 511; 2 934 471), it is a pale yellow liquid, with a fruity odour, b.p. 307°, n_D^{20} 1·5420. The technical grade has d_{20}^{20} 1·120. It is practically insoluble in water and in dilute alkali; miscible with petroleum oils, toluene, xylene, ethanol. It is non-corrosive and stable for long periods in drums; compatible with most insecticides.

Uses. It is an insect repellent effective against cockroaches, stable and horn flies, mosquitoes and gnats. Its main uses now are in pet sprays and in repellents for personal use in combination with other materials. It is no longer recommended for use on milk- or meat-producing animals.

Toxicology. The acute oral LD50 for rats is 2500 mg/kg; the acute dermal LD50 for rabbits is > 2000 mg/kg. In 90-d feeding trials rats receiving 20 000 mg/kg diet suffered no gross ill-effect.

Analysis. Product analysis is by GLC: details of the method are available from McLaughlin Gormley King Co.

Omethoate

$$O \quad\quad O$$
$$\| \quad\quad \|$$
$$(MeO)_2P.S.CH_2.C.NHMe$$

1OPO&O1&S1VM1 $C_5H_{12}NO_4PS$ (213·2)

Nomenclature and development. The common name omethoate is approved by BSI and ISO—exception Italy; also known as ENT 25776. The IUPAC name is *O,O*-dimethyl *S*-methylcarbamoylmethyl phosphorothioate, in *C.A.* usage *O,O*-dimethyl *S*-[2-(methylamino)-2-oxoethyl] phosphorothioate *[1113-02-6]* formerly *O,O*-dimethyl phosphorothioate *S*-ester with 2-mercapto-*N*-methylacetamide. It was first described as an insecticide by R. Santi & P. de Pietri-Tonelli, *Nature (London)*, 1959, **183**, 398, and was introduced in 1965 by Bayer AG under the code numbers 'Bayer 45432', 'S 6876', trade mark 'Folimat' and protected by DAS 1251 304.

Manufacture and properties. Produced either by the reaction of *O,O*-dimethyl hydrogen phosphorothioate with 2-chloro-*N*-methylacetamide or of methylamine with *S*-(methoxycarbonylmethyl) *O,O*-dimethyl phosphorothioate, it is a colourless to yellow oil, decomposing *c.* 135°, v.p. 2·5 x 10^{-5} mmHg at 20°, d_4^{20} 1·32, n_D^{20} 1·4987. It is miscible with water, acetone, ethanol and many hydrocarbons; slightly soluble in diethyl ether; almost insoluble in light petroleum. It is hydrolysed by alkali; 50% decomposition occurs at 24° in 2·5 d at pH 7.

Uses. Omethoate is a systemic acaricide and insecticide effective against aphids, woolly aphids, thrips, scale insects, caterpillars and beetles, especially on hops, and against wheat bulb fly. The usual rate is 50 g a.i./100 l.

Toxicology. The acute oral LD50 for rats is *c.* 50 mg/kg; the acute dermal LD50 (7 d) for rats is 700 mg/kg. Harmful to honey bees.

Formulations. These include soluble concentrates with a range of a.i. content; also concentrates for ULV application.

Analysis. Product analysis is by i.r. spectroscopy or by titration with potassium methoxide in ethylenediamine using azoviolet as indicator. Residues may be determined by GLC with FID (K. Wagner & H. Frehse, *Pflanzenschutz-Nachr. (Am. Ed.)*, 1976, **29**, 54) or by oxidation of the residue to phosphate which is measured colorimetrically by standard methods. Details of these methods are available from Bayer AG.

Oryzalin

ZSWR CNW ENW DN3&3

$C_{12}H_{18}N_4O_6S$ (346·4)

Nomenclature and development. The common name oryzalin is approved by BSI, ISO, ANSI and WSSA. The IUPAC name is 3,5-dinitro-N^4,N^4-dipropylsulphanilamide, in C.A. usage 4-(dipropylamino)-3,5-dinitrobenzenesulfonamide *[19044-88-3]* formerly 3,5-dinitro-N^4,N^4-dipropylsulfanilamide. It was introduced by Eli Lilly & Co. under the code number 'EL-119', trade marks 'Dirimal' and 'Surflan' and protected by USP 3367949. Its herbicidal properties were first described by J. V. Gramlich *et al.*, 1969 abstr. WSSA.

Manufacture and properties. Produced by the reaction of 4-dipropylamino-3,5-dinitro-benzenesulphonyl chloride and ammonium hydroxide, it is a yellow-orange crystalline solid, m.p. 141-142°. Its solubility at 25° is 2·4 mg/l water; readily soluble in polar organic solvents, *e.g.* acetone, ethanol, acetonitrile; only slightly soluble in benzene, xylene; insoluble in hexane.

Uses. Oryzalin is recommended for use, alone or in combination with other herbicides, as a pre-em. surface spray on soyabeans, cotton, groundnuts, winter rapeseed, sunflower and certain established crops at 1·0-2·0 kg a.i./ha.

Toxicology. The acute oral LD50 of the technical material is: for rats and gerbils > 10000 mg/kg; for cats and chickens 1000 mg/kg; for dogs > 1000 mg/kg. It caused a slight dermal but not eye irritation in rabbits and the dermal LD50 was > 2000 mg/kg. It had no effect on rats or dogs when fed for 90 d at 750 mg/kg diet. In teratology studies in rats there was no effect at 2250 mg/kg diet. The LC50 (96 h) for goldfish fingerlings is > 1·4 mg/l.

Formulation. A w.p. (750 g a.i./kg).

Analysis. Product and residue analysis is by TLC and GLC; particulars from Eli Lilly & Co. The a.i. is separated from the formulation by column chromatography and the effluent monitored spectrophotometrically. Residues are determined on the dimethyl derivative by GLC with ECD. See also O. D. Decker & W. S. Johnson, *Anal. Methods Pestic. Plant Growth Regul.,* 1976, **8,** 433.

Oxadiazon

T5NNVOJ BR BG DG EOY1&1& EX1&1&1 $C_{15}H_{18}Cl_2N_2O_3$ (345·2)

Nomenclature and development. The common name oxadiazon is approved by BSI, ISO, ANSI, WSSA and JMAF. The IUPAC name is 5-*tert*-butyl-3-(2,4-dichloro-5-isopropoxy-phenyl)-1,3,4-oxadiazol-2-one, in *C.A.* usage 3-[2,4-dichloro-5-(1-methylethoxy)phenyl]-5-(1,1-dimethylethyl)-1,3,4-oxadiazol-2(3*H*)-one *[19666-30-9]* formerly 2-*tert*-butyl-4-(2,4-dichloro-5-isopropoxyphenyl-$\triangle^2$-1,3,4-oxadiazolin-5-one. It was introduced in 1969 by Rhône-Poulenc Phytosanitaire under the code number RP 17 623, trade mark 'Ronstar' and protected by BP 1 110 500; USP 3 385 862. Its herbicidal properties were described by L. Burgaud *et al., Symp. New Herbic., 3rd,* 1969, p.201.

Manufacture and properties. Produced by the cyclisation, with carbonyl chloride, of 1-(2,4-dichloro-5-isopropoxyphenyl)-2-pivaloylhydrazine, which is obtained by acylation of the corresponding phenylhydrazine, it forms colourless crystals, m.p. *c.* 90°, v.p. $< 10^{-6}$ mmHg at 20°. Its solubility at 20° is: 0·7 mg/l water; 100 g/l ethanol, methanol; 600 g/l acetone, acetophenone, anisole; 1 kg/l benzene, toluene, chloroform. It is stable under normal storage conditions and non-corrosive.

Uses. Oxadiazon is a selective herbicide effective against mono- and di-cotyledonous weeds in rice at *c.* 1 kg/ha; in vineyards and orchards 2 kg/ha post-em. or 4 kg/ha pre-em.; on carnations 4 kg/ha. At these rates under temperate conditions 50% loss from soil occurs in 90-180 d.

Toxicology. The acute oral LD50 is: for rats >8000 mg/kg; for bobwhite quail 6000 mg/kg; for mallard duck 1000 mg/kg. The acute dermal LD50 for rats is >8000 mg/kg. In feeding trials rats and dogs receiving 25 mg/kg daily were not affected.

Formulations. These include e.c. (250 g a.i./l); granules (20 g/kg).

Analysis. Product and residue analysis is by GLC (J. Demoras *et al., Anal. Methods Pestic. Plant Growth Regul.,* 1973, **7**, 595).

Oxamyl

$$\underset{\underset{\displaystyle SMe}{|}}{Me_2N.\overset{\displaystyle O}{\overset{\|}{C}}.C}=N.O.\overset{\displaystyle O}{\overset{\|}{C}}.NHMe$$

1N1&VYS1&UNOVM1 $\hspace{6cm}$ $C_7H_{13}N_3O_3S$ (219·3)

Nomenclature and development. The common name oxamyl is approved by BSI, ANSI and JMAF and proposed by ISO. The IUPAC name is N,N-dimethyl-2-methyl-carbamoyloxyimino-2-(methylthio)acetamide formerly N,N-dimethyl-α-methylcarba-moyloxyimino-α-(methylthio)acetamide and S-methyl N',N'-dimethyl-N-(methylcarba-moyloxy)-1-thio-oxamimidate (I), in *C.A.* usage methyl 2-(dimethylamino)-N-[[(methyl-amino)carbonyl]oxy]-2-oxoethanimidothioate *[23135-22-0]* formerly (I). It was introduced in 1969 by E. I. du Pont de Nemours & Co. (Inc.) under the code number 'DPX 1410', it is now marketed under the trade mark 'Vydate' insecticide nematicide, protected by USP 3530220; 3658870.

Properties. It is a colourless crystalline solid, with a slight sulphurous odour, m.p. 100-102°, changing to a dimorphic form, m.p. 108-110°, v.p. 2·3 x 10^{-4} mmHg at 25°, 7·6 x 10^{-3} mmHg at 70°. Its solubility at 25° is: 280 g/kg water; 670 g/kg acetone; 330 g/kg ethanol; 110 g/kg propan-2-ol; 1·44 kg/kg methanol; 10 g/kg toluene. The aqueous solution is non-corrosive. The solid and most solutions are stable, but it decomposes to innocuous materials in natural waters and in soil. Aeration, sunlight, alkalinity and higher temperatures increase the rate of decomposition.

Uses. Oxamyl controls insects, mites and nematodes. A foliar spray 0·28-1·12 kg a.i./ha is effective by contact, moderately residual and controls a broad range of insect and mite species. Soil application in furrow at 1·12-3·36 kg/ha controls foliar-feeding insects and mites by systemic action. It is also a contact broad-range nematicide incorporated into soil pre-planting at 4·48-16·8 kg/ha. It is unique in functioning as a nematicide when applied to foliage of certain plants through downward translocation to the roots—redistribution of the compound through plants has been studied (C. A. Peterson *et al., Pestic. Biochem. Physiol.*, 1978, **8**, 1).

Toxicology. The acute oral LD50 is: for rats 5·4 mg a.i./kg, 8·9 mg a.i. (as liquid formulation)/kg; for coturnix quail 4·18 mg/kg. The acute dermal LD50 for male rabbits is 710 mg a.i. (as liquid formulation)/kg; it is not a skin sensitiser. There was no evidence of cumulative toxicity to male rats fed 2·4 mg/kg in 10 doses over a 14-d period. The LC50 (96 h) is: for bluegill 5·6 mg/l; for goldfish 27·5 mg/l; for rainbow trout 4·2 mg/l.

Formulations. These include: 'Vydate' L insecticide/nematicide, water-soluble liquid (240 g a.i./l); 'Vydate' G, granules (100 g/kg).

Analysis. Residues may be determined by GLC (R. F. Holt & H. L. Pease, *J. Agric. Food Chem.*, 1976, **24**, 263) or HPLC (J. E. Thean *et al., J. Assoc. Off. Anal. Chem.*, 1978, **61**, 15). See also R. F. Holt & R. G. Leitch, *Anal. Methods Pestic. Plant Growth Regul.*, 1978, **10**, 111.

Oxine-copper

$$C_{18}H_{12}CuN_2O_2 \quad (351 \cdot 9)$$

D566 1A L BND-CU-OJ C-& CD566 1A L BND-CU-OJ

Nomenclature and development. The common name oxine-copper is approved by BSI and ISO—exception France (oxyquinolinoléate de cuivre). The IUPAC name is bis(8-quinolinolato)copper (I) formerly known as cupric 8-quinolinoxide, in *C.A.* usage bis(8-quinolinolato-*N¹,O⁸*)copper *[10380-28-6]* formerly (**I**). It is a complex of copper and quinolin-8-ol, the latter has long been known by the trivial name oxine. It was introduced for crop protection in 1946 by D. Powell, *Phytopathology,* 1946, **36**, 572.

Manufacture and properties. Produced by the precipitation from solutions of copper salts with quinolin-8-ol, it is a greenish-yellow crystalline powder, stable $\leqslant 200°$, and non-volatile. It is insoluble in water, ethanol and common organic solvents; slightly soluble in pyridine. The co-ordinating complex is stable, chemically inert, and not degraded by u.v. light.

Uses. Oxine-copper is a protectant fungicide which is used on apples against *Venturia inaequalis* (D. Powell, *loc. cit.*), on soft fruit and nuts and market garden crops against *Anthracnose* and *Didymella* spp. at 50-70 g a.i./100 l. It is also used as a mixture with bitumen for sealing wounds and pruning cuts on trees. It is non-phytotoxic. It is also used for rendering fabrics mildew-proof (P. G. Benignus, *Ind. Eng. Chem.,* 1948, **40**, 1426; W. I. Illman, *Can. J. Res., Sec. F,* 1948, **26**, 311).

Toxicology. The acute oral LD50 for rats is *c.* 10 000 mg/kg; the intraperitoneal LD50 for mice is 67 mg/kg. It is non-irritating to the skin. The LC50 (48 h) for brown and rainbow trout is 0·2-0·3 mg/l (J. S. Alabaster, *Proc. Br. Weed Control Conf. 4th,* 1958, p.84). It is non-toxic to honey bees.

Formulations. These include: 'Quinolate 400', e.c. (400 g a.i./l); 'Arbrex 805', a mixture with bitumen.

Analysis. Product analysis is by reaction with sulphuric acid with standard AOAC Methods for copper content (*CIPAC Handbook,* 1970, **1**, 226). Residues in fabrics may be determined by digesting with dilute sulphuric acid, adjusting to pH 6 and colorimetry at 410 nm of a chloroform extract (A. Rose *et al., Am. Dyest. Rep.,* 1956, **45**, 362).

Oxycarboxin

T6O DSW BUTJ B1 CVMR $C_{12}H_{13}NO_4S$ (267·3)

Nomenclature and development. The common name oxycarboxin is approved by BSI, ISO, ANSI and JMAF. The IUPAC name is 2,3-dihydro-6-methyl-5-phenylcarbamoyl-1,4-oxathiin 4,4-dioxide, in *C.A.* usage 5,6-dihydro-2-methyl-*N*-phenyl-1,4-oxathiin-3-carboxamide 4,4-dioxide *[5259-88-1]* formerly 5,6-dihydro-2-methyl-1,4-oxathiin-3-carboxanilide 4,4-dioxide. It was introduced in 1966 by Uniroyal Inc. under the code number 'F 461', trade mark 'Plantvax' and protected by USP 3 399 214; 3 402 241; 3 454 391. Its fungicidal properties were described by B. von Schmeling & M. Kulka, *Science,* 1966, **152**, 659. The structure-activity relationships of carboxanilides are discussed by P. ten Haken & C. L. Dunn, *Proc. Br. Insectic. Fungic. Conf., 6th,* 1971, **2**, 453.

Manufacture and properties. Produced by oxidation of carboxin (p.86) with hydrogen peroxide, it forms colourless crystals, m.p. 127·5-130°. Its solubility at 25° is 1 g/l water; 360 g/kg acetone; 34 g/kg benzene; 2·23 kg/kg dimethyl sulphoxide; 30 g/kg ethanol; 70 g/kg methanol. It is compatible with all except highly acidic or alkaline pesticides.

Uses. Oxycarboxin is a systemic fungicide used for the treatment of rust diseases of vegetables, ornamentals (as w.p.), and foliar application to cereals at 200-400 g a.i. (as e.c.)/ha.

Toxicology. The acute oral LD50 for rats is 2000 mg/kg; the acute dermal LD50 for rabbits is > 16 000 mg/kg.

Formulations. These include: e.c. (200 g a.i./l); w.p. (750 g/kg).

Analysis. Product and residue analysis is by hydrolysis and determination of the aniline so formed, either by GLC (H. R. Siskin & J. E. Newell, *J. Agric. Food Chem.,* 1971, **19**, 738) or by diazotisation, coupling with 4-dimethylaminobenzaldehyde and colorimetric measurement of the resultant dye (J. R. Lane, *ibid.,* 1970, **18**, 409).

Oxydemeton-methyl

$$(MeO)_2P.SCH_2CH_2.S.Et$$

with two O double bonds above P and S.

OS2&2SPO&O1&O1 $C_6H_{15}O_4PS_2$ (246·3)

Nomenclature and development. The common name oxydemeton-methyl is approved by BSI and ISO—exception USSR (metilmercaptofosoksid); also known as ENT 24964. The IUPAC name is S-2-ethylsulphinylethyl O,O-dimethyl phosphorothioate, in *C.A.* usage S-[2-(ethylsulfinyl)ethyl] O,O-dimethyl phosphorothioate *[301-12-2].* It was introduced in 1960, after tests since 1956, by Bayer AG under the code numbers 'Bayer 21097', 'R 2170', the trade mark 'Metasystox-R' and protected by DBP 947368; USP 2963505.

Manufacture and properties. Produced by the oxidation of demeton-S-methyl (p.155) with hydrogen peroxide (DBP 947368) or with chlorine or bromine (DBP 949229), it is a clear amber-coloured liquid; m.p. < —10°, b.p. 106°/0·01 mmHg, d_4^{20} 1·289, n_D^{20} 1·5216. It is miscible with water; soluble in most organic solvents except light petroleum. It is hydrolysed in alkaline media.

Uses. Oxydemeton-methyl is a systemic and contact insecticide suitable for the control of sap-feeding insects and mites, with a range of action similar to that of demeton-S-methyl of which it is a metabolic product.

Toxicology. The acute oral LD50 for rats is 65-80 mg/kg; the acute intraperitoneal LD50 for rats is 20 mg/kg; the acute dermal LD50 for male rats is 250 mg/kg. In 84-d feeding trials rats receiving 20 mg/kg diet suffered a slight cholinesterase depression.

Formulations. These include: s.c. (500 g a.i./l, with water-miscible solvent); e.c. (250 g/l, with emulsifier); aerosol concentrate.

Analysis. Product analysis is either by TLC and colorimetric determination, after oxidation, of the zone from the plate by standard methods for phosphate (*CIPAC Handbook,* in press) or by reduction of the sulphoxide group with an excess of dititanium trisulphate and back titration with iron trichloride (details from Bayer AG). Residues in plants, soil or water may be determined, after oxidation to the corresponding sulphone (demeton-S-methyl sulphone), by GLC with FID (K. Wagner & J. S. Thornton, *Pflanzenschutz-Nachr.,* 1977, **30,** 1; J. H. van der Merwe & W. B. Taylor, *ibid.,* 1971, **24,** 259; M. C. Bowerman & M. Beroza, *J. Assoc. Off. Anal. Chem.,* 1969, **52,** 154). See also: *Anal. Methods Pestic. Plant Growth Regul.,* 1972, **6,** 432.

Paraquat

$$Me-\overset{+}{N}=\!\!\!\!\!\bigcirc\!\!\!\!\!-\!\!\!\!\!\bigcirc\!\!\!\!\!=\overset{+}{N}-Me \qquad 2A^{-}$$

T6KJ	A1	D-	DT6KJ	A1			*paraquat* $C_{12}H_{14}N_2$ (186·3)
T6KJ	A1	D-	DT6KJ	A1	&G	&G	*paraquat dichloride* $C_{12}H_{14}Cl_2N_2$ (257·2)

Nomenclature and development. The common name paraquat is approved by BSI, ISO, ANSI, WSSA and JMAF—exception Germany. The IUPAC name is 1,1'-dimethyl-4,4'-bipyridyldiylium ion, in *C.A.* usage 1,1'-dimethyl-4,4'-bipyridinium ion *[4685-14-7]*. It is normally formulated as the dichloride, Registry No. *[1910-42-5]*, though a di(methyl sulphate), *[2074-50-2]*, was at one time sold. Both salts were introduced in 1958 by ICI Ltd, Plant Protection Division under the protection of BP 813 531 and under the code numbers 'PP 148' and 'PP 910', respectively. Formulations containing the dichloride are sold under the trade marks 'Gramoxone', 'Dextrone X', 'Esgram' and 'Weedol' (which also contains diquat dibromide). Mixtures of paraquat with various residual herbicides are marketed as 'Dexuron', 'Tota-Col', 'Gramuron', 'Para-Col', 'Pathclear' and 'Gramonol'. The herbicidal activity of paraquat salts was first reported by R. C. Brian, *Nature (London)*, 1958, **181**, 446. Its properties have been reviewed (A. Calderbank, *Adv. Pest Control Res.*, 1968, **8**, 127).

Manufacture and properties. Produced by reacting aqueous methyl chloride with 4,4'-bipyridyl, which is obtained by treating pyridine with sodium in liquid ammonia giving a pyridyl radical which dimerises to tetrahydro-4,4'-bipyridyl and this oxidised by air, as colourless crystals of the dichloride, which decomposes *c.* 300°. It is very soluble in water; slightly in the short-chain alcohols; insoluble in hydrocarbons. These properties are shared by paraquat di(methyl sulphate) which is produced by the reaction of 4,4'-bipyridyl with dimethyl sulphate in water. Neither compound is isolated from the technical products which are > 95% pure when produced by the routes outlined above. Both salts are stable under acid conditions but are hydrolysed by alkali. They are in general compatible with non-alkaline aqueous solutions but are inactivated by inert clays and by anionic surfactants. The unformulated products are corrosive to common metals; when diluted they present no significant hazard to spray equipment. Paraquat is non-volatile and has no measurable vapour pressure.

Uses. Paraquat destroys green plant tissue by contact action with some translocation. It is rapidly inactivated on contact with soil. Uses include stubble cleaning (140-840 g a.i./ha); pasture renovation, *i.e.* the killing of unproductive grass in such a way that the ground can be resown without ploughing (140-2210 g/ha); inter-row weed control in vegetable crops (560-1120 g/ha); desiccation of various crops (140-1680 g/ha); weed control in plantation crops (280-560 g/sprayed ha).

Toxicology. The acute oral LD50 is: for rats 150 mg/kg; for mice 104 mg/kg; for sheep 70 mg/kg; for dogs 25-50 mg/kg; for hens 262 mg/kg. The acute dermal LD50 for rabbits is *c.* 236 mg/kg. Paraquat is absorbed through human skin only after prolonged exposure; shorter exposures can cause irritation and a delay in the healing of cuts and wounds. It is an irritant to the eyes, can cause temporary damage to the nails and, if inhaled, may cause nose bleeding. In 2-y feeding trials the 'no effect' level was: for dogs 34 mg/kg diet; for rats 170 mg/kg diet.

continued

Paraquat—*continued*

Formulations. These include paraquat dichloride as: aqueous concentrates (100-240 g paraquat/l); water-soluble granules (25 g paraquat/kg); mixtures (other herbicides + 100-120 g paraquat/l), and water-soluble granules (25 g paraquat + 25 g diquat/kg).

Analysis. Product analysis is by colorimetry at 600 nm of the colour produced with alkaline sodium dithionite (*CIPAC Handbook*, 1970, **1**, 547); impurities are determined by GLC (*ibid.*, 1979, **1A**, in press); FAO Specification (CP/50), (*Herbicides 1977*, pp.52, 54). Residues may be determined, after clean-up on an ion-exchange column, by reduction and colorimetry at 396 nm (A. Calderbank & S. H. Yuen, *Analyst (London)*, 1965, **90**, 99; T. H. Byast *et al., Agric. Res. Counc. (G.B.) Weed Res. Organ., Tech. Rep.*, No. 15 (2nd Ed.), p.35). See also: D. E. Pack, *Anal. Methods Pestic., Plant Growth Regul. Food Addit.*, 1967, **5**, 453; J. B. Leary, *Anal. Methods Pestic. Plant Growth Regul.*, 1978, **10**, 321. Details of methods are available from ICI Ltd, Plant Protection Division.

Parathion

$$\text{(EtO)}_2\overset{\displaystyle \overset{S}{\parallel}}{P}\text{-O-}\left\langle\!\!\!\bigcirc\!\!\!\right\rangle\text{-NO}_2$$

WNR   DOPS&O2&O2 $C_{10}H_{14}NO_5PS$ (291·3)

Nomenclature and development. The common name parathion is approved by BSI and ISO—exception USSR (thiophos); also known as ENT 15 108. The IUPAC name is *O,O*-diethyl *O*-4-nitrophenyl phosphorothioate (I), in *C.A.* usage (I) *[56-38-2]* formerly *O,O*-diethyl *O-p*-nitrophenyl phosphorothioate. It was developed by G. Schrader and first reported by H. Martin & H. Shaw, *BIOS, Final Report,* No. 1095, 1946. It was introduced in 1947 by the American Cyanamid Co. under the code number 'ACC 3422' and trade mark *'Thiophos',* but this firm no longer manufacture or market it; in 1948 by Bayer AG under the code number 'E-605' and trade marks 'Folidol' and 'Bladan'; other trade marks are 'Niran' (Monsanto Chemical Co.), 'Fosferno' (ICI Ltd, Plant Protection Division). It was protected by DBP 814 152; USP 1 893 018; 2 842 063.

Manufacture and properties. Produced by the condensation of *O,O*-diethyl phosphorochloridothioate with sodium 4-nitrophenoxide, it is a pale yellow liquid, b.p. 157-162°/0·6 mmHg, v.p. 3·78 x 10⁻⁵ mmHg at 20°, d_4^{25} 1·265, n_D^{25} 1·5370. Its solubility at 25° is 24 mg/l water; slightly soluble in petroleum oils; miscible with most organic solvents. The technical product is a brown liquid with a garlic-like odour. It is rapidly hydrolysed in alkaline media; at pH 5-6 and 25°, 1% in 62 d. On heating, it isomerises to the *O,S*-diethyl isomer.

Uses. Parathion is a non-systemic contact and stomach insecticide and acaricide with some fumigant action. It is non-phytotoxic except to some ornamentals and, under certain weather conditions, to pears and some apple varieties.

Toxicology. The acute oral LD50 is: for male rats 13 mg/kg; for female rats 3·6 mg/kg. The acute dermal LD50 is: for male rats 21 mg/kg; for female rats 6·8 mg/kg. Its toxicity is enhanced by a metabolic oxidation to diethyl 4-nitrophenyl phosphate.

Formulations. These include: w.p. and e.c. of various a.i. content; also dusts, smokes and aerosol concentrates.

Analysis. Product analysis is by GLC or HPLC (*CIPAC Handbook,* in press), by reduction followed by titration with sodium nitrite of the amino group so formed (AOAC Methods; *WHO Specifications for Pesticides,* 1961, 2nd Ed., p.70; FAO Specification (CP/32)) or by alkaline hydrolysis and spectrometry at 405 nm of the liberated 4-nitrophenol (*CIPAC Handbook,* 1970, **1,** 550). See also: J. C. Gage, *Analyst (London),* 1952, **77,** 123. Residues may be determined by GLC (E. Möllhoff, *Pflanzenschutz-Nachr. (Am. Ed.),* 1968, **21,** 331). See also: G. L. Sutherland & R. Miskus, *Anal. Methods Pestic., Plant Growth Regul. Food Addit.,* 1964, **2,** 321; *Anal. Methods Pestic. Plant Growth Regul.,* 1972, **6,** 321; D. E. Coffin, *Residue Rev.,* 1964, **7,** 61.

Parathion-methyl

$$\underset{(MeO)_2P.O-}{\overset{\overset{\textstyle S}{\|}}{}}\!\langle\bigcirc\rangle\!\!-NO_2$$

(MeO)$_2$P.O—⟨ ⟩—NO$_2$ with S double-bonded to P

WNR   DOPS&O1&O1 $C_8H_{10}NO_5PS$ (263·2)

Nomenclature and development. The common name parathion-methyl is approved by BSI and ISO—exceptions ESA (methyl parathion) and USSR (metaphos); also known as ENT 17 292. The IUPAC name is *O,O*-dimethyl *O*-4-nitrophenyl phosphorothioate (I), in *C.A.* usage (I) *[298-00-0]* formerly *O,O*-dimethyl *O-p*-nitrophenyl phosphorothioate. It was first described by G. Schrader, *Angew. Chem.* Monograph No. 52, 2nd Ed., 1952. It was introduced in 1949 by Bayer AG under the trade marks 'Folidol-M', 'Metacide', 'Bladan M', 'Nitrox 80' and formerly *'Dalf'* and protected by DBP 814 142.

Manufacture and properties. Produced by the reaction of *O,O*-dimethyl phosphoro-chloridothioate with sodium 4-nitrophenoxide, it is a colourless crystalline powder, m.p. 35-36°, v.p. 9·7 x 10^{-6} mmHg at 20°, d_4^{20} 1·358, n_D^{35} 1·5515. Its solubility at 25° is 55-60 mg/l water; slightly soluble in light petroleum and mineral oils; soluble in most other organic solvents. The technical product is a light to dark tan-coloured liquid *c.* 80% pure, f.p. *c.* 29°, d^{20} 1·20-1·22. It is hydrolysed by alkali and readily isomerises to the *O,S*-dimethyl analogue on heating. It is compatible with most other pesticides. It is a good methylating agent (G. Hilgetag, *Angew. Chem.*, 1959, **71**, 137).

Uses. Parathion-methyl is a non-systemic contact and stomach insecticide with some fumigant action. It is generally recommended at 15-25 g a.i./100 l. It is non-phytotoxic and non-persistent.

Toxicology. The acute oral LD50 is: for male rats 14 mg/kg; for female rats 24 mg/kg. The acute dermal LD50 for rats is 67 mg/kg. In 90-d feeding trials rats receiving 5 mg/kg diet showed no symptom of poisoning.

Formulations. These include: e.c., w.p. and dusts of various a.i. content.

Analysis. Product analysis is by GLC or HPLC (*CIPAC Handbook,* in press) or by hydrolysis to 4-nitrophenol which is determined colorimetrically at 405 nm (*ibid.,* 1970, **1**, 568). Residues may be determined by GLC (E. Möllhoff, *Pflanzenschutz-Nachr. (Am. Ed.),* 1968, **21**, 331).

Paris green

$$O$$
$$\|$$
$$(CH_3C.O)_2Cu.3Cu(AsO_2)_2$$

OV1   &OV1   &.CU4.AS-O2*6

$C_4H_6As_6Cu_4O_{16}$ (1014)

Nomenclature and development. The trivial name Paris green is normally used; also known as ENT 884, Emerald green, French green, Mitis green, Schweinfurtergrün. The IUPAC name is tetracopper bis(acetate) hexakis(arsenite), in *C.A.* usage bis(acetato)(hexametaarsenitotetracopper) *[12002-03-8]* (formerly *[1299-88-3]*). It was introduced *c.* 1867 for the control of *Leptinotarsa decemlineata* on potatoes.

Manufacture and properties. Produced by the interaction of sodium arsenite, copper sulphate and acetic acid, it is a green powder. It is sparingly soluble in water alone but, in combination with carbon dioxide, it is readily decomposed to give water-soluble and phytotoxic arsenic compounds.

Uses. Paris green is a stomach poison used as baits against slugs and other soil-dwelling pests and as a mosquito larvicide.

Toxicology. The acute oral LD50 for rats is 22 mg/kg. It is a violent poison to man when ingested and is apt to cause suppuration of open wounds.

Formulation. Powder.

Analysis. Product analysis for copper, total arsenic, water-soluble arsenic and acetate content is by standard methods (AOAC Methods; *MAFF Tech. Bull.,* No. 1, 1958, pp.61-63). Residues may be determined by the Gutzeit method (AOAC Methods).

Pebulate

$$O$$
$$\parallel$$
$$BuEtN.C.SPr$$

4N2&VS3 $C_{10}H_{21}NOS$ (203·3)

Nomenclature and development. The common name pebulate is approved by BSI, ISO, WSSA and JMAF. The IUPAC name is *S*-propyl butylethylthiocarbamate (I), in *C.A.* usage *S*-propyl butylethylcarbamothioate *[1114-71-2]* formerly (I). It was introduced in 1954 by Stauffer Chemical Co. under the code number 'R-2061', trade mark 'Tillam' and protected by USP 3 175 897. It was described by E. O. Burt, *Proc. South. Weed Conf., 12th,* 1959, p.19.

Manufacture and properties. Produced by the reaction of *N*-ethylbutylamine with carbonyl chloride to give butylethylcarbamoyl chloride, which is reacted with propane-1-thiol, it is a clear liquid, with an aromatic odour, b.p. 142·5°/20 mmHg, v.p. 6·8 x 10^{-2} mmHg at 30°, d^{20} 0·956, n_D^{30} 1·4750. Its solubility at 20° is 60 mg/l water; miscible with acetone, benzene, propan-2-ol, kerosine, methanol, toluene. It is stable and non-corrosive.

Uses. Pebulate is a pre-em. herbicide recommended for control of annual grasses, nut sedges and broad-leaved weeds in sugar beet, tomatoes and transplanted tobacco by soil incorporation at 4-6 kg a.i./ha. It is rapidly taken up by the roots, translocated throughout the plant and broken down into carbon dioxide.

Toxicology. The acute oral LD50 for rats is 1120 mg tech./kg; the acute dermal LD50 for rabbits is 4640 mg/kg. It is rapidly metabolised in animals; *c.* 50% of the compound administered to rats was expired as carbon dioxide in 3 d, *c.* 25% was excreted in the urine and 5% in the faeces. The LC50 (7 d) for bobwhite quail was 8400 mg/kg diet. The LC50 (96 h) for rainbow trout and bluegill is *c.* 7·4 mg/l.

Formulations. An e.c. (720 g a.i./l); granules (100 g/kg).

Analysis. Product analysis is by GLC (*CIPAC Handbook,* in press). Residues may be determined by GLC or by hydrolysis and conversion of the liberated *N*-ethylbutylamine to a copper dithiocarbamate complex which is measured colorimetrically: details of the methods are available from Stauffer Chemical Co. See also: G. G. Patchett *et al., Anal. Methods Pestic., Plant Growth Regul. Food Addit.,* 1964, **4**, 343; W. Y. Ja *et al., Anal. Methods Pestic. Plant Growth Regul.,* 1972, **6**, 698.

Pendimethalin

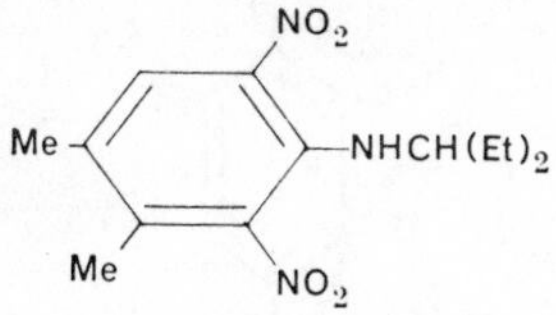

WNR B1 C1 ENW FMY2&2

$C_{13}H_{19}N_3O_4$ (281·3)

Nomenclature and development. The common name pendimethalin is approved by BSI, ANSI, New Zealand and by WSSA (who formerly used the name penoxalin). The IUPAC name is *N*-(1-ethylpropyl)-2,6-dinitro-3,4-xylidine, in *C.A.* usage *N*-(1-ethyl-propyl)-3,4-dimethyl-2,6-dinitrobenzenamine *[40487-42-1]*. It was introduced by American Cyanamid Co. under the code number 'AC 92 553', the trade marks 'Prowl', 'Stomp', and 'Herbadox' and protected by BelgP 787 939. Its herbicidal properties were described by P. L. Sprankle, *Proc. Br. Weed Control Conf., 12th,* 1974, **2**, 825.

Properties. It is an orange-yellow crystalline solid, m.p. 56-57°, v.p. 3×10^{-5} mmHg at 25°. Its solubility at 20° is 0·3 mg/l water; soluble in chlorinated hydrocarbons and aromatic solvents. It is stable to alkaline and acidic conditions, non-corrosive.

Uses. Pendimethalin is a selective herbicide effective against most annual grasses and several small-seeded annual broad-leaved weeds. It can be applied pre-em. after seeding in maize, cereals and rice or with shallow soil incorporation before seeding cotton, soyabeans, groundnuts and beans. It is also used to control suckers in tobacco.

Toxicology. The acute oral LD50 is: for albino rats 1050-1250 mg tech./kg; for albino mice 1340-1620 mg/kg; for beagle dogs > 5000 mg/kg. The acute dermal LD50 for albino rabbits is > 5000 mg/kg.

Formulations. These include: e.c. (330 and 500 g a.i./l); granules (30, 50 and 100 g/kg).

Analysis. Product analysis is by GLC (G. C. Wyckoff, *Anal. Methods Pestic. Plant Growth Regul.,* 1978, **10**, 461). Residues may be determined by GLC (A. Walker & W. Bond, *Pestic. Sci.,* 1977, **8**, 359; G. C. Wyckoff, *loc. cit.*). Details of analytical methods are available from American Cyanamid Co.

Pentachlorophenol

QR BG CG DG EG FG C_6HCl_5O (266·3)

Nomenclature and development. The IUPAC and *C.A.* name pentachlorophenol, Registry No. *[608-93-5]*, is accepted in lieu of a common name by BSI and ISO—exceptions JMAF and WSSA (PCP); also known as penchlorol. It was introduced *c.* 1936 as a timber preservative. Trade marks include: 'Dowicide EC7' (The Dow Chemical Co. Ltd), 'Monsanto Penta' for pentachlorophenol; *'Santobrite'* (Monsanto Chemical Co.), *'Dowicide G'* (The Dow Chemical Co. Ltd)—in both cases the product is no longer manufactured—and 'Cryptogil-Na' (Rhône-Poulenc Phytosanitaire) for the sodium salt.

Manufacture and properties. Produced by the catalytic chlorination of phenol (USP 2 131 259), it forms colourless crystals, with a phenolic odour, m.p. 191°, v.p. 0·12 mmHg at 100°, and volatile in steam. Its solubility at 30° is 20 mg/l; soluble in most organic solvents; slightly soluble in carbon tetrachloride, paraffins. The technical product is a dark grey powder or flakes, m.p. 187-189°. It is non-flammable. Though non-corrosive in the absence of moisture, its solutions in oil cause a deterioration of rubber, but synthetic rubbers may be used in equipment and protective clothing.

It is a weak acid, pK_a 4·71. The sodium salt, *C.A.* Registry No. *[131-52-2]*, forms a monohydrate, buff flakes, on crystallisation from water; its solubility at 25° is 330 g/l water, the solution having an alkaline reaction; insoluble in petroleum oils. The calcium and magnesium salts are also water-soluble.

Uses. Pentachlorophenol is used to control termites, to protect timber from fungal rots and wood-boring insects, as a pre-harvest defoliant and as a general herbicide. The sodium salt is used as a general disinfectant, *e.g.* for trays in mushroom houses. Its uses have been reviewed *(Pentachlorophenol)*.

Toxicology. The acute oral LD50 for rats is 210 mg/kg. In feeding trials dogs and rats receiving 3·9-10 mg/kg daily for 70-196 d suffered no fatality. It irritates mucous membranes and may provoke violent sneezing; the solid and aqueous solutions > 10 g/l cause skin irritation. The LC50 (48 h) for rainbow and brown trout is 0·17 mg sodium pentachlorophenoxide/l (J. S. Alabaster, *Proc. Br. Weed Control Conf.*, 4th, 1958, p.84).

Formulation. It is used as such; solutions in oil; technical sodium pentachlorophenoxide.

Analysis. Product analysis is by titration with alkali (*MAFF Tech. Bull.* No. 1, 1958, p.64). Residues may be determined by oxidation in benzene solutions with nitric and hydrochloric acids, the colour of the tetrachlorobenzoquinones being measured at 450 nm (*Monsanto Tech. Bull.* Me0-24), or by the formation with safranin-O of a coloured complex which is extracted into chloroform from buffered solution, and the colour intensity measured at 520-550 nm (W. T. Haskins, *Anal. Chem.*, 1951, **23**, 1672). See also: W. W. Kilgore & K. W. Cheng, *Anal. Methods Pestic., Plant Growth Regul. Food Addit.*, 1967, **5**, 313; *Anal. Methods Pestic. Plant Growth Regul.*, 1972, **6**, 581.

Pentanochlor

GR B1 EMVY3&1 $C_{13}H_{18}ClNO$ (293·7)

Nomenclature and development. The common name pentanochlor is approved by BSI and ISO—exceptions Canada, ANSI and WSSA (solan), JMAF (CMMP) and Turkey. The IUPAC name is 3'-chloro-2-methylvaler-*p*-toluidide also known as *N*-(3-chloro-*p*-tolyl)-2-methyl-valeramide, in *C.A.* usage *N*-(3-chloro-4-methylphenyl)-2-methylpentanamide *[2307-68-8]* formerly 3'-chloro-2-methyl-*p*-valerotoluidide. It was introduced by 1958 by the Agricultural Chemical Division of FMC Corp. under the code number 'FMC 4512' and protected by BP 869 169; USP 3 020 142, but this company no longer manufacture or sell it. Early work with it was described by D. H. Moore, *Proc. Northeast. Weed Control Conf.,* 1960, p.86; 1962, p.132.

Manufacture and properties. Produced by the reaction of 3-chloro-*p*-toluidine with 2-methylpentanoyl chloride in the presence of disodium carbonate, the technical product is a colourless to pale cream powder, m.p. 82-86° (pure 85-86°), d_{20}^{20} 1·106. Its solubility at room temperature is: 8-9 mg/l water; 460 g/kg di-isobutyl ketone; 550 g/kg isophorone; 520 g/kg 4-methylpentan-2-one; 410 g/kg pine oil; 200-300 g/kg xylene. It is stable to hydrolysis at room temperature and is non-corrosive.

Uses. Pentanochlor is a selective herbicide used post-em. to control annual weeds at < 4 kg a.i. in 500-1000 l water/ha in carrots, celery, parsley, strawberries, tomatoes and as a directed contact spray on carnations, chrysanthemums, roses, tomatoes, fruit and ornamental trees. It is also used pre-em. in combination with chlorpropham in narcissus and tulips in addition to many of the crops listed above. There is only a limited and short-lived residual activity in soil.

Toxicology. The acute oral LD50 for rats is > 10 000 mg/kg. In 140-d feeding trials rats receiving 20 000 mg/kg diet suffered no effect on body weight or survival, but histopathological changes were found in the liver. These effects were not observed at 2000 mg/kg diet.

Formulations. 'Herbon Solan' (Cropsafe), e.c. (400 g a.i./l); 'Herbon Brown' (Cropsafe), e.c. (pentanochlor + chlorpropham).

Analysis. Residues may be determined by hydrolysis to 3-chloro-*p*-toluidine which is estimated by GLC after clean-up.

Perfluidone

$$Ph-\overset{\overset{O}{\|}}{\underset{\underset{O}{\|}}{S}}-\text{benzene ring, with}-NH.\overset{\overset{O}{\|}}{\underset{\underset{O}{\|}}{S}}CF_3 \text{ and } Me$$

WSR&R   Cl   DMSWXFFF $C_{14}H_{12}F_3NO_4S_2$ (379·4)

Nomenclature and development. The common name perfluidone is approved by BSI, ANSI, New Zealand and WSSA, and proposed by ISO. The IUPAC name is 1,1,1-trifluoro-N-(4-phenylsulphonyl-o-tolyl)methanesulphonamide, in *C.A.* usage 1,1,1-trifluoro-N-[2-methyl-4-(phenylsulfonyl)phenyl]methanesulfonamide *[37924-13-3]*, also known as 1,1,1-trifluoro-4'-(phenylsulfonyl)methanesulfono-o-toluidide. It was introduced in 1971 by 3M Company under the code number 'MBR-8251', trade mark 'Destun' and protected by BP 1 306 564; BelgP 765 558. Its herbicidal properties were reported by W. A. Gentner, *Agric. Res., (Wash. D.C.)*, 1971, **20**(2), 5; *Weed Sci.*, 1973, **21**, 122.

Manufacture and properties. Produced by the sulphonylation of 4-phenylthio-o-toluidine with trifluoromethanesulphonyl fluoride followed by oxidation with hydrogen peroxide, it is a colourless crystalline solid, m.p. 142-144°, v.p. $< 1 \times 10^{-5}$ mmHg at 25°, pK_a 2·5. Its solubility at 22° is: 60 mg/l water; 750 g/l acetone; 11 g/l benzene; 162 g/l dichloromethane; 595 g/l methanol. It is stable to thermal degradation and to both acid and alkaline hydrolysis at 100°. In aqueous environments it is susceptible to degradation under u.v. radiation. Aqueous solutions and suspensions are mildly corrosive to metals on prolonged exposure.

Uses. Perfluidone is a selective herbicide which, at 1·1-4·5 kg a.i./ha, provides pre-em. control of *Cyperus esculentus,* many grass and several broad-leaved weed species in emerging cotton, established turf, container-grown ornamentals and transplanted tobacco. Other crops at various growth stages show a high degree of tolerance and include sugarcane seed pieces and ratoons, transplanted and established rice in water, and groundnuts at $\geqslant$ 4-leaf growth stage.

Toxicology. The acute oral LD50 is: for rats 633 mg tech./kg; for mice 920 mg/kg. The acute dermal LD50 for rabbits is > 4000 mg/kg; it is mildly irritating to the skin and irritating to the eyes of rabbits. The LC50 (96 h) is: for bluegill 318 mg/l; for rainbow trout 312 mg/l.

Formulations. These include: 'Destun', w.p. (500 g a.i./kg); 'Destun 4-S', solution (480 g/l); 'Destun 3·75 G' and 'Destun 5·0 G', granules (37·5 and 50 g/kg).

Analysis. Product and residue analysis is by GLC following methylation by diazomethane (C. D. Green, *Anal. Methods Pestic. Plant Growth Regul.,* 1978, **10**, 437).

Permethrin

L3TJ A1 A1 BVO1R COR&& C1UYGG $C_{21}H_{20}Cl_2O_3$ (391·3)

Nomenclature and development. The common name permethrin is approved by BSI, ISO and ANSI; also known as OMS 1821. The IUPAC name is 3-phenoxybenzyl (1*RS*)-*cis,trans*-3-(2,2-dichlorovinyl)-2,2-dimethylcyclopropanecarboxylate, in *C.A.* usage (3-phenoxyphenyl)methyl (1*RS*)-*cis,trans*-3-(2,2-dichloroethenyl)-2,2-dimethylcyclo-propanecarboxylate *[52645-53-1]*. The ratio of *cis/trans* isomers, which may vary with the manufacturing process, should be stated. It was discovered by M. Elliott *et al., Proc. Br. Insectic. Fungic. Conf., 7th,* 1973, **2,** 721; *Nature (London),* 1973, **246,** 169, under the code number 'NRDC 143' and protected by BP 1 413 491. It has been developed by FMC Corp. under the code number 'FMC 33 297'; by ICI Ltd, Plant Protection Division under the code number 'PP 557' and the trade marks 'Ambush', 'Ambushfog', 'Perthrine' and 'Kafil'; by Shell International Chemical Co. Ltd under the code number 'WL 43 479', the trade marks 'Talcord' (for agronomic use), 'Outflank' and 'Stockade' (for veterinary use); by the Wellcome Foundation under the trade marks 'Coopex', 'Qamlin', 'Stomoxin P', 'Perigen' and 'Stomoxin'; by Mitchell Cotts and by Sumitomo Chemical Co.

Manufacture and properties. Produced from (1*RS*)-*cis,trans*-3-(2,2-dichlorovinyl)-2,2-dimethylcyclopropanecarboxylic acid or its ethyl ester, and 3-phenoxybenzyl alcohol or the corresponding halide (*Synthetic Pyrethroids, ACS Symposium Series,* No. 42, 1977) the technical material is a yellowish-brown liquid which partially crystallises at ambient temperatures and is completely liquid ≥ 60°. Pure permethrin has m.p. 34-39°, b.p. 200°/0·01 mmHg, v.p. 3·4 x 10^{-7} mmHg at 25°, 1·1 x 10^{-6} mmHg at 80°. Its solubility at 20° is: < 0·1 mg tech./l water; > 450 g/l acetone, cyclohexanone, ethanol, xylene, chloroform. It is stable to heat; some photochemical degradation has been observed in laboratory studies (F. Barlow *et al., Pestic. Sci.,* 1977, **8,** 291) but field data indicate this does not adversely affect biological performance. It is more stable in acid than in alkaline media, optimum stability being *c.* pH 4.

Uses. Permethrin is a contact insecticide effective against a broad range of pests. It controls leaf- and fruit-eating lepidopterous and coleopterous pests in cotton at 100-150 g a.i./ha, fruit at 25-50 g/ha, vegetables at 40-70 g/ha, vines, tobacco and other crops at 50-200 g/ha. It has good residual activity on treated plants. It is effective against a wide range of animal ectoparasites, provides ≤ 60 d residual control of biting flies in animal housing at 200 mg a.i. (as e.c.)/m² wall or 30 mg a.i. (as w.p.)/m² wall and is effective as a wool preservative at 200 mg/kg wool (P. A. Duffield, *ibid.,* p.279). It provides ≥ 120 d control of cockroaches and other crawling insects at 100 mg a.i. (as w.p.)/m².

Toxicology. The acute oral LD50 varies with the *cis/trans* ratio of the sample, the carrier and conditions used in the study, typical values for a *cis/trans* ratio of *c.* 40:60 are: for rats 430-> 4000 mg/kg; for mice 540-2690 mg/kg; for chickens > 3000 mg/kg; for Japanese quail > 13 500 mg/kg. In 2-y feeding trials rats receiving 100 mg/kg diet showed no ill-effect. The LC50 (96 h) for rainbow trout is 9 µg/l. Metabolism in rats (M. Elliott *et al., J. Agric. Food Chem.,* 1976, **24,** 270; L. C. Gaughan *et al., ibid.,* 1977, **25,** 9) has been studied.

continued

Permethrin—*continued*

Formulations. These include: e.c. (100-500 g a.i./l), solution (50 g/l) for fogging, ULV concentrates (50 and 100 g/l), all for agronomic use; dusts (2·5-10 g/kg) and e.c. (25-200 g/l) for veterinary use; w.p. (100-500 g/kg) for industrial and public hygiene use. Many commercial formulations contain a *cis/trans* isomer ratio of 40-50 : 60-50 *m/m*, but formulations with a 25:75 ratio are also available.

Analysis. Product analysis is by GLC or HPLC. Residues may be determined by GLC with ECD (I. H. Williams, *Pestic. Sci.,* 1976, 7, 336). Details may be obtained from Shell International Chemical Co. Ltd.

Nomenclature and development. Petroleum oils are also known as mineral oils; refined grades have been called white oils. They consist largely of aliphatic hydrocarbons, both saturated and unsaturated, the content of the latter being reduced by refinement. Oils of higher distillation range came into use *c.* 1922 and highly refined oils, such as 'Volck' (Chevron Chemical Co.) *c.* 1924.

Manufacture and properties. Produced by the distillation and refinement of crude mineral oils, those used as pesticides generally distil $> 310°$. They may be classified by the proportion distilling at 335°, namely: 'light' (67-79%), 'medium' (40-49%), and 'heavy' (10-25%). Viscosity and density vary according to the geographical area from which the crude oils came; density rarely exceeds 0·92 g/cm³ at 15·5° (*cf.* tar oils).

Uses. Petroleum oils are effective against certain insects such as scales and red spider mites especially under glass on cucumbers, tomatoes and vines; they are ovicidal. Their use is limited by their phytotoxic properties; a semi-refined oil can be used as an ovicide on dormant trees; for foliage use a refined oil of narrow viscosity range and high content of unsulphonated residue is required. Such oils should not be applied to foliage bearing sulphur residues. An oil of lower distillation range containing some aromatic hydrocarbons is suitable for herbicidal use in umbelliferous crops.

Toxicology. Petroleum oils are relatively harmless to mammals, no toxicological problem having been encountered in practice.

Formulations. Stock emulsions; e.c. (solutions of a suitable surfactant in the oil).

Analysis Formulation analysis is by recovering the oil by extraction with diethyl ether from the dilute emulsion broken either by the addition of alkali and sodium chloride or, if an anionic emulsifier is used, by precipitation with a cationic surfactant. For the determination of the density, volatility or distillation range, unsulphonated residue on the oil so isolated, the methods prescribed for the particular specification should be followed (*MAFF Tech. Bull.,* No. 1, 1958, p.66; *CIPAC Handbook,* 1970, **1**, 582; *ibid.,* 1979, **1A,** in press).

Phenisopham

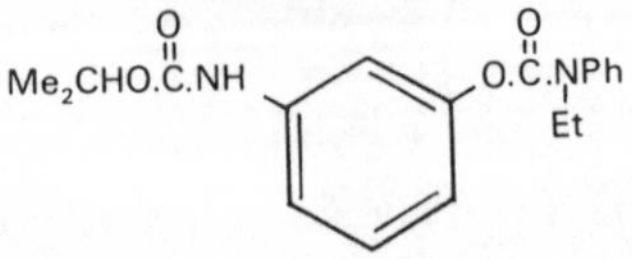

Me₂CHO.C.NH ⬡ O.C.NPh | Et

(structure)

2NR&VOR   CMVOY1&1

$C_{19}H_{22}N_2O_4$ (342.4)

Nomenclature and development. The common name phenisopham is approved by BSI and proposed by ISO. The IUPAC name is isopropyl 3-(N-ethyl-N-phenylcarbamoyloxy)-phenylcarbamate, in *C.A.* usage 3-[[(1-methylethoxy)carbonyl]amino]phenyl ethyl-phenylcarbamate *[57375-63-0]*. It was introduced in 1978 by Schering AG under the code number 'SN 58 132', trade mark 'Diconal' and protected by DBP 2 413 933. Its herbicidal effects were first described in Schering AG *Tech. Inf.* 1977.

Manufacture and properties. Produced by the reaction of isopropyl 3-hydroxyphenyl-carbamate with carbonyl chloride and N-ethylaniline, it is a colourless solid, m.p. 109-110°. It is insoluble in water; easily soluble in acetone and other polar organic solvents. It is unstable in alkaline conditions.

Uses. Phenisopham is a selective herbicide mainly used to control broad-leaved weeds in cotton. It acts mainly by contact but there also appears to be some activity through the soil. It should be applied as soon as most of the weeds have emerged and not later than the 2-4 true leaf stage. It is recommended at 1-2 kg a.i. in 300-400 l water/ha.

Toxicology. The acute oral LD50 is: for rats > 4000 mg/kg; for mice > 5000 mg/kg. The acute dermal LD50 for rabbits is > 1000 mg/kg.

Formulation. An e.c. (150 g a.i./l).

Analysis. Product analysis is by TLC and comparison with standards. Residues may be determined by hydrolysis to N-ethylaniline which is brominated and measured by GLC with ECD. Particulars available from Schering AG.

Phenmedipham

MeO.CO.NH—⟨phenyl ring⟩—O.CO.NH—⟨phenyl ring⟩—Me

lOVMR COVMR Cl $C_{16}H_{16}N_2O_4$ (300·3)

Nomenclature and development. The common name phenmedipham is approved by BSI, ISO, ANSI, WSSA and JMAF: The IUPAC name is methyl 3-*m*-tolylcarbamoyloxy-phenylcarbamate, formerly 3-methoxycarbonylaminophenyl *m*-tolylcarbamate, in *C.A.* usage 3-[(methoxycarbonyl)amino]phenyl (3-methylphenyl)carbamate *[13684-63-4]* formerly methyl *m*-hydroxycarbanilate *m*-methylcarbanilate. It was introduced in 1968 by Schering AG under the code number 'Schering 38 584', trade mark 'Betanal' and protected by BP 1 127050. It was described by F. Arndt & C. Kötter, *Abstr. Int. Congr. Plant Prot., 6th, Vienna,* 1967, p.433.

Manufacture and properties. Produced by the reaction of 3-methoxycarbonylaminophenol with *m*-tolyl isocyanate, it forms colourless crystals, m.p. 143-144° and of very low volatility. Its solubility at room temperature is: 3 mg/l water; *c.* 200 g/kg acetone, cyclo-hexanone; *c.* 50 g/kg methanol; 20 g/kg chloroform; 2·5 g/kg benzene; *c.* 500 mg/kg hexane. The technical product, m.p. 140-144°, v.p. 10^{-11} mmHg at 25°, is > 95% pure. No changes were observed when held for 6 d at 50°; hydrolysis at 30° in buffer/methanol solution gave 50% loss in *c.* 135 d at pH 7 and 10 min at pH 9.

Uses. Phenmedipham is a post-em. herbicide used for weeding beet crops, in particular sugar beet, at 1 kg/a.i. in 200-300 l water/ha after the emergence of most of the weeds and before they develop > 2-4 true leaves. It is absorbed through the leaves and has little action via the soil and roots. In soil, 71-86% of the amount determined 1 d after treatment was degraded in 90 d.

Toxicology. The acute oral LD50 is: for rats and mice > 8000 mg/kg; for dogs and guinea-pigs > 4000 mg/kg; for chickens > 3000 mg/kg. The acute dermal LD50 for rats is > 4000 mg/kg. In 120-d feeding trials rats receiving 125, 250 and 500 mg a.i./kg daily survived but there was a reduced food intake dependent on the dose.

Formulations. 'Betanal' (157 g a.i./l); in the UK 'Betanal E' (120 g/l).

Analysis. Product analysis is by bromometric titration (*CIPAC Handbook,* in press) or by quantitative TLC using a chromatogram spectrophotometer (K. Kossman & N. A. Jenny, *Anal. Methods Pestic. Plant Growth Regul.,* 1973, **7**, 611). Residues may be determined by hydrolysis to *m*-toluidine which is purified and either brominated and measured by GLC with ECD or measured colorimetrically after diazotisation and coupling with *N*-(1-naphthyl)ethylenediamine (M. Kossmann, *Weed Res.,* 1970, **10**, 340).

Phenothrin

Me$_2$C=CH— ... —C.OCH$_2$— ... —OPh

Me—

Me

L3TJ A1 A1 BVO1R COR&& C1UY1&1   C$_{23}$H$_{26}$O$_3$ (350·5)

Nomenclature and development. The common name phenothrin is approved by BSI and New Zealand and proposed by ISO. The IUPAC name is 3-phenoxybenzyl (1*RS*)-*cis,trans*-chrysanthemate, also known as 3-phenoxybenzyl (1*RS*)-*cis,trans*-2,2-dimethyl-3-(2-methylprop-1-enyl)cyclopropanecarboxylate, in *C.A.* usage (3-phenoxyphenyl)methyl 2,2-dimethyl-3-(2-methyl-1-propenyl)cyclopropanecarboxylate *[26002-80-2]*. It was introduced in 1973 by Sumitomo Chemical Co. Ltd under the code number 'S-2539' and protected by JapaneseP 618627; 631916; USP 3666789. It was first described by K. Fujimoto *et al., Agric. Biol. Chem.,* 1973, **37**, 2681. A mixture, sometimes called d-phenothrin, *C.A.* Registry No. *[26002-80-2]*, rich in the (1*R*)-*cis*- *[51186-88-0]* and (1*R*)-*trans*- *[26046-85-5]* isomers, *cis/trans* ratio 20 : 80 *m/m*, was introduced by Sumitomo Chemical Co. Ltd under the code number 'S-2539 Forte' and trade mark 'Sumithrin'; also known as ENT 27 972, OMS 1810.

Manufacture and properties. Produced by the esterification of 3-phenoxybenzyl alcohol with chrysanthemoyl chloride, phenothrin is a colourless liquid, n_D^{25} 1·5478, d_{25}^{25} 1·058. The (1*R*)-*cis,trans*-isomeric mixture is a colourless liquid, n_D^{25} 1·5482, d_{25}^{25} 1·061. Both products have v.p. 1·64 mmHg at 200°, 17·2 mmHg at 250°. Phenothrin's solubility at 30° is 2 mg/l water; miscible with most organic solvents. It is stable under irradiation, in most organic solvents and on inorganic mineral diluents.

Uses. Phenothrin is a non-systemic insecticide, effective by contact and as a stomach poison, used to control common insects that endanger public health.

Toxicology. The acute oral LD50 is: for rats and mice > 500 mg phenothrin/kg; > 10000 mg (1*R*)-*cis,trans*-isomeric mixture/kg. The LC50 (48 h) for goldfish is 0·25-0·5 mg phenothrin/l, the LC50 (96 h) for rainbow trout is 0·0167 mg (1*R*)-*cis,trans*-isomeric mixture/l.

Formulations. These include, with or without other pyrethroids or pyrethroid synergists: water- and oil-based aerosol concentrates; e.c.; dusts.

Analysis. Product analysis is by GLC. Residues may be determined by GLC with ECD. Details of these methods are available from Sumitomo Chemical Co. Ltd.

Phenthoate

$$(MeO)_2P.S.CH.C.OEt$$

with structure showing $\overset{S}{\underset{}{\|}}$ and $\overset{O}{\underset{}{\|}}$ above P and C, and Ph below CH

2OVYR&SPS&O1&O1

$C_{12}H_{17}O_4PS_2$ (320·4)

Nomenclature and development. The common name phenthoate is approved by BSI and ISO—exception JMAF (PAP); also known as ENT 27386, OMS 1075. The IUPAC name is *S*-α-ethoxycarbonylbenzyl *O,O*-dimethyl phosphorodithioate, in *C.A.* usage ethyl α-[(dimethoxyphosphinothioyl)thio]benzeneacetate *[2597-03-7]* formerly ethyl mercaptophenylacetate *S*-ester with *O,O*-dimethyl phosphorodithioate. It was introduced in 1961 by Montecatini S.p.A. (now Montedison S.p.A) under the code number 'L 561', trade marks 'Cidial' and 'Elsan' and protected by BP 834814; USP 2947662.

Manufacture and properties. Produced by the action of sodium *O,O*-dimethyl phosphorodithioate with ethyl α-bromophenylacetate, it is a colourless crystalline solid, with an aromatic odour, m.p. 17-18°, v.p. 4 x 10^{-5} mmHg at 40°, d_4^{20} 1·226, n_D^{20} 1·5550. Its solubility at 24° is 11 mg/l water. The technical product, 90-92% pure, is a reddish-yellow oil. Its solubility at 20° is: 200 mg/l water; 120 g/l hexane; > 200 g/l 1,2-dimethoxyethane; 100-170 g/l petroleum spirit. It is stable in acid and neutral media, but in buffer solution at pH 7 *c.* 25% is hydrolysed in 20 d.

Uses. Phenthoate is a non-systemic insecticide, with contact and stomach action. Applied at 0·5-1·0 kg a.i/ha, it protects cotton, rice, citrus, fruit, vegetables and other crops from Lepidoptera, jassids, aphids and soft scales. It is also effective against mosquito larvae and adults. It may be phytotoxic to some grape, peach and fig varieties and may discolour red-skinned apple varieties.

Toxicology. The acute oral LD50 is: for rats 300-400 mg tech./kg; for mice 350-400 mg/kg; for hares 72 mg/kg; for pheasants 218 mg/kg; for quail 300 mg/kg. In dermal tests on rats application of 4800 mg/kg for 4 h to the skins caused no toxic symptom. In 1·67-y trials the highest non-toxic dose for rats was *c.* 0·5 mg tech./kg daily. The LC100 (96 h) is: for goldfish 4·5 mg/l; for minnows 3·4 mg/l; for mosquito fish 0·3 mg/l; for guppies 2-3 mg/l. The LD50 for honey bees is 0·12 µg/bee.

Formulations. For agricultural use: 'Cidial 50L', liquid (500 g tech./kg); 'Cidial Oil', (50 g tech. + 800 g mineral oil); 'Cidial WP 40', w.p. (400 g/kg); 'Cidial D', dust (30 g/kg); 'Cidial Granules', granules (20 g/kg); 'Cidial AS', liquid (850 g/kg); 'Cidial ULV', ULV concentrate (1 kg/kg). For mosquito control: 'Tanone 50L', liquid (500 g/kg), 'Tanone AS', liquid (850 g/kg); 'Tanone ULV', ULV concentrate (1 kg/kg).

Analysis. Product analysis is by GLC or by TLC with colorimetric estimation of phosphorus in the appropriate spots by standard methods. Residues on apples or citrus fruits may be determined by GLC. Details of these methods are available from Montedison S.p.A. See also: B. Bazzi *et al., Anal. Methods Pestic. Plant Growth Regul.,* 1976, **8,** 159.

Phenylmercury acetate

$$O$$
$$\|$$
$$CH_3.C.O.Hg.Ph$$

1VO-HG-R $C_8H_8HgO_2$ (336·7)

Nomenclature and development. The IUPAC name phenylmercury acetate is accepted by BSI and ISO in lieu of a common name—exception JMAF (PMA). In *C.A.* usage it is called (acetato-*O*)phenylmercury *[62-38-4]*, formerly (acetato)phenylmercury. The use of organomercury compounds as pesticides dates from 1914 when E. Riehm, *Zentralbl. Bakteriol. Parasitenkd. Infektionskr. Hyg., Abt. 2*, 1914, **40**, 424, reported that phenylmercury chloride was effective against bunt of wheat. The first use of phenylmercury acetate as a herbicide was against crabgrass (J. A. De France, *Greenkeepers Rep.*, 1947, **15**(1), 30). Trade marks include: 'Tag HL 331' (Chevron Chemical Co.); 'Agrosan' D, 'Ceresol' and 'Harvesan Plus' (ICI Ltd, Plant Protection Division); 'Mist-O-Matic' mercury liquid seed treatment (Murphy Chemical Ltd); 'Leytosan' seed dressing (Steetley Chemicals Ltd—formerly F. W. Berk Ltd); 'Maysan' (May & Baker Ltd).

Manufacture and properties. Produced by the interaction of benzene and mercury diacetate (K. A. Kobe & P. F. Lueth, *Ind. Eng. Chem.*, 1942, **34**, 309), it forms colourless crystals, m.p. 149-153°, v.p. 9 x 10⁻⁶ mmHg at 35°. Its solubility at room temperature is 4·37 g/l water; soluble in acetone, benzene, ethanol; very soluble in 2-(2-ethoxyethoxy)ethanol. It is stable.

Uses. Phenylmercuric acetate is a powerful eradicant fungicide mainly used as a cereal seed dressing, sometimes combined with insecticides or other fungicides. It is a selective herbicide used on lawns for the control of crabgrass.

Toxicology. It is highly toxic to mammals (D. Hunter *et al., Q. J. Med.*, 1940, **9**, 193). Levels as low as 0·5 mg mercury/kg diet produced renal lesions (O. G. Fitzhugh *et al., Arch. Ind. Hyg.*, 1950, **2**, 433). There is some vapour hazard, so conditions under which seed is dressed are rigidly controlled by law.

Formulations. These include: powder or liquid seed treatments (10-300 g mercury/kg); other components include carboxin, gamma-HCH.

Analysis. Product and formulation analysis is by CIPAC methods (*CIPAC Handbook*, 1970, **1**, 516). Residues may be determined by digestion in concentrated nitric and sulphuric acids and measurement of mercury colorimetrically by dithizone: AOAC Methods. See also: *EPPO, Recommended Methods, Series A*, No.37.

416

Phenylmercury dimethyldithiocarbamate

$$\underset{\text{Me}_2\text{N.C.S.Hg.Ph}}{\overset{\overset{\text{S}}{\|}}{}}$$

1N1&YUS&S-HG-R $C_9H_{11}HgNS_2$ (397·9)

Nomenclature and development. The IUPAC name is phenylmercury dimethyldithio-carbamate, in *C.A.* usage (dimethylcarbamodithioato-*S,S*)phenylmercury *[32407-99-1]* formerly (dimethyldithiocarbamato)phenylmercury. It was introduced as a general fungicide by Berk Chemicals (now Steetley Chemicals Ltd) under the trade mark 'Phelam'. The use of organomercury compounds as pesticides dates from the discovery of the effect of phenylmercury chloride against bunt of wheat (E. Riehm, *Zentralbl. Bakteriol. Parasitenkd. Infektionskr. Hyg., Abt. 2,* 1914, **40,** 424).

Manufacture and properties. Produced by precipitation from a solution of phenylmercury acetate (p.416), phenylmercury dimethyldithiocarbamate is a fine, colourless to grey-white powder, m.p. 175°, v.p. 8×10^{-7} mmHg at 35°. It is stable at 180°, subliming slowly. Its solubility at 20° is: 6 mg/l water; 150 g/l aniline; 50 g/l cyclohexanone; 50 g/l pyridine; readily soluble in chloroform; sparingly soluble in acetic acid, alcohols, diethyl ether, tetrahydrofuran. It is stable in water $\leqslant$ pH 12.

Uses. Phenylmercury dimethyldithiocarbamate combines the fungicidal activities of phenylmercury compounds and dimethyldithiocarbamates and has the added advantages of low volatility and high stability. It is used to eradicate *Venturia* spp. on most varieties of apples and pears at 1·5-3·0 g (as mercury)/100 l HV. A few varieties are damaged by mercury compounds.

Toxicology. The acute oral LD50 for rats is 120 mg/kg. Like all arylmercury compounds it is highly toxic to mammals, but its low vapour pressure results in a much reduced hazard from the vapour.

Formulation. 'Phelam', w.p. (25 g a.i./kg; $\equiv$ 15·1 g mercury/kg) dispensed with wetting agent in water-soluble sachets, 'Solupacks', to minimise handling risks.

Analysis. Product analysis is by the CIPAC method (*CIPAC Handbook,* 1979, **1A,** in press). Residues may be determined by digestion in nitric and sulphuric acids and estimation of the mercury colorimetrically with dithizone: AOAC Methods.

417

Phenylmercury nitrate

Ph.Hg.NO₃ Ph.Hg.OH

WNO&-HG-R &Q-HG-R $C_{12}H_{11}Hg_2NO_4$ (634·4)

Nomenclature and development. The traditional chemical name phenylmercury nitrate is accepted by BSI and ISO in lieu of a common name though the material used commercially is the basic nitrate with formula given above. The IUPAC name is di(phenylmercury) hydroxide nitrate, in *C.A.* usage hydroxy(nitrato)diphenyldimercury *[8003-05-2]* (formerly *[6059-33-1]*). Phenylmercury salts have long been known as bactericides and fungicides (E. Riehm, *Zentralbl. Bakteriol. Parasitenkd. Infektionskr. Hyg., Abt. 2,* 1914, **40,** 424).

Manufacture and properties. Produced by the interaction of phenylmercury hydroxide and dilute nitric acid, the phenylmercury nitrate of commerce is a colourless powder, m.p. *c.* 188° (decomp.). Its solubility at 20° is: 600 mg/kg water; 5 g/kg acetone; 3 g/kg benzene; 1 g/kg ethanol; 9 g/kg methanol. The technical grade has a mercury content of 63·4% *m/m.*

Uses. Phenylmercury nitrate inhibits the growth of many bacteria and fungi at concentrations of 3-40 mg/l and has various pharmaceutical uses including the control of 'athlete's foot' and as stabiliser for opthalmic preparations. Its agricultural uses include the eradication of *Venturia* spp. on apples and pears at 1·5-3·0 g mercury/100 l HV though some varieties are sensitive to mercury. Seed dressings, usually combined with an insecticide or another fungicide, are used for cereals, sugar beet, fodder beet and mangolds.

Toxicology. The acute intravenous LD50 for mice is 27 mg/kg. It is highly toxic to mammals; some vapour hazard.

Formulations. These include: phenylmercury nitrate 2½% Steetley, w.p. (25 g a.i./kg; ≡ 15·8 g mercury/kg); 'Murcocide' liquid (15 g mercury/kg) with captan; 'Harvesan Plus' seed dressing (10 g mercury/kg) with gamma-HCH.

Analysis. Product analysis is by the CIPAC method (*CIPAC Handbook,* 1979, **1A,** in press). Residues may be determined by digestion in nitric and sulphuric acids and estimation of mercury colorimetrically by dithizone: AOAC Methods.

2-Phenylphenol

QR BR $C_{12}H_{10}O$ (170·2)

Nomenclature and development. The IUPAC name 2-phenylphenol is accepted by BSI and ISO in lieu of a common name; also known as biphenyl-2-ol, 2-hydroxybiphenyl and *o*-phenylphenol. In *C.A.* usage [1,1'-biphenyl]-2-ol *[90-43-7]* formerly 2-biphenylol. It was introduced in 1936 by R. G. Tompkins, *Rep. Food Investigation Board 1936,* p.149, for the treatment of fruit wrappers to prevent rotting of the produce.

Manufacture and properties. Produced in the reaction of sodium hydroxide with chlorobenzene, it forms colourless to pinkish crystals, of a mild odour, m.p. 57°, b.p. 286°, volatile in steam, d_{25}^{25} 1·217. Its solubility at 25° is 0·7 g/kg water; soluble in most organic solvents. It forms salts of which those of the alkali metals are water-soluble; the sodium salt, *C.A.* Registry No. *[132-27-4],* crystallises as a tetrahydrate, solubility *c.* 1·1 kg/kg water, giving a solution of pH 12·0-13·5 at 35°.

Uses. 2-Phenylphenol is a powerful disinfectant and fungicide for the impregnation of fruit wrappers and for the disinfection of seed boxes. The tendency to scald citrus fruits is reduced by the addition of hexamine. It is also applied during the dormant period to control apple canker.

Toxicology. The acute oral LD50 for white rats is 2480 mg/kg. In 2-y feeding trials rats receiving 2 g/kg diet showed no ill-effect (H. C. Hodge *et al., J. Pharmacol. Exp. Ther.,* 1952, **104,** 202). It may cause skin irritation.

Formulations. These include: 'Dowicide 1', 2-phenylphenol; 'Dowicide A', sodium 2-phenylphenoxide; 'Nectryl', canker paint (Stanhope).

Analysis. Residues in fruit wrappers may be determined by extraction with acetone or diethyl ether and, after purification, colorimetry at 450 nm after reaction with titanium disulphate (P. H. Caulfield & R. J. Robinson, *Anal. Chem.,* 1953, **25,** 982). Residues in citrus peel may be extracted by steam distillation, clean-up by liquid-liquid partition, and colorimetry after coupling with diazotised sulphanilic acid (R. G. Tompkins & F. A. Isherwood, *Analyst (London),* 1945, **70,** 330; *J. Assoc. Off. Agric. Chem.,* 1957 **40,** 238).

Phorate

$$\text{(EtO)}_2\overset{\overset{\text{S}}{\|}}{\text{P}}\text{.S.CH}_2\text{.SEt}$$

2S1SPS&O2&O2 $C_7H_{17}O_2PS_3$ (260·4)

Nomenclature and development. The common name phorate is approved by BSI, ISO and ANSI—exception USSR (timet); also known as ENT 24042. The IUPAC and *C.A.* name is *O,O*-diethyl *S*-ethylthiomethyl phosphorodithioate, Registry No. *[298-02-2]*. It was introduced in 1954 by the American Cyanamid Co. under the code number 'EI 3911', the trade mark 'Thimet' and protected by USP 2586655; 2596076; 2970080; and by 2759010 (Bayer AG).

Manufacture and properties. Produced by the reaction of *O,O*-diethyl hydrogen phosphorodithioate with formaldehyde followed by the addition of ethanethiol, it is a clear mobile liquid, b.p. 118-120°/0·8 mmHg, f.p. < —15°, v.p. 8·4 x 10^{-4} mmHg at 20°, n_D^{25} 1·5349. The technical grade, d^{25} 1·167, is > 90% pure. Its solubility at room temperature is 50 mg/l water; miscible with carbon tetrachloride, dioxane, vegetable oils, xylene, alcohols, ethers and esters. It is stable at room temperature. Environmental stability is optimum in the range pH 5-7. Highly acidic (pH < 2) or alkaline (pH > 9) media promote hydrolytic decomposition at rates dependent upon the temperature and pH.

Uses. Phorate is a systemic and contact insecticide and acaricide used to protect crops, primarily root and field crops, cotton, brassicas and coffee, from sucking and biting insects, mites and certain nematodes. Also as a soil insecticide on maize and sugar beet. In plants and animals it is metabolically-oxidised at both the thioether linkage and the co-ordinated sulphur yielding the sulphoxide and sulphone and their phosphorothioate analogues; as both the parent compound and the oxidation products are readily hydrolysed only a small proportion of the sulphone results (J. S. Bowman & J. E. Casida, *J. Agric. Food Chem.*, 1957, **5**, 192; R. L. Metcalf *et al.*, *J. Econ. Entomol.*, 1957, **50**, 338). Behaviour in soils is similar but the sulphones can persist under some conditions (D. L. Suett, *Pestic Sci.*, 1975, **6**, 385).

Toxicology. The acute oral LD50 for rats is 1·6-3·7 mg tech./kg. The acute dermal LD50 is: for rats 2·5-6·2 mg/kg; for guinea-pigs 20-30 mg/kg; values for granular formulations depend on a.i. content, carrier, test method and animal species—typical values include: for male rats 137 mg a.i. (as 5% granule)/kg, 98 mg a.i. (as 10% granule)/kg); for male rabbits 93-245 mg a.i. (as 5% granule)/kg, 116 mg a.i. (as 10% granule)/kg. In 90-d feeding trials rats receiving 6 mg tech./kg diet showed no ill-effect other than depression of cholinesterase levels.

Formulations. These include: e.c. of various a.i. content including 'Thimet LC-8' (960 g tech./l) also (200 and 250 g a.i./l); granules (50, 100 and 150 g/kg).

Analysis. Product analysis is by i.r. spectroscopy at 8·32 μm (*CIPAC Handbook,* in press). Residues of phorate and its oxidation products may be determined by GLC (D. L. Suett, *Pestic. Sci.*, 1971, **2**, 105; 1975, **6**, 385). Details of methods are available from the American Cyanamid Co. See also: G. L. Sutherland *et al., Anal. Methods Pestic., Plant Growth Regul. Food Addit.,* 1964, **2**, 487; J. G. Boyd, *Anal. Methods Pestic. Plant Growth Regul.,* 1972, **6**, 493.

Phosalone

T56 BNVOJ B1SPS&O2&O2 GG $\qquad$ $C_{12}H_{15}ClNO_4PS_2$ (367·8)

Nomenclature and development. The common name phosalone is approved by BSI, ISO, ANSI and JMAF—exception USSR (benzphos); also known as ENT 27 163. The IUPAC name is *S*-6-chloro-2,3-dihydro-2-oxobenzoxazol-3-ylmethyl *O,O*-diethyl phosphoro-dithioate, in *C.A.* usage *S*-[6-chloro-2-oxo-3(2*H*)-benzoxazolyl)methyl] *O,O*-diethyl phosphorodithioate *[2310-17-0]* formerly *O,O*-diethyl phosphorodithioate *S*-ester with 6-chloro-3-(mercaptomethyl)-2-benzoxazolinone. It was introduced in 1963 by Rhône-Poulenc Phytosanitaire under the code number 'RP 11974', trade mark 'Zolone' and protected by BP 1 005 372; BelgP 609 209: FP 1 482 025. Its insecticidal properties were first reported by J. Desmoras *et al., Phytiatr. Phytopharm.*, 1963, **12**, 199.

Manufacture and properties. Produced by the condensation of sodium *O,O*-diethyl phosphorodithioate with 6-chloro-3-chloromethyl-3*H*-benzoxazol-2-one, the latter being obtained by the interaction of urea with 2-aminophenol to give 3*H*-benzoxazol-2-one which is chlorinated to give the 6-chloro analogue and then chloromethylated, it forms colourless crystals, with a slight garlic odour, m.p. 48°, of negligible v.p. at room temperature. Its solubility at room temperature is 10 mg/l water; sparingly soluble in cyclohexane, light petroleum; soluble in acetone, acetonitrile, benzene, chloroform, dioxane, ethanol, methanol, toluene, xylene. It is stable under normal storage conditions, compatible with most other pesticides and non-corrosive.

Uses. Phosalone is a non-systemic insecticide and acaricide used on deciduous tree fruits at 300-600 g a.i./100 l against codling moth, tortrix moths, apple maggot, oriental fruit moth, plum curculio, pear psylla, aphids and red spider mites; on field and market garden crops at 400-800 g/ha; for the control of bollworms, aphids, jassids, thrips and red spider mites on cotton; tuber moths and aphids on potatoes; pollen beetles and weevils on rape. It persists on plants for about 14 h being metabolised to the corresponding phosphorothioate which is rapidly hydrolysed.

Toxicology. The acute oral LD50 is: for male rats 120-170 mg/kg; for mice 180 mg/kg; for guinea-pigs 380 mg/kg; for pheasants 290 mg/kg. The acute dermal LD50 is: for rats 1500 mg/kg; for rabbits > 1000 mg/kg. In 2-y feeding trials rats receiving 250 mg/kg diet and dogs 290 mg/kg diet suffered no ill-effect. Rates of 700 g/ha were without hazard to honey bees, provided the worker bees were not actively foraging at the time of spraying.

Formulations. These include: e.c. 'Zolone liquide' (350 g a.i./l), in Japan 'Rubitox' (350 g/l), in UK 'Zolone' (May & Baker Ltd) (330 g/l) and in Australia (300 g/l); w.p. (300 g/kg); dusts (25 and 40 g/kg). Various formulations known as 'Zolone DT' for use in cotton contain phosalone plus DDT.

Analysis. Product and residue analysis is by GLC: J. Desmoras *et al., Anal. Methods Pestic. Plant Growth Regul.*, 1973, **7**, 385.

Phosfolan

$$(EtO)_2\overset{\overset{O}{\|}}{P}-N= \text{(1,3-dithiolane ring)}$$

T5SYSTJ   BUNPO&O2&O2 $C_7H_{14}NO_3PS_2$ (255·3)

Nomenclature and development. The common name phosfolan is approved by BSI and ISO, in France (pholan); also known as ENT 25830. The IUPAC name is diethyl 1,3-dithiolan-2-ylidenephosphoramidate (I), formerly 2-(diethoxyphosphinylimino)-1,3-dithiolan, in *C.A.* usage (I) *[947-02-4]* formerly *P,P*-diethyl cyclic ethylene ester of phosphonodithioimidocarbonic acid. It was introduced in 1963 by the American Cyanamid Co. under the code number 'EI 47031', trade mark 'Cyolane', 'Cyolan', 'Cyalane' and 'Cylan', and protected by BP 974 138; FP 1 327 386.

Manufacture and properties. Produced by the reaction of 2-imino-1,3-dithiolane with diethyl phosphorochloridate, it is a colourless to yellow solid, m.p. 37-45°, b.p. 115-118°/ 1×10^{-3} mmHg. It is soluble in water, acetone, benzene, ethanol, cyclohexane, toluene; slightly soluble in diethyl ether; sparingly soluble in hexane. The aqueous solution is stable under neutral and slightly acidic conditions but is hydrolysed by alkali or acid (pH> 9 or < 2).

Uses. Phosfolan is a systemic insecticide for use against sucking insects, mites and lepidopterous larvae. It controls *Prodenia* spp. and *Spodoptera littoralis* on cotton at 0·75-1·0 kg/ha; whiteflies and jassids at 0·5 kg/ha; cabbage looper at 1 kg/ha; also onion thrips at 30 g/100 l. It is non-persistent in soil and, in plants and animals, is metabolised at the N-P bond to less toxic, water-soluble compounds.

Toxicology. The acute oral LD50 for male rats is 8·9 mg/kg. The acute dermal LD50 for guinea-pigs is 54 mg tech./kg; acute dermal LD50's for granular formulations range from 24 mg a.i. (10% granule)/kg to > 100 g a.i. (2% granule)/kg. In 90-d feeding trials dogs receiving 1 mg/kg daily showed no clinical symptom.

Formulations. An e.c. (250 g a.i./l); granules (20-50 g/kg).

Analysis. Residues may be determined by GLC (details are available from the American Cyanamid Co.) or by colorimetry (R. C. Blinn & J. E. Boyd, *J. Assoc. Off. Agric. Chem.*, 1964, **47**, 1106). See also: N. R. Pasarela & E. J. Orloski, *Anal. Methods Pestic. Plant Growth Regul.*, 1973, **7**, 231.

Phosmet

$$(MeO)_2\overset{\overset{\displaystyle S}{\|}}{P}.S.CH_2.N$$

T56   BVNVJ   C1SPS&O1&O1 $C_{11}H_{12}NO_4PS_2$ (317·3)

Nomenclature and development. The common name phosmet is approved by BSI, ISO and ESA—exceptions USSR (phtalofos), JMAF (PMP); also known as ENT 25 705. The IUPAC name is O,O-dimethyl S-phthalimidomethyl phosphorodithioate, in *C.A.* usage S-[(1,3-dihydro-1,3-dioxo-$2H$-isoindol-2-yl)methyl] O,O-dimethyl phosphoro-dithioate *[732-11-6]* formerly O,O-dimethyl phosphorodithioate S-ester with N-(mer-captomethyl)phthalimide. It was introduced in 1966 by Stauffer Chemical Co. under the code number 'R-1504', trade mark 'Imidan' and protected by USP 2 767 194. Its insecticidal properties were reported by B. A. Butt & J. C. Keller, *J. Econ. Entomol.*, 1961, **54**, 813.

Manufacture and properties. Produced by condensing sodium O,O-dimethyl phos-phorodithioate with N-chloromethylphthalimide, the latter being obtained from phthalimide and formaldehyde and conversion to the chloride, it is an off-white crystalline solid, with an offensive odour, m.p. 71·9°, v.p. 1 x 10^{-3} mmHg at 50°; it decomposes below its b.p. The technical product, m.p. 66·5-69·5°, is 95-98% pure. Its solubility at 25° is: 25 mg/l water; 100 g/kg acetone, dichloromethane, 4-methylpent-3-en-2-one, butanone, xylene. In buffered aqueous solutions of 20 mg/l at room temperature, 50% hydrolysis occurs in 13 d at pH 4·5, < 12 h at pH 7, < 4 h at pH 8·3. It is compatible with other pesticides except under alkaline conditions and is only slightly corrosive.

Uses. Phosmet is a non-systemic acaricide and insecticide, used on top fruit (*e.g.* apples, pears, peaches, apricots, cherries), citrus, grapes, potatoes and in forestry at rates (0·5-1·0 kg a.i./ha) such that it is safe for a range of predators of mites and therefore useful in integrated control programmes.

Toxicology. The acute oral LD50 for rats is 230-299 mg/kg; the acute dermal LD50 for albino rabbits is > 3160 mg/kg. In 2-y feeding trials the 'no effect' level for rats and dogs was 40 mg/kg diet. It is readily degraded both in the environment and in laboratory animals.

Formulations. These include: e.c. (202 and 295 g a.i./kg); w.p. (500 g/kg). Storage above 45° may lead to decomposition.

Analysis. Analysis is by GLC: details are available from Stauffer Chemical Co. See also: G. H. Batchelder *et al.*, *Anal. Methods Pestic., Plant Growth Regul. Food Addit.*, 1967, **5**, 257; J. E. Barney *et al.*, *Anal. Methods Pestic. Plant Growth Regul.*, 1972, **6**, 408.

Phosphamidon

(E)

(Z)

2N2&VYGUY1&OPO&O1&O1 $C_{10}H_{19}ClNO_5P$ (299·7)

Nomenclature and development. The common name phosphamidon is approved by BSI, ISO and ANSI—exceptions Italy and Turkey; also known as ENT 25 515. The IUPAC name is 2-chloro-2-diethylcarbamoyl-1-methylvinyl dimethyl phosphate, in *C.A.* usage 2-chloro-3-(diethylamino)-1-methyl-3-oxo-1-propenyl dimethyl phosphate *[13171-21-6]* ((Z)- + (E)-isomers, see below), formerly dimethyl phosphate ester with 2-chloro-N,N-diethyl-3-hydroxycrotonamide. Its insecticidal properties were first described by F. Bachmann & J. Meierhans, *Bull. Cent. Int. Antiparasit.,* 1956, Nov. p.18. It was introduced in 1956 by Ciba AG (now Ciba-Geigy AG) under the code number 'Ciba 570', trade mark 'Dimecron' and protected by BP 829 576; BelgP 552 284; USP 2 908 605. Its properties have been reviewed (*Residue Rev.,* 1971, **31**).

Manufacture and properties. Produced by reacting trimethyl phosphite in chlorobenzene with 2,2-dichloro-N,N-diethyl-3-oxobutyramide, which is prepared from sulphuryl chloride and N,N-diethyl-3-oxobutyramide, it is a pale yellow liquid, with a faint odour, b.p. 162°/1·5 mmHg, v.p. 2·5 x 10^{-5} mmHg at 20°, d_4^{25} 1·3132, n_D^{20} 1·4721. It is soluble in water and most organic solvents except paraffins.

Isomerism. Phosphamidon exists as a mixture of 70% *m/m* (Z)-isomer (β-isomer), *C.A.* Registry No. *[23783-98-4]*, and 30% (E)-isomer (α-isomer), Registry No. *[297-99-4]*, which are indistinguishable chemically, but may be separated by GLC or countercurrent distribution (Ciba AG, 1964; *Residue Rev.,* 1971, **37**, 1; D. L. Bull *et al., J. Econ. Entomol.,* 1967, **60**, 332; K. M. S. Sundaram & P. G. Davis, *Chem. Control Res. Inst., Ottawa, Inf. Rep.* CC-X, 1974, CC-X-66). The (Z)-isomer has the greater insecticidal activity.

Uses. Phosphamidon is a systemic insecticide rapidly absorbed by the plant; it has little contact action. It is effective against sap-feeding insects at 300-600 g/ha, and other pests including rice and sugarcane stem borers and rice leaf beetles at 500-1000 g/ha. It is non-phytotoxic except to some cherry varieties and sorghum varieties related to Red Swazi. In plants 50% loss occurs in *c.* 2 d (R. Schuppen, *Chim. Ind. (Paris),* 1961, **85**, 421).

Toxicology. The acute oral LD50 for rats is 17-30 mg/kg; the acute dermal LD50 for rats is 374-530 mg/kg. In 2-y feeding trials the 'no effect' level was: for rats 1·25 mg/kg daily; for dogs 0·1 mg/kg daily. It is highly toxic to birds and toxic to honey bees.

Formulations. These include: 'Dimecron 20', solution (200 g/l); 'Dimecron 50' (500 g/l); 'Dimecron 100' (1 kg/l); 'Dimecron' Ulvair 250.

Analysis. Product analysis is by GLC with internal standard (A. A. Carlstrom, *J. Assoc. Off. Anal. Chem.,* 1972, **55**, 1331) or by reaction with iodine under specified conditions to differentiate from impurities (R. Anlicker *et al., Helv. Chim. Acta.,* 1961, **44**, 1622). Residues may be determined by GLC (K. M. S. Sundaram & P. G. Davis, *loc. cit.*) or by chromatography and phosphorus determination by standard colorimetric methods (R. Anlicker & R. E. Menzer, *J. Agric. Food Chem.,* 1963, **11**, 291).

Phoxim

$$\underset{(EtO)_2P.O.N.=C.Ph}{\overset{S \qquad CN}{\overset{\|}{} \qquad \overset{|}{}}}$$

NCYR&UNOPS&O2&O2 $\qquad\qquad\qquad\qquad\qquad\qquad$ $C_{12}H_{15}N_2O_3PS$ $\quad$ (298·3)

Nomenclature and development. The common name phoxim is approved by BSI and ISO. The IUPAC name is *O,O*-diethyl α-cyanobenzylideneamino-oxyphosphonothioate, formerly α-(diethoxyphosphinothioyloxyimino)phenylacetonitrile, in *C.A.* usage α-[[(diethoxyphosphinothioyl)oxy]imino]benzeneacetonitrile *[14816-18-3]* formerly *O,O*-diethyl phenylglyoxylonitrile oxime phosphorothioate. It was introduced in 1969 by Bayer AG under the code numbers 'Bay 5621' and 'Bayer 77488', the trade marks 'Baython' and 'Volaton' and protected by BelgP 678139; DBP 1238902. Its insecticidal properties were described by A. Wybou & I. Hammann, *Meded. Rijksfac. Landbouwwet. Gent.*, 1968, **33**, 817.

Manufacture and properties. Produced by the reaction of *O,O*-diethyl phosphorochlorido-thioate with α-cyanobenzaldehyde oxime, it is a yellowish liquid, m.p. 5-6°, n^{20} 1·5405, d_4^{20} 1·176. Its solubility at 20° is 7 mg/l water; soluble in alcohols, ketones, aromatic hydrocarbons; less soluble in light petroleum. The technical product is a reddish oil at room temperature. It cannot be distilled without decomposition, but is stable to water and to acid media; 50% decomposition occurs at room temperature, in 170 min at pH 11·6, and 700 h at pH 7.0.

Uses. Phoxim is an insecticide effective against a broad range of insects, used particularly to control pests of stored products or man. Soil applications have controlled dipterous larvae at 50 mg a.i./plant, and rootworms and wireworms at 5 kg/ha (P. Villeroy & P. Poucharesse, *Pflanzenschutz-Nachr. (Am. Ed.)*, 1974, **27**, 284; G. Zoebelein, *ibid.*, 1975, **28**, 162, 178). ULV foliar applications have controlled grasshoppers. It is of brief persistence and has no systemic action. Photochemical breakdown and metabolism in cotton plants involve isomerisation to *O,O*-diethyl α-cyanobenzylideneaminothiophos-phonate, TEPP (p.498) being among the other products (G. Dräger, *ibid.*, 1971, **24**, 239).

Toxicology. The acute oral LD50 is: for rats 1976-2170 mg/kg; for mice 1935-2340 mg/kg; for female guinea-pigs *c.* 600 mg/kg; for female cats and dogs 250-500 mg/kg; for female rabbits 250-375 mg/kg. The dermal LD50 for male rats is 1000 mg/kg. In 90-d feeding trials rats receiving 50 mg/kg diet showed no symptom apart from a mild depression of cholinesterase activity; at 150 mg/kg diet this depression was accompanied by an increase in liver weight and kidney weight (only in female rats), otherwise no symptom was observed. The LC50 is: for trout and carp 0·1-1·0 mg/l; for goldfish 1-10 mg/l. Toxic to honey bees by contact and vapour action.

Formulations. These include: 'Volaton' for agricultural uses, e.c. (500 g a.i./l), granules (50 and 100 g/kg), seed dressing (200 g/kg), concentrate for ULV application; 'Baython' for use against pests of man and of stored products, e.c. (100, 200 and 500 g/l), w.p. (500 g/kg), dust (30 g/kg), cold fog solution (50 g/l).

Analysis. Residues may be determined by GLC (*idem, ibid.*, 1969, **22**, 301).

Picloram

$$NH_2$$

(structural diagram of picloram: pyridine ring with NH_2 at position 4, Cl at positions 3, 5, 6, and COOH at position 2)

T6NJ BVQ CG DZ EG FG $C_6H_3Cl_3N_2O_2$ (241·5)

Nomenclature and development. The common name picloram is approved by BSI, ISO, ANSI, WSSA and JMAF. The IUPAC name is 4-amino-3,5,6-trichloropyridine-2-carboxylic acid (I) or 4-amino-3,5,6-trichloropicolinic acid (II), in *C.A.* usage (I) *[1918-02-1]* formerly (II). Its herbicidal properties were first described by J. W. Hamaker *et al., Science,* 1963, **141,** 363, and it was introduced in 1963 by The Dow Chemical Co. under the trade mark 'Tordon' and protected by USP 3 285 925.

Manufacture and properties. Produced by the chlorination of 2-methylpyridine, hydrolysis and reaction with ammonia, it is a colourless powder, with a chlorine-like odour, decomposing *c.* 215° without melting, v.p. 6·16 x 10^{-7} mmHg at 35°. Its solubility at 25° is: 430 mg/l water; 19·8 g/l acetone; 5·5 g/l propan-2-ol; 0·6 g/l dichloromethane. It is acidic, pK_a 3·6, and forms water-soluble alkali metal and amine salts, *e.g.* picloram-potassium, *C.A.* Registry No. *[2545-60-0]* (*[15162-68-2]*, potassium content unstated), solubility at 25°: 400 g/l water.

Uses. Picloram and its salts are systemic herbicides producing epinasty and leaf curling, and are rapidly absorbed by leaves and roots and translocated, accumulating in new growth. Most broad-leaved crops, except crucifers, are sensitive; most grasses are resistant. For the control of annual weeds, rates as low as 17 g a.e./ha are effective; for deep-rooted perennials on non-crop land it is recommended alone at 2·2-3·3 kg a.e./ha or, in combination with 2,4-D, at 0·3-1·8 kg picloram/ha; for brush control in combination with 2,4,5-T or 2,4-D at 0·3-2·4 kg picloram/ha or pellets at 2·2-9·5 kg picloram/ha. At the higher rates 50% loss from soil varied from 30-330 d. Rate of loss is faster from warm, moist soils and is relatively slow in cold, dry soils.

Toxicology. The acute oral LD50 is: for rats 8200 mg/kg; for mice 2000-4000 mg/kg; for rabbits *c.* 2000 mg/kg; for guinea-pigs *c.* 3000 mg/kg; for sheep > 1000 mg/kg; for cattle > 750 mg/kg. The acute dermal LD50 for rabbits is > 4000 mg/kg; no serious hazard from eye or skin contact was found. In feeding trials there was no ill-effect on sheep, pigs, cattle or poultry. Toxicity to fish and aquatic animals is low (A. J. Watson & M. G. Wiltse, *Biokemia,* 1963, **2,** 11).

Formulations. These include 'Tordon 22K' a potassium salt concentrate (240 g a.e./l); 'Tordon 101 Mixture' (65 g picloram + 240 g 2,4-D/l—both as tri-isopropylamine salts); 'Tordon 225E Mixture' (120 g picloram, as tri-isopropanolamine salt + 120 g 2,4,5-T, as propyleneglycol butyl ether (2-butoxy-1-methylethyl) ester/l); 'Tordon 155 Mixture' (120 g a.e. picloram-iso-octyl ester + 480 g a.e. 2,4,5-T—as the propyleneglycol butyl ethyl ether esters/l); 'Tordon Beads' (20 g picloram-potassium/kg) on disodium tetraborate carrier; 'Tordon 10E Pellets' (100 g picloram-potassium/kg) on clay.

Analysis. Product analysis, including mixed 2,4-D and 2,4,5-T type formulations, is conveniently made by conversion to the methyl ester, which is estimated by GLC (P. A. Hargreaves & S. H. Rapkins, *Pestic. Sci.,* 1976, **7,** 515) or by HPLC (*CIPAC Handbook*, in press). Residues in plants and soil may be determined by GLC (G. L. Bjerke *et al., J. Agric Food Chem.,* 1967, **15,** 469; T. H. Byast *et al., Agric. Res. Counc. (G.B.) Weed Res. Organ., Tech. Rep.,* No. 15 (2nd Ed.), p.33). See also: J. R. Ramsey, *Anal. Methods Pestic., Plant Growth Regul. Food Addit.,* 1967, **5,** 537; *Anal. Methods Pestic. Plant Growth Regul.,* 1972, **6,** 700.

Pimaricin

$$C_{33}H_{47}NO_{13} \quad (665 \cdot 7)$$

T F3-24-6 A AO GO KVO IU OU QU SU UUTJ BQ DQ M1
C&VQ D&Q WO- BT6OTJ CQ DZ EQ F1

Nomenclature and development. Pimaricin is an antifungal antibiotic isolated from cultures of *Streptomyces natalensis*—it is identical with the antibiotics tennecetin and natamycin (BPC). Its structure was shown to be, in *C.A.* usage, 22-[(3-amino-3,6-dideoxy-β-D-mannopyranosyl)oxy]-1,3,26-trihydroxy-12-methyl-10-oxo-6,11,28-trioxatricyclo-[22.3.1.0^{5,7}]octacosa-8,14,16,18,20-pentaene-25-carboxylic acid *[7681-93-8]* formerly 16-[3-amino-3,6-dideoxy-β-D-mannopyranosyl)oxy]-18,20,24-trihydroxy-6-methyl-4,22-dioxo-5,27-dioxabicyclo[24.1.0]heptacosa-2,8,10,12,14-pentaene-19-carboxylic acid (B. T. Golding *et al., Tetrahedron Lett.*, 1966, p.3551; W. E. Meyer, *Chem. Commun.*, 1968, p.470; G. Gaudiano *et al., Chim. Ind. (Milan)*, 1966, **48,** 1327). It was introduced in 1955 as a fungicide by Gist-Brocades N.V. under the trade mark 'Delvolan' and protected by BP 712547; 844289; USP 3892850. Other trade marks are applied for its medical ('Pimafucin') and food additive ('Delvocid', 'Delvopos') uses.

Manufacture and properties. Produced by fermentation of *S. natalensis* and *S. chattanoogensis*, it is isolated as pale yellow crystals which decompose *c.* 200°. Its solubility at 20-22° is: 4·1 g/l water; 97 g/l methanol; 5·5 g/l ethanol; 50 mg/l benzene; 100 mg/l petroleum spirit; 850 mg/l acetone; > 200 g/l dimethyl sulphoxide (J. R. Marsh & P. J. Weiss, *J. Assoc. Off. Anal. Chem.*, 1967, **50,** 457). It forms water-soluble salts with acid or alkali. It is light-sensitive but otherwise stable when dry.

Uses. Pimaricim is used as a fungicide to control diseases of bulbs, especially basal rot of daffodils at 200 mg a.i./l—preferably combined with hot-water treatment of the bulbs.

Toxicology. The acute oral LD50 for rats is 2730-4670 mg/kg (G. J. Levinskas, *Toxicol. Appl. Pharmacol.*, 1966, **8,** 97).

Formulation. 'Delvolan, w.p. (68 g a.i./kg).

Analysis. Product analysis is either by bioassay with a suitable microorganism, and confirmed by the effect of the enzyme pimaricinase, or by u.v. spectrometry at 295·5, 303 and 311 nm in methanolic acetic acid. Residues may be determined by bioassay.

Pindone

L56 BV DV CHJ CVX1&1&1 $\qquad$ $C_{14}H_{14}O_3$ (230·3)

Nomenclature and development. The common name pindone is approved by BSI and ISO —exceptions France (pivaldione), Turkey (pival) and Portugal. The IUPAC name is 2-pivaloylindan-1,3-dione, in *C.A.* usage 2-(2,2-dimethyl-1-oxopropyl)-1*H*-indene-1,3(2*H*)-dione *[83-26-1]* formerly 2-pivaloyl-1,3-indandione. Its insecticidal properties were discovered in 1942 by L. B. Kilgore *et al., Ind. Eng. Chem.,* 1942, **34,** 494. It was introduced by Kilgore Chemical Co. under the trade marks 'Pivalyl valone', 'Pival', 'Pivalyn' and protected by USP 2 310 494.

Manufacture and properties. Produced by the condensation of diethyl phthalate with 3,3-dimethylbutanone, using sodium in benzene as the condensing agent, it is a yellow crystalline solid, m.p. 108·5-110·5°. Its solubility at 25° is 18 mg/l water; soluble in aqueous alkali or ammonia to give bright yellow salts, and in most organic solvents.

Uses. Pindone, originally suggested as a substitute for the greater part of the pyrethrins in pyrethrum sprays, is used as an anticoagulant rodenticide in baits containing 250 mg/kg.

Toxicology. The acute LD50 by injection for rats is *c.* 50 mg/kg, but it is more toxic when given in small daily doses of 15-35 mg/kg. Dogs are killed by daily doses of 2·5 mg/kg (J. R. Beauregard *et al., J. Agric. Food Chem.,* 1955, **3,** 124; J. P. Saunders *et al., ibid.,* p.762).

Formulations. These include: 'Pival' powder (5 g a.i./kg) for use in rat baits; 'Pivalyn', pindone-sodium with chelating agent.

Analysis. Analysis is by colorimetry at 283 nm (J. B. La Clair, *J. Assoc. Off. Agric. Chem.,* 1955, **38,** 299). All 2-acylindan-1,3-diones give an intense yellow-red colour with iron trichloride.

Piperonyl butoxide

T56 BO DO CHJ G3 H1O2O2O4 $C_{19}H_{30}O_5$ (338·4)

Nomenclature and development. The trivial name piperonyl butoxide is accepted in lieu of a common name by BSI, ISO and BPC; also known as ENT 14250. The IUPAC name is 5-[2-(2-butoxyethoxy)ethoxymethyl]-6-propyl-1,3-benzodioxole (I), in *C.A.* usage (I) *[51-03-6]* formerly α-[2-(2-butoxyethoxy)ethoxy]-4,5-methylenedioxy-2-propyltoluene. It was developed as a pyrethrum synergist by H. Wachs, *Science,* 1947, **105**, 530, and is protected by USP 2 485 681; 2 550 737.

Manufacture and properties. Produced by condensing 5-chloromethyl-6-propyl-1,3-benzodioxole with sodium 2-(2-butoxyethoxy)ethoxide, the technical product comprises ≥ 85% *m/m* piperonyl butoxide and ≤ 15% *m/m* related compounds, and is a pale yellow oil, b.p. 180°/1 mmHg, d^{25} 1.05-1.07, n_D^{20} 1·497-1·512, n_D^{25} 1·493-1·495. It is stable to light, resistant to hydrolysis and non-corrosive.

Uses. Piperonyl butoxide is a synergist for the pyrethrins and related insecticides.

Toxicology. The acute oral LD50 for rats and rabbits is *c.* 7500 mg/kg. It is non-carcinogenic and the safe human tolerance for chronic ingestion is estimated at 42 mg/kg diet (M. P. Sarles & W. B. Vandergrift, *Am. J. Trop. Hyg. Med.,* 1952, **1**, 862; J. R. M. Innes. *J. Nat. Cancer Res.,* 1969, **42**, 1101).

Formulations. It is used in conjunction with pyrethrins in ratios of 5 : 1-20 : 1, usually 8 : 1 *m/m,* in solutions, as aerosol concentrates, emulsions or dusts.

Analysis. Product analysis is by GLC with internal standard (*British Pharmacopoeia Vet.,* 1977, p.64) or by colorimetry at 625-635 nm after reaction with tannic acid in a mixture of phosphoric and acetic acids (R. A. Jones *et al., J. Assoc. Off. Agric. Chem.,* 1952, **35**, 771: AOAC Methods: *CIPAC Handbook,* 1979, **1A**, in press). See also: J. J. Velenovsky, *Anal. Methods Pestic., Plant Growth Regul. Food Addit.,* 1964, **2**, 393; *Anal. Methods Pestic. Plant Growth Regul.,* 1972, **6**, 458.

Piperophos

T6NTJ AV1SPS&O3&O3 B1

$C_{14}H_{28}NO_3PS_2$ (353·5)

Nomenclature and development. The common name piperophos is approved by BSI and ISO. The IUPAC name is *S*-2-methylpiperidinocarbonylmethyl *O,O*-dipropyl phosphorodithioate, in *C.A.* usage *S*-[2-(2-methyl-1-piperidinyl)-2-oxoethyl] *O,O*-dipropyl phosphorodithioate *[24151-93-7]* formerly *O,O*-dipropyl phosphorodithioate *S*-ester with 1-mercaptoacetyl-2-pipecoline. It was introduced in 1969 by Ciba-Geigy AG as an experimental herbicide under the code number 'C 19490', trade mark 'Rilof' and protected by BP 1255946. It was first described by D. H. Green & L. Ebner, *Proc. Br. Weed. Control Conf., 11th,* 1972, p.822.

Properties. It is an oil at room temperature and on heating decomposes without boiling. Its solubility at 20° is 25 mg/l water; miscible with most organic solvents.

Uses. Piperophos is a selective herbicide active against annual grasses and sedges in rice. It is marketed in combination with dimethametryn under the trade mark 'Avirosan', for the control of both mono- and di-cotyledonous weeds. Both herbicides are taken up by young plants through roots, coleoptiles and leaves. In tropical regions piperophos is preferably applied in mixture with a hormone-type weedkiller.

Toxicology. The acute oral LD50 for rats is 324 mg/kg; the acute dermal LD50 for rats is > 2150 mg/kg.

Formulations. These include: 'Rilof', e.c. (500 g a.i./l); 'Avirosan 3G', 'Avirosan 5G', granules; 'Avirosan 500 EC', e.c. (4:1 mixtures of piperophos + dimethametryn).

Analysis. Product analysis is by GLC with internal standard. Residues may be determined, after column chromatographic clean-up, by GLC with TID. Details of both methods are available from Ciba-Geigy AG.

Piproctanyl bromide

$$CH_2=CHCH_2 \overset{+}{\underset{\underset{Br^-}{|}}{N}} CH_2CH_2\overset{\overset{Me}{|}}{CH}-(CH_2)_3CHMe_2$$

T6KTJ A2Y1&3Y1&1 A2U1   *piproctanyl* $C_{18}H_{36}N$ (266·5)

T6KTJ A2Y1&3Y1&1 A2U1 &E   *piproctanyl bromide* $C_{18}H_{36}BrN$ (346·4)

Nomenclature and development. The common name piproctanyl is approved by BSI and proposed by ISO. The IUPAC name is 1-allyl-1-(3,7-dimethyloctyl)piperidinium, in *C.A.* usage 1-(3,7-dimethyloctyl)-1-(2-propenyl)piperidinium. The bromide salt, piproctanyl bromide, *C.A.* Registry No. *[56717-11-4]*, was introduced by R. Maag Ltd in 1976 as a plant growth regulator under the code number 'Ro 06-0761', the trade marks 'Alden' and 'Stemtrol' and protected by DOP 2 459 129 (to Hoffman-La Roche AG). Its properties were first described by G. A. Hüppi *et al., Experientia,* 1976, **32**, 37.

Properties. It is a pale yellow wax, m.p. 75°. It is highly soluble in water and alcohols. It is non-corrosive in aqueous solution, insensitive to light and stable at temperatures $\leqslant 100°$, and $\geqslant 3$ y under normal conditions.

Uses. Piproctanyl bromide is a plant growth regulator which reduces internodal elongation, reducing plant height, producing stronger stems and penduncles and a deeper green foliage. It is taken up through leaves and roots but is not readily translocated between shoots. A surfactant is included in sprays. It is used on chrysanthemums at 75-150 mg a.i./l, the spray concentration depends on cultivar. It is also used on *Begonia elatior, Calceolaria rugosa, Fuchsia hybrida* and *Petunia* spp; other species are under study.

Toxicology. The acute oral LD50 is: for rats 820-990 mg a.i./kg; for mice 182 mg/kg. The acute dermal LD50 for rats is 115-240 mg/kg; a solution (30 g/l acetone) caused no irritation to the skin of guinea-pigs; rabbits' eyes were not affected by 3 g/l water. The acute inhalation LC50 in rats is 1·5 g/m³ air. In 90-d feeding trials there was no significant effect for rats receiving 150 mg/kg daily or dogs 25 mg/kg daily. The LC50 (8 d) for bobwhite quail and mallard duck was > 10000 mg/kg diet. The LC50 (96 h) is: for rainbow trout 12·7 mg/l; for bluegill 62 mg/l.

Formulations. 'Alden' and 'Stemtrol' are water-miscible concentrates (50 g a.i./l, plus surfactant).

Analysis. Product analysis is by reaction with sodium thiophenolate, forming 1-(3,7-dimethyloctyl)piperidine and allyl phenyl sulphide, the latter being determined by GLC with 2-methylnaphthalene as internal standard. Residues may be determined in the same way after clean-up by liquid-liquid partition.

Pirimicarb

T6N CNJ BN1&1 DOVN1&1 E1 F1 $C_{11}H_{18}N_4O_2$ (238·3)

Nomenclature and development. The common name pirimicarb is approved by BSI, ISO and ANSI, in France (pyrimicarbe); also known as ENT 27766. The IUPAC name is 2-dimethylamino-5,6-dimethylpyrimidin-4-yl dimethylcarbamate, in *C.A.* usage 2-(dimethylamino)-5,6-dimethyl-4-pyrimidinyl dimethylcarbamate *[23103-98-2]*. It was introduced in 1969 by ICI Ltd, Plant Protection Division as an insecticide under the code number 'PP 062' and subsequently the trade marks 'Pirimor' and 'Aphox' and protected by BP 1 181 657. It was first described by F. L. C. Baranyovits & R. Ghosh, *Chem. Ind. (London),* 1969, p.1018.

Manufacture and properties. Produced either by condensing 2-dimethylamino-5,6-dimethylpyrimidin-4-one with dimethylcarbamoyl chloride and a base, or from the pirimidinone, carbonyl chloride and dimethylamine, it is a colourless solid, m.p. 90·5°, v.p. 3 x 10^{-5} mmHg at 30°. Its solubility at 25° is 2·7 g/l water; soluble in most organic solvents. It is decomposed by prolonged boiling with acids or alkali. Aqueous solutions are unstable to light. It forms well defined crystalline salts with acids, and they are readily soluble in water; the hydrochloride is deliquescent. It is non-corrosive to normal materials used in spray equipment.

Uses. Pirimicarb is a selective aphicide, being effective against organophosphorus-resistant aphid strains. It is fast acting and has fumigant and translaminar properties; is taken up by roots and translocated in the xylem system. It has neither acaricidal nor fungicidal properties, and can be used to control aphids in integrated control programmes.

Toxicology. The acute oral LD50 is: for rats 147 mg/kg; for mice 107 mg/kg; for poultry 25-50 mg/kg; for dogs 100-200 mg/kg. Daily applications of 500 mg/kg for 24 h to rabbit skin over a 14-d period produced no toxic symptom. A solution (5 g tech./l) was not an irritant to rabbit eyes. Rats exposed for 6 h/d (5 d/week) for 21 d, to air which had been passed over technical pirimicarb at room temperature developed no toxic sign, nor was there inhibition of cholinesterase. In 2-y feeding trials the 'no effect' level was: for dogs 1·8 mg/kg daily; for rats 250 mg/kg diet ($\equiv$ 12·5 mg/kg daily).

Formulations. These include: e.c. (80 g a.i./l), w.p. (500 g/kg), dispersible grains (500 g/kg) and smoke generators (10%) all for use in agriculture and horticulture; an aerosol concentrate for home garden use.

Analysis. Product analysis is by GLC (J. E. Bagness & W. G. Sharples, *Analyst (London),* 1974, **99**, 225). Residues may be determined by GLC with FID; a colorimetric method is also available. Details of these methods are available from ICI Ltd, Plant Protection Division. See also: D. J. W. Bullock, *Anal. Methods Pestic. Plant Growth Regul.,* 1973, **7**, 399.

Pirimiphos-ethyl

Me⎯⎯⎯O·P·(OEt)₂

T6N CNJ BN2&2 DOPS&O2&O2 F1 $C_{13}H_{24}N_3O_3PS$ (333·4)

Nomenclature and development. The common name pirimiphos-ethyl is approved by BSI, ISO and ANSI—in France (pyrimiphos-éthyl). The IUPAC name is O-2-diethylamino-6-methylpyrimidin-4-yl O,O-diethyl phosphorothioate, in *C.A.* usage O-[2-(diethylamino)-6-methyl-4-pyrimidinyl] O,O-diethyl phosphorothioate [23505-41-1]. It was introduced in 1971 by ICI Ltd, Plant Protection Division under the code number 'PP 211', and subsequently the trade marks 'Pirimicid' and 'Fernex' as an insecticide protected by BP 1 019 227; BP 1 205 000.

Manufacture and properties. Produced by condensing 1,1-diethylguanidine with ethyl 3-oxybutyrate and subsequent reaction with O,O-diethyl phosphorochloridothioate, it is a straw-coloured liquid, v.p. 2·9 x 10⁻⁴ mmHg at 25°, d^{20} 1·14, n_D^{25} 1·520. Decomposition begins > 130° so no b.p. is available. Its solubility at 30° is < 1 mg/l water; miscible with, or very soluble in most organic solvents. It is stable ⩾ 5 d at 80° and ⩾ 1 y at room temperature. It is corrosive to iron and unprotected tinplate.

Uses. Pirimiphos-ethyl is a broad-range insecticide effective against dipterous and coleopterous pests living in the soil or on the soil surface. It is relatively non-phytotoxic and may be used as a seed dressing, sometimes in combination with fungicides. It is also used in compost to prevent infestation of mushrooms by sciarids and phorids.

Toxicology. The acute oral LD50 is: for rats 140-200 mg/kg; for cats 25-50 mg/kg; for guinea-pigs 50-100 mg/kg. The acute dermal LD50 for rats is 1000-2000 mg/kg; no irritation followed the application of 100 mg/kg to shorn rabbit skin for 24-h periods on alternate days over a 10-d period; it is not a sensitiser. In 90-d feeding trials the 'no effect' level was: for rats 1·6 mg/kg diet (≡ 0·08 mg/kg daily); for dogs 0·2 mg/kg daily. For rats receiving 27 mg/kg diet (≡ 1·6 mg/kg daily) and dogs 2 mg/kg daily the only effect was on the cholinesterase levels. From inhalation tests on rats, the maximum allowable concentration is calculated to be 14 mg/m³.

Formulations. These include: encapsulated (200 g a.i./l); e.c. (250 and 500 g/l); granules (50 and 100 g/kg); 'Pirimicid' seed treatment (a mixture of drazoxolon + pirimiphos-ethyl).

Analysis. Product analysis is by GLC using an internal standard (J. E. Bagness & W. G. Sharples, *Analyst (London),* 1974, **99,** 225), or by u.v. spectrometry at 298 nm in methanol, after clean-up and TLC of the extract (S. H. Yuen, *ibid.,* 1976, **101,** 533). Residues may be determined by GLC with FPD or FTD; a colorimetric method is also available. Details of these methods are available from ICI Ltd, Plant Protection Division. See also: D. J. W. Bullock, *Anal. Methods Pestic. Plant Growth Regul.,* 1976, **8,** 171.

Pirimiphos-methyl

T6N CNJ BN2&2 DOPS&O1&O1 F1 $C_{11}H_{20}N_3O_3PS$ (305·3)

Nomenclature and development. The common name pirimiphos-methyl is approved by BSI, ISO and ANSI—in France (pyrimiphos-méthyl). The IUPAC name is O-2-diethyl-amino-6-methylpyrimidin-4-yl O,O-dimethyl phosphorothioate, in C.A. usage O-[2-(diethylamino)-6-methyl-4-pyrimidinyl] O,O-dimethyl phosphorothioate [29232-93-7]. It was introduced in 1970 by ICI Ltd, Plant Protection Division as an insecticide under the code number 'PP 511' and subsequently the trade marks 'Actellic', 'Actellifog', 'Silo San' and 'Blex' and protected by BP 1 019 227; BP 1 204 552.

Manufacture and properties. Produced by condensing 1,1-diethylguanidine with ethyl 3-oxobutyrate and subsequent reaction with O,O-dimethyl phosphorochloridothioate, it is a straw-coloured liquid, v.p. c. 1 x 10^{-4} mmHg at 30°, d^{30} 1·157, n_D^{25} 1·527. Its solubility at 30° is c. 5 mg/l water; miscible with, or very soluble in, most organic solvents. It is hydrolysed by concentrated acids and alkalies and does not corrode brass, stainless steel, nylon, polythene or aluminium; slightly corrosive to unprotected tinplate and mild steel.

Uses. Pirimiphos-methyl is a fast-acting insecticide and acaricide with both contact and fumigant action, able to penetrate leaf tissue to give a translaminar action. Its range of activity includes many crop pests and it is also used against pests of stored products, in public health, domestic and amenity use.

Toxicology. The acute oral LD50 is: for female rats 2050 mg/kg; for male mice 1180 mg/kg; for female guinea-pigs 1000-2000 mg/kg; for male rabbits 1150-2300 mg/kg; for female cats 575-1150 mg/kg; for hens 30-60 mg/kg. The acute dermal LD50 for rabbits is > 2000 mg/kg. In 90-d feeding trials the 'no effect' level was: for dogs 20 mg/kg diet (≡ 0·5 mg/kg daily); for rats 8 mg/kg diet (≡ 0·4 mg/kg daily) at 360 mg/kg diet the only effect was on cholinesterase levels. Rats were not affected when exposed to saturated vapour (40 mg/m³) for 6 h/d for 5 d/week over a period of 21 d. The LC50 (24 h) for mirror carp is 1·6 mg/l; the LC50 (48 h) is 1·4 mg/l.

Formulations. These include: e.c. (80, 250 and 500 g a.i./l); ULV concentrate (500 g/l); encapsulated (200 g/l); dust (20 g/kg); solvent-free formulations (900 g/kg); smoke generator; domestic sprays (pirimiphos-methyl + synergised pyrethrins).

Analysis. Product analysis is by GLC with internal standard (J. E. Bagness & W. G. Sharples, *Analyst (London),* 1974, **99**, 225) or by u.v. spectroscopy at 298 nm in methanol, after clean-up and TLC of the extract (S. H. Yuen, *ibid.,* 1976, **101**, 533). Residues may be determined by GLC with FPD or FTD, a colorimetric method is also available. See also: D. J. W. Bullock, *Anal. Methods Pestic. Plant Growth Regul.,* 1976, **8**, 181. Details are also available from ICI Ltd, Plant Protection Division.

Polyoxin B

$$C_{17}H_{25}N_5O_{13} \quad (507\cdot4)$$

T6NVMVJ E1Q A- BT5OTJ CQ DQ EYVQMVYZYQYQ1OVZ

Nomenclature and development. The common name polyoxins is approved by JMAF. They comprise a series of 13 antibiotics (polyoxin A to polyoxin M) of similar physical and chemical characteristics and isolated from culture solutions of *Streptomyces cacaoi*. In *C.A.* usage polyoxin B is 5-[[2-amino-5-*O*-(aminocarbonyl)-2-deoxy-L-xylonoyl]amino]-1,5-dideoxy-1-[3,4-dihydro-5-(hydroxymethyl)-2,4-dioxo-1(2*H*)-pyrimidinyl]-β-D-allofuranuronic acid *[19396-06-6]* formerly 5-(2-amino-2-deoxy-L-xylonamido)-1-[3,4-dihydro-5-hydroxymethyl-2,4-dioxo-1(2*H*)-pyrimidinyl]allofuranuronic acid β-D-monocarbamate. It was isolated in 1963 by S. Suzuki *et al.*, *J. Antibiot. Ser. A.*, 1965, **18**, 131. It was introduced by Nihon Nohyaku Ltd and by Kumiai Chemical Industry Co. Ltd.

Manufacture and properties. Produced by fermentation of *S. cacaoi* it is an amorphous powder, m.p. 160°. It is soluble in water; insoluble in acetone, benzene, chloroform, ethanol, hexane and methanol. It is stable over the range pH 1-8.

Uses. Polyoxin B is an effective fungicide at 50-100 mg a.i./l for use against *Alternaria* spp., causing an abnormal germination, and loss of phytopathogenicity and impeding hyphal development. It is also effective against *Botrytis* spp., *Pellicularia sasaki* of rice and several diseases of fruit, *e.g. Podosphaera leucotricha* on apple, *Fulvia fulva* on tomato, and cucumber scab. It is non-phytotoxic and compatible with other pesticides except those highly alkaline in reaction.

Polyoxin A is effective against brown spot of rice, polyoxin D against *Pellicularia sasaki* of rice.

Toxicology. Oral doses of 1500 mg polyoxin B/kg caused no deaths in mice, nor did intravenous injections of 800 mg/kg. It is non-irritant to mucous membranes or skin. Japanese killifish are unaffected by 100 mg/l for 72 h.

Formulation. A w.p. (100 g polyoxin B/l).

Prochloraz

T5N CNJ AVN3&2OR BG DG FG $C_{15}H_{16}Cl_3N_3O_2$ (376.7)

Nomenclature and development. The common name prochloraz is approved by BSI and proposed by ISO. The IUPAC name is N-propyl-N-[2-(2,4,6-trichlorophenoxy)ethyl]-imidazole-1-carboxamide or 1-{N-propyl-N-[2-(2,4,6-trichlorophenoxy)ethyl]} carbamoylimidazole, in *C.A.* usage N-propyl-N-[2-(2,4,6-trichlorophenoxy)ethyl]-1H-imidazole-1-carboxamide *[67747-09-5]*. It was introduced by The Boots Co. Ltd as an experimental fungicide under the code number 'BTS 40542' and protected by BP 1460772; USP 3991071; 4080462. Its properties were first reported by R. J. Birchmore *et al., Proc. 1977 Br. Crop Prot. Conf., Pests Dis.,* 1977, **2**, 543.

Manufacture and properties. Produced by reacting 2,4,6-trichlorophenol successively with 1,2-dibromoethane, propylamine, carbonyl chloride and imidazole, it is a colourless crystalline solid, m.p. 38.5-41.0°, v.p. 6×10^{-10} mmHg at 20°. Its solubility at 23° is 47.5 mg/l water; approximate values at 25° are: 2.5 kg/l chloroform, diethyl ether, toluene, xylene; 3.5 kg/l acetone. It is stable in water (50% loss in 1808 d (extrapolated) at 20° and pH 7) but is unstable to concentrated acid or alkali and to sunlight. The technical material, *c.* 97% pure, is a golden brown liquid that tends to solidify on cooling.

Uses. Prochloraz is a protectant and eradicant fungicide effective against a wide range of plant pathogens affecting field crops, fruit, vegetables, turf and ornamentals. For field crops it is used either as a seed treatment at 200-400 mg a.i./l or as a foliar spray at 0.3-1.0 kg/ha. Typical rates for pre-harvest spray on fruit and vegetable crops are 20-50 g/100 l, and as a post-harvest dip against storage rots 250-1000 mg/l.

Toxicology. The acute oral LD50 is: for rats *c.* 1600 mg/kg; for mice 2400 mg/kg; for mallard duck 3132 mg/kg. The acute dermal LD50 for rats is > 5000 mg/kg. The LC50 (5 d) for mallard duck was > 10000 mg/kg diet. The LC50 (96 h) for rainbow trout is 1 mg/l.

Formulations. These include: e.c. (250 g a.i./l) for cereals; w.p. (250 g/kg) for other crops; seed dressings, liquid (100 g/l) and powder (100 g/kg).

Analysis. Product analysis is by HPLC and GLC.

Procymidone

T35 DVNVTJ A1 Cl ER CG EG $\qquad$ $C_{13}H_{11}Cl_2NO_2$ (284·1)

Nomenclature and development. The common name procymidone is approved by BSI and proposed by ISO. The IUPAC name is *N*-(3,5-dichlorophenyl)-1,2-dimethylcyclo-propane-1,2-dicarboximide (I), in *C.A.* usage 3-(3,5-dichlorophenyl)-1,5-dimethyl-3-azabicyclo[3.1.0]hexane-2,4-dione *[32809-16-8]* formerly (I). It was introduced in 1969 by the Sumitomo Chemical Co. under the code number 'S-7131', trade marks 'Sumisclex' or 'Sumilex' and protected by BP 1298261; USP 3903090. Its usefulness as a fungicide for plant disease control was described by Y. Hisada *et al., J. Pestic. Sci.,* 1976, **1,** 145.

Manufacture and properties. Produced by the reaction of 3,5-dichloraniline with *cis*-1,2-dimethylcyclopropane-1,2-dicarboxylic acid, it is a colourless crystalline solid, m.p. 166°, v.p. 1·32 x 10⁻⁴ mmHg at 25°. It is practically insoluble in water; poorly soluble in alcohols; soluble in most organic solvents.

Uses. Procymidone is a preventive, curative and persistent fungicide with moderate systemic activity, effective against *Sclerotinia, Botrytis* and *Cochliobolus* spp. affecting fruits, vegetables, field crops and ornamentals. It is usually applied at 500-1000 g a.i./ha. It shows promise in controlling storage rots of fruits and vegetables.

Toxicology. The acute oral LD50 is: for rats 6800-7700 mg tech./kg; for mice 7800-9100 mg/kg. Neither skin nor eye irritation was observed in rabbits. The LC50 (48 h) for carp is > 10 mg/l.

Formulation. A w.p. (500 g a.i./kg).

Analysis. Product analysis is by GLC.

Profenofos

GR CE FOPO&S3&O2 $\qquad$ $C_{11}H_{15}BrClO_3PS$ (373·6)

Nomenclature and development. The common name profenofos is approved by BSI, ANSI and New Zealand, and proposed by ISO. The IUPAC and *C.A.* name is *O*-(4-bromo-2-chlorophenyl) *O*-ethyl *S*-propyl phosphorothioate, Registry No. *[41198-08-7]*. It was developed by Ciba-Geigy AG as an insecticide under the code number 'CGA 15324', trade mark 'Curacron' and protected by BelgP 798937. It was described by F. Buholzer, *Proc. Br. Insectic. Fungic. Conf., 8th*, 1975, **2**, 659.

Manufacture and properties. Produced by the reaction of 4-bromo-2-chlorophenol with *O*-ethyl *S*-propyl phosphorochloridothioate, it is a pale yellow liquid, b.p. 110°/1 x 10^{-3} mmHg, v.p. *c.* 10^{-5} mmHg at 20°. Its solubility at 20° is 20 mg/l water; miscible with most organic solvents.

Uses. Profenofos is a non-systemic broad spectrum insecticide for use against insect pests and mites on cotton and vegetables. It has contact and stomach action. Rates of application for sucking insects and mites are 250-500 g a.i./ha; for chewing insects 400-1200 g/ha.

Toxicology. The acute oral LD50 for rats is 358 mg/kg; the acute dermal LD50 for rats is *c.* 3300 mg/kg. It is highly toxic to birds and fish.

Formulations. These include: e.c. 'Curacron' 500 EC (500 g a.i./l); 400 EC (400 g/l); 720 EC (720 g/l); 250 ULV (250 g/l); granules (50 g/kg).

Analysis. Product analysis is by GLC with internal standard. Residues may be determined, after column chromatographic clean up, by GLC with TID. Details of these methods are available from Ciba-Geigy AG.

Profluralin

L3TJ A1N3&R BNW FNW DXFFF $C_{14}H_{16}F_3N_3O_4$ (347·3)

Nomenclature and development. The common name profluralin is approved by BSI, ISO, ANSI and WSSA. The IUPAC name is N-(cyclopropylmethyl)-α,α,α-trifluoro-2,6-dinitro-N-propyl-p-toluidine (I), formerly N-cyclopropylmethyl-2,6-dinitro-N-propyl-4-trifluoro-methylaniline, in *C.A.* usage N-(cyclopropylmethyl)-2,6-dinitro-N-propyl-4-(trifluoro-methyl)benzenamine *[26399-36-0]* formerly (I). It was introduced in 1970 by Ciba-Geigy AG as an experimental herbicide under the code number 'CGA 10832', trade mark 'Tolban' (in the USA) and protected by USP 3 546 295. It was first described by T. D. Taylor *et al., Annu. Meet. Weed Sci. Soc. Am.,* 1973, Abstr. 169.

Properties. It is a yellow-orange crystalline solid, m.p. 32°. Its solubility at 20° is 0·1 mg/l water; soluble with most organic solvents.

Uses. Profluralin is a soil-incorporated herbicide used pre-planting at 0·75-1·5 kg a.i./ha for annual and perennial weed and grass control in cotton, soyabeans and many other crops.

Toxicology. The acute oral LD50 for rats is *c.* 10000 mg/kg; the acute dermal LD50 for rats is > 3170 mg/kg. It is highly toxic to fish and toxic to honey bees.

Formulation. An e.c. (500 g a.i./l).

Analysis. Product analysis is by GLC with internal standard. Residues may be determined, after column chromatographic clean-up, by GLC with MCD. Details of these methods are available from Ciba-Geigy AG. See also: R. A. Kahrs, *Anal. Methods Pestic. Plant Growth Regul.,* 1978, **10**, 451.

Proglinazine-ethyl

T6N CN ENJ BMY1&1 DM1VQ FG   *proglinazine* $C_8H_{12}ClN_5O_2$ (245·6)

T6N CN ENJ BMY1&1 DM1VO2 FG   *proglinazine-ethyl* $C_{10}H_{16}ClN_5O_2$ (273·7)

Nomenclature and development. The common name proglinazine is approved by BSI and proposed by ISO. The IUPAC name is *N*-(4-chloro-6-isopropylamino-1,3,5-triazin-2-yl)-glycine, in *C.A.* usage *N*-[4-chloro-6-(1-methylethylamino)-1,3,5-triazine-2-yl]glycine *[68228-20-6]* formerly *N*-[4-chloro-6-(isopropylamino)-*s*-triazin-2-yl]glycine. The ethyl ester, Registry No. *[68228-18-2]*, was introduced in 1972 by Nitrokémia Ipartelepek as a herbicide under the code number 'MG-O7'.

Manufacture and properties. Produced by the reaction of trichloro-1,3,5-triazine with 1 equivalent of ethyl glycinate hydrochloride, followed by 1 equivalent of isopropylamine in the presence of an acid-binding agent, it forms colourless crystals, m.p. 110-112°. Its solubility at 25° is 750 mg/l water; soluble in organic solvents. It is stable in neutral, weakly acidic or weakly basic media at room temperature, but on heating it is hydrolysed by acid or alkali to the corresponding herbicidally inactive triazin-2-ol. It is compatible with other pesticides when used at normal rates and is non-corrosive. It is rapidly decomposed in soil.

Uses. It is used as a pre-em. herbicide in maize at 4 kg a.i./ha, and is especially effective for the control of dicotyledonous weed seedlings.

Toxicology. The acute oral LD50 is: for rats and mice > 8000 mg/kg; for guinea-pigs 857-923 mg/kg; for rabbits > 3000 mg/kg. The acute intraperitoneal LD50 is: for rats 829-891 mg/kg; for mice 720-1080 mg/kg. The acute percutaneous LD50 is: for rats > 1500 mg/kg; for rabbits > 4000 mg/kg. It is not a dermal or eye irritant to rabbits; the sensitising action for guinea-pigs is moderate.

Formulation. A w.p. (500 g a.i./kg).

Analysis. Product analysis is by GLC. Residues may be determined by GLC with TID.

Promecarb

MeNH.CO.O— [chemical structure: benzene ring with Me (methyl) group and Pr^i (isopropyl) group]

Me

Pr^i

1Y1&R   C1   EOVM1 $C_{12}H_{17}NO_2$ (207·3)

Nomenclature and development. The common name promecarb is approved by BSI, ISO and JMAF; also known as ENT 27300a, OMS 716. The IUPAC name is 5-isopropyl-*m*-tolyl methylcarbamate, formerly 3-isopropyl-5-methylphenyl methylcarbamate, in *C.A.* usage 3-methyl-5-(1-methylethyl)phenyl methylcarbamate *[2631-37-0]* formerly *m*-cym-5-yl methylcarbamate. It was introduced in 1965 as an experimental insecticide by Schering AG under the code number 'Schering 34615', trade mark 'Carbamult' and protected by DBP 1156272; BP 913707; USP 3167472. It was first described by A. Formigoni & G. P. Bellini at the Congresso Internazionale Degli Antiparassitari, Naples, March 1965 and by A. Jäger, *Z. Angew. Entomol.,* 1966, **58,** 188.

Manufacture and properties. Produced by the reaction of 5-isopropyl-*m*-cresol with methyl isocyanate, it is a colourless crystalline solid, m.p. 87-88°, b.p. 117°/0·01 mmHg, v.p. 3 x 10^{-5} mmHg at 25°. Its solubility at room temperature is: 92 mg/l water; 100-200 g/l carbon tetrachloride, xylene; 200-400 g/l cyclohexanol, cyclohexanone, propan-2-ol, methanol; 400-600 g/l acetone, dimethylformamide, 1,2-dichloroethane. The technical product is > 98% pure. No changes were observed on storage for 140 h at 50°; 50% loss occurs at 37° in 30 h at pH 7, 5·7 h at pH 9.

Uses. Promecarb is a non-systemic contact insecticide effective against coleopterous pests such as post-embryonic stages of *Leptinotarsa decemlineata* at 450 g a.i./ha, against lepidopterous pests and leaf miners of fruit at 50-100 g/100 l.

Toxicology. The acute oral LD50 for rats is 61-90 mg (in maize germ oil)/kg, 70-108 mg (in acacia gum suspension)/kg. The dermal LD50 for rats and rabbits is > 1000 mg a.i. (as 50% w.p.)/kg. In 1·5-y feeding trials rats receiving 5 mg a.i./kg (5 d/week) suffered no ill-effect.

Formulations. These include: e.c. (250 g a.i./l); w.p. (375 and 500 g/kg); dust (50 g/kg).

Analysis. Product analysis is by alkaline hydrolysis, the methylamine liberated being distilled into acid, followed by back titration. Residues may be determined by GLC measurement of 5-isopropyl-*m*-cresol produced on hydrolysis.

Prometon

T6N CN ENJ BO1 DMY1&1 FMY1&1 $C_{10}H_{19}N_5O$ (225·3)

Nomenclature and development. The common name prometon is approved by BSI, ISO, ANSI and WSSA—exceptions Canada and Germany. The IUPAC name is 2,4-bis-(isopropylamino)-6-methoxy-1,3,5-triazine, formerly 2,4-di(isopropylamino)-6-methoxy-1,3,5-triazine, in *C.A.* usage 6-methoxy-*N*,*N'*-bis(1-methylethyl)-1,3,5-triazine-2,4-diamine *[1610-18-0]*, formerly 2,4-bis(isopropylamino)-6-methoxy-*s*-triazine. Its herbicidal properties were reported by H. Gysin & E. Knüsli, *Adv. Pest Control Res.,* 1960, **3**, 289. It was introduced in 1959 by J. R. Geigy S.A. (now Ciba-Geigy AG) under the code number 'G 31435', trade marks 'Primatol' and 'Pramitol' and protected by SwissP 337019; 340372; BP 814948; USP 2909420.

Manufacture and properties. Produced by reacting propazine (p.448) with methanol in the presence of 1 molecular proportion of sodium hydroxide, it is a colourless crystalline solid, m.p. 91-92°, v.p. 2·3 x 10^{-6} mmHg at 20°. Its solubility at 20° is 620 mg/l water; readily soluble in most organic solvents. It is distillable and is stable under neutral and slightly acidic or alkaline conditions, but is hydrolysed by concentrated acid or alkali to the herbicidally-inactive 4,6-bis(isopropylamino)-1,3,5-triazin-2-ol. It is compatible with most other pesticides when used at normal rates and is non-corrosive under normal conditions of use.

Uses. Prometon is a non-selective herbicide used for the control of most annual and perennial broad-leaved and grass weeds on non-crop areas at 10-20 kg a.i./ha and may be applied to the ground before laying asphalt.

Toxicology. The acute oral LD50 for rats is 3000 mg/kg; the acute dermal LD50 for rabbits is > 2000 mg/kg. The sub-acute oral LD50 (28 d) for rats was 1000 mg/kg daily. It has a negligible toxicity to wildlife.

Formulations. These include, in the USA only: 'Pramitol 25 E', e.c. (250 g a.i./l); 'Primatol 80 WP', w.p. (800 g/kg).

Analysis. Product analysis is by GLC with an internal standard (K. Hofberg *et al., J. Assoc. Off. Anal. Chem.,* 1973, **56**, 586; *CIPAC Handbook,* 1979, **1A**, in press), or by titration with perchloric acid in acetic acid. Residues may be determined by TLC and GLC; A. M. Mattson *et al., J. Agric. Food Chem.,* 1965, **13**, 120; K. Ramsteiner *et al., J. Assoc. Off. Anal. Chem.,* 1974, **57**, 192. Details of unpublished methods are available from Ciba-Geigy AG. See also: E. Knüsli, *Anal. Methods Pestic., Plant Growth Regul. Food Addit.,* 1964, **4**, 171; *Anal. Methods Pestic. Plant Growth Regul.,* 1972, **6**, 679.

Prometryne

(Prometryn)

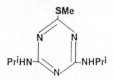

T6N CN ENJ BS1 DMY1&1 FMY1&1 $C_{10}H_{19}N_5S$ (241·4)

Nomenclature and development. The common name prometryn is approved by ISO, ANSI and WSSA—exceptions BSI and JMAF (prometryne) and France (prométryne). The IUPAC name is 2,4-bis(isopropylamino)-6-methylthio-1,3,5-triazine, formerly 2,4-di(isopropylamino)-6-methylthio-1,3,5-triazine, in *C.A.* usage *N,N'*-bis(1-methylethyl)-6-(methylthio)-1,3,5-triazine-2,4-diamine *[7287-19-6]* formerly 2,4-bis(isopropylamino)-6-(methylthio)-s-triazine. Its herbicidal properties were first described by H. Gysin, *Chem. Ind. (London),* 1962, p.1393. It was introduced in 1962 by J. R. Geigy S.A. (now Ciba-Geigy AG) under the code number 'G 34161' trade marks 'Gesagard', 'Caparol' (in the USA), and protected by SwissP 337019; BP 814948; USP 2909420.

Manufacture and properties. Produced either by reacting propazine (p.448) with methanethiol in the presence of 1 molecular proportion of sodium hydroxide, or by reacting 4,6-bis(isopropylamino)-1,3,5-triazine-2-thiol with a methylating agent in the presence of sodium hydroxide, it is a colourless crystalline solid, m.p. 118-120°, v.p. 1·0 x 10^{-6} mmHg at 20°. Its solubility at 20° is 40 mg/l water; readily soluble in organic solvents.

Uses. Prometryne is a herbicide used either pre- or post-em. for selective weed control in cotton, peas, carrots, celery, potatoes, sunflowers and onions at 1·0-1·5 kg a.i./ha pre-em. and 0·5-1·0 kg/ha post-em. At the higher rates, it persists in soils for 30-90 d.

Toxicology. The acute oral LD50 for rats is 3150-5233 mg/kg; the acute dermal LD50 for rats is > 3100 mg/kg. In 2-y feeding trials the 'no effect' level for rats was 1250 mg/kg diet. It is slightly toxic to fish.

Formulations. These include: 'Gesagard 50', w.p. (500 g a.i./kg); 'Gesagard 500 FW', flowable (500 g/l); 'Gesagard 80'/'Caparol 80', w.p. (800 g/kg). Also in combination with other triazines, and with metolachlor ('Codal 400 EC').

Analysis. Product analysis is by GLC with internal standard (*CIPAC Handbook,* 1979, **1A,** in press; R. T. Murphy *et al., J. Assoc. Off. Anal. Chem.,* 1971, **54,** 703); or by titration with perchloric acid in acetic acid (H. P. Bosshardt *et al., ibid.,* 1973, **56,** 749). Residues may be determined by GLC (K. Ramsteiner *et al., ibid.,* 1974, **57,** 192; T. H. Byast *et al., Agric. Res. Counc. (G.B.) Weed Res. Organ., Tech Rep.,* No. 15 (2nd Ed.), p.40). Details are available from Ciba-Geigy AG. See also: E. Knüsli, *Anal. Methods Pestic., Plant Growth Regul. Food Addit.,* 1964, **4,** 179; *Anal. Methods Pestic. Plant Growth Regul.,* 1972, **6,** 680.

Propachlor

$$\text{ClCH}_2.\overset{\displaystyle O}{\overset{\displaystyle \|}{C}}.N.\text{CHMe}_2$$
$$\overset{\displaystyle |}{\text{Ph}}$$

G1VNR&Y1&1 $C_{11}H_{14}ClNO$ (211·7)

Nomenclature and development. The common name propachlor is approved by BSI, ISO and WSSA. The IUPAC name is 2-chloro-N-isopropylacetanilide (I), formerly α-chloro-N-isopropylacetanilide, in *C.A.* usage 2-chloro-N-(1-methylethyl)-N-phenylacetamide *[1918-16-7]* formerly (I). It was introduced in 1965 as a herbicide by Monsanto Chemical Co. under the code number 'CP 31393', trade mark 'Ramrod' and protected by USP 2863752. Its herbicidal activity was first described by D. D. Baird *et al.,* at the North Cent. Weed Control Conference, 1964.

Manufacture and properties. Produced by the reaction of chloroacetyl chloride and N-isopropylaniline, it is a light tan solid, m.p. 67-76°, v.p. 0·03 mmHg at 110°. Its solubility at 20° is 700 mg/l water; readily soluble in most organic solvents, except aliphatic hydrocarbons.

Uses. Propachlor is a pre-em. herbicide effective against annual grasses and certain broad-leaved weeds in maize, cotton, soyabeans, sugarcane, groundnuts and vegetable crops including brassicas, onions, peas and beans, at 3·5-5·0 kg a.i./ha. At these rates it persists in soil for 28-42 d.

Toxicology. The acute oral LD50 for rats is 780 mg a.i. (as w.p.)/kg; the acute dermal LD50 for rabbits is 380 mg (as 10·4% suspension in water)/kg; a moderate eye irritant. In 90-d feeding trials dogs receiving 133·3 mg/kg daily gained weight normally; rats receiving 133·3 mg/kg daily showed no effect other than a 9% reduction in weight gain.

Formulations. A w.p. (650 g a.i./kg); granules (200 g/kg).

Analysis. Product analysis is by GLC. Residues may be determined, after column chromatographic clean-up, by GLC with ECD.

Propamocarb

$$O$$
$$\|$$
$$Me_2N.CH_2CH_2CH_2.NH.C.O.Pr$$

3OVM3N1&1

3OVM3N1&1 &GH

propamocarb $C_9H_{20}N_2O_2$ (188·3)

propamocarb hydrochloride $C_9H_{21}ClN_2O_2$ (224·7)

Nomenclature and development. The common name propamocarb is approved by BSI and proposed by ISO. The IUPAC and *C.A.* name is propyl 3-(dimethylamino)propyl-carbamate, Registry No. *[67338-66-3]*. Its hydrochloride, *C.A.* Registry No. *[67338-67-4]*, was introduced by Schering AG under the code number 'SN 66752', trade mark 'Previcur N' and protected by DBP 1 567 169; 1 643 040. Its fungicidal properties were first described by E. A. Pieroh *et al., Meded. Fac. Landbouwwet. Rijksuniv. Gent,* 1978, **43,** in press.

Manufacture and properties. Produced by the reaction of 3-aminopropyl(dimethyl)amine with propyl chloroformate, propamocarb is a colourless crystalline solid, m.p. 45-55°, v.p. 6×10^{-3} mmHg at 25°. Its hydrochloride is very hygroscopic with solubility at 23°: > 500 g/l water, methanol; > 450 g/l dichloromethane; 23 g/l ethyl acetate; < 100 mg/l toluene, hexane.

Uses. Propamocarb hydrochloride is a soil-applied systemic fungicide, but is also suitable for use as a dip treatment (bulbs and tubers) and as a seed protectant. Its action is specific against Phycomycetes, and its spectrum includes the following genera of fungi: *Aphanomyces, Bremia, Peronospora, Phytophthora, Pseudoperonospora* and *Pythium*. It is recommended as a preventive treatment.

Toxicology. The acute oral LD50 of the hydrochloride is: for rats 8600 mg/kg; for mice 1600-2000 mg/kg. The acute dermal LD50 for rats and rabbits is $> 5·0$ ml (70% solution)/kg.

Formulation. 'Previcur N', aqueous solution (722 g propamocarb hydrochloride/l).

Analysis. Product analysis is by potentiometric titration with perchloric acid in a non-aqueous medium. Residues may be determined, after clean-up by liquid-liquid partition, by GLC with nitrogen-specific FID. Particulars available from Schering AG.

Propanil

GR BG DMV2

$C_9H_9Cl_2NO$ (218·1)

Nomenclature and development. The common name propanil is approved by BSI, ISO, WSSA and JMAF—exceptions Austria and Germany. The IUPAC name is 3′,4′-dichloro-propionanilide (I), formerly *N*-(3,4-dichlorophenyl)propionamide, in *C.A.* usage *N*-(3,4-dichlorophenyl)propanamide *[709-98-8]* formerly (I). Its herbicidal properties were described in *Proc. South. Weed Control Conf.*, 1960, p.20 and in *Rice J.*, 1971, **64**, 14. It was introduced in 1960 by Rohm & Haas Co. under the code number 'FW-734' and the trade mark 'Stam F-34'. Subsequently it was introduced by Bayer AG under the trade mark 'Surcopur' and was formerly manufactured by Monsanto Chemical Co. under the trade mark *'Rogue'*. It is protected by DAS 1039779; BP 903766.

Manufacture and properties. Produced by the reaction of 3,4-dichloroaniline with propionic acid in the presence of thionyl chloride, it is a colourless solid, m.p. 92-93°, v.p. 9 x 10^{-5} mmHg at 60°. Its solubility at room temperature is 225 mg/l water; at 25°: 540 g/kg ethanol; 600 g/kg isophorone. The technical product is a brown crystalline solid, m.p. 88-91°. It is stable in emulsion concentrates, but is hydrolysed in acid and alkaline media to 3,4-dichloroaniline and propionic acid. The formulations should be packed in steel drums coated with 'Unichrome B-124-17' or glass-lined; polyethylene coatings are not suitable.

Uses. Propanil is a contact herbicide recommended for post-em. use in rice and potatoes at 1-4 kg/ha, at which dose its persistence in soils is brief. If applied to plants treated with organophosphorus insecticides, severe phytotoxicity may appear.

Toxicology. The acute oral LD50 for rats is 1285-1483 mg tech./kg; the acute dermal LD50 for rabbits is 7080 mg (in corn oil)/kg. The LC50 (48 h) for mirror carp is 0·42 mg/l; for goldfish 0·35 mg/l; for Japanese killifish 0·55 mg/l; thus indicating that contamination of streams must be avoided.

Formulations. These include: 'Stam F-34', e.c. (360 g a.i./l); 'Surcopur', e.c. (350 g/l); ULV (600 g/l).

Analysis. Product analysis is by hydrolysis in boiling alkaline ethylene glycol and titration of the liberated 3,4-dichloroaniline with nitrite. Residues may be determined by GLC (*Anal. Methods Pestic. Plant Growth Regul.*, 1972, **6**, 692) or colorimetrically by hydrolysis and diazotising the 3,4-dichloroaniline formed and coupling with *N*-(1-naphthyl)ethylenediamine (C. F. Gordon *et al.*, *Anal. Methods Pestic., Plant Growth Regul. Food Addit.*, 1964, **4**, 235).

Propargite

L6TJ AOSO&O2UU1 BOR DX1&1&1 $C_{19}H_{26}O_4S$ (350·5)

Nomenclature and development. The common name propargite is approved by BSI, ISO and ANSI; also known as ENT 27 226. The IUPAC name is 2-(4-*tert*-butylphenoxy)-cyclohexyl prop-2-ynyl sulphite, in *C.A.* usage 2-[4-(1,1-dimethylethyl)phenoxy]-cyclohexyl 2-propynyl sulfite *[2312-35-8]* formerly 2-(*p-tert*-butylphenoxy)cyclohexyl 2-propynyl sulfite. It was introduced in 1964 by Uniroyal Inc. as an acaricide under the code number 'DO 14', trade mark 'Omite' and protected by USP 3 272 854; 3 463 859.

Manufacture and properties. Produced by reacting 4-*tert*-butylphenol with 1,2-epoxy-cyclohexane followed by treatment with thionyl chloride giving the chlorosulphinate which is reacted with propargyl alcohol in the presence of pyridine, the technical product, > 85% a.i., is a dark amber viscous liquid, d^{25} 1·085-1·115. It is practically insoluble in water; soluble in most organic solvents.

Uses. Propargite is an acaricide effective for the control of phytophagous mites on fruit, hops, walnuts and other crops at 1·05-5·5 kg a.i. in 750-1900 l/ha.

Toxicology. The acute oral LD50 for rats is 2200 mg tech./kg; the acute dermal LD50 for rabbits is > 3000 mg/kg. It has not the suspected carcinogenic properties which have limited the use of the related 2-(4-*tert*-butylphenoxy)-1-methylethyl 2-chloroethyl sulphite (S. S. Sternberg *et al., Cancer,* 1960, **13,** 780).

Formulations. These include: e.c. 'Omite 6E' (680 g a.i./l), 'Omite 57E' (570 g/l), 'Comite' (730 g/l); w.p. 'Omite 30W' (300 g/kg); dust, 'Omite 4D' (40 g/kg).

Analysis. Product analysis is by i.r. spectroscopy in carbon disulphide solution at 3·03 μm. Residues may be determined by GLC (J. Devine & H. R. Siskin, *J. Agric. Food Chem.,* 1972, **20,** 59). See also: G. M. Stone, *Anal. Methods Pestic Plant Growth Regul.,* 1973, **7,** 355.

Propazine

T6N CN ENJ BMY1&1 DMY1&1 FG $C_9H_{16}ClN_5$ (229·7)

Nomenclature and development. The common name propazine is approved by BSI, ISO, ANSI, WSSA and JMAF—exceptions Canada, Germany and Sweden. The IUPAC name is 2-chloro-4,6-bis(isopropylamino)-1,3,5-triazine formerly 2-chloro-4,6-di(iso-propylamino)-1,3,5-triazine, in *C.A.* usage 6-chloro-*N,N'*-bis(1-methylethyl)-1,3,5-triazine-2,4-diamine *[139-40-2]* formerly 2-chloro-4,6-bis(isopropylamino)-*s*-triazine. Its herbicidal properties were first reported by H. Gysin & E. Knüsli, *Proc. Int. Congr. Crop Protect., 4th,* 1957, **1**, 549. It was introduced in 1960 by J. R. Geigy S.A. (now Ciba-Geigy AG) under the code number 'G 30 028', trade marks 'Gesamil' and 'Milogard' and protected by SwissP 329 277; BP 814 947; USP 2 891 855.

Manufacture and properties. Produced by the reaction of trichloro-1,3,5-triazine with 2 molecular proportions of isopropylamine in the presence of an acid acceptor, it forms colourless crystals, m.p. 212-214°, v.p. 2·9 x 10⁻⁸ mmHg at 20°. Its solubility at 20° is 5 mg/l water; sparingly soluble in organic solvents.

Uses. Propazine is a pre-em. herbicide recommended for the control of broad-leaved and grass weeds in sorghum and umbelliferous crops at 0·5-3·0 kg a.i./ha. It is metabolised in tolerant plants to the corresponding 2-hydroxy compound.

Toxicology. The acute oral LD50 for rats is 7700 mg/kg; the acute dermal LD50 for rats is > 3100 mg/kg. In 0·5-y feeding trials rats receiving 250 mg/kg daily suffered no observable ill-effect. It is slightly toxic to fish and birds.

Formulations. These include: 'Gesamil 50WP', w.p. (500 g/kg); and in the USA 'Milogard 80W', w.p. (800 g/kg).

Analysis. Product analysis is by GLC with internal standard (*CIPAC Handbook,* 1979, **1A**, in press). Residues may be determined by TLC and GLC (A. M. Mattson *et al., J. Agric. Food Chem.,* 1965, **13**, 120; K. Ramsteiner *et al., J. Assoc. Off. Anal. Chem.,* 1974, **57**, 192; T. H. Byast *et al., Agric. Res. Counc. (G.B.) Weed Res. Organ., Tech. Rep.,* No. 15 (2nd Ed.), p.40). Particulars of these methods are available from Ciba-Geigy AG. See also: K. Knüsli, *Anal. Methods Pestic., Plant Growth Regul. Food Addit.,* 1964, **4**, 187; *Anal. Methods Pestic. Plant Growth Regul.,* 1972, **6**, 234, 679.

Propetamphos

$$
\begin{array}{c}
\quad\quad\quad\quad O \\
\quad\quad\quad\quad \| \\
S \quad H-C-C\;OPr^{\,i} \\
\| \quad\quad\quad\quad \| \\
MeO-P.O-C-Me \\
| \\
EtNH
\end{array}
$$

2MPS&O1&OY1&U1VOY1&1 $C_{10}H_{20}NO_4PS$ (281·3)

Nomenclature and development. The common name propetamphos is approved by BSI and New Zealand and proposed by ISO. The IUPAC name is (*E*)-*O*-2-isopropoxy-carbonyl-1-methylvinyl *O*-methyl ethylphosphoramidothioate, in *C.A.* usage 1-methyl-ethyl (*E*)-3-[[(ethylamino)methoxyphosphinothioyl]oxy]-2-butenoate *[31218-83-4]*. It was introduced in 1969 by Sandoz AG under the code number 'SAN 52 139I', the trade mark 'Safrotin' and protected by DBP 2035 103; SwissP 526 585; BelgP 753 579. Its biological properties were first described by J. P. Leber, *Proc. Int. IUPAC Congr., 2nd,* 1974, **1**, 381.

Manufacture and properties. Produced either by the reaction of isopropyl 3-oxobutyrate with thiophosphoryl chloride in the presence of catalyst and acid-binding agent, and subsequent condensation with methanol and ethylamine, or by the reaction of *O*-methyl phosphorodichloridothioate with isopropyl 3-oxobutyrate and ethylamine, it is a yellowish liquid, b.p. 87-89°/5 x 10^{-3} mmHg, n_D^{20} 1·495. Its solubility at 24° is 110 mg/l water; soluble in most organic solvents. It is stable, 50% loss occurring at 20° in > 5 y (extrapolated), and in aqueous buffered solutions at 24° in 44 d at pH 5, in 47 d at pH 7 and 37 d at pH 9.

Uses. Propetamphos is a contact insecticide with stomach activity, effective against household and public health pests, notably cockroaches, flies and mosquitoes; also against cattle ticks.

Toxicology. The acute oral LD50 and acute dermal LD50 for male albino rats are 82 mg/kg and 2300 mg/kg, respectively.

Formulations. These include: for professional use, e.c. 'Safrotin' (500 g a.i./kg), 'Safrotin P20 2P' (200 g/kg); for household use, 'Safrotin Aerosol' (20 g propetamphos + 5 g dichlorvos/kg), 'Safrotin Liquid' (10 g propetamphos + 5 g dichlorvos/kg), 'Safrotin Powder' (20 g/kg).

Analysis. Product analysis is by GLC or by paper chromatography with subsequent phosphorus determination by standard colorimetric methods. Residues may be determined by GLC. Details are available from Sandoz AG.

Propham

$$O$$
$$\|$$
$$Ph.NH.C.OCHMe_2$$

1Y1&OVMR $C_{10}H_{13}NO_2$ (179·2)

Nomenclature and development. The common name propham is approved by BSI, ISO and WSSA—exception USSR (IFC); sometimes known as IPC. The IUPAC name is isopropyl phenylcarbamate, in *C.A.* usage 1-methylethyl phenylcarbamate *[122-42-9]* formerly isopropyl carbanilate. It was selected by W. G. Templeman & W. A. Sexton, *Nature (London),* 1945, **156,** 630; *Proc. Roy. Soc.,* 1946, **133B,** 480, as the most active of a series of arylcarbamates as plant growth substances. It is protected by BP 574995.

Manufacture and properties. Produced by the reaction of aniline with isopropyl chloroformate in the presence of sodium hydroxide, it forms colourless crystals, m.p. 87·0-87·6°. Its solubility at 20-25° has been variously cited as 32, 100 and 250 mg/l water; soluble in most organic solvents. The technical product is 99% pure, m.p. 86·5-87·5°. It is stable < 100°, is compatible with other herbicides and is non-corrosive.

Uses. Propham is used mainly for the control of annual grass weeds in peas and beet at 2·3-5·0 kg a.i./ha, or in mixture with other herbicides. It is absorbed by the roots and not through the leaves.

Toxicology. The acute oral LD50 for rats is 5000 mg/kg. In 30-d feeding trials rats receiving 10000 mg/kg diet suffered no ill-effect and there was no evidence of carcinogenic activity (W. C. Heuper, *Ind. Med.,* 1952, **21,** 71). No effect was observed on fish exposed to 5 mg/l.

Formulations. These include: w.p.; e.c. (in combination with fenuron and chlorpropham).

Analysis. Product analysis is by acid hydrolysis which liberates carbon dioxide that is absorbed in sodium hydroxide solution and determined titrimetrically (*CIPAC Handbook,* 1970, **1,** 593). Residues may be determined by acid hydrolysis, clean-up of the liberated aniline which is estimated by GLC after conversion to a suitable derivative (L. N. Gard & C. E. Ferguson, *Anal. Methods Pestic., Plant Growth Regul. Food Addit.,* 1964, **4,** 139; *Anal. Methods Pestic. Plant Growth Regul.,* 1972, **6,** 657).

Propineb

$$\left[\begin{array}{c} \overset{S}{\underset{\parallel}{}} \quad \overset{Me}{\underset{\mid}{}} \quad \overset{S}{\underset{\parallel}{}} \\ - \text{S.C.NH.CH}_2.\text{CH.NH.C.S.Zn} - \end{array} \right]_x$$

SUYS&MY1&1MYS&US &-ZN- $(C_5H_8N_2S_4Zn)_x$ $(289{\cdot}8)x$

Nomenclature and development. The common name propineb is approved by BSI, ISO and JMAF. The IUPAC name is polymeric zinc propylenebis(dithiocarbamate), in *C.A.* usage [[(1-methyl-1,2-ethanediyl)bis[carbamodithioato]](2-)]zinc homopolymer *[9016-72-2]* formerly [propylenebis[dithiocarbamato]]zinc. It was introduced in 1962 by Bayer AG under the code numbers 'Bayer 46 131' and 'LH 30/Z', trade mark 'Antracol' and protected by BelgP 611 960. Its fungicidal properties were described by H. Goeldner, *Pflanzenschutz-Nachr. (Am. Ed.),* 1963, **16,** 49.

Manufacture and properties. Produced by the precipitation of water-soluble propylenebis(dithiocarbamate)s with solutions of zinc salts, it is a white to yellowish powder which decomposes $> 160°$ and, at *c.* 300°, only a slight residue remains. It is practically insoluble in all common solvents. It is stable in cool, dry storage but is decomposed in strongly alkaline or acid media. Its degradation on apples and grapes has been studied (K. Vogeler *et al., ibid.,* 1977, **30,** 72).

Uses. Propineb is a protective fungicide with a long residual activity; suitable for the control of *Pseudoperonospora humuli* on hops, *Phytophthora infestans* on potatoes and tomatoes and *Venturia inaequalis* on apples. It has some inhibitory action on powdery mildews and on red spider mites.

Toxicology. The acute oral LD50 for male rats is 8500 mg/kg; the acute dermal LD50 (7 h) is > 1000 mg/kg. In 2-y feeding trials no ill-effect was caused in rats receiving 50 mg/kg diet. It is harmless to honey bees.

Formulations. These include: w.p. (650-750 g a.i./kg); dusts of various a.i. content; 'Antracol', w.p. (700 g/kg).

Analysis. Product analysis is by titration with iodine, after hydrolysis and conversion of the liberated carbon disulphide to dithiocarbonate (D. G. Clarke *et al., Anal. Chem.,* 1951, **23,** 1842). Residues may be determined by spectroscopy of a copper dithiocarbamate complex formed with the carbon disulphide produced on decomposition of the residue with acid (T. E. Cullen, *ibid.,* 1964, **36,** 221; G. E. Keppel, *J. Assoc. Off. Anal. Chem.,* 1969, **52,** 162; K. Vogeler, *Pflanzenschutz-Nachr. (Am. Ed),* 1967, **20,** 525) or by polarography *(idem., ibid.).*

Propoxur

1Y1&OR   BOVM1 $C_{11}H_{15}NO_3$ (209·2)

Nomenclature and development. The common name propoxur is approved by BSI and ISO —exception JMAF (PHC); also known as ENT 25671, OMS 33, the BSI common name *arprocarb* was originally used. The IUPAC name is 2-isopropoxyphenyl methylcarbamate, in *C.A.* usage 2-(1-methylethoxy)phenyl methylcarbamate *[114-26-1]* formerly *o*-isopropoxyphenyl methylcarbamate. It was introduced in 1959 by Bayer AG under the code numbers 'Bayer 39007', '58 12 315', the trade marks 'Baygon', 'Blattanex' and 'Unden' and protected by USP 3 111 539 and DBP 1 108 202. Its insecticidal properties were described by G. Unterstenhöfer, *Meded. Landbouwhogesch. Opzoekingsstn. Gent,* 1963, **28,** 758.

Manufacture and properties. Produced by the reaction of methyl isocyanate with 2-isopropoxyphenol, prepared from pyrocatechol and 2-chloropropane, it is a colourless crystalline powder, with a faint characteristic odour, m.p. 84-87°, v.p. 0·01 mmHg at 120°. Its solubility at 20° is *c.* 2 g/l water; soluble in most organic solvents. It is unstable in highly alkaline media, with 50% loss at 20° in 40 min at pH 10.

Uses. Propoxur is a non-systemic insecticide with rapid knock-down, effective against jassids, bugs, aphids, flies, mosquitoes, cockroaches and other household pests including ants and millipedes. It is non-phytotoxic for recommended uses and rates.

Toxicology. The acute oral LD50 is: for rats 90-128 mg/kg; for male mice 100-109 mg/kg; for male guinea-pigs 40 mg/kg; for red-winged blackbirds 2-6 mg/kg; for starlings 15-20 mg/kg. The dermal LD50 for male rats is 800-1000 mg/kg. In 2-y feeding trials male and female rats receiving 250 mg a.i./kg diet showed no ill-effect; at 750 mg/kg diet the liver weight of female rats increased, otherwise there was no ill-effect. It is highly toxic to honey bees.

Formulations. These include: e.c., w.p., dusts, granules, pressurised sprays, smokes and baits of different a.i. concentrations. 'Baygon' and 'Blattanex' are the trade marks for products used against household and public health pests; 'Unden' or 'Undene' for agricultural use.

Analysis. Product analysis is by hydrolysis and titration of the methylamine liberated (*CIPAC Handbook,* 1979, **1A,** in press), or by u.v. spectroscopy after alkaline hydrolysis to 2-isopropoxyphenol measuring the difference between spectra in alkali and acid solutions at 292 nm, or by i.r. spectroscopy (C. A. Anderson, *Anal. Methods Pestic. Plant Growth Regul.,* 1973, **7,** 163). Residues may be determined by GLC *(idem., ibid.)* or, after hydrolysis, by colorimetric estimation of 2-isopropoxyphenol (H. Niessen & H. Frehse, *Pflanzenschutz-Nachr. (Am. Ed.),* 1964, **17,** 25; P. Bracha, *J. Agric. Food Chem.,* 1964, **12,** 461).

Propyzamide

GR   CG   EVMX1&1&1UU1 $C_{12}H_{11}Cl_2NO$ (256·1)

Nomenclature and development. The common name propyzamide is approved by BSI, New Zealand and JMAF and proposed by ISO—exception WSSA (pronamide). The IUPAC and *C.A.* name is 3,5-dichloro-*N*-(1,1-dimethylpropynyl)benzamide, Registry No. *[23950-58-5]*. It was introduced in 1965 by the Rohm & Haas Co. under the code number 'RH 315', trade mark 'Kerb' and protected by BP 1 209 068; USP 3 534 098; 3 640 699.

Manufacture and properties. Produced by reacting 3,5-dichlorobenzoyl chloride with 1,1-dimethylpropynylamine, it is a colourless solid, m.p. 155-156°, v.p. 8·5 x 10⁻⁵ mmHg at 25°. Its solubility at 25° is 15 mg/l water; soluble in many aliphatic and aromatic solvents. The technical product is 94-95% pure. It should not be mixed with other types of pesticide but may be combined with certain other herbicides.

Uses. Propyzamide is a selective herbicide for post-em. use in new and established crops of lucerne and other small-seeded legumes grown for fodder, and for pre-em. treatment of lettuces, related leafy crops and many species of trees, ornamental shrubs, top fruit and some species of soft fruit (depending on season and soil type). It is also effective in controlling *Poa annua* in Bermuda grass, Zoysia and certain other specific turf species. Rates vary from 0·56-2·2 kg a.i./ha, depending on weed species, environmental conditions and the residual activity needed.

Toxicology. The acute oral LD50 is: for male rats 8350 mg tech./kg, for female rats 5620 mg/kg; the acute dermal LD50 for rabbits is > 3160 mg/kg. The w.p. formulation is only mildly irritating to the eyes and skin.

Formulations. 'Kerb 50-W', w.p. (500 g a.i./kg); granules (40 g/kg).

Analysis. Product analysis is by GLC. Residues may be determined by heating with methanolic sulphuric acid and measuring by GLC the methyl 3,5-dichlorobenzoate formed (I. L. Adler *et al., J. Assoc. Off. Anal. Chem.*, 1972, **55,** 802; *idem., Anal. Methods Pestic. Plant Growth Regul.*, 1976, **8,** 443).

Prothiocarb

$$\text{Me}_2\text{NCH}_2\text{CH}_2\text{CH}_2\text{NH}.\overset{\overset{\displaystyle O}{\|}}{C}.\text{SEt}$$

2SVM3N1&1   *prothiocarb* $C_8H_{18}N_2OS$ (190·3)

2SVM3N1&1 &GH   *prothiocarb hydrochloride* $C_8H_{19}ClN_2OS$ (226·8)

Nomenclature and development. The common name prothiocarb is approved by BSI and New Zealand and proposed by ISO. The IUPAC name is *S*-ethyl *N*-(3-dimethyl-aminopropyl)thiocarbamate, in *C.A.* usage *S*-ethyl [3-(dimethylamino)propyl]-carbamothioate, *[19622-08-3]*. Its hydrochloride, *C.A.* Registry No. *[19622-19-6]*, was introduced in 1974 by Schering AG under the code number 'SN 41 703', trade marks 'Previcur' and 'Dynone' (UK and Republic of South Africa) and protected by DBP 1 567 169. Its fungicidal properties were first described by M. G. Bastiaansen *et al.*, *Meded. Fac. Landbouwwet. Rijksuniv. Gent,* 1974, **39,** 1019.

Manufacture and properties. Produced by the reaction of *N,N*-dimethylpropane-1,3-diamine with *S*-ethyl chlorothioformate, giving the hydrochloride, as a colourless, odourless crystalline solid, m.p. 120-121°. It is hygroscopic; its solubility at 23° is: 890 g/l water; 680 g/l methanol; 100 g/l chloroform; < 150 mg/l benzene, hexane. The technical grade has a strong odour.

Uses. Prothiocarb hydrochloride is a soil-applied systemic fungicide with a specific action against Phycomycetes, *e.g. Phytophthora, Pythium* spp. It is taken up by the roots and translocated to the aerial parts. It is only recommended for ornamental crops, mainly as a protective fungicide but, under certain conditions, it has shown a curative action.

Toxicology. The acute oral LD50 of the hydrochloride is: for rats 1300 mg/kg; for mice 600-1200 mg/kg. The acute dermal LD50 is: for rats > 1470 mg a.i. (as w.p.)/kg; for rabbits > 980 mg a.i. (as w.p.)/kg.

Formulation. 'Previcur S70', aqueous solution (745 g prothiocarb hydrochloride/l).

Analysis. Product analysis is by argentometric titration of the ethanethiol liberated by hydrolysis. Residues may be determined by the fluorimetric measurement of the *N,N*-dimethylpropane-1,3-diamine liberated by alkaline hydrolysis. Particulars available from Schering AG.

Prothiofos

GR CG DOPS&S3&O2 $C_{11}H_{15}Cl_2O_2PS_2$ (345·2)

Nomenclature and development. The common name prothiofos is approved by BSI and New Zealand and proposed by ISO. The IUPAC and *C.A.* name is *O*-(2,4-dichloro-phenyl) *O*-ethyl *S*-propyl phosphorodithioate, Registry No. *[34643-46-4]*. The biological properties were first described by A. Kadamatsu, *Jpn. Pestic. Inf.,* 1976, (26), p.14. It was introduced in 1975 by Bayer AG under the code number 'BAY NTN 8629', the trade mark 'Tokuthion' and protected by DOS 2 111 414.

Properties. It is a colourless liquid, b.p. 125-128°/0·1 mmHg, v.p. < 7·5 x 10^{-6} mmHg at 20°, d_4^{20} 1·3. Its solubility at 20° is *c.* 1·7 mg/kg water; completely miscible with cyclohexanone, toluene.

Uses. Prothiofos is an insecticide for use against leaf-eating caterpillars. Generally recommended in vegetables at 50-75 g a.i./100 l

Toxicology. The acute oral LD50 for male rats is 925-966 mg/kg; the acute dermal LD50 for male rats is > 1·0 ml (1300 mg)/kg.

Formulations. An e.c. (500 g/l); w.p. (400 g/kg).

Analysis. Product analysis is by GLC. Residue analysis is by GLC, E. Möllhoff, *Pflanzenschutz-Nachr. (Am. Ed.),* 1975, **28,** 382.

Prothoate

$$\text{(EtO)}_2\overset{\overset{\displaystyle S}{\|}}{P}.S.CH_2\overset{\overset{\displaystyle O}{\|}}{C}NH.CHMe_2$$

2OPS&O2&S1VMY1&1

$C_9H_{20}NO_3PS_2$ (285·4)

Nomenclature and development. The common name prothoate is approved by BSI and ISO—exception Sweden; also known as ENT 24 652. The IUPAC name is O,O-diethyl S-isopropylcarbamoylmethyl phosphorodithioate, in *C.A.* usage O,O-diethyl S-[2-(1-methylethyl)amino-2-oxoethyl] phosphorodithioate *[2275-18-5]* formerly O,O-diethyl phosphorodithioate S-ester with N-isopropyl-2-mercaptoacetamide. It is one of a group of compounds protected in 1948 by the American Cyanamid Co., who used the code number 'E.I. 18 682', under USP 2494 283. It was introduced in 1956 as an insecticide by Montecatini S.p.A. (now Montedison S.p.A.) under the code number 'L 343', trade mark 'Fac' and protected by BP 791 824. It was first described in *Ital. Agric.*, 1955, **99**, 747.

Manufacture and properties. Produced either by the interaction of sodium O,O-diethyl phosphorodithioate with 2-chloro-N-isopropylacetamide, or by treating with isopropylamine the reaction product of sodium O,O-diethyl phosphorodithioate and phenyl chloroacetate (DBP 1076 662, C. H. Boehringer Sohn), it is a colourless crystalline solid, with a camphor-like odour, m.p. 28·5°, v.p. 1 x 10^{-4} mmHg at 40°, d^{32} 1·151, n_D^{32} 1·5128. Its solubility at 20° is 2·5 g/l water. The technical product is an amber to yellow semi-solid solidifying at 21-24°, miscible at 20° with most organic solvents; < 10 g/kg glycerol; < 20 g/kg light petroleum; < 30 g/kg cyclohexane, hexane and higher petroleum fractions. It is stable in neutral, moderately acid or slightly alkaline media, but it is decomposed in *c.* 48 h at 50° and pH 9·2.

Uses. Prothoate is a systemic acaricide and insecticide and is used at 20-30 g a.i./100 l (300-500 g/ha) for the protection of fruit, citrus and vegetable crops from Tetranychid and some Eryophyid mites and some insects, notably aphids, Tingidae, Thysanoptera and Psyllidae.

Toxicology. The acute oral LD50 is: for rats 8·0-8·9 mg tech./kg; for mice 19·8-20·3 mg/kg (R. Schuppon, *Chim. Ind. (France)*, 1961, **85**, 421). In 90-d feeding trials the highest rate producing no toxic effect was: for rats 0·5 mg/kg daily; for mice 1 mg/kg daily. The 'no effect' level (10 d) was: for goldfish 6-8 mg/l; for mosquito fish 2-3 mg/l; the toxic level (96 h) was: for goldfish 50-70 mg/l; for mosquito fish 7-8 mg/l.

Formulations. These include: 'Fac 20', liquid (200 g tech./kg); 'Fac 40', liquid (400 g/kg); 'Fac 40WP', w.p. (400 g/kg); 'Fac P', solid (30 g/kg); 'Fac Granules', granules (50 g/kg).

Analysis. Product analysis is by GLC, TLC or a volumetric method based on the determination of the amide nitrogen. Residues on crops are determined, after column chromatographic clean-up, by GLC (B. Bazzi *et al.*, *Pestic. Sci.*, 1974, **5**, 511). Details of the methods are available from Montedison S.p.A. See also: B. Bazzi, *Anal. Methods Pestic. Plant Growth Regul.*, 1976, **8**, 213.

456

Pyracarbolid

T6O BUTJ B1 CVMR $C_{13}H_{15}NO_2$ (217·3)

Nomenclature and development. The common name pyracarbolid is approved by BSI and ISO. The IUPAC name is 3,4-dihydro-6-methyl-2*H*-pyran-5-carboxanilide, formerly 3,4-dihydro-6-methylpyran-5-carboxanilide (I), in *C.A.* usage 3,4-dihydro-6-methyl-*N*-phenyl-2*H*-pyran-5-carboxamide *[24691-76-7]* formerly (I). It was introduced by Hoechst AG under the code numbers 'Hoe 13 764 OF' (formerly 'Hoe 2989', 'Hoe 6052', 'Hoe 6053'), the trade mark 'Sicarol' and protected by DBP 1 668 790. Its fungicidal properties were first described by H. Stingle *et al., Int. Congr. Plant Prot., 7th, Paris.,* 1970, p.205 (Abstr.); B. Jank & F. Grossman, *Pestic. Sci.,* 1971, **2**, 43.

Properties. It is a colourless solid, m.p. 110-111°, v.p. 7·66 x 10⁻³ mmHg. Its solubility at 40° is 0·6 g/l water; soluble in a wide range of organic solvents; only slightly in benzene, xylene, cyclohexane. It is stable to light and heat.

Uses. Pyracarbolid is a systemic fungicide, effective against Basidiomycetes. It controls rust (Uredinales), smut (Ustilaginales), and damping off disease *(Rhizoctonia solani).* It is well absorbed by plants via roots and leaves. It is used in cereal, coffee, tea and bean crops. Rates and concentrations vary considerably depending on disease/crop combination and formulation.

Toxicology. The acute oral LD50 for female rats is > 15 000 mg (in starch mucilage)/kg; the acute dermal LD50 for female rats is > 1000 mg/kg; the acute intraperitoneal LD50 for female rats is 1600 mg/kg. In 2-y feeding trials the 'no effect' level was: for rats 400 mg/kg diet; for dogs 1000 mg/kg diet. The LC50 (96 h) is: for carp 42·3 mg/l; for rainbow trout 45·5 mg/l.

Formulations. These include: w.p. (500 g a.i./kg); dispersion (137 g/l); dry seed dressing (750 g/kg).

Analysis. Product analysis is by titration. Residues may be determined by a colorimetric method. Details are available from Hoechst AG.

Pyrazophos

$$EtOC(=O) \quad \text{...pyrazolo pyrimidine ring...} \quad OP(OEt)_2$$

Me

T56 ANN FNJ COPS&O2&O2 G1 HVO2 $C_{14}H_{20}N_3O_5PS$ (373·4)

Nomenclature and development. The common name pyrazophos is approved by BSI and ISO. The IUPAC name *O*-6-ethoxycarbonyl-5-methylpyrazolo[1,5-*a*]pyrimidin-2-yl *O,O*-diethyl phosphorothioate, in *C.A.* usage ethyl 2-[(diethoxyphosphinothioyl)oxy]-5-methylpyrazolo[1,5-*a*]pyrimidine-6-carboxylate *[13457-18-6]* formerly ethyl 2-hydroxy-5-methylpyrazolo[1,5-*a*]pyrimidine-6-carboxylate *O*-ester with *O,O*-diethyl phosphorothioate. It was introduced in 1971 by Hoechst AG under the code number 'Hoe 02 873', the trade marks 'Afugan' and 'Curamil' and protected by DBP 1 545 790. Its fungicidal properties were described by F. M. Smit, *Meded. Rijksfac. Landbouwwet. Gent.*, 1969, **34**, 763; S. J. B. Hay, *Proc. Br. Insectic. Fungic. Conf., 6th,* 1971, **1**, 134.

Properties. It forms colourless crystals, m.p. 50-51°, v.p. (Antoine) 1·66 x 10⁻⁶ mmHg at 50°. Its solubility at 20° is 4·2 mg/l water; soluble in most organic solvents. It is stable ≥ 2 y under normal storage conditions but is decomposed by acid or alkali.

Uses. Pyrazophos is a systemic fungicide controlling powdery mildews on a wide range of crops at 10-30 g a.i./100 l, and on cereals at 500-700 g/ha. It has only a limited insecticidal and acaricidal activity. It is absorbed by foliage and green stems and translocated within the plant. When applied to the soil or as a seed dressing uptake by roots is insufficient for effective fungicidal action within the plant.

Toxicology. The acute oral LD50 is: for rats 151-632 mg tech./kg (depending on carrier and sex); for quail 395-480 mg (in sesame oil)/kg. The acute dermal LD50 for rats is > 2000 mg/kg. In 2-y feeding trials for rats the 'no effect' level was 5 mg/kg diet; a 4-generation test on rats showed no effect at 50 mg/kg diet. The LC50 (96 h at 22°) is: for guppies 0·8 mg a.i. (as e.c.)/l; for carp 4·5 mg a.i. (as e.c.)/l. It is non-toxic to honey bees at rates up to 1 g/l.

Formulations. These include: e.c. (295 g a.i./l); w.p. (300 g/kg) for HV spraying.

Analysis. Product and residue analysis is by GLC: details are available from Hoechst AG. See also J. Asshauer *et al., Anal. Methods Pestic. Plant Growth Regul.*, 1978, **10**, 237.

Pyrethrins or Pyrethrum

(i)   (ii)

Nomenclature. 'The pyrethrins' (*C.A.* Registry No. *[8003-34-7]*) is the collective trivial name accepted by BSI, ISO and ESA—in France (pyrèthres)—for the insecticidal compounds present in the flowers of *Pyrethrum cinerariaefolium* (once known as the Dalmatian Insect Flowers) and other species. They are esters of two cyclopropanecarboxylic acids with three ketonic alcohols. The acids are chrysanthemic acid (1*R*)-*trans*-2,2-dimethyl-3-(2-methylprop-1-enyl)cyclopropanecarboxylic acid (i, R = Me), in *C.A.* usage 2,2-dimethyl-3-(2-methyl-1-propenyl)cyclopropanecarboxylic acid; and pyrethric acid (1*R*)-*trans*-3-[(*E*)-2-methoxycarbonylprop-1-enyl]-2,2-dimethylcyclopropane-carboxylic acid (i, R = —CO.OMe), in *C.A.* usage 3-[(*E*)-3-methoxy-2-methyl-3-oxo-1-propenyl]-2,2-dimethylcyclopropanecarboxylic acid (formerly known as 3-carboxy-α,2,2-trimethylcyclopropanecarboxylic acid 1-methyl ester). The alcohols have (*S*)-configuration and (*Z*)-configuration in the unsaturated side-chain. They are: pyrethrolone (ii, R' = —CH₂.CH=CH.CH=CH₂); cinerolone (ii, R' = —CH₂.CH=CHMe); jasmolone (ii, R' = —CH₂.CH=CH.Et).

Esters of chrysanthemic acid have the designator, I, and of pyrethric acid, II. They are: pyrethrins, 2-methyl-4-oxo-3-[(*Z*)-penta-2,4-dienyl]cyclopent-2-enyl; cinerins, 3-[(*Z*)-but-2-enyl]-2-methyl-4-oxocyclopent-2-enyl; jasmolins, 2-methyl-4-oxo-3-[(*Z*)-pent-2-enyl]cyclopent-2-enyl.

Name	Wiswesser Line Notation		Molecular formula	Molecular weight	C.A. Registry No.
pyrethrin I	L5V BUTJ B2U2U1 C1 DOV-BL3TJ A1 A1 C1UY1&1		$C_{21}H_{28}O_3$	328·4	*[121-21-1]*
pyrethrin II	L5V BUTJ B2U2U1 C1 DOV-BL3TJ A1 A1 C1UY1&VO1		$C_{22}H_{28}O_5$	372·4	*[121-29-9]*
cinerin I	L5V BUTJ B2U2 C1 DOV-BL3TJ A1 A1 C1UY1&1		$C_{20}H_{28}O_3$	316·4	*[25402-06-6]*
cinerin II	L5V BUTJ B2U2 C1 DOV-BL3TJ A1 A1 C1UY1&VO1		$C_{21}H_{28}O_5$	360·4	*[121-20-0]*
jasmolin I	L5V BUTJ B2U3 C1 DOV-BL3TJ A1 A1 C1UY1&1		$C_{21}H_{30}O_3$	330·4	*[4466-14-2]*
jasmolin II	L5V BUTJ B2U3 C1 DOV-BL3TJ A1 A1 C1UY1&VO1		$C_{22}H_{30}O_5$	374·4	*[1172-63-0]*

Development and properties. The dried flower heads were introduced into Europe from western Asia *c.* 1820, but their use is now replaced by that of extracts of crops grown in Kenya, Tanzania and Ecuador. The ratio of pyrethrin : cinerin : jasmolin is generally 71 : 21 : 7; most commercial extracts contain 20-25% pyrethrins and are pale yellow, the plant waxes and pigments having been removed. They are unstable in light and are rapidly hydrolysed by alkali with loss of insecticidal properties.

continued

Uses. The pyrethrins are potent, non-systemic, contact insecticides causing a rapid paralysis or 'knockdown', death occurring at a later stage. Insecticidal activity is markedly increased by the addition of synergists, *e.g.* piperonyl butoxide.

Toxicology. The acute oral LD50 for rats is 584-900 mg/kg; the acute dermal LD50 for rats is >1500 mg/kg (J. C. Malone & N. C. Brown, *Pyrethrum Post,* 1968, **9**(3), 3). Constituents of the flowers may cause dermatitis to sensitised individuals (J. T. Martin & K. H. C. Hester, *Br. J. Dermatol.,* 1941, **53**, 127; F. E. Rickett *et al., Pestic. Sci.,* 1972, **3**, 57; 1973, **4**, 801) but are removed during the preparation of refined extracts. There is no evidence that synergists increase toxicity of the pyrethrins to mammals. Highly toxic to fish.

Formulations. Dusts of various a.i. content on non-alkaline carriers. For aerosol concentrates (up to 20 g/l) the extract is dissolved in refined odourless kerosine, occasionally with dichloromethane, plus suitable propellants. Synergists (usually piperonyl butoxide) and organophosphorus insecticides (especially diazinon) may be added to increase the kill.

Analysis. Commercial extracts of flowers may be analysed by GLC (D. B. McClellan, *Anal. Methods Pestic. Plant Growth Regul.,* 1972, **6**, 461; J. Sherma, *ibid.,* 1976, **8**, 225) or by the AOAC method, mercury reduction with Denigès reagent for pyrethrin I and titration of hydrolysed pyrethrin II. Due to interference by certain adjuvants, only approximate results on formulations may be obtained; an empirical factor and the determined pyrethrin I content may be used (*CIPAC Handbook,* 1970, **1**, 598; D. B. McClellan, *Anal. Methods Pestic., Plant Growth Regul. Food Addit.,* 1964, **2**, 399). Residues may be determined by colorimetry or GLC (D. B. McClellan, *loc. cit.*).

Pyridate

T6NNJ CR& DOVS8 FG $C_{19}H_{23}ClN_2O_2S$ (378·11)

Nomenclature and development. The common name pyridate is approved by BSI and proposed by ISO. The IUPAC name is 6-chloro-3-phenylpyridazin-4-yl *S*-octyl thiocarbonate, in *C.A.* usage *O*-(6-chloro-3-phenyl-4-pyridazinyl) *S*-octyl carbonothioate *[55512-33-9]*. It was developed by Chemie Linz AG as an experimental herbicide under the code number 'CL 11 344' and protected by AustrianP 326 409. Its biological properties were described by A. Diskus *et al., Proc. 1976 Br. Crop Prot. Conf.-Weeds,* 1976, **2**, 717.

Manufacture and properties. Produced by the reaction of 6-chloro-3-phenylpyridazin-4-ol with *S*-octyl chlorothioformate, it is a colourless crystalline solid, m.p. 27°. The technical material is a brown oil, b.p. > 220°/1·05 x 10^{-6} mmHg, v.p. 10^{-9} mmHg, d^{20} 1·15, n_D^{20} 1·568. It is practically insoluble in water; highly soluble in organic solvents.

Uses. Pyridate is a foliar-acting herbicide with contact activity on annual dicotyledonous plants, especially *Galium aparine* and *Amaranthus retroflexus* (atrazine-resistant biotypes), and some grassy weeds. It controls weeds selectively in cereals, maize, rice and other crops at 1·0-1·5 kg a.i./ha. Effectiveness depends on weed species and stage of weed development. There is a strong indication that the mode of action depends upon inhibition of the Hill reaction.

Toxicology. The acute oral LD50 is: for rats *c.* 2000 mg/kg; for 10-d-old pheasants and for 5-d-old Pekin ducks > 10000 mg/kg; for bobwhite quail 1500 mg/kg. The acute dermal LD50 for rabbits is > 3400 mg/kg; it is moderately irritant to rabbit skin; no irritant effect in rabbit eye mucosa. It produced evidence of sensitivity in guinea-pigs, but no symptom has been observed in human volunteers. The Ames test for mutagenicity was negative. The LC50 (96 h) is: for rainbow trout 81 mg/l; for bluegill 100 mg/l. It is non-toxic to honey bees.

Formulations. These include: w.p. (400 and 500 g a.i./kg); e.c. (700 g/l).

Analysis. Product analysis is by u.v. spectroscopy in morpholine or by HPLC. Residues may be determined by HPLC. Details of methods are available from Chemie Linz AG.

Quinalphos

T66 BN ENJ COPS&O2&O2 $C_{12}H_{15}N_2O_3PS$ (298·3)

Nomenclature and development. The common name quinalphos is approved by BSI, ISO and Canada—exceptions France and Germany (chinalphos); also known as ENT 27394. The IUPAC name is O,O-diethyl O-quinoxalin-2-yl phosphorothioate, in *C.A.* usage O,O-diethyl O-2-quinoxalinyl phosphorothioate *[13593-03-8]*. It was introduced by Bayer AG in 1969 under the code numbers 'Bayer 77049', trade mark 'Bayrusil' and protected by BelgP 681 443; DAS 1 545 817; and by Sandoz AG under the code numbers 'Sandoz 6538 e.c.' and 'Sandoz 6626 g', and the trade mark 'Ekalux'. Its insecticidal properties were first described by K. -J. Schmidt & L. Hammann, *Pflanzenschutz-Nachr. (Am. Ed.)*, 1969, **22**, 314.

Manufacture and properties. Produced by the reaction of o-phenylenediamine, chloroacetic acid and O,O-diethyl phosphorochloridothioate, it is a colourless crystalline solid, m.p. 31-32°, b.p. 142° (decomp.)/3 x 10^{-4} mmHg, v.p. 3·9 x 10^{-12} mmHg at 20°, d_4^{20} 1·235, n_D^{25} 1·5624 (supercooled melt). Its solubility at 24° is 22 mg/l water; soluble in xylene, diethyl ether, acetone, ethanol; slightly soluble in light petroleum. It is stable under ambient storage conditions when diluted in non-polar organic solvents and in the presence of stabilising agents. It is susceptible to hydrolysis, 50% loss occurring at 24°: in 56 d at pH 5, 40 d at pH 7, 30 d at pH 9.

Uses. Quinalphos is a contact and stomach insecticide and acaricide with good penetrative properties. It is used at 190-500 g a.i. (as e.c.)/ha against caterpillars on vegetables, groundnuts and cotton, also scales and caterpillars on fruit trees; at 250-500 g a.i. (as e.c.)/ha or 0·75-1·0 kg a.i. (as granules)/ha against the pest complex on rice. It is degraded in plants within a few days.

Toxicology. The acute oral LD50 and acute dermal LD50 for rats are 62-137 mg/kg and 1250-1400 mg/kg, respectively. In 90-d feeding trials rats receiving 10 mg/kg diet showed only a slight effect on serum and erythrocyte cholinesterase. It is dangerous to honey bees, the topical LC50 (24 h) is 1·6 mg/kg.

Formulations. These include: 'Ekalux', e.c. (250 g a.i./kg), ULV (300 g/kg), granules (50 g/kg), dust (15 g/kg); 'Bayrusil', e.c. (200 g/l).

Analysis. Product analysis is by TLC with elution of the spots and measurement by u.v. spectrometry, or by paper chromatography and subsequent combustion of the spots and phosphate determination by standard colorimetric methods—details are available from Sandoz AG. Residues may be determined by GLC (G. Dräger, *ibid.*, p.308).

Quinomethionate

(Chinomethionat)

T C566 BN DSVS HNJ K1 $C_{10}H_6N_2OS_2$ (234·3)

Nomenclature and development. The common name chinomethionat is approved by ISO—exceptions BSI (quinomethionate), Australia, USA and ESA (oxythioquinox), JMAF (quinoxalines); also known as ENT 25606. The IUPAC name is 6-methyl-1,3-dithiolo[4,5-*b*]quinoxalin-2-one (I) formerly *S,S*-(6-methylquinoxaline-2,3-diyl) dithiocarbonate, in *C.A.* usage (I), *[2439-01-2]* formerly *S,S*-(6-methyl-2,3-quinoxalinediyl) cyclic dithiocarbonate. Its biological properties were described by K. Sasse, *Hoefchen-Briefe (Engl. Ed.),* 1960, **13**, 197; K. Sasse *et al., Angew. Chem.,* 1960, **72**, 973. It was introduced in 1962 by Bayer AG under the code numbers 'Bayer 36205', 'Bayer Ss2074', the trade mark 'Morestan' and protected by DAS 1 100 372; BelgP 580478.

Manufacture and properties. Produced by the reaction of carbonyl chloride with 6-methylquinoxaline-2,3-dithiol, it forms yellow crystals, m.p. 169·8-170°, v.p. 2×10^{-7} mmHg at 20°. It is practically insoluble in water; its solubility is: 10 g/kg dimethylformamide; 18 g/kg cyclohexanone; 4 g/kg petroleum oils.

Uses. Quinomethionate is a selective non-systemic acaricide and fungicide specific to powdery mildews on fruits, vegetables and ornamentals at 7·5-12·5 g a.i./100 l.

Toxicology. The acute oral LD50 for rats is 2500-3000 mg/kg; the acute dermal LD50 (7 d) for rats is > 500 mg/kg. In 2-y feeding trials rats receiving 60 mg/kg diet suffered no ill-effect. It is non-toxic to honey bees.

Formulations. A w.p. (250 g a.i./kg); smokes; dusts.

Analysis. Product analysis is by u.v. spectroscopy in heptane at 262 nm. Residues may be determined by GLC (K. Vogeler & H. Niessen, *Pflanzenschutz-Nachr. (Am. Ed.),* 1967, **20**, 550), colorimetrically after conversion to a nickel complex of 6-methylquinoxaline-2,3-dithiol (H. Tietz *et al., ibid.,* 1962, **15**, 166; R. Havens *et al., J. Agric. Food Chem.,* 1964, **12**, 247) or by TLC (Y. Francoeur & V. Mallet, *J. Assoc. Off. Anal. Chem.,* 1976, **59**, 172). See also: C. A. Anderson, *Anal. Methods Pestic., Plant Growth Regul. Food Addit.,* 1967, **5**, 277.

Quinonamid

L66 BV EVJ CMVYGG DG $C_{12}H_6Cl_3NO_3$ (318·5)

Nomenclature and development. The common name quinonamid is approved by BSI and proposed by ISO—exception Germany (chinonamid). The IUPAC name is 2,2-dichloro-*N*-(3-chloronaphthoquinon-2-yl)acetamide, in *C.A.* usage 2,2-dichloro-*N*-(3-chloro-1,4-dihydro-1,4-dioxo-2-naphthalenyl)acetamide *[27541-88-4]* formerly 2,2-dichloro-*N*-(3-chloro-1,4-dihydro-1,4-dioxo-2-naphthyl)acetamide. It was developed by Hoechst AG, and introduced under the code number 'Hoe 2997', the trade mark 'Alginex', formerly *'Nosprasit'*, and protected by DBP 1 768 447. Its biological properties were described by P. Hartz *et al., Meded. Fac. Landbouwwet. Rijksuniv. Gent*, 1972, **37**, 699.

Manufacture and properties. Produced by the acylation of 2-amino-3-chloro-1,4-naphthoquinone by dichloroacetyl chloride, it forms yellow needles, m.p. 212-213°, v.p. 8·4 x 10^{-8} mmHg at 20°, 2·8 x 10^{-7} mmHg at 30°. Its solubility at 23° is 3·0 mg/l water at pH 4·6, 60 mg/l at pH 7; soluble in most organic solvents. It is decomposed in the presence of acid or alkali.

Uses. Quinonamid is effective against algae in the open as well as algae and mosses under glass. It can be used as a seed-dressing or spray for control of algae in paddy rice, as a dip for clay pots, and for treating benches, etc., in greenhouses.

Toxicology. The acute oral LD50 for rats is 11 700-15 000 mg/kg. In 90-d feeding trials the 'no effect' level for rats was 2000 mg/kg diet. The LC50 (24 h at 20°) for guppies is 5 mg/l.

Formulations. A w.p. (500 g a.i./kg); granules (100 g/kg).

Analysis. Product and residue analysis is by GLC: details are available from Hoechst AG.

Quintozene

WNR BG CG DG EG FG $C_6Cl_5NO_2$ (295·3)

Nomenclature and development. The common name quintozene is approved by BSI and ISO—exceptions Turkey (terrachlor), USSR (PKhNB) and JMAF (PCNB). The IUPAC and *C.A.* name is pentachloronitrobenzene, Registry No. *[82-68-8]*. It was introduced in the late 1930's as a fungicide by I. G. Farbenindustrie AG (now Bayer Ag, who no longer manufacture or sell it). Trade marks include 'Brassicol', 'Tritisan' (Hoechst AG); 'Folosan'; 'Terrachlor' (Olin Mathieson Chemical Corp.), and protected by DRP 682048.

Manufacture and properties. Produced by the chlorination of nitrobenzene at 60-70° with iodine as catalyst, it forms colourless needles, m.p. 146°, v.p. 0·0133 mmHg at 25°, d^{25} 1·718. It is practically insoluble in water; solubility at 25° *c.* 20 g/kg ethanol; soluble in benzene, carbon disulphide, chloroform. It is highly stable in soil and compatible with pesticides at < pH 7 and non-corrosive. The technical product is ≤98·5% pure, m.p. 142-145°.

Uses. Quintozene is a fungicide of specific use for seed and soil treatment, effective against *Tilletia caries* of wheat, *Botrytis, Rhizoctonia* and *Sclerotinia* spp. on brassicas, vegetable and ornamental crops.

Toxicology. The acute oral LD50 for rats is > 12 000 mg/kg. In 2-y feeding trials rats receiving 2500 mg/kg diet survived.

Formulations. These include: 'Brassicol', 'Tritisan' (200 g a.i./kg; in Canada 600 g/kg); 'Terrachlor', w.p. (750 g/kg); e.c. (240 g/l).

Analysis. Product analysis is by AOAC Methods (*CIPAC Handbook,* in press). Residues may be determined, after chromatographic clean-up, by reaction with potassium methoxide, the potassium nitrite liberated is used to diazotise 1-naphthylamine which is coupled to form a dye measured colorimetrically at 525 nm (H. J. Ackermann *et al., J. Agric. Food Chem.,* 1958, **6**, 747; A. R. Klein & R. J. Gajan, *J. Assoc. Off. Agric. Chem.,* 1961, **44**, 712). See also: M. J. Kolbezen, *Anal. Methods. Pestic., Plant Growth Regul. Food Addit.,* 1964, **3**, 127; *Anal. Methods Pestic. Plant Growth Regul.,* 1972, **6**, 577.

Red squill

scilliroside $C_{32}H_{44}O_{12}$ (620·7)

L   E5   B666   MUTJ   A1   E1   IQ   JQ
LOV1   F-   ET6OVJ&   OO-   BT6OTJ   CQ   DQ   EQ   F1Q

Nomenclature and development. The powdered bulbs of red squill, *Urginea (Scilla) maritima,* or an extract of the bulbs have long been used as a rat poison. The 2 varieties of *U. maritima,* red and white squill, contain cardiac glycosides but only the red squill is used against rats. The toxicity of fresh white squill to rats is lost on drying (H. Roques, *Bull. Fil. Soc. Biol., Paris,* 1942-44, pp.21, 22). The activity of red squill is thought to be protected by the red pigment but is lost on heating, hence the necessity for drying bulbs at < 80°.

The toxic principle, scilliroside was isolated by A. Stoll & J. Renz, *Helv. Chim. Acta,* 1942, **25,** 377; 1943, **26,** 648, and assigned the structure, in *C.A.* usage 3β,6β-6-acetyloxy-3-(β-D-glucopyranosyloxy)-8,14-dihydroxybufa-1,20,22-trienolide *[507-60-8]* formerly (3β-(β-D-glucopyranosyloxy)-3β,6β,8,14-tetrahydroxybufa-4,20,22-trienolide 6-acetate. The chemistry and toxicology of cardiac glycosides has been reviewed (A. Stoll, *Experientia,* 1954, **10,** 282).

Properties. Scilliroside forms hydrated prisms (from aqueous methanol) which lose *c.* 8% *m/m in vacuo* giving a hemihydrate, m.p. 168-170°, and decomposing at 200°, $[\alpha]_D^{20}$ —59 to —60° (methanol). It is sparingly soluble in: water, acetone, chloroform, ethyl acetate; soluble in: alcohols, ethylene glycol, dioxane, glacial acetic acid; practically insoluble in diethyl ether, petroleum spirit.

Uses. Extracts of red squill are used in baits to control rats. It is claimed that its specific toxicity to rats is due to the inability of the rodent to vomit, a reaction squill induces in other animals.

Toxicology. The acute oral LD50 of scilliroside is: for male rats 0·7 mg/kg; for female rats 0·43 mg/kg (A. Stoll & J. Renz, *loc. cit.*). Pigs and cats survived 16 mg/kg, fowls survived 400 mg/kg (S. A. Barnett *et al., J. Hyg.,* 1949, **47,** 431).

Analysis. Product analysis is by TLC, followed by elution and colorimetric determination after reaction with the Liebermann steroid reagent (details from Sandoz AG) or by separation on paper chromatography and, after elution, by photometric assay (M. Wichtl & L. Fuchs, *Arch. Pharm.,* 1962, **295,** 361).

466

Resmethrin

T5OJ B1R& D1OV- BL3TJ A1 A1 C1UY1&1 $C_{22}H_{26}O_3$ (338·4)

Nomenclature and development. The common name resmethrin is approved by BSI and ISO—exception France. The IUPAC name is 5-benzyl-3-furylmethyl (1*RS*)-*cis,trans*-chrysanthemate, in *C.A.* usage [5-phenylmethyl-3-furanyl]methyl 2,2-dimethyl-3-(2-methyl-1-propenyl)cyclopropanecarboxylate *[10453-86-8]* formerly 5-benzyl-3-furyl-methyl 2,2-dimethyl-3-(2-methylpropenyl)cyclopropanecarboxylate. This mixture of isomers, first described by M. Elliott *et al., Nature (London),* 1967, **213**, 493, has been called 'NRDC 104', 'SBP 1382', 'FMC 17370', 'benzyfuroline', and 'Chryson' (Sumitomo Chemical Co. Ltd.). The name cismethrin (NRDC 119) has been approved by BSI and ISO for the (1*R*)-*cis*-isomer, Registry No. *[35764-59-1]* (formerly *[31182-61-3]*); bioresmethrin, the (1*R*)-*trans*-isomer, is listed separately (p.43).

Manufacture and properties. Produced by esterification of 5-benzyl-3-furylmethanol (BP 1 168 797; 1 168 798; 1 168 799), is a colourless waxy solid, m.p. 43-48°, containing 20-30% of (1*RS*)-*cis*- and 80-70% of (1*RS*)-*trans*-isomers. All isomers are insoluble in water; soluble in all common organic solvents. Resmethrin is somewhat more stable than the pyrethrins, but is decomposed fairly rapidly on exposure to air and light (J. H. Fales *et al.,* CSMA Proceedings of the 55th Annual Meeting, December 1968, Washington, D.C.; A. B. Hadaway *et al., Bull. W. H. O.,* 1970, **42**, 387).

Uses. Resmethrin is a powerful contact insecticide effective against a wide range of insects (I. C. Brooks *et al., Soap Chem. Spec.,* 1969, **45**(3), 62). The toxicity to normal houseflies is 20 x that of natural pyrethrins (M. Elliott *et al., loc. cit.*), but these compounds are not synergised to any appreciable extent by pyrethrin synergists. Toxicity to plants is low.

Toxicology. The acute dermal LD50 for rats is *c.* 2000 mg/kg; dermal applications to rats of 3000 mg/kg caused no deaths. In 90-d feeding trials rats tolerated 3000 mg/kg diet without ill-effect. No teratogenic effect was seen in rats receiving 25 mg/kg daily, or in mice at 50 mg/kg daily.

Formulations. Resmethrin may be formulated with or without other pyrethroids and pyrethrin synergists in aerosol concentrates, and also as water-based sprays (*e.g.,* with tetramethrin, H. H. Incho, *Soap Chem. Spec.,* 1970, **46**(2), 37); e.c.; w.p.; concentrate for ULV application.

Analysis. Product analysis is by GLC with FID with an internal standard such as dicyclohexyl phthalate or di-octyl phthalate (B. B. Brown, *Anal. Methods Pestic. Plant Growth Regul.,* 1973, **7**, 441). The *cis*- and *trans*-isomers can be separated and estimated by HPLC (J. M. Zehner & R. A. Simonaitis, *J. Assoc. Off. Anal. Chem.,* 1976, **59**, 1101).

Rotenone

$C_{23}H_{22}O_6$ (394·4)

T G5 D6 B666 CV HO MO POT&TT&J IY1&U1 SO1 TO1

Nomenclature and development. The traditional name rotenone is accepted by BSI and ISO in lieu of a common name; also known as ENT 133. It is the main insecticidal compound in certain *Derris* and *Lonchocarpus* spp.; at one time called nicouline or yubatoxin. The roots of *Derris elliptica* are known as derris root, tuba-root or aker-tuba; those of *Lonchocarpus utilis, L. urucu* and *L. nicou* are known as barbasco, cube, haiari, nekoe, and timbo. The *C.A.* name is [2R-(2α,6aα,12aα)]-1,2,12,12a-tetrahydro-8,9-dimethoxy-2-(1-methylethenyl)[1]benzopyrano[3,4-*b*]furo[2,3-*h*][1]benzopyran-6(6aH)-one *[83-79-7]* formerly 1,2,12,12aα-tetrahydro-2a-isopropenyl-8,9-dimethoxy[1]benzopyrano[3,4-*b*]furo[2,3-*h*][1]benzopyran-6(6aH)-one.

Derris root has long been used as a fish poison and its insecticidal properties were known to the Chinese well before it was isolated in 1895 by E. Geoffrey, *Ann, Inst. Colon. (Marseilles)*, 1895, **2**, 1. Its structure was established in 1932 (E. B. LaForge *et al., Chem. Rev.*, 1933, **12**, 181).

Manufacture and properties. Isolated by crystallisation from carbon tetrachloride of extracts of roots of *Derris* and *Lonchocarpus* spp., it forms colourless crystals, m.p. 163°, a dimorphic form m.p. 181°, [α]$_D^{20}$ —231° (benzene). Its solubility at 100° is 15 mg/l water; it is slightly soluble in petroleum oils, carbon tetrachloride; soluble in polar organic solvents. It crystallises with solvent of crystallisation (H. A. Jones, *J. Am. Chem. Soc.*, 1931, **53**, 2738). It is readily racemised by alkali and readily oxidised, especially in the presence of light and alkali, to less insecticidal products, a change accelerated in certain solvents, *e.g.* pyridine.

Uses. Rotenone is a selective non-systemic insecticide with some acaricidal properties. It is of low persistence in spray or dust residues and is non-phytotoxic.

Toxicology. The acute oral LD50 is: for white rats 132-1500 mg/kg; for white mice 350 mg/kg. It is toxic to pigs (A. A. Kingscote *et al., Annu. Rep. Entomol. Soc. Ont.*, 1951, p.37). It is highly toxic to fish.

Formulations. These include: dusts of the ground root with a non-alkaline carrier; dusts prepared from extracts may be stabilised by a trace of phosphoric acid (R. S. Cahn *et al., J. Soc. Chem. Ind., London, Trans. Commun.*, 1945, **64**, 33).

Analysis. Product analysis is by colorimetry at 540 nm, after reaction with sulphuric acid plus nitrous acid or with alkaline sodium nitrite and sulphuric acid—and using an empirical correction factor; the purity of a benzene solution of rotenone may be checked by polarimetry (*CIPAC Handbook*, 1970, **1**, 610). Residues may be determined colorimetrically (C. R. Gross & C. M. Smith, *J. Assoc. Off. Agric. Chem.*, 1934, **17**, 336; L. D. Goodhue, *ibid.*, 1936, **19**, 118).

468

Ryanodine

$C_{25}H_{35}NO_9$ (493·6)

T5 G6556/BN/FO 5AABEL O LXOTJ AQ B1
CY1&1 CQ EQ F1 GQ J1 KQ NQ DOV- BT5MJ

Nomenclature and development. The trivial name ryania, *C.A.* Registry No. *[8047-13-0]*, is applied to the ground stemwood of *Ryania speciosa* (Flacourtiaceae), a shrub native to Trinidad and the Amazon basin. Its insecticidal properties were first reported by B. E. Pepper & L. A. Carruth, *J. Econ. Entomol.*, 1945, **38**, 59. It was introduced *c.* 1945 by S. B. Penick & Co., protected by USP 2400295. Ryanodine has been shown (K. Weisner *et al., Tetrahedron Lett.*, 1967, p.221; K. Weisner, *Adv. Org. Chem.*, 1972, **8**, 295) to have the structure in *C.A.* usage [3S-(3α,4β,4aS*,6a,6aα,7α,8β,8aα,8bβ,9β,9aα)]-octahydro-3,6a,9-trimethyl-7-(1-methylethyl)-6,9-methanobenzo[1,2]pentaleno[1,6-*bc*]furan-4,6,7,8,8a,8b,9a-heptol 3-(1*H*)-pyrrole-2-carboxylate, Registry No. *[15662-33-6]* (formerly *[15800-60-9]*).

Manufacture and properties. Produced by extracting the ground wood of *R. speciosa* with water, chloroform or methanol, the insecticidal compounds may be concentrated (E. F. Rogers *et al., J. Am. Chem. Soc.*, 1948, **70**, 3086) yielding ryanodine, crystals, m.p. 219-220° (decomp.) [α]$_D^{25}$ + 26° (methanol). It is soluble in water, ethanol, acetone, diethyl ether, chloroform; almost insoluble in benzene, light petroleum. It is neutral and does not react with the usual reagents for alkaloids.

Uses. Ryania is a selective stomach insecticide causing cessation of feeding and a slow death; ryanodine is *c.* 700 times as potent as the ground wood of *R. speciosa*.

Toxicology. The acute oral LD50 is: for white rats 750-1000 mg ryania/kg; for mice and rabbits 650 mg/kg; for dogs 750 mg/kg. In 150-d feeding trials guinea-pigs and chickens receiving 10000 mg ryania powder/kg diet suffered no toxic symptom (S. Kuna & R. E. Heal, *J. Pharmacol. Exp. Therap.*, 1948, **93**, 407).

Formulations. Powdered root stem and leaves of *R. speciosa* are mixed with an inert carrier; 'Ryno-tox' (800 mg ryanodine/kg).

Analysis. Product analysis is by u.v. spectroscopy at 270-271 nm after extraction and purification by liquid-liquid partition: details are available from S. B. Penick & Co.

Schradan

$$O \quad O$$
$$\parallel \quad \parallel$$
$$(Me_2N)_2P.O.P(NMe_2)_2$$

1N1&PO&N1&1&OPO&N1&1&N1&1 $C_8H_{24}N_4O_3P_2$ (286·3)

Nomenclature and development. The common name schradan is approved by BSI and ISO; sometimes known as OMPA. The IUPAC name is octamethylpyrophosphoric tetra-amide, in *C.A.* usage octamethyldiphosphoramide *[151-26-9]* formerly octamethylpyro-phosphoramide. Its systemic insecticidal properties were discovered in 1941 by G. Schrader & H. Kükenthal (cited by G. Schrader, *Die Entwicklung neuer insektizider Phosphorsäure-Ester,* 3rd Ed., p.88). Trade marks include *'Pestox 3'* (Fisons Ltd, who no longer market it), 'Sytam' (Murphy Chemical Ltd). Since 1968 it has been produced by Wacker-Chemie GmbH.

Manufacture and properties. Produced by the reaction of tetramethylphosphorodiamidic chloride (I) with ethyl tetramethylphosphorodiamidate (G. Schrader, *loc. cit.,*), by the reaction of (I) in pyridine or *N*-methyldibutylamine (BP 631 549; 652 981) or by heating (I) with an alkali metal hydroxide (M. Pianka, *J. Appl. Chem. (London),* 1955, **5,** 109), it is a colourless viscous liquid, b.p. 118-122°/0·3 mmHg, m.p. 14-20°, v.p. *c.* 1 x 10⁻³ mmHg at 25°, d^{25} 1·1343, n_D^{25} 1·4612. It is miscible with water and most organic solvents; slightly soluble in petroleum oils. It is readily extracted from aqueous solution by chloroform. It is stable to water and alkali, but hydrolysed under acid conditions to dimethylamine and orthophosphoric acid.

The technical product is a dark brown viscous liquid containing the amides of higher phosphoric acids of which decamethyltriphosphoric triamide predominates (25-50% in products obtained by the first 2 methods) $(Me_2N)_2PO.OPO.(NMe_2)OPO(NMe_2)_2$, b.p. 190-200°/0·5 mmHg, n_D^{25} 1·4660 (USP 2610 139).

Uses. Schradan is a systemic insecticide effective against sap-feeding insects and mites, but is relatively inert as a contact insecticide, used as a soil drench for transplanted trees. It is non-phytotoxic at insecticidal concentrations.

Toxicology. The acute oral LD50 is: for male rats 9·1 mg/kg, for female rats 42 mg/kg; the acute dermal LD50 is: for male rats 15 mg/kg, for female rats 44 mg/kg. In 1-y feeding trials rats receiving 30 mg/kg diet showed toxic symptoms at the beginning but not at the end of the period (J. M. Barnes & F. A. Denz, *Br. J. Ind. Med.,* 1954, **11,** 11).

Decamethyltriphosphoric triamide has similar systemic insecticidal properties to schradan but is reported to be of lower mammalian toxicity.

Formulations. Aqueous solution (300 g a.i./l); also anhydrous (750-800 g/l); or (600 g/l), with anhydrous surfactant.

Analysis. Product analysis is by GLC (details available from Wacker-Chemie GmbH) or by differential hydrolysis (*CIPAC Handbook,* 1970, **1,** 620). Residues may be determined, after extraction with chloroform and acid hydrolysis, by standard colorimetric methods for phosphoric acid (A. David *et al., J. Sci. Food Agric.,* 1951, **2,** 310), or by distilling off the dimethylamine which is measured colorimetrically (S. A. Hall, *Anal. Chem.,* 1951, **23,** 1866; D. F. Heath & J. Clough, *J. Agric. Food Chem.,* 1956, **4,** 230).

Secbumeton

T6N CN ENJ BO1 DMY2&1 FM2 $C_{10}H_{19}N_5O$ (225·3)

Nomenclature and development. The common name secbumeton is approved by BSI, ISO, ANSI and WSSA. The IUPAC name is 2-*sec*-butylamino-4-ethylamino-6-methoxy-1,3,5-triazine, in *C.A.* usage *N*-ethyl-6-methoxy-*N'*-(1-methylpropyl)-1,3,5-triazine-2,4-diamine *[26259-45-0]* formerly 2-(*sec*-butylamino)-4-(ethylamino)-6-methoxy-*s*-triazine. It was developed by J. R. Geigy S.A. and introduced by Ciba-Geigy AG under the code number 'GS 14254' and trade mark 'Etazine' (formerly *'Sumitol'* in USA). It was first described by A. Gast & E. Fankhauser, *Proc. Br. Weed Control Conf., 8th,* 1966, p.485.

Manufacture and properties. Produced by reacting 2-*sec*-butylamino-4-chloro-6-ethyl-amino-1,3,5-triazine with sodium hydroxide in methanol, it is a colourless powder, m.p. 86-88°, v.p. 7·3 x 10^{-6} mmHg at 20°. Its solubility at 20° is 600 mg/l water; readily soluble in organic solvents.

Uses. Secbumeton is a herbicide taken up by leaves and roots and controls mono- and di-cotyledonous weeds both annual and perennial. It is used in lucerne either at 1-3 kg a.i./ha or in combination with simazine. In combination with terbuthylazine it is suitable for non-cropped land.

Toxicology. The acute oral LD50 for rats is 2680 mg/kg. In 90-d feeding trials no toxic reaction was noted in rats receiving 2400 mg/kg diet, or dogs 40 mg/kg daily. It is of low toxicity to fish.

Formulations. These include: 'Etazine 50', w.p. (500 g a.i./kg); 'Etazine 3585', w.p. (350 g secbumeton + 150 g simazine/kg); 'Primatol 3588', w.p. (250 g secbumeton + 250 g terbuthylazine/kg).

Analysis. Product analysis is by GLC with internal standard or by titration with perchloric acid in acetic acid: details are available from Ciba-Geigy AG. Residues may be determined by GLC (T. H. Byast *et al., Agric. Res. Counc. (G.B.) Weed Res. Organ., Tech. Rep.,* No. 15 (2nd Ed.), p.40). See also: E. Knüsli, *Anal. Methods Pestic., Plant Growth Regul. Food Addit.,* 1964, **4,** 13.

Sesamex

O.CHMe.O.CH₂.CH₂.O.CH₂.CH₂.OEt

T56 BO DO CHJ GOY1&O2O2O2 $C_{15}H_{22}O_6$ (298·3)

Nomenclature and development. The common name sesamex is approved by ESA; also known as ENT 20 871. The IUPAC name is 5-{1-[2-(2-ethoxyethoxy)ethoxy]ethoxy}-1,3-benzodioxole (I), formerly 2-(1,3-benzodioxol-5-yloxy)-3,6,9-trioxaundecane, in *C.A.* usage (I) *[51-14-9]* formerly 4-[1-[2-(2-ethoxyethoxy)ethoxy]ethoxy]-1,2-methylenedioxy-benzene and acetaldehyde 2-(2-ethoxyethoxy)ethyl 3,4-methylenedioxyphenyl acetal. Its synergistic properties were described by M. Beroza, *J. Agric. Food Chem.,* 1956, **4,** 49. It was introduced in 1956 by Shulton Inc. under the trade mark 'Sesoxane'. It is no longer available commercially, but is a valuable research compound for studies on the mechanism of resistance to pyrethrins and some pyrethroids.

Manufacture and properties. Produced by the condensation of 3,6,9-trioxaundec-1-ene with benzodioxol-5-ol in the presence of hydrogen chloride, it is a straw-coloured liquid, with a faint odour, b.p. 137-141°/0·08 mmHg, n_D^{20} 1·491-1·493. It is readily soluble in kerosine and in dichlorodifluoromethane. It is unstable in sunlight and on powdered carriers.

Uses. Sesamex is a synergist for the pyrethrins and allethrin and is used in experimental studies.

Toxicology. The acute oral LD50 for rats is 2000-2270 mg/kg; the acute dermal LD50 for rabbits is > 9000 mg/kg.

472

Siduron

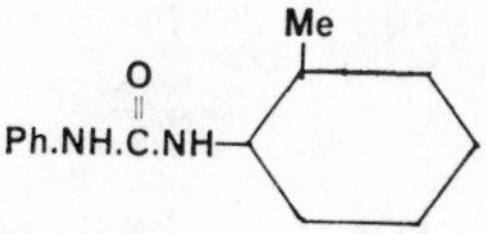

L6TJ   AMVMR&   B1

$C_{14}H_{20}N_2O$ (232·3)

Nomenclature and development. The common name siduron is approved by BSI, ISO, ANSI, WSSA and JMAF—exception Austria. The IUPAC name is 1-(2-methylcyclohexyl)-3-phenylurea (I), in *C.A.* usage *N*-(2-methylcyclohexyl)-*N'*-phenylurea *[1982-49-6]* formerly (I). Its herbicidal properties were described by R. W. Varner *et al., Proc. Br. Weed Control Conf., 7th,* 1964, p.38. It was introduced in 1964 by E. I. du Pont de Nemours & Co. (Inc.) under the code number 'Du Pont 1318', trade mark 'Tupersan' and protected by USP 3 309 192.

Properties. It is a colourless crystalline solid, m.p. 133-138°. Its solubility at 25° is: 18 mg/l water; > 100 g/kg *N,N*-dimethylacetamide, dimethylformamide, dichloromethane, isophorone. It is stable up to its m.p. and in water, but is slowly decomposed in acid and alkaline media. It is non-corrosive.

Uses. Siduron is a selective herbicide toxic to crabgrass and annual weed grasses, but tolerated by turf species, cereals and many broad-leaved crop plants. It is recommended for the treatment of grass at 2-6 kg a.i./ha for new seedlings, 8-12 kg/ha for established turf.

Toxicology. The acute oral LD50 for rats is 7500 mg a.i./kg; the maximum feasible dose (5500 mg/kg) applied to intact or abraded skin of rabbits caused no clinical sign of toxicity. In 96-d feeding trials no nutritional or clinical sign of toxicity was observed in albino rats receiving 5000-7500 mg/kg diet.

Formulation. A w.p. (500 g a.i./kg).

Analysis. Residues may be determined by colorimetry (R. L. Dalton & H. L. Pease, *J. Assoc. Off. Agric. Chem.,* 1962, **45**, 377).

Simazine

T6N CN ENJ BM2 DM2 FG $C_7H_{12}ClN_5$ (201·7)

Nomenclature and development. The common name simazine is approved by BSI, ISO, ANSI and WSSA—exceptions JMAF (CAT) and Turkey. The IUPAC name is 2-chloro-4,6-bis(ethylamino)-1,3,5-triazine formerly 2-chloro-4,6-di(ethylamino)-1,3,5-triazine, in *C.A.* usage 6-chloro-*N,N'*-diethyl-1,3,5-triazine-2,4-diamine *[122-34-9]* formerly 2-chloro-4,6-bis(ethylamino)-*s*-triazine. Its herbicidal properties were first described by A. Gast *et al., Experientia,* 1956, **12,** 146. It was introduced in 1956 by J. R. Geigy S.A. (now Ciba-Geigy AG) under code number 'G 27 692', trade marks 'Gesatop', 'Primatol' and, in the USA, 'Aquazine' and protected by SwissP 329 277; 342 784; BP 814 947; USP 2 891 855.

Manufacture and properties. Produced by the reaction of trichloro-1,3,5-triazine and two equivalents of ethylamine in the presence of an acid acceptor (W. M. Pearlman & C. K. Banks, *J. Am. Chem. Soc.,* 1948, **70,** 3726), it is a colourless crystalline solid, m.p. 225-227°, v.p. 6·1 x 10⁻⁹ mmHg at 20°. It solubility at 20° is 3·5 mg/l water; slightly soluble in most organic solvents. It is stable in neutral and slightly acidic or alkaline media, but is hydrolysed by more concentrated acids and alkalies to the herbicidally-inactive 4,6-bis(ethylamino)-1,3,5-triazin-2-ol, especially at higher temperatures. It is non-corrosive.

Uses. Simazine is a pre-em. herbicide recommended for the control of broad-leaved and grass weeds in deep-rooted crops such as citrus, deciduous fruits, olives, asparagus, grape vines, certain ornamentals, coffee, tea, cocoa and in broad beans and forest nurseries at 0·5-4·0 kg a.i./ha. A major use is on maize which can convert simazine to the inactive 2-hydroxy derivative. It may also be used for the non-selective control of herbaceous weeds at 5-20 kg/ha. It is also used as 'Aqualine' in the USA for the control of submerged vegetation and algae in farm ponds, fish hatcheries, etc.

Toxicology. The acute oral LD50 for rats is > 5000 mg/kg. In 2-y feeding trials rats receiving 100 mg/kg diet showed no toxic effect. It has a negligible toxicity to wildlife.

Formulations. These include: w.p. 'Gesatop 50' and 'Primatol S50' (500 g a.i./kg), 'Gesatop 80' and 'Primatol S80' (800 g/kg); flowable, 'Gesatop 500FW' (500 g a.i./l); also in mixture with other triazine herbicides and with aminotriazole.

Analysis. Product analysis is by GLC with internal standard (*CIPAC Handbook,* 1979, **1A,** in press) or by titration of ionic chloride liberated by treatment with morpholine (H. P. Bosshardt *et al., J. Assoc. Off. Anal. Chem.,* 1971, **54,** 749). Residues may be determined by GLC (K. Ramsteiner *et al., ibid.,* 1974, **57,** 192) or by HPLC (T. H. Byast *et al., Agric. Res. Counc. (G.B.) Weed Res. Organ., Tech. Rep.,* No. 15 (2nd Ed.) p.40). Details are available from Ciba-Geigy AG. See also: E. Knüsli *et al., Anal. Methods Pestic., Plant Growth Regul. Food Addit.,* 1964, **4,** 213; *Anal. Methods Pestic. Plant Growth Regul.,* 1972, **6,** 234, 691.

Simetryne

(Simetryn)

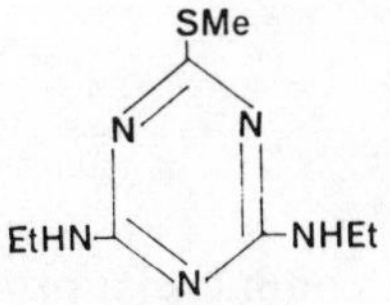

SMe

N═══N

EtHN—N—NHEt

T6N CN ENJ BS1 DM2 FM2 $C_8H_{15}N_5S$ (213·3)

Nomenclature and development. The common name simetryn is approved by ISO and WSSA; for BSI and JMAF (simetryne), in France (simétryne)—exceptions Canada and Germany. The IUPAC name is 2,4-bis(ethylamino)-6-methylthio-1,3,5-triazine formerly 2,4-di(ethylamino)-6-methylthio-1,3,5-triazine, in *C.A.* usage *N,N'*-diethyl-6-methylthio-1,3,5-triazine-2,4-diamine *[1014-70-6]* formerly 2,4-bis(ethylamino)-6-(methylthio)-*s*-triazine. It was first reported as a herbicide by J. R. Geigy S.A. (now Ciba-Geigy AG) under the code number G 32911 and protected by SwissP 337019. It is now marketed in Japan under the trade mark 'Gy-bon' by Nippon Kayaku Co., Hokko Chemical Industry Co. Ltd, Sankyo Co. Ltd and Nihon Nohyaku Co. Ltd.

Manufacture and properties. Produced by the reaction of simazine (p.474) with one equivalent of methanethiol in the presence of a base, it forms crystals, m.p. 82-83°. Its solubility at room temperature is 450 mg/l water.

Uses. Simetryne is used as a mixture with thiobencarb to control broad-leafed weeds in rice.

Toxicology. The acute oral LD50 for rats is 1830 mg/kg.

Formulations. These include: 'Saturn' (simetryne + thiobencarb).

Analysis. Product analysis is by titration (*CIPAC Handbook*, in press).

Nomenclature and development. The commercial product known as sodium arsenite was at one time thought to be sodium meta-arsenite, $NaAsO_2$, but is now considered to be a solid solution of the reactants rather than a definite compound. It has long been used for weed destruction and for poison baits against rodents.

Manufacture and properties. Produced by the controlled interaction of sodium hydroxide with diarsenic trioxide, it is a colourless to greyish powder, hygroscopic and readily soluble in water. It is decomposed by atmospheric carbon dioxide and containers must be air-tight.

Uses. Sodium arsenite is a powerful arsenical poison, sometimes used in ant baits, (30 g/kg sugar or honey solution), in rodent baits and as a non-selective herbicide.

Toxicology. The acute oral LD50 for small rodents is 10-50 mg/kg; it is apt to cause suppuration to open wounds.

Formulations. These include: solid ($\geqslant$ 780 g diarsenic trioxide/kg); solutions.

Analysis. Product analysis is by titration with iodine solution (*MAFF Tech. Bull.* No. 1, 1958, p.73). Residues may be determined by standard methods for arsenic (AOAC Methods).

Sodium chlorate

Na$^+$ [ClO$_3$]$^-$

. NA . . G-O3   ClNaO$_3$ (106·4)

Nomenclature and development. The IUPAC and *C.A.* name is sodium chlorate, Registry No. *[7775-09-9]*. It has been used as a weedkiller since *c.* 1910.

Manufacture and properties. Produced by the electrolysis of acidic concentrated solutions of sodium chloride at 80-90°, it is a colourless powder, m.p. 248°, decomposing with the evolution of oxygen at about 300°. Its solubility at 0° is 790 g/l water; soluble in ethanol, glycerol. It is a strong oxidising agent, reacting with organic materials in the presence of sunlight, a property creating a serious fire hazard, for example, with splashed clothing. It is somewhat corrosive to zinc and mild steel.

Uses. Sodium chlorate is a non-selective herbicide, used against established vegetation at 200-600 kg/ha and against annual weeds at 100-200 kg/ha. At 300 g/ha, it gives persistent control for *c.* 0·5 y but is leached from the soil by high rainfall.

Toxicology. The acute oral LD50 of sodium chlorate for rats is 1200 mg/kg (E. F. Edson, *Pharm. J.,* 1960, **185,** 361). It can cause local irritation to the skin and mucuous membranes and cases of poisoning in man, due to methaemoglobin production, have been reported following its use for oral hygiene (W. J. Cochrane & R. P. Smith, *Can. Med. Assoc. J.,* 1940, **42,** 23).

Formulations. Most of the marketed formulations incorporate a fire depressant: calcium chloride for liquid concentrates; borax, sodium chloride, sodium sulphate, trisodium phosphate in soluble powders. Combinations with organic herbicides such as atrazine, bromacil, 2,4-D, diuron, monuron are also available.

Analysis. Product analysis is either by reaction with acidified potassium bromide and potassium iodide followed by titration with sodium thiosulphate, or by adding an excess of an iron(II) salt and titration with potassium dichromate (*CIPAC Handbook,* 1970, **1,** 626; FAO Specification, see *Herbicides 1977,* p.41). Residues in soil may be determined after extraction with water, by colorimetry at 448 nm of the complex formed with *o*-tolidine [*beware;* carcinogenic] in hydrochloric acid (T. H. Byast *et al., Agric. Res. Counc. (G.B.) Weed Res. Organ., Tech. Rep.,* No. 15 (2nd Ed.), p.38).

Sodium cyanide

Na⁺ CN⁻

.NA..CN   CNNa   (49·01)

Nomenclature and development. The IUPAC and *C.A.* name is sodium cyanide, Registry No. *[143-33-9]*. It is marketed by ICI Ltd, Plant Protection Division as a rodenticide under the trade mark 'Cymag'.

Properties. It is a colourless solid, m.p. 564°. It is deliquescent and very soluble in water. It is hydrolysed and decomposed by carbon dioxide or acids, generating hydrogen cyanide (p.300) which is extremely toxic if inhaled. The technical grade is 95-98% pure.

Uses. Sodium cyanide is used to kill rabbits, rats, termites and nest-building ants at 11 g a.i./burrow. It is also used as a source of hydrogen cyanide for fumigating warehouses etc. Its use is restricted to trained personnel and it should be stored in cool, dry, well ventilated uninhabited places.

Toxicology. The acute oral LD50 for rats is 6·44 mg/kg. It is extremely toxic to humans; the antidote in case of poisoning is inhalation of pentyl (amyl) nitrite or intravenous injection of cobalt ethylenediaminetetraacetate (Co EDTA; 'Kelocyanor').

Formulation. 'Cymag', powder (400 g a.i./kg).

Analysis. Product analysis is by dissolving in alkali and titrating with silver nitrate (details available from ICI Ltd, Plant Protection Division).

Sodium fluoride

$$Na^+ \quad F^-$$

. NA . . F

FNa (41·99)

Nomenclature and development. The IUPAC and *C.A.* name is sodium fluoride, Registry No. *[7681-49-4]*. It has been used in baits against insects in stores.

Manufacture and properties. Produced by neutralisation of hydrofluoric acid or by the action of sodium carbonate on a fluorosilicate (USP 1 382 165), it is a colourless, non-volatile powder of high melting point, d 2·8. Its solubility at 18° is 42·2 g/kg water; slightly soluble in ethanol.

Uses. Sodium fluoride is a powerful stomach insecticide with some contact activity. As it is highly phytotoxic its use is limited to baits. It is also a timber preservative.

Toxicology. It is highly toxic to vertebrates, the lethal dose to man being 75-150 mg/kg.

Formulations. The commercial grade (93-99% pure), with a dye as warning of its poisonous properties, is used in the preparation of baits.

Analysis. Product analysis is by determination of the fluorine content by precipitation as lead chloride fluoride and estimation of chloride by the Volhard method (AOAC Methods).

Sodium fluoroacetate

$$FCH_2.\overset{\overset{\displaystyle O}{\|}}{C}.O^- \quad Na^+$$

OV1F &-NA- $\hspace{8cm}$ $C_2H_2FNaO_2$ (100·0)

Nomenclature and development. The IUPAC and *C.A.* name is sodium fluoroacetate, Registry No. *[62-74-8]*; also known as Compound 1080. It was suggested in the early 1940's as a rodenticide (E. R. Kalmbeck, *Science*, 1945, **102**, 232).

Manufacture and properties. Produced either by reaction of methyl chloroacetate with potassium fluoride at 220° and subsequent treatment with sodium hydroxide, or by the reaction at high pressures of carbon monoxide, hydrogen fluoride and formaldehyde to give fluoracetic acid which is converted to the sodium salt, it is a colourless, non-volatile powder, decomposing *c.* 200°. It is hygroscopic, very soluble in water; poorly soluble in ethanol, acetone, petroleum oils.

Uses. Sodium fluoroacetate is a potent mammalian poison used in baits as a rodenticide.

Toxicology. The acute oral LD50 for *Rattus norvegicus* is 0·22 mg/kg (S. H. Dieke & C. P. Richter, *U.S. Public Health Rep.,* 1946, **61,** 672). Its use is, in many countries, restricted to trained personnel. It is also a systemic insecticide, but is considered too toxic to mammals to be used for this purpose (W. A. L. David, *Nature (London),* 1950, **165,** 493).

Formulation. An aqueous solution, containing a dye as warning of its poisonous nature, for bait preparation.

Analysis. Product analysis is by conversion to sodium fluoride by digestion with metallic sodium and precipitation as lead chloride fluoride; sodium fluoride is a usual contaminant (*WHO Manual* (2nd Ed.), SRT/5).

Sodium metaborate

$$(NaBO_2)_x$$

. NA . . B-O2 $(BNaO_2)_x$

Nomenclature and development. The IUPAC name is sodium metaborate, in *C.A.* usage boric acid (HBO_2) sodium salt *[7775-19-1]*. It was introduced by the US Borax & Chemical Corp. as a non-selective herbicide protected by USP 3032405.

Manufacture and properties. Produced by the action of sodium hydroxide on disodium tetraborate, it forms colourless crystals which 'melt' in their water of crystallisation *c.* 55°; the anhydrous salt has m.p. 966°, d^{25} 1·90. Its solubility at 23° is 480 g/l water. It is stable and non-corrosive.

Uses. Sodium metaborate is used in admixture with sodium chlorate, as a non-selective herbicide on uncropped land and for the spot treatment of Johnson grass at 240-1950 kg/ha. It is rapidly leached in high rainfall areas.

Toxicology. The acute oral LD50 for rats is 2330 mg/kg; it is a moderate skin irritant.

Formulations. These include mixtures with sodium chlorate and with organic herbicides such as bromacil, *e.g.* 'Hibor' (US Borax & Chemical Corp.).

Analysis. Product analysis is by standard volumetric methods: see borax (p.47).

$$C_{21}H_{22}N_2O_2 \quad (334\cdot4)$$

T6 G656 B7 C6 E5 D 5ABCEF A& FX MNV
QO VN SU AHT&&TTTTJ

Nomenclature and development. The traditional name strychnine (I) is accepted in lieu of a common name by BSI and ISO. In *C.A.* usage strychnidin-10-one *[57-24-9]* formerly (I). The physiological activity of alkaloids of Nux-vomica has long been recognised and, of them, strychnine is the best known. Its synthesis was first accomplished by R. B. Woodward *et al., J. Am. Chem. Soc.,* 1954, **76**, 4749.

Manufacture and properties. Produced by extraction from the seeds of *Strychnos* spp. by benzene in the presence of calcium hydroxide, it is separated from the crude alkaloids by crystallisation from ethanol, and is a colourless crystalline powder, m.p. 270-280° (decomp.), $[\alpha]_D^{18}$ —139·3° (chloroform). Its solubility at room temperature is: 143 mg/l water; 6·7 g/l ethanol; 5·6 g/l benzene; 200 g/l chloroform; sparingly soluble diethyl ether, petroleum spirit.

It forms salts, the hydrochloride forms colourless prisms with 1·5-2 mol of water of crystallisation which are lost at 110°, and it is water-soluble. Strychnine sulphate forms colourless crystals with 5 mol of water of crystallisation lost at 110°, m.p. > 199°. Its solubility at 15° is 30 g/l water; soluble in ethanol; insoluble in diethyl ether.

Uses. Strychnine is an intense mammalian poison, uniquely effective against moles.

Toxicology. The lethal dose is: for rats 1-30 mg/kg; for man 30-60 mg/kg.

Formulations. Baits are usually prepared as coloured grains (5-10 g strychnine sulphate/kg).

Analysis. Product analysis in poisoned grain is by titration with acid after extraction of the bait with chloroform under alkaline conditions (AOAC Methods).

Sulfallate

$$\underset{\substack{ \\ }}{Et_2N.\overset{\displaystyle S}{\overset{\|}{C}}.S.CH_2\underset{\underset{\displaystyle Cl}{|}}{C}=CH_2}$$

2N2&YUS&S1YGU1 $C_8H_{14}ClNS_2$ (223·8)

Nomenclature and development. The common name sulfallate is approved by BSI and ISO—exception WSSA (CDEC). The IUPAC name is 2-chloroallyl diethyldithio-carbamate (I), in *C.A.* usage 2-chloro-2-propenyl diethylcarbamodithioate *[95-06-7]* formerly (I). It was introduced as a herbicide in 1954 by Monsanto Co. under the code number 'CP 4742' and trade mark 'Vegadex' and protected by USP 2 744 898; herbicides employing this compound are protected by USP 2 919 182.

Manufacture and properties. Produced by the reaction of 2,3-dichloropropene with sodium diethyldithiocarbamate, it is an amber oil, b.p. 128°/1 mmHg, v.p. 2·2 x 10^{-3} mmHg at 20°, d^{25} 1·088, n_D^{25} 1·5822. Its solubility at 25° is 92 mg/l water; soluble in most organic solvents. It is hydrolysed by alkali: 50% loss occurs in 47 d at pH 5, 30 d at pH 8.

Uses. Sulfallate is used as a pre-em. herbicide on a wide range of vegetable crops at 3-6 kg a.i./ha, being effective against annual weeds, but not established or vegetatively-propagating perennials. It is not absorbed by the leaves but is readily taken up by the roots. Applications of 4 kg/ha persist for 21-42 d.

Toxicology. The acute oral LD50 for rats is 850 mg/kg; it is somewhat irritant to the skin and eyes.

Formulations. An e.c. (480 g a.i./l); granules (200 g/kg).

Analysis. Product analysis is by total chlorine content. Residues may be determined by GLC (*Anal. Methods Pestic. Plant Growth Regul.,* 1972, **6,** 704) or by acid hydrolysis and measurement at 435 nm, after clean-up, as a copper bis(dithiocarbamate) complex (W. K. Lowen, *Anal. Chem.,* 1951, **23,** 1846). See also: R. A. Conkin & L. S. Gleason, *Anal. Methods Pestic., Plant Growth Regul. Food Addit.,* 1964, **4,** 249.

Sulfotep

$$\underset{(EtO)_2 \overset{\overset{S}{\|}}{P}.O.\overset{\overset{S}{\|}}{P}(OEt)_2}{}$$

2OPS&O2&OPS&O2&O2 $C_8H_{20}O_5P_2S_2$ (322·3)

Nomenclature and development. The common name sulfotep is approved by BSI and ISO —exception ESA (sulfotepp); also known as ENT 16273, dithio, dithione, thiotep. The IUPAC name is O,O,O',O'-tetraethyl dithiopyrophosphate, in *C.A.* usage thiodiphosphoric acid ([(HO)$_2$P(S)]$_2$O) tetraethyl ester *[3689-24-5]*, or tetraethyl thiodiphosphate formerly tetraethyl thiopyrophosphate. Its insecticidal properties were discovered in 1944 by G. Schrader & H. Kükenthal (cited by G. Schrader, *'Die Entwicklung neuer insektizider Phosphorsaure-Ester'*, 3rd Ed.). It was introduced *c.* 1947 by Bayer AG under the code number 'Bayer E 393', trade mark 'Bladafum' and protected by DBP 848 812. Another code number is 'ASP-47' (Victor Chemical Works).

Manufacture and properties. Produced by the action of sulphur on TEPP (p.498) (DBP 896 644) or by the interaction of O,O-diethyl phosphorochloridothioate with aqueous sodium carbonate in the presence of pyridine (A. D. F. Toy, *J. Am. Chem. Soc.,* 1951, **73**, 4670), it is a pale yellow mobile liquid, b.p. 136-139°/2 mmHg, v.p. $1·7 \times 10^{-4}$ mmHg at 20°, d_4^{25} 1·196, n_D^{25} 1·4753. Its solubility at room temperature is 25 mg/l water; miscible with most organic solvents including chloromethane. The technical product is a dark-coloured liquid, b.p. 131-135°/2 mmHg, n_D^{25} 1·4725. It is resistant to hydrolysis but corrosive to iron.

Uses. Sulfotep is a non-systemic insecticide with a wide range of action but of brief persistance on foliage.

Toxicology. The acute oral LD50 for rats is *c.* 5 mg/kg.

Formulation. 'Bladafum', for smoke generation.

Analysis. Product analysis is by i.r. spectroscopy (details available from Bayer AG). Residues may be determined by GLC (G. Dräger, *Pflanzenschutz-Nachr. (Am. Ed.)*, 1968, **21**, 359). See also: *Anal. Methods Pestic. Plant Growth Regul.*, 1972, **6**, 483.

Sulfoxide

$$\text{—CH}_2\text{CHMe.SO.(CH}_2)_7\text{Me}$$

T56 BO DO CHJ G1Y1&SO&8 $\qquad$ $C_{18}H_{28}O_3S$ (324·5)

Nomenclature and development. The common name sulfoxide is approved by ESA, there is no BSI or ISO name; also known as ENT 16 634. The IUPAC name is 1-methyl-2-(3,4-methylenedioxyphenyl)ethyl octyl sulphoxide, in *C.A.* usage 5-[2-(octyl-sulfinyl)propyl]-1,3-benzodioxole *[120-62-7]* formerly 1,2-methylenedioxy-4-[2-(octyl-sulfinyl)propyl]benzene. Its synergist properties were first described by M. E. Synerholm *et al., Contrib. Boyce Thompson Inst.,* 1947, **15,** 35. It was introduced in 1947 by S. B. Penick & Co., and protected by USP 2 486 445 (to Boyce Thompson Institute).

Manufacture and properties. Produced by the addition of octane-1-thiol to 1-(3,4-methylenedioxyphenyl)prop-1-ene followed by oxidation with hydrogen peroxide, the technical product ($\geqslant 88\%$ pure) is a brown liquid, of mild odour, which decomposes on attempted distillation, d_{25}^{25} 1·06-1·09, n_D^{25} 1·528-1·532. It is practically insoluble in water; miscible with most organic solvents; solubility in petroleum oils is 20-25 g/kg. It is stable under normal conditions.

Uses. Sulfoxide is a synergist for the pyrethrins and allethrin.

Toxicology. The acute oral LD50 for rats is 2000-2500 mg/kg; the acute dermal LD50 for rabbits is > 9000 mg/kg. In 1·25-y feeding trials no ill-effect was observed in rats receiving 2000 mg/kg diet.

Formulations. A component (10 g/l) with pyrethrins or allethrin (2 g a.i./l) in aerosol concentrates.

Analysis. Product analysis is by u.v. spectroscopy at 288 nm, after isolation by steam distillation (*WHO Insecticides,* 30 Rev.1, p.170).

Sulphur

$$S_x$$

. S*X   (32·06)x

Nomenclature and development. The IUPAC name sulphur, in *C.A.* usage sulfur *[7704-34-9]*, is accepted by ISO in lieu of a common name. It has long been in use as a pesticide.

Properties. It is a yellow solid, melting at 115° to a yellow mobile liquid which darkens and becomes viscous *c.*160°, v.p. 3·96 x 10⁻⁶ mmHg at 30·4°. It exists in allotropic forms including: rhombic, m.p. 112·8°; monoclinic, m.p. 119°. It is practically insoluble in water; slightly soluble in ethanol and diethyl ether; the crystalline forms are soluble in carbon disulphide whereas amorphous forms are not. It is compatible with most other pesticides, except petroleum oils.

Uses. Sulphur is a non-systemic direct and protective fungicide and acaricide. It is generally non-phytotoxic, except to certain varieties known as 'sulphur-shy'. It has been vaporised in glasshouses to control powdery mildews, care being taken to avoid the formation of the highly phytotoxic sulphur dioxide.

Toxicology. It is relatively non-toxic to mammals.

Formulations. Dusts, usually with inert material (⩽ 100 g/kg) to prevent electrostatic 'balling'; w.p.; finely-ground 'colloidal' suspensions.

Analysis. Product analysis is by reaction with sodium sulphite to form thiosulphate which is estimated volumetrically (*FAO Plant Prot. Bull.,* 1961, **9**(5), 80; *CIPAC Handbook,* 1970, **1,** 632). Residues may be determined using the Schönberg reagent (H. A. Ory *et al., Analyst (London),* 1957, **82,** 189).

Sulphuric acid

$$H_2SO_4$$

.. H2 . S-O4
$$H_2O_4S \quad (98.07)$$

Nomenclature and development. The IUPAC name sulphuric acid, in *C.A.* usage sulfuric acid *[7664-93-9]*, is accepted by ISO in lieu of a common name. Commercial grades were known as *brown oil of vitriol (BOV)*. It was first recommended as a weed-killer by C. D. Woods & J. M. Bartlett, *Bull. Me. Agric. Exp. Stn.,* No. 167, 1909.

Manufacture and properties. Produced by the catalytic oxidation of sulphur dioxide in the presence of water, the pure acid is a colourless viscous liquid, d 1.834. The commercial acid is a brown viscous liquid, d_4^{15} 1.675-1.710. It is extremely hygroscopic and reacts violently with water, evolving much heat; hence in dilution the concentrated acid should be added to water or ice. The concentrated acid attacks organic matter, is non-corrosive to lead and mild steel but the dilute acid attacks most metals.

Uses. Sulphuric acid was, at one time, used at 100 ml/l for the selective weeding of cereal crops and onions; also for pre-em. weed control in horticultural crops. It was also used at 100-120 ml/l or undiluted at 100-200 l/ha for the destruction of potato haulm 14 d before lifting. It can also be used as a pre-harvest desiccant for leguminous seed crops. The corrosive action on spray machinery tends to restrict its use to spray contractors.

Toxicology. The concentrated and dilute acids are very toxic to all forms of life; it attacks skin and clothing, and must be handled with great care.

Formulation. Commercial acid (1300 g sulphuric acid/l).

Analysis. Product analysis is usually by the determination of density and reference to standard tables or by titration with alkali.

Sulphuryl fluoride

$$F—\overset{\displaystyle O}{\underset{\displaystyle O}{\overset{\|}{\underset{\|}{S}}}}—F$$

WSFF F_2O_2S (102·1)

Nomenclature and development. The IUPAC name is sulphuryl fluoride, in *C.A.* usage sulfuryl fluoride *[2699-79-8]*. The insecticidal properties were first described by E. E. Kenaga, *J. Econ. Entomol.*, 1957, **40**, 1. It was introduced in 1957 by The Dow Chemical Co. under the trade mark 'Vikane' and protected by USP 2 875 127.

Manufacture and properties. Produced by the thermal decomposition of barium di(fluorosulphate), it is a colourless gas, b.p. —55·2°, v.p. 13 x 10^{+3} mmHg at 25°, critical temperature 96±20°. Its solubility at 25° is 750 mg/kg water; sparingly soluble in most organic solvents; miscible with methyl bromide. It is stable, non-corrosive and harmless to fabrics.

Uses. Sulphuryl fluoride is an insecticidal fumigant used for fumigating structures and wood products for the control of drywood termites and wood-infesting beetles. It has poor ovicidal effect, is phytotoxic but with little effect on the germination of weed and crop seeds. Its toxic effect may be due to formation of fluoride ion (R. W. Meikle *et al.*, *J. Agic. Food Chem.*, 1963, **11**, 226).

Toxicology. The adopted time-weighted average value for humans is 20 mg/m³ for 8 h/d, 5 d/week repeated exposure.

Formulations. It is used as the 99% *m/m* technical product.

Analysis. Atmospheres may be analysed by trapping in aqueous sodium hydroxide and titration of the sodium fluorosulphate produced or by a thermal conductivity meter (S. G. Heuser, *Anal. Chem.*, 1963, **35**, 1476). Details of methods for total fluoride residues are available from The Dow Chemical Co.

Sulprofos

3SPS&O2&OR DS1 $C_{12}H_{19}O_2PS_3$ (322·4)

Nomenclature and development. The common name sulprofos is approved by BSI and New Zealand and proposed by ISO. The IUPAC and *C.A.* name is *O*-ethyl *O*-(4-methyl-thiophenyl) *S*-propyl phosphorodithioate, Registry No. *[35400-46-4]*. It was introduced in 1976 by Bayer AG under the code number 'BAY NTN 9306' and the trade marks 'Bolstar' and 'Helothion'.

Properties. It is a colourless oil, b.p. 210°, v.p. < 7·5 x 10⁻⁷ mmHg at 20°, d_4^{20} 1·20. Its solubility at 20° is: < 5 mg/kg water; > 120 g/kg cyclohexanone; 400-600 g/kg propan-2-ol; > 1·2 kg/kg toluene.

Uses. Sulprofos is an insecticide with specific effect against Lepidoptera on cotton.

Toxicology. The acute oral LD50 for male rats is 304 mg/kg; the acute dermal LD50 for rats is > 1·0 ml (1200 mg)/kg.

Formulations. An e.c. (720 g/l); ULV (720 g/l).

Analysis. Product analysis is by GLC. Details of residue analysis are available from Bayer AG.

2,4,5-T

QV1OR BG DG EG $C_8H_5Cl_3O_3$ (255·5)

Nomenclature and development. The common name 2,4,5-T is approved by BSI, ISO, WSSA and JMAF—JMAF also apply it to the corresponding 2-butoxyethyl ester. The IUPAC and *C.A.* name is (2,4,5-trichlorophenoxy)acetic acid, Registry No. *[93-76-5]*. Its herbicidal properties were reported by C. L. Hamner & H. B. Tukey, *Science,* 1944, **100**, 154. It was introduced in 1944 by Amchem Products, Inc. under the trade mark 'Weedone'.

Manufacture and properties. It is produced by the condensation of sodium chloroacetate with 2,4,5-trichlorophenol, the latter being obtained by the action of alkali on 1,2,4,5-tetrachlorobenzene prepared by the action of chlorine on the trichlorobenzenes resulting from the dehydrochlorination of HCH (A. Galat, *J. Am. Chem. Soc.,* 1952, **74**, 3890). At high temperatures the action of alkali on 1,2,4,5-tetrachlorobenzene can produce some 2,3,7,8-tetrachlorodibenzo-*p*-dioxin. 2,4,5-T forms colourless crystals, m.p. 154-155°, d_{20}^{20} 1·80. Its solubility at 30° is 238 mg/kg water; at 50° 590 mg/kg ethanol, propan-2-ol. The technical grade is 98% pure, m.p. 150-151°. It is stable.

Its salts with alkali metals and amines are water-soluble, precipitation occurs with hard water in the absence of sequestering agents. Its triethanolamine salt has m.p. 113-115°, other salts used include the triethylammonium and the (*Z*)-(*N*-propyl)octadec-9-enammonium. 2,4,5-T esters are insoluble in water but soluble in oils. Typical examples include: 2-butoxyethyl ester; butyl ester (tech. grade solidifies at 20°); iso-octyl ester; propylene glycol butyl ether (2-butoxy-1-methylethyl) ester.

Uses. 2,4,5-T is used post-em. alone or with 2,4-D for the control of shrubs and trees. It is applied as a foliage, dormant shoot or basal bark spray. It is also used for girdling, injection or cut-stump treatment. It is absorbed through roots, foliage and bark. Esters of low volatility are used for ULV application.

Toxicology. The acute oral LD50 is: for rats 300 mg/kg; for dogs 100 mg/kg. In 90-d feeding trials no effect was observed on rats and dogs receiving 10 mg/kg daily, adverse symptoms occurred on dogs at 20 mg/kg daily (N. A. Drill & T. Hirayzka, *AMA Arch. Ind. Hyg.,* 1953, **7**, 61). In 1·5-y feeding trials no significant increase in tumours occurred in mice receiving 21·5 mg/kg daily. Hazards to wildlife are negligible at recommended rates. A contaminant, 2,3,7,8-tetrachlorodibenzo-*p*-dioxin, causes serious acne in man and produced foetal deaths in hamsters at 0·0091 mg/kg. Modern methods of manufacture of 2,4,5-T limit the amount of the contaminant to < 0·5 mg/kg. Teratogenic effects on mice and rats resulted from oral or subcutaneous administration of a sample containing 30 mg of the contaminant/kg 2,4,5-T (K. D. Courtney *et al., Science,* 1970, **168**, 864).

Formulations. Typical formulations include: 'Marks Brushwood Killer', e.c. (500 g a.e./l) as ester; 'Weedone' 2,4,5-T, e.c. (480 g a.e./l) as 2-butoxyethyl ester; 'Weedar' 2,4,5-T (480 g a.e./l), water-soluble concentrate of amine salts. Also mixtures with similar derivatives of 2,4-D.

continued

Analysis. Product analysis is by GLC of a suitable ester, by titration (*CIPAC Handbook*, 1970, **1**, 642; FAO Specification (CP/45), *Herbicides 1977*, p.32) or by i.r. or u.v. spectroscopy (AOAC Methods). Residues may be determined by GLC of the butyl ester (T. H. Byast *et al.*, *Agric. Res. Counc. (G.B.) Weed Res. Organ., Tech. Rep.*, No. 15 (2nd Ed.), p.31; H. E. Munro, *Pestic. Sci.*, 1977, **8**, 157). See also: R. P. Marquardt, *Anal. Methods Pestic., Plant Growth Regul. Food Addit.*, 1964, **4**, 247; *Anal. Methods Pestic. Plant Growth Regul.*, 1972, **6**, 702. For estimation of the 2,3,7,8-tetrachlorodibenzo-*p*-dioxin content see: E. A. Woolson *et al.*, *J. Agric. Food. Chem.*, 1972, **20**, 351; J. W. Edmunds *et al.*, *Pestic. Sci.*, 1973, **4**, 101.

Nomenclature and development. Tar oils are produced by the distillation of tars resulting from the high temperature carbonisation of coal and of coke oven and blast furnace tars. They consist mainly of aromatic hydrocarbons, but contain components soluble in aqueous alkali: the 'phenols' or 'tar acids'; and nitrogenous bases soluble in dilute mineral acids: 'tar bases'. Although they have been used for wood preservation since 1890, the introduction of the formulated products known as tar oil washes for crop protection, using the heavy creosote and anthracene oil ranges, dates from c. 1920.

Properties. The oils are brown to black liquids distilling from 230° to the temperature at which the residue is pitch, d^{15} 1·05-1·11. They are insoluble in water; soluble in organic solvents and in dimethyl sulphate.

Uses. Tar oils are toxic to the eggs of many insect species, particularly of aphids and psyllids. They are highly phytotoxic and their use on fruit trees is limited to the dormant season. They also kill moss and lichen on the tree trunks.

Toxicology. Liable to cause dermatitis to operators, especially in sunlight.

Formulations. An e.c. (solution of surfactants in the oil with higher phenols and petroleum oils as mutual solvents); stock emulsions (s.e.), concentrated emulsions using aqueous solutions of a suitable emulsifier.

Analysis. Product analysis is by breaking the dilute emulsion with alkali and sodium chloride and extracting the oil with diethyl ether; 'tar acids' are removed from the ether layers with alkali, 'tar bases' with acid; the neutral oil is recovered and its density, distillation range and solubility in dimethyl sulphate are determined (*MAFF. Tech. Bull.*, No. 1, 1958, p.81).

2,3,6-TBA

QVR BG CG FG $C_7H_3Cl_3O_2$ (225·5)

Nomenclature and development. The common name 2,3,6-TBA is approved by BSI, ISO and WSSA—exception JMAF (TCBA). The IUPAC and *C.A.* name is 2,3,6-trichlorobenzoic acid, Registry No. *[50-31-7]*. Its herbicidal activity was reported by H. J. Miller, *Weeds,* 1952, **1,** 185. It was introduced in 1954 by the Heyden Chemical Corp. under the code number 'HC-1281' and by E. I. du Pont de Nemours & Co. (Inc.) under the trade mark *'Trysben'* and protected by USP 2 848 470; 3 081 162. It is no longer manufactured or marketed by Amchem Products, Inc. or E. I. de Pont de Nemours & Co. (Inc.) but is available in the UK in mixtures with other herbicides.

Manufacture and properties. Produced by the chlorination of 2-chlorotoluene, giving a mixture of 2,3,6- and 2,4,5-trichlorotoluenes, which are then oxidised to the corresponding benzoic acids containing *c.* 60% 2,3,6-TBA, the technical product is a colourless to buff crystalline powder, m.p. 87-99°, v.p. 2·4 x 10^{-2} mmHg at 100°. Its solubility at 22° is 7·7 g/l water; readily soluble in most organic solvents. Pure 2,3,6-TBA has m.p. 125-126°. It forms water-soluble alkali metal and amine salts; the solubility at 25° is 440 g 2,3,6-TBA-sodium/kg water. 2,3,6-TBA is stable to light and at ≤ 60°. It is compatible with other hormone weedkillers.

Uses. 2,3,6-TBA is used post-em. in combination with other growth regulator herbicides in cereals and grass seed crops to control broad-leaved annual and perennial weeds including black bindweed, chickweed, cleavers, knotgrass, mayweeds and redshank.

Toxicology. The acute oral LD50 is: for rats 1500 mg/kg; for guinea-pigs and hens > 1500 mg/kg; for rabbits 600 mg/kg. The acute dermal LD50 for rats is > 1000 mg/kg. In 64-69-d feeding trials rats receiving 10 000 mg/kg diet suffered a minor disturbance of water metabolism, no trace of which was apparent at 1000 mg/kg diet. It is largely excreted unchanged in the urine.

Formulations. These include: 'Cambilene' (25 g 2,3,6-TBA + 19 g dicamba + 150 g mecoprop + 100 g MCPA/l), an aqueous concentrate of mixed sodium/potassium salts; 2,3,6-TBA + mecoprop (as mixed sodium/potassium salts).

Analysis. Product analysis is by GLC (*CIPAC Handbook,* in press). Residues may be determined by GLC after conversion to a suitable ester (J. J. Kirkland & H. L. Pease, *J. Agric. Food Chem.*, 1964, **12**, 468). Isomer determination has been studied (J. J. Kirkland, *Anal. Chem.*, 1961, **33**, 1502).

$$\underset{Cl_3C.C.O^-}{\overset{\overset{\displaystyle O}{\parallel}}{}} \quad Na^+$$

OVXGGG &-NA-   *sodium trichloroacetate* $C_2Cl_3NaO_2$ (185·4)

QVXGGG   *trichloroacetic acid* $C_2HCl_3O_2$ (163·4)

Nomenclature and development. The common name TCA is approved by BSI and ISO, with exceptions outlined below, for sodium trichloroacetate (IUPAC and *C.A.*), Registry No. *[650-51-1]*; in Australia, Canada, New Zealand, JMAF and WSSA, TCA applies to trichloroacetic acid (IUPAC and *C.A.*), Registry No. *[76-03-9]*, the salt being indicated (TCA-sodium); in France the chemical names are used. The sodium salt was introduced as a herbicide in 1947 by E. I. du Pont de Nemours & Co. (Inc.) (who no longer manufacture or market it) and by The Dow Chemical Co. Its herbicidal properties were first described by K. C. Barrons & A. J. Watson, *North Cent. Weed Control. Conf., Res. Rep.,* 1947, pp.43, 284.

Manufacture and properties. Trichloroacetic acid is produced by the oxidation of trichloroacetaldehyde by nitric or nitrous acids, or by the chlorination of acetic acid, it forms colourless hygroscopic crystals, m.p. 55-58°, b.p. 196-197°, d_4^{25} 1·62. Its solubility at 25° is 10 kg/l water; soluble in ethanol, diethyl ether. It tends to decompose to chloroform under alkaline conditions, but is stable in the absence of moisture; it is corrosive to iron, zinc and aluminium.

Technical sodium trichloroacetate is a yellowish, deliquescent powder. Its solubility at room temperature is 1·2 kg/l water; soluble in ethanol and many organic solvents. It is less corrosive than the acid and is compatible with other herbicides.

Uses. TCA is a pre-em. herbicide absorbed by the roots. It is used against: couch grass at 15-30 kg a.e./ha, in combination with certain tillage and cropping treatments; wild oats in sugar beet, peas and kale at 7·8 kg a.e./ha, prior to planting. Persistence in the soil is variable, 14 d- *c.* 90 d, dependent on soil type, moisture and temperature.

Toxicology. The acute oral LD50 is: for rats 3200-5000 mg sodium salt/kg, 400 mg acid/kg; for mice 5640 mg sodium salt/kg. The acid is extremely corrosive to skin and its sodium salt is a skin and eye irritant.

Formulations. Water-soluble solid (790-870 g a.e./kg) as sodium salt.

Analysis. Product analysis is by decarboxylation with sulphuric acid followed by titration of the excess of acid (*CIPAC Handbook,* 1970, **1**, 691). Residues may be determined by GLC with ECD, after conversion to the butyl ester (T. H. Byast *et al., Agric. Res. Counc. (G.B.) Weed Res. Organ., Tech. Rep.,* No. 15 (2nd Ed.), p.29).

Tebuthiuron

T5NN   DSJ   CX1&1&1   EN1&VM1   C$_9$H$_{16}$N$_4$OS   (228·3)

Nomenclature and development. The common name tebuthiuron is approved by BSI, ISO and WSSA. The IUPAC name is 1-(5-*tert*-butyl-1,3,4-thiadiazol-2-yl)-1,3-dimethylurea, in *C.A.* usage, *N*-[5-(1,1-dimethylethyl)-1,3,4-thiadiazol-2-yl]-*N,N'*-dimethylurea *[34014-18-1]*. It was introduced in 1974 by Eli Lilly & Co. under the code number 'EL-103', trade marks 'Spike' and 'Perflan' and protected by BP 1266172. Its herbicidal properties were described by J. F. Schwer, *Proc. Br. Weed Control Conf., 12th,* 1974, **2**, 847.

Manufacture and properties. Produced by the reaction of pivalic acid and 4-methyl-3-thiosemicarbazide in the presence of acid and reaction of the resulting intermediate with methyl isocyanate, it is a colourless solid, m.p. 161·5-164°, v.p. 2 x 10^{-6} mmHg at 25°. Its solubility at 25° is: 2·5 g/l water; 70 g/l acetone; 60 g/l acetonitrile; 6·1 g/l hexane; 170 g/l methanol; 60 g/l 2-methoxyethanol. It is photostable and non-corrosive to metals, polyethylene and spray equipment.

Uses. Tebuthiuron is a broad range herbicide for control of herbaceous and woody plants. Areas of use include: total vegetation control in non-crop areas, control of undesirable woody plants in pastures and rangeland, control of grass and broad-leaved weeds in sugarcane.

Toxicology. The acute oral LD50 is: for rats 644 mg/kg; for mice 579 mg/kg; for rabbits 286 mg/kg; for cats > 200 mg/kg; for dogs, bobwhite quail, mallard ducks and chickens > 500 mg/kg. Dermal applications of 200 mg/kg to rabbits produced no irritation nor did eye applications of 71 mg/kg cause any significant effect. In 90-d feeding trials to dogs, the 'no effect' level was 1000 mg/kg diet; no toxic effect was found in chickens fed for 30 d at 1000 mg/kg diet. Non-teratogenic in rats at 1800 mg/kg diet. Rats and mice fed for 2 y at 1600 mg/kg diet suffered no ill-effect. The LC50 (24 h) for trout and bluegill is > 160 mg/l; and for 96 h: for goldfish and fathead minnow > 160 mg/l; for trout 144 mg/l; and for bluegill 112 mg/l.

Formulations. These include: w.p. (800 g a.i./kg) and pellets (50, 100 and 400 mg/kg).

Analysis. Product analysis is by GLC using FID or by spectrophotometry. Residues are determined by GLC using FPD. Details may be obtained from Eli Lilly & Co.

Tecnazene

WNR BG CG EG FG $C_6HCl_4NO_2$ (260·9)

Nomenclature and development. The common name tecnazene is approved by BSI and ISO; also known as TCNB. The IUPAC and *C.A.* name is 1,2,4,5-tetrachloro-3-nitrobenzene, Registry No. *[117-18-0]*. It was introduced *c.*1946 by Bayer AG (who no longer manufacture or market it) under the trade marks 'Fusarex' and *'Folosan'*.

Manufacture and properties. Produced by the nitration of 1,2,4,5-tetrachlorobenzene (see 2,4,5-T, p.490), it forms colourless crystals, m.p. 99°, appreciably volatile at room temperature. It is practically insoluble in water; solubility at 25° *c.* 40 g/kg ethanol; readily soluble in benzene, carbon disulphide, chloroform.

Uses. Tecnazene is a fungicide selective for the control of *Fusarium caerulum* of potato tubers, and inhibiting the sprouting of seed potatoes (W. Brown, *Ann. Appl. Biol.,* 1947, **34**, 801; USP 2615801). Smoke formulations are used against *Botrytis* spp. on various glasshouse crops.

Toxicology. It has low toxicity to mammals. No ill-effect was noted in rats receiving 57 mg/kg daily or mice 215 mg/kg daily (G. A. H. Buttle & F. J. Dyer, *J. Pharm. Pharmacol.,* 1950, **2**, 371).

Formulations. Dusts (30-60 g a.i./kg); smokes, often in combination with gamma-HCH.

Analysis. Product analysis is by polarography (*CIPAC Handbook,* 1970, **1**, 663). Residues may be determined by polarography (J. G. Webster & J. A. Dawson, *Analyst (London),* 1952, **77**, 203) or by conversion by alkali to nitrite which is used to diazotise an amine and the diazonium salt coupled to form a dye that is estimated colorimetrically (T. Canback & J. Zajaczowska, *J. Pharm. Pharmacol.,* 1950, **2**, 545; D. J. Higgons & A. Toms, *J. Sci. Food Agric.,* 1957, **8**, 209).

Temephos

1OPS&O1&OR   DSR   DOPS&O1&O1 $C_{16}H_{20}O_6P_2S_3$ (466·5)

Nomenclature and development. The common name temephos is approved by BSI, ISO and ANSI; also known as ENT 27165. The IUPAC name is O,O,O',O'-tetramethyl O,O'-thiodi-*p*-phenylene bis(phosphorothioate) formerly O,O,O',O'-tetramethyl O,O'-thiodi-*p*-phenylene diphosphorothioate, in *C.A.* usage O,O'-(thiodi-4,1-phenylene) bis(O,O-dimethyl phosphorothioate) *[3383-96-8]* formerly O,O'-(thiodi-*p*-phenylene) O,O,O',O'-tetramethyl di(phosphorothioate). It was introduced in 1965 by American Cyanamid Co. under the code number 'AC 52160', the trade marks 'Abate', 'Abathion', 'Abat', 'Swebat', 'Nimitex' and 'Biothion' and protected by BelgP 648531; BP 1039238; USP 3317636.

Manufacture and properties. Produced by the condensation of O,O-dimethyl phosphoro-chloridothioate with 4,4'-thiodiphenol, which is obtained from phenol and thionyl chloride, it is a colourless crystalline solid, m.p. 30·0-30·5°. The technical product is a brown viscous liquid, 85-90% pure, n^{25} 1·586-1·588. It is almost insoluble in: water, hexane, methylcyclohexane; soluble in: acetonitrile, carbon tetrachloride, diethyl ether, 1,2-dichloroethane, toluene, lower alkyl ketones. It is stable at 25° and in natural fresh and saline waters; optimum stability is at pH 5-7, hydrolysis occurs at pH < 2 or pH > 9 at rates depending on the temperature and pH of the medium.

Uses. Temephos is used in public health programmes to control the larvae of mosquitoes, chironomid midges, blackfly (Simuliidae), biting midges (Ceratopogonidae) and moth and sand flies (Psychodidae). It is also effective in controlling human body lice, and fleas on dogs and cats. On crops, it is effective in controlling cutworms, thrips on citrus and Lygus bugs.

Toxicology. The acute oral LD50 for male rats is 8600 mg/kg, for female rats 1300 mg/kg. The acute dermal LD50 (24 h) is: for rabbits 1930 mg/kg; for rats > 4000 mg/kg. In 90-d feeding trials rats receiving 350 mg/kg diet showed no observable clinical effect. No toxic symptom was felt by humans receiving 256 mg/man for 5 d, or 64 mg/man for 28 d (R. L. Laws *et al., Arch. Environ. Health,* 1967, **14**, 289). It is relatively non-toxic to honey bees, birds and aquatic organisms (M. S. Mulla, *Mosq. News,* 1966, **26**(1), 87).

Formulations. These include: e.c. (200 and 500 g a.i./l); w.p. (500 g/kg); granules (10, 20 and 50 g/kg); dusting powder (20 g/kg).

Analysis. Residues may be determined by colorimetry (R. C. Blinn *et al., J. Agric. Food Chem.,* 1966, **14,** 152) or by GLC (N. R. Pasarela & E. J. Orloski, *Anal. Methods Pestic. Plant Growth Regul.,* 1973, **7**, 119).

$$O \quad O$$
$$\|\quad\|$$
$$(EtO)_2P.O.P(OEt)_2$$

2OPO&O2&OPO&O2&O2

$C_8H_{20}O_7P_2$ (290·2)

Nomenclature and development. The common name TEPP is approved by BSI and ISO— exception BPC (ethyl pyrophosphate); also known as ENT 18 771. The IUPAC name is tetraethyl pyrophosphate (I), in *C.A.* usage tetraethyl diphosphate *[107-49-3]* formerly (I). Its aphicidal properties were discovered in 1938 by G. Schrader & H. Kükenthal (cited by G. Schrader, *Die Entwicklung neuer insektizider Phosphorsäure-Ester,* 3rd. Ed.). In 1943 I. G. Farbenindustrie introduced a derivative thought to be hexaethyl tetraphosphate (hence HETP) but since shown to contain TEPP as the main active component. Trade marks include: *'Nifos T'* (Monsanto Chemical Co., who no longer manufacture or market the compound), 'Vapotone' (Chevron Chemical Co.).

Manufacture and properties. Produced by the hydrolysis of diethyl phosphorochloridate in the presence of pyridine or sodium hydrogen carbonate (A. D. F. Toy, *J. Am. Chem. Soc.,* 1948, **70**, 3882), it is a colourless hygroscopic liquid, b.p. 124°/1 mmHg, v.p. 1·55 x 10⁻⁴ mmHg at 20°, d_4^{20} 1·185, n_D^{20} 1·4196. It is miscible with water and most organic solvents; sparingly soluble in petroleum oils. The technical product is a dark amber-coloured mobile liquid, d_{25}^{25} *c.* 1·2. TEPP is rapidly hydrolysed by water, 50% decomposition occurring in 6·8 h at pH 7 and 25°. It decomposes at 170° with the evolution of ethylene; it is corrosive to most metals.

Uses. TEPP a non-systemic aphicide and acaricide of brief persistence.

Toxicology. The acute oral LD50 for rats is 1·12 mg/kg; the acute dermal LD50 for male rats is 2·4 mg/kg. It is rapidly metabolised by animals.

Formulation. Aerosol concentrate (solution in methyl chloride).

Analysis. Product analysis is by selective hydrolysis (AOAC Methods; *CIPAC Handbook,* 1970, **1**, 667). Residues may be determined by GLC (J. Crossley, *Anal. Methods Pestic. Plant Growth Regul.,* 1973, **7**, 471).

Terbacil

T6MVNVJ CX1&1&1 EG F1 $C_9H_{13}ClN_2O_2$ (216·7)

Nomenclature and development. The common name terbacil is approved by BSI, ANSI, WSSA, New Zealand and JMAF, and proposed by ISO. The IUPAC name is 3-*tert*-butyl-5-chloro-6-methyluracil (I), in *C.A.* usage 5-chloro-3-(1,1-dimethylethyl)-6-methyl-2,4(1*H*,3*H*)-pyrimidinedione *[5902-51-2]* formerly (I). The herbicidal activity of certain substituted uracils was reported by H. C. Bacha *et al., Science,* 1962, **137,** 537. Terbacil was introduced in 1966 by E. I. du Pont de Nemours & Co. (Inc.) under the code number 'Du Pont Herbicide 732', trade mark 'Sinbar' and protected by USP 3 235 357; BP 968 661; 968 663; 968 664; 968 665; 968 666; BelgP 625 897.

Manufacture and development. Produced by chlorination of 3-*tert*-butyl-6-methyluracil, it is a colourless crystalline solid, m.p. 175-177°. Its solubility at 25° is 710 mg/l water; soluble in cyclohexanone, dimethylformamide; moderately soluble in 4-methylpentan-2-one, butyl acetate, xylene. It is stable ≤ its m.p., below which it slowly sublimes. It is non-corrosive.

Uses. Terbacil is a herbicide, acting by inhibition of photosynthesis, absorbed mainly by the roots and used for the selective control of many annual and some perennial weeds in sugarcane, apples, peaches, citrus and mint at 1-4 kg a.i./ha and of perennials such as Bermuda grass and Johnson grass (in citrus only) at 4-8 kg/ha.

Toxicology. The acute oral LD50 for rats is > 5000 mg/kg. There was no clinical sign of toxicity to rabbits treated dermally at > 5000 mg/kg. In 2-y feeding trials the 'no effect' level for rats and dogs was 250 mg/kg diet.

Formulation. A w.p. (800 g a.i./kg).

Analysis. Product analysis by i.r. spectroscopy (details from E. I. du Pont de Nemours & Co. (Inc.)). Residues may be determined by GLC (H. L. Pease, *J. Agric. Food Chem.,* 1968, **16,** 54; T. H. Byast *et al., Agric. Res. Counc. (G.B.) Weed Res. Organ., Tech. Rep.,* No. 15 (2nd Ed.), p.47). See also: H. L. Pease *et al., Anal. Methods Pestic. Plant Growth Regul.,* 1978, **10,** 483).

Terbufos

$$\underset{\overset{\|}{(EtO)_2P.SCH_2.S.CMe_3}}{S}$$

2OPS&O2&S1SX1&1&1 $C_9H_{21}O_2PS_3$ (288·4)

Nomenclature and development. The common name terbufos is approved by BSI, ANSI and New Zealand, and proposed by ISO; also known as ENT 27 920. The IUPAC name is *S-tert*-butylthiomethyl *O,O*-diethyl phosphorodithioate, in *C.A.* usage *S*-[[(1,1-dimethyl-ethyl)thio]methyl] *O,O*-diethyl phosphorodithioate *[13071-79-9]*. It was introduced by American Cyanamid Co. under the code number 'AC 92 100' and trade mark 'Counter'. Its insecticidal properties were described by E. B. Fagan, *Proc. Br. Insectic. Fungic. Conf., 7th,* 1973, **2**, 695.

Properties. The technical product is a colourless to pale yellow liquid, ⩾85% pure, b.p. 69°/0·01 mmHg, m.p. —29·2°, v.p. 2·6 x 10^{-4} mmHg at 25°, 1 x 10^{-3} mmHg at 42°, d^{24} 1·105, flash point 88° (tag open cup). Its solubility at ordinary temperature is *c.* 10-15 mg/l water; soluble in acetone, alcohols, aromatic hydrocarbons, chlorinated hydrocarbons. It decomposes at > 120° or in the presence of acid (pH < 2) or alkali (pH > 9).

Uses. Terbufos has effective initial and residual activity against soil insects and other arthropods. Soil application of granules controls *Diabrotica* spp. larvae on maize, *Tetanops myopaeformis* on sugar beet, *Erioischia brassicae* on cabbages, onion maggot, wireworms, scutigerellids, millipedes and other soil-dwelling arthropods.

Toxicology. The acute oral LD50 is: for male albino rats 1·6 mg tech./kg; for female albino mice 5·4 mg/kg. The acute dermal LD50 is: for rabbits 1·0 mg tech./kg; for rats 7·4 mg tech./kg, and for 24-h contact 27·5 mg a.i. (as 10% granule)/kg, > 200 mg a.i. (as 2% granule)/kg. Feeding trials on rats (2 y), mice (1·5 y) and dogs (0·5 y) showed no adverse effect other than cholinesterase depression and associated syndrome.

Formulations. Granules (20-150 g a.i./kg).

Analysis. Product and residue analysis: details of GLC methods are available from American Cyanamid Co.

Terbumeton

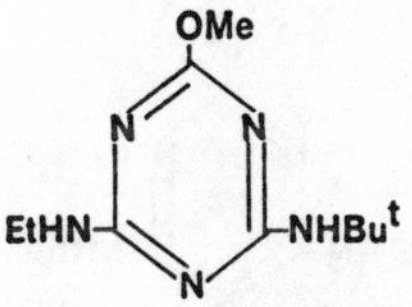

OMe

EtHN—[triazine ring]—NHBu^t

T6N CN ENJ BO1 DMX1&1&1 FM2 $C_{10}H_{19}N_5O$ (225·3)

Nomenclature and development. The common name terbumeton is approved by BSI and ISO. The IUPAC name is 2-*tert*-butylamino-4-ethylamino-6-methoxy-1,3,5-triazine, in *C.A.* usage *N*-(1,1-dimethylethyl)-*N'*-ethyl-6-methoxy-1,3,5-triazine-2,4-diamine *[33693-04-8]* formerly 2-(*tert*-butylamino)-4-(ethylamino)-6-methoxy-*s*-triazine. It was developed by J. R. Geigy S.A. (now Ciba-Geigy AG) under the code number 'GS 14 259'. It was introduced under the trade mark 'Caragard'. It was described by A. Gast & E. Fankhauser, *Proc. Br. Weed Control Conf., 8th,* 1966, p.485.

Manufacture and properties. Produced by the reaction of terbuthylazine (p.502) with methanol and sodium hydroxide, it is a colourless solid, m.p. 123-124°, v.p. $2\pm0.5 \times 10^{-6}$ mmHg at 20°. Its solubility at 20° is 130 mg/l water; soluble in organic solvents.

Uses. Terbumeton is a herbicide absorbed by leaves and roots and effective against grasses and broad-leaved weeds, both annual and perennial species. Mixed with terbuthylazine, it is used for post-em. weed control in citrus, apple orchards, vineyards and in forestry at 3-10 kg total a.i./ha.

Toxicology. The acute oral LD50 for rats is 483-651 mg/kg; the acute dermal LD50 for rats is > 3170 mg/kg. In 90-d feeding trials the 'no effect' level was: for rats 9 mg/kg daily; for dogs 22 mg/kg daily. It is slightly toxic to fish.

Formulations. 'Caragard 50', w.p. (500 g a.i./kg); also in combination with terbuthylazine.

Analysis. Product analysis is by GLC with internal standard or by titration with perchloric acid in acetic acid (details from Ciba-Geigy AG). Residues may be determined by GLC with MCD (K. Ramsteiner *et al., J. Assoc. Off. Anal. Chem.,* 1974, **57**, 192). See also: E. Knüsli, *Anal. Methods Pestic., Plant Growth Regul. Food Addit.,* 1964, **4**, 13.

Terbuthylazine

T6N CN ENJ BMX1&1&1 DM2 FG $\qquad$ $C_9H_{16}ClN_5$ (229·7)

Nomenclature and development. The common name terbuthylazine is approved by BSI, ISO, ANSI and WSSA. The IUPAC name is 2-*tert*-butylamino-4-chloro-6-ethylamino-1,3,5-triazine, in *C.A.* usage 6-chloro-*N*-(1,1-dimethylethyl)-*N'*-ethyl-1,3,5-triazine-2,4-diamine *[5915-41-3]* formerly 2-(*tert*-butylamino)-4-chloro-6-(ethylamino)-*s*-triazine. It was developed by J. R. Geigy S.A. (now Ciba-Geigy AG) under the code number 'GS 13529' and introduced by Ciba-Geigy AG under the trade mark 'Gardoprim'. It was described by A. Gast & E. Fankhauser, *Proc. Br. Weed Control Conf., 8th,* 1966, p.485.

Manufacture and properties. Produced by reacting trichloro-1,3,5-triazine in a first step with 1 equivalent of *tert*-butylamine and, in second step, with 1 equivalent of ethylamine in the presence of sodium hydroxide, it is a colourless solid, m.p. 177-179°, v.p. 1·12 x 10⁻⁶ mmHg at 20°. Its solubility at 20° is 5 mg/kg water; slightly soluble in organic solvents.

Uses. Terbuthylazine is a herbicide which controls a wide range of weeds, and is taken up mainly through the roots. It remains largely in the top soil. It is used pre-em. in sorghum at 1·2-1·8 kg a.i./ha; also for selective weed control in maize, vineyards, citrus and pod forests. Mixed with terbumeton it controls perennial weeds in established stands of apples, citrus and grapes. It may also be used for non-selective weed control when mixed with a methoxy-1,3,5-triazine. It is also used in combination with bromofenoxim at total a.i. rates of 0·8-1·25 kg/ha as a broad-spectrum broad-leaved herbicide in winter and spring cereals.

Toxicology. The acute oral LD50 for rats is 2000-2160 mg/kg; the acute dermal LD50 for rabbits is > 3000 mg/kg. In 90-d feeding trials the 'no effect' level for rats and dogs was estimated as 5 mg/kg daily. The toxic response to fish varies with the species from practically non-toxic to moderate.

Formulations. These include: 'Gardoprim 50', w.p. (500 g a.i./kg); 'Gardoprim 80' or 'Primatol 80', w.p. (800 g/kg); 'Gardoprim 500 FW', flowable (500 g/l). Also used in combination with terbumeton ('Caragard combi'), with bromofenoxim ('Faneron GB' or 'Mofix', 'Faneron combi' and 'Faneron multi'), with terbutryne ('Sorgoprim', 'Topogard').

Analysis. Product analysis is by GLC with internal standard (details from Ciba-Geigy AG). Residues may be determined by GLC (K. Ramsteiner *et al., J. Assoc. Off. Anal. Chem.,* 1974, **57**, 192). See also: E. Knüsli, *Anal. Methods Pestic., Plant Growth Regul. Food Addit.,* 1964, **4**, 13.

Terbutryne
(Terbutryn)

T6N CN ENJ BS1 DMX1&1&1 FM2 $C_{10}H_{19}N_5S$ (241·4)

Nomenclature and development. The common name terbutryn is approved by ISO, ANSI and WSSA—exceptions BSI and France (terbutryne). The IUPAC name is 2-*tert*-butylamino-4-ethylamino-6-methylthio-1,3,5-triazine, in *C.A.* usage *N*-(1,1-dimethylethyl)-*N'*-ethyl-6-(methylthio)-1,3,5-triazine-2,4-diamine *[886-50-0]* formerly 2-(*tert*-butylamino)-4-ethylamino-6-methylthio-*s*-triazine. It was introduced in 1966 by J. R. Geigy S.A. (now Ciba-Geigy AG) under the code number 'GS 14260', the trade marks 'Igran', 'Clarosan' and, for Great Britain only, 'Prebane' and protected by SwissP 393 344; BP 814 948; 978 249; USP 3 145 208. It was first described by A. Gast *et al., Proc. Symp. New Herbic., 2nd.,* 1965, p.305.

Manufacture and properties. Produced by reacting terbuthylazine (p.502) with methanethiol in the presence of an equivalent of sodium hydroxide or by reacting 4-*tert*-butylamino-6-ethylamino-1,3,5-triazine-2-thiol with a methylating agent in the presence of sodium hydroxide, it is a colourless powder, m.p. 104-105°, v.p. 9·6 x 10⁻⁷ mmHg at 20°. Its solubility at 20° is 25 mg/l water; readily soluble in most organic solvents.

Uses. Terbutryne is a selective herbicide for pre-em. use in winter cereals at 1-2 kg a.i./ha for the control of blackgrass and annual meadow grass. Among the autumn-germinating broad-leaved weeds controlled are chickweed, the mayweeds, speedwell, and poppies, but cleavers are rather resistant. Other pre-em. uses are in sunflower, sugarcane, potatoes and peas (for the last 2 crops it is generally used in mixture with terbuthylazine); post-em. directed on maize. Also used for control of algae and submerged vascular plants ('Clarosan').

Toxicology. The acute oral LD50 is: for rats 2000-2980 mg/kg; for mice 5000 mg/kg. The acute dermal LD50 for rats is > 2000 mg/kg. In 90-d feeding trials the 'no effect' level was: for rats 50 mg/kg daily; for dogs 40 mg/kg daily. It is moderately toxic to fish and of low toxicity to birds.

Formulations. These include: 'Igran 50', w.p. (500 g a.i./kg); 'Igran 80', w.p. (800 g/kg); 'Igran 500FW', flowable (500 g/l). Also used in combinations with atrazine ('Gesaprim combi') and with terbuthylazine ('Sorgoprim', 'Topogard') and with metobromuron ('Igrater').

Analysis. Product analysis is by GLC with internal standard (*CIPAC Handbook,* 1979, **1A**, in press; FAO Specification (CP/61); A. H. Hofberg *et al., J. Assoc. Off. Anal. Chem.,* 1973, **56**, 586), or by titration with perchloric acid in acetic acid. Residues may be determined by GLC (K. Ramsteiner *et al., ibid.,* 1974, **57**, 192; T. H. Byast *et al., Agric. Res. Counc. (G.B.) Weed Res. Organ., Tech. Rep.,* No. 15 (2nd Ed.), p. 40) or by HPLC (T. H. Byast *et al., loc. cit.*). See also: E. Knüsli, *Anal. Methods Pestic., Plant Growth Regul. Food Addit.,* 1964, **4**, 13.

Tetrachlorvinphos

(MeO)₂P—O ... C=C ... (Z)

(MeO)₂P—O ... C=C ... (E)

GR BG DG EYU1GOPO&O1&O1   C₁₀H₉Cl₄O₄P (366·0)

Nomenclature and development. The common name tetrachlorvinphos is approved by BSI and ISO—exception USA, ESA (stirofos); also known as ENT 25841. The IUPAC name is (Z)-2-chloro-1-(2,4,5-trichlorophenyl)vinyl dimethyl phosphate (I), in *C.A.* usage (Z)-2-chloro-1-(2,4,5-trichlorophenyl)ethenyl dimethyl phosphate *[22248-79-9]* formerly (I); the (E)-isomer has Registry No. *[22350-76-1]* and the mixed isomers *[961-11-5]*. It was introduced in 1966 by Shell Development Co. under the code number 'SD 8447', trade mark 'Gardona' (for crop protection use) and 'Rabond' (for veterinary use) and protected by USP 3 102 842. It was described by R. R. Whetstone *et al., J. Agric. Food Chem.*, 1966, **14**, 352.

Manufacture and properties. Produced by the reaction of triethyl phosphite with 2,2,2′,4′,5′-pentachloroacetophenone, the technical product, typically 98% (Z)-isomer, is a colourless crystalline solid, m.p. 95-97°. Its solubility at 20° is: 11 mg/l water; < 200 g/kg acetone; 400 g/kg chloroform, dichloromethane; < 150 g/kg xylene. It is stable ≤ 100°. It is slowly hydrolysed in water, 50% loss occurring at 50° in 1300 h at pH 3, in 1060 h at pH 7, in 80 h at pH 10·5. The pure (Z)-isomer has v.p. 4·2 x 10⁻⁸ mmHg at 20°.

Uses. Tetrachlorvinphos is a selective insecticide controlling: lepidopterous and dipterous pests of fruit at 25-75 g a.i./100 l; lepidopterous pests of rice at 250-600 g/ha, and of cotton and maize at 0·75-2·0 kg/ha; lepidopterous pests of vegetables at 0·25-1·0 kg/ha and of tobacco at 0·5-1·5 kg/ha. With certain exceptions, it does not exhibit high activity against hemipterous insects and other sucking pests and, because of rapid breakdown, is not effective against soil insects. It is used against flies in dairies and livestock barns and against pests of stored products. It is also used for pasture and forestry. Its metabolism and breakdown have been reviewed (K. I. Beynon *et al., Residue Rev.*, 1973, **47**, 55).

Toxicology. The acute oral LD50 is: for rats 4000-5000 mg/kg; for mice 2500-5000 mg/kg; for mallard duck and Chukar partridge > 2000 mg/kg. The acute percutaneous LD50 for rabbits is > 2500 mg/kg. In 2-y feeding trials the 'no effect' level was: for rats 125 mg/kg diet; for dogs 200 mg/kg diet. Reproduction studies on rats receiving 1000 mg/kg diet showed no adverse effect. Toxicity to fish is moderate to low.

Formulations. These include: e.c. (240 g a.i./l); w.p. (500 and 750 g/kg); granules (50 g/kg); suspension concentrate (700 g/l).

Analysis. Product analysis is by i.r. spectroscopy or GLC (details from Shell International Chemical Co.). Residues of tetrachlorvinphos and its major metabolites in plants may be determined by GLC (K. I. Beynon & A. N. Wright, *J. Sci. Food Agric.*, 1969, **20**, 250; K. I. Beynon *et al., Pestic. Sci.*, 1970, **1**, 250, 254, 259). See also: *Anal. Methods Pestic. Plant Growth Regul.*, 1973, **7**, 297.

Tetradifon

GR DSWR BG DG EG $C_{12}H_6Cl_4O_2S$ (356·0)

Nomenclature and development. The common name tetradifon is approved by BSI, ISO, ANSI and JMAF—exceptions Turkey and USSR (tedion) and Portugal; also known as ENT 23 737. The IUPAC name is 4-chlorophenyl 2,4,5-trichlorophenyl sulphone formerly 2,4,4′,5-tetrachlorodiphenyl sulphone, in *C.A.* usage 1,2,4-trichloro-5-[(4-chlorophenyl)-sulfonyl]benzene *[116-29-0]* formerly *p*-chlorophenyl 2,4,5-trichlorophenyl sulfone. Its acaricidal properties were first described by H. O. Huisman *et al., Nature (London),* 1955, **176,** 515. It was introduced in 1954 by Philips-Duphar B.V. under the code number 'V-18', the trade mark 'Tedion V-18' and protected by DutchP 81 359; USP 2 812 281.

Manufacture and properties. Produced by the Friedel Craft's reaction between 2,4,5-trichlorobenzenesulphonyl chloride and chlorobenzene (*idem, Rec. Trav. Chim. Pays Bas,* 1958, **77,** 103), it is a colourless crystalline solid, m.p. 148-149°, v.p. 2·4 x 10^{-10} mmHg at 20°. Its solubility at 50° is 200 mg/kg water; of low solubility in acetone and alcohols; more soluble in aromatic hydrocarbons, chloroform, dioxane. The technical product is ⩾ 94% pure, m.p. ⩾ 144°. It is resistant to hydrolysis by acid or alkali, is compatible with other pesticides and is non-corrosive.

Uses. Tetradifon is a non-systemic acaricide toxic to the eggs and all non-adult stages of phytophagous mites. It is recommended for application to top fruit, citrus, grapes, vegetables, ornamentals and nursery stock at 20 g a.i. (as w.p.)/100 l, tea at 16 g a.i. (as e.c.)/100 l, cotton at 240-400 g a.i. (as e.c.)/ha. At these rates it is non-phytotoxic and harmless to beneficial insects.

Toxicology. The acute oral LD50 for rats is > 14 700 mg/kg; the acute dermal LD50 for rabbits is 10 000 mg/kg; In 60-d feeding trials: rats receiving 500 mg/kg diet suffered no ill-effect; offspring of rats receiving 1000 mg/kg diet were normal.

Formulations. A w.p. (200 g a.i./kg); e.c. (80 g/l); also 'Childion', e.c. (400 g dicofol + 125 g tetradifon/l).

Analysis. Product analysis is by GLC (W. R. Mitchell, *J. Assoc. Off. Anal. Chem.,* 1976, **59,** 209; *CIPAC Handbook,* in press). Residues may be determined by GLC (G. A. Root, *Bull. Environ. Contam. Toxicol.,* 1967, **2,** 274). See also: C. C. Cassil & J. Yaffe, *Anal. Methods Pestic., Plant Growth Regul. Food Addit.,* 1964, **2,** 473; *Anal. Methods Pestic. Plant Growth Regul.,* 1972, **6,** 488; A van Rossum, *ibid.,* 1978, **10,** 119.

Tetramethrin

T56 BVNV&TJ C1OV- BL3TJ A1 A1 C1UY1&1 $C_{19}H_{25}NO_4$ (331·4)

Nomenclature and development. The common name tetramethrin is approved by BSI, ISO and ANSI; also known as phthalthrin. The IUPAC name is 3,4,5,6-tetrahydro-phthalimidomethyl (1*RS*)-*cis,trans*-chrysanthemate, in *C.A.* usage (1,3,4,5,6,7-hexahydro-1,3-dioxo-2*H*-isoindol-2-yl)methyl 2,2-dimethyl-3-(2-methyl-1-propenyl)cyclopropanecarboxylate *[7696-12-0]* formerly 2,2-dimethyl-3-(2-methylpropenyl)cyclopropanecarboxylic acid ester with *N*-(hydroxymethyl)-1-cyclohexane-1,2-dicarboximide. It was introduced in 1965 by Sumitomo Chemical Co. Ltd and by the Agricultural Chemical Division of FMC Corp. (who no longer manufacture or market it) under the code number 'FMC 9260', the trade mark 'Neo-Pynamin' and protected by JapanP 453929; 462108; USP 3 268 398.

Manufacture and properties. Produced by the reaction of tetrahydrophthalimidomethanol with chrysanthemoyl chloride, it is a colourless crystalline solid with a pyrethrum-like odour. The technical product has m.p. 65-80°, b.p. 185-190°/0·1 mmHg, v.p. 3·5 x 10^{-8} mmHg at 20°, d_{20}^{20} 1·109. Its solubility at 25° is: 400 g/kg acetone, toluene; 500 g/kg benzene; 45 g/kg ethanol. It is stable under normal conditions; there was no loss of biological activity after storage at 50° for 0·5 y.

Uses. Tetramethrin is a contact insecticide with a strong 'knockdown' action on flies, mosquitoes and other pests of public health.

Toxicology. The acute oral LD50 is: for rats > 4640 mg/kg; for mice 5200 mg/kg; for chickens 5000 mg/kg. The acute dermal LD50 for mice is > 15000 mg/kg. In 90-d feeding trials rats receiving ≤ 2000 mg/kg diet suffered no effect. The LC50 (48 h) for carp is 0·18 mg/l. Its metabolic fate in mammals has been studied (J. Miyamoto *et al., Agric. Biol. Chem.,* 1968, **32,** 628).

Formulations. These include: concentrates for the preparation of oil-based and water-based pressurised formulations; space and residual sprays containing tetramethrin with other insecticides and with piperonyl butoxide; dusts.

Analysis. Product analysis is by GLC, or by u.v. spectroscopy, after purification by TLC (J. Miyamoto, *Anal. Methods Pestic. Plant Growth Regul.,* 1973, **7,** 345).

O,O,O',O'-Tetrapropyl dithiopyrophosphate

$$\left[(PrO)_2 \overset{\overset{\displaystyle S}{\|}}{P} - \right]_2 O$$

3OPS&O3&OPS&O3&O3 $C_{12}H_{28}O_5P_2S_2$ (378·4)

Nomenclature and development. There is no common name; known as ENT 16894. The IUPAC name is O,O,O',O'-tetrapropyl dithiopyrophosphate, in *C.A.* usage tetrapropyl thiodiphosphate *[3244-90-4]* formerly tetrapropyl thiopyrophosphate. It was introduced in 1951 by Stauffer Chemical Co. under the code number 'ASP-51', trade mark 'Aspon' and protected by USP 2663722. It was described by A. D. F. Toy, *J. Am. Chem. Soc.,* 1951, **73**, 4670.

Manufacture and properties. Produced from O,O-dipropyl phosphorochloridothioate (DBP 848812, Bayer AG), the technical product is a straw to amber-coloured liquid, with a faintly aromatic odour, f.p. < 5°, b.p. 104°/0·01 mmHg, flash point 149° (open cup), d_{20}^{20} 1·119-1·123, n_D^{21} 1·4710. Its solubility at room temperature is 1·6 g/l water; sparingly soluble in light petroleum; miscible with most other organic solvents. It is stable with insignificant hydrolysis in water at room temperature; stable for at least 24 h at 100° but decomposes, without explosive hazard, at 149°. Discolouration and physical changes may occur on prolonged contact with metal, and it is corrosive to steel.

Uses. It is a non-systemic insecticide particularly effective for the control of chinch bugs in turf at 28·6 kg a.i./ha. It is persistent in soil.

Toxicology. The acute oral LD50 for male albino rats is 891-1700 mg/kg; the acute dermal LD50 for albino rabbits is 1838 mg a.i. (as e.c.)/kg. In 90-d feeding trials at sub-lethal rates rats showed depression of red blood cell cholinesterase.

Formulations. 'Aspon', e.c. (480 g a.i./l); granules (50 g/kg).

Analysis. Analysis is by GLC: details are available from Stauffer Chemical Co.

Tetrasul

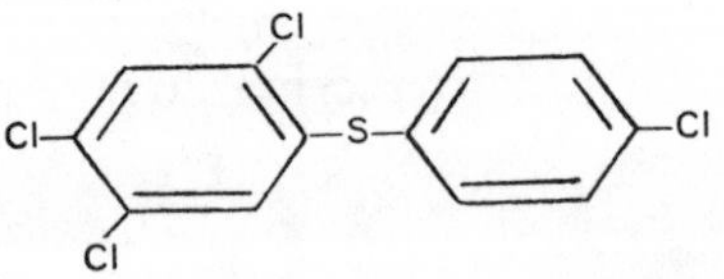

GR DSR BG DG EG

$C_{12}H_6Cl_4S$ (324·1)

Nomenclature and development. The common name tetrasul is approved by BSI and ISO —exceptions Canada, Germany and Italy; also known as ENT 27115. The IUPAC name is 4-chlorophenyl 2,4,5-trichlorophenyl sulphide formerly 2,4,4′,5-tetrachlorodiphenyl sulphide, in *C.A.* usage 1,2,4-trichloro-5-[4-(chlorophenyl)thio]benzene *[2227-13-6]* formerly *p*-chlorophenyl 2,4,5-trichlorophenyl sulfide. Its acaricidal properties were first described by J. Meltzer & F. C. Dietvoorst, *Proc. Int. Congr. Crop Protect., IVth,* 1957, **1**, 669. It was introduced in 1957 by Philips-Duphar B.V. under the code number 'V-101', trade mark 'Animert V-101' and protected by DutchP 94329; USP 3054719.

Manufacture and properties. Produced by the reaction of the sodium salt of 4-chloro-(thiophenol) with 1,2,4,5-tetrachlorobenzene, it is a colourless crystalline solid, m.p. 87·3-87·7°, v.p. 7·5 x 10^{-7} mmHg at 20°. It is slightly soluble in water; moderately in acetone, diethyl ether; soluble in benzene, chloroform. It is stable under normal conditions but should be protected against prolonged exposure to sunlight; it is oxidised to its sulphone, tetradifon (p.505). It is compatible with most other pesticides and is non-corrosive. The technical grade is a yellow-brown to cream solid, > 77·5% pure, m.p. 75-85°.

Uses. Tetrasul is a non-systemic acaricide, highly toxic to eggs and non-adult stages of phytophagous mites. It is recommended for use on apples, pears and cucumbers at 36 g a.i./100 l at the time when the winter eggs are hatching. At these concentrations it is non-phytotoxic. As it is highly selective it does not pose a hazard to beneficial insects or to wildlife.

Toxicology. The acute oral LD50 is: for female rats 6810 mg/kg; for female mice 5010 mg/kg; for female guinea-pigs 8800 mg/kg. The acute dermal LD50 for rabbits is > 2000 mg/kg.

Formulations. An e.c. (180 g a.i./l); w.p. (180 g/kg).

Analysis. Product and residue analysis is by GLC (L. R. Mitchell, *J. Assoc. Off. Anal. Chem.,* 1976, **59**, 209).

Thiabendazole

T56 BM DNJ C- ET5N CSJ $C_{10}H_7N_3S$ (201·2)

Nomenclature and development. The common name thiabendazole is approved by BSI, ISO, BPC and JMAF; also known as TBZ. The IUPAC name is 2-(thiazol-4-yl)benzimidazole, in *C.A.* usage 2-(4-thiazolyl)-1*H*-benzimidazole *[148-79-8]* formerly 2-(4-thiazolyl)-benzimidazole. It was introduced by Merck & Co. Inc. in 1962 as an anthelmintic and in 1968 as an agricultural fungicide under the code number 'MK-360' and the trade marks 'Mertect', 'Tecto' and 'Storite'. Its antifungal properties were reported by H. J. Robinson *et al., J. Invest. Dermatol.,* 1964, **42**, 479; T. Staron & C. Allard, *Phytiatr.-Phytopharm.,* 1964, **13**, 163, and its systemic properties in plants by D. C. Erwin *et al., Phytopathology,* 1968, **58**, 860.

Manufacture and properties. Produced by the condensation in polyphosphoric acid of thiazole-4-carboxamide with *o*-phenylenediamine (H. D. Brown *et al., J. Am. Chem. Soc.,* 1961, **83**, 1764; USP 3 017 415; V. J. Grenda *et al., J. Org. Chem.,* 1965, **30**, 259), it is a colourless powder, m.p. 304-305°, non-volatile at room temperature but sublimes when heated strongly to 310°. Its solubility is: at 25° *c.* 10 g/l water at pH 2, < 50 mg/l water at pH 5-12, > 50 mg/l water at pH 12, 4·2 g/l acetone, 7·9 g/l ethanol, 2·1 g/l ethyl acetate; at room temperature 230 mg/l benzene, 80 mg/l chloroform, 9·3 g/l methanol, 39 g/l dimethylformamide, 80 g/l dimethyl sulphoxide. It is stable under normal conditions to hydrolysis, light and heat.

Uses. Thiabendazole is a fungicide controlling diseases of the following crops: asparagus, avocado, banana, cabbage, celery, cherry, citrus, cotton, certain cucurbits, grapes, mushrooms, onions (and garlic), ornamentals (bulbs, corms and flowers), pome fruit, potatoes, rice, soyabean, sugar beet, sweet potato, tobacco, tomatoes, turf and wheat. Pathogenic fungi controlled include species of *Aspergillus, Botrytis, Ceratocystis, Cercospora, Colletotrichum, Diaporthe, Fusarium, Giberella, Gloesporium, Oospora, Penicillium, Phoma, Rhizoctonia, Sclerotinia, Septoria,* and *Verticillium.* It is also effective at 0·2-5·0 g a.i./l for the post-harvest treatment of fruit and vegetables to control storage diseases (D. S. Meredith, *Proc. 1977 Br. Crop Prot. Conf.—Pests Dis.,* 1977, **1**, 179). It is also used as an anthelmintic in human and veterinary medicine.

Toxicology. The acute oral LD50 is: for rats 3300 mg a.i./kg; for mice 3810 mg/kg; for rabbits 3850 mg/kg. No observable clinical effect followed chronic inhalation at 70 mg/m³. In 2-y feeding trials rats receiving 40 mg/kg daily showed no evidence of ill-effect.

Formulations. These include w.p. (400, 600 and 900 g a.i./kg); flowable suspension (450 g/l); fumigation tablets (7 g a.i.).

Analysis. Product analysis is by u.v. spectroscopy at 302 nm. Residues in food crops, after extraction and clean-up, are determined fluorimetrically (excitation at 302 nm, emission at 360 nm). See also: J. S. Wood, *Anal. Methods Pestic. Plant Growth Regul.,* 1976, **8**, 299.

Thiazafluron

T5NN DSJ CXFFF EN1&VM1 $C_6H_7F_3N_4OS$ (240·2)

Nomenclature and development. The common name thiazafluron is approved by BSI and ISO. The IUPAC name is 1,3-dimethyl-1-(5-trifluoromethyl-1,3,4-thiadiazol-2-yl)urea (I), in *C.A.* usage *N,N'*-dimethyl-*N*-[5-(trifluoromethyl)-1,3,4-thiadiazol-2-yl]urea *[25366-23-8]* formerly (I). It was introduced in 1972 by Ciba Geigy AG as an experimental herbicide under the code number 'GS 29 696' the trade mark 'Erbotan' and protected by SwissP 488 723; 493 988; BP 1 254 468. It was first described by G. Müller *et al., C. R. Journ. Etud. Herbic. Conf. COLUMA 7th,* 1973, p.32.

Properties. It is a crystalline solid, m.p. 136-137°, v.p. 2 x 10⁻⁶ mmHg at 20°. Its solubility at 20° is 2·1 g/l water; soluble in polar organic solvents.

Uses. Thiazafluron is a non-selective herbicide used pre- and post-em. for industrial weed control. It is effective against most annual and perennial mono- and dicotyledonous weeds. It is active mainly through plant roots, foliar activity is poor. The dosage range is 2-8 kg a.i./ha in moist and temperate and 6-12 kg/ha in warm and dry regions. Mixtures with triazines or other herbicides are used for special weed problems.

Toxicology. The acute oral LD50 is: for rats 278 mg/kg; for mice 630 mg/kg. The acute dermal LD50 for rats is > 2150 mg/kg. It is slightly toxic to birds.

Formulations. 'Erbotan' 50WP, 'Erbotan' 80WP, w.p. (500 and 800 g a.i./kg); 'Erbotan' 10G, granules (100 g/kg).

Analysis. Product analysis is by non-aqueous titration with tetrabutylammonium hydroxide in dimethylformamide. Residues in soil may be determined by extraction with methanol, a column chromatographic clean-up and GLC with estimation by a sulphur- or nitrogen-specific detector. Details of these methods are available from Ciba-Geigy AG.

Thidiazuron

T5NNSJ DMVMR $C_9H_8N_4OS$ (220·2)

Nomenclature and development. The common name thidiazuron is approved by BSI, ANSI and New Zealand and proposed by ISO. The IUPAC name is 1-phenyl-3-(1,2,3-thiadiazol-5-yl)urea, in *C.A.* usage *N*-phenyl-*N'*-(1,2,3-thiadiazol-5-yl)urea *[41118-83-6]*. It was introduced in 1976 by Schering AG under the code number 'SN 49 537', trade mark 'Dropp' and protected by DBP 2 506 690. Its plant growth regulating effects were first described by F. Arndt *et al., Plant Physiol.,* 1976, **57,** Supplement p.99.

Manufacture and properties. Produced by the reaction of 1,2,3-thiadiazol-5-ylamine with phenyl isothiocyanate, it is a colourless crystalline solid, m.p. 213°(decomp.), negligible v.p. at 25°. Its solubility at 23° is: < 0·05 g/l water; > 500 g/l dimethyl sulphoxide, dimethylformamide; 21 g/l cyclohexane; 4·5 g/l methanol; 8 g/l acetone; 0·8 g/l ethyl acetate. It is stable at 23° between pH 5-9.

Uses. Thidiazuron is a plant growth regulator, effective even at low dosages. It is used at 50-200 g a.i./ha for the defoliation of cotton, the leaves dropping when still green.

Toxicology. The acute oral LD50 is: for rats > 4000 mg/kg; for mice > 5000 mg/kg. The acute dermal LD50 for rats and rabbits is > 1000 mg/kg.

Formulations. A w.p. (500 g a.i./kg). Flowable and e.c. formulations are under investigation.

Analysis. Product analysis is by HPCL using 1,3-di-(*m*-tolyl)urea as internal standard. Residues may be determined, after suitable clean-up, by reversed phase HPLC with u.v. detection or by hydrolysis and determination of aniline, after bromination, by GLC with ECD. Details of methods are available from Schering AG.

Thiobencarb

GR D1SVN2&2 $C_{12}H_{16}ClNOS$ (257·8)

Nomenclature and development. The common name thiobencarb is approved by BSI, ANSI, New Zealand and WSSA and proposed by ISO—exception JMAF (benthiocarb). The IUPAC name is *S*-4-chlorobenzyl diethylthiocarbamate (I), in *C.A.* usage *S*-[(4-chlorophenyl)methyl] diethylcarbamothioate *[28249-77-6]* formerly (I). It was introduced by Kumiai Chemical Industry Co. Ltd under the code number 'B-3015' and the trade marks 'Saturn', and 'Bolero' and by Chevron Chemical Co. under the latter.

Manufacture and properties. Produced by reacting carbonyl sulphide with diethylamine and then with 4-chlorobenzyl chloride, it is a pale yellow liquid, m.p. 3·3°, b.p. 126-129°/8 x 10^{-3} mmHg, d^{20} 1·145-1·180. Its solubility at 20° is *c.* 30 mg/l water; soluble in most organic solvents. It is stable under acid and alkaline conditions. The technical product is *c.* 93% pure.

Uses. Thiobencarb is a herbicide of use in rice. For direct-seeded rice the concentrate is applied at 3-6 kg a.i./ha to the surface of the water 3-5 d before or 5-10 d after sowing. For transplanted rice it is applied at 3-6 kg a.i./ha to the water 3-7 d after transplanting. It also shows promise for the control of weeds and annual grasses in cotton, soyabeans, groundnuts, potatoes, green beans, beets and tomatoes.

Toxicology. The acute oral LD50 is: for rats 1300 mg/kg; for mice 560 mg/kg. The acute dermal LD50 for rats is 2900 mg/kg. The LC50 (48 h) for carp is 3·6 mg/l.

Formulations. These include: e.c. (500 g/kg), 'Bolero' 4 e.c. and 8 e.c. (480 and 960 g/l); 'Bolero' 5% and 10% granules (50 and 100 g/kg); 'Saturn', in combination with simetryne.

Analysis. Product and residue analysis is by GLC with FID and benzyl benzoate as internal standard (K. Kejima *et al., U.S.A.-Japan Seminar Envir. Toxicol. Pestic.*, 1971; K. Ishikawa *et al., Agric. Biol. Chem.*, 1971, **35**, 1161; S. K. De Datta *et al., Weed Res.*, 1971, **11**, 41; Y. Ishii, *Jpn. Pestic. Inf.*, 1974, No. 19, p.21).

Thiocyclam

T6SSSTJ EN1&1   *thiocyclam* $C_5H_{11}NS_3$ (181·3)

T6SSSTJ EN1&1 &QVVQ   *thiocyclam hydrogen oxalate* $C_7H_{13}NO_4S_3$ (271·4)

Nomenclature and development. The common name thiocyclam is approved by BSI and New Zealand and proposed by ISO. The IUPAC name is *N,N*-dimethyl-1,2,3-trithian-5-ylamine, in *C.A.* usage *N,N*-dimethyl-1,2,3-trithian-5-amine *[31895-21-3]*. Its hydrogen oxalate, Registry No. *[31895-22-4]*, was introduced in 1975 by Sandoz AG under the code number 'SAN 155 I', trade marks 'Evisect' and 'Evisekt' and protected by DOS 2 039 555. Its insecticidal properties were described by W. Berg & H. G. Knutti, *Proc. Br. Insectic. Fungi. Conf., 8th*, 1975, **2**, 683.

Manufacture and properties. Produced by the reaction of 2-dimethylaminotrimethylene-bis(sodium thiosulphonate) with aqueous disodium sulphide, the reaction mixture is extracted with an organic solvent and thiocyclam hydrogen oxalate precipitated by adding oxalic acid. The hydrogen oxalate is a colourless crystalline solid, m.p. 125-128° (decomp.). Its solubility at 23° is: 84 g/l water; < 10 g/l acetone, diethyl ether, ethanol, xylene; practically insoluble in kerosine. It is stable under ambient storage conditions (protected from light); 50% decomposition occurs at 22-24° in buffered aqueous solutions (40 mg/l) in: 42 d at pH 5, and 11 d at pH 7-9.

Uses. Thiocyclam is a selective insecticide acting as stomach poison and by contact with a 7-14 d residual activity and capable of acropetal translocation. It is effective against lepidopterous and coleopterous pests, particularly *Leptinotarsa decemlineata* larvae on potatoes at 250-400 g a.i./ha.

Toxicology. The acute oral LD50 is: for male rats 310 mg a.i./kg; for male mice 273 mg/kg. In 90-d feeding trials the 'no effect' level was: for rats 100 mg/kg diet; for dogs 75 mg/kg diet. The LC50 (96 h) for carp is 1·03 mg/l. It is moderately toxic to honey bees.

Formulations. 'Evisect' s.p., w.s.p. (900 g tech./kg); 'Evisect', granules (50 g tech./kg).

Analysis. Product analysis is by HPLC with u.v. detection or by TLC with u.v. spectrometry of the eluted substance. Residues may be determined by GLC. Details may be obtained from Sandoz AG.

Thiofanox

$$
\begin{array}{c}
\text{O} \\
\| \\
\text{NOCNHMe} \\
\| \\
\text{But} \!-\! \text{C}.\text{CH}_2.\text{SMe}
\end{array}
$$

1X1&1&Y1S1&UNOVM1 $C_9H_{18}N_2O_2S$ (218·3)

Nomenclature and development. The common name thiofanox is approved by BSI and ANSI and proposed by ISO. The IUPAC name is 3,3-dimethyl-1-(methylthio)butanone *O*-methylcarbamoyloxime, in *C.A.* usage 3,3-dimethyl-1-(methylthio)-2-butanone *O*-[(methylamino)carbonyl]oxime *[39196-18-4]*. It was introduced in 1971 by Diamond Shamrock Chemical Co. under the code number 'DS 15 647' and trade mark 'Dacamox'. Its insecticidal properties were described by R. L. Schauer, *Proc. Br. Insectic. Fungic. Conf., 7th*, 1973, **2**, 713.

Properties. It is a colourless solid, with a pungent odour, m.p. 56·5-57·5°, v.p. 1·7 x 10⁻⁴ mmHg at 25°. Its solubility at 22° is: 5·2 g/l water; very soluble in chlorinated and aromatic hydrocarbons, ketones and apolar solvents; sparingly soluble in aliphatic hydrocarbons. It is stable at normal storage temperatures and reasonably stable to hydrolysis < 30° at pH 5-9.

Uses. Thiofanox is a systemic insecticide and acaricide, used as a soil or seed treatment and giving protection from foliage-feeding and soil pests. In-furrow applications of: 0·4-1·0 kg a.i./ha to sugar beet controlled *Aphis fabae, Myzus persicae, Pegomya betae, Chaetocnema concinna* and *Atomaria linearis*; 1-3 kg/ha to potatoes controlled *Macrosiphum* spp., *Empoasca fabae, Epitrix* spp. and *Leptinotarsa decemlineata*.

The relationship between chemical structure and biological activity of analogues has been examined. (T. A. Magee & L. E. Limpel, *J. Agric. Food. Chem.*, 1977, **25**, 1376) and the degradation of thiofanox studied (W. T Chin *et al., ibid.*, 1976, **24**, 1001, 1071).

Toxicology. The acute oral LD50 is: for albino rats 8·5 mg/kg; for mallard duck 109 mg/kg; for bobwhite quail 43 mg/kg. The acute dermal LD50 for albino rabbits is 39 mg/kg. In 90-d feeding trials: the weight gain of rats receiving 100 mg/kg diet was not affected; the 'no effect' level for beagle dogs was 1·0 mg/kg daily. The LC50 (96 h) is: for bluegill 0·33 mg/l; for rainbow trout 0·13 mg/l.

Formulations. These include: granules, 'Dacamox 5G', 'Dacamox 10G', 'Dacamox 15G' (50, 100 and 150 g a.i./kg, respectively); seed treatment, 'Dacamox ST' (429 g/l).

Analysis. Residues, which are likely to include thiofanox, its sulphide and sulphone, are extracted and oxidised to the sulphone, which is determined by GLC giving the total residue (*idem, ibid.*, 1975, **23**, 963; M. B. Szalkowski *et al., J. Chromatogr.*, 1976, **128**, 426). Details are available from Diamond Shamrock Corp.

Thiometon

$$\underset{\parallel}{\overset{S}{(MeO)_2P.S.CH_2CH_2.SEt}}$$

2S2SPS&O1&O1 $C_6H_{15}O_2PS_3$ (246·3)

Nomenclature and development. The common name thiometon is approved by BSI, ISO and JMAF—exceptions France (dithiométhon), USSR (M-81), West Germany, Portugal and Turkey. The IUPAC name is *S*-2-ethylthioethyl *O,O*-dimethyl phosphorodithioate (I), in *C.A.* usage *S*-[2-(ethylthio)ethyl] *O,O*-dimethyl phosphorodithioate *[640-15-3]* formerly (I). It was first prepared in 1952 by W. Lorenz & G. Schrader (DBP 917 668) and independently by K. Lutz *et al.* (SwissP 319 579). It was introduced in 1953 under the code number 'Bayer 23 129' (Bayer AG, who no longer manufacture or market it) and the trade mark 'Ekatin' (Sandoz AG).

Manufacture and properties. Produced either by the reaction of 2-chloroethyl ethyl sulphide with sodium *O,O*-dimethyl phosphorodithioate, or by the reaction of 2-ethylthioethanol with *O,O*-dimethyl hydrogen phosphorodithioate and *p*-toluenesulphonyl chloride in the presence of an acid binding agent, it is a colourless oil, with a characteristic odour, b.p. 110°/0·1 mmHg, v.p. 2 x 10^{-4} mmHg at 20°, d_4^{20} 1·209, n_D^{20} 1·5515. Its solubility at 25° is 200 mg/l water; soluble in most organic solvents; sparingly soluble in light petroleum. The pure compound is unstable, but it is stable in non-polar solvents. It is hydrolysed in aqueous solution.

Uses. Thiometon is a systemic insecticide and acaricide which controls sucking insects, mainly aphids, and mites on most crops. It is metabolised by plants to the water-soluble sulphoxide and sulphone and the phosphorothioate analogues (related to demeton-S-methyl) (R. L. Metcalf & T. R. Fukuto, *J. Agric. Food Chem.*, 1954, **2**, 732).

Toxicology. The acute oral LD50 and the acute dermal LD50 for rats are 120-130 mg/kg and > 1000 mg/kg respectively.

Formulations. These include: 'Ekatin', e.c. (245 g a.i./l, with added blue dye and anti-odour adjuvant); 'Ekatin ULV', concentrate (125 g/l); 'Ekatin' WF, e.c. (250 g thiometon + 100 g parathion/kg); 'Ekatin WF ULV', concentrate (150 g thiometon + 60 g parathion/l).

Analysis. Product analysis is by i.r. spectroscopy or by paper chromatography followed by combustion and determination of phosphate by standard colorimetric methods (*CIPAC Handbook*, 1979, **1A**, in press), or by GLC. Residues may be determined by GLC or, after suitable clean-up, by conversion to phosphate and measurement by standard colorimetric methods. See also: M. Wisson *et al.*, *Anal. Methods Pestic. Plant Growth Regul.*, 1976, **8**, 239.

Thionazin

T6N DNJ BOPS&O2&O2 $C_8H_{13}N_2O_3PS$ (248·2)

Nomenclature and development. The common name thionazin is approved by BSI and proposed by ISO; also known as ENT 25 580. The IUPAC name is *O,O*-diethyl *O*-pyrazin-2-yl phosphorothioate, in *C.A.* usage *O,O*-diethyl *O*-2-pyrazinyl phosphorothioate *[297-97-2]*. It was introduced in 1958 by American Cyanamid Co. as 'Experimental Nematocide 18 133' the trade marks 'Nemafos', 'Zinophos' and, in the USA only, 'Cynem' and protected by USP 2 918 468; 2 938 831; 3 091 614.

Manufacture and properties. Produced by the reaction of *O,O*-diethyl phosphorochloridothioate with the sodium salt of pyrazin-2-ol in an aqueous medium, it is a colourless to pale yellow liquid, m.p. —1·67°, v.p. 3 x 10^{-3} mmHg at 30°, d^{25}1·207, n_D^{25} 1·5080. Its solubility at 27° is 1·14 g/l water; miscible with most organic solvents. The technical product is a dark brown liquid *c.* 90% pure, d^{25} 1·204-1·210. It is readily hydrolysed by alkali.

Uses. Thionazin is a soil insecticide and nematicide effective against a number of plant-parasitic as well as free-living nematodes, including those attacking bulbs, buds, leaves and roots, as well as soil pests such as symphylids and root maggots and foliar insects such as aphids and leaf miners. It is of short persistence (M. J. Way & N. E. A. Scopes, *Ann. Appl. Biol.,* 1965, **55,** 340). When incorporated in mushroom compost at spawning it is effective against mushroom flies.

Toxicology. The acute oral LD50 and acute dermal LD50 for rats are 12 mg/kg and 11 mg/kg, respectively. In 90-d feeding trials, rats receiving 25 mg/kg diet for 30 d and 50 mg/kg diet for the remaining 60 d showed a moderate depression in growth rate, but no abnormal behavioural reactions.

Formulations. An e.c. (460 g. a.i./l); granules (50 and 100 g/kg).

Analysis. Product analysis is by hydrolysis to produce the sodium salt of pyrazin-2-ol measured fluorimetrically (excitation at 315 nm, emission at 375 nm) (U. Kiigemagi *et al., J. Agric. Food Chem.,* 1963, **11,** 293). Residues may be determined by GLC: D. R. Coahran, *Bull. Environ. Contam. Toxicol.,* 1966, **1,** 208.

Thiophanate

2OVMYUS&MR BMYUS&MVO2 $C_{14}H_{18}N_4O_4S_2$ (370·4)

Nomenclature and development. The common name thiophanate is approved by BSI, ISO, JMAF and BPC—exception France (thiophanate-éthyl). The IUPAC name is diethyl 4,4'-(o-phenylene)bis(3-thioallophanate) (I) formerly 1,2-di-(3-ethoxycarbonyl-2-thioureido)benzene, in *C.A.* usage diethyl [1,2-phenylenebis(iminocarbonothioyl)]-biscarbamate *[23564-06-9]* formerly (I). It was introduced in 1969 by Nippon Soda Co. Ltd under the code number 'NF 35', the trade marks 'Topsin' and 'Cercobin' and protected by DBP 1 930 540. The methyl homologue, see thiophanate-methyl (p.518), is now more widely used. The fungicidal properties of these 2 compounds were described by K. Ishii, *Abstr. Int. Congr. Plant Prot., 7th, Paris,* 1970, p.200; E. Aelbers, *Meded. Fac. Landbouwwet. Rijksuniv., Gent,* 1971, **36,** 126. Thiophanate is also used an an anthelmintic, 'Nemafax'.

Manufacture and properties. Produced by the condensation of o-phenylenediamine with ethyl isothiocyanatoformate, obtained by reaction of ethyl chloroformate with potassium thiocyanate, it is a stable colourless crystalline solid, m.p. 195° (decomp.). It is almost insoluble in water; sparingly soluble in most organic solvents. With aqueous alkali it forms unstable solutions of salts and with divalent transition metal ions, *e.g.* copper, forms a complex.

Uses. Thiophanate is a systemic fungicide with a broad range of action, effective against *Venturia* spp. on apple and pear crops, powdery mildews, *Botrytis* spp. and *Sclerotinia* spp. on various crops. In plants it is converted into ethyl benzimidazol-2-ylcarbamate.

Toxicology. The acute oral and acute percutaneous toxicity to mice, rats, guinea-pigs and rabbits is > 10 000 mg/kg.

Formulations. A w.p. (500 g a.i./kg).

Analysis. Product analysis is by u.v. spectrometry at 260 nm.

1OVMYUS&MR   BMYUS&MVO1 $C_{12}H_{14}N_4O_4S_2$ (342·4)

Nomenclature and development. The common name thiophanate-methyl is approved by BSI, ISO, ANSI and JMAF. The IUPAC name is dimethyl 4,4'-(o-phenylene)bis(3-thioallophanate) (I) formerly 1,2-di-(3-methoxycarbonyl-2-thioureido)benzene, in *C.A.* usage dimethyl [1,2-phenylenebis(iminocarbonothioyl)]biscarbamate *[23564-05-8]* formerly (I). It was introduced in 1970-71 as a replacement for thiophanate (p.517) by the Nippon Soda Co. Ltd under the code number 'NF 44' and trade marks 'Topsin' Methyl, 'Cercobin' Methyl, 'Mildothane' (May & Baker Ltd), 'Cycosin' (American Cyanamid Co.). Thiophanate-methyl has replaced thiophanate in many uses as it has greater fungicidal activity and greater systemic movement. Their fungicidal properties were described by K. Ishii, *Abstr. Int. Congr. Plant Prot., 7th, Paris,* 1970, p.200; E. Aelbers, *Meded. Fac. Landbouwwet. Rijksuniv., Gent,* 1971, **36,** 126.

Manufacture and properties. Produced by the condensation of o-phenylenediamine with methyl isothiocyanatoformate, obtained by reaction of methyl chloroformate with potassium thiocyanate, it is a stable colourless crystalline solid, m.p. 172° (decomp.). Its solubility in water is very low; sparingly soluble in most organic solvents. It forms complexes with copper salts. It is compatible with other agricultural chemicals which are neither highly alkaline nor contain copper. Thiophanate-methyl has been shown to break down on prolonged storage of aqueous suspensions and in plant tissues to carbendazim (p.80). It, therefore, has activities similar to the fungicides in this group of compounds.

Uses. Thiophanate-methyl is a fungicide used at 30-50 g a.i./100 l and is effective against a wide range of fungal pathogens including: *Venturia* spp. on apples and pears; *Mycosphaerella musicola* on bananas; powdery mildews *(Podosphaera* spp., *Erysiphe* spp. and *Sphaerotheca fuliginea)* on apples, pears, small grains, vines and cucurbits; *Piricularia oryzae* on rice; leaf spot diseases caused by *Cercospora* spp., *Botrytis* spp. and *Sclerotinia* spp. on various crops.

Toxicology. The acute dermal LD50 is: for rats > 6000 mg/kg; for mice > 3000 mg/kg. The acute dermal LD50 for mice, rats, guinea-pigs and rabbits is > 10000 mg/kg.

Formulations. These include: w.p. (500 and 700 g a.i./kg); mixtures with other fungicides.

Analysis. Product analysis is by u.v. spectrometry at 269 nm. Residues may be determined by colorimetry (V. L. Miller *et al., J. Assoc. Off. Anal. Chem.,* 1977, **60,** 1154).

Thiram

$$\underset{Me_2N.C.S.S.C.NMe_2}{\overset{\displaystyle S \quad\quad S}{\overset{\displaystyle \| \quad\quad \|}{}}}$$

1N1&YUS&SSYUS&N1&1 $C_6H_{12}N_2S_4$ (240·4)

Nomenclature and development. The common name thiram is approved by BSI and ISO—exceptions USSR (TMTD) and JMAF (thiuram); also known as ENT 987. The IUPAC name is tetramethylthiuram disulphide formerly bis(dimethylthiocarbamoyl) disulphide (I), in *C.A.* usage tetramethylthioperoxydicarbonic diamide $[(Me_2N)C(S)]_2S_2$ *[137-26-8]* formerly (I). Its fungicidal properties were described by W. H. Tisdale & A. L. Flenner, *Ind. Eng. Chem.,* 1942, **34,** 501. It was introduced in 1931 by E. I. du Pont de Nemours & Co. (Inc.) under the trade marks 'Arasan' and 'Tersan' and protected by USP 1972961; and by Bayer AG under the trade mark 'Pomarsol' and protected by DRP 642532 (I. G. Farbenindustrie AG); other trade marks include 'Ferna-Col', 'Fernasan' (ICI Ltd, Plant Protection Division).

Manufacture and properties. Produced by the oxidation, using hydrogen peroxide or iodine, of sodium dimethyldithiocarbamate, which is obtained by the interaction of carbon disulphide and dimethylamine in the presence of sodium hydroxide, it forms colourless crystals, m.p. 155-156°, v.p. negligible at room temperature, d^{20} 1·29. Its solubility at room temperature is *c.* 30 mg/l water; slightly soluble in ethanol, diethyl ether; soluble in acetone, chloroform.

Uses. Thiram is a protective fungicide suitable for application to foliage or fruit to control: *Botrytis* spp. on soft fruit, vegetables and ornamentals; *Bremia lactucae* on lettuces; rusts on ornamentals; *Venturia pirina* on pears. Also as a seed treatment, sometimes with added insecticide or other fungicide, to control 'damping off' diseases of vegetables, maize and ornamentals. It is non-phytotoxic.

Toxicology. The acute oral LD50 for mammals is 375-865 mg/kg. It may cause skin irritation. Hens receiving 35 mg/kg diet suffered a severe drop in egg production (P. E. Waibel *et al., Science.* 1955, **121,** 401). At high doses it is repellent to mice in the field. The related tetraethylthiuram disulphide, disulfiram ('Antabuse'), blocks the in-vivo oxidation of ethanol at the acetaldehyde stage and has been used for the treatment of alcoholism.

Formulations. These include: w.p. ($\leqslant 800$ g a.i./kg); colloidal suspension, 'Ferna-Col'; seed treatments, 'Fernasan S' (slurry), 'Benlate T' (benomyl + thiram), also mixtures with HCH.

Analysis. Product analysis is either by reduction in alkaline solution followed by acid hydrolysis of the dithiocarbamate produced and colorimetric estimation of the carbon disulphide liberated, using standard methods; or by reaction with acetic acid and zinc oxide followed by distillation of the dimethylamine which is estimated by acid-base titration (*CIPAC Handbook,* 1970, **1,** 672). Residues may be determined by conversion to a dimethyldithiocarbamate copper complex (AOAC Methods), or by microcoulometry (L. C. Butler & D. C. Staiff, *J. Agric. Food Chem.,* 1978, **26,** 295).

Tolylfluanid

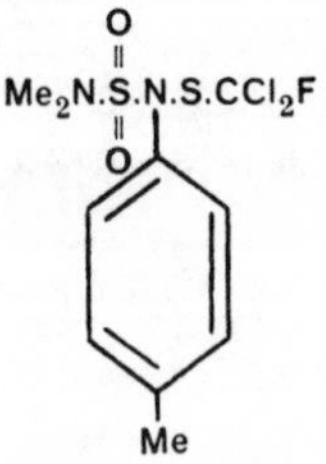

$$O$$
$$\|$$
$$Me_2N.S.N.S.CCl_2F$$
$$\|$$
$$O$$

Me

GXGFSNR D1&SWN1&1 $C_{10}H_{13}Cl_2FN_2O_2S_2$ (347·2)

Nomenclature and development. The common name tolylfluanid is approved by BSI and ISO. The IUPAC name is *N*-dichlorofluoromethylthio-*N',N'*-dimethyl-*N*-p-tolylsulphamide formerly *N*-dichlorofluoromethanesulphenyl-*N',N'*-dimethyl-*N*-p-tolylsulphamide, in *C.A.* usage 1,1-dichloro-*N*-[(dimethylamino)sulfonyl]-1-fluoro-*N*-(4-methylphenyl)methane-sulfenamide *[737-27-1]* formerly *N*-[(dichlorofluoromethyl)thio]-*N',N'*-dimethyl-*N*-p-tolylsulfamide. It was introduced in 1971 under the code number 'Bayer 49 854', trade mark 'Euparen M' and protected by DBP 1 193 498. The properties of tolylfluanid and related compounds, *e.g.* dichlofluanid (p.168), were described by E. Kühle *et al., Angew. Chem.,* 1964, **76,** 807.

Properties. It is a colourless to pale yellow powder, m.p. 95-97°, v.p. 10^{-5} mmHg at 45°. Its solubility at room temperature is: 4 g/l water; 46 g/l methanol; 570 g/l benzene; 230 g/l xylene.

Uses. Tolylfluanid is a protective fungicide with a broad range of activity (H. Kaspers & F. Grewe, *Abstr. Int. Congr. Crop Prot., 6th,* Vienna, 1967, p.345). It is mainly used in deciduous fruit crops and, when used regularly for control of *Venturia* spp., there is usually no need for additional steps to control *Phodosphaera leucotricha* on apples; or red spider mites (W. Kolbe, *Pflanzenschutz-Nachr. (Am. Ed.),* 1972, **25,** 123; R. Wäckers & C. van der Berge, *ibid.,* p.163).

Toxicology. The acute oral LD50 is: for rats > 1000 mg/kg; for male mice > 1100 mg/kg; for female guinea-pigs 250-500 mg/kg; for male rabbits 500 mg/kg; for female canaries 1000 mg/kg. The dermal LD50 for male rats is > 500 mg/kg. In 90-d feeding trials rats receiving 1000 mg/kg diet showed no effect.

Formulation. A w.p. (500 g a.i./kg).

Analysis. Product and residue analysis is by GLC, details are available from Bayer AG.

m-Tolyl methylcarbamate

1MVOR C1   C₉H₁₁NO₂ (165·2)

Nomenclature and development. The common name MTMC is approved by JMAF. The IUPAC name is *m*-tolyl methylcarbamate (I), in *C.A.* usage 3-methylphenyl methyl-carbamate *[1129-41-5]* formerly (I). It was introduced as an insecticide by the Nihon Nohyaku Co. Ltd under the code number 'C-3' and trade mark 'Tsumacide', and by the Sumitomo Chemical Co. Ltd under the trade mark 'Metacrate'.

Manufacture and properties. Produced either by the condensation of *m*-tolyl chloroformate with methylamine or by the reaction of *m*-cresol with methyl isocyanate, it is a colourless crystalline solid, m.p. 76-77°. Its solubility at 30° is 2·6 g/l water; soluble in most organic solvents. The technical grade is *c.* 95% pure.

Uses. *m*-Tolyl methylcarbamate is a systemic insecticide, also active in the vapour phase, mainly used to control leaf-hoppers on rice.

Toxicology. The acute oral LD50 for mice is 268 mg/kg; the acute dermal LD50 for rats is 6000 mg/kg.

Formulations. These include: e.c. (300 g a.i./l); w.p. (500 g/kg); dust (15 g/kg).

Analysis. Product analysis is by GLC. Residues may be determined, after conversion to 2,4-dinitrophenyl *m*-tolyl ether, by GLC with ECD (J. Miyamoto *et al., J. Pestic. Sci.,* 1978, **3,** 119).

3-*o*-Tolyloxypyridazine

T6NNJ COR B1 $C_{11}H_{10}N_2O$ (186·2)

Nomenclature and development. The common name credazine is approved by JMAF. The IUPAC name is 3-*o*-tolyloxypyridazine (I), in *C.A.* usage 3-(2-methylphenoxy)pyridazine *[14491-59-9]* formerly (I). It was introduced in 1970 by Sankyo Chemical Ltd under the code numbers 'H722', 'SW-6701' and 'SW-6721', the trade mark 'Kusakira' and protected by JapaneseP 509 596. Its use as a herbicide was first described by T. Jojima *et al., Agric. Biol. Chem.,* 1968, **32,** 1376; 1969, **33,** 96.

Properties. It forms colourless needles, m.p. 78°. Its solubility at room temperature is 2 g/l water; readily soluble in organic solvents. It is stable to light, heat and humidity and is non-corrosive to iron and stainless steel.

Uses. 3-*o*-Tolyloxypyridazine is a selective herbicide for soil application and is used, in Japan, for the control of annual grasses and some broad-leaved weeds in tomatoes, pimentoes and strawberries at 2-3 kg a.i./ha. It is moderately persistent in soil (S. Yoshimura *et al., Zasso Kenkyu,* 1969, No. 7, p.84). In plants the major metabolites are (2*H*)-pyridazin-3-one and 3-(1-β-D-glucosyloxy)pyridazine (M. Nakagawa *et al., Agric. Biol. Chem.,* 1971, **25,** 764).

Toxicology. The acute oral LD50 is: for rats 3090 mg/kg; for mice 569 mg/kg. The acute dermal LD50 for mice is > 10000 mg/kg. In 90-d feeding trials the 'no effect' level was: for rats 16·5 mg/kg daily; for mice 42 mg/kg daily. The LC50 (48 h) for carp is 62 mg/l.

Formulation. A w.p. (500 g a.i./kg).

Analysis. Product analysis is by GLC. Residues are determined by GLC after separation by TLC.

Triadimefon

T5NN DNJ AYOR DG&VX1&1&1 $\qquad$ $C_{14}H_{16}ClN_3O_2$ (293·8)

Nomenclature and development. The common name triadimefon is approved by BSI and proposed by ISO—exception Canada. The IUPAC name is 1-(4-chlorophenoxy)-3,3-dimethyl-1-(1,2,4-triazol-1-yl)butanone, in *C.A.* usage 1-(4-chlorophenoxy)-3,3-dimethyl-1-(1*H*-1,2,4-triazol-1-yl)-2-butanone *[43121-43-3]*. It was introduced in 1974 by Bayer AG under the code number 'BAY MEB 6447', trade mark 'Bayleton' and protected by BelgP 793867; USP 3912752. Its fungicidal activity was first reported by P. E. Frohberger, *Mitt. Biol. Bundesanst. Land- Forstwirtsch. Berlin-Dahlem,* 1973, **151,** 61; F. Grewe & K. H. Büchel, *ibid.,* p.208.

Properties. It is a colourless solid, m.p. 82·3°, v.p. $< 7.5 \times 10^{-10}$ mmHg at 20°. Its solubility at 20° is: 260 mg/l water; 0·6-1·2 kg/kg cyclohexanone; 1·2 kg/kg dichloromethane; 200-400 g/kg propan-2-ol; 400-600 g/kg toluene. It is stable at pH 1 and 13.

Uses. Triadimefon is a systemic fungicide with protective and curative action. It is effective against mildews and rusts attacking vegetables, cereals, coffee, stone fruit, grapes and ornamentals at 2·5-6·2 g a.i./100 l (H. Buchenauer, *Pflanzenschutz-Nachr. (Am. Ed.),* 1976, **29,** 266; R. Siebert, *ibid.,* p.303; W. Kolbe, *ibid.,* p.310). Its mode of action has been investigated (H. Buchenauer, *ibid.,* p.281; *idem., Pestic. Sci.,* 1978, **9,** 497) and the formation of the corresponding alcohol, the fungicide triadimenol, demonstrated in plants (T. Clark *et al., ibid.,* p.503; H. Buchenauer, *Pflanzenschutz-Nachr. (Am. Ed.),* 1976, **29,** 266).

Toxicology. The acute oral LD50 is: for rats 363-568 mg/kg; for mice 989-1071 mg/kg; for Japanese quail 1750-2500 mg/kg. The acute dermal LD50 for rats is > 1000 mg/kg. In 90-d feeding trials the 'no effect' level was: for rats 2000 mg/kg diet; for dogs 600 mg/kg diet. The LC50 (96 h) for goldfish is 10-50 mg/l. It is non-toxic to honey bees.

Formulations. These include: w.p. (50 and 250 g a.i./kg); e.c. (100 g/l); dust (10 g/kg).

Analysis. Product analysis is by i.r. spectroscopy. Residues of triadimefon and the corresponding alcohol (triadimenol) may be determined by GLC (W. Sprecht, *ibid.,* 1977, **30,** 55).

Tri-allate

$$O \atop (Me_2CH)_2N.C.S.CH_2C=CCl_2 \atop Cl$$

GYGUYG1SVNY1&1&Y1&1 $C_{10}H_{16}Cl_3NOS$ (304·7)

Nomenclature and development. The common name tri-allate is approved by BSI and ISO; by WSSA and France (triallate). The IUPAC name is *S*-2,3,3-trichloroallyl di-isopropylthiocarbamate (I), in *C.A.* usage *S*-(2,3,3-trichloro-2-propenyl) bis(1-methyl-ethyl)carbamothioate *[2303-17-5]* formerly (I). It was introduced in 1961 by Monsanto Co. under the code number 'CP 23 426', trade marks 'Avadex BW' and 'Fargo' and protected by USP 3 330 821. Herbicides employing this compound are protected by USP 3 330 642. Its herbicidal properties were described by G. Friesen, *Res. Proc. Nat. Weed Comm., Can.,* 1960.

Manufacture and properties. Produced by the condensation of di-isopropylamine, carbonyl sulphide and 1,1,2,3-tetrachloroprop-1-ene in the presence of a base, it is an oil, m.p. 29-30°, b.p. *c.* 148-149°/9 mmHg. Its solubility at 25° is 4 mg/l water; soluble in most organic solvents. It is non-flammable and non-corrosive.

Uses. Tri-allate is a herbicide of particular value for the control of wild oats in cereals and peas. It is incorporated into the soil at 1·0-1·5 kg a.i./ha as a pre-sowing or pre-em. treatment. It is also used in granular form for pre- or post-em. control of wild oats.

Toxicology. The acute oral LD50 for rats is 1675-2165 mg/kg; oral doses of 20 000 mg/kg caused no symptom in dogs. The dermal LD50 for rabbits is 2225-4050 mg/kg. No irritation was caused to human skin.

Formulations. These include: e.c. (400 and 480 g a.i./l); granules (100 g/kg).

Analysis. Product analysis is by GLC. Residues in soil, grain or straw may be determined by GLC with ECD (T. H. Byast *et al., Agric. Res. Counc. (G.B.) Weed Res. Organ., Tech. Rep.,* No. 15 (2nd Ed.), p.17).

Triamiphos

T5NN DNJ APO&N1&1&N1&1 CR& EZ $C_{12}H_{19}N_6OP$ (294·3)

Nomenclature and development. The common name triamiphos is approved by BSI and ISO; also known as ENT 27223. The IUPAC name is *P*-5-amino-3-phenyl-1,2,4-triazol-1-yl-*N,N,N',N'*-tetramethylphosphonic diamide, formerly 5-amino-1-(bisdimethylamino-phosphinyl)-3-phenyl-1,2,4-triazole, in *C.A.* usage *P*-(5-amino-3-phenyl-1*H*-1,2,4-triazol-1-yl)-*N,N,N',N'*-tetramethylphosphonic diamide *[1031-47-6]*. It was first described by B. G. van den Bos *et al., Rec. Trav. Chim. Pays-Bas,* 1960, **79**, 807. It was introduced in 1960 by Philips-Duphar B.V. under the code number 'WP 155', trade mark 'Wepsyn 155' and protected by DutchP 109 510; USP 3 121 090; 3 220 922.

Manufacture and properties. Produced by the reaction of the sodium salt of 5-phenyl-1,2,4-triazol-3-ylamine with *N,N,N',N'*-tetramethylphosphorodiamidic chloride, it is a colourless solid, m.p. 167-168°. Its solubility at 20° is 250 mg/l water; moderately to easily soluble in most organic solvents. It is stable at room temperature under neutral or slightly alkaline conditions but is readily hydrolysed by concentrated mineral acids. The technical product is ⩾ 95% pure, m.p. 166-170°. It is non-corrosive and compatible with common w.p. formulations of other pesticides.

Uses. Triamiphos is a fungicide for powdery mildew control and shows some systemic activity; it also has systemic insecticidal and acaricidal properties. Rates for powdery mildew control include: for apples 25 g a.i. (as w.p.)/100 l every 10 d; for roses 25 g a.i. (as water-miscible)/100 l. At these concentrations, it is non-phytotoxic and presents no hazard to wild life.

Toxicology. The acute oral LD50 for male rats is 20 mg/kg; the acute dermal LD50 for rabbits is 1500-3000 mg/kg. Honey bees tolerate 10 mg/kg diet.

Formulations. A w.p. (250 g a.i./kg); water-miscible (100 g/l).

Analysis. Product and residue analysis is by the colorimetric determination of 5-phenyl-1,2,4-triazol-3-ylamine formed on acid hydrolysis.

Triazophos

$$\text{(EtO)}_2\overset{\overset{\text{S}}{\|}}{\text{P}}\text{·O}$$

T5NN DNJ AR& COPS&O2&O2 $C_{12}H_{16}N_3O_3PS$ (313·3)

Nomenclature and development. The common name triazophos is approved by BSI and ISO. The IUPAC name is *O,O*-diethyl *O*-1-phenyl-1,2,4-triazol-3-yl phosphorothioate, in *C.A.* usage *O,O*-diethyl *O*-(1-phenyl-1*H*-1,2,4-triazol-3-yl) phosphorothioate *[24017-47-8]*. It was introduced in 1970 by Hoechst AG under the code number 'Hoe 2960', trade mark 'Hostathion'. It was described by M. Vulic *et al., Abstr. Int. Congr. Plant Prot., 7th,* Paris, 1970, p.123; S. J. B. Hay, *Proc. Br. Insectic. Fungic. Conf., 6th,* 1971, **2,** 597.

Manufacture and properties. Produced by the reaction of 1-phenyl-1,2,4-triazol-3-ol with *O,O*-diethyl phosphorochloridothioate, it is a light brown-yellowish liquid, m.p. 0-5°, v.p.2·9 x 10^{-6} mmHg at 30°. Its solubility at 23° is 39 mg/l water; soluble in most organic solvents. The technical grade is ≥ 92% pure but is available only in diluted form.

Uses. Triazophos is a broad spectrum insecticide and acaricide with some nematicidal properties. It controls aphids on fruit at 75-125 g a.i./100 l, on cereals at 320-600 g a.i. (40% e.c.)/ha. It is also used to control Lepidoptera on fruit and vegetables. When incorporated at 1-2 kg a.i. (40% e.c.)/ha in the soil prior to planting, it controls *Agrotis* spp. and other cutworms. Although it can penetrate plant tissue, it has no systemic activity.

Toxicology. The acute oral LD50 is: for rats 66 mg tech./kg; for dogs 320 mg/kg. The acute dermal LD50 for rats is 1100 mg/kg. In 2-y feeding trials rats receiving 1 mg/kg diet and dogs 0·3 mg/kg diet, the only effect noted was blood serum cholinesterase inhibition. The LC50 (96 h) is: for carp 5·6 mg/l; for golden orfe 11 mg/l.

Formulations. These include: e.c. (246 and 420 g a.i./l); granules (20 and 50 g/kg); ULV concentrates (242 and 408 g/l); w.p. (300 g/kg).

Analysis. Product and residue analysis is by GLC, details are available from Hoechst AG. See also: W. G. Thier *et al., Anal. Methods Pestic. Plant Growth Regul.,* 1978, **10,** 127.

S,S,S-Tributyl phosphorotrithioate

(BuS)$_3$PO

4SPO&S4&S4 $C_{12}H_{27}OPS_3$ (314·5)

Nomenclature and development. The IUPAC and *C.A.* name is *S,S,S*-tributyl phosphoro-trithioate, Registry No. *[78-48-8]*. It was tested as a defoliant by the Ethyl Corp. and introduced in 1956 by the Chemagro Corp. under the code number 'Chemagro B-1776', trade mark 'Def Defoliant' and protected by USP 2 943 107; 2 965 467.

Manufacture and properties. Produced by the reaction of butane-1-thiol on phosphoroyl chloride in the presence of a base, it is a pale yellow liquid, with a mercaptan-like odour, b.p. 150°/0·3 mmHg, m.p. < —25°, d^{20} 1·06, n_D^{25} 1·532. It is practically insoluble in water; soluble in most organic solvents, including chlorinated hydrocarbons. It is relatively stable to heat and to acids but is slowly hydrolysed under alkaline conditions.

Uses. It is strongly phytotoxic and is recommended for the defoliation of cotton: for complete defoliation 1·25-2·0 kg a.i./ha, for bottom defoliation 1·0-1·5 kg/ha.

Toxicology. The acute oral LD50 for female rats is 325 mg/kg (S. D. Murphy & K. P. DuBois, *AMA Arch. Ind. Health*, 1959, **20**, 161); the acute dermal LD50 for male rats is 850 mg/kg. In 84-d feeding trials dogs receiving 25 mg/kg diet showed no ill-effect.

Formulation. Spray concentrate (700 g a.i./l).

Analysis. Product analysis is by i.r. spectroscopy at 8·3 µm. Residues in cotton seed may be determined by GLC (R. F. Thomas & T. H. Harris, *J. Agric. Food Chem.*, 1965, **13**, 505). Details of methods are available from Chemagro Corp. See also: D. MacDougall, *Anal. Methods Pestic., Plant Growth Regul. Food Addit.*, 1964, **4**, 89; *Anal. Methods Pestic. Plant Growth Regul.*, 1972, **6**, 627.

Tributyl phosphorotrithioite

(BuS)₃P

4SPS4&S4 $\qquad$ $C_{12}H_{27}PS_3$ (298·5)

Nomenclature and development. The IUPAC and *C.A.* name is tributyl phosphorotrithioite, Registry No. *[150-50-5]*. It was introduced in 1957 by the Mobil Chemical Co. under the trade mark 'Folex' and, in the USA, 'Merphos' and protected by USP 2 955 803.

Manufacture and properties. Produced by the reaction of butane-1-thiol and phosphorus trichloride in a solvent, it is a colourless to pale yellow liquid, b.p. 115-134°/0·08 mmHg, d_4^{20} 0·99-1·01, n_D^{25} 1·55. It is sparingly soluble in water; very soluble in most organic solvents. The technical product is $\geqslant 95\%$ pure.

Uses. It is used to defoliate cotton at 1·2-2·5 kg a.i./ha and can induce leaf abscission in some other plants, such as roses and hydrangeas.

Toxicology. The acute oral LD50 for male albino rats is 1272 mg/kg; the acute dermal LD50 for albino rabbits is > 4600 mg/kg. In 90-d feeding trials dogs and rats receiving 750 mg a.i./kg diet showed depression of cholinesterase levels but no other effect on pathology or histology.

Formulation. 'Folex Cotton Defoliant' (720 g a.i./l).

Analysis. Product analysis is by GLC with internal standard. Residues may be determined by GLC with MCD as sulphur.

2,2,2-Trichloro-1-(3,4-dichlorophenyl)ethyl acetate

GXGGYOV1&R CG DG $C_{10}H_7Cl_5O_2$ (336.4)

Nomenclature and development. The common name is not yet agreed. The IUPAC name is 2,2,2-trichloro-1-(3,4-dichlorophenyl)ethyl acetate, in *C.A.* usage 3,4-dichloro-α-(trichloromethyl)benzenemethanol acetate *[51366-25-7]* formerly 3,4-dichloro-α-(trichloromethyl)benzyl alcohol acetate. It was introduced in 1974 by Bayer AG as an insecticide for public health use under the code number 'BAY MEB 6046', the trade mark 'Baygon MEB' and protected by DBP 2 110 056.

Properties. It is a colourless crystalline solid, m.p. 84.5°, v.p. 1.1×10^{-7} mmHg at 20°. Its solubility at 20° is 50 mg/l water; > 600 g/kg cyclohexanone; < 10 g/kg propan-2-ol.

Uses. It is an insecticide, recommended for use against pests such as flies, mosquitoes and clothes moths.

Toxicology. The acute oral LD50 for rats is > 10 000 mg/kg; the acute dermal LD50 for rats is > 1000 mg/kg.

Formulations. Aerosol concentrates (20 g/l), in combination with dichlorvos; oil-based spray (10 g/l), in combination with dichlorvos.

Analysis. Product analysis is by GLC.

Trichloronate

(Trichloronat)

GR BG DG EOPS&2&O2 $C_{10}H_{12}Cl_3O_2PS$ (333·6)

Nomenclature and development. The common name trichloronat is approved by ISO—exception BSI (trichloronate); also known as ENT 25 712. The IUPAC and *C.A.* name is *O*-ethyl *O*-2,4,5-trichlorophenyl ethylphosphonothioate, Registry No. *[327-98-0]*. It was introduced in 1960 by Bayer AG under the code numbers 'Bayer 37 289' and 'S 4400', the trade marks 'Agrisil', 'Agritox', 'Phytosol' and protected by DBP 1 099 530. Its insecticidal properties were described by R. O. Drummond, *J. Econ. Entomol.*, 1963, **56**, 831; B. C. Pass, *ibid.*, 1964, **57**, 1002.

Manufacture and properties. Produced by the condensation of 2,4,5-trichlorophenol with *O*-ethyl ethylphosphonochloridothioate, it is an amber-coloured liquid, b.p. 108°/0·01 mmHg, d_4^{20} 1·365. Its solubility at 20° is 50 mg/l water; soluble in acetone, ethanol, aromatic solvents, kerosine, chlorinated hydrocarbons. It is hydrolysed by alkali.

Uses. Trichloronate is a non-systemic insecticide recommended for the control of wireworms, root maggots and other soil-dwelling insects.

Toxicology. The acute oral LD50 is: for rats 16-37·5 mg/kg; for rabbits 25-50 mg/kg; for chickens 45 mg/kg. The acute dermal LD50 for male rats is 135-341 mg/kg. In 2-y feeding trials rats receiving 3 mg/kg diet showed no symptom of poisoning.

Formulations. Granules, e.c. and seed dressings of various a.i. content.

Analysis. Product analysis is by u.v. spectroscopy after alkaline hydrolysis to 2,4,5-trichlorophenol, measuring the difference between alkali and acid at 313 nm in methanolic-aqueous buffer solutions. Residues may be determined by GLC (E. Möllhoff, *Pflanzenschutz-Nachr. (Am. Ed.)*, 1968, **21**, 331). Details of both methods may be obtained from Bayer AG.

Trichlorphon

(Trichlorfon)

$$\underset{\underset{OH}{|}}{(MeO)_2.\overset{\overset{O}{||}}{P}.CH.CCl_3}$$

GXGGYQPO&O1&O1 $C_4H_8Cl_3O_4P$ (257·4)

Nomenclature and development. The common name trichlorfon is approved by ISO—exceptions BSI (trichlorphon), Turkey (dipterex), USSR (chlorophos) and JMAF (DEP); also known as ENT 19 763. The IUPAC and *C.A.* name is dimethyl 2,2,2-trichloro-1-hydroxyethylphosphonate, Registry No. *[52-68-6]*. It was first prepared by W. Lorenz. It was introduced in 1952 by Bayer AG under the code numbers 'Bayer 15 922', 'Bayer L 13/59', trade marks 'Dipterex', 'Neguvon', 'Tugon' and protected by USP 2 701 225. 'Dylox' is the trade mark registered by the Chemagro Division. Its biological properties were described by G. Unterstenhöfer, *Anz. Schaedlingskd.*, 1957, **30**, 7.

Manufacture and properties. Produced by the condensation of dimethyl hydrogen phosphite with trichloroacetaldehyde, it is a colourless crystalline powder, m.p. 83-84°, v.p. 7·8 x 10^{-6} mmHg at 20°, d_4^{20} 1·73, n_D^{20} (100 g/l aqueous solution) 1·3439. Its solubility at 25° is 154 g/l water; soluble in benzene, ethanol and most chlorinated hydrocarbons; insoluble in petroleum oils; poorly soluble in diethyl ether, carbon tetrachloride. Molecular weight determinations give a figure twice that quoted above (W. Lorenz *et al., J. Am. Chem. Soc.*, 1955, **77**, 2554). It is stable at room temperature, but is decomposed by water at higher temperatures and at pH < 5·5 to form dichlorvos (p.180).

Uses. Trichlorphon is a contact and stomach insecticide with penetrant action recommended for agricultural uses against lepidopterous larvae and fruit flies at 75-120 g a.i./100 l, for controlling household pests especially flies (with special formulations) and for the control of ectoparasites of domestic animals. Its activity is attributed to its metabolic conversion to dichlorvos (R. L. Metcalf *et al., J. Econ. Entomol.*, 1959, **52**, 44).

Toxicology. The acute oral LD50 and the acute dermal LD50 for rats are 560-630 mg/kg and > 2000 mg/kg, respectively. In 2-y feeding trials rats receiving 500 mg/kg diet suffered no ill-effect.

Formulations. These include: w.p. (500 g a.i./kg); s.p. (500, 800 and 950 g/kg); s.c. (500 g/l); ULV concentrate (250, 500 and 750 g/l); dust (50 g/kg); granules (25 and 50 g/kg); 'Tugon Stable Spray' (300 g/l); 'Tugon Fly Bait' (10 g/kg); 'Nevugon' preparations for veterinary use.

Analysis. Product analysis is either by polarography (P. A. Giang & R. L. Caswell, *J. Agric. Food Chem.*, 1957, **5**, 753), or by potentiometric titration of the chloride ion resulting from hydrolysis with ethanolic ethanolamine (*CIPAC Handbook*, 1970, **1**, 684; FAO Specification (CP/51)). Residues may be determined by GLC (R. J. Anderson *et al., J. Agric. Food Chem.*, 1966, **14**, 508). Details of these and other methods may be obtained from Bayer AG. See also: D. MacDougall, *Anal. Methods Pestic., Plant Growth Regul. Food Addit.*, 1964, **2**, 199; *Anal. Methods Pestic. Plant Growth Regul.*, 1972, **6**, 387.

Triclopyr

T6NJ BO1VQ CG EG FG $\qquad$ *triclopyr* $C_7H_4Cl_3NO_3$ (256·5)

T6NJ BO1VO2O4 CG EG FG $\quad$ *triclopyr-(2-butoxyethyl)* $C_{13}H_{16}Cl_3NO_4$ (356·6)

Nomenclature and development. The common name triclopyr is approved by BSI, ANSI, New Zealand and WSSA and proposed by ISO. The IUPAC name is 3,5,6-trichloro-2-pyridyloxyacetic acid, in *C.A.* usage 3,5,6-trichloro-2-pyridinyloxyacetic acid *[55335-06-3]*. It was introduced in 1970 by The Dow Chemical Co. under the code number 'DOWCO 233', and the trade mark 'Garlon'.

Manufacture and properties. Produced by the reaction of 3,5,6-trichloropyridine-2-ol with chloroacetic acid, it forms a fluffy colourless solid, m.p. 148-150°, v.p. 1·26 x 10^{-6} mmHg at 25°, pKa 2·68. Its solubility at 25° is: 440 mg/l water; 989 g/kg acetone; 307 g/kg octan-1-ol; poorly soluble in other common organic solvents. It is stable under normal storage conditions and to hydrolysis, but subject to photolysis with 50% loss in < 12 h.

Uses. Triclopyr is a systemic herbicide which is rapidly absorbed by both foliage and roots and translocated throughout the plant. In susceptible plants it induces auxin-type responses. Most broad-leaved plants are susceptible, while Gramineae are usually unaffected at normal application rates. It shows considerable promise for the control of certain 2,4-D resistant weeds in cereals, and for the control of problem broad-leaved vegetation in industrial areas and rangeland. It is also of promise for the control of broad-leaved weeds and brush in conifer reafforestation programmes. It undergoes relatively rapid microbial degradation in soil. Under non-leaching conditions, and those favouring microbial activity, 50% loss averages 46 d.

Toxicology. The acute oral LD50 is: for rats 713 mg/kg; for rabbits 550 mg/kg; for guinea-pigs 310 mg/kg. The acute percutaneous LD50 for rabbits is > 2000 mg/kg. It is a mild eye irritant, and essentially non-irritating to the skin. Toxicity to birds and fish is low.

Formulations. 'Garlon 4E' (480 g a.e./l) as the 2-butoxyethyl ester (*C.A.* Registry No. *[64470-88-8]*).

Analysis. Product and residue analysis: details of methods are available from The Dow Chemical Co.

Tricyclazole

T B556 BN DNN GSJ L1 $C_9H_7N_3S$ (189·2)

Nomenclature and development. The common name tricyclazole is approved by BSI, and New Zealand, and proposed by ISO. Its IUPAC name is 5-methyl-[1,2,4]-triazolo[3,4-*b*]-benzothiazole, in *C.A.* usage 5-methyl-1,2,4-triazolo[3,4-*b*]benzothiazole *[41814-78-2]*. It was introduced in 1975 by Eli Lilly and Co. under the code number 'EL-291', trade marks 'Beam', 'Bim', and 'Blascide', and protected by BP 1419121. It was described by J. D. Froyd *et al., Phytopathology,* 1976, **66**, 1135.

Manufacture and properties. Produced by the cyclisation of 2-hydrazino-4-methylbenzo-thiazole with formic acid, it is a crystalline solid, m.p. 187-188°, v.p. 2×10^{-7} mmHg at 25°. Its solubility at 25° is 1·6 g/l water; soluble in acetone, acetonitrile, chloroform, dichloromethane, ethanol, methanol.

Uses. Tricyclazole is a fungicide which has given effective control of *Pyricularia oryzae* in transplanted and direct-seeded rice by flat drench (1·5-2·4 g a.i./flat), root soak (0·5-2·0 g/m² of nursery bed) and foliar application (150-400 g/ha) methods. One or 2 applications by one or more of these methods give a season-long control of the disease.

Toxicology. The acute oral LD50 is: for adult mice $\geqslant 244$ mg/kg; for rats $\geqslant 290$ mg/kg; for bobwhite quail and mallard duck > 100 mg/kg; for beagle dogs > 50 mg/kg. It induced no skin irritation but caused some eye irritation in rabbits. The LC50 (96 h) is: for goldfish fingerlings $\geqslant 13·5$ mg/l; for bluegill fingerlings $\geqslant 1·96$ mg/l; for rainbow trout fingerlings $\geqslant 1·62$ mg/l.

Formulations. A w.p. (200 and 750 g a.i./kg); dust (10 g/kg); granules (30 g/kg).

Analysis. Product analysis is by GLC with FID after extraction of the compound from the formulation with chloroform. Residues in plant tissue are determined by refluxing in sulphuric acid and further extracting in ethyl acetate. Tricyclazole and its alcohol metabolite are separated on an alumina column and determined by GLC using FPD, the metabolite being first converted to the trimethylsilyl derivative. Detailed analytical procedures and other information may be obtained from Eli Lilly & Co.

Tridemorph

$$\text{C}_{13}\text{H}_{27}-\text{N} \quad \text{(morpholine ring)} \quad \text{O}, \quad \text{Me, Me}$$

T6N DOTJ A13 C1 E1 $C_{19}H_{39}NO$ (297·5)

Nomenclature and development. The common name tridemorph is approved by BSI and ISO. The IUPAC and *C.A.* name is 2,6-dimethyl-4-tridecylmorpholine, Registry No. *[24602-86-6]*. It was introduced in 1969 by BASF AG under the code number 'BASF 220' and the trade mark 'Calixin'. Its fungicidal properties were described by J. Kradel *et al., Proc. Br. Insecticide. Fungic. Conf., 5th,* 1969, **1**, 16; E. -H. Pommer *et al., ibid.,* 1969, **2**, 347.

Properties. It is a colourless oil, with a slight amine-like odour, b.p. 134°/0·5 mmHg, v.p. 3 x 10^{-4} mmHg at 20°. It is miscible with water, acetone, benzene, chloroform, cyclohexane, ethanol, olive oil. It is stable in unopened containers for $\geqslant 2$ y. It is compatible with most cereal herbicides.

Uses. Tridemorph is an eradicant fungicide with a systemic action, being absorbed through foliage and roots giving some protective action. It has controlled cereal mildews when applied between the 5-leaf and jointing stages of growth at 525 g a.i. in $\geqslant 250$ l/ha, often in combination with a hormone weedkiller or with chlormequat chloride. This treatment gives protection for 21-28 d when a second application may be made. Results on winter and spring barley indicate safety, but scorch has been observed on some winter wheat varieties under certain climatic conditions.

Toxicology. The acute oral toxicity is: for rats 825 mg a.i. (as e.c.)/kg; for rabbits 562 mg a.i. (as e.c.)/kg. The acute dermal LD50 for rabbits is *c.* 1350 mg a.i. (as e.c.)/kg. Rats survived a single 8-h exposure to air saturated with the compound at 20°.

Formulation. 'Calixin', e.c. (750 g/l).

Analysis. Product analysis is by extraction with chloroform followed by potentiometric titration with perchloric acid in glacial acetic acid. Residues in cereal straw and soil may be determined by extraction with acetone and, after clean-up by steam-distillation, reaction with methyl orange forming a chloroform-soluble dye, which is decomposed by acid and the liberated methyl orange measured colorimetrically.

Trietazine

T6N CN ENJ BN2&2 DM2 FG $\qquad$ $C_9H_{16}ClN_5$ (229·7)

Nomenclature and development. The common name trietazine is approved by BSI, ISO, ANSI and JMAF—exception India. The IUPAC name is 2-chloro-4-diethylamino-6-ethylamino-1,3,5-triazine, in *C.A.* usage 6-chloro-*N,N,N'*-triethyl-1,3,5-triazine-2,4-diamine *[1912-26-1]* formerly 2-chloro-4-diethylamino-6-ethylamino-*s*-triazine. Its herbicidal properties were described by H. Gysin & E. Knüsli, *Proc. Br. Weed Control Conf., 4th,* 1958, p.225. It was discovered by J. R. Geigy S.A. (now Ciba-Geigy AG) and given the code number 'G 27 901' and was first introduced commercially in 1972 by Fisons Ltd under the code number 'NC 1667', in combination with other herbicides. Its use as a herbicide was protected by SwissP 329 277; BP 814947; USP 2 819 855.

Manufacture and properties. Produced by the reaction of trichloro-1,3,5-triazine with 1 molecular proportion of ethylamine and then with 1 of diethylamine, it is a crystalline solid, m.p. 100-101°. Its solubility at 25° is: 20 mg/l water; 170 g/l acetone; 200 g/l benzene; > 500 g/l chloroform; 100 g/l dioxane; 30 g/l ethanol. It is stable to air and water, non-corrosive and compatible with other herbicides with which it is likely to be mixed.

Uses. Trietazine is a herbicide that is taken up by roots and foliage and inhibits the Hill reaction. It is used with linuron for weed control in potatoes, and with simazine in peas at 1·6-4·5 kg/ha according to crop and soil type.

Toxicology. The acute oral LD50 is: for rats 2830->4000 mg/kg; for quail 800 mg/kg. Application (24 h) of 1000 mg/kg to the skin of rats produced no ill-effect. In 90-d feeding trials rats receiving 16 mg/kg diet were unaffected. The LC50 (24 h) for guppies is 5·5 mg/l.

Formulations. These include: w.p. (linuron + trietazine), 'Bronox' (Fisons Ltd), 'Furol' (Murphy Chemical Ltd); w.p. (simazine + trietazine), 'Remtal' (Fisons Ltd), 'Aventox' (Murphy Chemical Ltd).

Analysis. Residues may be determined in a range of crops by extraction with methanol and, after clean-up by chromatography on alumina, GLC with FID.

Trifenmorph

$$Ph_3C-N \bigcirc O$$

T6N DOTJ AXR&R&R $C_{23}H_{23}NO$ (329·4)

Nomenclature and development. The common name trifenmorph is approved by BSI and ISO—exception France (triphenmorphe). The IUPAC name is 4-(triphenylmethyl)-morpholine (I), also known as 4-tritylmorpholine (II) in *C.A.* usage (I) *[1420-06-0]* formerly (II). Its molluscicidal properties were first described by C. B. C. Boyce *et al.*, *Nature (London)*, 1966, **210**, 1140. It was developed by Shell Research Ltd under the code number 'WL 8008' and the trade mark 'Frescon'.

Properties. It is a colourless crystalline solid which melts at 176-178° then re-solidifies, melting again at 185-187°, v.p. 1·4 x 10^{-7} mmHg at 20°. The technical grade is 90-95% pure, melting at 150-170° and again at 170-185°. Its solubility at 20° is: 0·02 mg/l water; 300 g/l carbon tetrachloride; 450 g/l chloroform; 255 g/l tetrachloroethylene. It is stable to heat and alkalies, but is hydrolysed by mild acid conditions to morpholine and triphenylmethanol. It is non-corrosive. Slight decomposition occurs in u.v. light.

Uses. Trifenmorph is a molluscicide effective against aquatic and semi-aquatic snails. It is recommended for application to irrigation and other moving water systems by a drip-feed technique at 0·03-0·10 mg/l. It is used for static water at *c.* 1·0-2·0 mg/l. It is also recommended for application to *Lymnaea truncatula* habitats at 500 g/ha to control fascioliasis. On plants, in soil and in water, trifenmorph is broken down initially to triphenylmethanol (K. I. Beynon *et al., Pestic. Sci.,* 1972, **3**, 689).

Toxicology. The acute oral LD50 is: for rats 446-2200 mg/kg; for mice 700-4800 mg/kg. The acute percutaneous LD50 for rats is > 1000 mg/kg. In 90-d feeding trials the 'no effect' level for rats was 100 mg/kg diet. Fish (*Barbus* and *Tilapia* spp.) survived a 10-d exposure to 0·025 mg/l; aquatic micro-flora and -fauna were unaffected at 0·2 mg/l.

Formulation. An e.c. (165 g a.i./l).

Analysis. Product analysis is by non-aqueous titration with a perchloric-acetic acid mixture (*Pesticides used in Public Health,* p.248) or by HPLC. Residues may be determined colorimetrically with cyclohexane-sulphuric acid (K. I. Beynon & G. R. Thomas, *Bull, WHO,* 1967, **37**, 47); a method suitable for use under field conditions is available (K. I. Beynon & A. N. Wright, *Pestic. Sci.,* 1975, **6**, 515). Details of all these methods are available from Shell International Chemical Co. Ltd.

Trifluralin

FXFFR CNW ENW DN3&3 $C_{13}H_{16}F_3N_3O_4$ (335·5)

Nomenclature and development. The common name trifluralin is approved by BSI, ISO, ANSI, WSSA and JMAF. The IUPAC name is α,α,α-trifluoro-2,6-dinitro-*N,N*-dipropyl-*p*-toluidine (I) formerly 2,6-dinitro-*N,N*-dipropyl-4-trifluoromethylaniline, in *C.A.* usage 2,6-dinitro-*N,N*-dipropyl-4-(trifluoromethyl)benzenamine *[1582-09-8]* formerly (I). Its herbicidal properties were first described by E. F. Alder *et al., North Cent. Weed Control Conf.,* 1960, p.23, and it was introduced in 1960 by Eli Lilly & Co., under the code number 'L-36 352', trade mark 'Treflan' and protected by USP 3 257 190.

Manufacture and properties. Produced by the nitration of 4-chloro-α,α,α-trifluorotoluene and reaction with dipropylamine, it is an orange crystalline solid, m.p. 48·5-49°, v.p. 1·03 x 10^{-4} mmHg at 25°. Its solubility at 27° is: < 1 mg/l water; 400 g/kg acetone; 580 g/kg xylene. The technical product is ⩾95% pure, m.p. > 42°. It is stable, though susceptible to photochemical decomposition, is compatible with most other pesticides and is non-corrosive.

Uses. Trifluralin is a pre-em. herbicide with little post-em. activity. When incorporated in the soil, it is effective for the control of annual grasses and broad-leaved weeds in cotton, forage legumes, established sugar beet and tomatoes, beans, groundnuts, brassicas, soyabeans, sunflowers, transplanted peppers, vineyards, orchards and ornamentals at rates of 0·5-1·0 kg a.i./ha. A combination with linuron controls broad-leaved and grass weeds when applied pre-em. to the soil surface in winter cereals. Trifluralin plus 2,4-D is used as a post-planting herbicide in transplanted rice. It shows a loss in soil of 85-90% in 0·5-1·0 y (G. W. Probst *et al., J. Agric. Food Chem.,* 1967, **15,** 592).

Toxicology. The acute oral LD50 is: for rats > 10000 mg/kg; for mice 500 mg/kg; for dogs, rabbits and chickens > 2000 mg/kg. Skin applications of 2000 mg/kg caused neither toxicity nor irritation to rabbits. In 2-y feeding trials rats receiving 2000 mg/kg diet and dogs 1000 mg/kg diet suffered no ill-effect. In reproduction studies, the safe level for rats was 200 mg/kg diet and for dogs 1000 mg/kg diet. In teratology studies there was no effect with: rats at 2000 mg/kg diet; dogs at 1000 mg/kg diet; rabbits at 1000 mg/kg bodyweight daily. The LC50 (96 h) to bluegill fingerlings is 0·089 mg/l.

Formulations. These include: e.c. 'Treflan' (480 g a.i./l), 'Trinulan' or 'Chandor' (240 g trifluralin + 120 g linuron/l), 'Neepex' (120 g trifluralin + 120 g napropamide/l); 'Treflan-R', spreadable (100 g trifluralin + 159 g 2,4-D-isopropyl/l); granules 'Treflan' (50 g/kg).

Analysis. Trifluralin is identified by GLC; FAO Specification (CP/64). Product analysis involves separation by column chromatography and spectrophotometric estimation; see also *CIPAC Handbook,* 1979, **1A,** in press. Residues are determined by GLC with ECD. See also: T. H. Byast *et. al., Agric. Res. Counc. (G.B.) Weed Res. Organ., Tech. Rep.,* No. 15 (2nd Ed.), p.11; J. B. Tepe & R. E. Scroggs, *Anal. Methods Pestic., Plant Growth Regul. Food Addit.,* 1967, **5,** 527; *Anal. Methods Pestic. Plant Growth Regul.,* 1972, **6,** 703. Particulars are available from Eli Lilly & Co.

Triforine

Cl₃C.CH.NH.CHO

$Cl_3C.CH.NH.CHO$

(structure: piperazine ring with two N-substituents)

$Cl_3C.CH.NH.CHO$

T6N DNTJ AYMVHXGGG DYMVHXGGG $C_{10}H_{14}Cl_6N_4O_2$ (435·0)

Nomenclature and development. The common name triforine is approved by BSI, ISO and ANSI. The IUPAC name is 1,4-bis(2,2,2-trichloro-1-formamidoethyl)piperazine previously known as 1,1'-piperazine-1,4-diyldi-[N-(2,2,2-trichloroethyl)formamide], in *C.A.* usage N,N'-[1,4-piperazinediylbis(2,2,2-trichloroethylidene)]bisformamide *[26644-46-2]*. This compound, first synthesised by C. H. Boehringer Sohn in 1967 (DOS 1901421), was introduced in 1969 by Cela GmbH under the code number 'Cela W524'. Its fungicidal properties were described by P. Schicke & K. H. Veen, *Proc. Br. Insectic. Fungic. Conf., 5th,* 1969, **2**, 569.

Manufacture and properties. Produced by the reaction of piperazine with N-(1,2,2,2-tetrachloroethyl)formamide, it forms colourless crystals, m.p. 155°, v.p. 2 x 10⁻⁷ mmHg at 25°. Its solubility at room temperature is 27-29 mg/l water; it is of low solubility in acetone, benzene, carbon tetrachloride, chloroform, dichloromethane, light petroleum; slightly soluble in dioxane or cyclohexanone; soluble in tetrahydrofuran; readily soluble in dimethylformamide, dimethyl sulphoxide or 1-methyl-2-pyrrolidone.

It is rapidly decomposed to trichloroacetaldehyde and piperazine salts by concentrated H_2SO_4 or HCl; slowly decomposed to chloroform and piperazine by strong alkalies. It is non-persistent in soil. The pathways of its photochemical and metabolic breakdown have been established: H. Buchenauer, *Pestic. Sci.,* 1975, **6**, 553; A. Fuchs *et al., ibid.,* 1976, **7**, 115, 127; S. Darda *et al., ibid.,* 1977, **8**, 183; S. Darda, *ibid.,* p.173; J. P. Rouchard *et al., ibid.,* p.65, 1978, **9**, 74, 139, 587; J. P. Rouchard, *Bull. Environ. Contam. Toxicol.,* 1977, **18**, 184.

Uses. Triforine is a systemic fungicide effective against: powdery mildew, scab and other diseases of fruit and berries at 20-25 g/100 l; powdery mildew, rust and black spot on ornamentals at 15 g/100 l; powdery mildew and other diseases of vegetables at 25 g/100 l; powdery mildew and other leaf diseases on cereals at 200-250 g/ha, rust on cereals at 300 g/ha. It is active against storage diseases of fruit and suppresses red spider mite activity.

Toxicology. The acute oral LD50 for rats, mice and quail is > 6000 mg/kg. It is non-toxic to guppies at 50 mg/l; non-toxic to honey bees.

Formulation. 'Saprol', e.c. (200 g a.i./l); 'Nimrod T', e.c. (62·5 g bupirimate + 62·5 g triforine/l); 'Funginex'.

Analysis. Product and residue analysis is by polarography or GLC (after degradation); particulars from Celamerck. See also R. Darskus & D. Eichler, *Anal. Methods Pestic. Plant Growth Regul.,* 1978, **10**, 243.

Undecan-2-one

$$\begin{array}{c} O \\ \| \\ Me(CH_2)_8.C.Me \end{array}$$

9V1 $C_{11}H_{22}O$ (170·3)

Nomenclature and development. The IUPAC name is undecan-2-one, in *C.A.* usage 2-undecanone *[112-12-9]*; also known as methyl nonyl ketone. It was introduced in 1966 by the McLaughlin Gormley King Co. under the name 'MGK Dog and Cat Repellent' and protected by USP 3 474 176.

Properties. Pure undecan-2-one has m.p. 11-13°, b.p. 231·5-232·5°; the technical product, 95% pure, is a clear oil at room temperature, b.p. 223°, flash point 90° (closed cup), d^{20} 0·789-0·819, n_D^{25} 1·4015. It is insoluble in water; soluble in most organic solvents.

Uses. Undecan-2-one is a repellent to dogs and cats and is used in the training of pets to break objectionable habits, including the soiling of flower beds and evergreens; claw sharpenings on furniture and trees. It is phytotoxic to foliage and to soft growth but it does not appear to damage ordinary lawn grasses nor hardy woody perennials when applied to trunk or stems. Its effectiveness as a repellent lasts *c.* 24 h.

Toxicology. The acute oral LD50 for rats is 2500-5000 mg a.i. (as 500 g/l propan-2-ol)/kg; the acute dermal LD50 for rabbits is > 4700 mg a.i. (as 500 g/l propan-2-ol)/kg. A solution of 1 g a.i./l saline did not sensitise guinea-pigs.

Formulations. These include: e.c. (660 g a.i./kg); pressurised spray (20 g/kg); granules (20 g/kg).

Analysis. Analysis is by GLC (particulars from the McLaughlin Gormley King Co.) or by a titrimetric method using hydroxylamine (L. D. Metcalfe & A. A. Schmitz, *Anal. Chem.*, 1955, **27**, 138).

Urea

$$\underset{H_2N.\overset{O}{\overset{\|}{C}}.NH_2}{}$$

ZVZ   CH$_4$N$_2$O (60·06)

Nomenclature and development. The IUPAC and *C.A.* name is urea, Registry No. *[57-13-6]*. It has long been used as a nitrogenous fertiliser being converted to ammonia and ultimately nitrate by bacteria in the soil. It is also a standard treatment to prevent attack of conifer stumps by the fungus *Fomes annosus*.

Manufacture and properties. Produced by hydrolysis of cyanamide or by the reaction of ammonia on carbon dioxide, it is a crystalline solid, m.p. 132·7°, d_4^{18} 1·32. Its solubility is: 1 kg/l water; 50 g/l ethanol; 167 g/l methanol; 500 g/l glycerol; almost insoluble in chloroform, diethyl ether. It is a weak base, forming salts with acids.

Uses. Urea is used as a solution (20 g/l water with marker dye) which is painted on conifer stumps to prevent infection by *F. annosus* (*Forestry Commission Leaflet,* No. 5, 1974).

Toxicology. It is essentially non-toxic to mammals.

Formulation. A w.s.p. (technical grade with marker dye).

Analysis. Product analysis is by acid hydrolysis, the resultant solution made alkaline and the liberated ammonia distilled into acid, the excess of which is estimated by titration with alkali. (AOAC Methods).

Validamycin A

$C_{20}H_{35}NO_{13}$ (497·5)

T6OTJ B1Q CQ DQ EQ FO- CL6TJ A1Q BQ DQ EM-
CL6UTJ A1Q DQ EQ FQ

Nomenclature and development. The common name validamycin A is approved by JMAF and the Japanese Antibiotics Research Association. The IUPAC name is N-[(1S)-(1,4,6/5)-3-hydroxymethyl-4,5,6-trihydroxycyclohex-2-enyl]-[O-β-D-glucopyranosyl-(1→3)]-(1S)-(1,2,4/3,5)-2,3,4-trihydroxy-5-hydroxymethylcyclohexylamine, in *C.A.* usage [1S-(1α,4α,5β,6α)]-1,5,6-trideoxy-3-O-β-D-glucopyranosyl-5-(hydroxymethyl)-1-[[4,5,6-trihydroxy-3-(hydroxymethyl)-2-cyclohexen-1-yl]amino]-D-chiro-inositol *[37248-47-8]*. It was introduced in 1972 by Takeda Chemical Industries Ltd under the trade marks 'Validacin' and 'Valimon'. It was described by T. Iwasa *et al., J. Antibiot.,* 1970, **23,** 595; 1971, **24,** 107, 119 and its chemical structure established by S. Horii *et al., ibid.,* 1972, **25,** 48.

Manufacture and properties. Produced by the fermentation of *Streptomyces hygroscopicus* var. *limoneus* nov. var., it is a colourless powder decomposing over a wide temperature range *c.* 130°, $[\alpha]_D^{24}$ +110° (water), and weakly basic, pK_a 6·0. It is very soluble in water; soluble in methanol, dimethylformamide, dimethyl sulphoxide; sparingly soluble in acetone, ethanol. It is stable at room temperature, in neutral and alkaline solution, but slightly unstable under acidic conditions. It is compatible with most pesticides and is not corrosive to steel.

Uses. Validamycin A is fungistatic against a narrow range of fungi and is used for the control of *Pellicularia (Corticium) sasakii* of rice at 45-90 g a.i./ha, of black scurf of potatoes and the damping-off of vegetable seedlings caused by *Rhizoctonia solani* (O. Wakae & K. Matsuura, *Rev. Plant Prot. Res.,* 1975, **8,** 81; M. Bakkeren *et al., Proc. 1977 Br. Crop Prot. Conf.—Pests Dis.* 1977, **2,** 541). It is non-phytotoxic to most of the crops tested.

Toxicology. The acute oral LD50 for rats and mice is > 20 000 mg/kg; dermal applications of a solution (500 g/l) caused no irritation to rabbits. In 90-d feeding trials no effect was produced in: rats receiving 1000 mg/kg daily; mice receiving 2000 mg/kg daily.

Formulations. These include: solution (30 g a.i./l); dust (3 g/kg).

Analysis. Product and residue analysis is by GLC of a trimethylsilyl derivative with FID detection (K. Nishi & K. Konishi, *Anal. Methods Pestic. Plant Growth Regul.,* 1976, **8,** 309).

Vamidothion

$$\text{(MeO)}_2\text{P.S.CH}_2\text{CH}_2\text{.S.CH.C.NHMe}$$

with the structural representation showing O (double bond) on the P, Me on the CH, and O (double bond) on the C.

1OPO&O1&S2SY1&VM1 $C_8H_{18}NO_4PS_2$ (287·3)

Nomenclature and development. The common name vamidothion is approved by BSI, ISO and JMAF; also known as ENT 26613. The IUPAC name is *O,O*-dimethyl *S*-2-(1-methylcarbamoylethylthio)ethyl phosphorothioate, in *C.A.* usage *O,O*-dimethyl *S*-[2-[[1-methyl-2-(methylamino)-2-oxoethyl]thio]ethyl] phosphorothioate *[2275-23-2]* formerly *O,O*-dimethyl phosphorothioate *S*-ester with 2-[(2-mercaptoethyl)thio]-*N*-methylpropionamide. It was introduced in 1961 by Rhône-Poulenc Phytosanitaire as an insecticide-acaricide under the code numbers 'RP 10465' and 'NPH 83', trade mark 'Kilval' (Rhône-Poulenc, May & Baker Ltd) and protected by BP 872823; BelgP 575106; USP 2943974. Its properties were first described by J. Dermoras *et al., Phytiatr.-Phytopharm.,* 1962, **11,** 107.

Manufacture and properties. Produced by the condensation of ammonium *O,O*-diethyl phosphorothioate with 2-(2-chloroethylthio)-*N*-methylpropionamide, which is obtained by the reaction of 2-mercaptoethanol with ethyl 2-chloropropionate and the resulting ethyl 2-(2-hydroxyethylthio)propionate treated successively with methylamine and thionyl chloride. The pure compound forms colourless needles, m.p. 46-48°. The technical product is an off-white waxy solid, m.p. *c.* 40°, of negligible v.p. at 20°. Its solubility is 4 kg/l water; very soluble in most organic solvents; practically insoluble in light petroleum, cyclohexane. The technical product and, to a lesser extent, the pure compound undergo slight decomposition at room temperature, a decomposition arrested in certain solvents such as anisole or butanone. It is compatible with most pesticides and is non-corrosive.

Uses. Vamidothion is a systemic insecticide and acaricide of high persistence value against *Eriosoma lanigerum;* used for the control of sap-feeding insects and mites on apples, pears, peaches, plums, hops, rice, cotton, etc., at 37-50 g a.i./100 l. It is metabolised in plants to the corresponding sulphoxide which is of similar activity to vamidothion but of greater persistence.

Toxicology. The acute oral LD50 is: for rats 64-105 mg/kg; for mice 34-57 mg/kg; for pheasants 35 mg/kg. The acute dermal LD50 is: for mice 1460 mg/kg; for rabbits 1160 mg/kg. The oral toxicity of the sulphoxide is about half that of vamidothion. In 90-d feeding trials the growth rate of rats receiving 50 mg vamidothion/kg diet or 100 mg of its sulphoxide/kg diet was unaffected. Goldfish survived 14 d in water containing 10 mg vamidothion/l.

Formulations. 'Kilval', 'Kilvar' (in Japan) and 'Trucidor', solutions (400 g a.i./l).

Analysis. Product and residue analysis is by GLC (J. Demoras *et al., Anal. Methods Pestic. Plant Growth Regul.,* 1973, **7,** 479).

Vernolate

$$\underset{Pr_2N.C.SPr}{\overset{\overset{O}{\|}}{}}$$

3SVN3&3

$C_{10}H_{21}NOS$ (203·3)

Nomenclature and development. The common name vernolate is approved by BSI, New Zealand, JMAF and WSSA, and proposed by ISO. The IUPAC name is S-propyl dipropylthiocarbamate (I), in *C.A.* usage S-propyl dipropylcarbamothioate *[1929-77-7]* formerly (I). It is one of a series of thiolcarbamates introduced in 1954 by the Stauffer Chemical Co. and had the code number 'R-1607', the trade mark 'Vernam' and is protected by USP 2 913 327.

Manufacture and properties. Produced by the reaction of dipropylcarbamoyl chloride with propane-1-thiol, it is a clear liquid, with an aromatic odour, b.p. 150°/30 mmHg, v.p. 10·4 x 10⁻³ mmHg at 25°, d^{20} 0·954, n_D^{20} 1·4736. Its solubility at 21° is 107 mg/l water; miscible with common organic solvents such as kerosine, 4-methylpentan-2-one, xylene. It is stable and non-corrosive.

Uses. Vernolate is toxic to germinating broad-leaved and grassy weeds and is recommended for the pre-em. weeding of soyabeans and groundnuts at 1·5-3·0 kg a.i./ha followed by soil incorporation. Its persistence in soil is brief.

Toxicology. The acute oral LD50 for male rats is 1780 mg tech./kg; the acute dermal LD50 for rabbits is 4640 mg/kg.

Formulations. These include: e.c. (720 and 840 g a.i./l); granules (50 and 100 g/kg).

Analysis. Product analysis is by GLC. Residues may be determined colorimetrically by hydrolysis, separation of the dipropylamine and estimation as the copper(II) dithiocarbamate complex (G. G. Patchett & G. H. Batchelder, *Anal. Methods Pestic., Plant Growth Regul. Food Addit.*, 1967, **5**, 537; W. J. Ja, *Anal. Methods Pestic. Plant Growth Regul.*, 1972, **6**, 708).

Vinclozolin

T5OVNV EHJ CR CG EG& E1U1 E1 $C_{12}H_9Cl_2NO_3$ (286·1)

Nomenclature and development. The common name vinclozolin is approved by BSI and New Zealand and proposed by ISO. The IUPAC name is 3-(3,5-dichlorophenyl)-5-methyl-5-vinyloxazolidine-2,4-dione, in *C.A.* usage 3-(3,5-dichlorophenyl)-5-ethenyl-5-methyl-2,4-oxazolidinedione *[50471-44-8]*. It was introduced in 1975 by the BASF AG under the code number 'BAS 352F' and trade mark 'Ronilan'. Its fungicidal properties were reported by E. -H. Pommer & D. Mangold, *Meded. Fac. Landbouwwet. Rijksuniv. Gent,* 1975, **40,** 713; C. Hess & F. Löcher, *Proc. Br. Insectic. Fungic. Conf., 8th,* 1975, **2,** 693.

Properties. It is a colourless crystalline solid, m.p. 108°. Its solubility at 20° is 1 g/l water; 435 g/kg acetone; 146 g/kg benzene; 319 g/kg chloroform; 253 g/kg ethyl acetate. It is stable at room temperature, in water and 0·1M hydrochloric acid but is slowly hydrolysed in alkaline solution.

Uses. Vinclozolin is a selective fungicide effective against *Botrytis cinerea, Monilia* spp. and *Sclerotinia sclerotiorum* and is used on vines, fruit, hops, tomatoes and ornamentals. On strawberries, 3 sprays during the flowering period at 0·75-1·0 kg a.i./ha controlled *B. cinerea.*

Toxicology. The acute oral LD50 is: for rats 10000 mg tech./kg; for guinea-pigs *c.* 8000 mg/kg. In 90-d feeding trials the 'no effect' level was: for rats 450 mg/kg diet; for dogs 300 mg/kg diet. The LC50 (96 h) is: for guppies 32·5 mg a.i. (as w.p.)/l; for trout 52·5 mg a.i./l. It is non-toxic to honey bees and to earthworms.

Formulation. 'Ronilan', w.p. (500 g a.i./kg).

Analysis. Product analysis is by GLC, after separation by column chromatography. Residues in crops may be determined by alkaline hydrolysis with conversion of the 3,5-dichloroaniline liberated to a volatile derivative which is measured by GLC with ECD. Details are available from BASF AG.

Warfarin

T66 BOVJ DYR&1V1 EQ $C_{19}H_{16}O_4$ (308·4)

Nomenclature and development. The common name warfarin is approved by BSI, ISO and BPC—exceptions France (coumaféne), USSR (zoocoumarin), JMAF (coumarins—a term also applied to coumatetralyl) and The Netherlands. The IUPAC name is 4-hydroxy-3-(3-oxo-1-phenylbutyl)coumarin formerly 3-(α-acetonylbenzyl)-4-hydroxy-coumarin (I), in *C.A.* usage 4-hydroxy-3-(3-oxo-1-phenylbutyl)-2*H*-1-benzopyran-2-one *[81-81-2]* formerly (I). Its anticoagulant properties were discovered by K. P. Link *et al., J. Biol. Chem.,* 1944, **153**, 5 at the Wisconsin Alumni Research Foundation, where it was given the code number 'WARF 42' and protected by USP 2 427 578.

Manufacture and properties. Produced by the condensation of 4-hydroxycoumarin with 4-phenylbut-3-enone, the racemate forms colourless crystals, m.p. 159-161°. It is practically insoluble in water; readily soluble in acetone, dioxane; moderately soluble in alcohols. The enolic form is acidic, forming metal salts; the sodium salt being soluble in water, insoluble in organic solvents.

Uses. Warfarin is an anticoagulant rodenticide to which rats do not develop 'bait-shyness'. Rats are killed by 5 daily doses of 1 mg/kg. It is used at 50 mg a.i./kg bait material for the control of *Rattus norvegicus,* 250 mg/kg for *R. rattus* and mice, baiting is continued for 14 d. The racemate has been resolved, the (*S*)-isomer showing 7-fold greater rodenticidal activity than the (*R*)-isomer (B. D. West *et al., J. Am. Chem. Soc.,* 1961, **83**, 2676). The anticoagulant properties of warfarin (as 'Coumadin' or the sodium salt 'Marevan') are used in medicine to reduce the risk of post-operative thrombosis.

Toxicology. Death followed 5 daily doses of: 3 mg/kg for cats; 1 mg/kg for pigs; poultry are more resistant.

Formulations. Dust (10 g a.i./kg) for use in holes and runs; dust (1 and 5 g/kg) for admixture with suitable protein-rich bait.

Analysis. Product analysis is by u.v. spectroscopy in chloroform at 305·5 nm after column chromatography on alumina (W. Armstrong, *Chem. Ind. (London),* 1959, p.154; AOAC Methods; *CIPAC Handbook,* 1970, **1**, 696). See also: C. H. Schroeder & J. N. Eble, *Anal. Methods Pestic., Plant Growth Regul. Food Addit.,* 1964, **3**, 197; C. H. Schroeder & J. Sherma, *Anal. Methods Pestic. Plant Growth Regul.,* 1973, **7**, 677.

$$O$$
$$\parallel$$
$$O.C.NHMe$$

(structure: benzene ring with O.C.NHMe carbamate group at top position, Me and Me substituents at bottom positions)

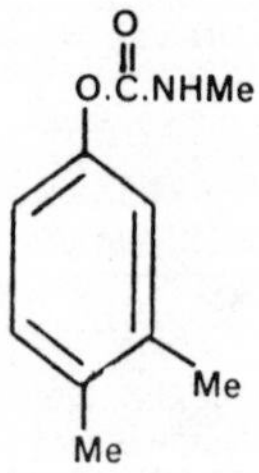

1MVOR   C1   D1 $C_{10}H_{13}NO_2$ (179·2)

Nomenclature and development. The common name MPMC is approved by JMAF. The IUPAC name is 3,4-xylyl methylcarbamate (I), in *C.A.* usage 3,4-dimethylphenyl methylcarbamate *[2425-10-7]* formerly (I). Its insecticidal properties were first described by R. L. Metcalf *et al., J. Econ. Entomol.,* 1963, **56,** 862. It was introduced in 1967 by the Sumitomo Chemical Co. Ltd under the code number 'S-1046' and trade mark 'Meobal'.

Manufacture and properties. Produced either by the condensation of 3,4-xylyl chloroformate with methylamine, or by the reaction of 3,4-xylenol with methyl isocyanate, it is a colourless crystalline solid, m.p. 79-80°, b.p. 123-130°/0·1 mmHg. Its solubility at 30° is 1·3 g/l water. It is readily hydrolysed in solutions of pH > 12 but is compatible with other than alkaline formulations. The technical product is *c.* 95% pure.

Uses. 3,4-Xylyl methylcarbamate is an insecticide used for the control of hoppers on rice and lepidopterous larvae on vegetables at *c.* 40 g a.i./100 l.

Toxicology. The acute oral LD50 for rats is 380 mg/kg; the acute dermal LD50 for mice is > 1500 mg/kg. The LC50 (48 h) for carp is 10 mg/l.

Formulations. These include: e.c. (300 g a.i./l); w.p. (500 g/kg); dust (20 g/kg); micro-granules (30 g/kg).

Analysis. Product analysis is by u.v. spectroscopy after purification by TLC. Residues may be determined, after conversion to 2,4-dinitrophenyl 3,4-xylyl ether, by GLC with ECD (J. Hiyamoto *et al., J. Pestic. Sci.,* 1978, **3,** 119).

3,5-Xylyl methylcarbamate

$$\text{O.CNHMe structure on 3,5-dimethylphenyl ring}$$

(structure: aromatic ring with O.CNHMe group at top bearing C=O, and Me groups at 3 and 5 positions)

1MVOR C1 E1 $C_{10}H_{13}NO_2$ (179·2)

Nomenclature and development. The common name XMC is approved by JMAF. The IUPAC name is 3,5-xylyl methylcarbamate (I), in *C.A.* usage 3,5-dimethylphenyl methylcarbamate *[2655-14-3]* formerly (I). It was introduced as an insecticide in 1968 by the Hokko Chemical Industry Co. Ltd and the Hodogaya Chemical Co. Ltd, under the code number 'H-69' and trade mark 'Macbal'.

Manufacture and properties. Produced either by the condensation of 3,5-xylyl chloroformate with methylamine or. by the reaction of 3,5-xylenol and methyl isocyanate, it forms colourless crystals, m.p. 99°. It is practically insoluble in water; soluble in most organic solvents including benzene, isophorone, cyclohexanone. The technical product is *c.* 97% pure.

Uses. 3,5-Xylyl methylcarbamate is a systemic insecticide recommended for use at 600 g a.i./ha against leafhoppers and planthoppers on rice.

Toxicology. The acute oral LD50 is: for rats 542 mg/kg; for rabbits 445 mg/kg. In 90-d feeding trials the 'no effect' level was for rats and mice 230 mg/kg daily. The LC50 (48 h) for carp is > 40 mg/l.

Formulations. These include: e.c. (200 g a.i./l); w.p. (500 g/kg); dust (20 g/kg); micro-granules (30 g/kg).

Analysis. Product analysis is by separation of the a.i. by TLC, recovery from the spot followed by hydrolysis to 3,5-xylenol which is measured by u.v. spectroscopy at 230 and 291 nm.

Zinc phosphide

$$Zn_3P_2$$

.ZN3.P2 P_2Zn_3 (258·1)

Nomenclature and development. The traditional chemical name zinc phosphide (I) is accepted by ISO and JMAF in lieu of a common name. The IUPAC name is trizinc diphosphide, in *C.A.* usage (I), Registry No. *[1314-84-7]*. It has long been in use as a rodent poison.

Manufacture and properties. Produced by the direct combination of zinc and phosphorus, it is a grey powder, *d* 4·55, of high m.p. and subliming when heated in the absence of oxygen. It is practically insoluble in water (with which it reacts), ethanol; soluble in benzene, carbon disulphide. It is stable when dry but decomposes slowly in moist air; it is decomposed violently by acids to produce phosphine which is a potent mammalian poison, and impurities in which render the gas spontaneously flammable. The technical product is 80-95% pure.

Uses. Zinc phosphide is a rodenticide, mainly restricted to use by trained personnel against rats, field mice and gophers. Baits (2·5 or 5·0 g/kg) are used against *Rattus rattus, R. norvegicus* and mice.

Toxicology. The acute oral LD50 for rats is 45·7 mg/kg. It is an intense poison to mammals and birds.

Formulations. Pastes (25 or 50 g a.i./kg) for bait preparation.

Analysis. Product analysis is by reaction with acid, the phosphine produced is either allowed to react with mercury dichloride and iodine, the excess of which is titrated with sodium thiosulphate (*CIPAC Handbook,* 1970, **1**, 703), or oxidised to phosphoric acid which is estimated by standard methods. (J. W. Elmore & F. R. Roth, *J. Assoc. Off. Agric. Chem.,* 1943, **26**, 559; 1947, **30**, 213; B. L. Griswold *et al, Anal. Chem.,* 1951, **23**, 192).

Zineb

$$\left[-\overset{\overset{\displaystyle S}{\|}}{S.C}.NH.CH_2CH_2NH.\overset{\overset{\displaystyle S}{\|}}{C}.S.Zn - \right]_x$$

SUYS&M2MYS&US &-ZN- $\hspace{5cm}$ $(C_4H_6N_2S_4Zn)x$ $\hspace{0.5cm}$ $(275\cdot8)x$

Nomenclature and development. The common name zineb is approved by BSI, ISO and JMAF—exception West Germany; also known as ENT 14874. The IUPAC name is zinc ethylenebis(dithiocarbamate), in *C.A.* usage [[1,2-ethanediylbis[carbamodithioato]]-(2-)]zinc complex *[12122-67-7]* formerly [ethylenebis[dithiocarbamato]]zinc. J. M. Heuberger & T. F. Manns, *Phytopathology,* 1943, **33,** 113, first reported that the addition of zinc sulphate improved the field performance of nabam (p.370). Zineb was then introduced by the Rohm & Haas Co. under the trade mark 'Dithane Z-78' and by E. I. du Pont de Nemours & Co. (Inc.) (who no longer manufacture or market it) under the trade mark *'Parzate',* protected by USP 2457674; 3050439.

Manufacture and properties. Produced by the precipitation from aqueous nabam on addition of soluble zinc salts, it is a light-coloured powder, which decomposes without melting, v.p. negligible at room temperature. Its solubility at room temperature is *c.* 10 mg/l water; soluble in carbon disulphide, pyridine. It is somewhat unstable to light, heat and moisture. When precipitated from concentrated solution a polymer is formed which is of lower fungicidal activity. Etem (see p.239) is one of its oxidation products (R. A. Ludwig *et al., Can. J. Bot.,* 1955, **33,** 42).

Uses. Zineb is a fungicide used to protect foliage and fruit of a wide range of crops against diseases such as potato and tomato blight, *Botrytis* spp., downy mildews and rusts. It is generally non-phytotoxic except to zinc-sensitive varieties.

Toxicology. The acute oral LD50 for rats is > 5200 mg/kg. Cases of irritation of skin and the mucous membranes have been reported.

Formulations. These include: w.p. (700 and 750 g a.i./kg); dust ('Dithane' dust).

Analysis. Product analysis is by acid hydrolysis, converting the carbon disulphide evolved into dithiocarbonate which is estimated by titration with iodine (*CIPAC Handbook,* 1970, **1,** 706; 1979, **1A,** in press). Residues may be determined by acid hydrolysis converting the carbon disulphide evolved into a copper/amine complex which is estimated colorimetrically (H. L. Pease, *J. Assoc. Off. Agric. Chem.,* 1957, **30,** 1113; G. E. Keppel, *J. Assoc. Off. Anal. Chem.,* 1969, **52,** 162). See also: W. K. Lowen & H. L. Pease, *Anal. Methods Pestic., Plant Growth Regul. Food Addit.,* 1964, **3,** 69.

Zineb-ethylenebis(thiuram disulphide) mixed precipitation

$$\left[\begin{array}{l} (-\mathrm{S.CS.NH.CH_2CH_2NH.CS.S}-)^{--} \; \mathrm{Zn(NH_3)_2^{++}} \big]_3 \\ (-\mathrm{S.CS.NH.CH_2CH_2NH.CS.S}-) \end{array} \right]_x$$

$$x > 1$$

Nomenclature and development. The common name metiram is approved by New Zealand and JMAF for a product, defined by the manufacturers as a mixed precipitation consisting of the ammonia complex of zinc ethylenebis(dithiocarbamate) and poly[ethylenebis(thiuram disulphide)]; this common name is no longer accepted by BSI (and was not adopted by ISO) because the product appears to be a mixture rather than a complex. It is indexed under this name by *C.A.*, Registry No. *[9006-42-2]*. It was introduced in 1958 by BASF AG under the trade mark 'Polyram' and protected by BP 840211; USP 3248400; the Agricultural Chemical Division of FMC Corp. used the code number 'FMC 9102'.

Manufacture and properties. Produced by the joint oxidation and precipitation of ethylenebis(dithiocarbamic acid) with a soluble zinc salt, it is a yellowish powder which begins to decompose *c.* 140°. It is practically insoluble in water, organic solvents such as acetone, benzene, ethanol; soluble, with decomposition, in pyridine. It is unstable under strongly acidic or alkaline conditions and tank mixes with certain insecticides, such as malathion or diazinon, it should be applied immediately after preparation; it is compatible with parathion but not with parathion oil sprays.

Uses. It is a non-systemic fungicide used for foliage application to control *Phytophthora infestans* on potatoes, *Pseudoperonospora humuli* on hops, *Botrytis tulipae* on tulips. It is also used in combination with tridemorph to control *Puccinia striiformis* on barley and winter wheat and with nitrothal-isopropyl to control powdery mildew on top fruit. It is non-phytotoxic at spray concentrations.

Toxicology. The acute oral LD50 is: for rats > 10000 mg/kg; for male mice > 5400 mg/kg; for female guinea-pigs 2400-4800 mg/kg. In feeding trials: rats receiving 10000 mg/kg diet, but not those 1000 mg/kg diet, for 14 d showed an increase in thyroid weight and decreased uptake of iodide; no ill-effect was observed for dogs receiving 45 mg/kg daily for 90 d, or 7·5 mg/kg daily for 1·92 y. The LC50 (and exposure period) for harlequin fish is 32 mg/l (24 h) and 17 mg/l (48 h).

Formulations. These include: 'Polyram' and 'Polyram Combi', w.p. (800 g/kg); dusts of various strength; combinations with other fungicides, *e.g.* 'Pallinal', w.p. (mixture with nitrothal-isopropyl).

Analysis. Product analysis is by acid hydrolysis, the carbon disulphide liberated is absorbed in methanolic potassium hydroxide and the resulting dithiocarbonate estimated by titration with iodine. Residues may be determined by reaction of the carbon disulphide, formed on hydrolysis, with a copper-amine reagent and the resulting dithiocarbamate complex estimated colorimetrically (D. G. Clarke *et al., Anal. Chem.,* 1951, **23,** 1842).

Ziram

$$\left[\begin{array}{c} \overset{\displaystyle S}{\underset{\displaystyle \|}{}} \\ Me_2N\,C.S^- \end{array} \right]_2 \ \ Zn^{2+}$$

SUYS&N1&1   2   &-ZN- $C_6H_{12}N_2S_4Zn$ (305·8)

Nomenclature and development. The common name ziram is approved by BSI, ISO and JMAF—exception West Germany; also known as ENT 988. The IUPAC name is zinc dimethyldithiocarbamate, in *C.A.* usage (*T*-4)-bis(dimethyldithiocarbamato)zinc *[137-30-4]*. It was introduced as a fungicide in the early 1930's by E. I. du Pont de Nemours & Co. (Inc.) (who no longer manufacture or market it) under the trade marks *'Milbam', 'Zerlate',* by Schering AG under the trade mark 'Fuklasin' and by Aagrunol as a rodent repellent under the trade mark 'Aaprotect'.

Manufacture and properties. Produced as a precipitate by the addition of an aqueous solution of a zinc salt to one of sodium dimethyldithiocarbamate, it is a colourless powder, m.p. 240°, v.p. negligible at room temperature. Its solubility at 25° is 65 mg/l water; slightly soluble in ethanol, diethyl ether; moderately soluble in acetone; soluble in dilute alkali, chloroform, carbon disulphide. The technical product has m.p. 240-244°. It is stable under normal conditions but decomposed by acids. It is compatible with other pesticides except copper and mercury compounds.

Uses. Ziram is a protective fungicide for use on fruit and vegetable crops against *Alternaria* and *Septoria* spp. It is non-phytotoxic except to zinc-sensitive plants. It is also used as a repellent to birds and rodents.

Toxicology. The acute oral LD50 for rats is 1400 mg/kg; it may cause irritation of the skin and mucous membranes. In 1-y feeding trials rats receiving 5 mg a.i./kg daily showed no effect, neither did weanling rats receiving 100 mg/kg diet for 30 d.

Formulations. A w.p. (900 g a.i./kg); 'Aaprotect', repellent paste (370 g/kg + sticker).

Analysis. Product analysis is by acid hydrolysis, the carbon disulphide liberated being absorbed in methanolic potassium hydroxide and the resulting potassium *O*-methyl dithiocarbonate estimated by titration with iodine (*CIPAC Handbook,* 1970, **1**, 716). Residues may be determined by acid hydrolysis followed by colorimetry of the complex produced by the liberated carbon disulphide with copper salts and amines (W. K. Lowen & H. L. Pease, *Anal. Methods Pestic., Plant Growth Regul. Food Addit.,* 1964, **3**, 69).

Appendix I

SUPERSEDED COMPOUNDS

The compounds in the following list, though at one time marketed or widely reported, are believed to be currently of little commercial interest in the subject areas covered by the *Pesticide Manual*. Some are still widely used but for different purposes. The following information is given, if available: common names (BSI, ISO, WSSA and JMAF, but *not* exceptions); chemical names (IUPAC and *C.A.*) and *C.A.* Registry No.; other trivial names; code numbers; trade marks (and firm(s)); molecular formula; Wiswesser Line-Notation. For those compounds described in earlier editions of the *Pesticide Manual* the last edition and page numbers on which they were reported are added.

Sometimes conflicting reports have been received from various countries about the current status of a given pesticide. The Editor will be pleased to receive details about any of the compounds in this list which are currently in commercial use and are likely to remain so in the foreseeable future.

(1) S-2-Acetamidoethyl O,O-dimethyl phosphorodithioate (IUPAC); S-[2-(acetylamino)-ethyl] O,O-dimethyl phosphorodithioate *[13265-60-6]* (*C.A.*); DAEP (JMAF); 'Amiphos' (Nippon Soda Co. Ltd); $C_6H_{14}NO_3PS_2$; 1VM2SPS&O1&O1.

(2)   Acrylonitrile (IUPAC; acceptable BSI, ISO); 2-propenenitrile *[107-13-1]* (*C.A.*); vinyl cyanide; ENT 54; 'Ventox' (American Cyanamid Co.; Degesch AG); C_3H_3N; NC1U1; Ed. **5**, p.3.

(3)   Allidochlor (BSI, ISO), CDAA (WSSA, JAMF); N,N-diallyl-2-chloroacetamide (IUPAC), 2-chloro-N,N-di-2-propenylacetamide *[93-71-0]* (*C.A.*); 'CP 6343', 'Randox' (Monsanto Co.); $C_8H_{12}ClNO$; G1VN2U1&2U1; Ed. **5**, p.8.

(4)   Allyxycarb (BSI, ISO), APC (JMAF); 4-diallylamino-3,5-xylyl methylcarbamate (IUPAC), 4-(di-2-propenylamino)-3,5-dimethylphenyl methylcarbamate *[6392-46-7]* (*C.A.*); 'Bayer 50282', 'A 546', 'Hydrol' (Bayer AG); $C_{16}H_{22}N_2O_2$; 1U2N2U1&R B1 F1 DOVM1; Ed. **4**, p.10.

(5)   Amidithion (BSI, ISO); S-(N-2-methoxyethylcarbamoylmethyl) O,O-dimethyl phosphorodithioate (IUPAC), S-[2-[(2-methoxyethyl)amino]-2-oxoethyl] O,O-dimethyl phosphorodithioate *[919-76-6]* (*C.A.*); ENT 27160; 'Ciba 2446', 'Thiocron' (Ciba-Geigy AG); $C_7H_{16}NO_4PS_2$; 1OPS&O1&S1VM2O1; Ed. **2**, p.16.

(6)   Amiton (BSI); S-2-(diethylamino)ethyl O,O-diethyl phosphorothioate (IUPAC, *C.A.*) *[78-53-5]*; ENT 24980-X; 'Tetram' (ICI Ltd, Plant Protection Division); $C_{10}H_{24}NO_3PS$; 2OPO&O2&S2N2&2.

(7)   Anabasine (JMAF); (S)-3-(piperidin-2-yl)pyridine (IUPAC), (S)-3-(2-piperidinyl)pyridine *[494-52-0]* (*C.A.*); racemate known as 'Neonicotine'; $C_{10}H_{14}N_2$; T6NJ C-BT6MTJ.

(8)   Anilazine (BSI, ISO), triazine (JMAF); 2,4-dichloro-6-(2-chloroanilino)-1,3,5-triazine (IUPAC), 4,5-dichloro-N-(2-chlorophenyl)-1,3,5-triazin-2-amine *[101-05-3]* (*C.A.*); 2,4-dichloro-6-(o-chloroanilino)-s-triazine; ENT 26058; 'B-622' (Ethyl Corp.), 'Dyrene' (Chemagro), 'Triazine' (Nippon Soda Co. Ltd); $C_9H_5Cl_3N_4$; T6N CN ENJ BMR BG& DG FG; Ed. **5**, p.19.

(9)   Athidathion (BSI, ISO); O,O-diethyl S-5-methoxy-2-oxo-1,3,4-thiadiazol-3-ylmethyl phosphorodithioate (IUPAC), O,O-diethyl S-[(5-methoxy-2-oxo-1,3,4-thiadiazol-3(2H)-yl)methyl] phosphorodithioate *[19691-80-6]* (*C.A.*); 'G 13006' (Ciba-Geigy AG); $C_8H_{15}N_2O_4PS_3$; T5NNVSJ B1SPS&O2&O2 EO1.

(10)   Atraton (BSI, ISO); 2-ethylamino-4-isopropylamino-6-methoxy-1,3,5-triazine (IUPAC), N-ethyl-6-methoxy-N'-(1-methylethyl)-1,3,5-triazine-2,4-diamine *[1610-17-9]* (*C.A.*); 2-ethylamino-4-isopropylamino-6-methoxy-s-triazine; 'G 32293', 'Gesatamin' (Ciba-Geigy AG); $C_9H_{17}N_5O$; T6N CN ENJ BO1 DMY1&1 FM2.

553

(1) 2-Azido-4-*tert*-butylamino-6-ethylamino-1,3,5-triazine (IUPAC), 6-azido-*N*-(1,1-dimethylethyl)-*N'*-ethyl-1,3,5-triazine-2,4-diamine *[2854-70-8]* (*C.A.*); 2-azido-4-*tert*-butylamino-6-ethylamino-*s*-triazine; WL 9385 (Shell Research Ltd); $C_9H_{16}N_8$; T6N CN ENJ BNNN DMX1&1&1 FM2.

(2) Azobenzene (IUPAC, accepted BSI, ISO); diphenyldiazene *[103-33-3]* (*C.A.*); ENT 14 611; $C_{12}H_{10}N_2$; RNUNR; Ed. **5**, p.27.

(3) Azothoate (BSI, ISO); *O*-4-(4-chlorophenylazo)phenyl *O,O*-dimethyl phosphorothioate (IUPAC, *C.A.*) *[5834-96-8]*; 'L 1058', 'Slam C' (Montecatini Edison S.p.A.); $C_{14}H_{14}ClN_2O_3PS$; GR DNUNR DOPS&O1&O1.

(4) Barium carbonate (IUPAC, *C.A.*) *[513-77-9]*; $CBaO_3$; .BA..C-O3; Ed. **5**, p.28.

(5) Benquinox (BSI, ISO); 1,4-benzoquinone 1-benzoylhydrazone 4-oxime (IUPAC), benzoic acid (4-hydroxyimino-2,5-cyclohexadien-1-ylidene)hydrazide *[495-73-8]* (*C.A.*); COBH; 'Bayer 15080', 'Ceredon' (Bayer AG); $C_{13}H_{11}N_3O_2$; L6Y DYJ AUNQ DUNMVR; Ed. **4**, p.35.

(6) Benzadox (BSI, WSSA); benzamido-oxyacetic acid (!UPAC), [(benzoylamino)oxy]acetic acid *[5251-93-4]* (*C.A.*); 'S 6173', 'Topcide' (Gulf Oil Corp.), 'MC 0035' (Murphy Chemical Ltd); $C_9H_9NO_4$; QV1OMVR; Ed. **4**, p.38.

(7) Benzipram (WSSA); *N*-benzyl-*N*-isopropyl-3',5'-dimethylbenzamide (IUPAC), 3,5-dimethyl-*N*-(1-methylethyl)-*N*-(phenylmethyl)benzamide *[35256-86-1]* (*C.A.*); *N*-benzyl-*N*-isopropyl-3,5-dimethylbenzamide; $C_{19}H_{23}NO$; 1Y1&N1R&VR C1 E1.

(8) *S*-Benzyl *O*-ethyl phenylphosphonothioate (IUPAC), *O*-ethyl *S*-phenylmethyl phenylphosphonothioate *[21722-85-0]* (*C.A.*); ESBP (JMAF); 'Inezin' (Nissan Chemical Industries Ltd); $C_{15}H_{17}O_2PS$; 2OPO&R&S1R.

(9) 5-[Bis[2-(2-butoxyethoxy)ethoxy]methyl]-1,3-benzodioxole (IUPAC, *C.A.*) *[5281-13-0]*; 1-bis[2-(2-butoxyethoxy)ethoxy]methyl-3,4-methylenedioxybenzene; piperonal bis[2-(2-butoxyethoxy)ethyl]acetal; 'Tropital' (McLaughlin Gormley King Co.); $C_{24}H_{40}O_8$; T56 BO DO CHJ GYO2O2O4&O2O2O4; Ed. **4**, p.166.

(10) Bis(2-chloroethyl) ether (IUPAC), 1,1'-oxybis[2-chloroethane] *[111-44-4]* (*C.A.*); di-(2-chloroethyl) ether; ENT 4505; $C_4H_8Cl_2O$; G2O2G; Ed. **4**, p.177.

(11) Bis(4-chlorophenoxy)methane (IUPAC), 1,1'-[methylenebis(oxy)]bis(4-chlorobenzene) *[555-89-5]* (*C.A.*); DCPM, oxythane; 'K-1875', 'Neotran' (The Dow Chemical Co.); $C_{13}H_{10}Cl_2O_2$; GR DO1OR DG; Ed. **1**, p.39.

(12) 1,1-Bis(4-chlorophenyl)-2-ethoxyethanol (IUPAC), 4-chloro-α-(4-chlorophenyl)-α-(ethoxymethyl)benzenemethanol *[6012-83-5]* (*C.A.*); 4,4'-dichloro-α-(ethoxymethyl)benzhydrol; 'Geigy 337', 'G 23645', 'Etoxinol' (Ciba-Geigy AG); $C_{16}H_{16}Cl_2O_2$; GR DXQR DG&1O2.

(13) 1,1-Bis(4-chlorophenyl)-2-nitrobutane + 1,1-bis(4-chlorophenyl)-2-nitropropane mixture (IUPAC); 1,1'-(2-nitrobutylidene)bis(4-chlorobenzene) *[117-26-0]* + 1,1'-(2-nitropropylidene)bis(4-chlorobenzene) *[117-27-1]* mixture *[8002-82-2]* (*C.A.*); ENT 18 066, (2:1 mixture); 'CS 708', 'Dilan' (mixture); 'CS 674A', 'Bulan' (butane); 'CS 645A', 'Prolan' (propane) (Commercial Solvents Corp.); $C_{16}H_{15}Cl_2NO_2$ + $C_{15}H_{13}Cl_2NO_2$; WNY2&YR DG&R DG + WNY1&YR DG&R DG; Ed. **5**, p.176.

(14) Bis(diethoxyphosphinothioyl)disulphide + bis(di-isopropoxyphosphinothioyl) disulphide 75:25 mixture (IUPAC); bis(diethoxyphosphinothioyl) disulfide *[2901-90-8]* + bis[di-(1-methylethoxy)phosphinothioyl] disulfide *[3031-21-8]* (*C.A.*); ENT 23 584; 'FMC 1137', 'Phostex' (FMC Corp.); $C_8H_{20}O_4P_2S_4$ + $C_{12}H_{28}O_4P_2S_4$; 2OPS&O2&S=SPS&O2&O2 + 1Y1&OPS&OY1&1&SSPS&OY1&1&OY1&1.

(1) 2,4-Bis(3-methoxypropylamino)-6-(methylthio)-1,3,5-triazine (IUPAC), *N,N'*-bis(3-methoxypropyl)-6-methylthio-1,3,5-triazine-2,4-diamine *[845-52-3]* (*C.A.*); 2,4-bis(3-methoxypropyl)-6-methylthio-*s*-triazine; MPMT: 'CP 17029', 'Lambast' (Monsanto Co.); $C_{12}H_{23}N_5O_2S$; T6N CN ENJ BS1 DM3O1 FM3O1.

(2) Bis(methylmercury) sulphate (IUPAC), bis(methylmercury) sulfate *[3810-81-9]* (*C.A.*); 'Cerewet', 'Aretan-nieuw' (Bayer AG); $C_2H_6Hg_2O_4S$; 1-HG-OSWO-HG-1; Ed. **2**, p.47.

(3) 1-Bromo-2-chloroethane (IUPAC, *C.A.*) *[107-04-0]*; ethylene chlorobromide; C_2H_4BrCl; G2E; Ed. **4**, p.255.

(4) 3-Bromo-1-chloroprop-1-ene (IUPAC) 3-bromo-1-chloro-1-propene *[3737-00-6]* (*C.A.*); 'CBP-55' (Shell Development Co.); C_3H_4BrCl; G1U2E.

(5) Brompyrazone (BSI), brompyrazon (ISO); 5-amino-4-bromo-2-phenylpyridazin-3-one (IUPAC), 5-amino-4-bromo-2-phenyl-3(2*H*)-pyridazinone *[3042-84-0]* (*C.A.*); 'Basanor' (BASF AG); $C_{10}H_8BrN_3O$; T6NNVJ BR& DE EZ; Ed. **5**, p.55.

(6) Butacarb (BSI, ISO); 3,5-di-*tert*-butylphenyl methylcarbamate (IUPAC), 3,5-bis(1,1-dimethylethyl)phenyl methylcarbamate *[2655-19-8]* (*C.A.*); 'RD 14639', 'BTS 14639' (The Boots Co. Ltd); $C_{16}H_{25}NO_2$; 1X1&1&R COVM1 EX1&1&1; Ed. **5**, p.59.

(7) Butonate (BSI, ISO); dimethyl 1-butyryloxy-2,2,2-trichloroethylphosphonate (IUPAC), 2,2,2-trichloro-1-(dimethoxyphosphinyl)ethyl butanoate *[126-22-7]* (*C.A.*); butyrate ester of dimethyl (2,2,2-trichloro-1-hydroxyethyl)phosphonate; ENT 20852; Prentiss Drug & Chemical Co.; $C_8H_{14}Cl_3O_5P$; GXGGYOV3&PO&O1&O1; Ed. **4**, p.61.

(8) 2-(2-Butoxyethoxy)ethyl thiocyanate (IUPAC, *C.A.*) *[112-56-1]*; butyl 'Carbitol' thiocyanate, butyl 'Carbitol' rhodanate; ENT 6; 'Lethane 384' (Rohm & Haas Co.); $C_9H_{17}NO_2S$; NCS2O2O4; Ed. **5**, p.62.

(9) Butoxypoly(propylene glycol) α-2-hydroxypropyl-ω-butoxypoly[oxy(1-methylethylene)] (IUPAC), α-butyl-ω-hydroxypoly[oxy(methyl-1,2-ethanediyl)] *[9003-13-8]* (*C.A.*); ENT 8286; 'Crag Fly Repellent' (Union Carbide Corp.); /*Y1&1O*/.

(10) 6-*tert*-Butyl-4,5-dihydro-3-isopropylisothiazolo[3,4-*d*]pyrimidin-4-one (IUPAC), 6-(1,1-dimethylethyl)-3-(1-methylethyl)isothiazolo[3,4-*d*]pyrimidin-4(5*H*)-one *[40915-86-4]* (*C.A*); 'FMC 19 873' (FMC Corp.); $C_{12}H_{17}N_3OS$; T56 BNS FVM INJ DY1&1 HX1&1&1; Ed. **4**, p.66.

(11) 6-*tert*-Butyl-4,5-dihydro-3-isopropylisoxazolo[3,4-*d*]pyrimidin-4-one (IUPAC), 6-(1,1-dimethylethyl)-3-(1-methylethyl)isoxazolo[3,4-*d*]pyrimidin-4(5*H*)-one *[38897-15-3]* (*C.A.*); 'FMC 23 486' (FMC Corp.); $C_{12}H_{17}N_3O_2$; T56 BNO FVM INJ DY1&1 HX1&1&1; Ed. **5**, p.66.

(12) 6-*tert*-Butyl-4,5-dihydro-3-isopropylisoxazolo[5,4-*d*]pyrimidin-4-one (IUPAC), 6-(1,1-dimethylethyl)-3-(1-methylethyl)isoxazolo[5,4-*d*]pyrimidin-4(5*H*)-one *[35258-87-8]* (*C.A.*); 'FMC 21844' (FMC Corp.); $C_{12}H_{17}N_3O_2$; T56 BON FVM INJ DY1&1 HX1&1&1; Ed. **4**, p.68.

(13) 6-*tert*-Butyl-4,5-dihydro-3-propylisoxazolo[5,4-*d*]pyrimidin-4-one (IUPAC), 6-(1,1-dimethylethyl)-3-propylisoxazolo[5,4-*d*]pyrimidin-4(5*H*)-one *[35260-91-4]* (*C.A.*); 'FMC 21861' (FMC Corp.); $C_{12}H_{17}N_3O_2$; T56 BON FVM INJ D3 HX1&1&1; Ed. **3**, p.75.

(14) 2-[2-(4-*tert*-Butylphenoxy)-1-methylethoxy]-1-methylethyl 2-chloroethyl sulphite (IUPAC), 2-chloroethyl 2-[2-[4-(1,1-dimethylethyl)phenoxy]-1-methylethoxy]-1-methylethyl sulfite *[3761-60-2]* (*C.A.*); 'OW9', 'Smite' (Uniroyal Inc.); $C_{18}H_{29}ClO_5S$; G2OSO&OY1&1O= Y1&1OR DX1&1&1.

(1) 2-(4-*tert*-Butylphenoxy)-1-methylethyl 2-chloroethyl sulphite (IUPAC), 2-chloroethyl 2-[4-(1,1-dimethylethyl)phenoxy]-1-methylethyl sulfite *[140-57-8]* (*C.A.*); ENT 16 519, aramite (JMAF); '88-R', 'Aramite' (Uniroyal Inc.); $C_{15}H_{23}ClO_4S$; G2OSO&OY1&1= OR DX1&1&1; Ed. **3,** p.73.

(2) Cadmium calcium copper zinc chromate sulphate (IUPAC), cadmium calcium copper zinc chromate sulfate *[12001-20-6]* (*C.A.*); 'Crag Turf Fungicide', 'Fungicide 531' (Union Carbide Corp.).

(3) Calcium cyanamide (IUPAC, *C.A.*) *[156-62-7]*; 'Cyanamid' (American Cyanamid Co.); $CCaN_2$; NCN &-CA-; Ed. **5,** p.72.

(4) Carbamorph (BSI, ISO); morpholinomethyl dimethyldithiocarbamate (IUPAC), *S*-morpholinomethyl dimethylcarbamodithioate *[31848-11-0]* (*C.A.*); 'MC 833' (Murphy Chemical Co.); $C_8H_{16}N_2OS_2$; T6N DOTJ A1SYUS&N1&1.

(5) Carbanolate (BSI, ISO); 6-chloro-3,4-xylyl methylcarbamate (IUPAC), 2-chloro-4,5-dimethylphenyl methylcarbamate *[671-04-5]* (*C.A.*); OMS 174; 'U 12927', 'Banol' (Upjohn Co.); $C_{10}H_{12}ClNO_2$; GR C1 D1 FOVM1.

(6) Chinosol, hydroxyquinoline sulfate (JMAF); bis(8-hydroxyquinolinium) sulphate (IUPAC); 8-quinolinol sulfate (2:1) *[134-31-6]* (*C.A.*); oxine sulphate; $C_{18}H_{16}N_2O_6S$; T66 BNJ JQ 2 &WSQQ; Ed. **5,** p.87.

(7) Chloraniformethan (BSI, ISO); *N*-[2,2,2-trichloro-1-(3,4-dichloroanilino)ethyl]formamide (IUPAC), *N*-[[2,2,2-trichloro-1-(3,4-dichlorophenyl)amino]ethyl]formamide *[20856-57-9]* (*C.A.*); 'Bayer 79 770', 'Imugan', 'Milfaron' (Bayer AG); $C_9H_7Cl_5N_2O$; GXGG= YMVHMR CG DG; Ed. **4,** p.91.

(8) Chloranil (accepted BSI, ISO); 2,3,5,6-tetrachloro-*p*-benzoquinone (IUPAC), 2,3,5,6-tetrachloro-2,5-cyclohexadiene-1,4-dione *[118-75-2]* (*C.A.*); ENT 3797; 'Spergon' (Uniroyal Chemicals); $C_6Cl_4O_2$; L6V DVJ BG CG EG FG; Ed. **5,** p.90.

(9) Chloranocryl (BSI, ISO), dicryl (ANSI, WSSA); 3',4'-dichloro-2-methylacrylanilide (IUPAC), *N*-(3,4-dichlorophenyl)-2-methyl-2-propenamide *[2164-09-2]* (*C.A.*); 'FMC 4556', 'Dicryl' (FMC Corp.); $C_{10}H_9Cl_2NO$; GR BG DMVY1&U1; Ed. **2,** p.85.

(10) Chlorazine (BSI); 2-chloro-4,6-bis(diethylamino)-1,3,5-triazine (IUPAC), 6-chloro-*N,N,N',N'*-tetraethyl-1,3,5-triazine-2,4-diamine *[580-48-3]* (*C.A.*); 2-chloro-4,6-bis(diethylamino)-*s*-triazine; 'G 30 031' (Ciba-Geigy AG); $C_{11}H_{20}ClN_5$; T6N CN ENJ BN2&2 DN2&2 FG.

(11) Chlorbenside (BSI, ISO); 4-chlorobenzyl 4-chlorophenyl sulphide (IUPAC), 1-chloro-4-[[(4-chlorophenyl)methyl]thio]benzene *[103-17-3]* (*C.A.*); *p*-chlorobenzyl *p*-chlorophenyl sulfide; ENT 20 696; 'HRS 860', 'RD 2195', 'Chlorparacide', 'Chlorsulphacide' (The Boots Co. Ltd); $C_{13}H_{10}Cl_2S$; GR DS1R DG; Ed. **3,** p.97.

(12) Chlorbicyclen (BSI, ISO); 1,2,3,4,7,7-hexachloro-5,6-bis(chloromethyl)-8,9,10-trinorborn-2-ene (IUPAC), 1,2,3,4,7,7-hexachloro-5,6-bis(chloromethyl)bicyclo[2.2.1]hept-2-ene (IUPAC, *C.A.*) *[2550-75-6]*; 'Hercules 426', 'Alodan' (Hercules Inc.); $C_9H_6Cl_8$; L55 A CUTJ AG AG BG CG DG EG F1G G1G.

(13) Chlordecone (BSI, ISO); decachloropentacyclo[5.2.1.0^{2,6}.0^{3,9}.0^{5,8}]decan-4-one (IUPAC), 1,1a,3,3a,4,5,5,5a,5b,6-decachloro-octahydro-1,3,4-metheno-2*H*-cyclobuta[*cd*]pentalen-2-one *[143-50-0]* (*C.A.*); ENT 16 391; 'GC-1189' 'Kepone' (Allied Chemical Corp.); $C_{10}Cl_{10}O$; L545 B4 C5 D 4ABCE J DVTJ AG BG CG EG FG GG HG HG IG JG; Ed. **5,** p.94.

(1) Chlorfensulphide (BSI, ISO), CPAS (JMAF); 4-chlorophenyl 2,4,5-trichlorophenylazo sulphide (IUPAC), [(4-chlorophenyl)thio](2,4,5-trichlorophenyl)diazene *[2274-74-0]* (*C.A.*); [(*p*-chlorophenyl)thio]-(2,4,5-trichlorophenyl)dimiide; present in 'Milbex' (Nippon Soda Co. Ltd); $C_{12}H_6Cl_4N_2S$; GR DSNUNR BG DG EG; Ed. **4**, p.101.

(2) Chlorflurazole (BSI, ISO); 4,5-dichloro-2-trifluoromethylbenzimidazole (IUPAC), 4,5-dichloro-2-(trifluoromethyl)-1*H*-benzimidazole *[3615-21-2]* (*C.A.*); 'NC 3363' (Fisons Ltd); $C_8H_3Cl_2F_3N_2$; T56 BM DNJ CXFFF FG GG.

(3) 2-Chloro-*N*-(2-cyanoethyl)acetamide (IUPAC, *C.A.*) *[17756-81-9]*; α-chloro-*N*-(2-cyanoethyl)acetamide; CECA (JMAF); 'Udonkor' (Nippon Soda Co. Ltd); $C_5H_7ClN_2O$; NC2MV1G.

(4) 1-Chloro-2,4-dinitronaphthalene (IUPAC, *C.A.*) *[2401-85-6]*; $C_{10}H_5ClN_2O_4$; L66J BG CNW ENW.

(5) Chloromebuform (BSI, ISO); N^1-butyl-N^2-(4-chloro-*o*-tolyl)-N^1-methylformamidine (IUPAC), *N*-butyl-*N'*-(4-chloro-2-methylphenyl)-*N*-methylmethanimidamide *[37407-77-5]* (*C.A.*); 'CGA 22 598' (Ciba-Geigy AG); $C_{13}H_{19}ClN_2$; GR Cl DNU1N4&1.

(6) *O*-2-Chloro-4-(methylthio)phenyl *O*-methyl ethylphosphoramidothioate (IUPAC, *C.A.*) *[54381-26-9]*; amidothioate (JMAF); 'Mitemate' (Nippon Kayaku Co. Ltd); $C_{10}H_{15}ClNO_2PS_2$; 2MPS&O1&OR BG DS1.

(7) *O*-3-Chloro-4-nitrophenyl *O,O*-dimethyl phosphorothioate (IUPAC, *C.A.*) *[500-28-7]*; 'Bayer 22/190', 'Chlorthion' (Bayer AG); $C_8H_9ClNO_5PS$; WNR BG DOPS&O1&O1.

(8) 1-Chloro-2-nitropropane (IUPAC, *C.A.*) *[2425-66-3]*; 'FMC 5916' 'Lanstan' (FMC Corp.); $C_3H_6ClNO_2$; WNY1&1G; Ed. **2**, p.104.

(9) 5-Chloro-4-phenyl-1,2-dithiol-3-one (IUPAC), 5-chloro-4-phenyl-3*H*-1,2-dithiol-3-one *[2425-05-0]* (*C.A.*); 'Hercules 3944' (Hercules Inc.); $C_9H_5ClOS_2$; T5SSVJ DR& EG; Ed. **3**, p.115.

(10) 3-(4-Chlorophenyl)-5-methyl-2-thioxothiazolidin-4-one (IUPAC), 3-(4-chlorophenyl)-5-methyl-2-thioxo-4-thiazolidinone *[6012-92-6]* (*C.A.*); 3-(4-chlorophenyl)-5-methyl-rhodanine; 'N 244' (Stauffer Chemical Co.); $C_{10}H_8ClNOS_2$; T5SYNV EHJ BUS CR DG& E1.

(11) 4-Chlorophenyl phenyl sulphone (IUPAC), 1-chloro-4-(phenylsulfonyl)benzene *[80-00-2]* (*C.A.*); *p*-chlorophenyl phenyl sulfone; 'R-242', 'Sulphenone' (Stauffer Chemical Co.); $C_{12}H_9ClO_2S$; WSR&R DG.

(12) 2-Chlorovinyl diethyl phosphate (IUPAC), 2-chloroethenyl diethyl phosphate *[311-47-7]* (*C.A.*); 'SD 1836', 'OS 1836' (Shell Chemical Co.); $C_6H_{12}ClO_4P$; G1U1OPO&O2&O2.

(13) 2-(4-Chloro-3,5-xylyloxy)ethanol (IUPAC), 2-[(4-chloro-3,5-dimethylphenyl)oxy]ethanol *[5825-79-6]* (*C.A.*); 'Experimental Chemotherapeutant 1182' (Union Carbide Corp.); $C_{10}H_{13}ClO_2$; Q2OR DG Cl E1.

(14) Chlorquinox (BSI, ISO); 5,6,7,8-tetrachloroquinoxaline (IUPAC, *C.A.*) *[3495-42-9]*; 'Lucel' (Fisons Ltd); $C_8H_2Cl_4N_2$; T66 BN ENJ GG HG IG JG; Ed. **5**, p.119.

(15) Copper bis(3-phenylsalicylate) (IUPAC), bis(2-hydroxy[1,1'-biphenyl]-3-carboxylato-O^2,O^0)copper *[5328-04-1]* (*C.A.*); bis(3-phenylsalicylato)copper(II); $C_{26}H_{18}CuO_6$; OVR BQ CR 2 &-CU-.

(16) Copper carbonate (basic); dicopper carbonate dihydroxide (IUPAC), [carbonato(2-)]di-hydroxydicopper *[12069-69-1]* (*C.A.*); malachite; $CH_2Cu_2O_5$; .CU2.C-O3.Q2; Ed. **5**, p.124.

(**1**) Copper sulphate pentahydrate (IUPAC), copper(2+) sulfate (1:1) *[7758-98-7]* (*C.A.*); blue vitriol; bluestone; blue copperas; $CuH_{10}O_9S$; .CU..S-O4 QH5; Ed. **5**, p.127.

(**2**) Copper zinc chromate (indefinite composition); 'Crag Fungicide 658', 'Experimental Fungicide 641' (Union Carbide Corp.).

(**3**) Coumaphos (BSI, ISO); *O*-3-chloro-4-methylcoumarin-7-yl *O,O*-diethyl phosphorothioate (IUPAC), *O*-(3-chloro-4-methyl-2-oxo-2*H*-benzopyran-7-yl) *O,O*-diethyl phosphorothioate *[56-72-4]* (*C.A.*); ENT 17 957; 'Bayer 21/199', 'Asuntol', 'Muscatox', 'Resitox' (Bayer AG): 'Co-Ral' (Baychem Corp.); $C_{14}H_{16}ClO_5PS$; T66 BOVJ DG E1 IOPS&O2&O2; Ed. **2**, p.120.

(**4**) Cupric hydrazinium sulphate, copper dihydrazinium disulphate (IUPAC), bis-(hydrazine)bis(hydrogen sulfato)copper *[33271-65-7]* (*C.A.*); copper dihydrazine disulfate; 'Mathieson 466', 'Omazene' (Olin Mathieson Chemical Corp.); $CuH_{10}N_4O_8S_2$; .Cu..ZZ.2.SO4*2.

(**5**) 2-Cyano-3-(2,4-dichlorophenyl)acrylic acid (IUPAC), 2-cyano-3-(2,4-dichlorophenyl)-2-propenoic acid *[6013-05-4]* (*C.A.*); 'Ethyl 214' (Ethyl Corp.); $C_{10}H_5Cl_2NO_2$; QVYCN&U1R BG DG.

(**6**) Cyanthoate (BSI, ISO); *S*-[*N*-(1-cyano-1-methylethyl)carbamoylmethyl] *O,O*-diethyl phosphorothioate (IUPAC), *S*-[2-[(1-cyano-1-methylethyl)amino]-2-oxoethyl] *O,O*-diethyl phosphorothioate *[3734-95-0]* (*C.A.*); 'M 1568', 'Tartan' (Montedison S.p.A.); $C_{10}H_{19}N_2O_4PS$; NCX1&1&MV1SPO&O2&O2; Ed. **5**, p.138.

(**7**) 3-(Cyclopent-2-enyl)-2-methyl-4-oxocyclopent-2-enyl (1*RS*)-*cis,trans*-chrysanthemate (IUPAC), 3-(2-cyclopenten-1-yl)-2-methyl-4-oxo-2-cyclopenten-1-yl (*RS*)-*cis,trans*-2,2-dimethyl-3-(2-methyl-1-propenyl)cyclopropanecarboxylate *[97-11-0]* (*C.A.*); cyclethrin; ENT 22952; American Cyanamid 43064; $C_{21}H_{28}O_3$; L5V BUTJ C1 B- CL5UTJ& DOV- BL3TJ A1 A1.

(**8**) Cypendazole (BSI, ISO); methyl 1-(5-cyanopentylcarbamoyl)benzimidazol-2-ylcarbamate (IUPAC), methyl [1-[(5-cyanopentyl)amino]carbonyl]-1*H*-benzimidazol-2-ylcarbamate *[28559-00-4]* (*C.A.*); 1-(5-cyanopentylcarbamoyl)benzimidazole-2-carbamate; 'DAM 18 654', 'Folcidin' (Bayer AG); $C_{16}H_{19}N_5O_3$; T56 BN DNJ BVM5CN CMVO1; Ed. **4**, p.145.

(**9**) Cyperquat (BSI, WSSA, proposed ISO); 1-methyl-4-phenylpyridinium ion (IUPAC, *C.A.*) *[48134-75-4]*; 'S 21634' (Gulf Oil Chemicals); $C_{12}H_{12}N$; T6KJ A1 DR.

(**10**) Cyprazine (BSI, ANSI, WSSA, proposed ISO); 2-chloro-4-cyclopropylamino-6-isopropylamino-1,3,5-triazine (IUPAC); 6-chloro-*N*-cyclopropyl-*N'*-(1-methylethyl)-1,3,5-triazine-2,4-diamine *[22936-86-3]* (*C.A.*); 2-chloro-4-cyclopropylamino-6-isopropylamino-*s*-triazine; 'S 6115', 'Outfox' (Gulf Oil Corp.); $C_9H_{14}ClN_5$; T6N CN ENJ BMY1&1 DG FM- AL3TJ; Ed. **4**, p.146.

(**11**) Cyprazole (BSI, WSSA, proposed ISO); *N*-[5-(2-chloro-1,1-dimethylethyl)-1,3,4-thiadiazol-2-yl]cyclopropanecarboxamide (IUPAC, *C.A.*) *[42089-03-2]*; 'S 19 073' (Gulf Oil Chemicals); $C_{10}H_{14}ClN_3OS$; T5NN DSJ CX1&1&1G EMV- AL3TJ.

(**12**) Cypromid (BSI, ISO, WSSA); 3',4'-dichlorocyclopropanecarboxanilide (IUPAC), *N*-(3,4-dichlorophenyl)cyclopropanecarboxamide *[2759-71-9]* (*C.A.*); 'S 6000' 'Clobber' (Gulf Oil Corp.); $C_{10}H_9Cl_2NO$; L3TJ AVMR CG DG; Ed. **3**, p.148.

(**13**) Delachlor (BSI, ISO); 2-chloro-*N*-(isobutoxymethyl)acet-2',6'-xylide (IUPAC); 2-chloro-*N*-(2,6-dimethylphenyl)-*N*-[(2-methylpropoxy)methyl]acetamide *[24353-58-0]* (*C.A.*); α-chloro-*N*-isobutoxymethylacet-2',6'-xylidide; 'CP 52 223', (Monsanto Co.); $C_{15}H_{22}ClNO_2$; 1Y1&1O1NV1GR B1 F1; Ed. **2**, p.100.

(1) 2,4-DEP (WSSA); mixture (*C.A.* Registry No. *[39420-34-3]*) of tris[2-(2,4-dichloro-phenoxy)ethyl] phosphite *[94-84-8]* + bis[2-(2,4-dichlorophenoxy)ethyl] phosphonate (IUPAC, *C.A.*); '3Y9', 'Falone' (Uniroyal Inc.); $C_{24}H_{21}Cl_6O_6P$ + $C_{16}H_{15}Cl_4O_5P$; GR CG DO2OPO2OR BG DG&O2OR BG DG + GR CG DO2OPHO&O2OR BG DG; Ed. **2**, p.143.

(2) Diamidafos (BSI, ANSI, proposed ISO); phenyl *N,N'*-dimethylphosphorodiamidate (IUPAC, *C.A.*) *[1754-58-1]*; 'DOWCO 169', 'Nellite' (The Dow Chemical Co.); $C_8H_{13}N_2O_2P$; 1MPO&OR&M1; Ed. **5**, p.413.

(3) Dibutyl adipate (IUPAC), dibutyl hexanedioate *[105-99-7]* (*C.A.*); 'Experimental Tick Repellent 3' (Union Carbide Corp.); $C_{14}H_{26}O_4$; 4OV4VO4.

(4) Dibutyl phthalate (IUPAC, accepted by BSI, ISO), dibutyl 1,2-benzenedicarboxylate *[84-74-2]* (*C.A.*); DBP; $C_{16}H_{22}O_4$; 4OVR BVO4; Ed. **5**, p.163.

(5) Dibutyl succinate (IUPAC), dibutyl butanedioate *[141-03-7]* (*C.A.*); *'Tabutrex'*, 'Tabatrex' (Glenn Chemical Co.); $C_{12}H_{22}O_4$; 4OV2VO4; Ed. **5**, p.164.

(6) Dicapthon (ANSI); *O*-(2-chloro-4-nitrophenyl) *O,O*-dimethyl phosphorothioate (IUPAC, *C.A.*) *[2463-84-5]*; ENT 17035, OMS 214; 'Experimental Insecticide 4124', 'Dicaptan' (American Cyanamid Co.); $C_8H_9ClNO_5PS$; WNR CG DOPS&O1&O1; Ed. **3**, p.169.

(7) Dichlone (BSI, ISO); 2,3-dichloro-1,4-naphthoquinone (IUPAC, 2,3-dichloro-1,4-naphthalenedione *[117-80-6]* (*C.A.*); ENT 3776; 'USR 604' 'Phygon' (Uniroyal Inc.); $C_{10}H_4Cl_2O_2$; L66 BV EVJ CG DG; Ed. **5**, p.169.

(8) Dichloralurea; 1,3-bis(2,2,2-trichloro-1-hydroxyethyl)urea (IUPAC, *C.A.*) *[116-52-9]*; DCU; 'Crag Herbicide 2' (Union Carbide Corp.); $C_5H_6Cl_6N_2O_3$; GXGGYQMVMYQXGGG; Ed. **1**, p.139.

(9) Dichlormate (BSI, ISO); 3,4-dichlorobenzyl methylcarbamate (IUPAC); 3,4-dichloro-benzenemethanol methylcarbamate *[1966-58-1]* (*C.A.*) (mixture with 2,3-dichloro-analogue *[62046-37-1]*); mixture 'UC 22 463' (pure 3,4-dichloro-isomer is 'UC 22 463A'), 'Rowmate' (Union Carbide Corp.), $C_9H_9Cl_2NO_2$; GR BG D1OVM1; Ed. **1**, p.141.

(10) *O*-(2,5-Dichloro-4-iodophenyl) *O*-ethyl ethylphosphonothioate (IUPAC, *C.A.*) *[25177-27-9]*; 'C 18 244' (Ciba-Geigy AG); $C_{10}H_{12}Cl_2IO_2PS$; IR BG EG DOPS&2&O2; Ed. **2**, p.231.

(11) 1,1-Dichloro-1-nitroethane (IUPAC, *C.A.*) *[594-72-9]*; 'Ethide' (Commercial Solvents Corp.); $C_2H_3Cl_2NO_2$; WNXGG1; Ed. **3**, p.177.

(12) 2,4-Dichlorophenyl benzenesulphonate (IUPAC), 2,4-dichlorophenyl benzenesulfonate *[97-16-5]* (*C.A.*); 'EM 293', 'Genite', 'Genitol' (Allied Chemical Corp.); $C_{12}H_8Cl_2O_3S$; WSR&OR BG DG; Ed. **4**, p.179.

(13) *N*-(3,5-Dichlorophenyl)succinimide (IUPAC), 1-(3,5-dichlorophenyl)-2,5-pyrrolidine-dione *[24096-53-5]* (*C.A.*); 'Ohric' (Sumitomo Chemical Co. Ltd); $C_{10}H_7Cl_2NO_2$; T5VNVJ BR CG EG.

(14) 1,3-Dichloro-1,1,3,3-tetrafluoropropane-2,2-diol (IUPAC, *C.A.*) *[993-57-8]*; 1,3-dichloro-1,1,3,3-tetrafluoroacetone hydrate (IUPAC), 1,3-dichloro-1,1,3,3-tetrafluoropropan-2-one *[121-21-9]* hydrate (*C.A.*); 'GC 9832' (Allied Chemical Corp.); $C_3H_2Cl_2F_4O_2$; GXFFXQQXGFF (diol), GXFFVXGFF &QH (hydrate); Ed. **4**, p.185.

(15) 3,4-Dichlorotetrahydrothiophene 1,1-dioxide (IUPAC, *C.A.*) *[3001-57-8]*; dichlorothiolane dioxide; 'PRD Experimental Nematicide' (Diamond Shamrock Chemical Co.); $C_4H_6Cl_2O_2S$; T5SWTJ CG DG.

(1) 2,2-Dichlorovinyl 2-(ethylsulphinyl)ethyl methyl phosphate (IUPAC), 2,2-dichloro-ethenyl 2-(ethylsulfinyl)ethyl methyl phosphate *[7076-53-1]* *(C.A.)*; 'Nexion 1378' **(Celamerck)**; $C_7H_{13}Cl_2O_5PS$; OS2&2OPO&O1&O1UYGG.

(2) Dichlozoline (BSI, ISO, JMAF); 3-(3,5-dichlorophenyl)-5,5-dimethyloxazolidine-2,4-dione (IUPAC), 3-(3,5-dichlorophenyl)-5,5-dimethyl-2,4-oxazolidinedione *[24201-58-9]* *(C.A.)*; 'Sclex' (Sumitomo Chemical Co. Ltd); $C_{11}H_9Cl_2NO_3$; T5OVNV EHJ CR CG EG& E1 E1.

(3) Dicyclohexylammonium 2-cyclohexyl-4,6-dinitrophenoxide (IUPAC); 2-cyclohexyl-4,6-dinitrophenol compound with dicyclohexylamine (1:1) *[317-83-9]* *(C.A.)*; ENT 30 828; 'DN 111' (The Dow Chemical Co.) *'Dynone II'* (Fisons Ltd); $C_{24}H_{37}N_3O_5$; L6TJ AR BQ CNW ENW &L6TJ AM- AL6TJ.

(4) *O,O*-Diethyl *O*-4-methylcoumarin-7-yl phosphorothioate (IUPAC), *O,O*-diethyl *O*-(4-methyl-2-oxo-2*H*-1-benzopyran-7-yl) phosphorothioate *[299-45-6]* *(C.A.)*; 'E 838', 'Potasan' (Bayer AG); $C_{14}H_{17}O_5PS$; T66 BOVJ E1 IOPS&O2&O2; Ed. **1**, p.158.

(5) Diethyl 5-methylpyrazol-3-yl phosphate (IUPAC), diethyl 5-methyl-1*H*-pyrazol-3-yl phosphate *[108-34-9]* *(C.A.)*; ENT 24 723; 'G 24 483', 'Pyrazoxon' (Ciba-Geigy AG); $C_8H_{15}N_2O_4P$; T5MNJ COPO&O2&O2 E1.

(6) *O,O*-Diethyl naphthalimido-oxyphosphonothioate (IUPAC), 2-[(diethoxyphosphino-thioyl)oxy-1*H*-benz[*de*]isoquinoline-1,3(2*H*)-dione *[2668-92-0]* *(C.A.)*; ENT 24 970; 'Bayer 22 408', 'S 125' (Bayer AG); $C_{16}H_{16}NO_5PS$; T666 1A M CVNVJ DOPS&O2&O2; Ed. **1**, p.159.

(7) *O,O*-Diethyl *O*-6-methyl-2-propylpyrimidin-4-yl phosphorothioate (IUPAC), *O,O*-diethyl *O*-[6-methyl-2-propyl-4-pyrimidinyl] phosphorothioate *[5826-91-5]* *(C.A.)*; 'G 24 622', 'Pirazinon' (Ciba-Geigy AG); $C_{12}H_{21}N_2O_3PS$; T6N CNJ B3 DOPS&O2&O2 F1.

(8) Dimethyl *cis*-bicyclo[2.2.1]hept-5-ene-2,3-dicarboxylate (IUPAC, *C.A.*) *[5826-73-3]*; dimethyl *cis*-norborn-5-ene-2,3-dicarboxylate; dimethylcarbate; 'Compound 3916', 'Dimelone'; 'NISY'; $C_{11}H_{14}O_4$; L55 A CUTJ FVO1 GVO1 &C.

(9) 5,5-Dimethyl-3-oxocyclohex-1-enyl dimethylcarbamate (IUPAC), 5,5-dimethyl-3-oxo-1-cyclohexen-1-yl dimethylcarbamate *[122-15-6]* *(C.A.)*; ENT 24 728; 'G 19 258', 'Dimetan' (Ciba-Geigy AG); $C_{11}H_{17}NO_3$; L6V BUTJ COVN1&1 E1 E1.

(10) 2-(4,5-Dimethyl-1,3-dioxolan-2-yl)phenyl methylcarbamate (IUPAC, *C.A.*) *[7122-04-5]*; ENT 27 410; 'C 10 015', 'Fondaren' (Ciba-Geigy AG); $C_{13}H_{17}NO_4$; T5O COTJ BR BOVM1& D1 E1; Ed. **3**, p.198.

(11) *O*-4-(Dimethylsulphamoyl)phenyl *O,O*-diethyl phosphorothioate (IUPAC), *O*-[4-[(dimethyl-amino)sulfonyl]phenyl] *O,O*-diethyl phosphorothioate *[3078-97-5]* *(C.A.)*; DSP (JMAF); 'Kaya-ace' (Nippon Soda Co. Ltd); $C_{12}H_{20}NO_5PS_2$; 2OPS&O2&OR DSWN1&1.

(12) Dimexan (BSI), dimexano (ISO); *O,O*-dimethyl dithiobis(thioformate) (IUPAC), dimethyl thioperoxydicarbonate *[1468-37-7]* *(C.A.)*; bis[methoxy(thiocarbonyl)] disulphide; 'Tri-PE' (Vondelingenplaat N.V.); $C_4H_6O_2S_4$; SUYO1&SSYUS&O1; Ed. **5**, p.206.

(13) Dinex (BSI, ISO); 2-cyclohexyl-4,6-dinitrophenol (IUPAC, *C.A.*) *[131-89-5]*; DN (JMAF); ENT 157; DNOCHP: 'DN1' (The Dow Chemical Co.); $C_{12}H_{14}N_2O_5$; L6TJ AR BQ CNW ENW.

(1) Dinocton is approved (BSI, ISO) to cover an isomeric reaction mixture comprising products that have been known as dinocton-4 and dinocton-6; see below.

(2) Dinocton-4; mixture *[32535-08-3]* including 2,6-dinitro-4-(1-propylpentyl)phenyl methyl carbonate *[6465-51-6]* + 4-(1-ethylhexyl)-2,6-dinitrophenyl methyl carbonate *[6465-60-7]* (IUPAC, *C.A.*); 'MC 1947' (Murphy Chemical Ltd); $C_{16}H_{22}N_2O_7$; 4Y3&R CNW ENW DOVO1 + 5Y2&R CNW ENW DOVO1; Ed. **4**, p.209.

(3) Dinocton-6; mixture *[32534-96-6]* including 2,4-dinitro-6-(1-propylpentyl)phenyl methyl carbonate *[19000-58-9]* + 2-(1-ethylhexyl)-4,6-dinitrophenyl methyl carbonate *[19000-52-3]* (IUPAC, *C.A.*); 'MC 1945' (Murphy Chemical Ltd); $C_{16}H_{22}N_2O_7$; 4Y3&R CNW ENW BOVO1 + 5Y2&R CNW ENW BOVO1; Ed. **4**, p.210.

(4) Dinoterb acetate (BSI, ISO); 2-*tert*-butyl-4,6-dinitrophenyl acetate (IUPAC), 2-(1,1-dimethylethyl)-4,6-dinitrophenyl acetate *[3204-27-1]* (*C.A.*); 'P 1108' (Murphy Chemical Ltd); $C_{12}H_{14}N_2O_6$; 1X1&1&R CNW ENW BOV1; Ed. **4**, p.214.

(5) Diphenyl sulphone (IUPAC, accepted BSI, ISO), 1,1'-sulfonylbis[benzene] *[127-63-9]* (*C.A.*); phenyl sulfone; DPS; $C_{12}H_{10}O_2S$; WSR&R; Ed. **4**, p.220.

(6) 2-(1,3-Dithiolan-2-yl)phenyl dimethylcarbamate (IUPAC, *C.A.*) *[21709-44-4]*; 'C 13 963' (Ciba-Geigy AG); $C_{12}H_{15}NO_2S_2$; T5S CSTJ BR BOVN1&1; Ed. **2**, p.209.

(7) DMPA (WSSA) *O*-2,4-dichlorophenyl *O*-methyl isopropylphosphoramidothioate (IUPAC), *O*-(2,4-dichlorophenyl) *O*-methyl (1-methylethyl)phosphoramidothioate *[299-85-4]* (*C.A.*); ENT 25 647, OMS 115; 'K 22 023', 'DOWCO 118', 'Zytron' (The Dow Chemical Co.); $C_{10}H_{14}Cl_2NO_2PS$; GR CG DOPS&O1&MY1&1; Ed. **2**, p.211.

(8) Endothion (BSI, ISO); *S*-5-methoxy-4-oxopyran-2-ylmethyl *O,O*-dimethyl phosphorothioate (IUPAC), *S*-[(5-methoxy-4-oxo-4*H*-pyran-2-yl)methyl] *O,O*-dimethyl phosphorothioate *[2778-04-3]* (*C.A.*); ENT 24 653; '7175 RP', 'Endocide' (Rhône Poulenc S.A.); 'AC 18 737' (American Cyanamid Co.); 'FMC 5767' (FMC Corp.); $C_9H_{13}O_6PS$; T6O DVJ B1SPO&O1&O1 EO1; Ed. **5**, p.234.

(9) Epofenonane (BSI, proposed ISO); 6,7-epoxy-3-ethyl-7-methylnonyl 4-ethylphenyl ether (IUPAC), 2-ethyl-3-[3-ethyl-5-(4-ethylphenoxy)pentyl]-2-methyloxirane *[57342-02-6]* (*C.A.*); 'Ro 10-3108' (Hoffman-La Roche & Co.); $C_{20}H_{32}O_2$; T3OTJ B2 B1 C2Y2&2OR D2.

(10) Erbon (ISO, ANSI, WSSA); 2-(2,4,5-trichlorophenoxy)ethyl 2,2-dichloropropionate (IUPAC), 2-(2,4,5-trichlorophenoxy)ethyl 2,2-dichloropropanoate *[136-25-4]* (*C.A.*); 'Daron', 'Erbon' (The Dow Chemical Co.); $C_{11}H_9Cl_5O_3$; GXG1&VO2OR BG DG EG; Ed. **5**, p.239.

(11) Ethiolate (BSI, ISO); *S*-ethyl diethylthiocarbamate (IUPAC), *S*-ethyl diethylcarbamothioate *[2941-55-1]* (*C.A.*); 'Prefox' (Gulf Oil Chemicals Co.); $C_7H_{15}NOS$; 2SVN2&2; Ed. **4**, p.246.

(12) Ethoate-methyl (BSI, ISO); *S*-ethylcarbamoylmethyl *O,O*-dimethyl phosphorodithioate (IUPAC), *S*-[2-(ethylamino)-2-oxoethyl] *O,O*-dimethyl phosphorodithioate *[116-01-8]* (*C.A.*); ENT 25 506; 'B/77', 'Fitios' (Snia Viscosa); $C_6H_{14}NO_3PS_2$; 2MV1SPS&O1&O1; Ed. **5**, p.245.

(13) Ethylene glycol bis(trichloroacetate); ethylene bis(trichloroacetate) (IUPAC), 1,2-ethanediyl bis(trichloroacetate) *[2514-53-6]* (*C.A.*), 1,2-bis(trichloroacetoxy)ethane; 'Glytac' (Hooker Chemical Corp.); $C_6H_4Cl_6O_4$; GXGGVO2OVXGGG; Ed. **5**, p.252.

(14) Ethyleneurea; imidazolidin-2-one (IUPAC), 2-imidazolidinone *[120-93-4]* (*C.A.*); $C_3H_6N_2O$; T5MVMTJ; Ed. **1**, p.214.

Appendix I—*continued*

(1) 2-Ethyl-5-methyl-5-(2-methylbenzyloxy)-1,3-dioxane (IUPAC), 2-ethyl-5-methyl-5-(2-methylphenylmethoxy)-1,3-dioxane (*C.A.*), *[41126-10-6]* (*cis*-isomer), *[41814-30-4]* (*trans*-isomer); 'FMC 25 213' (FMC Corp.); $C_{15}H_{22}O_3$; T6O COTJ B2 EO1R B1& E1; Ed. **5**, p.254.

(2) *N*-(Ethylmercuri)-*p*-toluenesulphonanilide (IUPAC), ethyl(4-methyl-*N*-phenylbenzene-sulfonamidato-*N*)mercury *[517-16-8]* (*C.A.*); 'Ceresan M', 'Granosan M' (E. I. du Pont de Nemours & Co. (Inc.)); $C_{15}H_{17}HgNO_2S$; 2-HG-NR&SWR D1; Ed. **3**, p.252.

(3) 4-(Ethylthio)phenyl methylcarbamate (IUPAC, *C.A.*) *[18809-57-9]*; EMPC (JMAF): 'Toxamate' (Nippon Kayaku Co. Ltd); $C_{10}H_{13}NO_2S$; 2SR DOVM1.

(4) EXD (WSSA); *O,O*-diethyl dithiobis(thioformate) (IUPAC), diethyl thioperoxydi-carbonate *[502-55-6]* (*C.A.*); bis[ethoxy(thiocarbonyl)] disulphide; 'Herbisan' (Roberts Chemicals Inc.); 'Sulfasan' (Monsanto Chemical Co.); $C_6H_{10}O_2S_4$; SUYO2&SSYUS&O2; Ed. **5**, p.260.

(5) Fenazflor (BSI, ISO); phenyl 5,6-dichloro-2-trifluoromethylbenzimidazole-1-carboxylate (IUPAC), phenyl 5,6-dichloro-2-(trifluoromethyl)-1*H*-benzimidazole-1-carboxylate *[14255-88-0]* (*C.A.*); 'NC 5016', 'Lovozal' (Fisons Ltd); $C_{15}H_7Cl_2F_3N_2O_2$; T56 BN DNJ BVOR& CXFFF GG HG; Ed. **4**, p.267.

(6) Fenson (BSI, ISO); 4-chlorophenyl benzenesulphonate (IUPAC), 4-chlorophenyl benzenesulfonate *[80-38-6]* (*C.A.*); *p*-chlorophenyl benzenesulphonate; ENT 4585; CPBS; PCPBS: 'Murvesco' (Murphy Chemical Ltd); $C_{12}H_9ClO_3S$; WSR&OR DG; Ed. **3**, p.260. [Common name fénizon in France, where it is still in use].

(7) Fluenetil (BSI, ISO); 2-fluoroethyl 4-biphenylylacetate (IUPAC), 2-fluoroethyl [1,1'-bi-phenyl]-4-acetate *[4301-50-2]* (*C.A.*); 'M 2060', 'Lambrol' (Montecatini-Edison S.p.A.); $C_{16}H_{15}FO_2$; F2OV1R DR; Ed. **2**, p.254.

(8) Fluorbenside (BSI, ISO); 4-chlorobenzyl 4-fluorophenyl sulphide (IUPAC), 1-chloro-4-[[(4-fluorophenyl)thio]methyl]benzene *[405-30-1]* (*C.A.*); *p*-chlorobenzyl *p*-fluoro-phenyl sulfide; 'HRS 924', 'RD 2454', 'Fluorparacide', 'Fluorsulphacide' (The Boots Co. Ltd); $C_{13}H_{10}ClFS$; GR D1SR DF; Ed. **1**, p.235.

(9) 2-Fluoro-*N*-methyl-*N*-1-naphthylacetamide (IUPAC), 2-fluoro-*N*-methyl-*N*-(1-naphthalenyl)-acetamide *[5903-13-9]* (*C.A.*); MNFA (JMAF); ENT 27 403; 'Nissol' (Nippon Soda Co. Ltd); $C_{13}H_{12}FNO$; L66J BN1&V1F.

(10) Fluothiuron (BSI, proposed ISO); 3-[3-chloro-4-(chlorodifluoromethylthio)phenyl]-1,1-dimethylurea (IUPAC), *N'*-[3-chloro-4-[(chlorodifluoromethyl)thio]phenyl]-*N,N*-dimethylurea *[33439-45-1]* (*C.A.*); 'BAY KUE 2079A', 'Clearcide' (Bayer AG); $C_{10}H_{10}Cl_2F_2N_2OS$; GXFFSR BG DMVN1&1; Ed. **5**, p.107.

(11) Fospirate (BSI, ISO, ANSI); dimethyl 3,5,6-trichloro-2-pyridyl phosphate (IUPAC), dimethyl 3,5,6-trichloro-2-pyridinyl phosphate *[5598-52-7]* (*C.A.*); 'DOWCO 217' (The Dow Chemical Co.); $C_7H_7Cl_3NO_4P$; T6NJ BOPO&O1&O1 CG EG FG.

(12) 3-Furfuryl-2-methyl-4-oxocyclopent-2-enyl (1*RS*)-*cis,trans*-chrysanthemate (IUPAC), (*RS*)-3-(2-furanylmethyl)-2-methyl-4-oxo-2-cyclopenten-1-yl (1*RS*)-*cis,trans*-2,2-dimethyl-3-(2-methyl-1-propenyl)cyclopropanecarboxylate *[17080-02-3]* (formerly *[7076-49-5]*) (*C.A.*); furethrin; $C_{21}H_{26}O_4$; T5OJ B1- BL5V BUTJ C1 DOV-BL3TJ A1 A1 C1UY1&1; Ed. **2**, p.264.

(13) Gliotoxin; 2,3,5a,6-tetrahydro-6-hydroxy-3-(hydroxymethyl)-2-methyl-10*H*-3,10a-epidithiopyrazino[1,2-*a*]indole-1,4-dione *[67-99-2]* (*C.A.*); $C_{13}H_{14}N_2O_4S_2$; T C6 B566/JO A 2BJ O AVN JXSS NNV EU GU MHTT&&J DQ M1Q N1; Ed. **4**, p.287.

(1) Glyodin (ISO); 2-heptadecyl-2-imidazolinium acetate (IUPAC), 2-heptadecyl-3,4-dihydro-1H-imidazolyl acetate (1:1) *[556-22-9]* (*C.A.*); 'Crag Fungicide 341' (Union Carbide Corp.); $C_{22}H_{44}N_2O_2$; T5M CN BUTJ B17 &QV1; Ed. **1**, p.245.

(2) Griseofulvin (BSI, ISO, BPC, JMAF); 7-chloro-4,6-dimethoxycoumaran-3-one-2-spiro-1'-(2'-methoxy-6'-methylcyclohex-2'-en-4'-one) (IUPAC), (1'S)-*trans*-7-chloro-2',4,6-trimethoxy-6'-methylspiro[benzofuran-2(3H),1'-[2]cyclohexene]-3,4'-dione *[126-07-8]* (*C.A.*); (Glaxo Laboratories Ltd); $C_{17}H_{17}ClO_6$; T56 BOXVJ FO1 HO1 IG C-& DL6V DX BUTJ CO1 E1; Ed. **5**, p.290.

(3) Halacrinate (BSI, proposed ISO); 7-bromo-5-chloroquinolin-8-yl acrylate (IUPAC), 7-bromo-5-chloro-8-quinolinyl 2-propenoate *[34462-96-9]* (*C.A.*); 'CGA 30 599' (Ciba-Geigy AG); $C_{12}H_7BrClNO_2$; T66 BNJ GG IE JOV1U1; Ed. **5**, p.292.

(4) Haloxydine (BSI, ISO); 3,5-dichloro-2,6-difluoropyridin-4-ol (IUPAC), 3,5-dichloro-2,6-difluoro-4-pyridinol *[2693-61-0]* (*C.A.*); 'PP 493' (ICI Ltd, Plant Protection Division); $C_5HCl_2F_2NO$; T6NJ BG CF DQ EF FG.

(5) 2-Heptadecyl-1-(2-hydroxyethyl)imidazoline; 2-(2-heptadecyl-2-imidazolin-1-yl)ethanol (IUPAC), 2-heptadecyl-4,5-dihydro-1H-imidazole-1-ethanol *[95-19-2]* (*C.A.*), 2-heptadecyl-2-imidazoline-1-ethanol; 'Fungicide 337' (Union Carbide Corp.); $C_{22}H_{44}N_2O$; T5N CN AUTJ B17 C2Q.

(6) Hexachloroacetone (IUPAC), 1,1,1,3,3,3-hexachloropropan-2-one *[116-16-5]* (*C.A.*); 'GC 1106' 'HCA Weedkiller', 'Urox' (Allied Chemical Corp.); C_3Cl_6O; GXGGVXGGG; Ed. **5**, p.296.

(7) Hexafluoroacetone trihydrate; 1,1,1,3,3,3-hexafluoropropane-2,2-diol dihydrate (IUPAC, *C.A.*) *[993-58-8]*; 'GC 7887' (Allied Chemical Corp.); $C_3H_6F_6O_4$; FXFFXQQXFFF &QH &QH; Ed. **4**, p.295.

(8) 1,1,1,7,7,7-Hexafluoro-4-methyl-2,6-bis(trifluoromethyl)hept-3-ene-2,6-diol *[756-91-2]* mixture with 1,1,1,7,7,7-hexafluoro-4-methylene-2,6-bis(trifluoromethyl)heptane-2,6-diol *[16202-91-8]* (IUPAC, *C.A.*); 'ACD 10 614' (Allied Chemical Corp.); $C_{10}H_8F_{12}O_2$; FXFFXQXFFF1YU1&1XQXFFFXFFF + FXFFXQXFFF1Y1&U1XQXFFFXFFF; Ed. **4**, p.296.

(9) Hexaflurate (WSSA); potassium hexafluoroarsenate (IUPAC), potassium hexafluoroarsenate(1-) *[17029-22-0]* (*C.A.*); 'Nopalmate' (Pennwalt Corp.); AsF_6K; .KA .. AS-F6; Ed. **5**, p.298.

(10) 1-Hydroxy-1H-pyridine-2-thione (IUPAC), 1-hydroxy-2(1H)-pyridinethione *[1121-30-8]* (*C.A.*)-tautomeric with pyridine-2-thiol 1-oxide (IUPAC), 2-pyridinethiol 1-oxide *[1121-31-9]* (*C.A.*); 'Omadine' (Olin Mathieson Chemical Corp.); C_5H_5NOS; T6NYJ AQ BUS + T6NJ AO BSH; salts include: 'Omadine OM 1563' (zinc), 'Omadine OM 1564' (manganese), 'Omadine OM 1565' (iron); Ed. **2**, p.395.

(11) Ipazine (BSI, ISO, WSSA); 2-chloro-4-diethylamino-6-isopropylamino-1,3,5-triazine (IUPAC), 6-chloro-N,N-diethyl-N'-(1-methylethyl)-1,3,5-triazine-2,4-diamine *[1912-25-0]* (*C.A.*); 2-chloro-4-diethylamino-6-isopropylamino-s-triazine; 'G 30 031', 'Gesabal' (Ciba-Geigy AG); $C_{10}H_{18}ClN_5$; T6N CN ENJ BN2&2 DMY1&1 FG.

(12) Isazophos (BSI, proposed ISO); O-(5-chloro-1-isopropyl-1,2,4-triazol-3-yl) O,O-diethyl phosphorothioate (IUPAC), O-[5-chloro-1-(1-methylethyl)-1H-1,2,4-triazol-3-yl] O,O-diethyl phosphorothioate *[42509-80-8]* (*C.A.*); 'CGA 12 223', 'Miral' (Ciba-Geigy AG); $C_9H_{17}ClN_3O_3PS$; T5NN DNJ AY1&1 COPS&O2&O2 EG; Ed. **5**, p.308.

(1) Isobenzan (BSI, ISO); 1,3,4,5,6,7,8,8-octachloro-1,3,3a,4,7,7a-hexahydro-4,7-methanoisobenzofuran (IUPAC, *C.A.*) *[297-78-9]*; ENT 25 545; 'SD 4402', 'Telodrin' (Shell International Chemical Co.); $C_9H_4Cl_8O$; T C555 A EOTJ AG AG BG DG FG HG IG JG.

(2) Isobornyl thiocyanoacetate; 1,7,7-trimethylbicyclo[2.2.1]hept-2-yl thiocyanoacetate (IUPAC, *C.A.*) *[115-31-1]*; ENT 92; 'Thanite' (Hercules Inc.); $C_{13}H_{19}NO_2S$; L55 ATJ A1 A1 B1 COV1SCN; Ed. **5**, p.309.

(3) Isocil (BSI, ISO, WSSA); 5-bromo-3-isopropyl-6-methyluracil (IUPAC), 5-bromo-6-methyl-3-(1-methylethyl)-2,4(1*H*,3*H*)-pyrimidinedione *[314-42-1]*; 'Du Pont Herbicide 82', 'Hyvar' (E. I. du Pont de Nemours & Co. (Inc.)); $C_8H_{11}BrN_2O_2$; T6MVNVJ CY1&1 EE F1; Ed. **1**, p.260.

(4) Isonoruron (BSI, ISO); mixture of 3-(hexahydro-4,7-methanoindan-1-yl)-1,1-dimethylurea + 3-(hexahydro-4,7-methanoindan-2-yl)-1,1-dimethylurea (IUPAC), *[28346-65-8]* for *N*,*N*-dimethyl-*N'*-[1- + 2-(2,3,3a,4,5,6,7,7a-octahydro-4,7-methano-1*H*-indenyl)]-urea (*C.A.*); present in 'Basanor', 'Basfitox' (BASF AG); $C_{13}H_{22}N_2O$; L C555 ATJ DMVN1&1 + L C555 ATJ EMVN1&1; Ed. **5**, p.313.

(5) 1-Isopropyl-3-methylpyrazol-5-yl dimethylcarbamate (IUPAC), 3-methyl-1-(1-methylethyl)-1*H*-pyrazol-5-yl dimethylcarbamate *[119-38-0]*; ENT 19 060; 'G 23 611', 'Isolan' (Ciba-Geigy AG); $C_{10}H_{17}N_3O_2$; T5NNJ AY1&1 C1 EOVN1&1; Ed. **1**, p.262.

(6) 3-Isopropylphenyl methylcarbamate (IUPAC), 3-(1-methylethyl)phenyl methylcarbamate *[64-00-6]* (*C.A.*); ENT 25 500; 'Hercules 5727' (Hercules Inc.); 'UC 10 854' (Union Carbide Corp.); $C_{11}H_{15}NO_2$; 1Y1&R COVM1; Ed. **1**, p.263.

(7) Isopyrimol (BSI, ANSI, proposed ISO); 1-(4-chlorophenyl)-2-methyl-1-pyrimidin-5-ylpropan-1-ol (IUPAC), α-(4-chlorophenyl)-α-(1-methylethyl)-5-pyrimidinemethanol *[55283-69-7]*; $C_{14}H_{15}ClN_2O$; T6N CNJ EXQR DG&Y1&1.

(8) 2-Isovalerylindan-1,3-dione (IUPAC), 2-(3-methyl-1-oxobutyl)-1*H*-indene-1,3(2*H*)-dione *[83-28-3]* (*C.A.*); 'Valone' (Kilgore Chemical Co.); $C_{14}H_{14}O_3$; L56 BV DV CHJ CV1Y1&1; Ed. **5**, p.318.

(9) Karbutilate (BSI, ISO, ANSI, WSSA); 3-(3,3-dimethylureido)phenyl *tert*-butylcarbamate (IUPAC), 3-[[(dimethylamino)carbonyl]amino]phenyl] (1,1-dimethylethyl)carbamate *[4849-32-5]* (*C.A.*); 'FMC 11092', 'Tandex' (FMC Corp.); $C_{14}H_{21}N_3O_3$; 1X1&1&MVOR CMVN1&1; Ed. **5**, p.319.

(10) Lirimfos (BSI, proposed ISO); *O*-6-ethoxy-2-isopropylpyrimidin-4-yl *O*,*O*-dimethyl phosphorothioate (IUPAC), *O*-[6-ethoxy-2-(1-methylethyl)-4-pyrimidinyl] *O*,*O*-dimethyl phosphorothioate *[38260-63-8]* (*C.A.*); 'SAN 201 I' (Sandoz AG); $C_{11}H_{19}N_2O_4PS$; T6N CNJ BY1&1 DOPS&O1&O1 FO2.

(11) Lythidathion (BSI, ISO); *S*-(5-ethoxy-2,3-dihydro-2-oxo-1,3,4-thiadiazol-3-ylmethyl) *O*,*O*-dimethyl phosphorodithioate (IUPAC), *S*-[5-ethoxy-2-oxo-1,3,4-thiadiazol-3(2*H*)-ylmethyl] *O*,*O*-dimethyl phosphorodithioate *[2669-32-1]* (*C.A.*); 'GS 12968' (Ciba-Geigy AG), 'NC 2962' (Fisons Ltd); $C_7H_{13}N_2O_4PS_3$; T5NNVSJ B1SPS&O1&O1 EO2.

(12) Mebenil (BSI, ISO); *o*-toluanilide (IUPAC), 2-methyl-*N*-phenylbenzamide *[7055-03-0]* (*C.A.*); 'BAS 3050' (BASF AG); $C_{14}H_{13}NO$; 1R BVMR; Ed. **5**, p.332.

(13) Mecarbinzide (BSI), mecarbinzid (ISO); methyl 1-(2-methylthioethylcarbamoyl)-benzimidazol-2-ylcarbamate (IUPAC), methyl 1-[[(2-methylthioethyl)amino]carbonyl]-1*H*-benzimidazol-2-ylcarbamate *[27386-64-7]* (*C.A.*); methyl 1-(2-methylthioethyl-carbamoyl)benzimidazole-2-carbamate; (BASF AG); $C_{13}H_{16}N_4O_3S$; T56 BN DNJ BVM2S1 CMVO1.

Appendix I—*continued*

(1) Mecarphon (BSI, ISO); *S*-(*N*-methoxycarbonyl-*N*-methylcarbamoylmethyl) *O*-methyl methylphosphonodithioate (IUPAC), methyl 3,7-dimethyl-6-oxo-2-oxa-4-thia-7-aza-3-phosphaoctan-8-oate 3-sulfide *[29173-31-7]* (*C.A.*); methyl [[methoxy(methyl-phosphinothioyl)thio]acetyl]methylcarbamate; 'MC 2420' (Murphy Chemical Ltd); $C_7H_{14}NO_4PS_2$; 1OVN1&V1SPS&1&O1; Ed. **4**, p.329.

(2) Medinoterb acetate (BSI, ISO); 6-*tert*-butyl-2,4-dinitro-*m*-tolyl acetate (IUPAC); 6-(1,1-dimethylethyl)-3-methyl-2,4-dinitrophenyl acetate *[2487-01-6]* (*C.A.*); 'P 1488', 'MC 1488' (Murphy Chemical Ltd); $C_{13}H_{16}N_2O_6$; 1X1&1&R D1 CNW ENW BOV1; Ed. **5**, p.335.

(3) Mesoprazine (BSI, proposed ISO); 2-chloro-4-isopropylamino-6-(3-methoxypropylamino)-1,3,5-triazine (IUPAC), 6-chloro-*N*-(3-methoxypropyl)-*N'*-(1-methylethyl)-1,3,5-triazine-2,4-diamine *[1824-09-5]* (*C.A.*); 2-chloro-4-isopropylamino-6-(3-methoxy-propylamino)-*s*-triazine; 'G 34 698' (Ciba-Geigy AG); $C_{10}H_{18}ClN_5O$; T6N CN ENJ BMY1&1 DM3O1 FG.

(4) Metflurazone (BSI), metflurazon (proposed ISO); 4-chloro-5-dimethylamino-2-(α,α,α-trifluoro-*m*-tolyl)pyridazin-3-one (IUPAC), 4-chloro-5-(dimethylamino)-2-[(3-trifluoromethyl)phenyl]-3(2*H*)-pyridazinone *[23576-23-0]* (*C.A.*); 'SAN 6706 H' (Sandoz AG); $C_{13}H_{11}ClF_3N_3O$; T6NNVJ BR CXFFF& DG EN1&1.

(5) Methacrifos (BSI, proposed ISO); *O*-2-methoxycarbonylprop-1-enyl *O,O*-dimethyl phosphorothioate (IUPAC), methyl 3-[(dimethoxyphosphinothioyl)oxy]-2-methyl-2-propenoate *[30864-28-9]* (*C.A.*); 'CGA 20 168' (Ciba-Geigy AG); $C_7H_{13}O_5PS$; 1OVY1&U1PS&O1&O1.

(6) Methanesulphonyl fluoride (IUPAC), methanesulfonyl fluoride *[558-25-8]* (*C.A.*); MSF; 'Fumette' (Bayer AG); CH_3FO_2S; WSF1.

(7) Methiuron (BSI, ISO); 1,1-dimethyl-3-*m*-tolyl-2-thiourea (IUPAC), *N,N*-dimethyl-*N'*-(3-methylphenyl)thiourea *[21540-35-2]* (*C.A.*); 'MH 090' (Yorkshire Tar Distillers Ltd); $C_{10}H_{14}N_2S$; 1N1&YUS&MR Cl; Ed. **2**, p.306.

(8) Methometon (BSI, ISO); 2-methoxy-4,6-bis(3-methoxypropylamino)-1,3,5-triazine (IUPAC), 6-methoxy-*N,N'*-bis(3-methoxypropyl)-1,3,5-triazine-2,4-diamine *[1771-07-9]* (*C.A.*); 2-methoxy-4,6-bis(3-methoxypropylamino)-*s*-triazine; 'G 34 690' (Ciba-Geigy AG); $C_{12}H_{23}N_5O_3$; T6N CN ENJ BO1 DM3O1 FM3O1.

(9) Methylarsenic sulphide (IUPAC), methylthioxoarsine *[2533-82-6]* (*C.A.*); MAS (JMAF); 'Rhizoctol', 'Urbasulf' (Bayer AG); CH_3AsS; S-AS-1; Ed. **4**, p.349.

(10) Methylarsinediyl bis(dimethyldithiocarbamate) (IUPAC), dimethylcarbamodithioic acid bis(anhydrosulfide) with methylarsonodithious acid *[2445-07-0]* (*C.A.*); 'Urbacid', 'Monzet' (Bayer AG); $C_7H_{15}AsN_2S_4$; 1N1&YUS&S-AS-1&SYUS&N1&1; Ed. **5**, p.356.

(11) *S*-Methyl *N*-(cabamoyloxy)thioacetimidate (IUPAC), methyl *N*-[(aminocarbonyl)oxy]-ethanimidothioate *[16960-39-7]* (*C.A.*); ENT 27 411; 'EI 1642' (E. I. du Pont de Nemours & Co. (Inc.)); $C_4H_8N_2O_2S$; ZVONUY1&S1; Ed. **3**, p.341.

(12) Methylmercury dicyandiamide (accepted BSI); 3-(methylmercurio)guanidinocarbonitrile (IUPAC), (cyanoguanidinato-*N'*)methylmercury *[502-39-6]* (*C.A.*); 'Panogen' (Panogen Inc.); $C_3H_6HgN_4$; NCMYUM&M-HG-1; Ed. **5**, p.361.

(13) Methylmercury 3,4,5,6,7,7-hexachloro-1,2,3,6-tetrahydro-3,6-endomethanophthalimide (IUPAC), (4,5,6,7,8,8-hexachloro-1,3,3a,4,7,7a-hexahydro-1,3-dioxo-4,7-methano-2*H*-isoindol-2-yl)methylmercury *[5902-79-4]* (*C.A.*); 1,4,5,6,7,7-hexachloro-5-norbornene-2,3-dicarboximidato)methylmercury, *N*-methylmercuri-1,4,5,6,7,7-hexachlorobicyclo[2.2.1]-hept-5-ene-2,3-dicarboximide; 'Memmi' (Velsicol Corp.); $C_{10}H_5Cl_6HgNO_2$; T C555 A DVNV IUTJ AG AG BG E-HG-1 HG IG JG.

(1) Methyl 2-naphthyloxyacetate (IUPAC; accepted BSI, ISO), methyl 2-naphthalenyloxyacetate *[1929-87-9]* (*C.A.*); methyl 2-naphthoxyacetate; methyl β-naphthoxyacetate; 'Schering 34 927', 'Kamillemittel Schering' (Schering AG); $C_{13}H_{12}O_3$; L66J CO1VO1; Ed. **2**, p.318.

(2) 3-Methyl-1-phenylpyrazol-5-yl dimethylcarbamate (IUPAC), 3-methyl-1-phenyl-1*H*-pyrazol-5-yl dimethylcarbamate *[87-47-8]* (*C.A.*); ENT 17 588; 'G 22 008', 'Pyrolan' (Ciba-Geigy AG); $C_{13}H_{15}N_3O_2$; T5NNJ AR& C1 EOVN1&1.

(3) 2-(*N*-Methyl-*N*-prop-2-ynylamino)phenyl methylcarbamate (IUPAC), 2-(methyl-2-propynylamino)phenyl methylcarbamate *[23504-07-6]* (*C.A.*); 'C 17 018' (Ciba-Giegy AG); $C_{12}H_{14}N_2O_2$; 1UU2N1&R BOVM1; Ed. **3**, p.339.

(4) 4-(*N*-Methyl-*N*-prop-2-ynylamino)-3,5-xylyl methylcarbamate (IUPAC), 3,5-dimethyl-4-(methyl-2-propynylamino)phenyl methylcarbamate *[23623-49-6]* (*C.A.*); 'C 20 132' (Ciba-Geigy AG); $C_{14}H_{18}N_2O_2$; 1UU2N1&R B1 F1 DOVM1; Ed. **2**, p.320.

(5) Methyl 2,3,5,6-tetrachloro-*N*-methoxy-*N*-methylterephthalamate (IUPAC), methyl 2,3,5,6-tetrachloro-4-[methoxy(methylamino)carbonyl]benzoate *[14419-01-3]* (*C.A.*); 'OCS 21 693' (Velsicol Corp.); $C_{11}H_9Cl_4NO_4$; 1OVR BG CG EG FG DVN1&O1; Ed. **4**, p.354.

(6) Mexacarbate (BSI, ISO, ANSI); 4-dimethylamino-3,5-xylyl methylcarbamate (IUPAC), 3,5-dimethyl-4-(dimethylamino)phenyl methylcarbamate *[315-18-4]* (*C.A.*); ENT 25 766; 'DOWCO 139', 'Zectran' (The Dow Chemical Co); $C_{12}H_{18}N_2O_2$; 1N1&R B1 F1 DOVM1; Ed. **4**, p.359.

(7) Milneb (BSI, ANSI), thiadiazine (JMAF); 4,4',6,6'-tetramethyl-3,3'-ethylenebis(tetrahydro-1,3,5-thiadiazine-2-thione) (IUPAC), 3,3'-(1,2-ethanediyl)bis[tetrahydro-4,6-dimethyl-2*H*-1,3,5-thiadiazine-2-thione] *[3773-49-7]* (*C.A.*); 'Experimental Fungicide 328', 'Banlate' (E. I. du Pont de Nemours & Co. (Inc.)); $C_{12}H_{22}N_4S_4$; T6NYS EMTJ BUS D1 F1 A2- AT6NYS EMTJ BUS D1 F1.

(8) Mipafox (BSI, ISO); *N,N'*-bisisopropylphosphorodiamidic fluoride (IUPAC), *N,N'*-bis-(1-methylethyl)phosphorodiamidic fluoride *[371-86-8]* (*C.A.*); 'Isopestox', 'Pestox 15' (Fisons Ltd); $C_6H_{16}FN_2OP$; 1Y1&MPO&FMY1&1.

(9) Mirex (ESA); dodecachloropentacyclo[5.2.1.0^{2,6}.0^{3,9}.0^{5,8}]decane (IUPAC), 1,1a,2,2,3,3a,-4,5,5,5a,5b,6-dodecachlorooctahydro-1,3,4-metheno-1*H*-cyclobuta[*cd*]pentalene *[2385-85-5]* (*C.A.*); ENT 25 719; 'GC 1283' (Allied Chemical Corp.); $C_{10}Cl_{12}$; L545 B4 C5 D 4ABCE JTJ AG BG CG DG DG EG FG GG HG HG IG JG; Ed. **5**, p.368.

(10) Morfamquat dichloride (BSI, ISO); 1,1'-bis(3,5-dimethylmorpholinocarbonylmethyl)-4,4'-bipyridyldiylium dichloride (IUPAC), 1,1'-bis[2-(3,5-dimethyl-4-morpholinyl)-2-oxoethyl]-4,4'-bipyridinium dichloride *[4638-83-3]* (*C.A.*); 'PP 745', 'Morfoxone' (ICI Ltd, Plant Protection Division); $C_{26}H_{36}Cl_2N_4O_4$; T6N DOTJ B1 F1 AV1- AT6KJ D- DT6KJ A1V- AT6N DOTJ B1 F1 &G &G; Ed. **3**, p.355.

(11) Morphothion (BSI, ISO); *O,O*-dimethyl *S*-(morpholinocarbonylmethyl) phosphorodithioate (IUPAC), *O,O*-dimethyl *S*-[2-(4-morpholinyl)-2-oxoethyl] phosphorodithioate *[144-41-2]* (*C.A.*); 'Ekatin M' (Sandoz AG); $C_8H_{16}NO_4PS_2$; T6N DOTJ AV1SPS&O1&O1.

(12) Mucochloric anhydride; 5,5'-oxybis(3,4-dichloro-5*H*-furan-2-one) (IUPAC), 5,5'-oxybis[3,4-dichloro-2(5*H*)-furanone] *[4412-09-3]* (*C.A.*); bis(3,4-dichloro-2,5-dihydro-5-oxo-2-furanyl) ether; 'GC 2466' (Allied Chemical Corp.); $C_8H_2Cl_4O_5$; T5OV EHJ CG DG EO- ET5OV EHJ CG DG; Ed. **2**, p.334.

(**1**) *N*-(3-Nitrophenyl)itaconimide (IUPAC), 3-methylene-*N*-(3-nitrophenyl)pyrrolidine-2,5-dione *[4137-12-6]* (*C.A.*); 'B 720' (Uniroyal Inc.); $C_{11}H_8N_2O_4$; T5VNVY EHJ BR CNW& DU1.

(**2**) 4-(2-Nitroprop-1-enyl)phenyl thiocyanate (IUPAC), 4-(2-nitro-1-propenyl)phenyl thiocyanate *[950-00-5]* (*C.A.*); nitrostyrene (JMAF); 'Styrocide' (Nippon Kayaku Co. Ltd); $C_{10}H_8N_2O_2S$; WNY1&U1R DSCN.

(**3**) Nornicotine; 3-(pyrrolidin-2-yl)pyridine (IUPAC), 3-(2-pyrrolidinyl)pyridine *[494-97-3]* (*C.A.*); 2-(3-pyridyl)pyrrolidine; $C_9H_{12}N_2$; T6NJ C- BT5MTJ.

(**4**) Noruron (BSI, ISO), norea (ANSI, WSSA); 3-(hexahydro-4,7-methanoindan-5-yl)-1,1-dimethylurea (IUPAC), *N*,*N*-dimethyl-*N'*-(octahydro-4,7-methano-1*H*-inden-5-yl)-urea 3aα,4α,5α,7α,7aα-isomer *[18530-56-8]*, unstated steroechemistry *[2163-79-3]* (*C.A.*); 'Hercules 7531', 'Herban' (Hercules Inc.); $C_{13}H_{22}N_2O$; L C555 ATJ IMVN1&1; Ed. **3**, p.368.

(**5**) Octachlorocyclohex-2-en-1-one (IUPAC), 2,3,4,4,5,5,6,6-octachloro-2-cyclohexen-1-one *[4024-81-1]* (*C.A.*); 'OCH', 'Oktone' (P. F. Goodrich Chemical Co.); C_6Cl_8O; L6V BUTJ BG CG DG DG EG EG FG FG.

(**6**) Oxapyrazone-sodium (BSI), oxapyrazon-sodium (ISO); sodium *N*-(5-bromo-6-oxo-1-phenylpyridazin-4-yl)oxamate (IUPAC), sodium *N*-(5-bromo-6-oxo-1-phenyl-4(1*H*)-pyridazinyl)oxamate *[25316-56-7]*; BASF AG; $C_{12}H_7BrN_3NaO_4$; T6NNVJ BR& DE FMVVO &-NA-.

(**7**) Oxydisulfoton (BSI, ISO); *O*,*O*-diethyl *S*-2-ethylsulphinylethyl phosphorodithioate (IUPAC), *O*,*O*-diethyl *S*-[2-(ethylsulfinyl)ethyl] phosphorodithioate *[2497-07-6]* (*C.A.*); 'Bayer 23 323', 'S 309', 'L 16/184', 'Disyston-S' (Bayer AG); $C_8H_{19}O_3PS_3$; OS2&2SPS&O2&O2; Ed. **4**, p.389.

(**8**) Parafluron (BSI, ISO); 1,1-dimethyl-3-(α,α,α-trifluoro-*p*-tolyl)urea (IUPAC), *N*,*N*-dimethyl-*N'*-[4-(trifluoromethyl)phenyl]urea *[7159-99-1]* (*C.A.*); 'C 15935' (Ciba-Geigy AG); $C_{10}H_{11}F_3N_2O$; FXFFR DMVN1&1.

(**9**) Phenkapton (BSI, ISO); *S*-(2,5-dichlorophenylthiomethyl) *O*,*O*-diethyl phosphorodithioate (IUPAC, *C.A.*) *[2275-14-1]*; ENT 25585, 'G 28 029' (Ciba-Geigy AG); $C_{11}H_{15}Cl_2O_2PS_3$; GR DG BS1SPS&O2&O2.

(**10**) Phenobenzuron (BSI, ISO); 1-benzoyl-1-(3,4-dichlorophenyl)-3,3-dimethylurea (IUPAC), *N*-(3,4-dichlorophenyl)-*N*-[(dimethylamino)carbonyl]benzamide *[3134-12-1]* (*C.A.*); 'PP 65-25', 'Benzomarc' (Rhône Poulenc Phytosanitaire); $C_{16}H_{14}Cl_2N_2O_2$; GR BG DNVR&VN1&1; Ed. **5**, p.409.

(**11**) 2-Phenyl-3,1-benzoxazin-4-one (IUPAC), 2-phenyl-4*H*-3,1-benzoxazin-4-one *[1022-46-4]* (*C.A.*); 'H-170', 'Linurotox' (BASF AG); $C_{14}H_9NO_2$; T66 BVO ENJ DR; Ed. **5**, p.412.

(**12**) Phosacetim; *O*,*O*-bis(4-chlorophenyl) *N*-acetimidoylphosphoramidothioate (IUPAC), *O*,*O*-bis(4-chlorophenyl) (1-iminoethyl)phosphoramidothioate *[4104-14-7]* (*C.A.*); 'Bayer 38819', 'Gophacide' (Bayer AG); $C_{14}H_{13}Cl_2N_2O_2PS$; MUY1&MPS&OR DG&OR DG; Ed. **3**, p.394.

(**13**) Piperonyl cyclonene; 5-(benzodioxol-5-yl)-3-hexylcyclohex-2-enone (IUPAC), 5-(5-benzo-1,3-dioxolyl)-3-hexyl-2-cyclohexen-1-one *[8022-12-4]* (*C.A.*); 3-hexyl-5-(3,4-methylenedioxyphenyl)-2-cyclohexen-1-one; $C_{19}H_{24}O_3$; T56 BO DO CHJ G-EL6V BUTJ C6.

(**14**) Pirimetaphos (BSI, ISO); 2-diethylamino-6-methylpyrimidin-4-yl methyl methyl-phosphoramidate (IUPAC), 2-(diethylamino)-6-methyl-4-pyrimidinyl methyl methylphosphoramidate *[31377-69-2]* (*C.A.*); 'SAN 52 135' (Sandoz AG); $C_{11}H_{21}N_4O_3P$; T6N CNJ BN2&2 DOPO&1&M1 F1.

Appendix I—*continued*

(1) Polychlorocyclopentadiene isomers; *C.A.* Registry No. *[8029-29-6]*; 'Bandane' (Velsicol Chemical Corp.).

(2) Potassium cyanate (IUPAC, *C.A.*) *[590-28-3]*; 'Aero' cyanate (American Cyanamid Co.); CKNO; .KA..OCN; Ed. **5**, p.432.

(3) Propyl isome (ESA); dipropyl 1,2,3,4-tetrahydro-3-methyl-6,7-methylenedioxynaphthalene-1,2-dicarboxylate (IUPAC), dipropyl 5,6,7,8-tetrahydro-7-methylnaphtho[2,3-*d*]-1,3-dioxole-5,6-dicarboxylate *[83-59-0]* (*C.A.*); dipropyl maleate isosafrole condensate; (S. B. Penick & Co.); $C_{20}H_{26}O_6$; T C566 DO FO EH&&TJ JVO3 KVO3 L1; Ed. **5**, p.446.

(4) Proxan-sodium (BSI, ISO); sodium *O*-isopropyl dithiocarbonate (IUPAC), sodium *O*-(1-methylethyl) carbonodithioate *[140-93-2]* (*C.A.*); sodium *O*-isopropyl xanthate; 'Good-rite n.i.x.' (B. F. Goodrich Chemical Co.); $C_4H_7NaOS_2$; SUYS&OY1&1 &-NA-; Ed. **1**, p.360.

(5) Pydanon (BSI, ISO) (±)-hexahydro-4-hydroxy-3,6-dioxopyridazin-4-ylacetic acid (IUPAC); hexahydro-4-hydroxy-3,6-dioxo-4-pyridazineacetic acid *[22571-07-9]* (*C.A.*); 'H 1244' (C. F. Spiess & Sohn); $C_8H_8N_2O_5$; T6VMMVTJ EQ E1VQ.

(6) Pyrichlor (WSSA); 2,3,5-trichloropyridin-4-ol (IUPAC), 2,3,5-trichloro-4-pyridinol *[1970-40-7]* (*C.A.*); 'Daxtron' (The Dow Chemical Co.); $C_5H_2Cl_3NO$; T6NJ BG CG DQ EG; Ed. **1**, p.363.

(7) Pyridinitril (BSI, ISO); 2,6-dichloro-4-phenylpyridine-3,5-dicarbonitrile (IUPAC), 2,6-dichloro-4-phenyl-3,5-pyridinedicarbonitrile *[1086-02-8]* (*C.A.*); 2,6-dichloro-3,5-dicyano-4-phenylpyridine; 'IT 3296', 'Ciluan' (E. Merck); $C_{13}H_5Cl_2N_3$; T6NJ BG CCN DR& ECN FG; Ed. **5**, p.454.

(8) Pyrimithate (BSI, BPC), pyrimitate (ISO); *O*-(2-dimethylamino-6-methylpyrimidin-4-yl) *O,O*-diethyl phosphorothioate (IUPAC), *O*-[2-(dimethylamino)-6-methyl-4-pyrimidinyl] *O,O*-diethyl phosphorothioate *[5221-49-8]* (*C.A.*); 'ICI 29 661', 'Diothyl' (ICI Ltd, Pharmaceuticals Division); $C_{11}H_{20}N_3O_3PS$; T6N CNJ BN1&1 DOPS&O2&O2 F1.

(9) Pyroxychlor (BSI, proposed ISO); 2-chloro-6-methoxy-4-trichloromethylpyridine (IUPAC, *C.A.*) *[7159-34-4]*; 'DOWCO 269', 'Nurelle', 'Lorvek' (The Dow Chemical Co.); $C_7H_5Cl_4NO$; T6NJ BO1 DXGGG FG.

(10) Quinacetol sulphate (BSI, ISO); bis(5-acetyl-8-hydroxyquinolinium) sulphate (IUPAC), 2-(8-hydroxy-5-quinolinyl)ethanone sulfate (2:1) *[57130-91-3]* (*C.A.*); 'G 20 072', 'Fongoren', 'Risoter' (Ciba-Geigy AG); $C_{22}H_{20}N_2O_8S$; T66 BNJ GV1 JQ 2 &WSQQ; Ed. **5**, p.455.

(11) Quinalphos-methyl (BSI, ISO); *O,O*-dimethyl *O*-quinoxalin-2-yl phosphorothioate (IUPAC), *O,O*-dimethyl *O*-(2-quinoxalinyl) phosphorothioate *[13593-08-3]* (*C.A.*); 'SAN 52 056 I' (Sandoz AG); $C_{10}H_{11}N_2O_3PS$; T66 BN ENJ COPS&O1&O1.

(12) Sabadilla; cevadilla; *Schoenocaulon officinale*; Ed. **5**, p.464.

(13) Salicylanilide (IUPAC; accepted BSI, ISO), 2-hydroxy-*N*-phenylbenzamide *[87-17-2]* (*C.A.*); 'Shirlan' (ICI Ltd); $C_{13}H_{11}NO_2$; QR BVMR; Ed. **5**, p.465.

(14) Sesamolin; 1-(1,3-benzodioxol-5-yl)-4-(1,3-benzodioxol-5-yloxy)tetrahydrofuro[3,4-*c*]-furan, 2-(1,3-benzodioxol-5-yl)-6-(1,3-benzodioxol-5-yloxy)-3,7-dioxabicyclo[3.3.0]-octane (IUPAC), [1*S*-(1α,3aα,4α,6aα)-5-[4-(1,3-benzodioxol-5-yloxy)tetrahydro-1*H*,-3*H*-furo[3,4-*c*]furan-1-yl]-1,3-benzodioxole *[526-07-8]* (*C.A.*); present in sesame oil from *Sesamum indicum* seeds; $C_{20}H_{18}O_7$; T56 BO DO CHJ GO- BT55 CO GOTJ F- GT56 BO DO CHJ; the related sesamin, **not** sesamolin, is described in Ed. **5**, p.469.

(1) Simeton (BSI); 2,4-bis(ethylamino)-6-methoxy-1,3,5-triazine (IUPAC), *N,N'*-diethyl-6-methoxy-1,3,5-triazine-2,4-diamine *[673-04-1]* (*C.A.*); 2,4-bis(ethylamino)-6-methoxy-*s*-triazine; 'G 30 044' (Ciba Geigy AG); $C_8H_{15}N_5O$; T6N CN ENJ BO1 DM2 FM2.

(2) Sodium (*Z*)-3-chloroacrylate (IUPAC), sodium (*Z*)-3-chloro-2-propenoate *[4312-97-4]* (*C.A.*); sodium *cis*-3-chloroacrylate; 'UC 20 299', 'Prep' (Union Carbide Corp.); $C_3H_2ClNaO_2$; OV1U1G &-NA-; Ed. **1**, p.386.

(3) Sodium selenate (IUPAC), selenic acid (H_2SeO_4) disodium salt *[13410-01-0]* (*C.A.*); Na_2O_4Se; .NA2.SE-O4.

(4) Streptomycin (BSI, ISO, BPC); 2,4-diguanidino-3,5,6-trihydroxycyclohexyl 5-deoxy-2-*O*-(2-deoxy-2-methylamino-α-L-glucopyranosyl)-3-*C*-formyl-β-L-lyxopentanofuranoside (IUPAC), *O*-2-deoxy-2-(methylamino)-α-L-glucopyranosyl-(1→2)-*O*-5-deoxy-3-*C*-formyl-α-L-lyxofuranosyl-(1→4)-*N,N'*-bis(aminoiminomethyl-D-streptamine *[57-92-1]* (*C.A.*); 'Spikespray' (ICI Ltd, Plant Protection Division), 'Agrimycin 100' (Chas. Pfizer Co., Inc.); $C_{21}H_{39}N_7O_{12}$; T6OTJ B1Q CQ DQ EM1 FO- DT5OTJ B1 CVH CQ EO- AL6TJ BQ CQ DMYZUM EQ FMYZUM; Ed. **5**, p.478.

(5) Swep (ANSI, WSSA), MCC (JMAF); methyl 3,4-dichlorophenylcarbamate (IUPAC, *C.A.*) *[1918-18-9]*, methyl 3,4-dichlorocarbanilate; 'FMC 2995' (FMC Corp.); $C_8H_7Cl_2NO_2$; GR BG DMVO1; Ed. **5**, p.486.

(6) TDE (BSI, ISO); 1,1-dichloro-2,2-bis(4-chlorophenyl)ethane (IUPAC), 1,1'-(2,2-dichloroethylidene)bis[4-chlorobenzene] *[72-54-8]* (*C.A.*); 1,1-dichloro-2,2-bis(*p*-chlorophenyl)ethane; DDD; ENT 4225; 'Rhothane' (Rohm & Haas Co.); $C_{14}H_{10}Cl_4$; GYGYR DG&R DG; Ed. **3**, p.457. *Note:* TDE, though no longer manufactured commercially, is found in nature as a degradation product of DDT (p.150).

(7) Terbucarb (BSI, ISO); 2,6-di-*tert*-butyl-*p*-tolyl methylcarbamate (IUPAC), 2,6-bis(1,1-dimethylethyl)-4-methylphenyl methylcarbamate *[1918-11-2]* (*C.A.*); 'Hercules 9573', 'Azak' (Hercules Inc.); $C_{17}H_{27}NO_2$; 1X1&1&R C1 FOVM1 EX1&1&1; Ed. **4**, p.476.

(8) Terbuchlor (BSI, proposed ISO); *N*-butoxymethyl-6'-*tert*-butyl-2-chloroacet-*o*-toluidide (IUPAC), *N*-(butoxymethyl)-2-chloro-*N*-[2-(1,1-dimethylethyl)-6-methylphenyl]-acetamide *[4212-93-5]* (*C.A.*); 'MON 0358', 'CP 46 358' (Monsanto Co.); $C_{18}H_{28}ClNO_2$; G1VN1O4&R B1 FX1&1&1.

(9) Terpene polychlorinates; polychloroterpenes; chlorinated mixed terpenes, *C.A.* Registry No. *[8001-50-1]*; ENT 19 442; 'Compound 3961', 'Strobane' (B.F. Goodrich & Co.); Ed. **1**, p.410.

(10) Tetrachlorothiophene (IUPAC, *C.A.*) *[6012-97-1]*; ENT 25 764; TCTP; 'Penphene' (Pennwalt); C_4Cl_4S; T5SJ BG CG DG EG; Ed. **3**, p.464.

(11) Tetrahydro-5-methyl-6-thioxo-2*H*-1,3,5-thiadiazin-3-ylacetic acid (IUPAC), dihydro-5-methyl-6-thioxo-2*H*-1,3,5-thiadiazine-3(4*H*)-acetic acid *[3655-88-7]* (*C.A.*); 'Terracur' (Bayer AG); $C_6H_{10}N_2O_2S_2$; T6NYS ENTJ A1 BUS E1VQ; Ed. **1**, p.71.

(12) 2-Thiocyanatoethyl laurate (IUPAC), 2-thiocyanatoethyl dodecanoate *[301-11-1]* (*C.A.*); 'Lethane 60' (Rohm & Haas Co.); $C_{15}H_{27}NO_2S$; SCN2OV11.

(13) Thioquinox (BSI, ISO); 1,3-dithiolo[4,5-*b*]quinoxaline-2-thione (IUPAC, *C.A.*) *[93-75-4]*; quinoxaline-2,3-diyl trithiocarbonate, 2-thioxo-1,3-dithiolo[4,5-*b*]quinoxaline; ENT 25 579; 'Bayer 30686', 'Ss 1451', 'Eradex' (Bayer AG); $C_9H_4N_2S_3$; T C566 BN DSYS HNJ EUS; Ed. **4**, p.491.

(1) Triarimol (BSI, ISO, ANSI); 2,4-dichloro-α-(pyrimidin-5-yl)benzhydryl alcohol (IUPAC), α-(2,4-dichlorophenyl)-α-phenyl-5-pyrimidinemethanol *[26766-27-8]* (*C.A.*); 'EL 273', 'Trimidal' (Eli Lilly & Co.); $C_{17}H_{12}Cl_2N_2O$; T6N CNJ EXQR&R BG DG; Ed. **3**, p.481.

(2) Tricamba (BSI, ISO, ANSI, WSSA); 3,5,6-trichloro-*o*-anisic acid (IUPAC), 2,3,5-trichloro-6-methoxybenzoic acid *[2307-49-5]* (*C.A.*); 'Banvel T' (Velsicol Corp.); $C_8H_5Cl_3O_3$; QVR BG CG EG FO1.

(3) 4,5,7-Trichlorobenzo-2,1,3-thiadiazole (IUPAC, *C.A.*) *[1982-55-4]*; 'PH 40-21' (Philips Duphar N.V.); $C_6HCl_3N_2S$; T56 BNSNJ FG GG IG; Ed. **3**, p.483.

(4) Trichlorobenzyl chloride (IUPAC); trichloro(chloromethyl)benzene *[1344-32-7]* (formerly *[25429-36-1]*) (*C.A.*) (isomer composition not stated); ar,ar,ar,α-tetrachlorotoluene; TCBC; 'Randon-T' (Monsanto Co.); $C_7H_4Cl_4$; G1R XG XG XG; Ed. **4**, p.500.

(5) 1-(2,3,6-Trichlorobenzyloxy)propan-2-ol (IUPAC), 1-[(2,3,6-trichlorophenyl)methoxy]-2-propanol *[1861-44-5]* (*C.A.*); 'Tritac' (Hooker Chemical Corp.); $C_{10}H_{11}Cl_3O_2$; QY1&1O1R BG CG FG.

(6) Tris[1-dodecyl-3-methyl-2-phenylbenzimidazolium] hexacyanoferrate (IUPAC), tris[1-dodecyl-3-methyl-2-phenyl-1*H*-benzimidazolium] hexakis(cyano-*C*)ferrate (*C.A.*); tris[1-dodecyl-3-methyl-2-phenylbenzimidazolium] ferricyanide; 'Bayer 32 394', 'B 169 Ferricyanide', 'Fungilon' (Bayer AG); $C_{84}H_{111}FeN_{12}$; T56 BK DNJ B12 CR& D1 3 &.FE-CN6; Ed. **2**, p.213.

(7) Undec-10-enic acid (IUPAC), 10-undecenoic acid *[112-38-9]* (*C.A.*); undecyl-10-enic acid, 10-hendecenoic acid; $C_{11}H_{20}O_2$; QV9U1; Ed. **5**, p.531.

Appendix II

ABBREVIATIONS

The following abbreviations have been used, some being SI units.

ACS .. .. American Chemical Society

a.e. .. .. acid equivalent—active ingredient expressed in terms of the parent acid

AG .. .. Aktiengesellschaft (Company)

a.i. .. .. active ingredient

ANSI .. .. American National Standards Institute

AOAC .. Association of Official Analytical Chemists or, before 1966, Association of Official Agricultural Chemists

AustriaP .. Austrian Patent

BCPC .. .. British Crop Protection Council

BelgP .. .. Belgian Patent

BIOS .. .. British Intelligence Objectives Sub-Committee

BP .. .. British Patent

b.p. .. .. boiling point/at stated pressure

BPA .. .. British Patent Application

BPC .. .. British Pharmacopoeia Commission

BSI .. .. British Standards Institution

Bu .. .. butyl, $-(CH_2)_3CH_3$

Bu^i .. .. isobutyl, $-CH_2CH(CH_3)_2$

Bu^s .. secondary butyl, *sec*-butyl, $-\overset{\overset{\displaystyle CH_3}{|}}{C}HCH_2CH_3$

Bu^t .. .. tertiary butyl, *tert*-butyl, $-C(CH_3)_3$

B.V. .. .. Beperkt Vennootschap (Limited)

c. .. .. about

C.A. .. .. *Chemical Abstracts*

cf .. .. compare

CIPAC .. Collaborative International Pesticides Analytical Committee

Co.	.. ..	Company
Corp.	.. ..	Corporation
d	.. ..	day(s)
d_x^t	..	density of compound at t° compared to that of water at x°
DAFS	.. ..	Department of Agriculture and Fisheries for Scotland
DAS	.. ..	Deutsche AuslegeSchrift (an intermediate stage towards DBP which retains the same number)
DBP	.. ..	West German Patent (German Federal Republic)
DDRP	..	East German Patent (German Democratic Republic)
decomp.	..	with decomposition
DOS	.. ..	Deutsche OffenlegungSchrift (the first stage towards DBP which retains the same number)
d.p.	.. ..	dispersible powder
DRP	.. ..	German Patent (before 1945)
DutchP	..	Netherlands Patent
e.c.	.. ..	emulsifiable concentrate
ECD	.. ..	electron-capture detection
ed.	.. ..	editor
Ed.	.. ..	Edition
e.g.	.. ..	for example
EPA	.. ..	Environmental Protection Agency (of USA)
ESA	.. ..	Entomological Society of America
Et	.. ..	ethyl, $-CH_2CH_3$
et al.	.. ..	and other (authors)
EPPO	..	European and Mediterranean Plant Protection Organisation
EWRC	..	European Weed Research Council

FAO .. .. Food and Agricultural Organisation (of the United Nations)

FID .. .. flame-ionisation detection

FP .. .. French Patent

f.p. .. .. freezing point

FPD .. .. flame-photometric detection

FTP .. .. flame thermionic detection

g .. .. gram(s)

GC-MS .. combined gas chromatography—mass spectrometry

GLC .. .. gas-liquid chromatography

h hour(s)

ha hectare(s) (10^4 m²)

HMSO .. Her Majesty's Stationery Office

HPLC .. .. high pressure liquid chromatography (also known as high performance liquid chromatography)

HV .. .. High Volume (herbicide usage > 670 l/ha; insecticide and fungicide usage > 1120 l/ha for bushes and trees and > 670 l/ha for ground crops)

ibid. .. .. in the journal last mentioned

idem. .. .. by the author(s) last mentioned

i.e. .. .. that is

i.r. .. .. infrared

Inc. .. .. Incorporated

ISO .. .. International Standardization Organization

i.u. .. .. international unit (measure of activity of micro-organisms)

IUPAC .. International Union of Pure and Applied Chemistry

JapanP .. Japanese Patent

JapanPA .. Japanese Patent Application (Kokai)

JMAF .. .. Japanese Ministry for Agriculture and Forestry

kg .. .. kilogram(s)

l .. .. litre(s)

LC50 .. .. concentration required to kill 50% of the test animals

LD50 .. .. dose required to kill 50% of the test animals

Ltd .. .. Limited

LV .. .. Low Volume (herbicide usage 55-220 l/ha; insecticide and fungicide usage < 550 l/ha for bushes or trees and < 22 l/ha on ground crops)

m .. .. metre; as multiplyer 10^{-3} (0·001)

M .. .. molar

MAC .. .. maximum allowed concentration

MAFF .. Ministry of Agriculture Fisheries and Food (England and Wales)

MCD .. .. microcoulometric detection

Me .. .. methyl, —CH_3

mg .. .. milligram(s), 10^{-3} g, 0·001 g

mm .. .. millimetre(s), 10^{-3} m, 0·001 m

m/m .. .. proportion by mass

mmHg .. .. pressure equivalent to 1 mm of mercury (133·3 Pascals)

mol. wt .. molecular weight

m.p. .. .. melting point

n_D^t .. .. refractive index for the sodium D lines at a temperature of t° Celsius

nm .. .. nanometre(s), 10^{-9} m

NMR .. .. nuclear magnetic resonance

NRDC .. National Research and Development Corporation

N.V. .. .. Naamloze Vennotschap (Limited)

575

Appendix II—*continued*

o.m. .. .. organic matter

OMS .. .. Organisation mondiale de la Sante = WHO

op. cit. .. .. in the book cited

Pa .. .. Pascal

Ph .. .. phenyl, $—C_6H_5$ 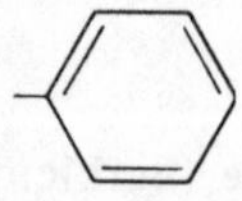

pH .. .. $—\log_{10}$ hydrogen ion concentration

pK_a .. .. $—\log_{10}$ acid dissociation constant

post-em. .. after emergence

Pr .. .. propyl, $—CH_2CH_2CH_3$

Pr^i .. .. isopropyl, $—CH(CH_3)_2$

pre-em. .. before emergence

RBC .. .. red blood corpuscles

r.h. .. .. relative humidity

s .. .. second(s)

S.A. .. .. Société Anonyme (Company)

s.p. .. .. soluble powder

sp. .. .. species (singular)

spp. .. .. species (plural)

S.p.A. .. .. Société par Actions (Company)

SwissP Swiss Patent

t .. .. tonne, 1000 kg

tech. .. .. technical grade

TID .. .. thermionic detection

TLC .. .. thin-layer chromatography

UK	.. ..	United Kingdom
ULV	.. ..	Ultra-low Volume (< 5 l/ha)
USA	.. ..	United States of America
USDA	.. ..	United States Department of Agriculture
USP	.. ..	United States Patent
USPharm.	..	United States Pharmacopoeia
u.v.	.. ..	ultraviolet
v.p.	.. ..	vapour pressure
WHO	.. ..	World Health Organisation (of the United Nations) = OMS
w.p.	.. ..	wettable powder
w.s.c.	.. ..	water-soluble concentrate
w.s.p.	.. ..	water-soluble powder
WSSA	.. ..	Weed Science Society of America
y	.. ..	year(s)
$[a]_D^t$	.. ..	specification rotation (degrees) for sodium D lines at temperature t° Celsius
μ	.. ..	multiplier for SI units (10^{-6})
μl	.. ..	microlitre (10^{-6} litre)
μg	.. ..	microgram (10^{-6} gram)
ρ	.. ..	density
—°	.. ..	temperature of t degrees Celsius (formerly Centigrade)
>	.. ..	greater than
≥	.. ..	greater than or equal to
<	.. ..	less than
≤	.. ..	less than or equal to

Scientific and common names of mammals, fish, birds and insects quoted in the Toxicology Section of entries in the *Pesticide Manual*.

Agelaius phoeniceus .... Blackbird, red-winged
Alectoris chukar .... Partridge, chukar
Anas platyrhynchos .... Duck, mallard
Anas platyrhynchos domesticus .... Duck, pekin
Apis mellifera .... Bee, honey

Barbel .... *Barbus barbus*
Barbel, southern .... *Barbus meridionalis*
Barbus barbus .... Barbel
Barbus meridionalis .... Barbel, southern
Bass, largemouth .... *Micropterus salmoides*
Bass, largemouth, black .... *Micropterus salmoides*
Bee, honey .... *Apis mellifera*
Blackbird, European .... *Turdus merula*
Blackbird, red-winged .... *Agelaius phoeniceus*
Bluegill ....*Lepomis macrochirus*
Bos taurus .... Cattle
Bullhead, black .... *Ictaluras melas*

Canary ....*Serinius canarius*
Canis familiaris .... Dog
Carassius auratus .... Goldfish
Carp ....*Cyprinus carpio*
Carp, Japanese
Carp, mirror ....*Cyprinus carpio*
Cat ....*Felis domestica*
Catfish .... Siluroids, covering several families
Catfish, channel ....*Ictalurus punctatus*
Cattle .... *Bos taurus*
Cavia porcellus .... Guinea-pig
Chicken .... *Gallus gallus*
Colinus virginianus .... Quail, bobwhite
Columba leucocephala .... Pigeon
Coturnix coturnix .... Quail, European
Coturnix coturnix japonica .... Quail, Japanese
Cow .... *Bos taurus*
Crassotrea virginica .... Oysters
Crayfish....................... *Procambarus acutus*
Cyprinodon variegatus .... Minnow, sheepshead
Cyprinus carpio .... Carp

Dog .... *Canis familiaris*
Dog, beagle
Dove, mourning .... *Zenaidura macroura*
Duck, mallard ....*Anas platyrhychos*
Duck, pekin ....*Anas platyrhychos domesticus*

Felis domestica .... Cat
Finfish ....*Fundulus heteroclitus*
Fowl, domestic .... *Gallus gallus*
Fundulus heteroclitus .... Finfish

Gallus gallus....................................... Chicken
Gambusia affinis .... Mosquito fish
Gammarus spp.................................... Shrimps
Goldfish ....*Carassius auratus*
Guinea-pig .... *Cavia porcellus*
Guppy......... *Poecilia reticulata—Libistes reticulatus*

Hamster, golden ....*Mesocricetus auratus*
Hare, European.....................*Lepus cuniculus*
Harlequin fish ....*Rasbora heteromorpha*
Hen .... *Gallus gallus*

Ictaluras melas ....Bullhead, black
Ictaluras punctatus ....Catfish, channel
Ide ....*Idus idus*
Idus idus ⎱   ⎰ Ide (silver) or
Idus melanotus ⎰   ⎱ Orfe (golden)

Killifish, Japanese ....*Oryzias latipes*

Leiostomus xanthurus .... Spot
Lepomis cyanellus ....Sunfish, green
Lepomis macrochirus .... Bluegill
Lepomis microlophus .... Sunfish, redear
Lepus cuniculus .... Hare, European
Libistes reticulatus .... Guppy

Macaca mulatta .... Monkey, rhesus
Mesocricetus auratus .... Hamster, golden
Mice, albino ....*Mus musculus*
Micropterus salmoides ....Bass, largemouth
Minnow, fathead .... *Pimephals promelas*
Minnow, sheepshead .... *Cyprinodon variegatus*
Monkey, rhesus.........................*Macaca mulatta*
Mosquito fish.......................... *Gambusia affinis*
Mouse, house ....*Mus musculus*
Mugil cephalus ....Mullet, black
Mugil curema ....Mullet, silver
Mullet, black ....*Mugil cephalus*
Mullet, silver .... *Mugil curema*
Mus musculus .... Mouse, house

Notemigonus crysoleucas .... Shiner, golden

Oncorhynchus kisutch .... Salmon, coho
Oncorhynchus tshawytscha .... Salmon, chinook
Orfe.......................................*Idus idus*
Orizias latipes
                    Ricefish, Japanese—Killifish, Japanese
Oryctolagus cuniculus .... Rabbit, European
Ovis aries .... Sheep
Oysters................................ *Crassotrea virginica*

Partridge, chukar ....*Alectoris chukar*
Partridge, European, grey .... *Perdix perdix*
Passer domesticus ....Sparrow, house
Passer italiae ....Sparrow
Perca flavescens .... Perch, yellow
Perch, yellow ....*Perca flavescens*
Perdix perdix .... Partridge, European, grey
Phasianus colchicus
               Pheasant, ring-neck (ring-necked)
Pheasant ....*Phasianus* spp.
Pheasant, ring-neck ⎫
Pheasant, ring-necked ⎬ *Phasianus colchicus*
Phoxinux laevis .... Minnow
Pig .... *Sus scrofa*
Pigeon..............................*Columba leucocephala*
Pimephals promelas ....Minnow, fathead
Poecilia reticulata .... Guppy
Procambarus acutus...............................Crayfish

Quail, bobwhite.....................*Colinus virginianus*
Quail, European ....*Coturnix coturnix*
Quail, Japanese ....*Coturnix coturnix japonica*

Rabbit..............................*Oryctolagus cunniculus*
Rasbora heteromorpha .... Harlequin fish
Rat..................................... *Rattus norvegicus*
Rat, brown .... *Rattus rattus*
Rattus norvegicus .... Rat
Rattus rattus................................. Rat, brown
Ricefish, Japanese ....*Oryzias latipes*

Salmo spp...Trout
Salmo gaidneri .... Trout, rainbow
Salmo trutta ....Trout, brown
Salmon, chinook.......... *Oncorhynchus tshawytscha*
Salmon, coho..................... *Oncorhynchus kisutch*
Salvelinus fontinalis .... Trout, brook
Serinius canarius ....Canary
Sheep..*Ovis aries*
Shiner, golden................. *Notemigonus crysoleucas*
Shrimp .... *Gammarus* spp.
Sparrow...................................... *Passer italiae*
Sparrow, house ....*Passer domesticus*
Spot.............................*Leiostomus xanthurus*
Starling, common .... *Sturnus vulgaris*
Sturnus vulgaris........................Starling, common
Sunfish ...... A general term covering several classes
Sunfish, bluegill..................*Lepomis macrochirus*
Sunfish, green ....*Lepomis cyanellus*
Sunfish, redear ....*Lepomis microlophus*
Sus scrofa .... Pig

Tilapia spp.
Trout .... *Salmo* spp.
Trout, brook.......................... *Salvelinus fontinalis*
Trout, brown .... *Salmo trutta*
Trout, rainbow ....*Salmo gairdneri*
Turdus merula ....Blackbird, European

Zenaidura macroura ....Dove, mourning

Appendix IV

ADDRESSES OF FIRMS MENTIONED IN TEXT

List of the main firms mentioned in the text as discoverers or producers of the compounds concerned. In general, the address of the firm's headquarters is given; the regional offices may be able to help with additional information but there is no guarantee of this.

Aagrunol B.V., Groningen, Netherlands.

Abbott Laboratories, Agriculture and Veterinary Division, North Chicago, Illinois 60064, USA.

Aldrich Chemical Company Incorporated, 940 West Street, Paul Avenue, Milwaukee, Wisconsin 53233, USA.

Allied Chemical Corporation (For former Agricultural Division, *see* Hopkins Agricultural Chemical Company).

Amchem Products Incorporated, Ambler, Pennsylvania 19002, USA. (Member of Union Carbide Corporation.)

American Cyanamid Company, Agricultural Division, P.O. Box 400, Princeton, New Jersey 08540, USA.

Atlas Products and Services Limited, Fraser Road, Erith, Kent DA5 1PN, England.

BASF AG, 6703 Limburgerhof/Pfalz, Birkenweg 2, Federal Republic of Germany.

BASF (UK) Limited, Agrochemical Division, Lady Lane, Hadleigh, Ipswich, Suffolk IP7 6BQ, England.

Bayer AG, PF-AT Beratung, 5090 Leverkusen Bayerwerk, Federal Republic of Germany.

The Boots Company Limited, Industrial Division, Agricultural Research, Lenton House, Nottingham NG7 2QD, England.

Borax (*see* United States Borax and Chemical Corporation).

Celamerck GmbH, 6507 Ingelheim am Rhein, Federal Republic of Germany.

Chemie Linz AG, Postfach 296, A-4021 Linz, Austria.

Chevron Chemical Company, 940 Hensley Steet, Richmond, California 94804, USA.

Ciba AG (now amalgamated as Ciba-Geigy AG).

Ciba-Geigy AG, CH-4002, Basle, Switzerland.

Cropsafe Limited, Salisbury Road, Downton, Salisbury, Wiltshire, England.

Crown Chemicals, Agricultural Division (*see* Hopkins Agricultural Chemical Company).

Diamond Alkali Company (*see* Diamond Shamrock Corporation).

Diamond Shamrock Corporation, Agricultural Chemicals Division, 1100 Superior Avenue, Cleveland, Ohio 44114, USA.

The Dow Chemical Company, Ag-Organics Department, Midland, Michigan 48640, USA.

E. I. Du Pont de Nemours and Company Incorporated, Biochemicals Department, Wilmington, Delaware 19898, USA.

Elanco Products (*see* (Eli) Lilly and Company).

Eli Lilly and Company (*see* (Eli) Lilly and Company).

Fisons Limited, Agrochemical Division, Chesterford Park Research Station, Near Saffron Walden, Essex CB10 1XL, England.

FMC Corporation, Agricultural Chemical Division, 100 Niagara Street, Middleport, New York 14105, USA.

J. R. Geigy S.A. (now amalgamated as Ciba-Geigy AG).

Gist-Brocades N.V., P.O. Box 1, Delft, Netherlands.

Gulf Oil Chemicals, 9009 W. 67th Street, Merriam, Kansas 66202, USA.

Hercules Incorporated, Agricultural Chemicals Laboratory, Wilmington, Delaware 19899, USA.

Hoechst AG, Landwirtschaft, Beratung, Lyoner Strasse 30, D-6 Frankfurt/Main 71, Federal Republic of Germany.

Hokko Chemical Industry Company Limited, Mitsui Bldg. No. 2, 24, Nihonbashi-Hongokucho, Chuo-ku, Tokyo, Japan.

Hooker Chemical Corporation, Industrial Chemical Division, P.O. Box 344, Niagara Falls, New York 14302, USA.

Hopkins Agricultural Chemical Company, P.O. Box 7532, Madison, Wisconsin 53707, USA. (Formerly Agricultural Division of Allied Chemical Corporation, bought by Crown Chemicals).

ICI (*see* Imperial Chemical Industries Limited).

I. G. Farbenindustrie—disbanded after 1945, forming what became BASF AG, Bayer AG and Hoechst AG.

Imperial Chemical Industries Limited, Pharmaceuticals Division, Alderley House, Alderley Park, Macclesfield, Cheshire SK10 4TF, England.

Imperial Chemical Industries Limited, Plant Protection Division, Fernhurst, Haslemere, Surrey GU27 3JE, England.

Janssen Pharmaceutica N.V., Turnhoutseweg 30, B-2340 Beerse, Belgium.

Kenogard A.B., Box 11033, S-100 61 Stockholm, Sweden.

Kumiai Chemical Industry Company Limited, 4-26, Ikenohata 1-Chome, Taitoh-ku, Tokyo 110, Japan.

Eli Lilly and Company, Lilly Research Laboratories, P.O. Box 708, Greenfield, Indianapolis, Indiana 46206, USA.

3M Company, Commercial Chemical Division, 3M Centre, St. Paul, Minnesota 55101, USA.

Dr. R. Maag Limited, Chemical Works, CH-8157 Dielsdorf, Switzerland.

Mallinckrodt Incorporated, Mallinckrodt and Second Street, St. Louis, Missouri 63147, USA.

A. H. Marks and Company Limited, Wyke Lane, Wyke, Bradford, West Yorkshire BD12 9EJ, England.

May and Baker Limited, Agrochemical Division, 37-39 Manor Road, Romford, Essex RM1 2TL, England.

McLaughlin, Gormley King and Company, 8810 Tenth Avenue, N. Minneapolis, Minnesota 55427, USA.

Merck Animal Health Division, Merck and Company Incorporated, Lincoln Avenue, Rahway, New Jersey 07065, USA.

MGK (*see* McLaughlin, Gormley King and Company).

Mitsubishi Chemical Industries Limited, Mitsubishi Bldg., 5-2, Marunouchi 2-Chome, Chiyoda-ku, Tokyo 100, Japan.

Mobil Chemical Company, P.O. Box 26683, Richmond, Virginia 23261, USA.

Monsanto Commercial Products Company, Agricultural Division, 800 N. Lindbergh Boulevard, St. Louis, Missouri 63166, USA.

Montecatini S.p.A. (*see* Montedison S.p.A.).

Montedison S.p.A., DIAG/C.R.A., Via Bonfadini 148, 20138 Milan, Italy.

Murphy Chemical Company, Wheathampstead, St. Albans, Hertfordshire AL4 8QU, England. (Member of the Dalgety Group).

Nihon Nohyaku Company Limited, Eitaro Bldg., 521, Nihonbashi, Chuo-ku, Tokyo, Japan.

Nippon Kayaku Company Limited, Kaijo Bldg., 121, Marunouchi, Chiyoda-ku, Tokyo, Japan.

Nippon Soda Company Limited, Shin-Ohtemachi Bldg., 3rd floor, 2-1. 2-Chrome, Ohtemachi, Chiyoda-ku, Tokyo, Japan.

Nissan Chemical Industries Limited, Kowa-Hitosubashi Bldg., 7-1, 3-Chrome, Kanda-Nishiki-Cho, Chiyoda-ku, Tokyo, Japan.

Nitrokémia Ipartelepek, 8184 Fúzfógyártelep, Hungary.

Olin Corporation, Agricultural Division, P.O. Box 991, Little Rock, Arkansas 72203, USA.

Pan Britannica Industries Limited, Britannica House, Waltham Cross, Hertfordshire EN8 7DY, England.

Penick Corporation, 1050 Wall Street West, Lyndhurst, New Jersey 07071, USA.

Pennwalt Corporation, P.O. Box 1297, Tacoma, Washington 98401, USA.

Philips-Duphar B.V., Crop Protection Division, Postbox 54, 's-Graveland, Netherlands.

Plant Protection Division ⎫
Plant Protection Limited ⎭ (*see* Imperial Chemical Industries Limited, Plant Protection Division).

Rentokil Limited, Felcourt, East Grinstead, Sussex RH19 2JY, England.

Rhône-Poulenc-Phytosanitaire, BP9163, Lyon 09-69263, Lyone Cedex 1, France.

Rohm and Haas Company, Independence Mall West, Philadelphia, Pennsylvania 19105, USA.

Roussel Uclaf, Division Agro-Veterinaire, 163 Avenue Gambetta, 75020 Paris, France.

Sandoz AG, Agrochemicals Division, Research, CH-4002 Basle, Switzerland.

Sankyo Company Limited, 17 10, 3 chome Ginza, Chuo-ku, Tokyo, Japan.

Schering AG, Agricultural Division, Postfach 650311, D 10000 Berlin 65, Federal Republic of Germany.

Shell International Chemical Company Limited (Ref. CAMK/343), Shell Centre, London SE1 7PG, England.

Sorex (London) Limited, Agricultural Research Department, Halebank Factory, Lower Road, Widnes, Cheshire WA8 8NS, England.

C. F. Spiess and Sohn, Chemische Fabrik, 6719 Kleinkarlbach Über Grünstadt/Pfalz, Federal Republic of Germany.

Stauffer Chemical Company, Agricultural Research Centre, P.O. Box 760, Mountain View, California 94042, USA.

Steetley Chemicals Manufacturing Division, Berk Limited, Canning Road, Stratford, London E15 3NX, England.

Sumitomo Chemical Company Limited, 15, 5-chome, Kitahama, higashi-ku, Osaka, Japan.

Synchemicals Limited, Agrochemicals and Garden Products, 44 Grange Walk, London SE1 3EN, England.

Takeda Chemical Industries Limited, 12-10, Nihonbashi 2-chome, Chuo-ku, Tokyo 103, Japan.

Union Carbide Agricultural Products Company Incorporated, 7825 Baymeadows Way, Jacksonville, Florida 32216, USA.

Uniroyal Chemical Division, Agricultural Chemicals, Research and Development, Amity Road, Connecticut 06525, USA.

The Upjohn Company, Agricultural Division, Plant Health Research and Development, 301 Henrietta Street, Kalamazoo, Michigan 49001, USA.

United States Borax and Chemical Corporation, U.S. Borax Research Corporation, 412 Crescent Way, Anaheim, California 92801, USA.

Velsicol Chemical Corporation, Commercial Development Department, 341 East Ohio Street, Chicago, Illinois 60611, USA.

Wacker-Chemi GmbH, Prinzregentenstrasse 22, D-800 München 22, Federal Republic of Germany.

Wellcome Research Laboratories (Berkhamsted), Berkhamsted Hill, Berkhamsted, Hertfordshire HP4 2QE, England.

Appendix V

DETAILS OF BOOKS MENTIONED IN THE MAIN SECTION OR APPENDIXES

Note: in several cases other volumes in a given series deal with compounds other than those described in detail in this edition of the *Manual* or with general methods or principles and are not listed below.

Advances in Chemistry Series, American Chemical Society.

Advances in Pest Control Research, ed. R. L. Metcalf. New York, Interscience. Volume **3,** 1960; Volume **8,** 1968.

Analysis of Insecticides and Acaricides, F. A. Gunther & R. G. Blinn. (*Chemical Analysis,* Volume **6**). New York, Interscience, 1955.

Analytical Methods for Pesticides, Plant Growth Regulators and Food Additives. Series editor, G. Zweig. New York, Academic Press. Volumes: **2,** 1964; **3,** 1964; **4,** 1964. (See also next entry).

Analytical Methods for Pesticides and Plant Growth Regulators. (cf. previous entry). Series editor, G. Zweig. New York, Academic Press. Volumes: **5,** ed. G. Zweig & J. Sherma, 1967; **6,** ed. G. Zweig & J. Sherma, 1972; **7,** ed. J. Sherma & G. Zweig, 1973; **8,** ed. G. Zweig & J. Sherma, 1976; **10,** ed. G. Zweig & J. Sherma, 1978. Further volumes are in preparation.

Chemistry and Action of Herbicide Antidotes, ed. F. C. Pallos & J. E. Casida. New York, Academic Press, 1978.

CIPAC Handbook, Volume **1,** ed. G. R. Raw. Harpenden, Collaborative International Pesticides Council Ltd, 1970. Volume **1A** is due to be published shortly (Cambridge, Heffer, 1979). Material has been assembled for later volumes and is quoted in the text as 'in press'.

Common Names for Pesticides and Other Agrochemicals, ISO 1750 (1979), in press. Geneva, International Organization for Standardization.

Degradation of Herbicides, ed. P. C. Kearney & D. D. Kaufmann. New York, Dekker, 1969.

Die Entwicklung neuer insektizider Phosphorsäure-Ester, (3rd Ed.), G. Schrader. Weinheim, Verlag Chemie GmbH, 1963.

1967 Evaluations of Some Pesticide Residues in Food. FAO/WHO Meeting Report, PL-1967-M-11-1; Food Additives 68.30. Rome, 1968.

Fumigation of Small Enclosures with Methyl Bromide. MAFF, 1972.

Fumigation of Soil and Compost under Gas-proof Sheeting with Methyl Bromide. MAFF, 1973.

Fumigations Using Methyl Bromide, Guidance Note. Health & Safety Executive, 1976.

Fumigation with Methyl Bromide under Gas-proof Sheets, (3rd Ed.), H. K. Heseltine & R. H. Thompson. MAFF, 1974.

Herbicides: Chemistry, Degradation and Mode of Action, (2nd revised Ed.), ed. P. C. Kearney & D. D. Kaufmann. New York, Dekker: Volume **1,** 1975; Volume **2,** 1976.

Insecticide and Fungicide Handbook for Crop Protection, (5th Ed.), ed. H. Martin & C. R. Worthing. Oxford, Blackwell Scientific Publications, 1976.

Lindane, ed. E. Ulmann. Freiburg, Verlag Karl Schillinger (1072). Supplement I (1974), Supplement II (1976).

Appendix V—*continued*

Official Methods of Analysis, (12th Ed.). Washington, Association of Official Analytical Chemists, 1975. Also amendments to methods, as described in *Journal of the Association of Official Analytical Chemists.*

Organic Insecticides: their Chemistry and Mode of Action, R. L. Metcalf. New York, Interscience, 1955.

[*Pathology of the Silkworm*], K. Ishikawa. Tokyo, Meinbundo (1902). (In Japanese).

Pentachlorophenol, ed. K. R. Rao. Proceedings of a Symposium, Pensacola, Florida, 1977. *(Environmental Science Research,* Volume **12**). Plenum Press, 1978.

Pesticide Analytical Manual: Methods for Individual Pesticide Residues. Washington, Food & Drug Administration, USA, 1971.

Pesticides Considered Not to Require Common Names. ISO 765 (1976); Geneva, International Organization for Standardization.

Recommended Common Names for Pesticides. BS 1831 : 1969, Amendment Slip No. 1 (1979); Supplement No. 1 (1970); Supplement No. 2 (1970); Supplement No. 3 (1974) and Amendment Slip No. 1 (1975). London, British Standards Institution.

Specifications for Pesticides.
Insecticides. (1st Rev.), Rome, WHO.
Organochlorine Insecticides 1973. Rome, FAO, 1973; (MAFF, 1973).
Pesticides Used in Public Health. Rome, WHO, 1973.
Herbicides 1977. Rome, FAO, 1977; (MAFF, 1977).

The Wiswesser Line-Formula Chemical Notation, (3rd Ed.), E. G. Smith & P. A. Baker. Cherry Hill, New Jersey, Chemical Information Management Inc., 1976.

NOTES ON THE WISWESSER LINE-FORMULA NOTATION (WLN)†

For most of the chemical compounds in this manual, the structures and molecular formulae are shown, together with the Wiswesser Line-Formula Notations. These notations are strings of symbols constructed by strict rules to provide a compact, unique and unambiguous description of the molecular structure in linear form. The notations can be used in manual and computer-based indexing and retrieval systems.

Although the principles of encoding structures into WLNs require some weeks of training, any chemist can quickly learn to decode most notations. The list of notation symbols opposite will aid this process and the reader may like to practise on the notations in this manual. As an example, consider monuron, whose WLN is— **GR DMVN1&1**

From the list opposite, it is seen that **G** is chlorine, **R** a benzene ring, **M** an NH-group, **V** a carbonyl, **N** a branched nitrogen, and **1** a C₁ alkyl chain. The **D** with a space in front indicates a ring position, and the **&** shows the end of the first alkyl chain. When these fragments are put together in the order shown by the notation, the structure is shown to be:

The notations for more complex cyclic compounds begin with a description of the ring system; *e.g.,* captan:

T56 BVNV GUTJ CSXGGG

The opening **T** indicates a heterocyclic ring system. The two numerals following show a 5-membered and 6-membered ring fused together. In such bicyclic systems, the ring positions are lettered in order from the fusion point round the smaller and then the larger ring. The letter **B** with a space in front indicates the ring position of the following **V** or carbonyl group; immediately adjacent in the ring is a nitrogen **N** and then another carbonyl **V**. The position of the unsaturation **U** is shown, and the second **T** means that the rings are otherwise saturated. The **J** closes the ring description. Finally the position and nature of the substituent group is shown: **S** stands for sulphur, **X** for a four-branched carbon, and the three **G** symbols for the three chlorines attached to it.

---

† For further introductory information on WLN and its applications see the books listed in Appendix V. Also: G. Palmer, *World Rev. Pest Control,* 1970, **9**, 128; *Chem. Br.,* 1970, **6**, 422.

or write to the Secretary of the Chemical Notation Association (UK), Mrs. J. Ash, Penfield House, Rodmersham, SITTINGBOURNE, Kent, England ME9 0QP.

THE WISWESSER LINE-FORMULA NOTATION SYMBOLS

All the international atomic symbols are used except K, U, V, W, Y, Cl, and Br. Two-letter atomic symbols in organic notations are enclosed between hyphens. Single letters preceded by a blank space indicate ring positions.

Numerals preceded by a space are multipliers of preceding notation suffixes or within ring signs L . . . J or T . . . J show the number of multicyclic points in the ring structure.

Numerals not preceded by a space show ring sizes if within the ring signs—elsewhere numerals show the length of internally saturated, unbranched alkyl chains and segments.

Single letters not preceded by a blank space have the following meanings:

A   Generic alkyl.

B   Boron atom.

C   Unbranched carbon atom multiply bonded to an atom other than carbon, or doubly bonded to two other carbon atoms.

D   Proposed symbol for a chelate bond and initial symbol of a chelate notation.

E   Bromine atom.

F   Fluorine atom.

G   Chlorine atom.

H   When preceded by a locant within ring signs, shows the position of a carbon atom bonded to four other atoms—elsewhere H means hydrogen atom.

I   Iodine atom.

J   Sign for the end of a ring description.

K   Nitrogen atom bonded to more than three other atoms.

L   First symbol of a carbocyclic ring notation.

M   Imino or imido -NH- group.

N   Nitrogen atom, hydrogen free, attached to no more than three other atoms.

O   Oxygen atom, hydrogen free; note that Ø represents the numeral zero.

P   Phosphorus atom.

Q   Hydroxyl group, -OH.

R   Benzene ring.

S   Sulphur atom.

T   First symbol of a heterocyclic ring notation- or within ring signs indicates a ring containing two or more carbon atoms each bonded to four other atoms.

U   Double bond; UU shows an acetylenic triple bond.

V   Carbonyl connective, -CO- (carbon attached to three other atoms).

W   Nonlinear (branching) dioxo group (as in $-NO_2$ or $-SO_2-$).

X   Carbon atom attached to four atoms other than hydrogen.

Y   Carbon atom attached to three atoms other than hydrogen or doubly bonded oxygen.

Z   Amino or amido $-NH_2$ group.

&   Punctuation mark showing the end of a side chain- or preceded by a space, sign of ionic salt, addition compound or suffixed information- or within ring signs indicates a ring not containing two or more carbon atoms that are bonded to four other atoms- or following a hyphen, shows certain spiro ring connections.

-   Separator or connective or other special uses.

/   Encloses polymer notations; precedes each non-consecutive locant pair.

*   (1) Points of attachment in polymer repeat units; (2) coincident atoms in polymer notations; (3) a multiplier symbol in inorganic notations.

.   Space-filling symbol for inorganic notations.

Ø   Zero

INDEX 1

Based on Wiswesser Line-Formula Notation

These entries include those in the main section and Appendix 1 (superseded compounds) and are listed in the hierarchial order:

'nothing'

.

&

-

other punctuation
letters in alphabetical order
digits Ø 1 2 3 4 5 6 7 8 9

Note: Because the comparison is made with each sign (or space) in turn the correct numeric order is 1, 1Ø, 11 etc., preceding 2 etc.

When possible WLNs are broken in this index at spaces between fragments. When this is not possible the sign = is placed at the end of the line, signifying that the following line runs immediately after it. The - sign has its usual significance in WLN (see Appendix VI). Ions (diquat) and salts (diquat dibromide, sodium fluoroacetate) are listed as such to agree with the WLNs and formulae in the text and Appendix I.

The italic figures in parentheses following page numbers refer to position of compound on page in Appendix I (superseded compounds), *viz.* 553(*3*) allidochlor.

589

Index 1—*continued*

Index 1—*continued*

Index 1—*continued*

INDEX 2

Based on Molecular Formulae

These entries include those in the Main Section and Appendix I (superseded compounds). Individual salts and esters mentioned only under the sections on formulations are not included. Ions (diquat) and salts (diquat dibromide, sodium fluoroacetate) are listed as such to agree with the formulae in the text. Compounds (zineb) believed to be polymeric are indexed as monomers.

The italic figures in parentheses following page numbers refer to position of compound on page in Appendix I (superseded compounds), *viz.* 553(*3*) allidochlor.

599

Index 2—*continued*

Index 2—*continued*

INDEX 3

Based on 'Code Numbers'

This index includes Codes used to identify pesticides. All the Codes are made up of an identifier and a serial number. It contains Codes used by official bodies (qualifier AI3-, AN4-, ENT, OMS etc.) and those used by discoverers, manufacturers and suppliers (indicated by single quotation marks ' '). Some less well known common names based on letters and numbers (e.g. M-81) are also included.

The index is arranged in order of the numerical component of the code. When more than one entry has the same numeral, separation is by the alphabetical qualifier. Numbers of the type 12AB34 or 12/34 are indexed in all possible ways, e.g. under 12, 34 and 1234.

The italic figures in parentheses following page numbers refer to position of compound on page in Appendix I (superseded compounds), *viz.* 553(*3*) allidochlor.

Index 3—*continued*

Index 3—*continued*

Index 3—*continued*

INDEX 4

Based on Chemical, Common and Trivial Names and Trade Marks

Numerical and alphabetical prefixes and infixes signifying positions in chemical names or stereochemistry have been omitted. Brackets [ ] and parentheses ( ) have been introduced for convenience in distinguishing between, *e.g.* chloro(methyl) (two groups) and (chloromethyl) (one group).

Trade marks have, as far as possible, been placed within single quotation marks ' ' and their use, even without this indication, is not to be considered to imply that they are not protected by law. No attempt has been made to distinguish between registered and unregistered trade marks. Qualifying numbers or letters are included only when these indicate additional active ingredient(s) in the product—not the concentration of a.i.

Pages on which the properties of the stated pesticides are given are listed in normal print; pages on which the compound is mentioned as an active component of a mixture are shown in *italics*; pages on which the compound is mentioned as a degradation product or source of another pesticide are shown in parentheses ( ).

The italic figures in parentheses following page numbers refer to position of compound on page in Appendix I (superseded compounds), *viz.* 553(*3*) allidochlor.

The order is strictly alphabetical, for instance: 2,4-D, dalapon, 2,4-D-ammonium. Readers should bear in mind national spelling variations, especially those between f/ph, t/th and the addition or deletion of a terminal 'e' in syllables such as ol(e), on(e) and yn(e).

Index 4—*continued*

625

Index 4—*continued*

NOTES

NOTES

NOTES

NOTES

NOTES

NOTES

NOTES